Extremum Seeking through Delays and PDEs

Advances in Design and Control

SIAM's Advances in Design and Control series consists of texts and monographs dealing with all areas of design and control and their applications. Topics of interest include shape optimization, multidisciplinary design, trajectory optimization, feedback, and optimal control. The series focuses on the mathematical and computational aspects of engineering design and control that are usable in a wide variety of scientific and engineering disciplines.

Series Volumes

Oliveira, Tiago Roux and Krstic, Miroslav, *Extremum Seeking through Delays and PDEs*
Kaszkurewicz, Eugenius and Bhaya, Amit, *Business Dynamics Models: Optimization-Based One Step Ahead Optimal Control*
Pedregal, Pablo, *A Variational Approach to Optimal Control of ODEs*
Garon, André and Delfour, Michel C., *Transfinite Interpolations and Eulerian/Lagrangian Dynamics*
Martinelli, Agostino, *Observability: A New Theory Based on the Group of Invariance*
Betts, John T., *Practical Methods for Optimal Control Using Nonlinear Programming, Third Edition*
Rodrigues, Luis, Samadi, Behzad, and Moarref, Miad, *Piecewise Affine Control: Continuous-Time, Sampled-Data, and Networked Systems*
Ferrara, A., Incremona, G. P., and Cucuzzella, C., *Advanced and Optimization Based Sliding Mode Control: Theory and Applications*
Morelli, A. and Smith, M., *Passive Network Synthesis: An Approach to Classification*
Özbay, Hitay, Gümüşsoy, Suat, Kashima, Kenji, and Yamamoto, Yutaka, *Frequency Domain Techniques for $\mathcal{H}_\infty$ Control of Distributed Parameter Systems*
Khalil, Hassan K., *High-Gain Observers in Nonlinear Feedback Control*
Bauso, Dario, *Game Theory with Engineering Applications*
Corless, M., King, C., Shorten, R., and Wirth, F., *AIMD Dynamics and Distributed Resource Allocation*
Walker, Shawn W., *The Shapes of Things: A Practical Guide to Differential Geometry and the Shape Derivative*
Michiels, Wim and Niculescu, Silviu-Iulian, *Stability, Control, and Computation for Time-Delay Systems: An Eigenvalue-Based Approach, Second Edition*
Narang-Siddarth, Anshu and Valasek, John, *Nonlinear Time Scale Systems in Standard and Nonstandard Forms: Analysis and Control*
Bekiaris-Liberis, Nikolaos and Krstic, Miroslav, *Nonlinear Control under Nonconstant Delays*
Osmolovskii, Nikolai P. and Maurer, Helmut, *Applications to Regular and Bang-Bang Control: Second-Order Necessary and Sufficient Optimality Conditions in Calculus of Variations and Optimal Control*
Biegler, Lorenz T., Campbell, Stephen L., and Mehrmann, Volker, eds., *Control and Optimization with Differential-Algebraic Constraints*
Delfour, M. C. and Zolésio, J.-P., *Shapes and Geometries: Metrics, Analysis, Differential Calculus, and Optimization, Second Edition*
Hovakimyan, Naira and Cao, Chengyu, *$\mathcal{L}_1$ Adaptive Control Theory: Guaranteed Robustness with Fast Adaptation*
Speyer, Jason L. and Jacobson, David H., *Primer on Optimal Control Theory*
Betts, John T., *Practical Methods for Optimal Control and Estimation Using Nonlinear Programming, Second Edition*
Shima, Tal and Rasmussen, Steven, eds., *UAV Cooperative Decision and Control: Challenges and Practical Approaches*
Speyer, Jason L. and Chung, Walter H., *Stochastic Processes, Estimation, and Control*
Krstic, Miroslav and Smyshlyaev, Andrey, *Boundary Control of PDEs: A Course on Backstepping Designs*
Ito, Kazufumi and Kunisch, Karl, *Lagrange Multiplier Approach to Variational Problems and Applications*
Xue, Dingyü, Chen, YangQuan, and Atherton, Derek P., *Linear Feedback Control: Analysis and Design with MATLAB*
Hanson, Floyd B., *Applied Stochastic Processes and Control for Jump-Diffusions: Modeling, Analysis, and Computation*
Michiels, Wim and Niculescu, Silviu-Iulian, *Stability and Stabilization of Time-Delay Systems: An Eigenvalue-Based Approach*
Ioannou, Petros and Fidan, Barış, *Adaptive Control Tutorial*
Bhaya, Amit and Kaszkurewicz, Eugenius, *Control Perspectives on Numerical Algorithms and Matrix Problems*
Robinett III, Rush D., Wilson, David G., Eisler, G. Richard, and Hurtado, John E., *Applied Dynamic Programming for Optimization of Dynamical Systems*
Huang, J., *Nonlinear Output Regulation: Theory and Applications*
Haslinger, J. and Mäkinen, R. A. E., *Introduction to Shape Optimization: Theory, Approximation, and Computation*
Antoulas, Athanasios C., *Approximation of Large-Scale Dynamical Systems*
Gunzburger, Max D., *Perspectives in Flow Control and Optimization*
Delfour, M. C. and Zolésio, J.-P., *Shapes and Geometries: Analysis, Differential Calculus, and Optimization*
Betts, John T., *Practical Methods for Optimal Control Using Nonlinear Programming*
El Ghaoui, Laurent and Niculescu, Silviu-Iulian, eds., *Advances in Linear Matrix Inequality Methods in Control*
Helton, J. William and James, Matthew R., *Extending $\mathcal{H}^\infty$ Control to Nonlinear Systems: Control of Nonlinear Systems to Achieve Performance Objectives*

Extremum Seeking through Delays and PDEs

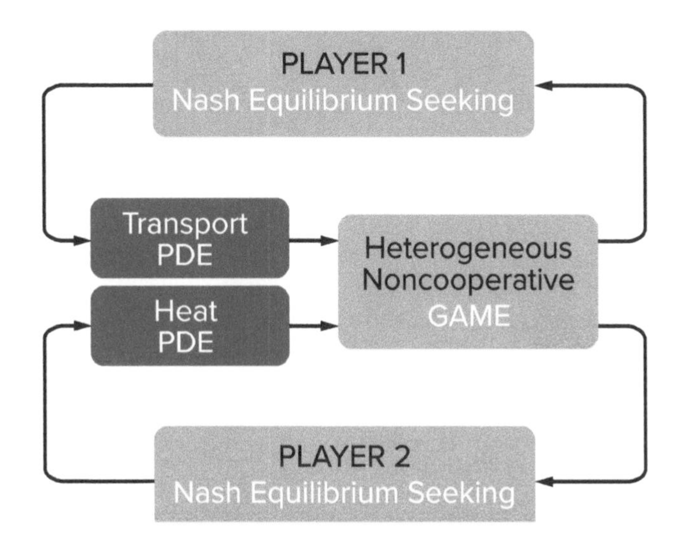

Tiago Roux Oliveira
State University of Rio de Janeiro
Rio de Janeiro, Brazil

Miroslav Krstic
University of California, San Diego
La Jolla, California

Society for Industrial and Applied Mathematics
Philadelphia

10 9 8 7 6 5 4 3 2 1

Royalties from the sale of this book are placed in a fund to help students attend SIAM meetings and other SIAM-related activities. This fund is administered by SIAM, and qualified individuals are encouraged to write directly to SIAM for guidelines.

Publications Director	Kivmars H. Bowling
Executive Editor	Elizabeth Greenspan
Managing Editor	Kelly Thomas
Production Editor	David Riegelhaupt
Copy Editor	Bruce Owens
Production Manager	Donna Witzleben
Production Coordinator	Cally A. Shrader
Compositor	Cheryl Hufnagle
Graphic Designer	Doug Smock

Library of Congress Control Number: 2022947933

Contents

Preface

The year in which this book appears is the centennial of the 1922 invention, in the context of maximizing the power transfer to an electric tram car [130], of the method for real-time model-free optimization called extremum seeking (ES). Forgotten after a few decades of practical use and attempts at theoretical study, ES returned to life when the stability of its operation for not only static input-output maps but for dynamic systems modeled by general nonlinear ODEs was proven in 2000 [129].

It is natural to ask "why limit the use and the theoretical advances of ES to ODE systems?" So many—if not the majority of physical systems—involve delays or are modeled by PDEs; why not pursue the application of ES in the presence of delays and for PDE systems?

With this book, which is a labor of the last eight years (since we began to work on ES through delays in 2014), we make an ambitious first venture into the vast space of possibilities of designing ES algorithms for infinite-dimensional systems.

We begin with systems of infinite dimension which are arguably the most common in control practice—delays at the input or due to computation, sometimes also referred to as "dead time." With predictor feedback designs for input delays, we compensate delays of arbitrary length in ES feedback loops and ensure convergence of the algorithms. Then we extend such results, using PDE backstepping, to various PDEs in the actuation or sensing pathways of unknown input-output maps. Finally, we study the use of delay-compensated and PDE-compensated ES algorithms by multiple players in noncooperative game settings and prove that users of such algorithms are assured that their actions will converge to a Nash equilibrium, namely, to the best possible strategy in the presence of other rational players, even though the players employing ES algorithms may not even be aware of the presence or the competing players, let alone know the competing players' actions or payoffs.

It is the exhaustive consideration of these three problems, and various extensions (multivariable maps, dynamical plants, nonconstant and uncertain delays, distributed delays, etc.), that comprise the three parts of this book and its approximately four hundred pages.

The field of ES has grown vast since the paper [129] re-awoke it in 2000. Over 10,000 papers have been published on this topic since and they currently emerge at the rate of about 1,000 per year. It is impossible to survey a literature of this size in either this preface or later in the book. But we do give a few highlights.

As said earlier, the idea of ES emerged in a French patent and article [130] on maximizing power transfer to a tram car. Soviet control engineers intensively studied and advanced this concept in the 1940s [107], before it was introduced in the 1950s in the U.S. by Draper and Li [56]. ES vanished from the research horizon in the 1960s and returned in 2000 with the stability proof in [129]. An advancement of the stability study from local to semi-global was provided in [217]. An extension to multivariable maps with the convergence rate being made uniform across all the input channels using a Newton-based ES approach was introduced in [73]. Stochastic ES algorithms and the stochastic averaging theory needed for their stability study were introduced in [142]. Performance improvements for parameter convergence is pursued in [82, 2, 80, 165].

The most popular ES approach relies on a small periodic excitation, usually sinusoidal, to disturb the parameters being tuned [129, 10, 123, 82, 2, 217, 219]. This approach generates an estimate of the gradient of the map, which is then being driven to zero by the ES algorithm, to find an extremum. Along with the perturbations employed to estimate the gradient, and in some cases the Hessian of the unknown map, come the mathematical methods needed for the convergence study—the averaging theory (periodic and stochastic) and the singular perturbation theory.

What Motivates ES through Delays? Actuator and sensor delays are some of the most common phenomena that arise in engineering practice. Long delays are as detrimental, if left uncompensated, to ES as they are to other feedback loops. Real-time optimization gives rise to significant delays, such as when post-processing of the plant's measured output translates into a considerable delay in generating the control input to be applied to the plant. Examples are the image processing that takes place in laser-based light sources for photolithography in semiconductor manufacturing [195, 38], or in various chemical and biochemical processes where analysis of samples takes place. For instance, the phase lag observed in batch cultures applied to bioreactors [216, 138] illustrates the delay phenomenon occurring in the biological optimizing process. These delays are typically known, constant, and relatively large.

The infinite-dimensional backstepping transformation for systems with input delays [119] has recently enabled the design of predictor feedback designs for delay compensation, the construction of explicit Lyapunov functionals for their stability analysis, and the quantification of their performance, robustness, and optimality. However, predictor feedback designs encounter a significant obstacle to their applicability in ES—they are model-based. The presence of an unknown nonlinear map in the feedback loop is an antithesis to model-based design. In this book we resolve this challenge and design algorithms that enable the estimation of the unknown Hessian of the map, which is needed for the implementation of predictor-based delay compensation in ES algorithms.

What Motivates ES through PDEs? Many complex industrial systems—perhaps the majority—are described by partial differential equations (PDEs) or even PDEs coupled with ODEs. These ODE-PDE cascades systems occur, for example, when sensors and actuators are not collocated and, particularly, in systems involving transport of heat and materials. Among several other applications are the screw extrusion processes, metal rolling-cutting processes, oil drilling, populations, transport delays in fuel/air ratio systems, production of manufacturing systems, and tubular bioreactors. The extension of the backstepping designs from delay-ODE cascades to PDE-ODE cascades [119] provides the tools for ES design in the presence of PDEs, along with the perturbation-based estimation of the Hessian of the map and the PDE motion planning-based design perturbation signals employed in extremum seeking.

What Motivates ES for Noncooperative Games under Delays and PDE Dynamics? Game theory provides a theoretical framework for conceiving social situations among competing players and using mathematical models of strategic interaction among rational decision-makers [72, 22]. Game-theoretic approaches to designing, modeling, and optimizing emerging engineering systems, biological behaviors, and mathematical finance make this topic of research an important tool in many fields with a wide range of applications [85, 8, 23], and with sustained advances over many decades [214, 191, 213, 238, 45, 17, 4].

One categorization of games is into *cooperative* and *noncooperative* [21]. A game is cooperative if the players are able to form binding commitments externally enforced (e.g., through contract law), resulting in collective payoffs. A game is noncooperative if players cannot form alliances or if all agreements need to be self-enforcing (e.g., through credible threats), focusing on predicting individual players' actions and payoffs and analyzing *Nash equilibria* [163]. Nash equilibrium is an outcome which, once achieved or reached, means no player can increase its payoff by changing decisions unilaterally [21].

The development of algorithms to achieve convergence to a Nash equilibrium has been a focus of researchers for several decades [133, 20]. Learning schemes for reaching a Nash equilibrium have also been studied [249]. But fully model-free seeking of Nash equilibria, in an online fashion and in the presence of plant dynamics, was not achieved until [70], where the players locally converge to their Nash strategies without the need for model information, information on the actions and payoffs of the other players, or even the awareness of the participation of other players.

Delays and/or PDEs in a game context arise in applications like network virtualization, software defined networks, cloud computing, the Internet of Things, context-aware networks, green communications, and security [85, 8, 7]. In particular, differential games with delays are dealt with (in a partly or fully model-based fashion) in [42, 157, 105, 61, 76, 187, 41]. PDE dynamics arise in the Black–Scholes model of behavior in financial markets [197, 132]. Hence, strong motivation exists for designs that ensure convergence to Nash equilibria in the presence of delays and PDEs. We provide them in this book.

What Does the Book Cover? The book comprises three parts. Part I deals with delays, Part II with PDEs, and Part III with games.

Part I contains results for single-input and multi-input maps with distinct delays, point delays and distributed delays, constant and varying delays, and known and uncertain delays, as well as for seeking an extremum of a map and of any of the map's derivatives.

We first design (multivariable) ES algorithms which incorporate compensation of delays by means of predictor feedback. Predictor feedback requires a known model. In extremum seeking problems, the model—and, particularly, the Hessian of the map—is unknown. To deal with this challenge, we do not estimate the Hessian parametrically, as is done in [82, 2, 217, 219, 80, 81, 165, 155]. Instead, we design a perturbation-based estimator of the Hessian [73, 166], which is incorporated into our design of a predictor for a gradient extremum seekers. In addition, we also develop predictor-based Newton ES designs in which the predictor does not require the Hessian but the ES algorithm requires an *inverse* of the Hessian, which we estimate using a Riccati (quadratic) ODE. With such a predictor-based Newton ES design we make the convergence rate independent of the unknown parameters of the map and make the convergence rate assignable by the user.

While designing ES algorithms with predictors is the main practical benefit of this book, and this design is of non-negligible complexity, it is in the analysis that a greater share of the book's innovation resides. We introduce a carefully crafted stability analysis process—a template of sorts—with precisely sequenced steps: (1) representation of the delay as a transport PDE; (2) representation of the closed-loop system in a form in which the perturbation signals are present and the parameter estimation error is delayed; (3) derivation of the average PDE-ODE model; (4) application of a backstepping transformation to the averaged PDE-ODE model; (5) introduction of a Lyapunov–Krasovskii functional for the backstepping target system and a stability computation for this functional; (6) generation of an exponential stability estimate for the average PDE-ODE system in the original variables, using the inverse backstepping transformation; (7) application of the averaging theorem for delay systems, by Hale and Lunel, to obtain exponential stability of a small periodic orbit near the extremum; (8) derivation of the asymptotic estimates for the convergence of the output of the unknown map to the vicinity of its extremum. This process will serve the needs of future researchers who deal with ES under delays and PDEs.

In Part II we introduce ES algorithms for PDEs preceding unknown maps. We consider both parabolic PDEs (the heat equation and the reaction-advection-diffusion equation) and hyperbolic PDEs (the wave equation and the delay-wave cascade). The compensation of the PDE in the ES algorithm is approached using the extension of the backstepping designs from delay-ODE cascades to PDE-ODE cascades [119].

For estimating the gradient and the Hessian, the perturbation signal cannot be simply advanced in time, as is the case with input delays. The perturbation signal is designed using a motion planning approach or solving the trajectory generation problem for PDEs, as sketched in the textbook [128].

On the analysis side, the averaging results for delay systems, which we employ in Part I, do not suffice for PDEs. More general results by Hale and Lunel [83] are employed for averaging of PDEs.

Multivariable PDE problems are particularly challenging because of the coupling among them that results from the non-diagonal character, in general, of the unknown Hessian. We remove the need to deal with multiple coupled PDEs by employing the Newton ES approach and its associated inversion of the unknown Hessian using a Riccati filter ODE of the Hessian's estimate (which we generate using demodulation signals designed by PDE motion planning). The estimate of the Hessian's inverse diagonalizes and decouples the PDE system, permitting us to employ standard PDE backstepping compensators for single PDEs.

In Part III we deal with ES algorithms that are engaged in a noncooperative game interaction amongst each other. No algorithm is aware of the other algorithms running. No algorithm employs the information on the actions applied by the other algorithms or the payoffs obtained by the other algorithms. And no player has the knowledge of the analytical form of even his own payoff function, let alone of the payoff functions of the other players. Only in one particular scenario—of a cooperative game with input delays—do we permit information exchange among the players, as well as cooperation in compensating each others' input delays.

Diagonal dominance of the Hessian and other small-gain relationships are standard in iterative algorithms for seeking Nash equilibria, even when the full modeling information is available, which is the typical assumption. In our ES algorithms that compensate for the PDE dynamics, much more complex coupling arises—among the PDEs themselves, and even through the boundary conditions, which is the most challenging form of coupling as the input operator for an input acting through the boundary condition is unbounded. To resolve such coupling relationships, we use the latest forms of Small-Gain Theorems for PDEs in [99]. The small-gain approach to studying stability of coupled PDEs is "agnostic" to the PDE class and allows us to handle even PDEs from distinct classes appearing concurrently in a noncooperative game.

Whom Is the Book For? As we indicated earlier, the last two decades have seen over 10,000 publications on extremum seeking, with about 1,000 papers appearing each year on topics of ES, Nash equilibrium seeking, and source seeking. One could estimate the community of active ES researchers in several thousands. They work in many areas of engineering, in applied mathematics, and in physics. Their applications range from robotics, to fluids, to energy systems, to various areas of manufacturing, to automotive systems, to traffic, to chemical and bioreactors, to neuromuscular electrical simulation, and to lasers and charged particle accelerators. Many of these researchers can benefit from the exposure, through this book, to the advances in ES designs to delays and PDE systems.

The book is not only an extension of ES to delay and PDEs. It is also—and conversely—an extension of control of delay systems and PDEs to ES. The PDE control community is rapidly growing as a result of the emergence of PDE backstepping, which supplies the user with controllers with explicit gains for PDEs whose levels of open-loop instability are unrestricted. The delay systems community is even larger and with a much longer history. Many of these researchers can deploy their experiences in model-free and online optimization problems through the ES approach, suitably integrated with their techniques.

A key ingredient in the analysis of stability of ES algorithms is averaging theory, sometimes combined with singular perturbation theory. The design of ES algorithms often lurches ahead of the existing analysis tools. It is common that designs are possible, perform powerfully in simulations, and guarantee stability of an average system, but an averaging theorem is lacking to infer stability of the original system. Vast opportunities exist for applied mathematicians, particularly for specialists in delay differential equations, PDEs, and stochastic differential equations, to develop advances in averaging theory which would serve as a mathematical foundation for further useful advances in extremum seeking.

May we be excused for mentioning, for the third time, the rate at which the area of extremum seeking is expanding—about 1,000 papers per year. Significant graduate instructional need and opportunity exists for a field of this size. The literature on ES probably far exceeds the size of the literature in adaptive control in the mid-1990s, at which time courses in adaptive control were common in many, if not most, strong graduate programs in control engineering. The instructor of a course on extremum seeking has several research monographs at her disposal to use in assembling course material, including [10, 142, 29]. However, each of these books tells only a small fraction of the overall story. While the present book is not crafted as a textbook, and it certainly lacks homework sets and exercises, it does contain both accessible and relatively comprehensive material on ES, starting with Chapter 1. It is also the most up-to-date book on ES and the most complete on the basic capabilities of ES. The only topic that is glaringly missing is *source seeking* for nonholonomic vehicles seeking signal sources in GPS-denied environments. For this material, which is one of the core capabilities of ES as it directly leverages the vehicle kinematics in the algorithm design, with no possibility that other static optimization approaches could reproduce such a capability, the instructor would have to rely on papers like [58] and the earlier [43].

With the explosion of interest in learning algorithms, for all sorts of purposes including optimization and recently even for control, it would be regrettable if it were overlooked that ES is also a learning-based approach (in the minimalistic, most efficient sense of learning), also a model-free approach, and also a data-based approach. The real-time capabilities, as well as the rigorous convergence guarantees, and even convergence rate assignment capabilities, which ES possesses, are what machine learning and reinforcement learning algorithm can only dream of. It is the authors' hope that some of the researchers and students who are pursuing learning-based capabilities will take note of what this books offers, especially in terms of the possibility of blending model-free learning and optimization with model-based compensation of physical (and social) processes such as transport and diffusion.

The previous knowledge that the reader is assumed to have in order to be able to read this book is the basic graduate-level background on differential equations and calculus. All required notions, such as Lyapunov stability, the basic functional analysis inequalities, asymptotic theory such as Averaging, and Small-Gain Theorems, are incorporated in the appendices for the reader's convenience.

Finally, while the book's material is of interest to both applied mathematicians and engineers, the exposition is optimized for the background and the style that is customary for engineers.

Acknowledgments. We would like to thank Daisuke Tsubakino, Damir Rusiti, Mamadou Diagne, Jan Feiling, Shumon Koga, Huan Yu, Nikolaos Bekiaris-Liberis, Victor Hugo Pereira Rodrigues, Paulo Cesar Souza da Silva, Paulo Paz, Denis César Ferreira, and Maurício Linhares Galvão for their generous collaboration and valuable contribution in several parts of this book.

We are grateful to Iasson Karafyllis and Tamer Basar for their inspiring and pioneering contributions to the Input-to-State Stability analysis for PDEs and noncooperative games, respectively.

The work of Miroslav Krstic was supported by the National Science Foundation and the Air Force Office of Scientific Research.

Tiago Roux Oliveira thanks the Brazilian funding agencies CAPES, CNPq, and FAPERJ for supporting this research.

Tiago is thankful for his parents Vera Lúcia and Delci, his sister Mirela, and his in-laws Tania Mara and José Francisco.

For all the hours sacrificed and for all the continuous support and encouragement, Tiago dedicates this book to his beautiful wife Deborah. Special thanks go to Trufa and Polenguinho, Tiago's little dogs, for their companionship on this journey.

Tiago's gratitude and love go to all of them.

Rio de Janeiro – RJ, Brazil
La Jolla – CA, USA
May, 2022

Tiago Roux Oliveira
Miroslav Krstic

Chapter 1

Fundamentals of Extremum Seeking

Extremum seeking (ES) is a method for real-time non-model-based optimization. Though ES was invented in 1922, the "turn of the 21st century" has been its golden age, both in terms of the development of theory and in terms of its adoption in industry and in fields outside of control engineering. This chapter supplies an overview of the basic gradient- and Newton-based versions of ES with periodic and stochastic perturbation signals.

1.1 ▪ The Basic Idea of Extremum Seeking

Many versions of ES exist, with various approaches to their stability study [129], [217], [142]. The most common version employs perturbation signals for the purpose of estimating the gradient of the unknown map that is being optimized. To understand the basic idea of extremum seeking, it is best to first consider the case of a static single-input map of the quadratic form

$$f(\theta) = f^* + \frac{f''}{2}(\theta - \theta^*)^2, \tag{1.1}$$

where f^*, f'' and θ^*, are all unknown, as shown in Figure 1.1.

Three different thetas appear in Figure 1.1: θ^* is the unknown optimizer of the map, $\hat{\theta}(t)$ is the real-time estimate of θ^*, and $\theta(t)$ is the actual input into the map. The actual input $\theta(t)$ is based on the estimate $\hat{\theta}(t)$ but is perturbed by the signal $a\sin(\omega t)$ for the purpose of estimating the unknown gradient $f'' \cdot (\theta - \theta^*)$ of the map $f(\theta)$ in (1.1). The sinusoid is only one choice for a perturbation signal—many other perturbations, from square waves to stochastic noise, can be used in lieu of sinusoids, provided they are of zero mean. The estimate $\hat{\theta}(t)$ is generated with the integrator k/s with the adaptation gain k controlling the speed of estimation.

The ES algorithm is successful if the error between the estimate $\hat{\theta}(t)$ and the unknown θ^*, namely the signal

$$\tilde{\theta}(t) = \hat{\theta}(t) - \theta^*, \tag{1.2}$$

converges towards zero or some small neighborhood of zero as $t \to +\infty$. Based on Figure 1.1, the estimate $\hat{\theta}(t)$ is governed by the differential equation $\dot{\hat{\theta}} = k\sin(\omega t) f(\theta)$, which means that the estimation error is governed by

$$\frac{d\tilde{\theta}}{dt} = k\, a \sin(\omega t)\left[f^* + \frac{f''}{2}\left(\tilde{\theta} + a\sin(\omega t)\right)^2\right]. \tag{1.3}$$

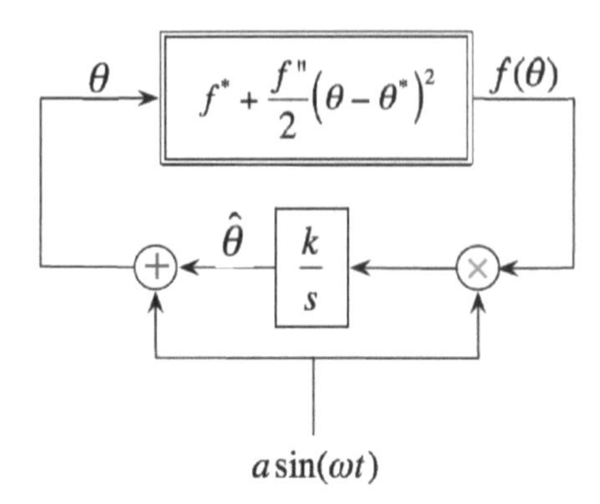

Figure 1.1. *The simplest perturbation-based extremum seeking scheme for a quadratic single-input map $f(\theta)$ in (1.1). The user has to only know the sign of f'', namely, whether the quadratic map has a maximum or a minimum, and has to choose the adaptation gain k such that* $\mathrm{sgn}(k) = -\mathrm{sgn}\left(f''\right)$. *The user has to also choose the frequency ω as relatively large compared to a, k and f''.*

Expanding the right-hand side, one obtains

$$\begin{aligned}\frac{d\tilde{\theta}(t)}{dt} =& k\,a\,f^* \underbrace{\sin(\omega t)}_{\text{mean}=0} + k\,a^3\frac{f''}{2}\underbrace{\sin^3(\omega t)}_{\text{mean}=0}\\ &+ k\,a\frac{f''}{2}\underbrace{\sin(\omega t)}_{\text{fast, mean}=0}\underbrace{\tilde{\theta}(t)^2}_{\text{slow}}\\ &+ k\,a^2 f''\underbrace{\sin^2(\omega t)}_{\text{fast, mean}=1/2}\underbrace{\tilde{\theta}(t)}_{\text{slow}}.\end{aligned} \tag{1.4}$$

A theoretically rigorous time-averaging procedure [108, Section 10.4] allows one to replace the above sinusoidal signals by their means, yielding the "average system" [108, p. 404]:

$$\frac{d\tilde{\theta}_{\mathrm{av}}}{dt} = \frac{\overbrace{k\,f''\,a^2}^{<0}}{2}\tilde{\theta}_{\mathrm{av}}, \tag{1.5}$$

which is exponentially stable. The averaging theory guarantees that there exists a sufficiently large ω such that, if the initial estimate $\hat{\theta}(0)$ is sufficiently close to the unknown θ^*, one has

$$|\theta(t) - \theta^*| \le |\theta(0) - \theta^*|\,e^{\frac{kf''a^2}{2}t} + \mathcal{O}\left(\frac{1}{\omega}\right) + \mathcal{O}(a) \quad \forall t \ge 0. \tag{1.6}$$

For the user, the inequality (1.6) guarantees that if a is chosen small and ω is chosen large, the input $\theta(t)$ exponentially converges to a small interval—of order $\mathcal{O}(\frac{1}{\omega}+a)$—around the unknown θ^* and, consequently, the output $f(\theta(t))$ converges to the vicinity of the optimal output f^*.

1.2 ▪ Extremum Seeking for Multivariable Static Maps

For static maps, ES extends in a straightforward manner from the single-input case shown in Figure 1.1 to the multi-input case shown in Figure 1.2.

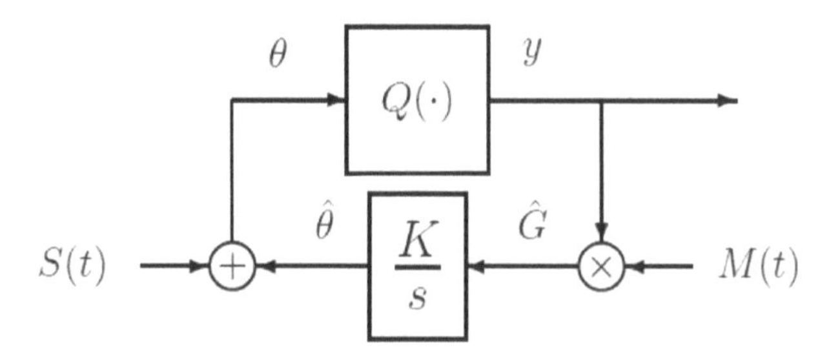

Figure 1.2. *Extremum seeking algorithm for a multivariable map $y = Q(\theta)$, where θ is the input vector $\theta = [\theta_1, \theta_2, \ldots, \theta_n]^T$. The algorithm employs the additive perturbation vector signal $S(t)$ given in* (1.7) *and the multiplicative demodulation vector signal $M(t)$ given in* (1.8).

The algorithm measures the scalar signal $y(t) = Q(\theta(t))$, where $Q(\cdot)$ is an unknown map whose input is the vector $\theta = [\theta_1, \theta_2, \ldots, \theta_n]^T$. The gradient is estimated with the help of the signals

$$S(t) = [a_1 \sin(\omega_1 t) \; \cdots \; a_n \sin(\omega_n t)]^T, \tag{1.7}$$

$$M(t) = \left[\frac{2}{a_1} \sin(\omega_1 t) \; \cdots \; \frac{2}{a_n} \sin(\omega_n t) \right]^T, \tag{1.8}$$

with nonzero perturbation amplitudes a_i, $i \in \{1, \ldots, n\}$, and with an adaptation gain matrix K that is diagonal. To guarantee convergence, the user should choose $\omega_i \neq \omega_j$. This is a key condition that differentiates the multi-input case from the single-input case. In addition, in the convergence analysis, the user should choose ω_i / ω_j as rational and $\omega_i + \omega_j \neq \omega_k$ for distinct i, j, and k.

If the unknown map is quadratic, namely,

$$Q(\theta) = Q^* + \frac{1}{2}(\theta - \theta^*)^T H (\theta - \theta^*), \tag{1.9}$$

the averaged system [108, p. 404] is

$$\dot{\tilde{\theta}}_{\text{av}} = K H \tilde{\theta}_{\text{av}}, \quad H = \text{Hessian}. \tag{1.10}$$

If, for example, the map $Q(\cdot)$ has a maximum that is locally quadratic (which implies $H = H^T < 0$), and if the user chooses the elements of the diagonal gain matrix K as positive, the ES algorithm is guaranteed to be at least locally convergent (for a compact set of initial conditions around θ^*). However, the convergence rate depends on the unknown Hessian H. This weakness of the gradient-based ES algorithm is removed with the Newton-based ES algorithm, discussed later on.

On the other hand, a stochastic version [142] of the algorithm in Figure 1.2 also exists, in which $S(t)$ and $M(t)$ are replaced by

$$S(\eta(t)) = [a_1 \sin(\eta_1(t)) \; \cdots \; a_n \sin(\eta_n(t))]^T, \tag{1.11}$$

$$M(\eta(t)) = \left[\frac{2}{a_1(1 - e^{-q_1^2})} \sin(\eta_1(t)) \; \cdots \; \frac{2}{a_n(1 - e^{-q_n^2})} \sin(\eta_n(t)) \right]^T, \tag{1.12}$$

where $\eta_i = \frac{q_i \sqrt{\epsilon_i}}{\epsilon_i s + 1} \left[\dot{W}_i \right]$ and $\left[\dot{W}_i \right]$ are independent unity-intensity white noise processes.

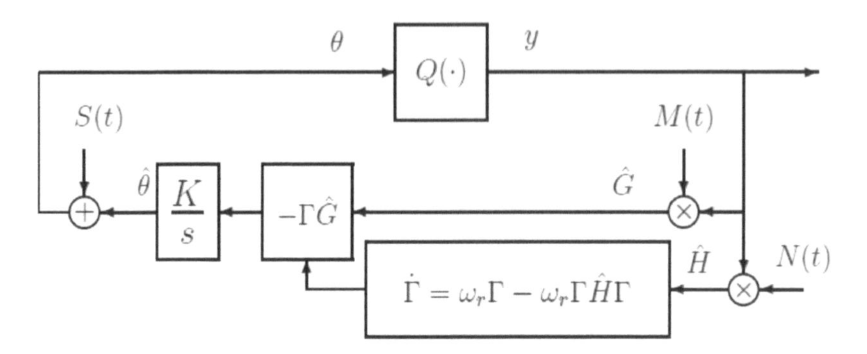

Figure 1.3. *A Newton-based ES algorithm for a static map. The multiplicative excitation $N(t)$ helps to generate the estimate of the Hessian $H = \frac{\partial^2 Q(\theta)}{\partial \theta^2}$ as $\hat{H}(t) = N(t)y(t)$. The Riccati matrix differential equation $\Gamma(t)$ generates an estimate of the Hessian's inverse matrix, avoiding matrix inversions of Hessian estimates that may be singular during the transient.*

1.3 ▪ Newton Extremum Seeking Algorithm for Static Maps

A Newton version of the ES algorithm, shown in Figure 1.3, ensures that the convergence rate be user-assignable rather than being dependent on the unknown Hessian of the map.

The elements of the demodulating matrix $N(t)$ for generating the estimate of the Hessian are given by [73]

$$N_{ii}(t) = \frac{16}{a_i^2}\left(\sin^2(\omega_i t) - \frac{1}{2}\right), \tag{1.13}$$

$$N_{ij}(t) = \frac{4}{a_i a_j}\sin(\omega_i t)\sin(\omega_j t), \qquad i \neq j, \qquad i,j \in \{1,\ldots,n\}. \tag{1.14}$$

For a quadratic map, the averaged system [108, p. 404] in the error variables $\tilde{\theta} = \hat{\theta} - \theta^*$, $\tilde{\Gamma} = \Gamma - H^{-1}$ is

$$\frac{d\tilde{\theta}_{\text{av}}}{dt} = -K\tilde{\theta}_{\text{av}} - K\underbrace{\tilde{\Gamma}_{\text{av}} H \tilde{\theta}_{\text{av}}}_{\text{quadratic}}, \tag{1.15}$$

$$\frac{d\tilde{\Gamma}_{\text{av}}}{dt} = -\omega_r\tilde{\Gamma}_{\text{av}} - \omega_r\underbrace{\tilde{\Gamma}_{\text{av}} H \tilde{\Gamma}_{\text{av}}}_{\text{quadratic}}. \tag{1.16}$$

Since the eigenvalues of K and $\omega_r > 0$ are independent of the unknown H, the (local) convergence rate is user-assignable.

1.4 ▪ Maximizing Map Sensitivity and Higher Derivatives via Newton Extremum Seeking

There is a common thread amongst the existing ES literature. The unknown input-to-output map available for (noisy) measurement is almost always the unknown objective function meant to be optimized. In [229] and [230], Vinther et al. expressed that a desirable operating point of their refrigeration system can be found in a non-model-based manner.

However, unlike typical ES problems, the function available for measurement has sigmoid-like properties and the desired operating point is the maximum slope. This is our motivating example to generalize the ES problem further: maximization (minimization) of an unknown map's higher derivatives through measurements of the map. For those interested in system identification, maximizing the sensitivity of an input-to-output function can be useful in improving the signal-to-noise ratio. Minimizing parameter sensitivity can be useful when designing feed-forward controllers. This approach may be useful in applications where the quantity of interest is not available for direct measurement, but some other quantity related through an antiderivative relationship is available, e.g., biological systems and materials properties. As long as sufficient knowledge of the output to objective function relationship is known, the ES scheme can still estimate the required derivatives of the objective function.

In [151], the emphasis for maximizing higher derivatives is on choosing proper demodulation signals for a fixed choice of perturbation signal. Through careful selection of the perturbation and demodulation signals, the same average system can be obtained with stochastic or deterministic methods, where the stability of the average system needs only be addressed once.

Without loss of generality, let us consider the maximization of nth derivative of the output

$$y(t) = h(\theta(t)), \tag{1.17}$$

using Newton-based ES, where the maximizing value of θ is denoted by θ^*. We state our optimization problem as follows:

$$\max_{\theta\in\mathbb{R}} h^{(n)}(\theta), \tag{1.18}$$

with nonlinear map $h(\cdot)$ satisfying the next assumption.

Assumption 1.1. *Let $h^{(n)}(\cdot)$ be the nth derivative of a smooth function $h(\cdot)$: $\mathbb{R}\to\mathbb{R}$. Let us also define*

$$\Theta_{max} = \{\theta | h^{(n+1)}(\theta) = 0, \quad h^{(n+2)}(\theta) < 0\} \tag{1.19}$$

to be a collection of maxima where $h^{(n)}$ is locally concave. We assume that $\Theta_{max} \neq \emptyset$ and $\exists\theta^* \in \Theta_{max}$.

As in the previous sections, $\hat{\theta}$ is the estimate of the optimizer, and

$$\tilde{\theta}(t) = \hat{\theta}(t) - \theta^* \tag{1.20}$$

is the estimation error. Analogously, the error dynamics and the control law can be written as

$$\dot{\tilde{\theta}}(t) = \dot{\hat{\theta}}(t) := U(t), \quad U(t) = -\gamma(t)\widehat{h^{(n+1)}}(t), \tag{1.21}$$

with $\gamma(t)$ and $\widehat{h^{(n+1)}}(t)$ defined in what follows. We have

$$\dot{\gamma} = k_R\gamma(1-\gamma\widehat{h^{(n+2)}}), \quad k_R > 0, \tag{1.22}$$

where (1.22) is a differential Riccati equation. Equation (1.22) will be used again to generate an estimate of the Hessian's inverse [73] according to the following error transformation:

$$\tilde{\gamma}(t) = \gamma(t) - \frac{1}{h^{(n+2)}(\theta^*)}. \tag{1.23}$$

Rearranging the equations given in [151], we can write

$$\theta(t) = \hat{\theta}(t) + a\sin(\omega t), \tag{1.24}$$

$$\Upsilon_j(t) = C_j \sin\left(j\omega t + \frac{\pi}{4}(1+(-1)^j)\right), \tag{1.25}$$

$$C_j = \frac{2^j j!}{a^j}(-1)^F, \tag{1.26}$$

$$F = \frac{j - \left|\sin\left(\frac{j\pi}{2}\right)\right|}{2}. \tag{1.27}$$

Then, we define the additive dither signal as in the classical approach

$$S(t) = a\sin(\omega t). \tag{1.28}$$

However, the demodulation signals $M(t)$ and $N(t)$ are replaced by $\Upsilon_{(n+1)}(t)$ and $\Upsilon_{(n+2)}(t)$, respectively, so that the estimates

$$\widehat{h^{(n+1)}}(t) = \Upsilon_{(n+1)}(t)y(t), \tag{1.29}$$

$$\widehat{h^{(n+2)}}(t) = \Upsilon_{(n+2)}(t)y(t) \tag{1.30}$$

can be obtained [151]. Basically, $\widehat{h^{(j)}} = y\Upsilon_j$ are used to estimate the delayed gradient ($j = n+1$) and the Hessian ($j = n+2$) of $h^{(n)}$. For $n = 0$, the demodulation signals $\Upsilon_1(t) = M(t) = \frac{2}{a}\sin(\omega t)$ and $\Upsilon_2(t) = N(t) = -\frac{8}{a^2}\cos(2\omega t)$ are simply the scalar versions of (1.8) and (1.13), respectively.

1.5 ▪ Extremum Seeking for Dynamic Systems

ES extends in a relatively straightforward manner from static maps to dynamic systems, provided the dynamics are stable and the algorithm's parameters are chosen so that the algorithm's dynamics are slower than those of the plant.

1.5.1 ▪ Gradient Algorithm for Dynamic Systems

The gradient-based ES algorithm is shown in Figure 1.4. If the dynamics are stable and the user employs parameters in the ES algorithm which make the algorithm dynamics slower than the dynamics of the plant [129], the parametric convergence is guaranteed (at least locally). The high-pass filter $s/(s+\omega_h)$ and the low-pass filter $\omega_l/(s+\omega_l)$, with cutoff frequencies $\omega_h > 0$ and $\omega_l > 0$, are useful in the implementation to reduce the adverse effect of the perturbation signals on the asymptotic performance but they are not needed in the stability analysis.

The technical conditions for convergence in the presence of dynamics are that the equilibria $x = l(\theta)$ of the system

$$\dot{x} = f(x, \alpha(x,\theta)), \tag{1.31}$$

where $\alpha(x,\theta)$ is the control law of an internal feedback loop, are locally exponentially stable uniformly in θ and that, given the output map $y = h(x)$, there exists at least one $\theta^* \in \mathbb{R}^n$ such that

$$\frac{\partial}{\partial\theta}(h\circ l)(\theta^*) = 0 \quad \text{and} \quad \frac{\partial^2}{\partial\theta^2}(h\circ l)(\theta^*) = H < 0, \tag{1.32}$$

with $H = H^T$. Without loss of generality, it was assumed the maximization problem ($H < 0$).

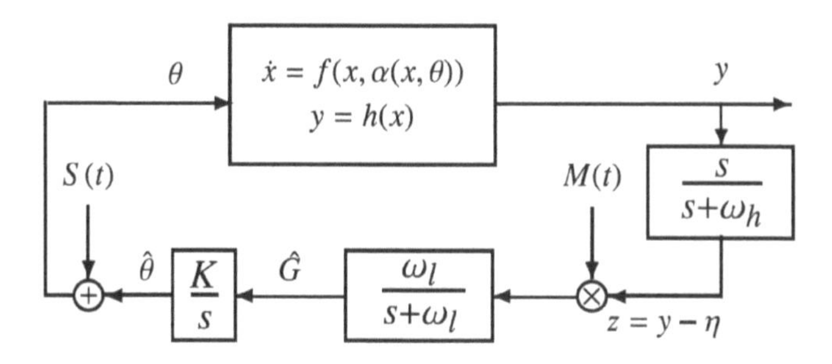

Figure 1.4. *The gradient-based ES algorithm in the presence of dynamics with an equilibrium map $\theta \to y$ that satisfies the same conditions as in the static case.*

The stability analysis in the presence of dynamics employs both averaging and singular perturbations, in a specific order [129]. The design guidelines for the selection of the algorithm's parameters follow the analysis. The guidelines ensure that the plant's dynamics are on a fast time scale, the perturbations are on a medium time scale, and the ES algorithm is on a slow time scale. In order to avoid unnecessary repetitions, we will not show the complete sequence of steps involving averaging and singular perturbations for the gradient-based ES scheme, but we will give the full derivations in the next section for the Newton-based ES algorithm.

1.5.2 ▪ Newton Algorithm for Dynamic Systems

Consider a general multi-input-single-output (MISO) nonlinear model

$$\dot{x} = f(x,u), \tag{1.33}$$

$$y = h(x), \tag{1.34}$$

where $x \in \mathbb{R}^m$ is the state, $u \in \mathbb{R}^n$ is the input, $y \in \mathbb{R}$ is the output, and $f : \mathbb{R}^m \times \mathbb{R}^n \to \mathbb{R}^m$ and $h : \mathbb{R}^m \to \mathbb{R}$ are both smooth (continuous and differentiable). Suppose that we know a smooth control law $u = \alpha(x,\theta)$ parameterized by a vector parameter $\theta \in \mathbb{R}^n$. The closed-loop system $\dot{x} = f(x,\alpha(x,\theta))$ then has equilibria parameterized by θ. We make the following assumptions about the closed-loop system, as in [129].

Assumption 1.2. *There exists a smooth function $l : \mathbb{R}^n \to \mathbb{R}^m$ such that $f(x,\alpha(x,\theta)) = 0$ if and only if $x = l(\theta)$.*

Assumption 1.3. *For each $\theta \in \mathbb{R}^n$, the equilibrium $x = l(\theta)$ of the system $\dot{x} = f(x,\alpha(x,\theta))$ is locally exponentially stable with decay and overshoot constants uniform in θ.*

Assumption 1.4. *There exists $\theta^* \in \mathbb{R}^n$ such that*

$$\frac{\partial}{\partial\theta}(h \circ l)(\theta^*) = 0, \tag{1.35}$$

$$\frac{\partial^2}{\partial\theta^2}(h \circ l)(\theta^*) = H < 0, \quad H = H^T. \tag{1.36}$$

Our objective is to develop a feedback mechanism which maximizes the steady-state value of y but without requiring the knowledge of either θ^* or the functions h and l. The gradient-based ES design that achieves this objective, suitably adapted from [129] to the multivariable

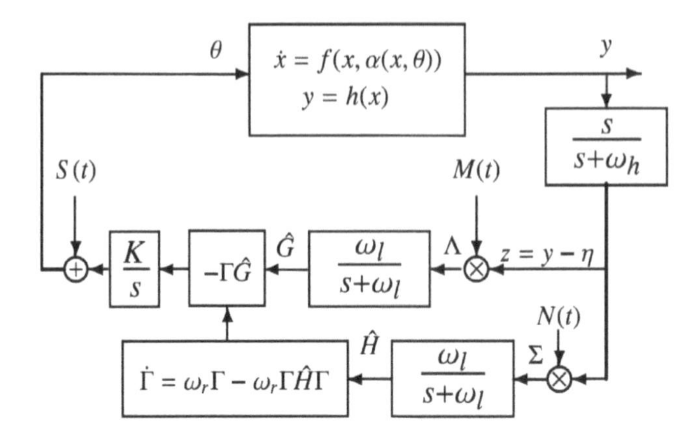

Figure 1.5. *The Newton-based ES algorithm in the presence of dynamics with an equilibrium map $\theta \to y$ that satisfies the same conditions as in the static case.*

case, is shown schematically in Figure 1.4. Parallel to this, we present the generalized scheme for multivariable Newton-based ES as shown in Figure 1.5.

The perturbation signals are defined by (1.7), (1.8), (1.13), and (1.14). The probing frequencies ω_i's, the filter coefficients ω_h, ω_l, and ω_r, and the adaptation gain K are selected as

$$\omega_i = \omega\omega_i' = \mathcal{O}(\omega), \quad i \in \{1,2,\ldots,n\}, \tag{1.37}$$

$$\omega_h = \omega\omega_H = \omega\delta\omega_H' = \mathcal{O}(\omega\delta), \tag{1.38}$$

$$\omega_l = \omega\omega_L = \omega\delta\omega_L' = \mathcal{O}(\omega\delta), \tag{1.39}$$

$$\omega_r = \omega\omega_R = \omega\delta\omega_R' = \mathcal{O}(\omega\delta), \tag{1.40}$$

$$K = \omega K' = \omega\delta K'' = \mathcal{O}(\omega\delta), \tag{1.41}$$

where ω and δ are small positive constants, ω_i' is a rational number, ω_H', ω_L', and ω_R' are $\mathcal{O}(1)$ positive constants, K'' is an $n \times n$ diagonal matrix with $\mathcal{O}(1)$ positive elements, and $K' = \delta K''$.

The analysis of [129, 10] shows that, in the gradient-based scheme, for "sufficiently small" ω and $|a|$, where $a = [a_1\, a_2\, \cdots\, a_n]^T$, and sufficiently small δ, which imply small filter cutoff frequencies ω_h and ω_l, the states $(x, \hat{\theta})$ of the closed-loop system exponentially converge to an $\mathcal{O}(\omega + \delta + |a|)$-neighborhood of $(l(\theta^*), \theta^*)$ and the output y converges to an $\mathcal{O}(\omega + \delta + |a|)$-neighborhood of the optimum output $y^* = (h \circ l)(\theta^*)$.

In Section 1.5.4 we prove that the average value of the signal $\sum(t)$ (in Figure 1.5) over period

$$\Pi := 2\pi \times \text{LCM}\{1/\omega_i\} \tag{1.42}$$

is close enough to the actual value of the Hessian, under specific conditions on ω, δ, and a, where LCM stands for the least common multiple. Since we are integrating over a finite time period, and we set a priori the phase of the periodic perturbation signals equal to zero, it is possible to exclude condition $\omega_i \neq \omega_j + \omega_k$. The probing frequencies need to satisfy

$$\omega_i' \notin \left\{\omega_j', \frac{1}{2}(\omega_j' + \omega_k'), \omega_j' + 2\omega_k', \omega_j' + \omega_k' \pm \omega_l'\right\} \tag{1.43}$$

for all distinct i, j, k, and l. As we will see in Section 1.5.4, ignoring the conditions (1.43) is shifting the estimate of the parameter away from its true value and leading to inaccurate estimates of the gradient vector and Hessian matrix.

1.5.3 ▪ Stability of the Closed-Loop System with the Newton-Based Extremum Seeking Algorithm

We summarize the closed-loop system in Figure 1.5 as

$$\frac{d}{dt}\begin{bmatrix} x \\ \tilde{\theta} \\ \hat{G} \\ \tilde{\Gamma} \\ \tilde{H} \\ \tilde{\eta} \end{bmatrix} = \begin{bmatrix} f(x,\alpha(x,\theta^*+\tilde{\theta}+S(t))) \\ -K(\tilde{\Gamma}+H^{-1})\hat{G} \\ -\omega_l\hat{G}+\omega_l(y-h\circ l(\theta^*)-\tilde{\eta})M(t) \\ \omega_r(\tilde{\Gamma}+H^{-1})(I-(\tilde{H}+H)(\tilde{\Gamma}+H^{-1})) \\ -\omega_l\tilde{H}-\omega_l H+\omega_l(y-h\circ l(\theta^*)-\tilde{\eta})N(t) \\ -\omega_h\tilde{\eta}+\omega_h(y-h\circ l(\theta^*)) \end{bmatrix}, \tag{1.44}$$

where I is the identity matrix with appropriate dimensions.

To conduct a stability analysis we have introduced error variables $\tilde{\theta}=\hat{\theta}-\theta^*$, $\theta=\hat{\theta}+S(t)$, $\tilde{\eta}=\eta-h\circ l(\theta^*)$, $\tilde{\Gamma}=\Gamma-H^{-1}$, and $\tilde{H}=\hat{H}-H$, where η is governed by

$$\dot{\eta}=-\omega_h\eta+\omega_h y. \tag{1.45}$$

We perform a slight abuse of notation by stacking matrix quantities $\tilde{\Gamma}$ and $\tilde{H}$ along with vector quantities, as alternative notational choices would be more cumbersome.

Our main stability result is stated in the following theorem.

Theorem 1.1. *Consider the feedback system* (1.44) *under Assumptions* 1.2 *to* 1.4. *There exists* $\overline{\omega}>0$ *and for any* $\omega\in(0,\overline{\omega})$ *there exist* $\overline{\delta}$, $\overline{a}>0$ *such that for the given* ω *and any* $|a|\in(0,\overline{a})$ *and* $\delta\in(0,\overline{\delta})$ *there exists a neighborhood of the point* $(x,\hat{\theta},\hat{G},\Gamma,\hat{H},\eta)=(l(\theta^*),\theta^*,0,H^{-1},H,h\circ l(\theta^*))$ *such that any solution of the system* (1.44) *initialized on this neighborhood, exponentially converges to an* $\mathcal{O}(\omega+\delta+|a|)$*-neighborhood of that point. Furthermore,* $y(t)$ *converges to an* $\mathcal{O}(\omega+\delta+|a|)$*-neighborhood of* $h\circ l(\theta^*)$.

To prepare for the proof of Theorem 1.1, which is given in Sections 1.5.4 and 1.5.5, we summarize the system (1.44) in the time scale $\tau=\omega t$ as

$$\omega\frac{dx}{d\tau}=f(x,\alpha(x,\theta^*+\tilde{\theta}+\overline{S}(\tau))), \tag{1.46}$$

$$\frac{d}{d\tau}\begin{bmatrix} \tilde{\theta} \\ \hat{G} \\ \tilde{\Gamma} \\ \tilde{H} \\ \tilde{\eta} \end{bmatrix} = \delta\begin{bmatrix} -K''(\tilde{\Gamma}+H^{-1})\hat{G} \\ -\omega_L'\hat{G}+\omega_L'(y-h\circ l(\theta^*)-\tilde{\eta})\overline{M}(\tau) \\ \omega_R'(\tilde{\Gamma}+H^{-1})(I-(\tilde{H}+H)(\tilde{\Gamma}+H^{-1})) \\ -\omega_L'(\tilde{H}+H)+\omega_L'(y-h\circ l(\theta^*)-\tilde{\eta})\overline{N}(\tau) \\ -\omega_H'\tilde{\eta}+\omega_H'(y-h\circ l(\theta^*)) \end{bmatrix}, \tag{1.47}$$

where $\overline{S}(\tau)=S(t/\omega)$, $\overline{M}(\tau)=M(t/\omega)$, and $\overline{N}(\tau)=N(t/\omega)$.

1.5.4 ▪ Averaging Analysis

The first step in our analysis is to study the system in Figure 1.5. We "freeze" x in (1.46) at its equilibrium value $x=l(\theta^*+\tilde{\theta}+\overline{S}(\tau))$ and substitute it into (1.47), getting the *reduced system*

$$\frac{d}{d\tau}\begin{bmatrix} \tilde{\theta}_r \\ \hat{G}_r \\ \tilde{\Gamma}_r \\ \tilde{H}_r \\ \tilde{\eta}_r \end{bmatrix} = \delta\begin{bmatrix} -K''(\tilde{\Gamma}_r+H^{-1})\hat{G}_r \\ -\omega_L'\hat{G}_r+\omega_L'(v(\tilde{\theta}_r+\overline{S}(\tau))-\tilde{\eta}_r)\overline{M}(\tau) \\ \omega_R'(\tilde{\Gamma}_r+H^{-1})(I-(\tilde{H}_r+H)(\tilde{\Gamma}_r+H^{-1})) \\ -\omega_L'\tilde{H}_r-\omega_L'H+\omega_L'(v(\tilde{\theta}_r+\overline{S}(\tau))-\tilde{\eta}_r)\overline{N}(\tau) \\ -\omega_H'\tilde{\eta}_r+\omega_H'v(\tilde{\theta}_r+\overline{S}(\tau)) \end{bmatrix}, \tag{1.48}$$

where $v(z) = h \circ (\theta^* + z) - h \circ l(\theta^*)$. In view of Assumption 1.4, $v(0) = 0$, $\partial v(0)/\partial z = 0$, and $\partial^2 v(0)/\partial z^2 = H < 0$.

To prove the overall stability of (1.44), first we show that the reduced system (1.48) has a unique exponentially stable periodic solution around its equilibrium.

Theorem 1.2. *Consider system* (1.48) *under Assumption* 1.4. *There exist* $\overline{\delta},\ \overline{a} > 0$ *such that for all* $\delta \in (0,\ \overline{\delta})$ *and* $|a| \in (0, \overline{a})$ *system* (1.48) *has a unique exponentially stable periodic solution* $(\tilde{\theta}_r^\Pi(\tau),\ \hat{G}_r^\Pi(\tau), \tilde{\Gamma}_r^\Pi(\tau),\ \tilde{H}_r^\Pi(\tau),\ \tilde{\eta}_r^\Pi(\tau))$ *of period* Π *defined in* (1.42) *and this solution satisfies*

$$\left|\tilde{\theta}_{r,i}^{\Pi}(\tau) - \sum_{j=1}^{n} c_{jj}^i a_j^2\right| \leq \mathcal{O}(\delta + |a|^3), \tag{1.49}$$

$$\left|\hat{G}_r^\Pi(\tau)\right| \leq \mathcal{O}(\delta), \tag{1.50}$$

$$\left|\tilde{\Gamma}_r^\Pi(\tau) + \sum_{i=1}^{n}\sum_{j=1}^{n} H^{-1} W^i H^{-1} c_{jj}^i a_j^2\right| \leq \mathcal{O}(\delta + |a|^3), \tag{1.51}$$

$$\left|\tilde{H}_r^\Pi(\tau) - \sum_{i=1}^{n}\sum_{j=1}^{n} W^i c_{jj}^i a_j^2\right| \leq \mathcal{O}(\delta + |a|^3), \tag{1.52}$$

$$\left|\tilde{\eta}_r^\Pi(\tau) - \frac{1}{4}\sum_{i=1}^{n} H_{ii} a_i^2\right| \leq \mathcal{O}(\delta + |a|^4) \tag{1.53}$$

for all $\tau \geq 0$, *where*

$$\begin{bmatrix} c_{jj}^1 \\ \vdots \\ c_{jj}^{i-1} \\ c_{jj}^i \\ c_{jj}^{i+1} \\ \vdots \\ c_{jj}^n \end{bmatrix} = -\frac{1}{12} H^{-1} \begin{bmatrix} \frac{\partial^3 v}{\partial z_j \partial z_1^2}(0) \\ \vdots \\ \frac{\partial^3 v}{\partial z_j \partial z_{j-1}^2}(0) \\ \frac{3}{2}\frac{\partial^3 v}{\partial z_j^3}(0) \\ \frac{\partial^3 v}{\partial z_j \partial z_{j+1}^2}(0) \\ \vdots \\ \frac{\partial^3 v}{\partial z_j \partial z_n^2}(0) \end{bmatrix} \qquad \forall i,j \in \{1,2,\ldots,n\}, \tag{1.54}$$

$$(W^i)_{jk} = \frac{\partial^3 v(0)}{\partial z_i \partial z_j \partial z_k} \qquad \forall i,j,k \in \{1,2,\ldots,n\}. \tag{1.55}$$

The proof of Theorem 1.2 is presented in [73].

1.5.5 ▪ Singular Perturbation Analysis

Now, we address the full system in Figure 1.5 whose state-space model is given by (1.46) and (1.47) in the time scale $\tau = \omega t$. To make the notation in our further analysis compact, we write (1.47) as

$$\frac{d\xi}{d\tau} = \delta E(\tau,\ x,\ \xi), \tag{1.56}$$

where $\xi = (\tilde{\theta}, \hat{G}, \tilde{\Gamma}, \tilde{H}, \tilde{\eta})$. By Theorem 1.2, there exists an exponentially stable periodic solution $\xi_r^{\Pi}(\tau)$ such that

$$\frac{d\xi_r^{\Pi}(\tau)}{d\tau} = \delta E(\tau, L(\tau, \xi_r^{\Pi}(\tau)), \xi_r^{\Pi}(\tau)), \tag{1.57}$$

where $L(\tau, \xi) = l(\theta^* + \tilde{\theta} + \overline{S}(\tau))$. To bring the system (1.46) and (1.56) into the standard singular perturbation form, we shift the state ξ using the transformation $\tilde{\xi} = \xi - \xi_r^{\Pi}(\tau)$ and get

$$\frac{d\tilde{\xi}}{d\tau} = \delta \tilde{E}(\tau, x, \tilde{\xi}), \tag{1.58}$$

$$\omega \frac{dx}{d\tau} = \tilde{F}(\tau, x, \tilde{\xi}), \tag{1.59}$$

where

$$\tilde{E}(\tau, x, \tilde{\xi}) = E(\tau, x, \tilde{\xi} + \xi_r^{\Pi}(\tau)) - E(\tau, L(\tau, \xi_r^{\Pi}(\tau)), \xi_r^{\Pi}(\tau)), \tag{1.60}$$

$$\tilde{F}(\tau, x, \tilde{\xi}) = f\left(x, \alpha(x, \tilde{\xi}_1 + \theta^* + \tilde{\theta}_r^{\Pi}(\tau) + \overline{S}(\tau))\right). \tag{1.61}$$

We note that $x = L(\tau, \tilde{\xi}_r + \tilde{\xi}_r^{\Pi}(\tau))$ is the quasi-steady state, and that the reduced model

$$\frac{d\tilde{\xi}_r}{d\tau} = \delta \tilde{E}(\tau, L(\tau, \tilde{\xi}_r + \xi_r^{\Pi}(\tau)), \tilde{\xi}_r + \xi_r^{\Pi}(\tau)) \tag{1.62}$$

has an equilibrium at the origin $\tilde{\xi}_r = 0$. This equilibrium has been shown in Section 1.5.4 to be exponentially stable for a small $|a|$.

To complete the singular perturbation analysis, we also study the *boundary layer model* (in the time scale $t - t_0 = \tau/\omega$):

$$\begin{aligned} \frac{dx_b}{d\tau} &= \tilde{F}(\tau, x_b + L(\tau, \tilde{\xi} + \xi_r^{\Pi}(\tau)), \tilde{\xi}) \\ &= f(x_b + l(\theta), \alpha(x_b + l(\theta), \theta)), \end{aligned} \tag{1.63}$$

where $\theta = \theta^* + \tilde{\theta} + \overline{S}(\tau)$ should be viewed as a parameter independent from the time variable t. Since $f(l(\theta), \alpha(l(\theta), \theta)) \equiv 0$, then $x_b \equiv 0$ is an equilibrium of (1.63). By Assumption 1.3, this equilibrium is locally exponentially stable uniformly in θ (and hence $l(\theta)$).

By combining exponential stability of the reduced model (1.62) with the exponential stability of the boundary layer model (1.63), using Tikhonov's theorem on the Infinite Interval (Theorem 9.4 in [108]), we conclude the following:

(a) The solution $\xi(\tau)$ of (1.56) is $\mathcal{O}(\omega)$-close to the solution $\xi_r(\tau)$ of (1.62), and therefore it exponentially converges to an $\mathcal{O}(\omega)$-neighborhood of the periodic solution $\xi_r^{\Pi}(\tau)$, which is $\mathcal{O}(\delta)$-close to the equilibrium $\xi_r^{a,e}$. This, in turn, implies that the solution $\tilde{\theta}(\tau)$ of (1.47) exponentially converges to an $\mathcal{O}(\omega + \delta)$-neighborhood of

$$\sum_{j=1}^{n} \left[c_{jj}^1 \, c_{jj}^2 \cdots c_{jj}^n \right]^T a_j^2 + \left[\mathcal{O}(|a|^3) \right]_{n \times 1}. \tag{1.64}$$

It follows then that $\theta(\tau) = \theta^* + \tilde{\theta}(\tau) + \overline{S}(\tau)$ exponentially converges to an $\mathcal{O}(\omega + \delta + |a|)$-neighborhood of θ^*.

(b) The solution $x(\tau)$ of (1.59) satisfies

$$x(\tau) - l(\theta^* + \tilde{\theta}_r(\tau) + \overline{S}(\tau)) - x_b(t) = \mathcal{O}(\omega), \tag{1.65}$$

where $\tilde{\theta}_r(\tau)$ is the solution of the reduced model (1.48) and $x_b(t)$ is the solution of the boundary layer model (1.63).

From (1.65), we get

$$x(\tau) - l(\theta^*) = \mathcal{O}(\omega) + l(\theta^* + \tilde{\theta}_r(\tau) + \overline{S}(t)) - l(\theta^*) + x_b(t). \tag{1.66}$$

Since $\tilde{\theta}_r(\tau)$ exponentially converges to the periodic solution $\tilde{\theta}_r^{\Pi}(\tau)$, which is $\mathcal{O}(\delta)$-close to the average equilibrium (1.64), and since the solution $x_b(t)$ of (1.63) is exponentially decaying, then by (1.66), $x(\tau) - l(\theta^*)$ exponentially converges to an $\mathcal{O}(\omega + \delta + |a|)$-neighborhood of zero. Consequently, $y = h(x)$ exponentially converges to an $\mathcal{O}(\omega + \delta + |a|)$-neighborhood of its maximal equilibrium value $h \circ l(\theta^*)$.

This completes the proof of Theorem 1.1.

1.6 ▪ Nash Equilibrium Seeking in Noncooperative Games

To introduce the Nash seeking algorithm originally presented in [70], we first consider a specific two-player noncooperative game which, for example, may represent two firms competing for profit in a duopoly market structure. Common duopoly examples include the soft drink companies, Coca-Cola and Pepsi, and the commercial aircraft companies, Boeing and Airbus. We present a duopoly price game in this section for motivational purposes before proving convergence to the Nash equilibrium when N players with quadratic payoff functions employ the Nash seeking strategy in Section 1.6.2.

1.6.1 ▪ Two-Player Game

Let Players P1 and P2 represent two firms that produce the same good, have dominant control over a market, and compete for profit by setting their prices u_1 and u_2, respectively. The profit of each firm is the product of the number of units sold and the profit per unit, which is the difference between the sale price and the marginal or manufacturing cost of the product. In mathematical terms, the profits are modeled by

$$J_i(t) = s_i(t)\big(u_i(t) - m_i\big), \tag{1.67}$$

where s_i is the number of sales, m_i the marginal cost, and $i \in \{1,2\}$ for P1 and P2. Intuitively, the profit of each firm will be low if it either sets the price very low, since the profit per unit sold will be low, or if it sets the price too high, since then consumers will buy the other firm's product. The maximum profit is to be expected to lie somewhere in the middle of the price range, and it crucially depends on the price level set by the other firm.

To model the market behavior, we assume a simple, but quite realistic model, where for whatever reason, the consumer prefers the product of P1, but is willing to buy the product of P2 if its price u_2 is sufficiently lower than the price u_1. Hence, we model the sales for each firm as

$$s_1(t) = S_d - s_2(t), \tag{1.68}$$

$$s_2(t) = \frac{1}{p}\big(u_1(t) - u_2(t)\big), \tag{1.69}$$

where the total consumer demand S_d is held fixed for simplicity, the preference of the consumer for P1 is quantified by $p > 0$, and the inequalities $u_1 > u_2$ and $(u_1 - u_2)/p < S_d$ are assumed to hold.

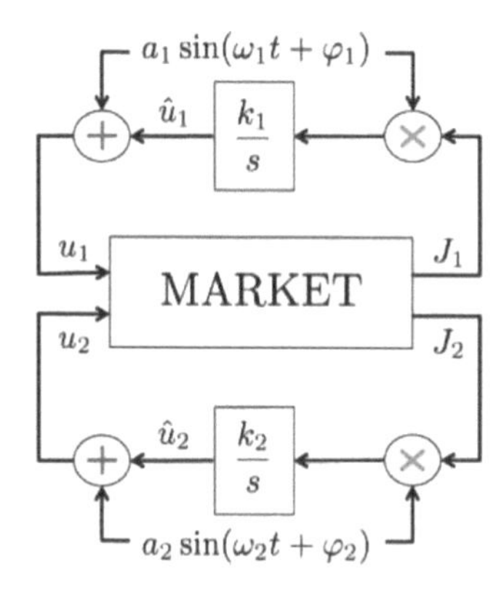

Figure 1.6. *Deterministic Nash seeking schemes applied by players in a duopoly market structure.*

Substituting (1.68) and (1.69) into (1.67) yields expressions for the profits $J_1(u_1, u_2)$ and $J_2(u_1, u_2)$ that are both quadratic functions of the prices u_1 and u_2, namely,

$$J_1 = \frac{-u_1^2 + u_1 u_2 + (m_1 + S_d p)u_1 - m_1 u_2 - S_d p m_1}{p}, \tag{1.70}$$

$$J_2 = \frac{-u_2^2 + u_1 u_2 + m_2 u_1 + m_2 u_2}{p}, \tag{1.71}$$

and thus the Nash equilibrium is easily determined to be

$$u_1^* = \frac{1}{3}(2m_1 + m_2 + 2S_d p), \tag{1.72}$$

$$u_2^* = \frac{1}{3}(m_1 + 2m_2 + S_d p). \tag{1.73}$$

To make sure the constraints $u_1 > u_2$, $(u_1 - u_2)/p < S_d$ are satisfied by the Nash equilibrium, we assume that $m_1 - m_2$ lies in the interval $(-S_d p, 2S_d p)$. If $m_1 = m_2$, this condition is automatically satisfied.

For completeness, we provide here the definition of a Nash equilibrium $u^* = [u_1^*, \ldots, u_N^*]^T$ in an N-player game:

$$J_i(u_i^*, u_{-i}^*) \geq J_i(u_i, u_{-i}^*) \qquad \forall u_i \in U_i,\ i \in \{1, \ldots, N\}, \tag{1.74}$$

where J_i is the payoff function of Player i, u_i is its action, U_i is its action set, and u_{-i} denotes the actions of the other players. Hence, no player has an incentive to unilaterally deviate its action from u^*. In the duopoly example, $U_1 = U_2 = \mathbb{R}_+$, where $\mathbb{R}_+$ denotes the set of positive real numbers.

To attain the Nash strategies (1.72)–(1.73) without any knowledge of modeling information, such as the consumer's preference p, the total demand S_d, or the other firm's marginal cost or price, the firms implement a non-model based real-time optimization strategy, e.g., deterministic extremum seeking with sinusoidal perturbations, to set their price levels. Specifically, P1 and P2 set their prices, u_1 and u_2 respectively, according to the time-varying strategy (Figure 1.6):

$$\dot{\hat{u}}_i(t) = k_i \mu_i(t) J_i(t), \tag{1.75}$$

$$u_i(t) = \hat{u}_i(t) + \mu_i(t), \tag{1.76}$$

where $\mu_i(t) = a_i \sin(\omega_i t + \varphi_i)$, k_i, a_i, $\omega_i > 0$, and $i \in \{1,2\}$. Further, the frequencies are of the form

$$\omega_i = \omega \overline{\omega}_i, \tag{1.77}$$

where ω is a positive real number and $\overline{\omega}_i$ is a positive rational number. This form is convenient for the convergence analysis performed in Section 1.6.2.

In contrast, the firms are also guaranteed to converge to the Nash equilibrium when employing the standard parallel action update scheme [21, Proposition 4.1]

$$u_1^{(k+1)} = \frac{1}{2}\left(u_2^{(k)} + m_1 + S_d p\right), \tag{1.78}$$

$$u_2^{(k+1)} = \frac{1}{2}\left(u_1^{(k)} + m_2\right), \tag{1.79}$$

which requires each firm to know both its own marginal cost and the other firm's price at the previous step of the iteration, and also requires P1 to know the total demand S_d and the consumer preference parameter p. In essence, P1 must know nearly all the relevant modeling information. When using the extremum seeking algorithm (1.75)–(1.76), the firms only need to measure the value of their own payoff functions, J_1 and J_2. Convergence of (1.78)–(1.79) is global, whereas the convergence of the Nash seeking strategy for this example can be proved to be semi-global, following [217], or locally, by applying the theory of averaging [108]. We establish local results in this section since we consider non-quadratic payoff functions in Section 1.6.3. We do, however, state a nonlocal result for static games with quadratic payoff functions using the theory found in [222]. For detailed analysis of the nonlocal convergence of extremum seeking controllers applied to general convex systems, the reader is referred to [217].

1.6.2 ▪ N-Player Games with Quadratic Payoff Functions

We now generalize the duopoly example in Section 1.6.1 to static noncooperative games with N players that wish to maximize their quadratic payoff functions. We prove convergence to a neighborhood of the Nash equilibrium when the players employ the Nash seeking strategy (1.75)–(1.76).

First, we consider games with general quadratic payoff functions. Specifically, the payoff function of player i is of the form

$$J_i(t) = \frac{1}{2}\sum_{j=1}^{N}\sum_{k=1}^{N} D^i_{jk} u_j(t) u_k(t) + \sum_{j=1}^{N} d^i_j u_j(t) + c^i, \tag{1.80}$$

where the action of Player j is $u_j \in U_j = \mathbb{R}$, D^i_{jk}, d^i_j and c^i are constants, $D^i_{ii} < 0$, and $D^i_{jk} = D^i_{kj}$. Quadratic games of this form are studied in [21, Section 4.6], where Proposition 4.6 states that the N-player game with payoff functions (1.80) admits a Nash equilibrium $u^* = [u_1^*, \ldots, u_N^*]^T$ if and only if

$$D^i_{ii} u_i^* + \sum_{j \neq i} D^i_{ij} u_j^* + d^i_i = 0, \qquad i \in \{1, \ldots, N\}, \tag{1.81}$$

admits a solution. Rewritten in matrix form, we have $Du^* = -d$, where

$$D \triangleq \begin{bmatrix} D^1_{11} & D^1_{12} & \cdots & D^1_{1N} \\ D^2_{21} & D^2_{22} & & \\ \vdots & & \ddots & \\ D^N_{N1} & & & D^N_{NN} \end{bmatrix}, \qquad d = \begin{bmatrix} d^1_1 \\ d^2_2 \\ \vdots \\ d^N_N \end{bmatrix} \tag{1.82}$$

and u^* is unique if D is invertible. We make the following stronger assumption concerning this matrix.

Assumption 1.5. *The matrix D given by* (1.82) *is strictly diagonally dominant, i.e.,*

$$\sum_{j\neq i}^{N} \left|D^i_{ij}\right| < \left|D^i_{ii}\right|, \qquad i \in \{1,\dots,N\}. \tag{1.83}$$

By Assumption 1.5, the Nash equilibrium u^* exists and is unique since strictly diagonally dominant matrices are nonsingular by the Levy–Desplanques theorem [221], [90]. To attain u^* stably in real time, without any modeling information, each Player i employs the extremum seeking strategy (1.75)–(1.76).

Theorem 1.3. *Consider the system* (1.75)–(1.76) *with* (1.80) *under Assumption* 1.5 *for an N-player game, where $\omega_i \neq \omega_j$, $2\omega_i \neq \omega_j$, and $\omega_i \neq \omega_j + \omega_k$ for all distinct i, j, $k \in \{1,\dots,N\}$, and where ω_i/ω_j is rational for all i, $j \in \{1,\dots,N\}$. There exist ω^*, $M, m > 0$ such that for all $\omega > \omega^*$, if $\left|\Delta(0)\right|$ is sufficiently small, then for all $t \geq 0$,*

$$\left|\Delta(t)\right| \leq Me^{-mt}\left|\Delta(0)\right| + \mathcal{O}\left(\frac{1}{\omega} + \max_i a_i\right), \tag{1.84}$$

where $\Delta(t) = [u_1(t) - u_1^*, \dots, u_N(t) - u_N^*]^T$ and $|\cdot|$ denotes the Euclidean norm.

Proof. Denote the relative Nash equilibrium error as

$$\begin{aligned} \tilde{u}_i(t) &= u_i(t) - \mu_i(t) - u_i^* \\ &= \hat{u}_i(t) - u_i^*. \end{aligned} \tag{1.85}$$

By substituting (1.80) into (1.75)–(1.76), we get the error system

$$\begin{aligned} \dot{\tilde{u}}_i(t) = k_i\mu_i(t)\Bigg(&\frac{1}{2}\sum_{j=1}^{N}\sum_{k=1}^{N} D^i_{jk}\Big(\tilde{u}_j(t) + u_j^* + \mu_j(t)\Big)\Big(\tilde{u}_k(t) + u_k^* + \mu_k(t)\Big) \\ &+ \sum_{j=1}^{N} d^i_j\Big(\tilde{u}_j(t) + u_j^* + \mu_j(t)\Big) + c^i\Bigg). \end{aligned} \tag{1.86}$$

Let $\tau = \omega t$, where ω is the positive real number in (1.77). Rewriting (1.86) in the time scale τ and rearranging terms yields

$$\begin{aligned} \frac{d\tilde{u}_i(\tau)}{d\tau} = \frac{k_i\mu_i(\tau)}{2\omega} \times \Bigg[&\sum_{j=1}^{N}\sum_{k=1}^{N} D^i_{jk}(\tilde{u}_j(\tau) + u_j^*)(\tilde{u}_k(\tau) + u_k^*) \\ &+ 2\sum_{j=1}^{N}\sum_{k=1}^{N} D^i_{jk}(\tilde{u}_j(\tau) + u_j^*)\mu_k(\tau) + \sum_{j=1}^{N}\sum_{k=1}^{N} D^i_{jk}\mu_j(\tau)\mu_k(\tau) \\ &+ 2\sum_{j=1}^{N} d^i_j(\tilde{u}_j(\tau) + u_j^* + \mu_j(\tau)) + 2c^i\Bigg] \\ = \frac{1}{\omega} f_i\Big(\tau, \tilde{u}_1, \dots, \tilde{u}_N, \frac{1}{\omega}\Big), \end{aligned} \tag{1.87}$$

where $\mu_i(\tau) = a_i \sin(\overline{\omega}_i \tau + \varphi_i)$ and $\overline{\omega}_i$ is a rational number. Hence, the error system (1.87) is periodic with period

$$T := 2\pi \times \text{LCM}\{1/\overline{\omega}_1, \ldots, 1/\overline{\omega}_N\}, \tag{1.88}$$

where LCM denotes the least common multiple. With $1/\omega$ as a small parameter, (1.87) admits the application of the averaging theory [108] for stability analysis. The average error system can be shown to be

$$\begin{aligned}\frac{d\tilde{u}_i^{\text{av}}}{d\tau} &= \frac{1}{\omega}\left(\frac{1}{T}\int_0^T f_i(\tau, \tilde{u}_1^{\text{av}}, \ldots, \tilde{u}_N^{\text{av}}, 0)d\tau\right) \\ &= \frac{1}{2\omega} k_i a_i^2 \sum_{j=1}^{N} D_{ij}^i \tilde{u}_N^{\text{av}}(\tau),\end{aligned} \tag{1.89}$$

which in matrix form is $d\tilde{u}^{\text{av}}/d\tau = A\tilde{u}^{\text{av}}$, where

$$A = \frac{1}{2\omega}\begin{bmatrix} \kappa_1^1 D_{11} & \kappa_1^1 D_{12} & \cdots & \kappa_1^1 D_{1N} \\ \kappa_2^2 D_{21} & \kappa_2^2 D_{22} & & \\ \vdots & & \ddots & \\ \kappa_N D_{N1}^N & & & \kappa_N D_{NN}^N \end{bmatrix} \tag{1.90}$$

and $\kappa_i = k_i a_i^2$, $i \in \{1, \ldots, N\}$. The details of computing (1.89), which require that $\omega_i \neq \omega_j$, $2\omega_i \neq \omega_j$, and $\omega_i \neq \omega_j + \omega_k$ for all distinct $i, j, k \in \{1, \ldots, N\}$, are shown in [70, Appendix A].

From the Gershgorin Circle Theorem [90, Theorem 6.1.1], we have $\lambda(A) \subseteq \bigcup_{i=1}^{N} \rho_i$, where $\lambda(A)$ denotes the spectrum of A and ρ_i is a Gershgorin disc:

$$\rho_i = \frac{k_i a_i^2}{2\omega}\left\{ z \in \mathbb{C} \middle| \, |z - D_{ii}^i| < \sum_{j \neq i} |D_{ii}^i| \right\}. \tag{1.91}$$

Since $D_{ii}^i < 0$ and D is strictly diagonally dominant, the union of the Gershgorin discs lies strictly in the left half of the complex plane, and we conclude that $\text{Re}\{\lambda\} < 0$ for all $\lambda \in \lambda(A)$. Thus, given any matrix $Q = Q^T > 0$, there exists a matrix $P = P^T > 0$ satisfying the Lyapunov equation $PA + A^T P = -Q$.

Using $V(\tau) = (\tilde{u}^{\text{av}}(\tau))^T P \tilde{u}^{\text{av}}(\tau)$ as a Lyapunov function, we obtain

$$\dot{V} = -(\tilde{u}^{\text{av}})^T Q \tilde{u}^{\text{av}} \leq -\lambda_{\min}(Q)|\tilde{u}^{\text{av}}|^2. \tag{1.92}$$

Noting that V satisfies the bounds, $\lambda_{\min}(P)|\tilde{u}^{\text{av}}(\tau)|^2 \leq V(\tau) \leq \lambda_{\max}(P)|\tilde{u}^{\text{av}}(\tau)|^2$, and applying the Comparison Lemma [108] gives

$$|\tilde{u}^{\text{av}}(\tau)| \leq M e^{-m\tau/\omega}|\tilde{u}^{\text{av}}(0)|, \tag{1.93}$$

where

$$M = \sqrt{\frac{\lambda_{\max}(P)}{\lambda_{\min}(P)}}, \quad m = \frac{\omega \lambda_{\min}(Q)}{2\lambda_{\max}(P)}. \tag{1.94}$$

From (1.93) and [108, Theorem 10.4], we obtain $|\tilde{u}(\tau)| \leq M e^{-m\tau/\omega}|\tilde{u}(0)| + \mathcal{O}(1/\omega)$, provided $\tilde{u}(0)$ is sufficiently close to $\tilde{u}^{\text{av}}(0)$. Reverting to the time scale t and noting that $u_i(t) - u_i^* = \tilde{u}_i(t) + \mu_i(t) = \tilde{u}_i(t) + \mathcal{O}(\max_i a_i)$ completes the proof. □

From the proof, we see that the convergence result holds if (1.90) is Hurwitz, which does not require the strict diagonal dominance assumption. However, Assumption 1.5 allows convergence to hold for k_i, $a_i > 0$, whereas merely assuming (1.90) is Hurwitz would create a potentially intricate dependence on the unknown game model and the selection of the parameters k_i, a_i. Also, while we have considered only the case where the action variables of the players are scalars, the results equally apply to the vector case, namely, $u_i \in \mathbb{R}^n$, by simply considering each different component of a player's action variable to be controlled by a different (virtual) player. In this case, the payoff functions of all virtual players corresponding to Player i will be the same.

Even though u^* is unique for quadratic payoffs, Theorem 1.3 is local due to our use of standard local averaging theory. From the theory in [222], we have the following nonlocal result.

Corollary 1.1. *Consider N players with quadratic payoff functions* (1.80) *that implement the Nash seeking strategy* (1.75)–(1.76) *with frequencies satisfying the inequalities stated in Theorem* 1.3. *Then, the Nash equilibrium u^* is semi-globally practically asymptotically stable.*

Proof. In the proof of Theorem 1.3, the average error system (1.89) is shown to be globally asymptotically stable. By [222, Theorem 2], with the error system (1.87) satisfying the theorem's conditions, the origin of (1.87) is semi-globally practically asymptotically stable. □

For more details on semi-global convergence with extremum seeking controllers, the reader is referred to [217].

1.6.3 ▪ N-Player Games with Non-quadratic Payoff Functions

Now consider a more general noncooperative game with N players and a dynamic mapping from the players' actions u_i to their payoff values J_i. Each player attempts to maximize the steady-state value of its payoff. Specifically, we consider a general nonlinear model

$$\dot{x} = f(x,u), \tag{1.95}$$

$$J_i = h_i(x), \quad i \in \{1,\dots,N\}, \tag{1.96}$$

where $x \in \mathbb{R}^n$ is the state, $u \in \mathbb{R}^N$ is a vector of the players' actions, u_i is the action of player i, $J_i \in \mathbb{R}$ is its payoff value, $f : \mathbb{R}^n \times \mathbb{R}^N \to \mathbb{R}^n$ and $h_i : \mathbb{R}^n \to \mathbb{R}$ are smooth, and h_i is a possibly non-quadratic function. Oligopoly games may possess nonlinear demand and cost functions [162], which motivate the inclusion of the dynamic system (1.95) in the game structure and the consideration of non-quadratic payoff functions. For this scenario, we pursue local convergence results since the payoff functions h_i may be non-quadratic and multiple-isolated Nash equilibria may exist. If the payoff functions are quadratic, semi-global practical stability can be achieved following the results of [217].

We make the following assumptions about this N-player game.

Assumption 1.6. *There exists a smooth function $l : \mathbb{R}^N \to \mathbb{R}^n$ such that*

$$f(x,u) = 0 \quad \text{if and only if} \quad x = l(u). \tag{1.97}$$

Assumption 1.7. *For each $u \in \mathbb{R}^N$, the equilibrium $x = l(u)$ of* (1.95) *is locally exponentially stable.*

Hence, we assume that for all actions, the nonlinear dynamic system is locally exponentially stable. We can even relax the requirement that this assumption holds for each $u \in \mathbb{R}^N$ as we need to be only concerned with the action sets of the players, namely, $u \in U = U_1 \times \cdots \times U_N \subset \mathbb{R}^N$. For notational convenience, we use this more restrictive case.

The following assumptions are central to the Nash seeking scheme as they ensure that at least one stable Nash equilibrium exists at steady state.

Assumption 1.8. *There exists at least one, possibly multiple, isolated stable Nash equilibria $u^* = [u_1^*, \dots, u_n^*]$ such that, for all $i \in \{1, \dots, N\}$,*

$$\frac{\partial (h_i \circ l)}{\partial u_i}(u^*) = 0,$$

$$\frac{\partial^2 (h_i \circ l)}{\partial u_i^2}(u^*) < 0. \tag{1.98}$$

Assumption 1.9. *The matrix*

$$\Lambda = \begin{bmatrix} \frac{\partial^2 (h_1 \circ l)(u^*)}{\partial u_1^2} & \frac{\partial^2 (h_1 \circ l)(u^*)}{\partial u_1 \partial u_2} & \dots & \frac{\partial^2 (h_1 \circ l)(u^*)}{\partial u_i \partial u_N} \\ \frac{\partial^2 (h_2 \circ l)(u^*)}{\partial u_1 \partial u_2} & \frac{\partial^2 (h_2 \circ l)(u^*)}{\partial u_2^2} & & \\ \vdots & & \ddots & \\ \frac{\partial^2 (h_N \circ l)(u^*)}{\partial u_1 \partial u_N} & & & \frac{\partial^2 (h_N \circ l)(u^*)}{\partial u_N^2} \end{bmatrix} \tag{1.99}$$

is strictly diagonally dominant and hence nonsingular.

By Assumptions 1.8 and 1.9, Λ is Hurwitz.

As with the games considered in Sections 1.6.1 and 1.6.2, each player converges to a neighborhood of u^* by implementing the extremum seeking strategy (1.75)–(1.76) to evolve its action u_i according to the measured value of its payoff J_i . Unlike the previous games, however, we select the parameters $k_i = \varepsilon \omega K_i = \mathcal{O}(\varepsilon \omega)$, where ε, ω are small, positive constants and ω is related to the players' frequencies by (1.77). Intuitively, ω is small since the players' actions should evolve more slowly than the dynamic system, creating an overall system with two time scales. In contrast, our earlier analysis assumed $1/\omega$ to be small, which can be seen as the limiting case where the dynamic system is infinitely fast and allows ω to be large.

Formulating the error system clarifies why these parameter selections are made. The error relative to the Nash equilibrium is denoted by (1.85), which in the time scale $\tau = \omega t$ leads to

$$\omega \frac{dx}{d\tau} = f(x, u^* + \tilde{u} + \mu(\tau)), \tag{1.100}$$

$$\frac{d\tilde{u}_i}{d\tau} = \varepsilon K_i \mu_i(\tau) h_i(x), \qquad i \in \{1, \dots, N\}, \tag{1.101}$$

where $\tilde{u} = [\tilde{u}_1, \dots, \tilde{u}_N]$, $\mu(\tau) = [\mu_1(\tau), \dots, \mu_N(\tau)]$, and $\mu_i(\tau) = a_i sin(\overline{\omega}_i \tau + \varphi_i)$. The system (1.100)–(1.101) is in the standard singular perturbation form with ω as a small parameter. Since ε is also small, we analyze (1.100)–(1.101) using the averaging theory for the quasi-steady state of (1.100), followed by the use of the singular perturbation theory for the full system.

1.6.4 ▪ Averaging Analysis

For the averaging analysis, we first "freeze" x in (1.100) at its quasi-steady state $x = l(u^* + \tilde{u} + \mu(\tau))$, which we substitute into (1.101) to obtain the "reduced system"

$$\frac{d\tilde{u}_i}{d\tau} = \varepsilon K_i \mu_i(\tau)(h_i \circ l)(u^* + \tilde{u} + \mu(\tau)). \tag{1.102}$$

This system is in the form to apply averaging theory [108] and leads to the following result.

Theorem 1.4. *Consider the system* (1.102) *for an N-player game under Assumptions* 1.8 *and* 1.9, *where* $\overline{\omega}_i \neq \overline{\omega}_j$, $\overline{\omega}_i \neq \overline{\omega}_j + \overline{\omega}_k$, $2\overline{\omega}_i \neq \overline{\omega}_j + \overline{\omega}_k$, *and* $\overline{\omega}_i \neq 2\overline{\omega}_j + \overline{\omega}_k$ *for all distinct* $i, j, k \in \{1, \ldots, N\}$ *and* $\overline{\omega}_i$ *is rational for all* $i \in \{1, \ldots, N\}$. *There exist parameters* $\Xi, \xi > 0$ *and* ε^*, a^* *such that, for all* $\varepsilon \in (0, \varepsilon^*)$ *and* $a_i \in (0, a^*)$, *if* $|\Theta(0)|$ *is sufficiently small, then for all* $\tau > 0$,

$$|\Theta(\tau)| \leq \Xi e^{-\xi\tau}|\Theta(0)| + \mathcal{O}\Big(\varepsilon + \max_i a_i^3\Big), \tag{1.103}$$

where $\Theta(\tau) = \Big[\tilde{u}_1(\tau) - \sum_{j=1}^N c_{jj}^1 a_j^2, \ldots, \tilde{u}_N(\tau) - \sum_{j=1}^N c_{jj}^N a_j^2\Big]^T$, *and*

$$\boldsymbol{c}_j = -\frac{1}{4}\Lambda^{-}1\boldsymbol{g}_j, \tag{1.104}$$

$$\boldsymbol{c}_j \triangleq \Big[c_{jj}^1, \ldots, c_{jj}^{j-1}, c_{jj}^j, c_{jj}^{j+1}, \ldots, c_{jj}^N\Big]^T, \tag{1.105}$$

$$\boldsymbol{g}_j \triangleq \Big[g_j^1, \ldots, g_j^{j-1}, g_j^j, g_j^{j+1}, \ldots, g_j^N\Big]^T, \tag{1.106}$$

$$g_j^i = \begin{cases} \dfrac{\partial^3 (h_k \circ l)(u^*)}{\partial u_k \partial u_j^2} & \text{if } i \neq j, \\ \dfrac{1}{2}\dfrac{\partial^3 (h_j \circ l)(u^*)}{\partial u_j^3} & \text{if } i = j. \end{cases} \tag{1.107}$$

Proof. As already noted, the form of (1.102) allows for the application of averaging theory, which yields the average error system

$$\frac{d\tilde{u}_i^{\text{av}}}{d\tau} = \varepsilon\left(\frac{K_i}{T}\int_0^T \mu_i(\tau)(h_i \circ l)(u^* + \tilde{u}^{\text{av}} + \mu(\tau))d\tau\right). \tag{1.108}$$

The equilibrium $\tilde{u}^e = [\tilde{u}_1^e, \ldots, \tilde{u}_N^e]$ of (1.108) satisfies

$$0 = \frac{1}{T}\int_0^T \mu_i(\tau)(h_i \circ l)(u^* + \tilde{u}^e + \mu(\tau))d\tau \tag{1.109}$$

for all $i \in \{1, \ldots, N\}$, and we postulate that $\tilde{u}^e$ has the form

$$\tilde{u}_i^e = \sum_{j=1}^N b_j^i a_j + \sum_{j=1}^N \sum_{k \geq j}^N c_{jk}^i a_j a_k + \mathcal{O}\Big(\max_i a_i^3\Big). \tag{1.110}$$

By approximating $(h_i \circ l)$ about u^* in (1.109) with a Taylor polynomial and substituting (1.110), the unknown coefficients b_j^i and c_{jk}^i can be determined.

To capture the effect of higher-order derivatives on the average error system's equilibrium, we use the Taylor polynomial approximation [9], which requires $(h_i \circ l)$ to be $k+1$ times differentiable

$$\begin{aligned} (h_i \circ l)(u^* + \tilde{u}^e + \mu(\tau)) &= \sum_{|\alpha|=0}^k \frac{D^\alpha (h_i \circ l)(u^*)}{\alpha!}(\tilde{u}^e + \mu(\tau))^\alpha + \sum_{|\alpha|=k+1} \frac{D^\alpha (h_i \circ l)(\zeta)}{\alpha!}(\tilde{u}^e + \mu(\tau))^\alpha \\ &= \sum_{|\alpha|=0}^k \frac{D^\alpha (h_i \circ l)(u^*)}{\alpha!}(\tilde{u}^e + \mu(\tau))^\alpha + \mathcal{O}\Big(\max_i a_i^{k+1}\Big), \end{aligned} \tag{1.111}$$

where ζ is a point on the line segment that connects the points u^* and $u^* + \tilde{u}^e + \mu(\tau)$. In (1.111), we have used multi-index notation, namely, $\alpha = (\alpha_1, \ldots, \alpha_N)$, $|\alpha| = \alpha_1 + \cdots + \alpha_N$, $\alpha! = \alpha_1! \cdots \alpha_N!$, $u^\alpha = u_1^{\alpha_1} \cdots u_N^{\alpha_N}$, and $D^\alpha (h_i \circ l) = \partial^{|\alpha|} (h_i \circ l) / \partial u_1^{\alpha_1} \cdots \partial u_N^{\alpha_N}$. The second term on the last line of (1.111) follows by substituting the postulated form of $\tilde{u}^e$ (1.110).

For this analysis, we choose $k = 3$ to capture the effect of the third-order derivative on the system as a representative case. Higher-order estimates of the bias can be pursued if the third order derivative is zero. Substituting (1.111) into (1.109) and computing the average of each term gives

$$
\begin{aligned}
0 = \frac{a_i^2}{2} \times \Bigg[& \tilde{u}_i^e \frac{\partial^2 (h_i \circ l)(u^*)}{\partial u_i^2} + \sum_{j \neq i}^{N} \tilde{u}_j^e \frac{\partial^2 (h \circ l)(u^*)}{\partial u_i \partial u_j} + \left(\frac{1}{2} (\tilde{u}_i^e)^2 + \frac{a_i^2}{8} \right) \frac{\partial^3 (h \circ l)(u^*)}{\partial u_i^3} \\
& + \tilde{u}_i^e \sum_{j \neq i}^{N} \tilde{u}_i^e \frac{\partial^3 (h_i \circ l)(u^*)}{\partial u_i^2 \partial u_j} + \sum_{j \neq i}^{N} \left(\frac{1}{2} (\tilde{u}_j^e)^2 + \frac{a_j^2}{4} \right) \frac{\partial^3 (h \circ l)(u^*)}{\partial u_i \partial u_j^2} + \sum_{j \neq i}^{N} \sum_{\substack{k > j \\ k \neq i}}^{N} \tilde{u}_j^e \tilde{u}_k^e \frac{\partial^3 (h \circ l)(u^*)}{\partial u_i \partial u_j \partial u_k} \Bigg] \\
& + \mathcal{O}\left(\max_i a_i^5 \right),
\end{aligned}
\tag{1.112}
$$

where we have noted (1.98), utilized (1.110), and computed the integrals, as shown in Appendices A and B of [70]. Substituting (1.110) into (1.112) and matching first-order powers of a_i gives

$$
\begin{bmatrix} 0 \\ \vdots \\ 0 \end{bmatrix} = a_1 \Lambda \begin{bmatrix} b_1^1 \\ \vdots \\ b_1^N \end{bmatrix} + \cdots + a_N \Lambda \begin{bmatrix} b_1^1 \\ \vdots \\ b_N^N \end{bmatrix},
\tag{1.113}
$$

which implies that $b_j^i = 0$ for all i, j since Λ is nonsingular by Assumption 1.9. Similarly, matching second-order terms of a_i, and substituting $b_j^i = 0$ to simplify the resulting expressions, yields

$$
\begin{bmatrix} 0 \\ \vdots \\ 0 \end{bmatrix} = \sum_{j=1}^{N} \sum_{k > j}^{N} a_j a_k \Lambda \begin{bmatrix} c_{jk}^1 \\ \vdots \\ c_{jk}^N \end{bmatrix} + \sum_{j=1}^{N} a_j^2 \left(\Lambda \begin{bmatrix} c_{jj}^1 \\ \vdots \\ c_{jj}^N \end{bmatrix} + \frac{1}{4} \mathbf{g}_j \right),
$$

where $\mathbf{g}_j$ is defined in (1.106). Thus, $c_{jk}^i = 0$ for all i, j, k when $j \neq k$ and c_{jj}^i is given by (1.105) for $i, j \in \{1, \ldots, N\}$. The equilibrium of the average system is then

$$
\tilde{u}_i^e = \sum_{j=1}^{N} c_{jj}^i a_j^2 + \mathcal{O}\left(\max_i a_i^3 \right).
\tag{1.114}
$$

By again utilizing a Taylor polynomial approximation, one can show that the Jacobian $\Psi^{\mathrm{av}} = [\psi_{i,j}]_{N \times N}$ of (1.108) at $\tilde{u}^e$ has elements given by

$$
\begin{aligned}
\psi_{i,j} &= \varepsilon \frac{K_i}{T} \int_0^T \mu_i(\tau) \frac{\partial (h_i \circ l)}{\partial u_j} (u^* + \tilde{u}^e + \mu(\tau)) d\tau \\
&= \frac{1}{2} \varepsilon K_i a_i^2 \frac{\partial^2 h_i \circ l}{\partial u_i \partial u_j} (u^*) + \mathcal{O}\left(\varepsilon \max_i a_i^3 \right).
\end{aligned}
\tag{1.115}
$$

By Assumptions 1.8 and 1.9, Ψ^{av} is Hurwitz for sufficiently small a_i, which implies that the equilibrium (1.114) of the average error system (1.108) is locally exponentially stable; i.e., there exist constants $\Xi, \xi > 0$ such that $|\tilde{u}^{\mathrm{av}}(\tau) - \tilde{u}^e| \leq \Xi e^{-\xi\tau}|\tilde{u}^{\mathrm{av}}(0) - \tilde{u}^e|$, which with [108, Theorem 10.4] implies

$$|\tilde{u}(\tau) - \tilde{u}^e| \leq \Xi e^{-\xi\tau}|\tilde{u}(0) - \tilde{u}^e| + \mathcal{O}(\varepsilon) \tag{1.116}$$

provided $\tilde{u}(0)$ is sufficiently close to $\tilde{u}^{\mathrm{av}}(0)$. Defining $\Theta(\tau)$ as in Theorem 1.4 completes the proof. □

From Theorem 1.4, we see that u of reduced system (1.102) converges to a region that is biased away from the Nash equilibrium u^*. This bias is in proportion to the perturbation magnitudes a_i and the third derivatives of the payoff functions, which are captured by the coefficients c^i_{jj}. Specifically, $\hat{u}_i$ of the reduced system converges to $u_i^* + \sum_{j=1}^N c^i_{jj} a_j^2 + \mathcal{O}\left(\varepsilon + \max_i a_i^3\right)$ as $t \to \infty$.

Theorem 1.4 can be viewed as a generalization of Theorem 1.3, but with a focus on the error system to highlight the effect of the payoff functions' non-quadratic terms on the players' convergence. This emphasis is needed because u_i of the reduced system converges to an $\mathcal{O}(\varepsilon + \max_i a_i)$-neighborhood of u_i^*, as in the quadratic payoff case, since $u_i = \hat{u}_i + \mu_i$.

1.6.5 ▪ Singular Perturbation Analysis

We analyze the full system (1.100)–(1.101) in the time scale $\tau = \omega t$ using singular perturbation theory [108]. In Section 1.6.4, we analyzed the reduced model (1.102), and now we must study the boundary layer model to state our convergence result. First, however, we translate the equilibrium of the reduced model to the origin by defining $z_i = \tilde{u}_i - \tilde{u}_i^p$, where by [108, Theorem 10.4], $\tilde{u}_i^p$ is a unique, exponentially stable, T-periodic solution $\tilde{u}^p = [\tilde{u}_1^p, \ldots, \tilde{u}_N^p]$ such that

$$\frac{d\tilde{u}_i^p}{d\tau} = \varepsilon K_i \mu_i(\tau)(h_i \circ l)(u^* + \tilde{u}^p + \mu(\tau)). \tag{1.117}$$

In this new coordinate system, we have for $i \in \{1, \ldots, N\}$

$$\frac{dz_i}{d\tau} = \varepsilon K_i \mu_i(\tau) \times \left[h_i(x) - (h_i \circ l)(u^* + \tilde{u}^p + \mu(\tau))\right], \tag{1.118}$$

$$\omega \frac{dx}{d\tau} = f(x, u^* + z + \tilde{u}^p + \mu(\tau)), \tag{1.119}$$

which from Assumption 1.6 has the quasi-steady state $x = l(u^* + z + \tilde{u}^p + \mu(\tau))$, and consequently, the reduced model in the new coordinates is

$$\frac{dz_i}{d\tau} = \varepsilon K_i \mu_i(\tau)\left[(h_i \circ l)(u^* + z + \tilde{u}^p + \mu(\tau)) - (h_i \circ l)(u^* + \tilde{u}^p + \mu(\tau))\right], \tag{1.120}$$

which has an equilibrium at $z = 0$ that is exponentially stable for sufficiently small a_i.

To formulate the boundary layer model, let $y = x - l(u^* + z + \tilde{u}^p + \mu(\tau))$, and then in the time scale $t = \tau/\omega$, we have

$$\begin{aligned} \frac{dy}{dt} &= f(y + l(u^* + z + \tilde{u}^p + \mu(\tau)), u^* + z + \tilde{u}^p + \mu(\tau)) \\ &= f(y + l(u), u), \end{aligned} \tag{1.121}$$

where $u = u^* + \tilde{u} + \mu(\tau)$ should be viewed as a parameter independent of the time variable t. Since $f(l(u), u) = 0, y = 0$ is an equilibrium of (1.121) and is exponentially stable by Assumption 1.7.

With ω as a singular perturbation parameter, we apply Tikhonov's Theorem on the Infinite Interval [108, Theorem 11.2] to (1.118)–(1.119), which requires the origin to be an exponentially stable equilibrium point of both the reduced model (1.120) and the boundary layer model (1.121) and leads to the following:

- the solution $z(\tau)$ of (1.118) is $\mathcal{O}(\omega)$-close to the solution $\bar{z}(\tau)$ of the reduced model (1.120), so
- the solution $\tilde{u}(\tau)$ of (1.101) converges exponentially to an $\mathcal{O}(\omega)$-neighborhood of the T-periodic solution $\tilde{u}^p(\tau)$, and
- the T-periodic solution $\tilde{u}^p(\tau)$ is $\mathcal{O}(\varepsilon)$-close to the equilibrium $\tilde{u}^e$.

Hence, as $t \to \infty$, $\tilde{u}(\tau)$ converges to an $\mathcal{O}(\omega+\varepsilon)$-neighborhood of

$$\tilde{u}^e = \left[\sum_{j=1}^{N} c_{jj}^1 a_j^2, \ldots, \sum_{j=1}^{N} c_{jj}^N a_j^2\right] + \mathcal{O}(\max_i a_i^3).$$

Since $u(\tau) - u^* = \tilde{u}(\tau) + \mu(\tau) = \tilde{u}(\tau) + \mathcal{O}(\max_i a_i), u(\tau)$ converges to an $\mathcal{O}(\omega+\varepsilon+\max_i a_i)$-neighborhood of u^*.

Also from Tikhonov's Theorem on the Infinite Interval, the solution $x(\tau)$ of (1.119), which is the same as the solution of (1.100), satisfies

$$x(\tau) - l(u^* + \tilde{u}^r(\tau) + \mu(\tau)) - y(t) = \mathcal{O}(\omega), \tag{1.122}$$

where $\tilde{u}^r(\tau)$ is the solution of the reduced model (1.102) and $y(t)$ is the solution of the boundary layer model (1.121). Rearranging terms and subtracting $l(u^*)$ from both sides yields

$$x(\tau) - l(u^*) = \mathcal{O}(\omega) + l(u^* + \tilde{u}^r(\tau) + \mu(\tau)) - l(u^*) + y(t). \tag{1.123}$$

After noting that

- $\tilde{u}^r(\tau)$ converges exponentially to $\tilde{u}^p(\tau)$, which is $\mathcal{O}(\varepsilon)$-close to the equilibrium $\tilde{u}^e$,
- $\mu(\tau)$ is $\mathcal{O}(\max_i a_i)$, and
- $y(t)$ is exponentially decaying,

we conclude that $x(\tau) - l(u^*)$ exponentially converges to an $\mathcal{O}(\omega+\varepsilon+\max_i a_i)$-neighborhood of the origin. Thus, $J_i = h_i(x)$ exponentially converges to an $\mathcal{O}(\omega+\varepsilon+\max_i a_i)$-neighborhood of the payoff value $(h_i \circ l)(u^*)$.

We summarize with the following theorem and Remark 1.1.

Theorem 1.5. *Consider the system* (1.95)–(1.96) *with* (1.75)–(1.76) *for an N-player game under Assumptions* 1.6–1.9, *where $\omega_i \neq \omega_j, \omega_i \neq \omega_j + \omega_k, 2\omega_i \neq \omega_j + \omega_k$, and $\omega_i \neq 2\omega_j + \omega_k$ for all distinct $i,j,k \in \{1,\ldots,N\}$ and where ω_i/ω_j is rational for all $i,j \in \{1,\ldots,N\}$. There exists $\omega^* > 0$ and for any $\omega \in (0,\omega^*)$ there exist $\varepsilon^*, a^* > 0$ such that for the given ω and any $\varepsilon \in (0,\varepsilon^*)$ and $\max_i a_i \in (0,a^*)$, the solution $(x(t), u_1(t), \ldots, u_N(t))$ converges exponentially to an $\mathcal{O}(\omega+\varepsilon+\max_i a_i)$-neighborhood of the point $(l(u^*), u_1^*, \ldots, u_n^*)$, provided the initial conditions are sufficiently close to this point.*

Remark 1.1. Note that the material in this section contains the following special cases:

- Multivariable static extremum seeking of Section 1.2, as the corollary of Theorem 1.4, when $h_1 = \cdots = h_n$,

- Multivariable dynamic extremum seeking of Section 1.5, as the corollary of Theorem 1.5, when $h_1 = \cdots = h_n$, and
- Static Nash equilibrium seeking in Theorem 1.4.

1.7 ▪ Finite-Dimensional Systems (ODEs) vs. Infinite-Dimensional Systems (PDEs)

As briefly discussed so far, ES is a powerful method for solving optimization problems without the knowledge of the operating map, using only the measurements of the output of the map. Tackling similar problems as evolutionary/genetic algorithms, ES was invented half a century earlier within the early control community and is well suited for real-time implementation on plants with significant dynamics. Modern ES algorithms, developed since 2000, are capable of guaranteeing stability, and even prescribed rates of convergence in spite of the plant model and the performance index function being unknown. Since the publication of the first proof of stability of extremum seeking [129], thousands of papers have been published on this topic, presenting further theoretical developments and applications of ES.

Some fundamental ES results, including deterministic ES algorithms, ES for noncooperative games, and extensions of ES from gradient- to Newton-based updates, were developed. A proof that expands the validity of extremum seeking from local to semi-global stability was published in [217]. The book [142] presents stochastic versions of the algorithms, where the sinusoids are replaced by filtered white noise perturbation signals. Many hundreds of applications of ES have also emerged since 2000, such as source seeking for autonomous vehicles in GPS-denied environments, MPPT for solar and wind energy sources, liquid tin droplet targeting by lasers in semiconductor photolithography, and a Mars Rover application.

In almost one century since its first applications and more than two decades since its formal convergence proof, the extremum seeking algorithm has been recognized as one of the most important model-free real-time optimization tools. However, until recently extremum seeking has been restricted to dynamic systems represented by connections of Ordinary Differential Equations (ODEs) and nonlinear convex maps with unknown extremum points.

In this context, the next chapters of this book present the first results on the theory and design of extremum seeking strategies for systems governed by Partial Differential Equations (PDEs). The main ideas for the design of the Gradient-Newton methods and the stability analysis for infinite-dimensional systems will be discussed considering a wide class of parabolic and hyperbolic PDEs: delay equations, wave equation, and reaction-advection-diffusion models. Moreover, engineering applications are presented, including problems of noncooperative games, neuromuscular electrical stimulation, biological reactors, oil-drilling systems, and flow-traffic control for urban mobility.

1.8 ▪ Organization of the Book

The book is structured as follows.

1. Part I - Extremum Seeking for Time-Delay Systems

   In **Chapters 2** and **3**, novel predictor designs with perturbation-based estimate of the Hessian and the Hessian's inverse are introduced to cope with input-output delays in the control loop of gradient-based and Newton-based extremum seeking controllers, respectively. The resulting approach preserves exponential stability and convergence of the system output to a small neighborhood of the extremum point, despite the presence of actuator and sensor delays. A rigorous proof is given in terms of backstepping transformation and averaging analysis in infinite dimension. The convergence rate of the Newton-based scheme addressed in Chapter 3 is independent of the unknown Hessian, whereas the convergence of the gradient method in Chapter 2 is dictated by it. In other words, the delay compensation can now be achieved with an arbitrarily assigned convergence rate, improving the performance of the previous controller.

In **Chapter 4**, we derive inverse optimality results for extremum seeking feedback with the low-pass filtered modification of the predictor-based feedback for delay compensation proposed in Chapter 2. Extremum seeking is studied with Laypunov tools and has a control input whose cost can be optimized over infinite time. We establish the stability robustness to varying the filter pole to large values, recovering in the limit the basic, unfiltered predictor-based feedback. This robustness property might be intuitively expected from a singular perturbation idea, though an off-the-shelf theorem for establishing this property would be highly unlikely to be found in the literature, due to the infinite dimensionality and the special hybrid (ODE-PDE) structure of the system at hand. Although the delay case as well as the gradient-based design are the main focus throughout the chapter, the inverse optimality can also be extended for the delay-free case and Newton-based approach.

In **Chapter 5**, the extension of the Newton-based extremum seeking (Chapter 3) to the real-time optimization for higher derivatives of unknown dynamic maps with delayed output signal is presented using stochastic perturbations rather than deterministic ones. The richness of the stochastic dithers and demodulation signals improve the closed-loop responses by generating a faster adaptation and search for the extremum point.

In **Chapter 6**, the robustness to delay uncertainties of the Newton-based extremum seeking for higher-derivatives maps via prediction feedback is discussed. We show that the maximum delay mismatch is a function of the dither frequency and the order of the derivative that is subject to optimization. Moreover, the resulting approach of using predictor feedback to compensate the nominal (known) delay still guarantees exponential stability and convergence of the system output to a small neighborhood of the extremum point despite the sufficiently small delay mismatch.

In **Chapter 7**, extremum seeking controllers are developed for multivariable real-time optimization in the presence of actuator and sensor delays. The control scheme introduced there for output delay compensation uses prediction feedback with perturbation-based estimate of the Hessian or its inverse associated with an adequate choice of the demodulation signals. Two different predictor designs for gradient-based and Newton-based extremum seeking control are formulated. The generalization for multi-input-single-output maps with distinct input delays is also addressed. While the Newton-based version for distinct input delays can be obtained via the conventional backstepping method, the Gradient extremum seeking is based on a novel successive backstepping-like state transformation. The predictor in the Newton case is much simpler than in the Gradient case. The Newton algorithm effectively "diagonalizes" the map and allows "decentralized" predictors for each control channel, whereas the Gradient algorithm has to perform prediction of the cross-coupling of the channels. In addition, an alternative gradient-based extremum seeking scheme is developed using the reduction approach rather than backstepping transformations, which results in a simpler multiparameter boundary control law and preserves the delay-independent convergence properties of the previous methods.

In **Chapter 8**, we provide a new predictor-feedback design for gradient-based extremum seeking in order to address arbitrarily large variable delays in the output (input) signals of an unknown locally quadratic static map. The predictor control law with distributed terms of the delay is developed with perturbation-based estimates for the Gradient and Hessian of the nonlinear map to be optimized. Numerical implementation aspects of the proposed predictor are also addressed. We present all the steps of the proof for local exponential convergence to a small neighborhood of the extremum point using analysis tools of backstepping transformation and averaging in infinite dimensions. The treatment of time- and state-dependent delays is also discussed as well as the possibility of designing extremum seeking boundary control laws by means of representative PDE-PDE cascades of hyperbolic types. Numerical results are performed to evaluate the sensitivity of the results with respect to measurement noise and the rate of variation of the delay. The application of the proposed scheme in the neuromuscular electrical stimulation (NMES) problem is also discussed.

In **Chapter 9**, we propose a single-parameter extremum seeking scheme for a static map in the presence of distributed delays. To compensate the distributed delays, we first introduce new perturbation and probing signals which become conventional sinusoidal signals after delay. Then, the extremum seeking scheme is developed based on the predictor feedback control law. The effectiveness of the proposed scheme is shown theoretically and numerically. Extensions of the proposed approach to multiparameter cases and Newton-based scheme are also addressed there.

2. Part II - Extremum Seeking for PDE Systems

In **Chapter 10**, we present a design and stability analysis of extremum seeking for static maps with known actuation dynamics governed by diffusion PDEs. The average-based controller to compensate the actuation dynamics is designed via the backstepping method and fed with the perturbation-based gradient and Hessian estimates of the static map. The additive perturbation signal also takes the actuation dynamics into account and compensates that. Local exponential stability and convergence to a small neighborhood of the extremum point are shown, extending the theory of extremum seeking to cover also a wider class of infinite-dimensional systems. Finally, the generalization for the Stefan problem (diffusion PDE with moving boundary rather than fixed domain) is also discussed.

In **Chapter 11**, we present and prove local stability of the proposed gradient extremum seeking algorithm based on boundary control for actuation dynamics governed by reaction-advection-diffusion PDEs. The proposed method maximizes in real time the output of an unknown static nonlinear map employing perturbation-based estimates of its gradient and Hessian. The resulting approach guarantees exponential convergence of the system input to a small neighborhood around the maximizer point where extremum occurs, despite the presence of the PDE. A rigorous proof is given in terms of backstepping transformation and averaging analysis in infinite dimensions for locally quadratic objective functions. The proposed approach has also broad applicability in practice since the presence of PDE models is often listed as a major limiting factor in the application of extremum seeking controllers in some practical situations. As a further contribution, we considered the extension to scalar Newton-based extremum seeking designs in order to alleviate the scaling issues associated with gradient descent algorithms and yield significant improvements in transient performance and faster convergence rates. The proposed method maximizes arbitrary nth derivatives of an unknown static map. The only available measurement is from the map output itself and not of its derivatives.

In **Chapter 12**, we provide a complete analysis of the gradient-based extremum seeking feedback for wave PDE compensation in different design scenarios, including Neumann–Dirichlet actuation forms, anti-stable wave PDEs, and delay-wave PDE cascades. The wave PDE dynamics must be known, whereas no knowledge is required for the map, that is, the parameters of the map were assumed to be unknown. The boundary control law is proposed in order to counteract the wave actuator dynamics was designed via backstepping methodology by combining state-of-the-art techniques in the field of PDE boundary control and averaging-based estimates of the map's gradient and Hessian. The additive perturbation-dither signal is designed in order to compensate the effects of the wave dynamics as well. Local stability and ultimate convergence to a small neighborhood of the unknown extremum point are proved as well.

In **Chapter 13**, we address the design and analysis of multivariable Newton-based extremum seeking for static locally quadratic maps subject to actuation dynamics governed by diffusion partial differential equations (PDEs). Multi-input systems with distinct diffusion coefficients in each individual input channel are dealt with. The phase compensation of the dither signals is handled as a trajectory generation problem and the inclusion of a multivariable diffusion feedback controller with a perturbation-based (averaging-based) estimate of the Hessian's inverse allows one to obtain local exponential convergence results to a small neighborhood of the optimal point. The stability analysis is carried out using backstepping transformation and averaging in infinite dimensions, capturing the infinite-dimensional state due to the diffusion PDEs. In addition, the generalization of the results for different classes of parabolic (reaction-advection-diffusion) PDEs, wave equations, and/or first-order hyperbolic (transport-dominated) PDEs is also discussed. As for the ordinary differential equations (ODEs) case, the proposed Newton approach removes the dependence of the algorithm's convergence rate on the unknown Hessian of the nonlinear map to be optimized, being user-assignable unlike the Gradient algorithm.

In **Chapter 14**, we develop an ES boundary control approach (Gradient and Newton-based versions) in order to compensate cascades of PDEs in the input of locally quadratic static maps with unknown parameters. The topic of ES for PDE-PDE cascades is in its infancy. Its first comprehensive coverage is the aim of this chapter. We start with the representation of a wave PDE as a feedback loop of two transport delay equations. We achieve stabilization for such a wave equation plant. The result of this delay-delay feedback loop helps us to resolve the Datko problem [47] for ES in the sense that the proposed ES boundary control stabilizes the wave equation system in the presence of an arbitrarily long delay, not only in the presence of a small delay. We do derive the feedback laws and make statements of closed-loop stability through an input-to-state stability analysis of coupled PDEs (with and without small gains in the loops), averaging in infinite dimensions and the associated estimates for the backstepping transformations between the plant and the target system. Local exponential convergence to a small neighborhood of the extremum is guaranteed.

3. Part III - Noncooperative Games under Delays and PDE Dynamics

In **Chapter 15**, we develop a non-model-based approach via extremum seeking and predictor feedback to find, in a distributed way, the Nash equilibria of noncooperative games with N players with unknown quadratic payoff functions and time-delayed actions, under two different information sharing scenarios—cooperative and noncooperative. In the noncooperative scenario, a player can stably attain its Nash equilibrium by measuring only the value of its payoff function (no other information about the game is needed), while in the cooperative scenario the players must share part of the information about the Hessian matrix estimated by each of them. Local stability and convergence are guaranteed by means of averaging theory in infinite dimensions, Lyapunov functionals, and/or a small-gain analysis for ODE-PDE loops. Such approaches have the potential for many applications. One possibility is in electronic markets, where players negotiate prices in real time as the supply and demand fluctuates, such as households in a smart electric grid, and, in addition, time delays appear naturally for different reasons or can be even artificially introduced to perturb the overall large-scale system. Numerical simulations conducted for a two-player game under constant delays support our theoretical results.

In **Chapter 16**, we introduce a non-model-based approach via extremum seeking and boundary control to find, in a distributed way, the Nash equilibria of noncooperative duopoly games with unknown quadratic payoff functions and the players acting through heat PDE dynamics. A player can stably attain its Nash equilibrium by measuring only the value of its payoff function (no other information about the game is needed). Local stability and convergence are guaranteed by means of averaging theory in infinite dimensions and a small-gain analysis for ODE-PDE loops.

In **Chapter 17**, we propose a non-model-based approach via extremum seeking and boundary control to find, in a decentralized fashion, the Nash equilibrium of noncooperative duopoly games with unknown quadratic payoff functions and with players acting through heterogeneous transport-heat PDE dynamics. Nash equilibrium is achieved for such a class of hyperbolic-parabolic PDEs with the players having access only to their own payoff functions via averaging theory in infinite dimensions and a small-gain analysis for cascades of ordinary and partial differential equations of different nature.

1.9 ▪ Notation, Norms, and Terminology

The 2-norm of the state vector $X(t)$ for a finite-dimensional system described by an Ordinary Differential Equation (ODE) is denoted by single bars, $|X(t)|$. In contrast, norms of functions (of x) are denoted by double bars. By default, $\|\cdot\|$ denotes the spatial $L_2[0,D]$-norm, i.e., $\|\cdot\| = \|\cdot\|_{L_2[0,D]}$, where we drop the index $L_2([0,D])$ if not otherwise specified. Since the state variable $u(x,t)$ of the infinite-dimensional system governed by a Partial Differential Equation (PDE) is a function of two arguments, we should emphasize that taking a norm in one of the variables makes the norm a function of the other variable, as adopted in [119]. For example, the $L_2[0,D]$-norm of $u(x,t)$ in $x \in [0,D]$ is $\|u(t)\| = \left(\int_0^D u^2(x,t)dx\right)^{1/2}$, whereas the $L_\infty[0,D]$-norm is defined by $\|u(t)\|_{L_\infty[0,D]} = \|u(t)\|_\infty = \sup_{x\in[0,D]} |u(x,t)|$. Moreover, the $\mathcal{H}_1$-norm is given by $\|u(t)\|^2_{\mathcal{H}_1} = \|u(t)\|^2_{L_2} + \|u_x(t)\|^2_{L_2}$.

We denote the partial derivatives of a function $u(x,t)$ as $\partial_x u(x,t) = \partial u(x,t)/\partial x$, $\partial_t u(x,t) = \partial u(x,t)/\partial t$. We conveniently use the compact forms $u_x(x,t)$ and $u_t(x,t)$ for the former and the latter, respectively.

While the transport PDE has only one boundary condition, second-order (in space) PDEs like the heat equation and the wave equation have two boundary conditions, one at $x=0$ and the other at $x=D$. When neither of the two conditions is of the homogeneous Dirichlet type (it may be the Neumann type, or the mixed/Robin type, or the damping/anti-damping type), the norm on the actuation/sensor dynamics has to be defined slightly differently. The norm in that case has to include a boundary value of the state at $x=0$ or $x=D$, such as, for example, $\left(u^2(0,t)+\int_0^D[u_x^2(x,t)+u_t^2(x,t)]dx\right)^{1/2}$ in the case of a wave equation with a Neumann boundary condition on the "free end" and with Neumann actuation on the controlled end. The reader should note that since the wave equation is not only second-order in x but also second-order in t, the state of the PDE includes both the "shear" variable $u_x(x,t)$ and the velocity variable $u_t(x,t)$. Likewise, the norm of the system includes both the potential and the kinetic energy.

Consider a generic nonlinear system $\dot{x} = f(t,x,\epsilon)$, where $x \in \mathbb{R}^n$, $f(t,x,\epsilon)$ is periodic in t with period T, i.e., $f(t+T,x,\epsilon) = f(t,x,\epsilon)$. Thus, for $\epsilon > 0$ sufficiently small, we can obtain its average model given by $\dot{x}_{\rm av} = f_{\rm av}(x)$, with $f_{\rm av}(x) = 1/T\int_0^T f(\tau,x,0)d\tau$, where $x_{\rm av}(t)$ denotes the average version of the state $x(t)$ [108].

As defined in [108], a vector function $f(t,\epsilon) \in \mathbb{R}^n$ is said to be of order $\mathcal{O}(\epsilon)$ over an interval $[t_1,t_2]$ if $\exists k,\bar{\epsilon}: |f(t,\epsilon)| \leq k\epsilon \;\forall \epsilon \in [0,\bar{\epsilon}]$ and $\forall t \in [t_1,t_2]$. In most cases we give no estimation of the constants k and $\bar{\epsilon}$. Then $\mathcal{O}(\epsilon)$ can be interpreted as an order-of-magnitude relation for sufficiently small ϵ.

The term "s" stands either for the Laplace variable or the differential operator "d/dt", according to the context. For a transfer function $H_0(s)$ with a generic input u, pure convolution $h_0(t) * u(t)$, with $h_0(t)$ being the impulse response of $H_0(s)$, is also denoted by $H_0(s)u$, as done in [92]. The maximum and minimum eigenvalues of a generic quadratic matrix A are denoted by $\lambda_{\max}(A)$ and $\lambda_{\min}(A)$, respectively.

The definition of the Input-to-State Stability (ISS) for ODE-based as well as for PDE-based systems are assumed to be as provided in [212] and [99], respectively.

Let $A \subseteq \mathbb{R}^n$ be an open set. By $C^0(A;\Omega)$, we denote the class of continuous functions on A, which take values in $\Omega \subseteq \mathbb{R}^m$. By $C^k(A;\Omega)$, where $k \geq 1$ is an integer, we denote the class of functions on $A \subseteq \mathbb{R}^n$ with continuous derivatives of order k, which take values in $\Omega \subseteq \mathbb{R}^m$. In addition, $C([a,b];\mathbb{R}^n)$ is the Banach space of continuous functions mapping the interval $[a,b]$ into $\mathbb{R}^n$; see [84, Chapter 2]. Alternatively, $C^n(\mathcal{X})$ denotes a n-times continuously differentiable function on the domain $\mathcal{X}$. In addition, $\mathbb{R}_+$ stands for the domain of positive real numbers including 0.

According to [84, 68], we assume the usual definitions for any delayed-system $\dot{x}(t) = f(t,x_t)$, $t \geq t_0$ and $x(t_0+\Theta) = \xi(\Theta)$, $\Theta \in [-D_{\max},0]$, where t_0 is an arbitrary initial time instant $t_0 \geq 0$, $x(t) \in \mathbb{R}^N$ is the state vector, $D_{\max} > 0$ is the maximum time delay allowed, the history function of the delayed state is given by $x_t(\Theta) = x(t+\Theta) \in C([-D_{\max},0];\mathbb{R}^N)$, and the functional initial condition ξ is also assumed to be continuous on $[-D_{\max},0]$. Without loss of generality, we consider $t_0 = 0$ throughout the chapters.

1.10 ▪ Mathematical Background

The purpose of this section is to provide some familiarity to the reader with the methods of averaging, semigroup theory, infinite-dimensional backstepping and Input-to-State Stability for PDEs, which are used for the analysis throughout the book chapters.

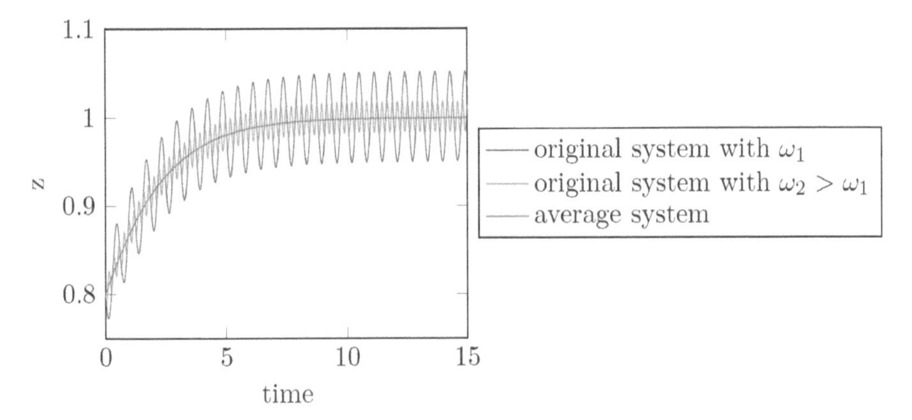

Figure 1.7. *Example of the averaging method* [108] *with the system dynamics* $\dot{z} = z\sin^2(\omega t) - 0.5z^2$ *and the average system dynamics* $\dot{z} = 0.5(z - z^2)$, *converging to the equilibrium* $z = 1$ *and its neighborhood, respectively.*

1.10.1 ▪ Averaging Theory

Averaging is one of the core elements of ES and its stability proof [129]. This method approximates the solution of a periodic nonlinear system by the solution of a so-called *average system*, obtained by averaging the original system with respect to a high frequency, as shown in Figure 1.7 and summarized in the following theorem.

Theorem 1.6 (Averaging for ODE systems [108]). *Consider the original system*

$$\dot{z} = f(\omega t, z), \quad z(0) = z_0\,. \tag{1.124}$$

Let $f(\omega t, z)$ *and its partial derivatives with respect to* z *up to the second order be continuous and bounded for* $(\omega t, z) \in [0, \infty) \times D_0$, *for every compact set* $D_0 \subset D$, *where* $D \subset \mathbb{R}^n$ *is the domain. Suppose f is T-periodic with* $T > 0$ *in* ωt, *i.e.,* $f(s, z) = f(s + T, z)$ *for all* $s \in [0, \infty)$. *Consider the average system*

$$\dot{z}_{\text{av}} = f_{\text{av}}(z_{\text{av}}), \quad z_{\text{av}}(0) = z_{\text{av},0} \tag{1.125}$$

with

$$f_{\text{av}}(z_{\text{av}}) = \frac{1}{T}\int_0^T f(\tau, z_{\text{av}})d\tau. \tag{1.126}$$

Suppose that $z_{\text{av}} = 0 \in D$ *is an exponentially stable equilibrium point of the average system* (1.125), $\Omega \subset D$ *is a compact subset of its region of attraction,* $z_{\text{av}}(0) \in \Omega$, *and* $z(0) - z_{\text{av}}(0) = \mathcal{O}(1/\omega)$. *Then there exists some* $\bar{\omega} > 0$ *such that for all* $\omega > \bar{\omega}$

$$\|z(t) - z_{\text{av}}(t)\|_{\mathbb{R}^n} \leq \mathcal{O}(1/\omega) \quad \forall\, t \in [0, \infty). \tag{1.127}$$

Furthermore, (1.124) *has a unique exponentially stable, T-periodic solution* $\bar{z}(t, 1/\omega)$ *with the property* $\|\bar{z}(t, 1/\omega)\|_{\mathbb{R}^n} \leq \mathcal{O}(1/\omega)$.

Note that $\bar{z}(t, 1/\omega)$ is the approximation of the solution of (1.124), while considering the solution of the average system (1.125); see [108]. By analyzing an average system and showing that it is exponentially stable, one can apply the theorem given above. Thus, the original system

(1.124) is exponentially stable and the solution behaves like (1.127). In case of ES, the analysis of the average closed-loop system is much easier as the analysis of the original closed-loop system, as we will see later.

Actuation dynamics governed by PDEs represent an infinite-dimensional system, which cause justification issues of the transformation in the derivation of the averaging theory [108], [83]. Therefore, Theorem 1.6 has to be extended to averaging for infinite-dimensional systems, where the main result can be found in [83] and is conveniently summarized in the following theorem.

Theorem 1.7 (Averaging for infinite-dimensional systems [83]). *Consider the following infinite-dimensional system, defined in the Banach space $\mathcal{X}$:*

$$\dot{z}(t) = Az(t) + F(\omega t, z), \tag{1.128}$$

with $z(0) = z_0 \in \mathcal{X}$, $A : \mathcal{D}(A) \to \mathcal{X}$ generates a continuous semigroup $T_A(t)$ with the property $\|T_A(t)\| \leq Me^{\mu t}$ for some M, μ. Moreover the nonlinearity $F : \mathbb{R}_+ \times \mathcal{X} \to \mathcal{X}$ with $t \mapsto F(t,z)$ is Fréchet differentiable in z, strongly continuous and periodic in t uniformly with respect to z in the compact subset of $\mathcal{X}$. Additionally the following hypothesis **(H)** *has to be satisfied:*

If $h : [s, \infty) \to \mathcal{X}$ is norm continuous, then

(i) $\int_s^t T_A(t-\tau)h(\tau) \in \mathcal{D}(A)$ *for* $s \leq t$;

(ii) $\|A \int_s^t T_A(t-\tau)h(\tau)d\tau\| \leq Me^{\mu t} \sup_{s \leq \tau \leq t} \|h(\tau)\|$.

Along with (1.128)*, the average system*

$$\dot{z}_{\text{av}} = Az_{\text{av}} + F_0(z) \quad \text{with} \quad F_0 = \lim_{T \to \infty} \frac{1}{T} \int_0^T F(\tau, z) d\tau \tag{1.129}$$

is considered. Suppose that $z_{\text{av}} = 0 \in D \subset \mathcal{X}$ is an exponentially stable equilibrium point of the average system (1.129)*. Then for some $\bar{\omega} > 0$ and $\omega > \bar{\omega}$,*

(a) *there exists a unique exponentially stable periodic solution $t \mapsto \bar{z}(t, 1/\omega)$, continuous in t and $1/\omega$, with $\|\bar{z}(t, 1/\omega)\| \leq \mathcal{O}(1/\omega)$ for $t > 0$;*

(b) $\|z_0 - z_{\text{av}}(0)\| < \mathcal{O}(1/\omega)$*, the solution estimate of* (1.128)*, is given by*

$$\|z(t) - z_{\text{av}}(t)\| \leq \mathcal{O}(1/\omega), \quad t > 0; \tag{1.130}$$

(c) *and for $\|z_0\| < \mathcal{O}(1/\omega)$ and by the stable manifold theorem, it holds that*

$$\|z(t) - \bar{z}(t, 1/\omega)\| \leq Ce^{-\gamma t}, \quad t > 0, \tag{1.131}$$

for some $C, \gamma > 0$.

See also Appendixes A and B concerning averaging theorems for a general class of FDEs and PDEs, respectively.

Remark 1.2. As in [83], hypothesis **(H)**, a smoothness property of the operator A, is trivially satisfied if A in (1.128) and (1.129), respectively, generates an analytic semigroup. The definition of analytic semigroups can be found in Section 1.10.2.

As the above theorem and Remark 1.2 state, the original closed-loop system (1.128) has to satisfy some assumptions to apply this theorem on an exponentially stable infinite-dimensional average system to conclude the exponential stability of the original system. Especially, one has to define a suitable infinite-dimensional state vector of the system and then show that A generates an analytic semigroup. We emphasize that this theorem is the key step for the stability proof of the closed-loop PDE systems.

1.10.2 ▪ Analytic Semigroups

As seen in the section before, linear PDEs can be expressed as so-called *evolutionary equations* [145]. Therefore, consider the PDE state $u(x,t)$, $x \in [0,1]$ as a function which belongs to a Banach space $\mathcal{X}$, i.e., $z(t) := u(\cdot,t)$ with $z(t) \in \mathcal{X}$. To describe a solution of an evolutionary equation the semigroup theory is used. We introduce a special class of semigroups, namely analytic semigroups in this section. Let us first consider the Cauchy problem in the space $\mathbb{R}$ ($z(t) \in \mathbb{R}$):

$$\dot{z}(t) = Az(t) + f(t), \quad z(0) = z_0. \tag{1.132}$$

The solution of (1.132) can be simply calculated with the well-known variation-of-constants formula to

$$z(t) = e^{At} z_0 + \int_0^t e^{A(t-\tau)} f(\tau) d\tau. \tag{1.133}$$

Now, consider the Cauchy problem (1.132) defined in a Banach space $\mathcal{X}$ (with $z(t) \in \mathcal{X}$), thus an infinite-dimensional problem. To describe the solution of this infinite-dimensional problem, the theory of semigroups [190] is used. We say that the operator A generates the semigroup $T_A(t) = (e^{At})_{t \geq 0}$, such that the solution of the infinite-dimensional Cauchy problem is given as

$$z(t) = T_A(t) z_0 + \int_0^t T_A(t-\tau) f(\tau) d\tau. \tag{1.134}$$

The generated semigroup is called analytic if the conditions as stated in the next definition are satisfied.

Definition 1.1 ([62]). *Let $\mathcal{X}$ be a Banach space and Σ_δ a sector of angle δ in $\mathbb{C}$. A family of operators $(T(t))_{t \in \Sigma_\delta \cup \{0\}} \subset \mathcal{L}(\mathcal{X})$ is called an analytic semigroup (of angle $\delta \in (0,\pi/2]$) if*

(i) *$T(0) = I$ and $T(t_1 + t_2) = T(t_1)T(t_2)$ for all $t_1, t_2 \in \Sigma_\delta$;*

(ii) *the map $t \mapsto T(t)$ is analytic in Σ_δ;*

(iii) *$\lim_{\Sigma_{\delta'} \ni t \to 0} T(t)z = z$ for all $z \in \mathcal{X}$ and $0 < \delta' < \delta$.*

In the publications by Henry Lunardi [145] and Engel et al. [62], several examples for analytic semigroups generators can be found, such as the Laplace operator

$$\Delta\varphi = \frac{\partial^2 \varphi}{\partial x^2} \tag{1.135}$$

with domain

$$D(\Delta)=\left\{\varphi\in\mathcal{X}:\varphi,\frac{d}{dx}\varphi\in\mathcal{X}\ \text{ a.c.},\ \frac{d^2}{dx^2}\varphi\in\mathcal{X},\ \frac{d}{dx}\varphi(0)=0,\ \varphi(1)=0\right\}, \tag{1.136}$$

where a.c. means *absolute continuous*. In this book, we are dealing with operator matrices, i.e., matrices which consist of bounded and infinite-dimensional operators as elements. Therefore, the next theorem states the conditions for each operator of the matrix such that the operator matrix generates an analytic semigroup.

Theorem 1.8 ([161]). *Consider the operator matrix*

$$\boldsymbol{A}:=\begin{bmatrix} A & B \\ 0 & D \end{bmatrix} \tag{1.137}$$

with $\boldsymbol{A}: D(\boldsymbol{A})=D(A)\times(D(D)\cap D(B))\to\mathcal{E}\times\mathcal{F}$*, where* A,B,D *are linear operators and* $\mathcal{E},\mathcal{F}$ *are suitable Banach spaces. If* A *and* D *generate analytic semigroups on* $\mathcal{E}$ *and* $\mathcal{F}$*, respectively, and* B *is* D*-bounded, then* $\boldsymbol{A}$ *is the generator of an analytic semigroup, where* D*-bounded is defined as*

$$\exists\,\epsilon_1\geq 0,\ \epsilon_2\geq 0,\ s.t.\ \|Bz\|\leq\epsilon_1\|Dz\|+\epsilon_2\|z\|,\quad \textit{with}\quad D(D)\subset D(B). \tag{1.138}$$

1.10.3 ▪ Boundary Control and Backstepping for PDEs

Two PDE control settings exist:

- "in domain" control (actuation penetrates inside the domain of the PDE system or is evenly distributed everywhere in the domain, likewise with sensing);
- "boundary" control (actuation and sensing are only through the boundary conditions).

In this sense, boundary control is physically more realistic because actuation and sensing are non-intrusive (think fluid flow where actuation is from the walls). For instance, "body force" actuation of electromagnetic type is also possible but it has low control authority and its spatial distribution typically has a pattern that favors the near-wall region. Moreover, boundary control is a harder problem, because the "input operator" (the analog of the B matrix in the LTI finite-dimensional model $\dot{x}=Ax+Bu$) and the output operator (the analog of the C matrix in $y=Cx$) are unbounded operators.

Most books on control of PDEs either do not cover boundary control or dedicate only small fractions of their coverage to boundary control. This book is devoted exclusively to boundary control.

Backstepping for PDEs is a boundary control concept [128], where destabilizing effects that appear throughout the spatial domain $[0,D]$ can be eliminated. The main idea is to find a change of variables that transform the PDE system with boundary control (at $x=D$)

$$u_t(x,t)=D_u u(x,t),\ x\in[0,D], \tag{1.139}$$

$$R_u u(x,t)=U(t) \tag{1.140}$$

into a *target system*

$$w_t(x,t)=D_w w(x,t),\ x\in[0,D], \tag{1.141}$$

$$R_w w(x,t)=0, \tag{1.142}$$

which has to be exponentially stable. The operators D_u, D_w are spatial differential operators and R_u, R_w are boundary operators. A standard coordinate transformation

$$w(x,t) = u(x,t) - \int_0^x k(x,y)u(y,t)dy \tag{1.143}$$

is proposed in [128] and preceding publications by Krstic and his coauthors, where $k(x,y)$ is the *gain kernel* and the integral in (1.143) is a Volterra integral transformation. Since the coordinate transformation (1.143) is invertible [128] the stability of the target system translates into stability of the original system. Hence, if the boundary controller is chosen as

$$U(t) = u(D,t) = \int_0^D k(x,y)u(y,t)dy \tag{1.144}$$

and the gain kernel $k(x,y)$ exists, the original system (1.139)–(1.140) with the controller (1.144) is exponentially stable. The gain kernel $k(x,y)$ has to satisfy the conditions which occur by substituting the transformation (1.143) into the target system (1.141)–(1.142) along with the original system equations (1.139)–(1.140). The triangular structure of the Volterra integral transformation and its relation to the finite-dimensional backstepping is highlighted in Remark 1.3.

Remark 1.3. There exists also *ODE backstepping*, especially integrator backstepping [125], where the feedback law is propagated "backwards" through a chain of integrators. The name backstepping for PDEs is caused by the triangular structure of the Volterra integral transformation similar to ODE backstepping.

1.10.4 ▪ Input-to-State Stability for PDEs

This subsection reviews the challenges for the extension of the Input-to-State Stability (ISS) property for systems described by Partial Differential Equations (PDEs).

ISS is a stability notion that has played a key role in the development of modern control theory. ISS combines internal and external stability notions and has been heavily used in the finite-dimensional case for the stability analysis of control systems as well as for the construction of feedback stabilizers. The present chapter is devoted to the extension of the ISS property to infinite-dimensional systems and more specifically, to control systems that involve PDEs. The methodologies that have been used in the literature for the derivation of ISS estimates are presented. Examples on ISS for PDEs are also provided.

Problems and Methods of ISS for PDEs

The extension of input-to-state stability (ISS) to systems described by partial differential equations (PDEs) is a challenging topic that has required extensive research efforts by many researchers. The main reason for the difficulty of such an extension is the fact that there are key features which are absent in finite-dimensional systems but play an important role for the study of ISS in PDE systems. Such key features are the following:

1. The nature of the disturbance inputs. There are two cases for PDE systems where a disturbance input may appear: a disturbance may appear in the PDE itself (distributed disturbance) and/or in the boundary conditions (boundary disturbance). The transformation of a boundary disturbance to a distributed disturbance does not solve the problem because in this way time derivatives of the disturbance appear in the PDE. This happens because the boundary input operator is unbounded. This is a fundamental obstacle which cannot be avoided.

2. Systems described by PDEs require various mathematical tools. For example, the mathematical theory and techniques for parabolic PDEs are completely different from the mathematical theory and techniques for hyperbolic PDEs. Moreover, there are PDEs which may not fall into one of these two categories (hyperbolic and parabolic).

3. When nonlinear PDEs are studied, the place where the nonlinearity appears is very important. PDEs with nonlinear differential operators are much more demanding than PDEs with linear differential operators and nonlinear reaction terms (semilinear PDEs).

4. A variety of spatial norms may be used for the solution of a PDE. However, the ISS property written in different state norms has a different meaning. For example, the ISS property in certain L_p-norm with $p < +\infty$ cannot give pointwise estimates of the solution. On the other hand, the ISS property in the sup-spatial norm can indeed give pointwise estimates of the state.

5. Different notions of a solution are used in the theory of PDEs. The notion of the solution of a PDE has a tremendous effect on the selection of the state space and the set of allowable disturbance inputs. So, if weak solutions are used for a PDE, then the state space is usually an L_p space with $p < +\infty$, and it makes no sense to talk about the ISS property in the sup-norm. However, it should be noticed that the state space does not necessarily determine the (spatial) norm in the ISS property: for example, if the state space is $\mathcal{H}_1(0,1)$, then the ISS property may be written in an $L_p(0,1)$-norm with $p \leq +\infty$.

6. While the main tool of proving the ISS property for finite-dimensional systems is the ISS Lyapunov function, there are important cases for PDEs where ISS Lyapunov functionals are not available (e.g., the heat equation with boundary disturbances and Dirichlet boundary conditions).

7. A PDE may or may not include nonlocal terms. Moreover, the nonlocal terms may appear in the boundary conditions of a PDE system. The latter case is particularly important for control theory, because it is exactly the case that appears when boundary feedback control is applied to a PDE system. However, it should be noticed that PDEs with nonlocal terms are not always covered by standard results of the theory of PDEs.

All the above features do not appear in systems described by ordinary differential equations (ODEs), and this has led the research community to create novel mathematical tools which can be used for the derivation of an ISS estimate for the solution of a PDE. So far research has mostly focused on 1-D linear or semilinear PDEs. The reader can consult the recent book [99] for a detailed description. More specifically, for hyperbolic PDEs researchers have used:

(a) ISS Lyapunov functionals, which usually provide ISS estimates in L_p-norms with $p < +\infty$, and

(b) the strong relation between hyperbolic PDEs and delay equations [25, 100, 101]. This relation is most useful in 1-D hyperbolic PDEs and provides ISS estimates in the sup-spatial norm.

For parabolic PDEs researchers have used numerically inspired methods which relate the solution of a PDE to the solution of a finite-dimensional discrete-time system. Such methods were used in [102, 103], where ISS estimates in various spatial state norms were derived (including the sup-spatial norm). Researchers have also used

(a) ISS Lyapunov functionals,

(b) eigenfunction expansions of the solution, which can be used for the L_2 spatial norm as well as for the spatial $\mathcal{H}_1$-norm,

(c) semigroup theory, which is particularly useful for linear PDEs, and

(d) monotonicity methods, i.e., methods that exploit the monotonicity of a particular PDE system [154].

Examples of Results of ISS for PDEs

As an example, consider the 1-D heat equation

$$\frac{\partial u}{\partial t}(t,x) = p\frac{\partial^2 u}{\partial x^2}(t,x) + f(t,x) \quad \forall (t,x) \in (0,+\infty)\times(0,1), \tag{1.145}$$

where $p > 0$ is a constant, subject to the boundary conditions

$$u(t,0) - d_0(t) = u(t,1) = 0 \tag{1.146}$$

with $d_0 : \mathbb{R}_+ \to \mathbb{R}$ being a given locally bounded boundary disturbance input and $f : \mathbb{R}_+ \times [0,1] \to \mathbb{R}$ being a locally bounded distributed input. Using the results in Chapters 5 and 6 of [99], we have that

- every solution $u \in C^0(\mathbb{R}_+; L_2(0,1))$ with $u \in C^1((0,+\infty)\times[0,1])$ satisfying $u[t] \in C^2([0,1])$, with $u[t]$ denoting the profile at certain time $t \geq 0$, i.e., $(u[t])(z) = u(t,z)$ for all $z \in [0,1]$, (1.146) for $t > 0$ and (1.145) satisfies the following ISS estimate for all $t \geq 0$:

$$||u[t]||_2 \leq \exp(-p\pi^2 t)||u[0]||_2 + \frac{1}{\sqrt{3}}\sup_{0<s<t}(|d_0(s)|) + \frac{1}{p\pi^2}\sup_{0<s<t}(||f[s]||_2); \tag{1.147}$$

- every solution $u \in C^0(\mathbb{R}_+ \times [0,1])$ with $u \in C^1((0,+\infty)\times[0,1])$ satisfying $u[t] \in C^2([0,1])$ for $t > 0$, $u[0] \in C^2([0,1])$, (1.146) for $t > 0$ and (1.145) satisfies the following ISS estimate for all $t \geq 0$:

$$||u[t]||_\infty \leq \sqrt{2}\max\left(\exp\left(-\frac{p\pi^2}{4}t\right)||u[0]||_\infty, \sup_{0\leq s\leq t}(|d_0(s)|)\right) + \frac{4\sqrt{2}}{p\pi^2}\sup_{0\leq s\leq t}(||f[s]||_\infty]), \tag{1.148}$$

where $||u[t]||_2 = \left(\int_0^1 u^2(t,x)dx\right)^{1/2}$ and $||u[t]||_\infty = \max\{|u(t,x)| : x \in [0,1]\}$.

The reader can notice the difference between estimates (1.147) and (1.148). As remarked above, the selection of different spatial (state) norms leads to completely different ISS estimates. It is worth mentioning that the notation adopted in [99] and repeated above is a little bit different from that assumed in this book, although they are completely equivalent.

Small-gain results are also useful for the derivation of ISS estimates for various kinds of PDE systems. In the book [99], small-gain methods were used for the study of ISS in a wide variety of PDE loops (i.e., interconnected PDEs). Moreover, PDE-ODE loops were also studied—for convenience, see Appendix C. It should be noticed that small-gain results rely on accurate estimations of the gain functions (or gain coefficients), and consequently it is usually the case that small-gain results cannot be used in conjunction with qualitative characterizations of the ISS property.

The effort for a complete qualitative characterization of ISS for PDEs continues. The reader should consult [153] and notice that many characterizations of the ISS property for ODEs are not valid for PDEs. It should be also noted that at this point in time, there has been very little research devoted to the Input-to-Output Stability (IOS) property.

It is expected that ISS theory for PDEs will enable researchers to solve important feedback design problems and observer design problems. ISS was recently used in [103] for boundary feedback design in 1-D parabolic semilinear PDEs with nonlinear reaction terms that satisfy a linear growth condition. Future research may address feedback stabilization problems for systems of PDEs as well as more complicated mathematical models, such as free-boundary parabolic PDEs. For instance, the ISS property for a boundary controlled Stefan problem was recently shown in [111].

Since PDEs are used for the mathematical modeling of many phenomena, it is also expected that ISS theory for PDEs will have important applications to all sciences. In this book, it is particularly explored in Chapters 14 to 17.

Part I

EXTREMUM SEEKING FOR TIME-DELAY SYSTEMS

Chapter 2

Gradient Extremum Seeking for Scalar Static Map with Delay

2.1 ▪ This Chapter as a Template for Understanding the Whole Book

Following the introductory Chapter 1, this is the first chapter through which the book makes a contribution to the literature on extremum seeking. This is also the cornerstone chapter of the book. The key ideas in the design and analysis are introduced here. The chapter is not general and all-encompassing. The later chapters are not specializations of this chapter but they are extensions to this chapter. If the reader understands the design and the proof process in this chapter, he will have a chance to understand the same elements in the subsequent chapters. Conversely, if the reader does not succeed in understanding the tools of this chapter, his chances to understand the later chapters are limited. All this means, in particular, is that if the reader has the time to read only one chapter of this book, it is this chapter.

In this chapter, we derive the control algorithm and present the stability analysis for scalar gradient extremum seeking (ES) in the presence of arbitrarily long input-output delays. In our design we employ a predictor, to compensate for the delay. Since a predictor is always model-based, we are faced with the problem that we need to perform a prediction of a system in which either the input map or the output map contains a dependency on the Hessian, which is unknown. Hence, to perform a model-based prediction, i.e., delay compensation, we need an estimate of the Hessian. Our estimate of the Hessian is generated similarly to the estimation of the Hessian in Newton-based extremum seeking in Section 1.3, namely, using a perturbation or demodulation signal $N(t)$.

In our proof, we guarantee the exponential stability and convergence to a small neighborhood of the unknown extremum point. This result is carried out using a backstepping transformation and averaging in infinite dimensions. The proof process is quite elaborate and it is crucial that the reader absorbs its steps and their order: (1) representation of the delay as a transport PDE; (2) representation of the closed-loop system in a form in which the perturbation signals are present and the parameter estimation error is delayed; (3) derivation of the average PDE-ODE model; (4) application of a backstepping transformation to the averaged PDE-ODE model; (5) introduction of a Lyapunov–Krasovskii functional for the backstepping target system and a stability computation for this functional; (6) generation of an exponential stability estimate for the average PDE-ODE system in the original variables, using the inverse backstepping transformation; (7) application of the averaging theorem for delay systems to obtain exponential stability of a small periodic orbit near the extremum; and (8) derivation of the asymptotic estimates for the convergence of the output of the unknown map to the vicinity of its extremum.

We start in Section 2.2 introducing the predictor design for delay compensation in scalar gradient ES. Exponential stability with explicit Lyapunov–Krasovskii functionals and the real-time convergence to a small neighborhood of the desired extremum are proved in Section 2.3. Section 2.4 presents numerical examples to illustrate the applicability of the proposed extremum seeking with delay compensation. Finally, Sections 2.5 and 2.6 conclude the chapter identifying some potential applications and open problems as well as discussing the literature review on the topic.

2.2 ▪ Extremum Seeking with Delays

Scalar ES considers applications in which the goal is to maximize (or minimize) the output $y \in \mathbb{R}$ of an unknown nonlinear static map $Q(\theta)$ by varying the input $\theta \in \mathbb{R}$.

Here, we additionally assume that there is a *constant and known* delay $D \geq 0$ in the actuation path or measurement system such that the measured output is given by

$$y(t) = Q(\theta(t-D)). \tag{2.1}$$

For notational clarity, we assume that our system is output-delayed in the following presentation and block diagrams. However, the extension of the results in this chapter to the input-delay case is straightforward since, for a static map, any input delay can be moved to the output. The case when input delays D_{in} and output delays D_{out} occur simultaneously could also be handled by declaring that the total delay to be compensated is $D = D_{\text{in}} + D_{\text{out}}$, with $D_{\text{in}}, D_{\text{out}} \geq 0$.

Without loss of generality, let us consider the maximum seeking problem such that the maximizing value of θ is denoted by θ^*. For the sake of simplicity, we also assume that the nonlinear map is quadratic, i.e.,

$$Q(\theta) = y^* + \frac{H}{2}(\theta - \theta^*)^2, \tag{2.2}$$

where, besides the constants $\theta^* \in \mathbb{R}$ and $y^* \in \mathbb{R}$ being unknown, the scalar $H < 0$ is the unknown Hessian of the static map.

By plugging (2.2) into (2.1), we obtain the *quadratic static map with delay*,

$$y(t) = y^* + \frac{H}{2}(\theta(t-D) - \theta^*)^2. \tag{2.3}$$

2.2.1 ▪ Probing and Demodulation Signals

Let $\hat{\theta}$ be the estimate of θ^* and

$$\tilde{\theta}(t) = \hat{\theta}(t) - \theta^* \tag{2.4}$$

be the *estimation error*. From Figure 2.1 and (2.4), the *error dynamics* can be written as

$$\dot{\tilde{\theta}}(t) = \dot{\hat{\theta}}(t) = U(t) \quad \text{and} \quad \dot{\tilde{\theta}}(t-D) = U(t-D). \tag{2.5}$$

Moreover, one has

$$G(t) = M(t)y(t), \tag{2.6}$$

$$\theta(t) = \hat{\theta}(t) + S(t), \tag{2.7}$$

where the perturbation signal and demodulation signal, respectively, are given by

$$S(t) = a\sin(\omega(t+D)), \tag{2.8}$$

$$M(t) = \frac{2}{a}\sin(\omega t) \tag{2.9}$$

with nonzero perturbation amplitude a and frequency ω.

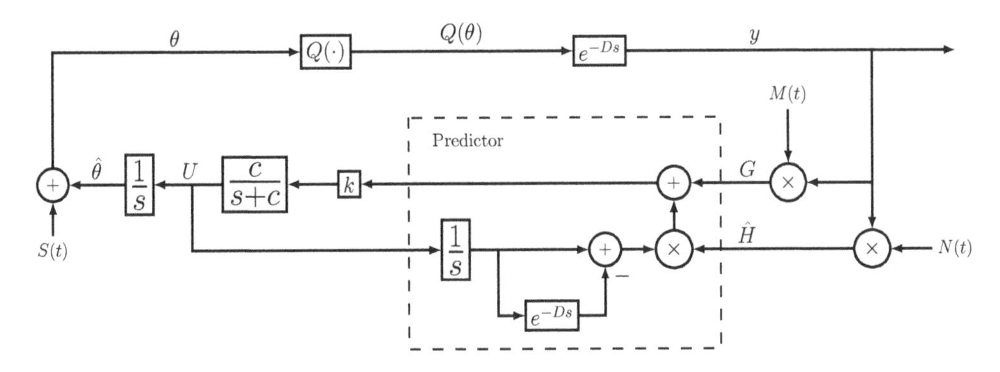

Figure 2.1. *Block diagram of the basic prediction scheme for output-delay compensation in gradient ES. The predictor feedback with a perturbation-based estimate of the Hessian obeys (2.22); the probing and demodulation signals are given, respectively, by $S(t) = a\sin(\omega(t+D))$ and $M(t) = \frac{2}{a}\sin(\omega t)$; and the demodulating signal $N(t) = -\frac{8}{a^2}\cos(2\omega t)$.*

Notice that (2.8) is different from the perturbation signal used in the standard gradient extremum seeking free of delays. Since the output delay can be transferred to the integrator output for analysis purposes (or, equivalently, to its input), the phase shift $+\omega D$ in (2.8) is applied to compensate the delay effect in the perturbation signal $S(t)$.

The signal

$$\hat{H}(t) = N(t)y(t) \tag{2.10}$$

is used to obtain an estimate of the unknown Hessian H, where the demodulating signal $N(t)$ is given by

$$N(t) = -\frac{8}{a^2}\cos(2\omega t). \tag{2.11}$$

In [73], it was proved that

$$\frac{1}{\Pi}\int_0^{\Pi} N(\sigma)y d\sigma = H, \quad \Pi = 2\pi/\omega \tag{2.12}$$

if a quadratic map as in (2.2) is considered. In other words, the average version of (2.10) is $\hat{H}_{\text{av}} = (Ny)_{\text{av}} = H$; i.e., the unknown Hessian is correctly estimated, on the average, using a product of available signals, the measured $y(t)$ and the given $N(t)$.

2.2.2 ▪ Predictor Feedback with a Perturbation-Based Estimate of the Hessian

By using the averaging analysis, we can verify that the average version of the signal (2.6) is

$$G_{\text{av}}(t) = H\tilde{\theta}_{\text{av}}(t-D). \tag{2.13}$$

From (2.5) and (2.13), the following average models are also obtained:

$$\dot{\tilde{\theta}}_{\text{av}}(t-D) = U_{\text{av}}(t-D), \tag{2.14}$$

$$\dot{G}_{\text{av}}(t) = HU_{\text{av}}(t-D), \tag{2.15}$$

where $U_{\text{av}} \in \mathbb{R}$ is the resulting average control for $U \in \mathbb{R}$.

In order to motivate the predictor feedback design, the idea here is to compensate for the delay by feeding back the future state $G(t+D)$, or $G_{\rm av}(t+D)$ in the equivalent average system.

Given any stabilizing gain $k>0$ for the undelayed system, our wish is to have a control that achieves

$$U_{\rm av}(t)=kG_{\rm av}(t+D)\quad \forall t\geq 0\,, \tag{2.16}$$

and this feedback law appears to be nonimplementable since it requires future values of the state. However, by applying the variation-of-constants formula to (2.15), we can express the future state as

$$G_{\rm av}(t+D)=G_{\rm av}(t)+H\int_{t}^{t+D}U_{\rm av}(\sigma-D)d\sigma\,, \tag{2.17}$$

where the current state $G_{\rm av}(t)$ is the initial condition. Shifting the time variable under the integral in (2.17), we obtain

$$G_{\rm av}(t+D)=G_{\rm av}(t)+H\int_{t-D}^{t}U_{\rm av}(\sigma)d\sigma\,, \tag{2.18}$$

which gives the future state $G_{\rm av}(t+D)$ in terms of the average control signal $U_{\rm av}(\sigma)$ from the past window $[t-D,t]$. It yields the following feedback law:

$$U_{\rm av}(t)=k\left[G_{\rm av}(t)+H\int_{t-D}^{t}U_{\rm av}(\sigma)d\sigma\right]. \tag{2.19}$$

Hence, from (2.18) and (2.19), the average feedback law (2.16) can be obtained indeed as desired. Consequently,

$$\dot{\hat{\theta}}_{\rm av}(t)=kG_{\rm av}(t+D)\quad \forall t\geq 0\,. \tag{2.20}$$

Therefore, from (2.13), one has

$$\frac{d\tilde{\theta}_{\rm av}(t)}{dt}=kH\tilde{\theta}_{\rm av}(t)\quad \forall t\geq D\,, \tag{2.21}$$

with an exponentially attractive equilibrium $\tilde{\theta}^{e}_{\rm av}=0$, since $k>0$ in the control design and $H<0$ by assumption. It means that the delay is perfectly compensated in D units of time, namely that the system evolves as if the delay were absent after $D\geq 0$.

In the next section, we show that the control objectives can still be achieved if a simple modification of the above basic predictor-based controller, which employs a low-pass filter, is applied. In this case, we propose the following infinite-dimensional and averaging-based predictor feedback in order to compensate the delay [115]

$$U(t)=\frac{c}{s+c}\left\{k\left[G(t)+\hat{H}(t)\int_{t-D}^{t}U(\tau)d\tau\right]\right\}, \tag{2.22}$$

where $c>0$ is sufficiently large, i.e., the predictor feedback is of the form of a low-pass filtering of the nonaverage version of (2.19). This low-pass filtering is particularly required in the stability analysis when the averaging theorem in infinite dimensions [83, 131], rewritten in Appendix A, is invoked. Note that we mix the time- and frequency-domain notation in (2.22) by using the braces $\{\cdot\}$ to denote that the transfer function acts as an operator on a time-domain function.

The predictor feedback (2.22) is infinite-dimensional because the integral involves the control history over the interval $[t-D,t]$. This feedback is also averaging-based (perturbation-based) because $\hat{H}$ is updated according to the estimate (2.10) of the unknown Hessian H, which satisfies the averaging property (2.12).

2.3 ▪ Stability Analysis

The main stability/convergence results in the L_2-norm for the closed-loop system are summarized in the next theorem for gradient ES in the presence of time delays.

Theorem 2.1. *Consider the closed-loop system in Figure* 2.1 *with delayed output* (2.3). *There exists* $c^* > 0$ *such that,* $\forall c \geq c^*$, $\exists\, \omega^*(c) > 0$ *such that,* $\forall \omega > \omega^*$, *the closed-loop delayed system* (2.5), (2.22), (2.6), *and* (2.10) *with state* $\tilde{\theta}(t-D)$, $U(\tau)$, $\forall \tau \in [t-D,t]$, *has a unique exponentially stable periodic solution in* t *of period* $\Pi = 2\pi/\omega$, *denoted by* $\tilde{\theta}^{\Pi}(t-D)$, $U^{\Pi}(\tau)$, $\forall \tau \in [t-D,t]$, *satisfying,* $\forall t \geq 0$,

$$\left(\left|\tilde{\theta}^{\Pi}(t-D)\right|^2+\left[U^{\Pi}(t)\right]^2+\int_{t-D}^{t}\left[U^{\Pi}(\tau)\right]^2 d\tau\right)^{1/2} \leq \mathcal{O}(1/\omega). \tag{2.23}$$

Furthermore,

$$\limsup_{t\to+\infty}|\theta(t)-\theta^*| = \mathcal{O}(a+1/\omega)\,, \tag{2.24}$$

$$\limsup_{t\to+\infty}|y(t)-y^*| = \mathcal{O}(a^2+1/\omega^2)\,. \tag{2.25}$$

Proof. The proof consists of the following eight steps (whose order is crucial).

Step 1: Transport PDE for Delay Representation

According to [119], the delay in (2.5) can be represented using a transport PDE as

$$\dot{\tilde{\theta}}(t-D) = u(0,t)\,, \tag{2.26}$$

$$u_t(x,t) = u_x(x,t)\,, \quad x \in [0,D]\,, \tag{2.27}$$

$$u(D,t) = U(t)\,, \tag{2.28}$$

where the solution of (2.27)–(2.28) is

$$u(x,t) = U(t+x-D)\,. \tag{2.29}$$

Step 2: Equations of the Closed-Loop System

First, substituting (2.8) into (2.7), we obtain

$$\theta(t) = \hat{\theta}(t) + a\sin(\omega(t+D))\,. \tag{2.30}$$

Now, plug (2.4) and (2.30) into (2.3) so that the output is given in terms of $\tilde{\theta}$:

$$y(t) = y^* + \frac{H}{2}(\tilde{\theta}(t-D)+a\sin(\omega t))^2\,. \tag{2.31}$$

By plugging (2.9) into (2.6) and (2.11) into (2.10) and representing the integrand in (2.22) using the transport PDE state, one has

$$U(t)=\frac{c}{s+c}\left\{k\left[G(t)+\hat{H}(t)\int_0^D u(\sigma,t)d\sigma\right]\right\}, \tag{2.32}$$

$$G(t)=\frac{2}{a}\sin(\omega t)y(t), \tag{2.33}$$

$$\hat{H}(t)=-\frac{8}{a^2}\cos(2\omega t)y(t). \tag{2.34}$$

Plug (2.31) into (2.33) and (2.34), and then the resulting (2.33) and (2.34) into (2.32). By extracting the common factor y in the resulting version of (2.32), one has

$$\begin{aligned}U(t)=\frac{c}{s+c}&\left\{k\left[y^*+\frac{H}{2}(\tilde{\theta}(t-D)+a\sin(\omega t))^2\right]\right.\\&\left.\times\left[\frac{2}{a}\sin(\omega t)-\frac{8}{a^2}\cos(2\omega t)\int_0^D u(\sigma,t)d\sigma\right]\right\}.\end{aligned} \tag{2.35}$$

By expanding the binome in (2.35), we obtain

$$\begin{aligned}U(t)=\frac{c}{s+c}&\left\{k\left[y^*+\frac{H}{2}\tilde{\theta}^2(t-D)\right.\right.\\&\left.+Ha\sin(\omega t)\tilde{\theta}(t-D)+\frac{a^2H}{2}\sin^2(\omega t)\right]\\&\left.\times\left[\frac{2}{a}\sin(\omega t)-\frac{8}{a^2}\cos(2\omega t)\int_0^D u(\sigma,t)d\sigma\right]\right\}.\end{aligned} \tag{2.36}$$

Finally, substituting (2.36) into (2.28), we can rewrite (2.26)–(2.28) as

$$\dot{\tilde{\theta}}(t-D)=u(0,t), \tag{2.37}$$

$$\partial_t u(x,t)=\partial_x u(x,t),\quad x\in[0,D], \tag{2.38}$$

$$\begin{aligned}u(D,t)=&\frac{c}{s+c}\left\{k\left[y^*+\frac{H}{2}\tilde{\theta}^2(t-D)\right.\right.\\&\left.+Ha\sin(\omega t)\tilde{\theta}(t-D)+\frac{a^2H}{2}\sin^2(\omega t)\right]\\&\left.\times\left[\frac{2}{a}\sin(\omega t)-\frac{8}{a^2}\cos(2\omega t)\int_0^D u(\sigma,t)d\sigma\right]\right\}\\=&\frac{c}{s+c}\left\{k\left[y^*\frac{2}{a}\sin(\omega t)-y^*\frac{8}{a^2}\cos(2\omega t)\int_0^D u(\sigma,t)d\sigma\right.\right.\\&+\frac{H}{a}\tilde{\theta}^2(t-D)\sin(\omega t)\\&-\frac{4H}{a^2}\tilde{\theta}^2(t-D)\cos(2\omega t)\int_0^D u(\sigma,t)d\sigma\\&+2H\sin^2(\omega t)\tilde{\theta}(t-D)\\&-\frac{8H}{a}\sin(\omega t)\tilde{\theta}(t-D)\cos(2\omega t)\int_0^D u(\sigma,t)d\sigma\\&\left.\left.+aH\sin^3(\omega t)-4H\sin^2(\omega t)\cos(2\omega t)\int_0^D u(\sigma,t)d\sigma\right]\right\}\end{aligned}$$

$$
\begin{aligned}
&= \frac{c}{s+c}\Bigg\{k\Bigg[y^*\frac{2}{a}\sin(\omega t) - y^*\frac{8}{a^2}\cos(2\omega t)\int_0^D u(\sigma,t)d\sigma \\
&+\frac{H}{a}\tilde{\theta}^2(t-D)\sin(\omega t) \\
&-\frac{4H}{a^2}\tilde{\theta}^2(t-D)\cos(2\omega t)\int_0^D u(\sigma,t)d\sigma \\
&+H\tilde{\theta}(t-D) - H\cos(2\omega t)\tilde{\theta}(t-D) \\
&-\frac{4H}{a}[\sin(3\omega t) - \sin(\omega t)]\tilde{\theta}(t-D)\int_0^D u(\sigma,t)d\sigma \\
&+\frac{3aH}{4}\sin(\omega t) - \frac{aH}{4}\sin(3\omega t) - 2H\cos(2\omega t)\int_0^D u(\sigma,t)d\sigma \\
&+[H + H\cos(4\omega t)]\int_0^D u(\sigma,t)d\sigma\Bigg]\Bigg\}.
\end{aligned} \tag{2.39}
$$

Step 3: Average Model of the Closed-Loop System

Now denoting

$$
\tilde{\vartheta}(t) = \tilde{\theta}(t-D), \tag{2.40}
$$

the average version of system (2.37)–(2.39) is

$$
\dot{\tilde{\vartheta}}_{\text{av}}(t) = u_{\text{av}}(0,t), \tag{2.41}
$$

$$
\partial_t u_{\text{av}}(x,t) = \partial_x u_{\text{av}}(x,t), \quad x \in [0,D], \tag{2.42}
$$

$$
\frac{d}{dt}u_{\text{av}}(D,t) = -cu_{\text{av}}(D,t) + ckH\left[\tilde{\vartheta}_{\text{av}}(t) + \int_0^D u_{\text{av}}(\sigma,t)d\sigma\right], \tag{2.43}
$$

where in the last line we have simply set all the averages of the sine and cosine functions of ω, 2ω, 3ω, and 4ω to zero in (2.39). Moreover, the filter $c/s+c$ is also represented in the state-space form. Analogously to (2.29), the solution of the transport PDE (2.42)–(2.43) is given by

$$
u_{\text{av}}(x,t) = U_{\text{av}}(t+x-D). \tag{2.44}
$$

Step 4: Backstepping Transformation, Its Inverse, and the Target System

Consider the infinite-dimensional backstepping transformation of the delay state

$$
w(x,t) = u_{\text{av}}(x,t) - kH\left[\tilde{\vartheta}_{\text{av}}(t) + \int_0^x u_{\text{av}}(\sigma,t)d\sigma\right], \tag{2.45}
$$

which maps the system (2.41)–(2.43) into the target system:

$$
\dot{\tilde{\vartheta}}_{\text{av}}(t) = kH\tilde{\vartheta}_{\text{av}}(t) + w(0,t), \tag{2.46}
$$

$$
w_t(x,t) = w_x(x,t), \quad x \in [0,D], \tag{2.47}
$$

$$
w(D,t) = -\frac{1}{c}\partial_t u_{\text{av}}(D,t). \tag{2.48}
$$

Using (2.45) for $x = D$ and the fact that $u_{\rm av}(D,t) = U_{\rm av}(t)$, from (2.48) we get (2.43), i.e.,

$$U_{\rm av}(t) = \frac{c}{s+c}\left\{kH\left[\tilde{\vartheta}_{\rm av}(t) + \int_0^D u_{\rm av}(\sigma,t)d\sigma\right]\right\}. \tag{2.49}$$

Let us now consider $w(D,t)$. It is easily seen that

$$w_t(D,t) = \partial_t u_{\rm av}(D,t) - kHu_{\rm av}(D,t), \tag{2.50}$$

where $\partial_t u_{\rm av}(D,t) = \dot{U}_{\rm av}(t)$. The inverse of the backstepping transformation (2.45) is given by

$$u_{\rm av}(x,t) = w(x,t) + kH\left[e^{kHx}\tilde{\vartheta}_{\rm av}(t) + \int_0^x e^{kH(x-\sigma)}w(\sigma,t)d\sigma\right]. \tag{2.51}$$

Plugging (2.48) and (2.51) into (2.50), we get

$$\begin{aligned} w_t(D,t) &= -cw(D,t) - kHw(D,t) \\ &\quad -(kH)^2\left[e^{kHD}\tilde{\vartheta}_{\rm av}(t) + \int_0^D e^{kH(D-\sigma)}w(\sigma,t)d\sigma\right]. \end{aligned} \tag{2.52}$$

Step 5: Lyapunov–Krasovskii Functional

Now, consider the following Lyapunov functional:

$$V(t) = \frac{\tilde{\vartheta}_{\rm av}^2(t)}{2} + \frac{a}{2}\int_0^D (1+x)w^2(x,t)dx + \frac{1}{2}w^2(D,t), \tag{2.53}$$

where the parameter $a > 0$ is to be chosen later. We have

$$\begin{aligned} \dot{V}(t) &= kH\tilde{\vartheta}_{\rm av}^2(t) + \tilde{\vartheta}_{\rm av}(t)w(0,t) \\ &\quad + a\int_0^D (1+x)w(x,t)w_x(x,t)dx + w(D,t)w_t(D,t) \\ &= kH\tilde{\vartheta}_{\rm av}^2(t) + \tilde{\vartheta}_{\rm av}(t)w(0,t) + \frac{a(1+D)}{2}w^2(D,t) \\ &\quad - \frac{a}{2}w^2(0,t) - \frac{a}{2}\int_0^D w^2(x,t)dx + w(D,t)w_t(D,t) \\ &\leq kH\tilde{\vartheta}_{\rm av}^2(t) + \frac{\tilde{\vartheta}_{\rm av}^2(t)}{2a} - \frac{a}{2}\int_0^D w^2(x,t)dx \\ &\quad + w(D,t)\left[w_t(D,t) + \frac{a(1+D)}{2}w(D,t)\right]. \end{aligned}$$

Recalling that $k > 0$ and $H < 0$, let us choose

$$a = -\frac{1}{kH}. \tag{2.54}$$

Then,

$$\dot{V}(t) \le \frac{kH}{2}\tilde{\vartheta}_{\rm av}^2(t) + \frac{1}{2kH}\int_0^D w^2(x,t)dx$$
$$+w(D,t)\left[w_t(D,t) - \frac{(1+D)}{2kH}w(D,t)\right]$$
$$= -\frac{1}{2a}\tilde{\vartheta}_{\rm av}^2(t) - \frac{a}{2}\int_0^D w^2(x,t)dx$$
$$+w(D,t)\left[w_t(D,t) + \frac{a(1+D)}{2}w(D,t)\right]. \tag{2.55}$$

Hereafter, we consider (2.55) along with (2.52). With a completion of squares, we obtain

$$\dot{V}(t) \le -\frac{1}{4a}\tilde{\vartheta}_{\rm av}^2(t) - \frac{a}{4}\int_0^D w^2(x,t)dx + a\left|(kH)^2 e^{kHD}\right|^2 w^2(D,t)$$
$$+\frac{1}{a}\left\|(kH)^2 e^{kH(D-\sigma)}\right\|^2 w^2(D,t)$$
$$+\left[\frac{a(1+D)}{2} - kH\right] w^2(D,t) - cw^2(D,t). \tag{2.56}$$

To obtain (2.56), we have used

$$-w(D,t)\left\langle (kH)^2 e^{kH(D-\sigma)}, w(\sigma,t)\right\rangle \le |w(D,t)|\left\|(kH)^2 e^{kH(D-\sigma)}\right\| \|w(t)\| \tag{2.57}$$
$$\le \frac{a}{4}\|w(t)\|^2 + \frac{1}{a}\left\|(kH)^2 e^{kH(D-\sigma)}\right\|^2 w^2(D,t),$$

where the first inequality is the Cauchy–Schwarz and the second one is the Young inequality; the notation $\langle\cdot,\cdot\rangle$ denotes the inner product in the spatial variable $\sigma \in [0,D]$, on which both terms $e^{kH(D-\sigma)}$ and $w(\sigma,t)$ depend; and $\|\cdot\|$ denotes the L_2-norm in σ. Then, from (2.56), we arrive at

$$\dot{V}(t) \le -\frac{1}{4a}\tilde{\vartheta}_{\rm av}^2(t) - \frac{a}{4(1+D)}\int_0^D (1+x)w^2(x,t)dx - (c-c^*)w^2(D,t), \tag{2.58}$$

where

$$c^* = \frac{a(1+D)}{2} - kH + a\left|(kH)^2 e^{kHD}\right|^2 + \frac{1}{a}\left\|(kH)^2 e^{kH(D-\sigma)}\right\|^2. \tag{2.59}$$

From (2.59), it is clear that an upper bound for c^* can be obtained from known lower and upper bounds of the unknown Hessian H. Hence, from (2.58), if c is chosen such that $c > c^*$, we obtain

$$\dot{V}(t) \le -\mu V(t) \tag{2.60}$$

for some $\mu > 0$. Thus, the closed-loop system is exponentially stable in the sense of the full-state norm

$$\left(|\tilde{\vartheta}_{\rm av}(t)|^2 + \int_0^D w^2(x,t)dx + w^2(D,t)\right)^{1/2}, \tag{2.61}$$

i.e., in the transformed variable $(\tilde{\vartheta}_{\rm av}, w)$.

Step 6: Exponential Stability Estimate (in the L_2-Norm) for the Average System (2.41)–(2.43)

To obtain exponential stability in the sense of the norm

$$\left(|\tilde{\vartheta}_{\mathrm{av}}(t)|^2+\int_0^D u_{\mathrm{av}}^2(x,t)dx+u_{\mathrm{av}}^2(D,t)\right)^{1/2},$$

we need to show there exist positive numbers α_1 and α_2 (see Appendix D for a complete computation) such that

$$\alpha_1\Psi(t)\leq V(t)\leq\alpha_2\Psi(t), \tag{2.62}$$

where $\Psi(t)\triangleq|\tilde{\vartheta}_{\mathrm{av}}(t)|^2+\int_0^D u_{\mathrm{av}}^2(x,t)dx+u_{\mathrm{av}}^2(D,t)$ or, equivalently,

$$\Psi(t)\triangleq|\tilde{\theta}_{\mathrm{av}}(t-D)|^2+\int_{t-D}^t U_{\mathrm{av}}^2(\tau)d\tau+U_{\mathrm{av}}^2(t), \tag{2.63}$$

using (2.40) and (2.44). This is straightforward to establish by using (2.45), (2.51), (2.53) and employing the Cauchy–Schwarz inequality and other calculations, as in the proof of Theorem 2.1 in [119]. Hence, with (2.60), we get

$$\Psi(t)\leq\frac{\alpha_2}{\alpha_1}e^{-\mu t}\Psi(0), \tag{2.64}$$

which completes the proof of exponential stability.

Step 7: Invoking the Averaging Theorem

First, note that the closed-loop system (2.5) and (2.22) can be rewritten as

$$\dot{\tilde{\theta}}(t-D)=U(t-D), \tag{2.65}$$

$$\dot{U}(t)=-cU(t)+c\left\{k\left[G(t)+\hat{H}(t)\int_{t-D}^t U(\tau)d\tau\right]\right\}, \tag{2.66}$$

where $z(t)=[\tilde{\theta}(t-D),U(t)]^T$ is the state vector. Moreover, from (2.6) and (2.10), one has

$$\dot{z}(t)=f(\omega t,z_t), \tag{2.67}$$

where $z_t(\Theta)=z(t+\Theta)$ for $-D\leq\Theta\leq0$ and f is an appropriate continuous functional, such that the averaging theorem by [83] and [131] in Appendix A can be directly applied considering $\omega=1/\epsilon$.

From (2.64), the origin of the average closed-loop system (2.41)–(2.43) with transport PDE for delay representation is exponentially stable. Then, according to the averaging theorem [83, 131] in Appendix A, for ω sufficiently large, (2.37)–(2.39), or equivalently (2.65)–(2.66), has a unique exponentially stable periodic solution around its equilibrium (origin) satisfying (2.23).

Step 8: Asymptotic Convergence to a Neighborhood of the Extremum (θ^*,y^*)

By using the change of variables (2.40) and then integrating both sides of (2.26) within the interval $[t,\sigma+D]$, we have

$$\tilde{\vartheta}(\sigma+D)=\tilde{\vartheta}(t)+\int_t^{\sigma+D}u(0,s)ds. \tag{2.68}$$

From (2.29), we can rewrite (2.68) in terms of U, namely

$$\tilde{\vartheta}(\sigma+D)=\tilde{\vartheta}(t)+\int_{t-D}^{\sigma}U(\tau)d\tau\,. \tag{2.69}$$

Now, note that

$$\tilde{\theta}(\sigma)=\tilde{\vartheta}(\sigma+D)\quad \forall\sigma\in[t-D,t]\,. \tag{2.70}$$

Hence,

$$\tilde{\theta}(\sigma)=\tilde{\theta}(t-D)+\int_{t-D}^{\sigma}U(\tau)d\tau\quad \forall\sigma\in[t-D,t]\,. \tag{2.71}$$

By applying the supremum norm in both sides of (2.71), we have

$$\begin{aligned}\sup_{t-D\leq\sigma\leq t}\left|\tilde{\theta}(\sigma)\right| &= \sup_{t-D\leq\sigma\leq t}\left|\tilde{\theta}(t-D)\right|+\sup_{t-D\leq\sigma\leq t}\left|\int_{t-D}^{\sigma}U(\tau)d\tau\right| \\ &\leq \sup_{t-D\leq\sigma\leq t}\left|\tilde{\theta}(t-D)\right|+\sup_{t-D\leq\sigma\leq t}\int_{t-D}^{t}\left|U(\tau)\right|d\tau \\ &\leq \left|\tilde{\theta}(t-D)\right|+\int_{t-D}^{t}\left|U(\tau)\right|d\tau \quad \text{(Cauchy–Schwarz)} \\ &\leq \left|\tilde{\theta}(t-D)\right|+\left(\int_{t-D}^{t}d\tau\right)^{1/2}\times\left(\int_{t-D}^{t}\left|U(\tau)\right|^{2}d\tau\right)^{1/2} \\ &\leq \left|\tilde{\theta}(t-D)\right|+\sqrt{D}\left(\int_{t-D}^{t}U^{2}(\tau)d\tau\right)^{1/2}\,. \end{aligned} \tag{2.72}$$

Now, it is easy to check

$$\left|\tilde{\theta}(t-D)\right|\leq\left(\left|\tilde{\theta}(t-D)\right|^{2}+\int_{t-D}^{t}U^{2}(\tau)d\tau\right)^{1/2}, \tag{2.73}$$

$$\left(\int_{t-D}^{t}U^{2}(\tau)d\tau\right)^{1/2}\leq\left(\left|\tilde{\theta}(t-D)\right|^{2}+\int_{t-D}^{t}U^{2}(\tau)d\tau\right)^{1/2}. \tag{2.74}$$

By using (2.73) and (2.74), one has

$$\begin{aligned}\left|\tilde{\theta}(t-D)\right|+\sqrt{D}\left(\int_{t-D}^{t}U^{2}(\tau)d\tau\right)^{1/2}\leq(1+\sqrt{D})\Bigg(&\left|\tilde{\theta}(t-D)\right|^{2} \\ &+\int_{t-D}^{t}U^{2}(\tau)d\tau\Bigg)^{1/2}. \end{aligned} \tag{2.75}$$

From (2.72), it is straightforward to conclude that

$$\sup_{t-D\leq\sigma\leq t}\left|\tilde{\theta}(\sigma)\right|\leq(1+\sqrt{D})\left(\left|\tilde{\theta}(t-D)\right|^{2}+\int_{t-D}^{t}U^{2}(\tau)d\tau\right)^{1/2} \tag{2.76}$$

and, consequently,

$$\left|\tilde{\theta}(t)\right|\leq(1+\sqrt{D})\left(\left|\tilde{\theta}(t-D)\right|^{2}+\int_{t-D}^{t}U^{2}(\tau)d\tau\right)^{1/2}. \tag{2.77}$$

Inequality (2.77) can be given in terms of the periodic solution $\tilde{\theta}^{\Pi}(t-D)$, $U^{\Pi}(\sigma)$ $\forall\sigma\in[t-D,t]$ as follows:

$$|\tilde{\theta}(t)|\leq(1+\sqrt{D})\left(\left|\tilde{\theta}(t-D)-\tilde{\theta}^{\Pi}(t-D)+\tilde{\theta}^{\Pi}(t-D)\right|^{2}\right.$$
$$\left.+\int_{t-D}^{t}\left[U(\tau)-U^{\Pi}(\tau)+U^{\Pi}(\tau)\right]^{2}d\tau\right)^{1/2}. \tag{2.78}$$

By applying Young's inequality and some algebra, the RHS of (2.78) and $|\tilde{\theta}(t)|$ can be majorized by

$$|\tilde{\theta}(t)|\leq\sqrt{2}\,(1+\sqrt{D})\left(\left|\tilde{\theta}(t-D)-\tilde{\theta}^{\Pi}(t-D)\right|^{2}+\left|\tilde{\theta}^{\Pi}(t-D)\right|^{2}\right.$$
$$\left.+\int_{t-D}^{t}\left[U(\tau)-U^{\Pi}(\tau)\right]^{2}d\tau+\int_{t-D}^{t}\left[U^{\Pi}(\tau)\right]^{2}d\tau\right)^{1/2}. \tag{2.79}$$

According to the averaging theorem [83, 131] (see Appendix A), we can conclude that

$$\tilde{\theta}(t-D)-\tilde{\theta}^{\Pi}(t-D)\rightarrow 0, \tag{2.80}$$
$$\int_{t-D}^{t}\left[U(\tau)-U^{\Pi}(\tau)\right]^{2}d\tau\rightarrow 0, \tag{2.81}$$

exponentially. Hence,

$$\limsup_{t\rightarrow+\infty}|\tilde{\theta}(t)|=\sqrt{2}\,(1+\sqrt{D})$$
$$\times\left(\left|\tilde{\theta}^{\Pi}(t-D)\right|^{2}+\int_{t-D}^{t}[U^{\Pi}(\tau)]^{2}d\tau\right)^{1/2}. \tag{2.82}$$

From (2.23) and (2.82), we can write

$$\limsup_{t\rightarrow+\infty}|\tilde{\theta}(t)|=\mathcal{O}(1/\omega). \tag{2.83}$$

From (2.4) and recalling that $\theta(t)=\hat{\theta}(t)+S(t)$ with $S(t)=a\sin(\omega(t+D))$, one has that

$$\theta(t)-\theta^{*}=\tilde{\theta}(t)+S(t). \tag{2.84}$$

Since the first term in the RHS of (2.84) is ultimately of order $\mathcal{O}(1/\omega)$ and the second term is of order $\mathcal{O}(a)$, then

$$\limsup_{t\rightarrow+\infty}|\theta(t)-\theta^{*}|=\mathcal{O}(a+1/\omega). \tag{2.85}$$

Finally, from (2.3) and (2.85), we get (2.25). □

2.4 ▪ Simulation with an Academic Example

In order to evaluate the proposed extremum seeking with delay compensation, the following static quadratic map is considered:

$$Q(\theta)=5-(\theta-2)^{2}, \tag{2.86}$$

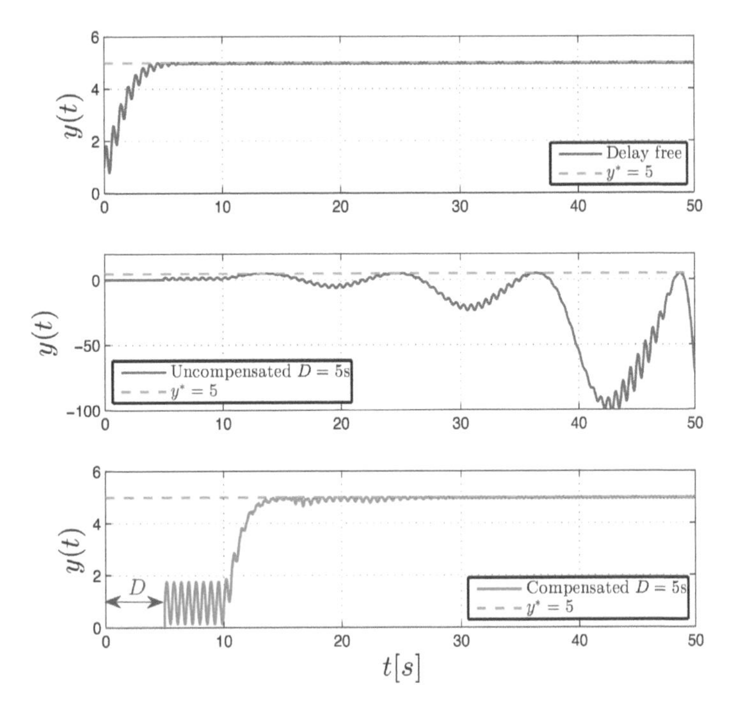

Figure 2.2. ***Gradient-based ES plus output delay (time response of*** $y(t)$***):*** (a) *basic ES works well without delays;* (b) *ES goes unstable in the presence of delays;* (c) *predictor fixes this.*

subject to an output delay of $D = 5\,$s. According to (2.86), the extremum point is $(\theta^*,\ y^*) = (2,\ 5)$ and the Hessian of the map is $H = -2$.

In our simulations, we use low-pass and washout filters with corner frequencies ω_h and ω_l to improve the controller performance as usual in extremum seeking design [129, 123]; see Figures 1.4 and 1.5 in Chapter 1.

In what follows, we present numerical simulations of the predictor (2.22), where $\hat{H}$ is given by (2.10) and $c = 20$. We perform our tests with the following parameters: $a = 0.2$, $\omega = 10$, $k = 0.2$, $\omega_h = \omega_l = 1$, and $\theta(0) = 0$.

Figure 2.2 shows the system output $y(t)$ in 3 situations: (a) free of output delays, (b) in the presence of output delay but without any delay compensation, and (c) with output-delay and predictor-based compensation.

Figure 2.3 presents relevant variables for ES. The red curves are shown when the proposed predictor is applied in comparison to the case free of delay (with blue curves).

It is remarkable how the predictor-based scheme searches and finds the maximum of the unknown map and how it completes the perturbation-based estimation of the unknown Hessian $H = -2$.

2.5 ▪ Application to Traffic Congestion Control with a Downstream Bottleneck

This section develops boundary control for freeway traffic with a downstream bottleneck [232]. Traffic on a freeway segment with capacity drop at the outlet of the segment is a common phenomenon that leads to traffic bottleneck problems.

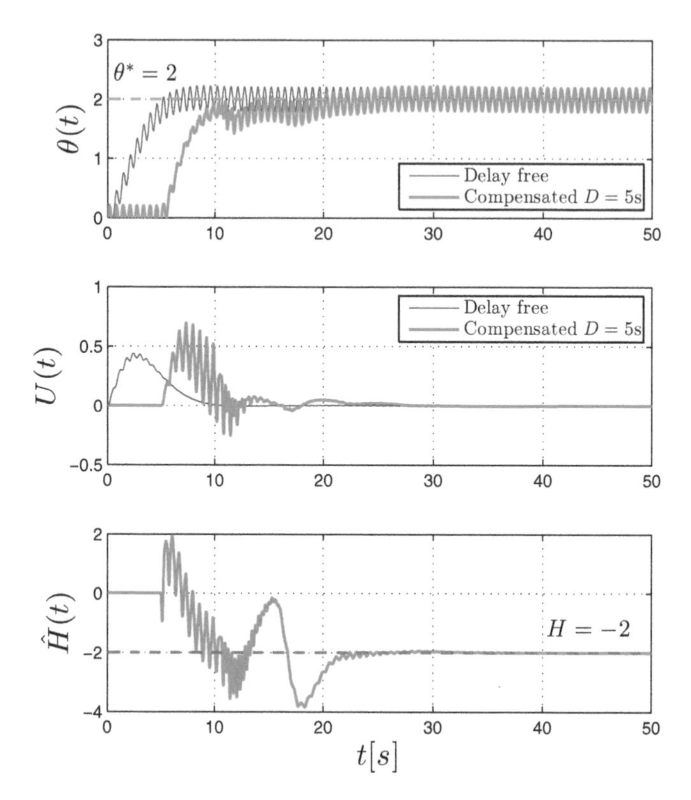

Figure 2.3. ***Gradient-based ES plus output delay:*** (a) *Time response of the parameter* $\theta(t)$*;* (b) *the control signal* $U(t)$*;* (c) *Hessian's estimate* $\hat{H}(t)$.

The capacity drop can be caused by lane-drop, hills, tunnels, bridges, or curvatures on the road. If incoming traffic flow remains unchanged, traffic congestion forms upstream of the bottleneck because the upstream traffic demand exceeds its capacity. Therefore, it is important to regulate the incoming traffic flow of the segment to avoid overloading the bottleneck area. Traffic densities on the freeway segment are described with the Lighthill–Whitham–Richards (LWR) macroscopic Partial Differential Equation (PDE) model. To mitigate the traffic congestion upstream of the bottleneck, incoming flow at the inlet of the freeway segment is controlled so that the optimal density that maximizes the outgoing flow is reached. The density and traffic flow relation at the bottleneck area, described with fundamental diagram, is considered to be unknown. We tackle this problem using Extremum Seeking (ES) Control with delay compensation for LWR PDE [232]. ES control, a non-model-based approach for real-time optimization, is adopted to find the optimal density for the unknown fundamental diagram. A predictor feedback control design is proposed to compensate the delay effect of traffic dynamics in the freeway segment.

2.5.1 ▪ Problem Statement

We consider a traffic congestion problem on a freeway-segment with lane-drop bottleneck downstream of the segment. The freeway segment upstream of the bottleneck and the lane-drop area are shown in Figure 2.4, which illustrates the clear Zone C and the bottleneck Zone B, respectively. The flow is conserved through the clear Zone C to the bottleneck Zone B. The local road capacity is changed due to the lane-drop in Zone B, which could be caused by working zones,

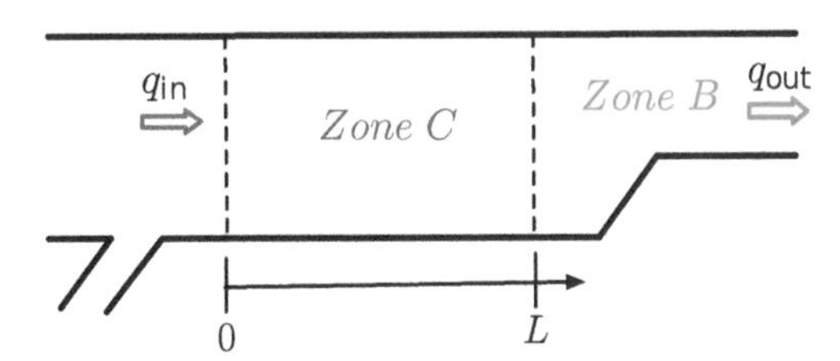

Figure 2.4. *Traffic on a freeway segment with lane-drop.*

accidents, or lane closures. To prevent the traffic in Zone B from overflowing its capacity and then causing congestion in the freeway segment, we aim to find out the optimal density ahead of Zone C that maximizes the outgoing flux of Zone B given an unknown density-flow relation. Traffic dynamics in Zone C is described with the macroscopic LWR traffic model for the aggregated values of traffic density.

Due to the reduction of lanes in Zone B, the fundamental diagram for the flow and density relation usually changes, which leads to a capacity drop in Zone B. The control objective is to find the optimal input density at the inlet of Zone C that drives the measurable output flux of Zone B to its unknown optimal value of an unknown fundamental diagram.

2.5.2 ▪ LWR Traffic Model

The traffic dynamics in Zone C upstream of Zone B is described with the first-order, hyperbolic LWR model. Traffic density $\rho(x,t)$ in Zone C is governed by the following nonlinear hyperbolic PDE, where $x \in [0,L]$, $t \in [0,\infty)$:

$$\partial_t \rho + \partial_x (Q_C(\rho)) = 0. \tag{2.87}$$

The fundamental diagram of traffic flow and density function $Q_C(\rho)$ is given by $Q_C(\rho) = \rho V(\rho)$, where traffic velocity follows an equilibrium velocity-density relation $V(\rho)$. There are different models to describe the flux and density relation. A basic and popular choice is Greenshield's model for $V(\rho)$, which is given by $V(\rho) = v_f \left(1 - \frac{\rho}{\rho_m}\right)$, where $v_f \in \mathbb{R}^+$ is defined as maximum velocity and $\rho_m \in \mathbb{R}^+$ is maximum density for Zone C [232]. Then the fundamental diagram of flow and density function $Q_C(\rho)$ is in a quadratic form of density,

$$Q_C(\rho) = -\frac{v_f}{\rho_m}\rho^2 + v_f \rho. \tag{2.88}$$

A critical value of density segregates the traffic into the free flow regime whose density is smaller than the critical value and the congested regime whose density is greater than the critical value. The critical density may be assumed as $\rho_c = \rho_m/2$ for (2.88) [232]. For the fundamental diagram calibrated with the freeway empirical data, the critical density usually appears at 20% of the maximum value of the density [51], [64].

In practice, the quadratic fundamental diagram sometimes does not fit well with traffic density-flow field data. There are several other equilibrium models, e.g., the Greenberg model, the Underwood model, and the diffusion model, for which the fundamental diagrams are nonlinear functions; see [232] and references therein. However, according to Taylor expansion, a second-order differentiable nonlinear function can be approximated as a quadratic function in the neighborhood of its extremum. The following assumption is made for the nonlinear fundamental diagram. The stability results derived in this section hold locally for the general form of fundamental diagram $Q(\rho)$ that satisfies the following assumption. Here we can adopt other density-flow relations for the fundamental diagram $\mathcal{Q}(\rho)$ but requiring Assumption 2.1 below to be satisfied.

Assumption 2.1. *The fundamental diagram $\mathcal{Q}(\rho)$ is a smooth function, and it holds that $\mathcal{Q}'(\rho_c) = 0$, $\mathcal{Q}''(\rho_c) < 0$.*

Under Assumption 2.1, the fundamental diagram can be approximated around the critical density ρ_c as follows: $\mathcal{Q}(\rho) = q_c + \frac{\mathcal{Q}''(\rho)}{2}(\rho(t) - \rho_c)^2$, where $q_c = \mathcal{Q}(\rho_c)$ is defined as the road capacity or maximum flow, with $\mathcal{Q}''(\rho) < 0$.

2.5.3 ▪ Lane-Drop Bottleneck Control Problem

Due to the reduction of the number of the lanes from Zone C to Zone B, we consider the equilibrium density-flow relation of Zone B as shown in Figure 2.5, as pointed out in [225]. There is a capacity drop ΔC of Q_B in Zone B compared to Q_C in Zone C after the congestion has formed upstream of the lane-drop area. The capacity drop caused by a sudden lane-drop is hard to measure in real time, and the traffic dynamics of Zone B are affected by the lane-changing and merging activities. Therefore we assume that the fundamental diagram $Q_B(\rho)$ of Zone B is unknown.

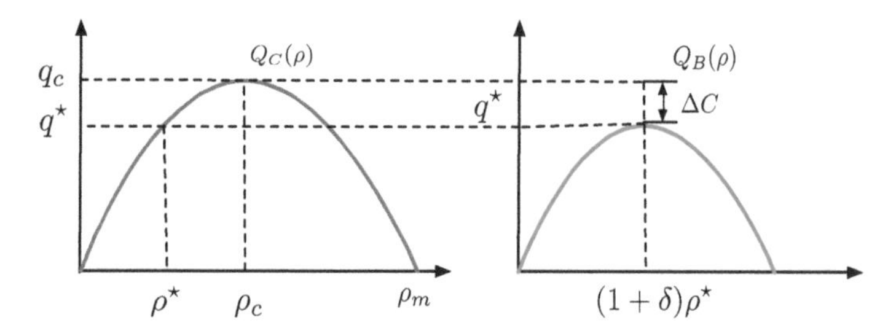

Figure 2.5. *Quadratic fundamental diagram for the clear Zone C and the bottleneck Zone B.*

In Figure 2.5, the capacity is

$$\Delta C = Q_C(\rho_c) - Q_B((1+\delta)\rho^\star), \tag{2.89}$$

$$q^\star = Q_C(\rho^\star) = Q_B((1+\delta)\rho^\star), \tag{2.90}$$

where ΔC is unknown. The $\rho^\star \in \mathbb{R}^+$ represents the optimal density that keeps Zone C in the free regime while $(1+\delta)\rho^\star$ reaches the critical density of Zone B so that the discharging flow rate reaches its maximum value $q^\star \in \mathbb{R}^+$. The ratio δ accounts for the density discontinuity before the outlet in Zone C and after the outlet in Zone B. We assume that ΔC and δ are unknown and therefore the optimal density and flow rate $(\rho^\star, q^\star)$ are unknown.

When there is a lane-drop bottleneck presenting downstream, the density at the outlet of Zone C is $\rho(L,t)$ governed by the PDE in (2.87) for $x \in [0,L]$, $t \in [0,\infty)$. The inlet boundary flow is $q_{\text{in}}(t) = Q_C(\rho(0,t))$. The output measurement of traffic flow in Zone B, $q_{\text{out}}(t)$, is given by $Q(\rho)$ with outlet density $\rho(L,t)$, $q_{\text{out}}(t) = Q(\rho(L,t))$, where the function $Q(\rho)$ of outlet boundary $x = L$ connecting Zone C and Zone B is defined as follows:

$$Q(\rho(L,t)) = \begin{cases} Q_C(\rho(L,t)), & \rho(L,t) < \rho^\star, \\ Q_C(\rho^\star) = q^\star = Q_B((1+\delta)\rho^\star), & \rho(L,t) = \rho^\star, \\ Q_B((1+\delta)\rho(L,t)), & \rho(L,t) > \rho^\star, \end{cases} \tag{2.91}$$

so that the flow is conserved through the boundary, entering from Zone C to Zone B. Note that when the optimal density $\rho^\star$ is reached, the flow rate at the outlet of Zone C and the input of Zone C reaches the equilibrium and its maximum value $q^\star$.

The control objective is to design the traffic flow input $q_{\text{in}}(t)$ so that the outgoing flow in lane-drop area Zone B, $q_{\text{out}}(t)$, is maximized. We aim to find out the optimal outlet density $\rho(L,t)=\rho^\star$ that maximizes $q_{\text{out}}(t)$ of Zone B and then use the PDE that describes the dynamics of traffic in Zone C to obtain the desirable flow input $q_{\text{in}}(t)$ from the inlet of Zone C. Here we approximate $q_{\text{out}}(t)$ with a function that satisfies Assumption 2.1 and $q_{\text{out}}(t)$ can be written as

$$q_{\text{out}}(t)=q^\star+\frac{H}{2}(\rho(L,t)-\rho^\star)^2, \tag{2.92}$$

where $H<0$ is the unknown Hessian of the approximated static map $q_{\text{out}}(t)$.

Note that we use a static fundamental diagram to model the traffic in the bottleneck Zone B. Therefore, the upstream propagating traffic waves from Zone B to Zone C cannot be captured by our model if Zone B is very congested. Since this result is focused on maximizing the discharging flow rate at the bottleneck area, the ES control seeks the optimal traffic density value in its neighborhood. In bottleneck Zone B, the closer the outlet traffic density $\rho(L,t)$ to the optimal value $\rho^\star$ where $Q'(\rho)=0$ is satisfied, the smaller the propagating characteristic speed of the traffic waves $Q'(\rho)$. Therefore, the spill-back traffic from Zone B to Zone C is negligible in our model.

In order to find the unknown optimal density at the bottleneck area, we design ES control for the unknown static map $Q(\rho)$ with actuation dynamics governed by a nonlinear hyperbolic PDE in (2.87). In the following section, we linearize the nonlinear PDE and the traffic dynamics can be represented by the delay effect for the control input design.

2.5.4 ▪ Linearized Reference Error System

We linearize the nonlinear LWR model around a constant reference density $\rho_r\in\mathbb{R}^+$, which is assumed to be close to the optimal density $\rho^\star$. Note that the reference density ρ_r is in the free regime of $Q(\rho)$ of Zone C and thus is smaller than the critical density ρ_c, and therefore the following is satisfied $\rho_r<\rho_c$. Define the reference error density as

$$\tilde{\rho}(x,t)=\rho(x,t)-\rho_r, \tag{2.93}$$

and reference flux q_r is $q_r=Q(\rho_r)>0$. By the governing equation (2.87) together with (2.88), the linearized reference error model is derived as

$$\partial_t\tilde{\rho}(x,t)+u\partial_x\tilde{\rho}(x,t)=0, \tag{2.94}$$

$$\tilde{\rho}(0,t)=\rho(0,t)-\rho_r, \tag{2.95}$$

where the constant transport speed u is given by $u=Q'(\rho)|_{\rho=\rho_r}=V(\rho_r)+\rho_r V'(\rho)|_{\rho=\rho_r}$. The equilibrium velocity-density relation $V(\rho)$ is a strictly decreasing function. The reference density ρ_r is in the left-half plane of the fundamental diagram $Q_c(\rho)$, which yields the following inequality for the propagation speed $u>0$. We define the input density as $\varrho(t)=\rho(0,t)$, and the linearized input at the inlet is

$$\tilde{\varrho}(t)=\varrho(t)-\rho_r. \tag{2.96}$$

The linearized error dynamics in (2.94), (2.95) is a transport PDE with an explicit solution for $t>\frac{x}{u}$ and thus is represented with input density $\tilde{\rho}(x,t)=\tilde{\varrho}\left(t-\frac{x}{u}\right)$. The density variation at the outlet is

$$\tilde{\rho}(L,t)=\tilde{\varrho}(t-D), \tag{2.97}$$

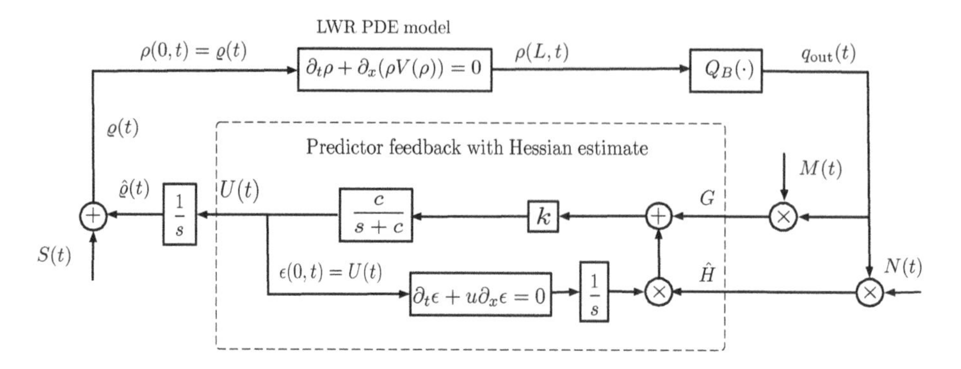

Figure 2.6. *Block diagram for implementation of ES control design for nonlinear LWR PDE model.*

where the time delay is $D=\frac{L}{u}$. Therefore, the density at the outlet is given by a delayed input density variation and the reference

$$\rho(L,t)=\rho_r+\tilde{\rho}(L,t). \tag{2.98}$$

Finally, substituting (2.97), (2.98) into the static map (2.92), we arrive at the following:

$$\begin{aligned} q_{\text{out}}(t)=&q^\star+\frac{H}{2}(\tilde{\varrho}(t-D)+\rho_r-\rho^\star)^2\\ =&q^\star+\frac{H}{2}\left(\varrho(t-D)-\rho^\star\right)^2. \end{aligned} \tag{2.99}$$

The control objective is to regulate the input $q_{\text{in}}(t)$ so that $\varrho(t-D)$ reaches to an unknown optimal $\rho^\star$ and the maximum of the uncertain quadratic flux-density map $q_{\text{out}}(t)$ can be achieved. We can apply the method of extremum seeking for the static map with delays of Section 2.2.2, originally developed in [179]. The extremum seeking control is designed for finding the extremum of the unknown map.

In practice, control of density at the inlet can be realized with a coordinated operation of a ramp metering and a VSL at the inlet, which is widely used in freeway traffic management [40, 87, 144, 245, 104, 233, 231]. The controlled density at the inlet is implemented by $\varrho(t)=\frac{q_{\text{in}}(t)}{v_c}$, where v_c is the speed limit implemented by VSL and $q_{\text{in}}(t)$ is actuated by an on-ramp metering upstream of the inlet. Note that the linearized model is valid at the optimal density $\rho^\star$ since the reference density is assumed to be chosen near the optimal value.

2.5.5 ▪ Online Optimization by Extremum Seeking Control

In this section, we present the design of extremum seeking control with delay by following analogously the procedure in [179]. The block diagram of the delay-compensated ES algorithm applied to the LWR PDE model is depicted in Figure 2.6.

Let $\hat{\varrho}(t)$ be the estimate of $\rho^\star$, and let $e(t)$ be the estimation error defined as

$$e(t)=\hat{\varrho}(t)-\rho^\star, \tag{2.100}$$

where $\hat{\varrho}(t)$ is an integrator of the predictor-based feedback signal $U(t)$ as $\dot{\hat{\varrho}}(t)=U(t)$. From Figure 2.6, the error dynamics can be written as

$$\dot{e}(t-D)=U(t-D), \tag{2.101}$$

given the delayed estimation error dynamics modeled by $\epsilon(x,t)=U(t-\frac{x}{u})$.

We introduce the additive perturbation $S(t)$,

$$S(t) = a\sin\left(\omega(t+D)\right), \tag{2.102}$$

and the multiplicative demodulation signals $(M(t), N(t))$ given by

$$M(t) = \frac{2}{a}\sin\left(\omega t\right), \quad N(t) = -\frac{8}{a^2}\cos\left(2\omega t\right), \tag{2.103}$$

where a and ω are amplitude and frequency of a slow periodic perturbation signal $a\sin(\omega t)$ introduced later. Using the demodulation signals, we calculate estimates of the gradient and Hessian of the cost function, denoted as $(G(t), \hat{H}(t))$,

$$G(t) = M(t)q_{\text{out}}(t), \quad \hat{H}(t) = N(t)q_{\text{out}}(t), \tag{2.104}$$

where $\hat{H}(t)$ is to estimate the unknown Hessian H. The averaging of $G(t)$ and $\hat{H}(t)$ yields that

$$G_{\text{av}}(t) = He_{\text{av}}(t-D), \quad \hat{H}_{\text{av}} = (Nq_{\text{out}})_{\text{av}} = H. \tag{2.105}$$

Taking the average of (2.101), we have $\dot{e}_{\text{av}}(t-D) = U_{\text{av}}(t-D)$, where $U_{\text{av}}(t)$ is the averaged value for $U(t)$ designed later. Substituting the above equation into (2.105) gives that

$$\dot{G}_{\text{av}}(t) = HU_{\text{av}}(t-D). \tag{2.106}$$

The motivation for predictor feedback design is to compensate for the delay by feeding back future states in the equivalent averaged system $G_{av}(t+D)$. Given an arbitrary control gain $k > 0$, we aim to design

$$U_{\text{av}}(t) = kG_{\text{av}}(t+D) \quad \forall t \geq 0, \tag{2.107}$$

which requires knowledge of future states. Therefore, we have the following by plugging (2.107) into (2.101):

$$\dot{e}_{\text{av}}(t) = U_{\text{av}}(t) = kHe_{\text{av}}(t) \quad \forall t \geq D. \tag{2.108}$$

Recalling that $k > 0, H < 0$, the equilibrium of the average system $e_{\text{av}}(t) = 0$ is exponentially stable. Applying the variation-of-constants formula $G_{\text{av}}(t+D) = G_{\text{av}}(t) + \hat{H}_{\text{av}}(t)\int_{t-D}^{t} U_{\text{av}}(\tau)d\tau$ and, from (2.107), one has

$$U_{\text{av}}(t) = k\left(G_{\text{av}}(t) + \hat{H}_{\text{av}}(t)\int_{t-D}^{t} U_{\text{av}}(\tau)d\tau\right), \tag{2.109}$$

which represents the future state $G_{\text{av}}(t+D)$ in (2.106) in terms of the average control signal $U_{\text{av}}(\tau)$ for $\tau \in [t-D, t]$. The control input is infinite-dimensional due to its use of history over the past D time units.

For the stability analysis in which the averaging theorem for infinite-dimensional systems is used, we employ a low-pass filter for the above basic predictor feedback controller and then derive an infinite-dimensional and averaging-based predictor feedback given by

$$U(t) = \mathscr{T}\left\{k\left(G(t) + \hat{H}(t)\int_{t-D}^{t} U(\tau)d\tau\right)\right\}, \tag{2.110}$$

where $k > 0$ is an arbitrary control gain, and the Hessian estimate $\hat{H}(t)$ is updated according to (2.104), satisfying the average property in (2.105). $\mathscr{T}\{\cdot\}$ is the low-pass filter operator defined by

$$\mathscr{T}\{\varphi(t)\} = \mathscr{L}^{-1}\left\{\frac{c}{s+c}\right\} * \varphi(t), \tag{2.111}$$

where $c \in \mathbb{R}^+$ is the corner frequency, $\mathscr{L}^{-1}$ is the inverse Laplace transformation, and $*$ is the convolution in time.

Hence, according to Theorem 2.1, we can conclude

$$\lim_{t\to+\infty} \sup|\varrho(t) - \rho^\star| = \mathcal{O}(a + 1/\omega), \tag{2.112}$$

$$\lim_{t\to+\infty} \sup|q_{\text{out}}(t) - q^\star| = \mathcal{O}(a^2 + 1/\omega^2). \tag{2.113}$$

2.6 ▪ Notes and References

In spite of the large number of publications on extremum seeking (ES) control [129, 123, 2, 217, 219, 166, 73, 142, 173, 181], no work in the literature has rigorously dealt with the problem of ES in the presence of delays. In the present chapter and [174], we give an answer to this question by considering scalar gradient-based ES under input-output delays.

The proposed solution based on prediction feedback with a perturbation-based estimate of the Hessian is introduced to tackle delays in the feedback loop with gradient extremum seeking controllers. The perturbation-based (averaging-based) approach to predictor design is employed due to the necessity of estimating the unknown second derivative (Hessian) [166, 73] of the nonlinear map to be optimized.

The resulting approach preserves exponential stability and convergence of the system output to a small neighborhood of the extremum point, despite the presence of actuator and sensor delays. The stability analysis is rigorously conducted with the help of a backstepping transformation [119] and averaging in infinite dimension [83, 131].

While the convergence rate of the gradient method is dictated by the unknown Hessian, the Newton-based scheme addressed in Chapter 3 (and in the companion paper [174]) is independent of it and, thus, the delay compensation can be achieved with an arbitrarily assigned convergence rate, improving the controller performance. The results presented here were given for scalar plants, but the extension to the multivariable case can be achieved by using our earlier methods for multiparameter extremum seeking, as discussed in the subsequent Chapter 7.

To illustrate the practical need for delay-compensated ES, on a rather exciting application problem, we employed ES to find an optimal density input for freeway traffic when there is a downstream bottleneck. To prevent the traffic flow in the bottleneck area from overflowing the road capacity and furthermore causing congestion upstream in the freeway segment, the incoming traffic density at the inlet of the freeway segment is regulated. The control design uses a delay compensation for the ES control because the upstream traffic is governed by the linearized LWR model, which is nothing but a pure delay. The optimal density and flow are achieved in the bottleneck area. This theoretical result was validated in [232] through numerical simulations with the control design being applied to the nonlinear LWR PDE model along with an unknown fundamental diagram.

Chapter 3

Newton-Based ES for Scalar Static Map with Delay

3.1 ▪ Assigning Convergence Rate in Spite of Unknown Hessian

The Hessian H of the unknown map is the key uncertainty, besides the optimizer θ^*, in extremum seeking. In Chapter 2, on gradient-based ES in the presence of a delay, in order to compensate for the delay, we had to estimate the unknown Hessian using the demodulation signal $N(t)$. However, we did not fully get rid of the effect of the Hessian—H did appear in the average system (2.46), does influence the dominant pole of this system, and, therefore, impacts the convergence rate under the ES algorithm.

While happy to have a guarantee of convergence, the user cannot be satisfied that the convergence rate, kH, in which k is the user's gain and H is the unknown Hessian, is unknown overall. The user would prefer having full control over the convergence rate and being able to arbitrarily assign it.

Assigning the convergence rate would be perhaps possible by replacing the scalar update gain k by a matrix update gain K and by including the inverse $\hat{H}(t)^{-1}$ of the Hessian estimate $\hat{H}(t) = N(t)y(t)$ into K, hoping that, in the limit, KH would be the negative of the identity matrix or, more generally, a user-chosen negative definite matrix.

However, while H is a definite matrix and, therefore, invertible, its estimate $\hat{H}(t)$ does not necessarily remain nonsingular for all time. In fact, in simulations, $\hat{H}(t)$ typically does lose rank for some time instants. So the direct inversion of $\hat{H}(t)$ is not an option.

One can further envision including, more conservatively, a division by a lower bound on $|H|$ in k. However, the point of ES is that it be applicable without any assumptions of a priori knowledge of the unknown map, except that the map has a maximum or a minimum. This is to say that $\operatorname{sgn} H$ must be known a priori but knowing a lower bound on $|H|$ is too much to ask in model-free optimization.

While, at the outset, it may, therefore, seem that it would take a miracle to make the convergence rate independent of the unknown H, this is, in fact, possible, using a Newton approach to extremum seeking. So, in this chapter, we derive an algorithm, and provide a convergence analysis, for scalar Newton-based extremum seeking in the presence of input-output delays.

A predictor different than the one for the gradient ES in Chapter 2 is developed in this chapter. While, for gradient ES, it suffices to estimate the Hessian H, for the predictor in the Newton-based approach it is necessary to also estimate the Hessian *inverse* H^{-1}.

Additionally, while the estimation of H requires a static operation—a multiplication by the demodulation signal $N(t)$—for estimating H^{-1}, it does not suffice to just take a reciprocal of the estimate of H since this estimate may be going through zero. Instead, a separate estimator of H^{-1} is designed which generates the estimate of H^{-1} using a differential equation of the Riccati type. This equation is, in fact, an ODE that governs the reciprocal of the estimate of H. With this ODE estimator of H^{-1}, it becomes possible to eliminate the unknown H from the convergence rate kH.

Exponential stability and convergence to a small neighborhood of the unknown extremum point are achieved by using a backstepping transformation and averaging in infinite dimension. Except for the inclusion of the linearized dynamics of the Hessian inverse in a few places, the proof process follows the outline given in Section 2.1.

In Section 3.2, we introduce the predictor design for delay compensation in scalar Newton-based ES. Exponential stability with explicit Lyapunov–Krasovskii functionals and the real-time convergence to a small neighborhood of the desired extremum are proved in Section 3.3. Section 3.4 presents simulation results to show the applicability of the proposed delay-compensated extremum seeking. Finally, Section 3.5 concludes the chapter with some notes and references.

3.2 ▪ Newton-Based Extremum Seeking for Delay Systems

As in Chapter 2, we are going to consider scalar ES applications in which the goal is to maximize (or minimize) the output $y \in \mathbb{R}$ of an unknown nonlinear static map $Q(\theta)$ by varying the input $\theta \in \mathbb{R}$. Here, we again assume that there is a *constant and known* delay $D \geq 0$ in the actuation path or measurement system such that the measured output is given by

$$y(t) = Q(\theta(t-D)). \tag{3.1}$$

For notational clarity, we assume that our system is output-delayed in the following presentation and block diagrams. However, the results in this chapter can also be extended to the input-delay case since any input delay can be moved to the output of the static map. The case when input delays D_{in} and output delays D_{out} occur simultaneously could be handled in a similar fashion by assuming that the total delay to be compensated for is $D = D_{\text{in}} + D_{\text{out}}$, with $D_{\text{in}}, D_{\text{out}} \geq 0$.

Without loss of generality, let us consider the maximum seeking problem such that the maximizing value of θ is denoted by θ^*. For the sake of simplicity, we also assume that the nonlinear map is quadratic, i.e.,

$$Q(\theta) = y^* + \frac{H}{2}(\theta - \theta^*)^2, \tag{3.2}$$

where besides the constants $\theta^* \in \mathbb{R}$ and $y^* \in \mathbb{R}$ being unknown, the scalar $H < 0$ is the unknown Hessian of the static map.

In Figure 3.1, we illustrate the proposed scalar version of the Newton-based ES based on predictor feedback for delay compensation.

3.2.1 ▪ Probing and Demodulation Signals

By plugging (3.2) into (3.1), we obtain the *quadratic static map with delay* of interest:

$$y(t) = y^* + \frac{H}{2}(\theta(t-D) - \theta^*)^2. \tag{3.3}$$

Let $\hat{\theta}$ be the estimate of θ^* and

$$\tilde{\theta}(t) = \hat{\theta}(t) - \theta^* \tag{3.4}$$

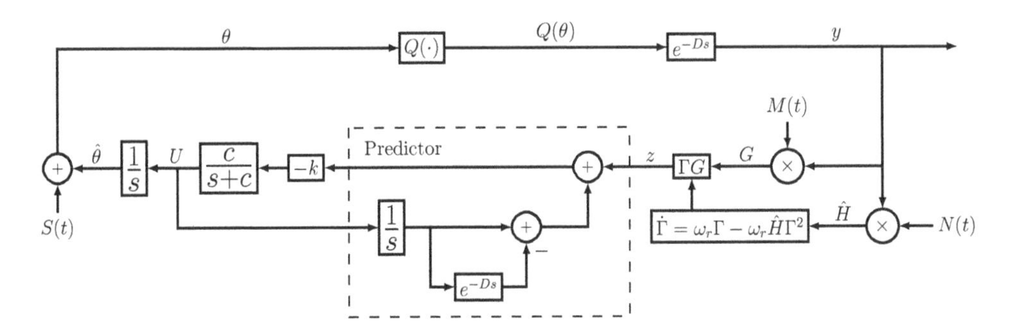

Figure 3.1. *Block diagram of the basic prediction scheme for output-delay compensation in Newton-based ES. The predictor feedback with a perturbation-based estimate of the Hessian's inverse obeys* (3.13), (3.14), *and* (3.26), *the additive dither is given by* $S(t) = a\sin(\omega(t+D))$, *and the demodulation signals* $M(t) = \frac{2}{a}\sin(\omega t)$ *and* $N(t) = -\frac{8}{a^2}\cos(2\omega t)$.

be the *estimation error*. From the block diagram of Figure 3.1, the *error dynamics* is written as

$$\dot{\tilde{\theta}}(t-D) = U(t-D). \tag{3.5}$$

Furthermore, we can also write

$$G(t) = M(t)y(t), \tag{3.6}$$
$$\theta(t) = \hat{\theta}(t) + S(t), \tag{3.7}$$

where the additive dither and the demodulation signal are given by

$$S(t) = a\sin(\omega(t+D)), \tag{3.8}$$
$$M(t) = \frac{2}{a}\sin(\omega t), \tag{3.9}$$

with nonzero perturbation amplitude a and frequency ω.

Notice that (3.8) is different from the additive dither signal used in the standard Newton extremum seeking free of delays. Since the output delay can be transferred to the integrator output for analysis purposes (or, equivalently, to its input), thus the phase shift $+\omega D$ is applied to compensate the delay effect in the dither signal $S(t)$.

The signal

$$\hat{H}(t) = N(t)y(t) \tag{3.10}$$

is applied to generate an estimate of the unknown Hessian H, where the demodulating signal $N(t)$ is given by

$$N(t) = -\frac{8}{a^2}\cos(2\omega t). \tag{3.11}$$

In [73], it was proved that

$$\frac{1}{\Pi}\int_0^{\Pi} N(\sigma)y d\sigma = H, \quad \Pi = 2\pi/\omega, \tag{3.12}$$

if a quadratic map as in (3.2) is considered. In other words, the averaging version $\hat{H}_{\text{av}} = (Ny)_{\text{av}} = H$.

Now, let us define the measurable signal

$$z(t)=\Gamma(t)G(t), \tag{3.13}$$

where $\Gamma(t)$ is updated by the following Riccati differential equation [73]:

$$\dot{\Gamma}=\omega_r\Gamma-\omega_r\hat{H}\Gamma^2, \tag{3.14}$$

with $\omega_r>0$ being a design constant. Equation (3.14) generates an estimate of the Hessian's inverse, avoiding inversions of the Hessian estimates which may be zero during the transient phase.

The estimation error of the Hessian's inverse can be defined as

$$\tilde{\Gamma}(t)=\Gamma(t)-H^{-1}, \tag{3.15}$$

and its dynamic equation can be written from (3.14) and (3.15) by

$$\dot{\tilde{\Gamma}}=\omega_r\left(\tilde{\Gamma}+H^{-1}\right)\left[1-\hat{H}\left(\tilde{\Gamma}+H^{-1}\right)\right]. \tag{3.16}$$

3.2.2 ▪ Predictor Feedback via Hessian's Inverse Estimation

By using the averaging analysis, we can verify from (3.6) and (3.13) that

$$z_{\mathrm{av}}(t)=\Gamma_{\mathrm{av}}(t)H\tilde{\theta}_{\mathrm{av}}(t-D). \tag{3.17}$$

From (3.15), equation (3.17) can be written in terms of $\tilde{\Gamma}_{\mathrm{av}}(t)=\Gamma_{\mathrm{av}}(t)-H^{-1}$ as

$$z_{\mathrm{av}}(t)=\tilde{\theta}_{\mathrm{av}}(t-D)+\tilde{\Gamma}_{\mathrm{av}}(t)H\tilde{\theta}_{\mathrm{av}}(t-D). \tag{3.18}$$

The second term in the right side of (3.18) is quadratic in $(\tilde{\Gamma}_{\mathrm{av}},\ \tilde{\theta}_{\mathrm{av}})$; thus, the linearization of $\Gamma_{\mathrm{av}}(t)$ at H^{-1} results in the linearized version of (3.17) given by

$$z_{\mathrm{av}}(t)=\tilde{\theta}_{\mathrm{av}}(t-D). \tag{3.19}$$

From (3.5) and (3.19), the following average models can be written:

$$\dot{\tilde{\theta}}_{\mathrm{av}}(t-D)=U_{\mathrm{av}}(t-D), \tag{3.20}$$

$$\dot{z}_{\mathrm{av}}(t)=U_{\mathrm{av}}(t-D), \tag{3.21}$$

where $U_{\mathrm{av}}\in\mathbb{R}$ is the resulting average control for $U\in\mathbb{R}$.

In order to motivate the predictor feedback design, the idea here is to compensate for the delay by feeding back the future state $z(t+D)$, or $z_{\mathrm{av}}(t+D)$ in the equivalent average system.

To obtain $z_{\mathrm{av}}(t+D)$ with the variation-of-constants formula to (3.21), the future state is written as

$$z_{\mathrm{av}}(t+D)=z_{\mathrm{av}}(t)+\int_{t-D}^{t}U_{\mathrm{av}}(\sigma)d\sigma \tag{3.22}$$

in terms of the control signal $U_{\mathrm{av}}(\sigma)$ from the past window $[t-D,t]$. Given any stabilizing gain $k>0$, the average control would be given by

$$U_{\mathrm{av}}(t)=-k\left[z_{\mathrm{av}}(t)+\int_{t-D}^{t}U_{\mathrm{av}}(\sigma)d\sigma\right], \tag{3.23}$$

resulting in the average control feedback

$$U_{\text{av}}(t) = -kz_{\text{av}}(t+D) \quad \forall t \geq 0, \tag{3.24}$$

as desired. Hence, the average system would be, $\forall t \geq D$,

$$\frac{d\tilde{\theta}_{\text{av}}(t)}{dt} = -k\tilde{\theta}_{\text{av}}(t) - k\tilde{\Gamma}_{\text{av}}(t+D)H\tilde{\theta}_{\text{av}}(t). \tag{3.25}$$

Since $k\tilde{\Gamma}_{\text{av}}H\tilde{\theta}_{\text{av}}$ is quadratic in $(\tilde{\Gamma}_{\text{av}}, \tilde{\theta}_{\text{av}})$, the linearization of the system (3.25) has all its eigenvalues determined by $-k$. The (local) exponential stability of the algorithm can be guaranteed with a convergence rate which is independent of the unknown Hessian H, being user-assignable.

In the next section, we show that the control objectives can still be achieved if a simple modification of the above basic predictor-based controller, which employs a low-pass filter, is applied. In this case, we propose the following infinite-dimensional and averaging-based predictor feedback in order to compensate the delay [115]:

$$U(t) = \frac{c}{s+c}\left\{-k\left[z(t) + \int_{t-D}^{t} U(\tau)d\tau\right]\right\}, \tag{3.26}$$

where $c > 0$ is sufficiently large, i.e., the predictor feedback is of the form of a low-pass filtering of the nonaverage version of (3.23). This low-pass filtering is particularly required in the stability analysis when the averaging theorem in infinite dimension [83, 131], included in Appendix A, is invoked. Note that we mix the time- and frequency-domain notation in (3.26) by using the braces $\{\cdot\}$ to denote that the transfer function acts as an operator on a time-domain function.

The predictor feedback (3.26) is infinite-dimensional because the integral involves the control history over the interval $[t-D, t]$ and is also averaging-based (perturbation-based) because z in (3.13) is updated according to the estimate Γ for the unknown Hessian's inverse H^{-1} given by (3.14), with $\hat{H}(t)$ in (3.10) satisfying the averaging property (3.12).

3.3 ▪ Stability Analysis

The main stability/convergence results for the closed-loop system are summarized in the next theorem for Newton-based ES in the presence of time delays.

Theorem 3.1. *Consider the block diagram of the closed-loop system in Figure* 3.1 *with output delays* (3.3)*. There exists $c^* > 0$ such that, $\forall c \geq c^*$, $\exists\, \omega^*(c) > 0$ such that, $\forall \omega > \omega^*$, the closed-loop delayed system* (3.5), (3.26), *with $z(t)$ in* (3.13), *$G(t)$ in* (3.6), *$\Gamma(t)$ in* (3.14), *and the state $\tilde{\Gamma}(t)$, $\tilde{\theta}(t-D)$, $U(\tau)$, $\forall \tau \in [t-D, t]$, has a unique locally exponentially stable periodic solution in t of period $\Pi = 2\pi/\omega$, denoted by $\tilde{\Gamma}^{\Pi}(t)$, $\tilde{\theta}^{\Pi}(t-D)$, $U^{\Pi}(\tau)$, $\forall \tau \in [t-D, t]$, satisfying, $\forall t \geq 0$,*

$$\left(\left|\tilde{\Gamma}^{\Pi}(t)\right|^2 + \left|\tilde{\theta}^{\Pi}(t-D)\right|^2 + \left[U^{\Pi}(t)\right]^2 + \int_{t-D}^{t}\left[U^{\Pi}(\tau)\right]^2 d\tau\right)^{1/2} \leq \mathcal{O}(1/\omega). \tag{3.27}$$

Furthermore,

$$\limsup_{t\to+\infty} |\theta(t) - \theta^*| = \mathcal{O}(a + 1/\omega), \tag{3.28}$$

$$\limsup_{t\to+\infty} |y(t) - y^*| = \mathcal{O}(a^2 + 1/\omega^2). \tag{3.29}$$

Proof. The proof is given in the following eight steps.

Step 1: Transport PDE for Delay Representation

According to [119], the delay in (3.5) can be represented using a transport PDE as

$$\dot{\tilde{\theta}}(t-D) = u(0,t), \tag{3.30}$$

$$u_t(x,t) = u_x(x,t), \quad x \in [0,D], \tag{3.31}$$

$$u(D,t) = U(t), \tag{3.32}$$

where the solution of (3.31)–(3.32) is

$$u(x,t) = U(t+x-D). \tag{3.33}$$

Step 2: Equations of the Closed-Loop System

First, substituting (3.8) into (3.7), we obtain

$$\theta(t) = \hat{\theta}(t) + a\sin(\omega(t+D)). \tag{3.34}$$

Now, plug (3.4) and (3.34) into (3.3) so that the output is given in terms of $\tilde{\theta}$:

$$y(t) = y^* + \frac{H}{2}(\tilde{\theta}(t-D) + a\sin(\omega t))^2. \tag{3.35}$$

Plug (3.9) into (3.6) and then the resulting (3.6) into (3.13). After that, by representing the integrand in (3.26) using the transport PDE state, one has

$$U(t) = \frac{c}{s+c}\left\{-k\left[z(t) + \int_0^D u(\sigma,t)d\sigma\right]\right\}, \tag{3.36}$$

$$z(t) = \Gamma(t)\frac{2}{a}\sin(\omega t)y(t). \tag{3.37}$$

By plugging (3.35) into (3.37) and then the resulting (3.37) into (3.36), one has

$$\begin{aligned} U(t) = \frac{c}{s+c}\Bigg\{&-k\left[y^* + \frac{H}{2}(\tilde{\theta}(t-D) + a\sin(\omega t))^2\right] \\ &\times\left[\Gamma(t)\frac{2}{a}\sin(\omega t)\right] - k\int_0^D u(\sigma,t)d\sigma\Bigg\}. \end{aligned} \tag{3.38}$$

By expanding the binome in (3.38), we obtain

$$\begin{aligned} U(t) = \frac{c}{s+c}\Bigg\{&-k\left[y^* + \frac{H}{2}\tilde{\theta}^2(t-D)\right. \\ &\left. + Ha\sin(\omega t)\tilde{\theta}(t-D) + \frac{a^2H}{2}\sin^2(\omega t)\right] \\ &\times\left[\Gamma(t)\frac{2}{a}\sin(\omega t)\right] - k\int_0^D u(\sigma,t)d\sigma\Bigg\}. \end{aligned} \tag{3.39}$$

Finally, substituting (3.39) into (3.32), we can rewrite (3.30)–(3.32) as

$$\dot{\tilde{\theta}}(t-D) = u(0,t), \tag{3.40}$$

$$\partial_t u(x,t) = \partial_x u(x,t), \quad x \in [0,D], \tag{3.41}$$

$$\begin{aligned} u(D,t) &= \frac{c}{s+c}\left\{-k\left[y^* + \frac{H}{2}\tilde{\theta}^2(t-D) + Ha\sin(\omega t)\tilde{\theta}(t-D)\right.\right. \\ &\quad \left.\left. + \frac{a^2H}{2}\sin^2(\omega t)\right] \times \left[\Gamma(t)\frac{2}{a}\sin(\omega t)\right] - k\int_0^D u(\sigma,t)d\sigma\right\} \\ &= \frac{c}{s+c}\left\{-k\left[\Gamma(t)y^*\frac{2}{a}\sin(\omega t) + \Gamma(t)\frac{H}{a}\tilde{\theta}^2(t-D)\sin(\omega t)\right.\right. \\ &\quad + 2\Gamma(t)H\sin^2(\omega t)\tilde{\theta}(t-D) + \Gamma(t)aH\sin^3(\omega t) \\ &\quad \left.\left. + \int_0^D u(\sigma,t)d\sigma\right]\right\} \\ &= \frac{c}{s+c}\left\{-k\left[\Gamma(t)y^*\frac{2}{a}\sin(\omega t) + \Gamma(t)\frac{H}{a}\tilde{\theta}^2(t-D)\sin(\omega t)\right.\right. \\ &\quad + \Gamma(t)H\tilde{\theta}(t-D) - \Gamma(t)H\cos(2\omega t)\tilde{\theta}(t-D) \\ &\quad + \frac{3aH}{4}\Gamma(t)\sin(\omega t) - \frac{aH}{4}\Gamma(t)\sin(3\omega t) \\ &\quad \left.\left. + \int_0^D u(\sigma,t)d\sigma\right]\right\}. \end{aligned} \tag{3.42}$$

Step 3: Average Model of the Closed-Loop System

The average version of system (3.40)–(3.42) is simply

$$\dot{\tilde{\theta}}_{\rm av}(t-D) = u_{\rm av}(0,t), \tag{3.43}$$

$$\partial_t u_{\rm av}(x,t) = \partial_x u_{\rm av}(x,t), \quad x \in [0,D], \tag{3.44}$$

$$u_{\rm av}(D,t) = \frac{c}{s+c}\left\{-k\left[\Gamma_{\rm av}(t)H\tilde{\theta}_{\rm av}(t-D) + \int_0^D u_{\rm av}(\sigma,t)d\sigma\right]\right\}. \tag{3.45}$$

From (3.17) and (3.18), we conclude the linearization of $\Gamma_{\rm av}(t)$ at H^{-1} results in the linearized version of (3.17) given by (3.19), i.e., $z_{\rm av}(t) = \tilde{\theta}_{\rm av}(t-D)$. Thus, the term $\Gamma_{\rm av}(t)H\tilde{\theta}_{\rm av}(t-D)$ in (3.45) can be replaced by $\tilde{\theta}_{\rm av}(t-D)$ in the linearized model.

Now, denoting

$$\tilde{\vartheta}(t) = \tilde{\theta}(t-D), \tag{3.46}$$

we have $\tilde{\vartheta}_{\rm av}(t) = z_{\rm av}(t) = \tilde{\theta}_{\rm av}(t-D)$ and the following linearized average version of system (3.40)–(3.42) can be obtained:

$$\dot{\tilde{\vartheta}}_{\rm av}(t) = u_{\rm av}(0,t), \tag{3.47}$$

$$\partial_t u_{\rm av}(x,t) = \partial_x u_{\rm av}(x,t), \quad x \in [0,D], \tag{3.48}$$

$$\frac{d}{dt}u_{\rm av}(D,t) = -cu_{\rm av}(D,t) - ck\left[\tilde{\vartheta}_{\rm av}(t) + \int_0^D u_{\rm av}(\sigma,t)d\sigma\right], \tag{3.49}$$

where the filter $c/s+c$ is also represented in the state-space form. The solution of the transport PDE (3.48)–(3.49) is given by

$$u_{\mathrm{av}}(x,t) = U_{\mathrm{av}}(t+x-D). \tag{3.50}$$

On the other hand, the average model for the Hessian's inverse estimation error in (3.16) is

$$\frac{d\tilde{\Gamma}_{\mathrm{av}}(t)}{dt} = -\omega_r \tilde{\Gamma}_{\mathrm{av}}(t) - \omega_r H \tilde{\Gamma}^2_{\mathrm{av}}(t), \tag{3.51}$$

and its linearized version is given by

$$\frac{d\tilde{\Gamma}_{\mathrm{av}}(t)}{dt} = -\omega_r \tilde{\Gamma}_{\mathrm{av}}(t). \tag{3.52}$$

Step 4: Backstepping Transformation, Its Inverse, and the Target System

Consider the infinite-dimensional backstepping transformation of the delay state

$$w(x,t) = u_{\mathrm{av}}(x,t) + k\left[\tilde{\vartheta}_{\mathrm{av}}(t) + \int_0^x u_{\mathrm{av}}(\sigma,t)d\sigma\right], \tag{3.53}$$

which maps the system (3.47)–(3.49) into the target system:

$$\dot{\tilde{\vartheta}}_{\mathrm{av}}(t) = -k\tilde{\vartheta}_{\mathrm{av}}(t) + w(0,t), \tag{3.54}$$

$$w_t(x,t) = w_x(x,t), \quad x \in [0,D], \tag{3.55}$$

$$w(D,t) = -\frac{1}{c}\partial_t u_{\mathrm{av}}(D,t). \tag{3.56}$$

Using (3.53) for $x = D$ and the fact that $u_{\mathrm{av}}(D,t) = U_{\mathrm{av}}(t)$, from (3.56) we get (3.49), i.e.,

$$U_{\mathrm{av}}(t) = \frac{c}{s+c}\left\{-k\left[\tilde{\vartheta}_{\mathrm{av}}(t) + \int_0^D u_{\mathrm{av}}(\sigma,t)d\sigma\right]\right\}. \tag{3.57}$$

It is easily seen that

$$w_t(D,t) = \partial_t u_{\mathrm{av}}(D,t) + k\left[u_{\mathrm{av}}(D,t) + \int_0^D u_{\mathrm{av}}(\sigma,t)d\sigma\right], \tag{3.58}$$

where $\partial_t u_{\mathrm{av}}(D,t) = \dot{U}_{\mathrm{av}}(t)$. The inverse of (3.53) is given by

$$u_{\mathrm{av}}(x,t) = w(x,t) - k\left[e^{-kx}\tilde{\vartheta}_{\mathrm{av}}(t) + \int_0^x e^{-k(x-\sigma)}w(\sigma,t)d\sigma\right]. \tag{3.59}$$

Plugging (3.59) into (3.58), after a lengthy calculation that involves a change of the order of integration in a double integral, we get

$$\begin{aligned} w_t(D,t) &= -cw(D,t) + kw(D,t) \\ &\quad -k^2\left[e^{-kD}\tilde{\vartheta}_{\mathrm{av}}(t) + \int_0^D e^{-k(D-\sigma)}w(\sigma,t)d\sigma\right]. \end{aligned} \tag{3.60}$$

Step 5: Lyapunov–Krasovskii Functional

Now, consider the following Lyapunov functional

$$V(t)=\frac{\tilde{\vartheta}_{\rm av}^2(t)}{2}+\frac{a}{2}\int_0^D(1+x)w^2(x,t)dx+\frac{1}{2}w^2(D,t), \tag{3.61}$$

where the parameter $a>0$ is to be chosen later. We have

$$\begin{aligned}\dot{V}(t)&=-k\tilde{\vartheta}_{\rm av}^2(t)+\tilde{\vartheta}_{\rm av}(t)w(0,t)+\frac{a(1+D)}{2}w^2(D,t)\\&\quad-\frac{a}{2}w^2(0,t)-\frac{a}{2}\int_0^D w^2(x,t)dx+w(D,t)w_t(D,t)\\&\le -k\tilde{\vartheta}_{\rm av}^2(t)+\frac{\tilde{\vartheta}_{\rm av}^2(t)}{2a}-\frac{a}{2}\int_0^D w^2(x,t)dx\\&\quad+w(D,t)\left[w_t(D,t)+\frac{a(1+D)}{2}w(D,t)\right].\end{aligned}$$

Recalling that $k>0$, let us choose

$$a=\frac{1}{k}. \tag{3.62}$$

Then,

$$\begin{aligned}\dot{V}(t)&\le-\frac{k}{2}\tilde{\vartheta}_{\rm av}^2(t)-\frac{1}{2k}\int_0^D w^2(x,t)dx\\&\quad+w(D,t)\left[w_t(D,t)+\frac{(1+D)}{2k}w(D,t)\right]\\&=-\frac{1}{2a}\tilde{\vartheta}_{\rm av}^2(t)-\frac{a}{2}\int_0^D w^2(x,t)dx\\&\quad+w(D,t)\left[w_t(D,t)+\frac{a(1+D)}{2}w(D,t)\right].\end{aligned} \tag{3.63}$$

Now we consider (3.63) along with (3.60). With a completion of squares, we obtain

$$\begin{aligned}\dot{V}(t)&\le-\frac{1}{4a}\tilde{\vartheta}_{\rm av}^2(t)-\frac{a}{4}\int_0^D w^2(x,t)dx+a\left|k^2e^{-kD}\right|^2w^2(D,t)\\&\quad+\frac{1}{a}\left\|k^2e^{-k(D-\sigma)}\right\|^2w^2(D,t)\\&\quad+\left[\frac{a(1+D)}{2}+k\right]w^2(D,t)-cw^2(D,t).\end{aligned} \tag{3.64}$$

To obtain (3.64), we have used

$$\begin{aligned}-w(D,t)\left\langle k^2e^{-k(D-\sigma)},w(t)\right\rangle&\le|w(D,t)|\left\|k^2e^{-k(D-\sigma)}\right\|\|w(t)\|\\&\le\frac{a}{4}\|w(t)\|^2+\frac{1}{a}\left\|k^2e^{-k(D-\sigma)}\right\|^2w^2(D,t),\end{aligned} \tag{3.65}$$

where the first inequality is the Cauchy–Schwarz and the second is Young's inequality; the notation $\langle\cdot,\cdot\rangle$ denotes the inner product in the spatial variable $\sigma\in[0,D]$, on which both $e^{-k(D-\sigma)}$

and $w(\sigma,t)$ depend; and $\|\cdot\|$ denotes the L_2-norm in σ. Then, from (3.64), we arrive at

$$\dot{V}(t) \leq -\frac{1}{4a}\tilde{\vartheta}_{\rm av}^2(t) - \frac{a}{4}\int_0^D w^2(x,t)dx - (c-c^*)w^2(D,t), \tag{3.66}$$

where

$$c^* = \frac{a(1+D)}{2} + k + a\left|k^2 e^{-kD}\right|^2 + \frac{1}{a}\left\|k^2 e^{-k(D-\sigma)}\right\|^2. \tag{3.67}$$

Hence, from (3.66), if c is chosen such that $c > c^*$, we obtain

$$\dot{V}(t) \leq -\mu V(t) \tag{3.68}$$

for some $\mu > 0$. Thus, the closed-loop system is exponentially stable in the sense of the full-state norm

$$\left(|\tilde{\vartheta}_{\rm av}(t)|^2 + \int_0^D w^2(x,t)dx + w^2(D,t)\right)^{1/2}, \tag{3.69}$$

i.e., in the transformed variable $(\tilde{\vartheta}_{\rm av}, w)$.

Step 6: Exponential Stability Estimate (in the L_2-Norm) for the Average System (3.47)–(3.49)

To obtain exponential stability in the sense of the norm

$$\left(|\tilde{\vartheta}_{\rm av}(t)|^2 + \int_0^D u_{\rm av}^2(x,t)dx + u_{\rm av}^2(D,t)\right)^{1/2},$$

we need to show that there exist positive numbers α_1 and α_2 such that

$$\alpha_1\Psi(t) \leq V(t) \leq \alpha_2\Psi(t), \tag{3.70}$$

where $\Psi(t) \triangleq |\tilde{\vartheta}_{\rm av}(t)|^2 + \int_0^D u_{\rm av}^2(x,t)dx + u_{\rm av}^2(D,t)$ or, equivalently,

$$\Psi(t) \triangleq |\tilde{\theta}_{\rm av}(t-D)|^2 + \int_{t-D}^t U_{\rm av}^2(\tau)d\tau + U_{\rm av}^2(t), \tag{3.71}$$

using (3.46) and (3.50).

This is straightforward to establish by using (3.53), (3.59), (3.61) and employing the Cauchy–Schwarz inequality and other calculations, as in the proof of Theorem 2.1 in [119]. Hence, with (3.68), we get

$$\Psi(t) \leq \frac{\alpha_2}{\alpha_1}e^{-\mu t}\Psi(0), \tag{3.72}$$

which completes the proof of exponential stability.

Step 7: Invoking the Averaging Theorem

First, note that the closed-loop system (3.5), (3.26), and (3.16) can be rewritten as

$$\dot{\hat{\theta}}(t-D) = U(t-D), \tag{3.73}$$

$$\dot{U}(t) = -cU(t) - ck\left[z(t) + \int_{t-D}^t U(\tau)d\tau\right], \tag{3.74}$$

$$\dot{\tilde{\Gamma}}(t) = \omega_r[\tilde{\Gamma}(t) + H^{-1}] \times [1 - \hat{H}(t)(\tilde{\Gamma}(t) + H^{-1})], \tag{3.75}$$

where $\xi(t) = [\tilde{\theta}(t-D), U(t), \tilde{\Gamma}(t)]^T$ is the state vector. Moreover, from (3.6), (3.10), and (3.13), one has

$$\dot{\xi}(t) = f(\omega t, \xi_t), \tag{3.76}$$

where $\xi_t(\Theta) = \xi(t+\Theta)$ for $-D \leq \Theta \leq 0$ and f is an appropriate continuous functional, such that the averaging theorem by [83] and [131] in Appendix A can be directly applied considering $\omega = 1/\epsilon$.

We can conclude that the equilibrium $\tilde{\Gamma}_{\rm av}(t) = 0$ of the linearized error system (3.52) is exponentially stable since $\omega_r > 0$. In addition, from (3.72), the origin of the average closed-loop system (3.47)–(3.49) with transport PDE for delay representation is also exponentially stable. Thus, there exist positive constants α and β such that all solutions satisfy

$$\Upsilon(t) \leq \alpha e^{-\beta t}\Upsilon(0) \quad \forall t \geq 0, \tag{3.77}$$

where

$$\Upsilon(t) \triangleq |\tilde{\Gamma}_{\rm av}(t)|^2 + |\tilde{\theta}_{\rm av}(t-D)|^2 + U_{\rm av}^2(t) + \int_{t-D}^{t} U_{\rm av}^2(\tau)d\tau. \tag{3.78}$$

Then, according to the averaging theorem [83, 131] rewritten in Appendix A, for ω sufficiently large, the system (3.73)–(3.75) has a unique locally exponentially stable periodic solution around its equilibrium (origin) satisfying (3.27).

Step 8: Asymptotic Convergence to a Neighborhood of the Extremum (θ^*, y^*)

Finally, to obtain (3.28) and (3.29), we only have to use the change of variables (3.46) and then integrate both sides of (3.43) within the interval $[t, \sigma + D]$ to get

$$\tilde{\vartheta}(\sigma + D) = \tilde{\vartheta}(t) + \int_{t}^{\sigma+D} u(0,s)ds. \tag{3.79}$$

After that, we have to reproduce exactly the developments presented in Step 8 of the proof of Theorem 2.1. □

3.4 ▪ Simulation Results

In order to evaluate the proposed delay-compensated extremum seeking, the following static quadratic map is considered:

$$Q(\theta) = 5 - 0.1(\theta - 2)^2, \tag{3.80}$$

subject to an output delay of $D = 5$s. According to (3.80), the extremum point is $(\theta^*,\ y^*) = (2,\ 5)$ and the Hessian of the map is $H = -0.2$.

As in the numerical simulations of Section 2.4, we use low-pass and washout filters with corner frequencies ω_h and ω_l as usual in extremum seeking designs [129, 123]; see Figures 1.4 and 1.5 in Chapter 1. On the other hand, unlike the numerical simulations of Section 2.4, when the nonlinear map $y = Q(\theta)$ is flat with the Hessian matrix satisfying the inequality $|H| < 1$, the convergence rate with the gradient ES in the neighborhood of the extremum is expected to become slow. While the convergence rate of the gradient method is dictated by the unknown Hessian, the Newton-based scheme is independent of that and thus the delay compensation can be achieved with an arbitrarily assigned convergence rate, improving the controller performance.

In what follows, we present numerical simulations of the predictor (3.26), where $c = 20$, and z is given by (3.13) with G in (3.6) and Γ in (3.14). We perform our tests with the following parameters: $a = 0.2$, $\omega = 10$, $k = 0.2$, $\omega_h = \omega_l = 1$, $\theta(0) = -5$, $\Gamma(0) = -1$, and $\omega_r = 0.1$.

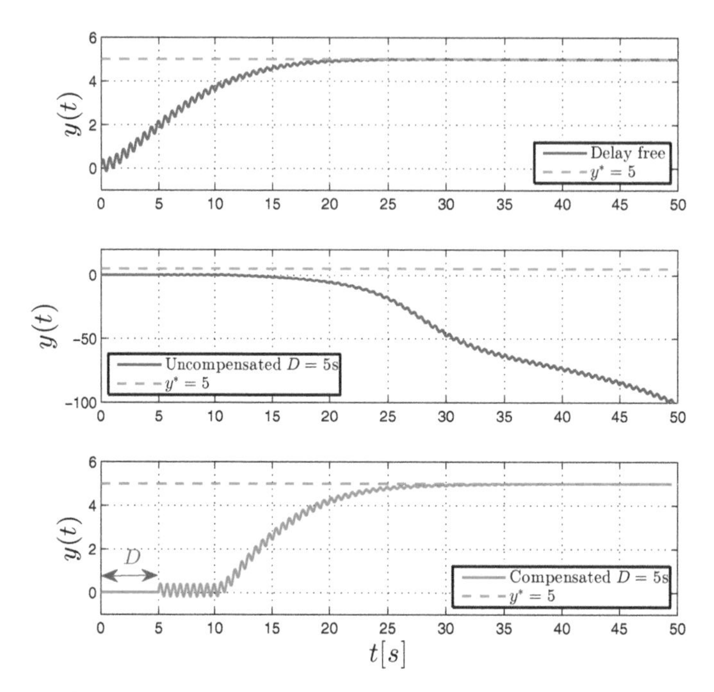

Figure 3.2. ***Newton-based ES plus output delay (time response of*** $y(t)$***):*** (a) *basic ES works well without delays;* (b) *ES goes unstable in the presence of delays;* (c) *predictor fixes this.*

Figure 3.2 shows the system output $y(t)$ in 3 situations: (a) free of output delays, (b) in the presence of output delay but without any delay compensation, and (c) with output-delay and predictor-based compensation.

Figure 3.3 presents relevant variables for ES. The red curves are shown when the developed predictor is applied in comparison to the free of delays case (with blue curves). The remarkable evolution of the new prediction scheme in searching the maximum and the Hessian's inverse $H^{-1} = -5$ is clear. This exact estimation allow us to cancel the Hessian H and thus guarantee convergence rates that can be arbitrarily assigned by the user.

In order to make a fair comparison, all common parameters used so far for the present delay-compensated Newton-based ES are exactly the same of the simulations at the gradient method introduced in [174]. Figure 3.4 illustrates the estimate of the maximum for both approaches. As expected, it is worth noting that the Newton algorithm converges faster to the extremum than the gradient scheme, even in the presence of delays.

3.5 ▪ Notes and References

Actuator and sensor delays are some of the most common phenomena that arise in control practice [119, 25]. Recently, the infinite-dimensional backstepping transformation [128] has shed light on the classical predictor feedback designs for delay compensation and provided the means for many extensions. In this chapter, we addressed the problem of real-time optimization under input-output delays by considering scalar Newton-based ES with predictor feedback.

The advantages of Newton-based over gradient ES in the absence of delays were deeply studied in [123, 73] and the discussion can be summarized in the fact that the former removes the dependence of the convergence rate of the algorithm on the unknown second derivative (Hessian)

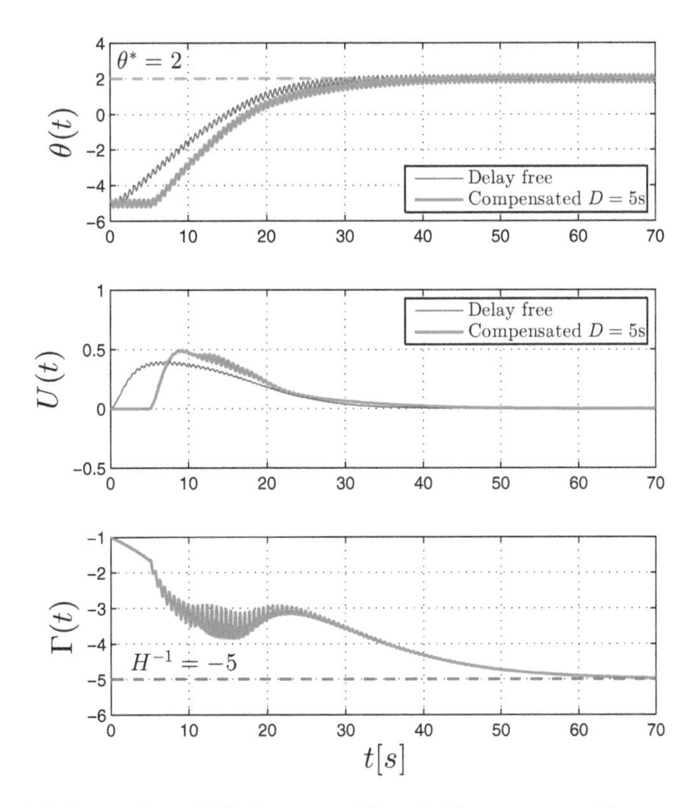

Figure 3.3. ***Newton-based ES plus output delay:*** (a) *Time response of the parameter* $\theta(t)$*;* (b) *the control signal* $U(t)$*;* (c) *Hessian's inverse estimate* $\Gamma(t)$.

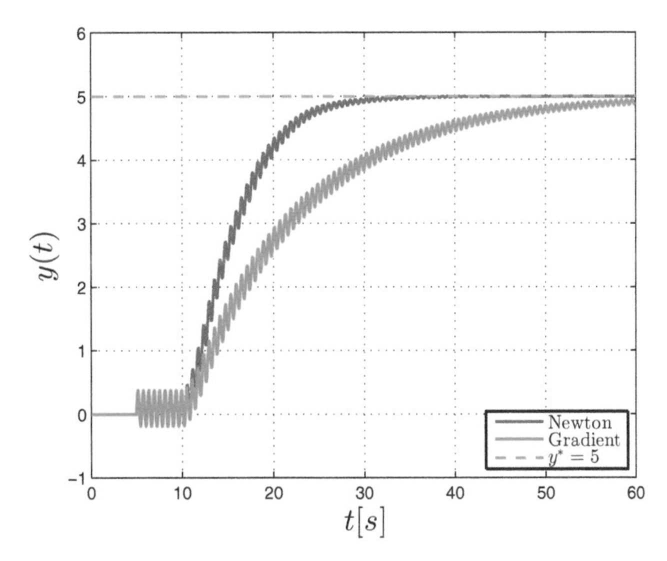

Figure 3.4. ***Newton-based ES versus Gradient ES, both under delays:*** *Time response of the output* $y(t)$ *subject to an output-delay of* $D = 5\,s$.

of the nonlinear map to be optimized, thus being arbitrarily assignable. The guarantee of this property, even in the presence of delays, is the aim achieved in the present chapter. When ignored, delay not only severely restricts the convergence rate but destabilizes the closed-loop system.

The solution to the aforementioned problem is obtained by employing a predictor feedback with a perturbation-based (averaging-based) estimate of the inverse of the unknown Hessian [166, 73]. The stability analysis is rigorously constructed via a backstepping transformation [119] and averaging theory in infinite dimension [83, 131], taking into account the entire system, i.e., including both the infinite-dimensional state of the time delay and the state of the estimator of the Hessian's inverse.

Chapter 4

Inverse Optimal ES under Delay

4.1 ▪ Inverse Optimality: What Is It and Why Aim for It?

Extremum seeking is *a method for optimization by control* but it is *not a method of optimal control.*

This sentence, if not read carefully, may seem like a contradiction. But it is not. Let us recall the distinction between optimization, which ES pursues, and optimal control.

ES pursues the optimization of a static map[1]—whether in the presence or in the absence of a dynamic system. Hence, all that matters in ES is the optimality as the time approaches infinity. The trajectory through which ES approaches an optimal steady state may be quick and straight, or slow, roundabout, and complex, but the trajectory is not what is being penalized or rewarded. Only the asymptotic value of the output is what is being minimized or maximized.

In optimal control, the asymptotic value of the output is not what is being optimized—this value must be taken to zero. It is, instead, the transient of the output, state, and the input that is being minimized, in a suitable temporal norm.

Hence, while ES is performing optimization, it is not necessarily optimal in the sense of optimal control, i.e., in the sense of the transient of its estimate, or of the update rate, being minimized.

This realization brings up the question whether this transient—towards the optimum—can itself be made optimal.

The answer to this question is, in general, negative. The reason for the impossibility of optimizing the ES transient is that optimal control is model based and, under ES, at least the output map is unknown.

However, we are still unwilling to give up on optimality, even if it is not achieved by deliberate minimization of a cost functional of the transient but is, instead, of a serendipitous nature.

This insistence on obtaining optimality, by accident if not by design, has its root in Kalman's question (paraphrased) "Is every stabilizing feedback law optimal?" Kalman's answer to this question, in his 1964 paper [96], was negative but, in providing this answer, he provided something very useful: a characterization of "inverse optimal controllers," i.e., all the controllers that happen to be optimal with respect to a cost that is meaningful (positive definite) though not chosen at will by the user.

Kalman's notion of inverse optimality has spread from LTI systems to nonlinear control [160, 206], ISS stabilization and differential games [126], stochastic nonlinear stabilization [48, 49], adaptive control [134], PDE control [209], and control of delay systems [115].

[1] or of a steady non-stationary motion, such as in the minimization of a limit cycle amplitude

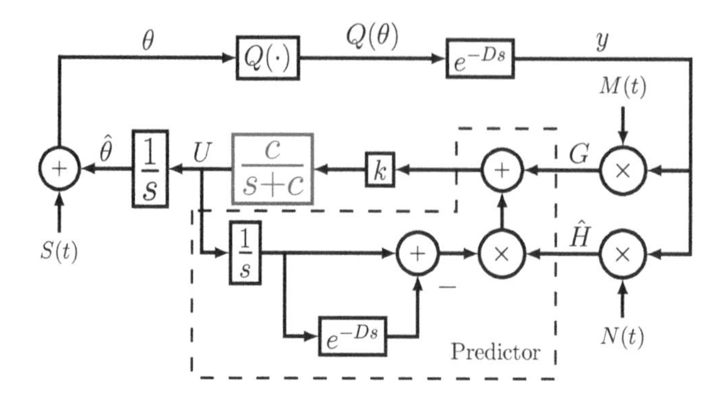

Figure 4.1. *Block diagram of the basic prediction scheme for output-delay compensation in gradient ES. The low-pass filter (in blue) is the key element employed to guarantee the inverse optimality for the closed-loop system.*

The success of inverse optimality in the context of control of delay systems, specifically, in the context of predictor feedback for systems with input delays [115], is what inspires our interest in inverse optimality of ES in the presence of delays.

In this chapter, we revisit the gradient-based extremum seeking algorithm of Chapter 2 in the presence of a delay. We show that the inclusion of a filter with a single sufficiently negative pole guarantees inverse optimality of the average ES system. The infinite-horizon cost functional that ES happens to minimize includes a positive definite cost on the ES parameter estimation error, the parameter estimation rate over a moving time window of length equal to the delay D, and the derivative of the parameter estimation error rate (i.e., parameter estimation "acceleration").

This "fast enough single-pole low-pass filter" is a part of the common ES implementation anyway for two reasons, one practical and one mathematical. First, the practical reason for including a low-pass filter is to attenuate higher-order perturbation terms of frequencies ω and higher, which arise due to injected perturbation and the (locally) quadratic nature of the map being optimized, as expanded in (1.4). Second, the mathematical reason for including the low-pass filter is to make the predictor-compensated ES feedback amenable to the application of the existing averaging theorem, by Hale and Lunel [83], for functional differential equations, which includes systems with delays on the state.

The same inverse optimality considerations can be conducted for the Newton-based extremum seeking approach presented in Chapter 3.

Results and simulations in this chapter illustrate the benefit of endowing ES with an inverse optimality property.

4.2 ▪ Extremum Seeking under Delay

The goal of scalar ES remains the same, i.e., to maximize (or minimize) the output $y \in \mathbb{R}$ of an unknown nonlinear static map $Q(\theta)$ by changing its input $\theta \in \mathbb{R}$. We further assume there is a *constant and known* delay $D \geq 0$ in the measurement system such that the delayed output signal is

$$y(t) = Q(\theta(t-D)). \tag{4.1}$$

For the sake of clarity, we illustrate the complete output-delayed system in the block diagram of Figure 4.1. Without loss of generality, we also consider the maximum seeking problem. For the sake of simplicity, we assume the nonlinear map to be optimized is quadratic as

$$Q(\theta) = y^* + \frac{H}{2}(\theta - \theta^*)^2, \tag{4.2}$$

with $\theta^* \in \mathbb{R}$, $y^* \in \mathbb{R}$, and $H \in \mathbb{R}$ being unknown scalar constants. The maximizer of the input parameter θ is given by θ^*, while y^* is the extremum point and $H < 0$ is the unknown Hessian of the map.

Substituting (4.2) into (4.1) leads us to the following quadratic static map under delay:

$$y(t) = y^* + \frac{H}{2}(\theta(t-D) - \theta^*)^2. \tag{4.3}$$

4.2.1 ▪ Probing and Demodulation Signals

Let us define the *estimation error*:

$$\tilde{\theta}(t) = \hat{\theta}(t) - \theta^*, \tag{4.4}$$

where $\hat{\theta}$ is the estimate of θ^*. According to Figure 4.1, one can write the *error dynamics* as

$$\dot{\tilde{\theta}}(t-D) = U(t-D). \tag{4.5}$$

In addition, one gets

$$G(t) = M(t)y(t), \quad \theta(t) = \hat{\theta}(t) + S(t), \tag{4.6}$$

with additive dither and demodulation signal given by

$$S(t) = a\sin(\omega(t+D)), \quad M(t) = \frac{2}{a}\sin(\omega t). \tag{4.7}$$

In (4.7), the amplitude a and frequency ω must be nonzero, assuming small and large values (as discussed later on), respectively.

The estimate of the unknown Hessian H is given by

$$\hat{H}(t) = N(t)y(t), \tag{4.8}$$

with the demodulating signal $N(t)$ being

$$N(t) = -\frac{8}{a^2}\cos(2\omega t). \tag{4.9}$$

From [73], it is possible to show

$$\frac{1}{\Pi}\int_0^{\Pi} N(\sigma)y\,d\sigma = H, \quad \Pi = 2\pi/\omega, \tag{4.10}$$

for a quadratic map, as assumed in (4.2). Basically, the expression (4.9) give us an averaging-based estimate of H by means of (4.10), i.e., $\hat{H}_{\rm av} = (Ny)_{\rm av} = H$.

4.2.2 ▪ Predictor Feedback with Averaging-Based Estimates

From the averaging analysis, we can obtain the average version of $G(t)$ in (4.6) as follows:

$$G_{\rm av}(t) = H\tilde{\theta}_{\rm av}(t-D)\,. \tag{4.11}$$

Moreover, from (4.5), we can also derive the average models

$$\dot{\tilde{\theta}}_{\rm av}(t-D) = U_{\rm av}(t-D)\,, \tag{4.12}$$
$$\dot{G}_{\rm av}(t) = HU_{\rm av}(t-D)\,, \tag{4.13}$$

with $U_{\rm av} \in \mathbb{R}$ being the average version of the control signal $U \in \mathbb{R}$.

As shown in [174, 179], by computing the variation-of-constants formula of (4.13), we can write the future state as

$$G_{\rm av}(t+D) = G_{\rm av}(t) + H\int_{t-D}^{t} U_{\rm av}(\sigma)d\sigma\,, \tag{4.14}$$

which is given in terms of the average control signal $U_{\rm av}(\sigma)$ from the past window $[t-D,t]$. It results in the next predictor-feedback law

$$U_{\rm av}(t) = k\left[G_{\rm av}(t) + H\int_{t-D}^{t} U_{\rm av}(\sigma)d\sigma\right], \quad k>0\,, \tag{4.15}$$

which is able to make the equilibrium $\tilde{\theta}^e_{\rm av} = 0$ of the closed-loop system (4.12) and (4.15) exponentially stable.

In the previous chapters, it was shown that the same stability objectives could be guaranteed if a simple modification of the above basic predictor-based controller (4.15) employing a low-pass filter [115] was applied.

As in Chapter 2, the following infinite-dimensional filtered predictor feedback, computed from its nonaverage version (4.15), is again considered:

$$U(t) = \frac{c}{s+c}\left\{k\left[G(t) + \hat{H}(t)\int_{t-D}^{t} U(\tau)d\tau\right]\right\}, \tag{4.16}$$

where $c > 0$ is a design constant chosen sufficiently large.

This low-pass filtering was particularly required in the stability analysis of our earlier publications [174, 179] when the averaging theorem in infinite dimensions was invoked [83].

In the next section, we demonstrate that the advantages of such a filtering option go beyond merely overcoming technical limitations in the averaging literature for the purpose of our analysis but that this filtering option improves the control performance of the closed-loop ES system, as rigorously justified through the concept of inverse optimality [96].

4.3 ▪ Inverse Optimal Design

In the formulation of the inverse optimality problem we consider $\dot{U}_{\rm av}(t)$ as the input to the system, whereas $U_{\rm av}(t)$ is still the actuated variable. Hence, our inverse optimal design will be implementable after integration in time, i.e., as dynamic feedback. Treating $\dot{U}_{\rm av}(t)$ as an input is the same as adding an integrator, which has been observed as being beneficial in the control design for delay systems in [93].

Theorem 4.1. *There exists c^* such that the average feedback system of* (4.5) *and* (4.16) *is exponentially stable in the sense of the norm*

$$\Psi(t) = \left(|\tilde{\theta}_{\rm av}(t-D)|^2 + \int_{t-D}^{t} U_{\rm av}(\tau)^2 d\tau + U_{\rm av}(t)^2 \right)^{1/2} \tag{4.17}$$

for all $c > c^*$. Furthermore, there exists $c^{**} > c^*$ such that for any $c \geq c^{**}$, the feedback (4.16) *minimizes the cost functional*

$$J = \int_0^\infty (\mathcal{L}(t) + \dot{U}_{\rm av}^2(t))dt, \tag{4.18}$$

where $\mathcal{L}(t)$ is a functional of $(\tilde{\theta}_{\rm av}(t-D), U(\tau))$, $\tau \in [t-D, t]$ and such that

$$\mathcal{L}(t) \geq \mu \Psi(t)^2 \tag{4.19}$$

for some $\mu(c) > 0$ with a property that $\mu(c) \to \infty$ as $c \to \infty$.

Proof. The proof is structured into **Step 1** to **Step 6**, analogously to what has been done in [179]. Some overlaps with respect to Chapter 2 (see also [174]) are incorporated for pedagogical reasons. Some minimal technical details must be given lest the reader miss the key novel points in **Step 6** regarding the inverse optimality result.

Step 1: Transport PDE for Delay Representation

Considering [119, p. 19], the delay in (4.5) is represented using a transport PDE such as

$$\dot{\tilde{\theta}}(t-D) = u(0,t), \tag{4.20}$$

$$u_t(x,t) = u_x(x,t), \quad x \in [0, D], \tag{4.21}$$

$$u(D,t) = U(t), \tag{4.22}$$

with the solution of (4.21)–(4.22) being

$$u(x,t) = U(t+x-D). \tag{4.23}$$

Step 2: Average Model of the Closed-Loop System

By denoting

$$\tilde{\vartheta}(t) := \tilde{\theta}(t-D), \qquad \tilde{\vartheta}_{\rm av}(t) := \tilde{\theta}_{\rm av}(t-D), \tag{4.24}$$

the average version of system (4.20)–(4.22), with $U(t)$ in (4.16), is given by

$$\dot{\tilde{\vartheta}}_{\rm av}(t) = u_{\rm av}(0,t), \tag{4.25}$$

$$\partial_t u_{\rm av}(x,t) = \partial_x u_{\rm av}(x,t), \quad x \in [0, D], \tag{4.26}$$

$$\frac{d}{dt} u_{\rm av}(D,t) = -c u_{\rm av}(D,t) + ckH \left[\tilde{\vartheta}_{\rm av}(t) + \int_0^D u_{\rm av}(\sigma,t) d\sigma \right], \tag{4.27}$$

where the filter $c/(s+c)$ in (4.16) was also represented in the state-space form. The solution of the transport PDE (4.26)–(4.27) is given by

$$u_{\rm av}(x,t) = U_{\rm av}(t+x-D). \tag{4.28}$$

Step 3: Backstepping Transformation, Its Inverse, and the Target System

Since we are not able to prove directly the stability for the average closed-loop system (4.25)–(4.27), we consider the infinite-dimensional backstepping transformation of the delay state

$$w(x,t) = u_{\rm av}(x,t) - kH\left[\tilde{\vartheta}_{\rm av}(t) + \int_0^x u_{\rm av}(\sigma,t)d\sigma\right], \tag{4.29}$$

with inverse given by

$$u_{\rm av}(x,t) = w(x,t) + kH\left[e^{kHx}\tilde{\vartheta}_{\rm av}(t) + \int_0^x e^{kH(x-\sigma)}w(\sigma,t)d\sigma\right]. \tag{4.30}$$

The transformation (4.29) maps the system (4.25)–(4.27) into the target system:

$$\dot{\tilde{\vartheta}}_{\rm av}(t) = kH\tilde{\vartheta}_{\rm av}(t) + w(0,t), \tag{4.31}$$
$$w_t(x,t) = w_x(x,t), \quad x \in [0,D], \tag{4.32}$$
$$w(D,t) = -\frac{1}{c}\partial_t u_{\rm av}(D,t). \tag{4.33}$$

Step 4: Lyapunov–Krasovskii Functional

Consider the following Lyapunov functional:

$$V(t) = \frac{\tilde{\vartheta}^2_{\rm av}(t)}{2} + \frac{a}{2}\int_0^D (1+x)w^2(x,t)dx + \frac{1}{2}w^2(D,t), \tag{4.34}$$

where the parameter $a = -\frac{1}{kH}$ and $kH < 0$. Computing the time derivative of (4.34) along with (4.31)–(4.33), we have

$$\begin{aligned}
\dot{V}(t) &= kH\tilde{\vartheta}^2_{\rm av}(t) + \tilde{\vartheta}_{\rm av}(t)w(0,t) \\
&\quad + a\int_0^D (1+x)w(x,t)w_x(x,t)dx + w(D,t)w_t(D,t) \\
&= kH\tilde{\vartheta}^2_{\rm av}(t) + \tilde{\vartheta}_{\rm av}(t)w(0,t) + \frac{a(1+D)}{2}w^2(D,t) \\
&\quad - \frac{a}{2}w^2(0,t) - \frac{a}{2}\int_0^D w^2(x,t)dx + w(D,t)w_t(D,t) \\
&\leq kH\tilde{\vartheta}^2_{\rm av}(t) + \frac{\tilde{\vartheta}^2_{\rm av}(t)}{2a} - \frac{a}{2}\int_0^D w^2(x,t)dx \\
&\quad + w(D,t)\left[w_t(D,t) + \frac{a(1+D)}{2}w(D,t)\right].
\end{aligned} \tag{4.35}$$

Now, following the same procedure given in Chapter 2 and [174], we get

$$\begin{aligned}
\dot{V}(t) &\leq -\frac{1}{4a}\tilde{\vartheta}^2_{\rm av}(t) - \frac{a}{4(1+D)}\int_0^D (1+x)w^2(x,t)dx \\
&\quad - (c-c^*)w^2(D,t),
\end{aligned} \tag{4.36}$$

where

$$c^* = \frac{a(1+D)}{2} - kH + a\left|(kH)^2 e^{kHD}\right|^2 + \frac{1}{a}\left\|(kH)^2 e^{kH(D-\sigma)}\right\|^2. \tag{4.37}$$

According to (4.37), an upper bound for c^* can be obtained from the known delay D as well as some lower and upper bounds of the Hessian H. Thus, from (4.36), if we chose c such that $c > c^*$, we arrive at

$$\dot{V}(t) \leq -\mu^* V(t) \tag{4.38}$$

for some $\mu^* > 0$. Hence, the closed-loop system is exponentially stable in the sense of the full-state norm

$$\left(|\tilde{\vartheta}_{\rm av}(t)|^2 + \int_0^D w^2(x,t)dx + w^2(D,t)\right)^{1/2}, \tag{4.39}$$

i.e., in the transformed variable $(\tilde{\vartheta}_{\rm av}, w)$.

Step 5: Average System Exponential Stability Estimate (in the L_2-Norm)

In order to ensure exponential stability for the average system (4.25)–(4.27) in the sense of the norm

$$\left(|\tilde{\vartheta}_{\rm av}(t)|^2 + \int_0^D u_{\rm av}^2(x,t)dx + u_{\rm av}^2(D,t)\right)^{1/2},$$

we need to show there exist constants $\alpha_1 > 0$ and $\alpha_2 > 0$ such that

$$\alpha_1 \Psi(t) \leq V(t) \leq \alpha_2 \Psi(t), \tag{4.40}$$

where $\Psi(t) := |\tilde{\vartheta}_{\rm av}(t)|^2 + \int_0^D u_{\rm av}^2(x,t)dx + u_{\rm av}^2(D,t)$ or, using (4.24) and (4.28),

$$\Psi(t) := |\tilde{\theta}_{\rm av}(t-D)|^2 + \int_{t-D}^t U_{\rm av}^2(\tau)d\tau + U_{\rm av}^2(t). \tag{4.41}$$

The inequality (4.40) can be directly established from (4.29), (4.30), (4.34) by using the Cauchy–Schwarz inequality and other calculations, such as in the proof of Theorem 2.1 in [119, p. 24]. Thus, taking into account (4.38), we obtain

$$\Psi(t) \leq \frac{\alpha_2}{\alpha_1} e^{-\mu^* t} \Psi(0), \tag{4.42}$$

which concludes the proof of exponential stability in the original variables $(\tilde{\vartheta}_{\rm av}, u_{\rm av})$.

Step 6: Inverse Optimality

Based on the proof of Theorem 6 in [209] and Theorem 2.8 in [124], we chose $c^{**} = 4c^*$, $c = 2c^*$ and define $\mathcal{L}(t)$ as

$$\begin{aligned}\mathcal{L}(t) &= -2c\dot{V}(t) + c(c-4c^*)w^2(D,t) \\ &\geq c\left(\frac{1}{2a}\tilde{\vartheta}_{\rm av}^2(t) + \frac{a}{2}\int_0^D w^2(x,t)dx + (c-2c^*)w^2(D,t)\right),\end{aligned} \tag{4.43}$$

where $\vartheta_{\rm av}(t) := \tilde{\theta}_{\rm av}(t-D)$, according to (4.24).

Using (4.29) for $x = D$ and the fact that $u_{\rm av}(D,t) = U_{\rm av}(t)$, from (4.33) we get (4.27). Let us now consider $w(D,t)$. From (4.29) and (4.30), it is easy to see that

$$w_t(D,t) = \partial_t u_{\rm av}(D,t) - kHu_{\rm av}(D,t), \tag{4.44}$$

where $\partial_t u_{\rm av}(D,t)=\dot{U}_{\rm av}(t)$. Plugging (4.33) and (4.30) into (4.44), we get

$$\begin{aligned} w_t(D,t) &= -cw(D,t) - kHw(D,t) \\ &\quad -(kH)^2\left[e^{kHD}\tilde{\vartheta}_{\rm av}(t) + \int_0^D e^{kH(D-\sigma)}w(\sigma,t)d\sigma\right]. \end{aligned} \tag{4.45}$$

By plugging (4.45) into the derivative of the Lyapunov functional (4.35), one has

$$\begin{aligned} \dot{V}(t) &= kH\tilde{\vartheta}_{\rm av}^2(t) + \tilde{\vartheta}_{\rm av}(t)w(0,t) + \frac{a(1+D)}{2}w^2(D,t) \\ &\quad -\frac{a}{2}w^2(0,t) - \frac{a}{2}\int_0^D w^2(x,t)dx - 2c^*w^2(D,t) \\ &\quad -kHw^2(D,t) - (kH)^2 w(D,t)e^{kHD}\tilde{\vartheta}_{\rm av}(t) \\ &\quad -(kH)^2 w(D,t)\int_0^D e^{kH(D-\sigma)}w(\sigma,t)d\sigma. \end{aligned} \tag{4.46}$$

Then, by applying (4.46) to (4.43), $\mathcal{L}(t)$ can be written as

$$\begin{aligned} \mathcal{L}(t) &= -2ckH\tilde{\vartheta}_{\rm av}^2(t) - 2c\tilde{\vartheta}_{\rm av}(t)w(0,t) - 2c\frac{a(1+D)}{2}w^2(D,t) \\ &\quad +caw^2(0,t) + ca\int_0^D w^2(x,t)dx + 2ckHw^2(D,t) \\ &\quad +2c(kH)^2 w(D,t)e^{kHD}\tilde{\vartheta}_{\rm av}(t) \\ &\quad +2c(kH)^2 w(D,t)\int_0^D e^{kH(D-\sigma)}w(\sigma,t)d\sigma + c^2w^2(D,t). \end{aligned} \tag{4.47}$$

On the other hand, substituting the average version of the system (4.25) into the target system (4.31), we obtain

$$u_{\rm av}(0,t) = kH\tilde{\vartheta}_{\rm av}(t) + w(0,t). \tag{4.48}$$

Rearranging (4.48) in order to isolate $w(0,t)$, we can write

$$w(0,t) = u_{\rm av}(0,t) - kH\tilde{\vartheta}_{\rm av}(t). \tag{4.49}$$

Then, plugging (4.49) into (4.47), and adding-subtracting the term $\gamma\tilde{\vartheta}_{\rm av}^2(t)$ (in blue) in the right-hand side of the resulting equation, lead us to

$$\begin{aligned} \mathcal{L}(t) &= c\Big(a(kH)^2\tilde{\vartheta}_{\rm av}^2(t) - 2(akH+1)u_{\rm av}(0,t)\tilde{\vartheta}_{\rm av}(t) \\ &\quad -a(1+D)w^2(D,t) - \gamma\tilde{\vartheta}_{\rm av}^2(t) + 2(kH)^2 w(D,t) \\ &\quad \times\left[e^{kHD}\tilde{\vartheta}_{\rm av}(t) + \int_0^D e^{kH(D-\sigma)}w(\sigma,t)d\sigma\right]\Big) \\ &\quad +au_{\rm av}^2(0,t) + \frac{a}{2}\int_0^D w^2(x,t)dx + w^2(D,t)(2c^* + 2kH) \\ &\quad +c\Big(\gamma\tilde{\vartheta}_{\rm av}^2(t) + \frac{a}{2}\int_0^D w^2(x,t)dx + (c-2c^*)w^2(D,t)\Big). \end{aligned} \tag{4.50}$$

Recalling that $a=-\frac{1}{kH}$, and replacing kH by $-\frac{1}{a}$ in (4.50), one has

$$\begin{aligned}\mathcal{L}(t)=&c\Big(\Big[\frac{1}{a}-\gamma\Big]\tilde{\vartheta}_{\text{av}}^2(t)\big(2c^*-a(1+D)-\frac{2}{a}\big)w^2(D,t)\\&+au_{\text{av}}^2(0,t)+\frac{a}{2}\int_0^D w^2(x,t)dx+\frac{2}{a^2}w(D,t)\\&\times\Big[e^{kHD}\tilde{\vartheta}_{\text{av}}(t)+\int_0^D e^{kH(D-\sigma)}w(\sigma,t)d\sigma\Big]\Big)\\&+c\big(\gamma\tilde{\vartheta}_{\text{av}}^2(t)+\frac{a}{2}\int_0^D w^2(x,t)dx+(c-2c^*)w^2(D,t)\big).\end{aligned}\tag{4.51}$$

After some mathematical manipulations, the term $\mathcal{L}(t)$ in (4.51) can be rewritten as

$$\begin{aligned}\mathcal{L}(t)=&\Upsilon(D,t)+c\big(\gamma\tilde{\vartheta}_{\text{av}}^2(t)+\frac{a}{2}\int_0^D w^2(x,t)dx\\&+(c-2c^*)w^2(D,t)\big),\end{aligned}\tag{4.52}$$

where $\Upsilon(D,t)$ is given by

$$\begin{aligned}\Upsilon(D,t)=&c\Big(\Big[\frac{1}{a}-\gamma\Big]\tilde{\vartheta}_{\text{av}}^2(t)+\big(2c^*-a(1+D)-\frac{2}{a}\big)w^2(D,t)\\&+au_{\text{av}}^2(0,t)+\frac{a}{2}\int_0^D w^2(\sigma,t)d\sigma+\frac{2}{a^2}w(D,t)e^{kHD}\tilde{\vartheta}_{\text{av}}(t)\\&+\frac{2}{a^2}w(D,t)\int_0^D e^{kH(D-\sigma)}w(\sigma,t)d\sigma\Big).\end{aligned}\tag{4.53}$$

In order to satisfy inequality (4.19), it is necessary to ensure $\Upsilon(D,t)\geq 0$. To ensure the latter condition, we will analyze the terms in (4.53) with undefined signs so that we can guarantee they are nonnegative. After adding and subtracting the terms $\frac{1}{a^2}[\tilde{\vartheta}_{\text{av}}^2+w^2(D,t)]$ and $\frac{2\sqrt{D}}{a^2}[w^2(D,t)+\int_0^D w^2(\sigma,t)d\sigma]$ (in blue and red) into (4.53), $\Upsilon(D,t)$ can be rewritten as

$$\begin{aligned}\Upsilon(D,t)=&c\Big(\Big[\frac{1}{a}-\frac{1}{a^2}-\gamma\Big]\tilde{\vartheta}_{\text{av}}^2(t)\\&+\big(2c^*-a(1+D)-\frac{2}{a}-\frac{1}{a^2}-\frac{2\sqrt{D}}{a^2}\big)w^2(D,t)\\&+au_{\text{av}}^2(0,t)+\Big[\frac{a}{2}-\frac{2\sqrt{D}}{a^2}\Big]\int_0^D w^2(\sigma,t)d\sigma\\&+\frac{2}{a^2}w(D,t)e^{kHD}\tilde{\vartheta}_{\text{av}}(t)+\frac{1}{a^2}w^2(D,t)+\frac{1}{a^2}\tilde{\vartheta}_{\text{av}}^2(t)\\&+\frac{2}{a^2}w(D,t)\int_0^D e^{kH(D-\sigma)}w(\sigma,t)d\sigma\\&+\frac{2\sqrt{D}}{a^2}w^2(D,t)+\frac{2\sqrt{D}}{a^2}\int_0^D w^2(\sigma,t)d\sigma\Big).\end{aligned}\tag{4.54}$$

By employing the Young and Cauchy–Schwarz inequalities, it is possible verify valid lower bounds for the terms which were added and subtracted in (4.54), so that

$$\frac{1}{a^2}w^2(D,t)+\frac{1}{a^2}\tilde{\vartheta}_{\text{av}}^2(t)\geq\frac{2}{a^2}\Big|w(D,t)e^{kHD}\tilde{\vartheta}_{\text{av}}(t)\Big|,\tag{4.55}$$

$$\frac{2\sqrt{D}}{a^2}\Big(w^2(D,t)+\int_0^D w^2(\sigma,t)d\sigma\Big)\geq\frac{2}{a^2}\Big|w(D,t)\int_0^D e^{kH(D-\sigma)}w(\sigma,t)d\sigma\Big|.\tag{4.56}$$

Analyzing $\Upsilon(D,t)$ in terms of the lower bounds in (4.55) and (4.56), we get

$$\begin{aligned}\Upsilon(D,t) \geq c\Big(&\Big[\frac{1}{a}-\frac{1}{a^2}-\gamma\Big]\tilde{\vartheta}_{\rm av}^2(t)\\&+\big(2c^*-a(1+D)-\frac{2}{a}-\frac{1}{a^2}-\frac{2\sqrt{D}}{a^2}\big)w^2(D,t)\\&+au_{\rm av}^2(0,t)+\Big[\frac{a}{2}-\frac{2\sqrt{D}}{a^2}\Big]\int_0^D w^2(\sigma,t)d\sigma\\&+\frac{2}{a^2}w(D,t)e^{kHD}\tilde{\vartheta}_{\rm av}(t)+\frac{2}{a^2}\left|w(D,t)e^{kHD}\tilde{\vartheta}_{\rm av}(t)\right|\\&+\frac{2}{a^2}w(D,t)\int_0^D e^{kH(D-\sigma)}w(\sigma,t)d\sigma\\&+\frac{2}{a^2}\left|w(D,t)\int_0^D e^{kH(D-\sigma)}w(\sigma,t)d\sigma\right|\Big).\end{aligned}\tag{4.57}$$

Then, to ensure $\Upsilon(D,t)\geq 0$ it is necessary to satisfy the following conditions:
1st Condition:

$$\frac{1}{a}-\frac{1}{a^2}-\gamma>0, \qquad \gamma<\frac{a-1}{a^2}.$$

2nd Condition:
Recalling that $c=2c^*$,

$$2c^*-a(1+D)-\frac{2}{a}-\frac{1}{a^2}-\frac{2\sqrt{D}}{a^2}>0,$$
$$c>a(1+D)+\frac{2}{a}+\frac{1}{a^2}+\frac{2\sqrt{D}}{a^2}.$$

3rd Condition:

$$\frac{a}{2}-\frac{2\sqrt{D}}{a^2}>0, \qquad a>\sqrt[3]{4\sqrt{D}}.$$

Therefore, considering $\mathcal{L}(t)$ given in (4.52) and $\Upsilon(D,t)$ given in (4.53), under the conditions above imposed for γ, a, and c, one can conclude $\Upsilon(D,t)\geq 0$ and

$$\mathcal{L}(t)\geq c\left(\frac{1}{2a}\tilde{\vartheta}_{\rm av}^2(t)+\frac{a}{2}\int_0^D w^2(x,t)dx+(c-2c^*)w^2(D,t)\right),$$

with $\gamma=\frac{1}{2a}$.

Hence, we have $\mathcal{L}(t)\geq \mu\Psi(t)^2$ for the same reason that (4.40) holds, completing the proof of inverse optimality. □

Corollary 4.1. *For the delay-free case ($D=0$), the ES feedback*

$$\dot{U}(t)=-cU(t)+ckG(t) \tag{4.58}$$

is inverse optimal and stabilizing in the norm

$$\bar{\Psi}(t)=\left(|\tilde{\theta}_{\rm av}(t-D)|^2+U_{\rm av}(t)^2\right)^{1/2}. \tag{4.59}$$

In particular, there exist $c^{**} > c^* > 0$ such that for any $c \geq c^{**}$, the feedback (4.58) *minimizes the cost functional*

$$\bar{J} = \int_0^\infty (\mathcal{L}(t) + \dot{U}_{\rm av}^2(t))dt, \tag{4.60}$$

where $\mathcal{L}(t)$ is a functional of $(\tilde{\theta}_{\rm av}(t), U_{\rm av}(t))$ and such that

$$\mathcal{L}(t) \geq \mu \bar{\Psi}(t)^2 \tag{4.61}$$

for some $\mu(c) > 0$ with a property that $\mu(c) \to \infty$ as $c \to \infty$.

The result above is obtained from Theorem 4.1 with the Lyapunov function $V(t) = \tilde{\theta}_{\rm av}^2(t)/2$ employed in the proof.

Remark 4.1. The feedback (4.15) is not inverse optimal; however, the feedback (4.16) is for any $c \in [c^{**}, \infty)$. Its optimality holds for a relevant cost functional, which is underbounded by the temporal $L_2[0, \infty)$ norm of the ODE state $\tilde{\vartheta}_{\rm av}(t)$, the norm of the average control $U_{\rm av}(t)$, as well as the norm of its derivative $\dot{U}_{\rm av}(t)$—in addition to $\int_{-D}^0 U_{\rm av}(\tau)^2 d\tau$, which is fixed because feedback has no influence on it. The controller (4.16) is stabilizing for $c = \infty$, namely, in its nominal form (4.15); however, since $\mu(\infty) = \infty$, it is not optimal with respect to a cost functional that includes a penalty on $\dot{U}_{\rm av}(t)$.

Remark 4.2. In our inverse optimality results of Theorem 4.1 and Corollary 4.1, we also want to minimize the update rate $\dot{U}_{\rm av}(t)$ of the filtered-predictor feedbacks (4.16) or (4.58) over the infinite interval [115] in order to improve the closed-loop performance in terms of transient responses and smooth control signals. However, we could study the inverse optimality of the average system with a different cost functional and the "input" in (4.18) or (4.60) not being the average update rate. For instance, the controller in [39, Theorem 3], in the context of stabilization rather than ES, does not employ a filter and the cost functional employs (only) the control instead of its derivative. Unfortunately, this result cannot be used for ES since time-varying delays are considered there, and we would not be able to apply the averaging theorem by [83] to complete the proof, as discussed in [182].

Finally, analogously to the Steps 6 and 7 performed for the proof of Theorem 2.1 in Chapter 2, we can invoke the averaging theorem in infinite dimensions by [83] (see Appendix A) and still conclude the results for constant delays, where the estimation errors $\theta(t) - \theta^*$ and $y(t) - y^*$ are ultimately of order $\mathcal{O}(a + 1/\omega)$ and $\mathcal{O}(a^2 + 1/\omega^2)$, respectively.

4.4 ▪ Numerical Simulations

In order to evaluate the effects of the inverse optimality for the ES feedback under delays, the quadratic map (4.1)–(4.2) is considered: $Q(\theta) = 5 - 0.1(\theta - 3)^2$, with an output delay of $D = 5$ s. The extremum point is $(\theta^*;\ y^*) = (3;\ 5)$ and the Hessian of the corresponding static map is $H = -0.2$. For the simulation tests, the following parameters were employed: $\omega = 10$ rad/s, $k = 0.8$, $\hat{\theta}(0) = -5$, and $a = 0.2$. The time constant of the low-pass filter $c = 40$ was chosen to satisfy Conditions 1–3 in **Step 6** of the proof of Theorem 4.1.

Figure 4.2 presents a numerical comparison between the ES fundamental variables with and without using the filter $\frac{c}{s+c}$ in feedback law (4.16). As it can be observed, in the first case where the inverse optimality is guaranteed, the input-output signals, $\theta(t)$ and $y(t)$, converge monotonically rather than swinging up and down, thus improving the transient responses.

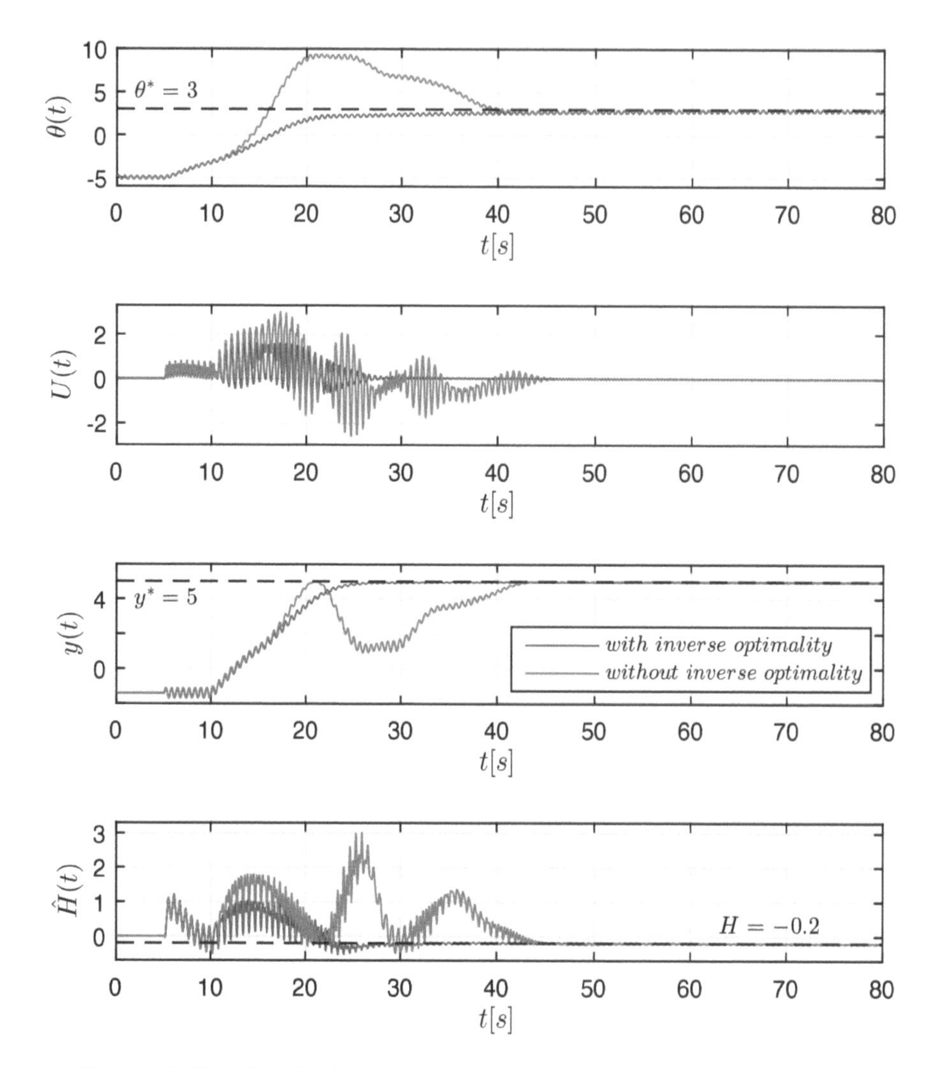

Figure 4.2. *ES with and without inverse optimality:* (a) *input signal* $\theta(t)$*;* (b) *control signal* $U(t)$*;* (c) *output signal* $y(t)$*;* (d) *Hessian estimate* $\hat{H}(t)$.

Most interestingly, inverse optimality (namely, low-pass filtering before the update) has the effect of calming down the estimation of the Hessian. As evident from Figure 4.2, the oscillations in the Hessian estimate $\hat{H}(t)$ are reduced as well as the amplitude of the control signal $U(t)$.

4.5 ▪ Notes and References

In the ES literature there are many publications applying high-pass and low-pass filters in order to improve the closed-loop system performance and to facilitate the tuning parameters [2, 219, 166, 73, 142]. However, they do not present any theoretical support that justifies the inclusion of such filters; on the contrary, only heuristic arguments are given.

In this chapter, the proof of inverse optimality and its influence on the extremum seeking feedback is discussed in the presence of delays (although the results are also valid in the case without delays). We show that the basic predictor feedback controller originally proposed in

[174, 179], when applied through a low-pass filter, is inverse optimal and we study its robustness to the low-pass filter time constant.

Inverse optimality was defined by [96] as follows: "Given a dynamic system and a known control law, find performance criteria (if any) for which this control law is optimum." Inverse optimality can be related to a Lyapunov concept, with a special control law. In this sense, the inverse optimality is guaranteed when a stabilizing controller is optimal for some criteria and, for a given Lyapunov function, it is possible to show the feedback law is optimal with respect to some cost function. In general, this functional includes a control input penalty (or in its derivative or rate) and has an infinite gain margin [115, 39].

Notice that, so far, classical ES was neither studied with Lyapunov tools nor has a control input been included in a cost that should be optimized over infinite time. In this chapter, we studied the inverse optimality of the average system, which we do via Lyapunov method, and we also minimize the update rate over the infinite interval.

Chapter 5

Stochastic ES for Higher-Derivative and Dynamic Maps with Delay

5.1 ▪ Seeking Inflection Points and Extrema of Higher Derivatives of Maps

The conventional use of ES is for seeking maxima and minima, namely, for seeking points where the first derivative of the map is zero. This is achieved through a careful choice of probing and demodulation signals which enable the estimation of the gradient (as well as the Hessian) of the unknown map.

But why stop at problems where the first derivative is driven to zero? Why stop at seeking minima and maxima? Why not seek also points where higher derivatives are zero, like inflexion points? We don't get into applications here but they certainly exist, including in thermal management systems.

Indeed, we shall not stop at seeking points where the first derivative is zero. We pursue the seeking of points where the second or any higher derivative is zero. But we cannot seek all such points at once. We can only seek individual critical points and we need to decide up front which algorithm we will employ, depending on a priori knowledge of what kind of critical points the map has.

The key to seeking points at which higher derivatives are zero is designing demodulation signals that allow us to estimate the respective higher derivatives. Once we have an estimate of the $n+1$st derivative of a function, feeding this estimate into a standard gradient-type ES algorithm will result in seeking the point where the nth derivative of the function is maximized or minimized. In particular, once we have an estimate of the second derivative (Hessian) of a function, feeding this estimate into a standard ES algorithm will result in seeking the point where the gradient is maximal or minimal, i.e., an inflection point.

In the presence of a delay, we know that the gradient-based seeking of maxima or minima requires also the estimation of the map's Hessian. So, when we seek points where the nth derivative of a map is zero, in addition to needing an estimate of the $n+1$st derivative, we also need an estimate of the map's $n+2$nd derivative.

This informs us regarding what types of higher-order demodulation signals we need. If we are employing a standard zero-mean perturbation S, we will need much higher order demodulation signals to estimate the $n+1$st and $n+2$nd derivatives, and these demodulation signals will need to incorporate a delay of S if the map's input or output are delayed. While these demodulation signals are too complicated to state in this introduction, we indicate here that they will be denoted by Υ_{n+1} and Υ_{n+2}.

But this is a rather special chapter of the book in which we pursue more than seeking extrema of higher derivatives. In this chapter we also present a stochastic version of delay-compensated ES. The chapter is to be regarded as a generalization of stochastic ES from seeking minima and maxima of a map to seeking maxima and minima of higher derivatives of a map.

Furthermore, this is one of the few chapters in which we pursue delay-compensated ES not just for static maps but for dynamic systems. In this chapter, the delay acts on the measured output of the dynamic system. We seek a maximum of a higher derivative of a map from an adjustable system parameter of the dynamic system to the equilibrium value of the output once the system reaches an equilibrium, namely, of a function that is a composition of the function relating the plant equilibrium with the plant parameter and the plant's output map.

Finally, since the delay compensation requires an estimation of the $n+2$nd derivative of the parameter-to-output map, we design a Newton version of a seeker of the maximum of the nth derivative of the parameter-to-output map.

Stochastic ES is a branch of extremum seeking research which involves a very different averaging theory, along with which go different probing and demodulation signals. Averaging is not conducted in time but relative to random realizations of the stochastic noise perturbation signals. Stochastic ES was introduced in the work of Liu and Krstic [140, 143]. Stochastic ES requires significant advances relative to the previous theorems of stochastic averaging and these ES-suitable stochastic averaging theorems can be found in the book [10] by Liu and Krstic.

Stochastic ES is a cousin of the "stochastic approximation" method. The differences include the facts that Stochastic ES does not pursue unbiased estimation through a tuning gain that decays to zero because the map might change later and, more importantly, that stochastic ES has been developed with guaranteed convergence properties for online application with dynamic systems, such as source seeking with nonholonomic unicycles [141].

Deterministic versions of the results in this chapter, with periodic probing and demodulation signals, do exist and can be found in [200] and in Chapter 6 of this book.

We incorporate a novel predictor feedback for delay compensation and show local exponential stability along with convergence to a small neighborhood of the unknown extremum point. For the purpose of the proof, we apply a backstepping transformation and averaging theory in infinite dimension for stochastic systems. Numerical simulations show that the stochastic version outperforms the periodic one in terms of faster convergence rates.

5.2 ▪ Problem Formulation

Different from the previous chapters, now we are interested in maximizing (w.l.o.g.) arbitrary derivatives of unknown nonlinear *dynamic maps*

$$\dot{\mathbf{x}} = \mathbf{f}(\mathbf{x},u), \tag{5.1}$$

$$y(t-D) = g(\mathbf{x}(t-D)), \tag{5.2}$$

under constant output delays, where $\mathbf{x}\in\mathbb{R}^m$ is the m-dimensional state vector; $u\in\mathbb{R}$ and $y\in\mathbb{R}$ represent the scalar input and output, respectively; and $\mathbf{f}\colon \mathbb{R}^m\times\mathbb{R}\to\mathbb{R}^m$ as well as $g\colon \mathbb{R}^m\to\mathbb{R}$ are smooth [129]. Establishing that a smooth control law $\alpha\colon \mathbb{R}^m\times\mathbb{R}\to\mathbb{R}$

$$u = \alpha(\mathbf{x},\theta) \tag{5.3}$$

is acting upon the plant, one obtains the closed-loop system

$$\dot{\mathbf{x}} = \mathbf{f}\big(\mathbf{x},\alpha(\mathbf{x},\theta)\big). \tag{5.4}$$

Its equilibria, characterized by the scalar parameter θ, are specified by the next assumptions [151].

Assumption 5.1. *Let* $\mathbf{l}\colon \mathbb{R} \to \mathbb{R}^m$ *be an existing, smooth vector field, such that only when the state vector* $\mathbf{x}$ *follows the assignment*

$$\mathbf{x} = \mathbf{l}(\theta), \tag{5.5}$$

the closed-loop system (5.4) *is in the equilibrium*

$$\mathbf{f}\big(\mathbf{x}, \alpha(\mathbf{x},\theta)\big) = \mathbf{0}. \tag{5.6}$$

Assumption 5.2. *For every value of the parameter* $\theta \in \mathbb{R}$, *the equilibrium* (5.6) *is exponentially stable with decay rate and overshoot constant uniform in* θ.

For $D \geq 0$, the delayed output y is

$$y(t-D) = g\Big(\mathbf{l}(\theta(t-D))\Big), \tag{5.7}$$

which is only true in the equilibrium (Assumptions 5.1 and 5.2). Input delays are not considered since, in the case of dynamic maps, they cannot be simply transformed to output delays.

Defining $\nu\colon \mathbb{R} \to \mathbb{R}$ as the composition of the scalar output function g in (5.2) and the state vector function $\mathbf{l}$ in (5.5)

$$\nu(\cdot) = (g \circ \mathbf{l})(\cdot), \tag{5.8}$$

with $n \in \mathbb{N}_0$, we formulate our optimization problem as

$$\max_{\theta\in\mathbb{R}} \nu^{(n)}(\theta(t-D)) := \max_{\theta\in\mathbb{R}} \frac{d^n \nu}{d\theta^n}(\theta(t-D)), \tag{5.9}$$

where the corresponding maximizing value is denoted by θ^*.

Assumption 5.3. *Consider the set*

$$\Theta_{\max} = \Big\{\theta \mid \nu^{(n+1)}(\theta) = 0, \quad \nu^{(n+2)}(\theta) < 0\Big\} \neq \emptyset \tag{5.10}$$

including all stationary points which are local maxima, i.e., locally concave. We assume that $\exists \theta^* \in \Theta_{\max}$.

5.3 ▪ Probing and Demodulation Signals

We only require the map g (5.2) itself to be measurable. Moreover, all constants, i.e., the dither parameters a and ω, the integrator gains

$$k_I = \varepsilon\omega k_I' = \mathcal{O}(\varepsilon\omega), \tag{5.11}$$

$$k_R = \varepsilon\omega k_R' = \mathcal{O}(\varepsilon\omega), \tag{5.12}$$

and the time scale separation ε, as well as the entailing gains k_I' and k_R', are user-assignable, having to be positive [151]. For the averaging theory we require $0 < \varepsilon \ll 1$. The analysis strategy leads to multiple time scales with the plant dynamics (5.4) being the fastest [166].

We employ a stochastic perturbation of the unknown map via the sinusoid of a Wiener process $\eta(t)$ about the boundary of a circle [151],

$$S(t) = a\sin(\eta(t)), \tag{5.13}$$

such that

$$\theta(t) = \hat{\theta}(t) + a\sin(\eta(t)), \tag{5.14}$$

where

$$\eta(t) = \omega\pi\left(1 + \sin(W_{\omega t})\right) \tag{5.15}$$

represents a homogeneous ergodic Markov process with invariant distribution, where $W_{\omega t}$ is a standard Brownian motion process (also referred to as the Wiener process) [142]. To satisfy the Markov property, there is no difference in future predictions based on just the current process state or with its full history. We then refer to a Markov process as a stochastic process which satisfies the Markov property with respect to its natural filtration.

The special demodulation signals

$$\Upsilon_k(t) = C_k \sin\left(k\eta(t) + \frac{\pi}{4}(1 + (-1)^k)\right), \tag{5.16}$$

with the normalizing gain

$$C_k = \frac{2^k k!}{a^k}(-1)^{\frac{k - \left|\sin\left(\frac{k\pi}{2}\right)\right|}{2}}, \tag{5.17}$$

allow a sufficiently precise estimate of the gradient ($k = n+1$) and Hessian ($k = n+2$) in an average sense.

By using Ito's stochastic chain rule [142], one obtains

$$d\eta = -\omega\frac{\pi}{2}\sin(W_{\omega t})dt + \omega\pi\cos(W_{\omega t})dW_{\omega t} \tag{5.18}$$

as the perturbational middle time scale [151].

With $\hat{\theta}$ being the best estimate of the maximizing value θ^*, and from the block diagram in Figure 5.1, we can write

$$\frac{d\hat{\theta}}{dt} = -\varepsilon\omega k_I' U(t), \tag{5.19}$$

where the control signal $U(t)$ is generated through our proposed predictor-based strategy in Section 5.5.

Furthermore, we use a Riccati filter [73]

$$\frac{d\gamma}{dt} = \varepsilon\omega k_R' \gamma(t)\left(1 - \gamma(t)\widehat{\nu^{(n+2)}}(t)\right) \tag{5.20}$$

to dynamically estimate the Hessian's inverse of $\nu^{(n)}(\theta)$. In addition, by setting the sign of the initial value $\mathrm{sgn}(\gamma(0)) = \mathrm{sgn}\left(\nu^{(n+2)}(\theta^*)\right)$ one can switch from a maximization to a minimization problem. Altogether, these differential parameter update equations (5.19) and (5.20) follow the slowest time scale.

The demodulated signals are defined as

$$\widehat{\nu^{(k)}}(t) = \Upsilon_k(t - D)y(t - D). \tag{5.21}$$

The estimates for the Gradient (1st derivative) and Hessian (2nd derivative) of the nth derivative map $\nu^{(n)}$ in (5.7)–(5.9) are obtained by means of (5.21) by setting $k = n+1$ and $k = n+2$, respectively. They are also delayed by D time units in order to cope with the delayed output y.

Lastly, we define the measurable signal

$$z(t) = \gamma(t)\widehat{\nu^{(n+1)}}(t), \tag{5.22}$$

where the estimate $\gamma(t)$ of the Hessian's inverse is updated according to (5.20), and $\widehat{\nu^{(n+1)}}(t)$ represents the gradient estimate according to (5.21) [200].

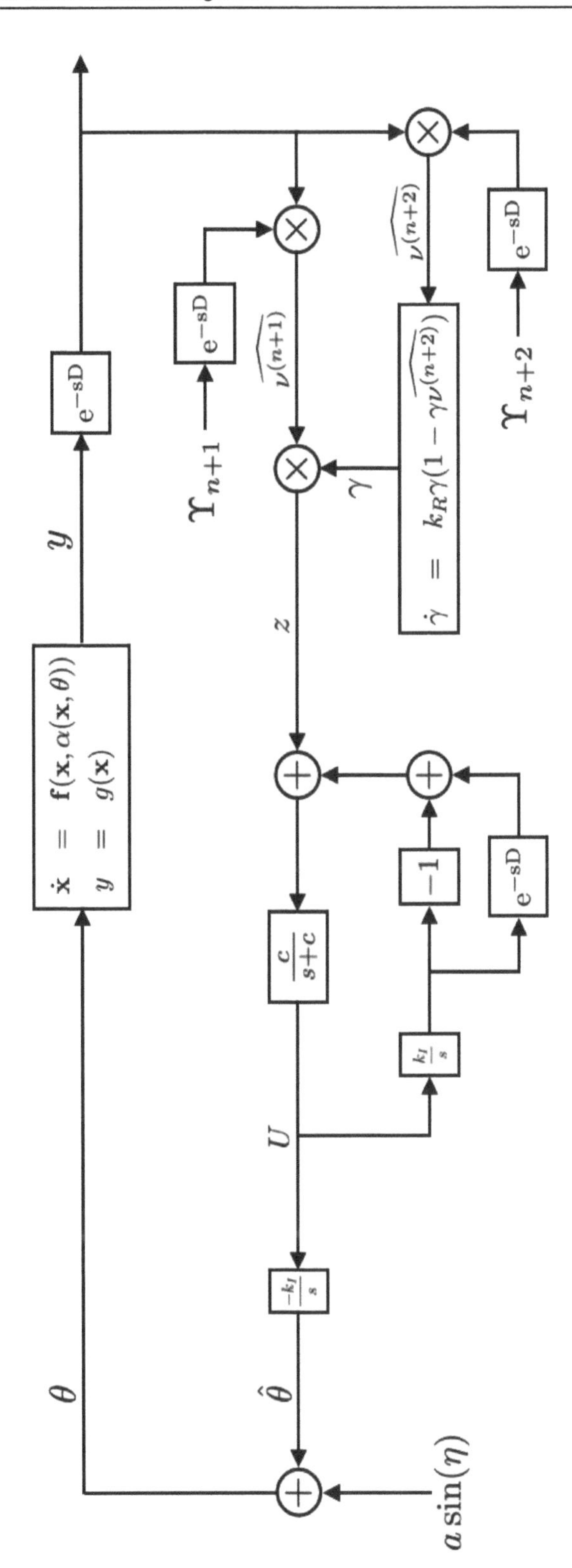

Figure 5.1. *Block diagram of the basic prediction scheme for time-delay compensation in Newton-based extremum seeking with stochastic perturbation and an unknown, nonlinear, dynamic map for maximizing arbitrarily higher derivatives.*

5.4 ▪ Averaging Analysis

Consider the error transformations

$$\tilde{\theta}(t) = \hat{\theta}(t) - \theta^* , \tag{5.23}$$

$$\tilde{\gamma}(t) = \gamma(t) - \gamma^* , \tag{5.24}$$

with the inverse of the Hessian

$$\gamma^* = \frac{1}{\nu^{(n+2)}(\theta^*)} \tag{5.25}$$

being the corresponding optimum value of γ.

Lemma 5.1. *For maximizing any higher derivative of the unknown, dynamic, nonlinear, and time-delayed map $y(t-D)$ satisfying Assumptions* 5.1–5.3*, the reduced, averaged, and linearized version of the measurable signal $z(t)$ as stated in* (5.22) *can be expressed as*

$$z_{\mathrm{r}}^{\mathrm{av}}(t) = \tilde{\theta}_{\mathrm{r}}^{\mathrm{av}}(t-D) \tag{5.26}$$

when using the stochastic perturbation $\eta(t)$ being a homogeneous ergodic Markov process, given by (5.14)–(5.15)*, and the estimates given by the demodulated signals* (5.21)*.*

Proof. For the uncompensated-delay case, where $U(t)$=$z(t)$, using (5.19) and (5.23) as well as (5.20) and (5.24), we can express the full-error dynamics as

$$\frac{d\tilde{\theta}}{dt} = -\varepsilon\omega k_I'\big(\tilde{\gamma}(t)+\gamma^*\big)\Upsilon_{n+1}(t-D)g\big(\mathbf{x}(t-D)\big), \tag{5.27}$$

$$\frac{d\tilde{\gamma}}{dt} = \varepsilon\omega k_R'\big(\tilde{\gamma}(t)+\gamma^*\big) \times\Big(1-\big(\tilde{\gamma}(t)+\gamma^*\big)\Upsilon_{n+2}(t-D)g\big(\mathbf{x}(t-D)\big)\Big). \tag{5.28}$$

After employing a singular perturbation reduction [151] in order to freeze the delayed-state vector $\mathbf{x}(t-D)$ in (5.4) at its quasi-steady state value

$$\mathbf{x}(t-D) = \mathbf{l}\Big(\tilde{\theta}(t-D)+a\sin\big(\eta(t-D)\big)\Big), \tag{5.29}$$

we obtain the reduced error system of the form

$$\frac{d\tilde{\theta}_{\mathrm{r}}}{dt} = -\varepsilon\omega k_I'\big(\tilde{\gamma}_{\mathrm{r}}(t)+\gamma^*\big) \times\Upsilon_{n+1}(t-D)\nu\Big(\tilde{\theta}_{\mathrm{r}}(t-D)+a\sin\big(\eta(t-D)\big)\Big), \tag{5.30}$$

$$\frac{d\tilde{\gamma}_r}{dt} = \varepsilon\omega k_R'\big(\tilde{\gamma}_{\mathrm{r}}(t)+\gamma^*\big)\Bigg(1-\big(\tilde{\gamma}_{\mathrm{r}}(t)+\gamma^*\big) \times\Upsilon_{n+2}(t-D)\nu\Big(\tilde{\theta}_{\mathrm{r}}(t-D)+a\sin\big(\eta(t-D)\big)\Big)\Bigg). \tag{5.31}$$

We assume the maximizing value θ^* to remain constant in the reduced system, since the parameter differential equations follow the slowest time scale. For sufficiently small ε, the reduced parameter error system is applicable to the theory of averaging for stochastic systems [106].

Now, for the averaging method, we replace the frozen quantities $\tilde{\theta}_{\mathrm{r}}$ and $\tilde{\gamma}_{\mathrm{r}}$ with autonomous values $\tilde{\theta}_{\mathrm{r}}^{\mathrm{av}}$ and $\tilde{\gamma}_{\mathrm{r}}^{\mathrm{av}}$, respectively. Then, comparing (5.30) and (5.31), we focus on the evaluation of the average

$$\xi_{\mathrm{r,k}}^{\mathrm{av}} = \mathrm{AVE}\left\{\widehat{\nu^{(k)}}\left(\tilde{\theta}_{\mathrm{r}}^{\mathrm{av}}(t-D)\right)\right\} \tag{5.32}$$

exploiting the ergodicity and invariant distribution. The operator $\mathrm{AVE}\left\{\cdot\right\}$ simply denotes the average computation in the RHS of (5.32). The average of the reduced-demodulated error signal $\xi_{\mathrm{r,k}}^{\mathrm{av}}$ (5.32) can be solved for any arbitrary $k \in \mathbb{N}$, and one obtains the expression [151]

$$\xi_{\mathrm{r,k}}^{\mathrm{av}} = \nu^{(k)}\left(\tilde{\theta}_{\mathrm{r}}^{\mathrm{av}}(t-D)\right) + \frac{\nu^{(k+2)}\left(\tilde{\theta}_{\mathrm{r}}^{\mathrm{av}}(t-D)\right)}{4(k+1)}a^2 + \mathcal{O}(a^4). \tag{5.33}$$

Consequently, we can now express the averaged-reduced error system as

$$\frac{d\tilde{\theta}_{\mathrm{r}}^{\mathrm{av}}}{dt} = -\varepsilon\omega k_I'\left(\tilde{\gamma}_{\mathrm{r}}^{\mathrm{av}}(t)+\gamma^*\right)\xi_{\mathrm{r,n+1}}^{\mathrm{av}}, \tag{5.34}$$

$$\frac{d\tilde{\gamma}_{\mathrm{r}}^{\mathrm{av}}}{dt} = \varepsilon\omega k_R'\left(\tilde{\gamma}_{\mathrm{r}}^{\mathrm{av}}(t)+\gamma^*\right)\left(1-\left(\tilde{\gamma}_{\mathrm{r}}^{\mathrm{av}}(t)+\gamma^*\right)\xi_{\mathrm{r,n+2}}^{\mathrm{av}}\right). \tag{5.35}$$

We use a local quadratic approximation of the objective function [200] at $\theta = \theta^*$, which yields

$$\nu^{(n)}(\theta) = Q^* + \frac{\nu^{(n+2)}(\theta^*)}{2}\left(\theta(t)-\theta^*\right)^2. \tag{5.36}$$

Thus, the equilibria of (5.34) and (5.35) become

$$\tilde{\theta}_r^{\mathrm{av,e}} = 0, \tag{5.37}$$

$$\tilde{\gamma}_r^{\mathrm{av,e}} = 0. \tag{5.38}$$

After plugging (5.36) into (5.33) for $k = n+1$, we obtain

$$\xi_{\mathrm{r,n+1}}^{\mathrm{av}} = \nu^{(n+2)}(\theta^*)\tilde{\theta}_{\mathrm{r}}^{\mathrm{av}}(t-D). \tag{5.39}$$

Finally, consider the averaged-reduced version of (5.22):

$$z_{\mathrm{r}}^{\mathrm{av}}(t) = \left(\tilde{\gamma}_{\mathrm{r}}^{\mathrm{av}}(t)+\gamma^*\right)\nu^{(n+2)}(\theta^*)\tilde{\theta}_{\mathrm{r}}^{\mathrm{av}}(t-D). \tag{5.40}$$

By the linearization at the desired extremum operating point corresponding to the error equilibrium (5.37)–(5.38) we get (5.26), which completes the proof of Lemma 5.1. □

5.5 ▪ Predictor-Feedback Design

The idea is to derive a control law which, taking (5.26) as the input, feeds back the future state into the equivalent averaged-reduced system:

$$U_{\mathrm{r}}^{\mathrm{av}}(t) = z_{\mathrm{r}}^{\mathrm{av}}(t+D) = \tilde{\theta}_{\mathrm{r}}^{\mathrm{av}}(t+D-D) = \tilde{\theta}_{\mathrm{r}}^{\mathrm{av}}(t). \tag{5.41}$$

Using (5.11), (5.19) and (5.23), we consider

$$\dot{\hat{\theta}}(t-D) = -k_I U(t-D) \tag{5.42}$$

as well as its shifted, average-reduced version

$$\dot{\hat{\theta}}_{\mathrm{r}}^{\mathrm{av}}(t) = -k_I U_{\mathrm{r}}^{\mathrm{av}}(t). \tag{5.43}$$

Delaying (5.43) by D time units and using (5.26), we obtain

$$\dot{z}_{\mathrm{r}}^{\mathrm{av}}(t) = -k_I U_{\mathrm{r}}^{\mathrm{av}}(t-D). \tag{5.44}$$

Now, we apply the Laplace transformation to (5.44)

$$sZ_{\mathrm{r}}^{\mathrm{av}}(s) - \underbrace{z_{\mathrm{r}}^{\mathrm{av}}(0)}_{=0} = -k_I \upsilon_{\mathrm{r}}^{\mathrm{av}}(s)e^{-sD} \tag{5.45}$$

$$\Leftrightarrow \; Z_{\mathrm{r}}^{\mathrm{av}}(s)e^{sD} = -k_I \upsilon_{\mathrm{r}}^{\mathrm{av}}(s)\frac{1}{s}, \tag{5.46}$$

with $Z_{\mathrm{r}}^{\mathrm{av}}(s)$ and $\upsilon_{\mathrm{r}}^{\mathrm{av}}(s)$ being the respective transforms of $z_{\mathrm{r}}^{\mathrm{av}}(t)$ and $U_{\mathrm{r}}^{\mathrm{av}}(t)$, respectively. Now, we are able to express the future state as

$$z_{\mathrm{r}}^{\mathrm{av}}(t+D) = -k_I \int_0^t U_{\mathrm{r}}^{\mathrm{av}}(\tau)d\tau. \tag{5.47}$$

Since we are only interested in predicting the change over the next D time units, (5.47) is rewritten into

$$\begin{aligned} z_{\mathrm{r}}^{\mathrm{av}}(t+D) &= z_{\mathrm{r}}^{\mathrm{av}}(t) + z_{\mathrm{r}}^{\mathrm{av}}(t+D) - z_{\mathrm{r}}^{\mathrm{av}}(t) \\ &= z_{\mathrm{r}}^{\mathrm{av}}(t) - k_I \int_0^t U_{\mathrm{r}}^{\mathrm{av}}(\tau)d\tau + k_I \int_0^{t-D} U_{\mathrm{r}}^{\mathrm{av}}(\tau)d\tau \\ &= z_{\mathrm{r}}^{\mathrm{av}}(t) - k_I \int_{t-D}^t U_{\mathrm{r}}^{\mathrm{av}}(\tau)d\tau. \end{aligned} \tag{5.48}$$

Therefore, from (5.41), we arrive at the following expression for our predictor-based controller $\forall t \geq D$:

$$U_{\mathrm{r}}^{\mathrm{av}}(t) = z_{\mathrm{r}}^{\mathrm{av}}(t) - k_I \int_{t-D}^t U_{\mathrm{r}}^{\mathrm{av}}(\tau)d\tau. \tag{5.49}$$

Now, we propose the usage of the nonaverage and unreduced version of (5.49),

$$U(t) = \frac{c}{s+c}\left\{ z(t) - k_I \int_{t-D}^t U(\tau)d\tau \right\}, \tag{5.50}$$

by employing an additional low-pass filter with a sufficiently large $c > 0$. The braces $\{\cdot\}$ in (5.50) are used to indicate the effect of the frequency-domain transfer function of the filter as an operator on the according time-domain signal.

5.6 ▪ Stability Analysis

In the following, utilizing the approaches of [179], [200] we prove the stability of the stochastic perturbation-based ES control system from Section 5.3 applying the predictor (5.50). The operators $\mathrm{E}\{\cdot\}$ and $\mathrm{P}\{\cdot\}$ denote the expectation and the probability of a signal, respectively.

Theorem 5.1. *There exists a low-pass filter constant $c^* > 0$ such that, $\forall c \geq c^*$, the reduced closed-loop system described by* (5.19), (5.20), *and* (5.50) *with error signals* (5.23)–(5.24) *and states $\tilde{\theta}_{\mathrm{r}}(t-D)$, $U_{\mathrm{r}}(s)$, $\tilde{\gamma}_{\mathrm{r}}(t)$ has a unique, locally exponentially stable solution $\forall t \geq 0, \forall s \in [t-D, t]$ which satisfies the following upper bound for the expectation:*

$$\begin{aligned} &E\left\{ |\tilde{\gamma}_{\mathrm{r}}(t)|^2 + |\tilde{\theta}_{\mathrm{r}}(t-D)|^2 + U_{\mathrm{r}}^2(t) + \int_{t-D}^t U_{\mathrm{r}}^2(s)ds \right\} \\ &\qquad \leq a_1 e^{-a_2 t \varepsilon} |\varphi(\delta)|^2 + \Delta(\varepsilon), \end{aligned} \tag{5.51}$$

with $\Delta(\varepsilon) > 0$ such that $\Delta(\varepsilon) \to 0$ as $\varepsilon \to 0$, some positive constants a_1, a_2, and any continuous three-dimensional vector function $\varphi(t) \in \mathbf{C}_3$ as defined in [106] *with initial condition*

$$\varphi(\delta) = \begin{bmatrix} \tilde{\theta}_{\mathrm{r}}(\delta) \\ U_{\mathrm{r}}(\delta) \\ \tilde{\gamma}_{\mathrm{r}}(\delta) \end{bmatrix}, \quad -D \le \delta \le 0, \tag{5.52}$$

and history function $\varphi_t(\delta) = \varphi(t+\delta)$. Moreover, there exist constants $r > 0$, $M > 0$, $\lambda > 0$, $\Delta_1 > 0$ and function $T(\varepsilon) : (0,1) \to \mathbb{N}$ such that for any initial condition $|\varphi(\delta)| < r$ and any $\Delta_1 > 0$ the error vector norm $|\varphi(t)|$ converges below a residual value Δ_1 exponentially fast almost surely (a.s.) and in probability

$$\liminf_{\varepsilon \to 0} \left\{ \forall t \ge 0 : |\varphi(t)| > M|\varphi(\delta)|e^{-\lambda t} + \Delta_1 \right\} = \infty, \text{ a.s.}, \tag{5.53}$$

$$\lim_{\varepsilon \to 0} P \left\{ |\varphi(t)| \le M|\varphi(\delta)|e^{-\lambda t} + \Delta_1, \forall t \in [0, T(\varepsilon)] \right\} = 1, \tag{5.54}$$

with $\lim_{\varepsilon \to 0} T(\varepsilon) = \infty$. Finally, by choosing $\Delta_1(\varepsilon) = \mathcal{O}(\varepsilon)$, we can conclude

$$\lim_{\varepsilon \to 0} P \left\{ \limsup_{t \to \infty} |\theta_r(t) - \theta^*| \right\} = \mathcal{O}(a + \varepsilon). \tag{5.55}$$

The proof of Theorem 5.1 is made according to the steps described in Sections 5.6.1 to 5.6.5.

5.6.1 ▪ Reduced ODE-PDE Average System

Using (5.11), (5.19), and (5.23), we consider

$$\dot{\tilde{\theta}}(t-D) = -k_I U(t-D). \tag{5.56}$$

We define

$$u(x,t) = U_{\mathrm{r}}(t + x - D), \tag{5.57}$$

where t is time, D is the delay, and x the spatial variable enabling the following representation for the reduced version of (5.56) as an ODE-PDE system:

$$\dot{\tilde{\theta}}_{\mathrm{r}}(t-D) = -k_I u(0,t), \tag{5.58}$$

$$\partial_t u(x,t) = \partial_x u(x,t), \quad x \in [0, D], \tag{5.59}$$

$$u(D,t) = U_{\mathrm{r}}(t). \tag{5.60}$$

Consequently, (5.57) represents the solution of the PDE subsystem (5.59)–(5.60) above, with (5.58) being the ODE subsystem.

Employing a singular perturbation reduction [151] in order to freeze the delayed state vector $\mathbf{x}(t-D)$ in (5.4) at its quasi-steady state value

$$\mathbf{x}(t-D) = \mathbf{1}\Big(\tilde{\theta}(t-D) + a\sin\big(\eta(t-D)\big)\Big), \tag{5.61}$$

we derive an expression for the delayed and reduced output $y_{\mathrm{r}}(t-D)$ in terms of $\tilde{\theta}_{\mathrm{r}}(t-D)$ by plugging (5.8), (5.5), and then (5.61) into (5.2):

$$y_{\mathrm{r}}(t-D) = \nu\Big(\tilde{\theta}_{\mathrm{r}}(t-D) + a\sin\big(\eta(t-D)\big)\Big). \tag{5.62}$$

Then we rewrite the integrand of the reduced version of our predictor (5.50) using (5.57) to

$$U_{\mathrm{r}}(t) = \frac{c}{s+c}\left\{ z_{\mathrm{r}}(t) - k_I \int_0^D u(\tau,t)d\tau \right\}. \tag{5.63}$$

Following the same steps as in [200], using (5.63) we derive the averaged version of the reduced ODE-PDE system

$$\dot{\tilde{\theta}}_{\rm r}^{\rm av}(t-D) = -k_I u^{\rm av}(0,t)\,, \tag{5.64}$$

$$\partial_t u^{\rm av}(x,t) = \partial_x u^{\rm av}(x,t), \quad x \in [0,D]\,, \tag{5.65}$$

$$u^{\rm av}(D,t) = \frac{c}{s+c}\left\{ z_{\rm r}^{\rm av}(t) - k_I \int_0^D u^{\rm av}(\tau,t)d\tau \right\}. \tag{5.66}$$

5.6.2 ▪ Exponential Stability in the Sense of the Full-State Norm

Next, we consider the following infinite-dimensional backstepping transformation of the delay state

$$w(x,t) = u^{\rm av}(x,t) - \left(\tilde{\vartheta}_{\rm r}^{\rm av}(t) - k_I \int_0^x u^{\rm av}(\tau,t)d\tau\right), \tag{5.67}$$

mapping the local version the system (5.64)–(5.66) into

$$\dot{\tilde{\vartheta}}_{\rm r}^{\rm av}(t) = -k_I\left(\tilde{\vartheta}_{\rm r}^{\rm av}(t) + w(0,t)\right), \tag{5.68}$$

$$\partial_t w(x,t) = \partial_x w(x,t), \quad x \in [0,D]\,, \tag{5.69}$$

$$\partial_t u^{\rm av}(D,t) = -cw(D,t)\,, \tag{5.70}$$

where we have defined $\tilde{\vartheta}_{\rm r}^{\rm av}(t) := \tilde{\theta}_{\rm r}^{\rm av}(t-D)$. Using (5.64), we partially derive the transformed state $w(x,t)$ in (5.67) with respect to time t for $x = D$:

$$\partial_t w(D,t) = \partial_t u^{\rm av}(D,t) + k_I u^{\rm av}(D,t)\,. \tag{5.71}$$

Furthermore, consider the inverse transformation of (5.67)

$$u^{\rm av}(x,t) = w(x,t) + e^{-k_I x}\tilde{\vartheta}_{\rm r}^{\rm av}(t) - k_I \int_0^x e^{-k_I(x-\tau)} w(\tau,t)d\tau\,. \tag{5.72}$$

After plugging (5.70) and (5.72) into (5.71), we obtain

$$\begin{aligned} \partial_t w(D,t) = {} & -cw(D,t) + k_I w(D,t) + k_I e^{-k_I D}\tilde{\vartheta}_{\rm r}^{\rm av}(t) \\ & - k_I^2 \int_0^D e^{-k_I(D-\tau)} w(\tau,t)d\tau\,. \end{aligned} \tag{5.73}$$

Given the following Lyapunov–Krasovskii functional

$$V(t) = \frac{\left(\tilde{\vartheta}_{\rm r}^{\rm av}(t)\right)^2}{2} + \frac{b}{2}\int_0^D (1+x)w^2(x,t)dx + \frac{1}{2}w^2(D,t), \tag{5.74}$$

as already done in the previous chapter and in [200], for $b = 1/k_I$ and $c > 0$ in (41) sufficiently large, one can show that

$$\dot{V}(t) \leq -\mu V(t) \tag{5.75}$$

is guaranteed for some $\mu > 0$. Thus, the closed-loop system is exponentially stable in the sense of the full-state norm

$$\sqrt{|\tilde{\vartheta}_{\rm r}^{\rm av}(t)|^2 + w^2(D,t) + \int_0^D w^2(x,t)dx}\,, \tag{5.76}$$

i.e., in the transformed variable $(\tilde{\vartheta}_{\rm r}^{\rm av}, w)$.

5.6.3 ▪ Exponential Stability Estimate of the Averaged-Reduced System

To obtain exponential stability in the sense of the norm

$$\sqrt{|\tilde{\theta}_{\mathrm{r}}^{\mathrm{av}}(t-D)|^2 + \left(U_{\mathrm{r}}^{\mathrm{av}}(t)\right)^2 + \int_{t-D}^{t} \left(U_{\mathrm{r}}^{\mathrm{av}}(\tau)\right)^2 d\tau}\,, \tag{5.77}$$

we need to show

$$\alpha_1 \Psi(t) \leq V(t) \leq \alpha_2 \Psi(t) \tag{5.78}$$

for α_1 and α_2 being appropriate positive numbers and

$$\Psi(t) \triangleq |\tilde{\theta}_{\mathrm{r}}^{\mathrm{av}}(t-D)|^2 + \left(U_{\mathrm{r}}^{\mathrm{av}}(t)\right)^2 + \int_{t-D}^{t} \left(U_{\mathrm{r}}^{\mathrm{av}}(\tau)\right)^2 d\tau\,. \tag{5.79}$$

Ultimately, one obtains

$$\Psi(t) \leq \frac{\alpha_2}{\alpha_1} e^{-\mu t} \Psi(0)\,, \tag{5.80}$$

which completes the proof of exponential stability for the averaged-reduced system.

5.6.4 ▪ Invoking the Averaging Theorem

Using (5.28), (5.50), and (5.56), while considering in (5.11), (5.12), (5.22) as well as (5.16) and (5.21), we obtain

$$\frac{d}{dt}\begin{bmatrix} \tilde{\theta}(t-D) \\ U(t) \\ \tilde{\gamma}(t) \end{bmatrix} = \begin{bmatrix} 0 \\ -cU(t) \\ 0 \end{bmatrix} + \varepsilon \begin{bmatrix} -\omega k_I' U(t-D) \\ -c\left((\tilde{\gamma}(t)+\gamma^*)\frac{1}{\varepsilon}\widehat{\nu^{(n+1)}}(t) - \omega k_I' \int\limits_{t-D}^{t} U(\tau)d\tau\right) \\ \omega k_R'(\tilde{\gamma}(t)+\gamma^*)\left(1-(\tilde{\gamma}(t)+\gamma^*)\widehat{\nu^{(n+2)}}(t)\right) \end{bmatrix} \tag{5.81}$$

with

$$\widehat{\nu^{(k)}}(t) = \frac{2^k k!}{a^k}(-1)^{\lfloor \frac{k}{2} \rfloor} y(t-D)\sin\left(k\eta(t-D) + \frac{\pi}{4}\left(1+(-1)^k\right)\right). \tag{5.82}$$

Now, we define the state vector

$$\mathbf{u}^{\varepsilon}(t) = \begin{bmatrix} \tilde{\theta}(t-D) \\ U(t) \\ \tilde{\gamma}(t) \end{bmatrix}, \tag{5.83}$$

which allows us to generally express (5.81) in the form of the three-dimensional stochastic functional differential equation

$$\frac{d}{dt}\mathbf{u}^{\varepsilon}(t) = G(\mathbf{u}_t^{\varepsilon}) + \varepsilon F(t, \mathbf{u}_t^{\varepsilon}, \eta(t), \varepsilon)\,. \tag{5.84}$$

Therefore, since $\eta(t)$ is a homogeneous ergodic Markov process with invariant measure $\mu(d\eta)$ in some domain Y satisfying the property of exponential ergodicity and $\mathbf{u}_t^{\varepsilon}(\delta) = \mathbf{u}^{\varepsilon}(t+\delta)$ for $-D \leq \delta \leq 0$, $G\colon \mathbf{C}_3([-D,0]) \to \mathbb{R}^3$ as well as the Lipschitz $F\colon \mathbb{R}_+ \times \mathbf{C}_3([-D,0]) \times Y \times [0,1) \to \mathbb{R}^3$, with $F(t,0,\eta,\varepsilon) = 0$, being continuous mappings, we can apply the averaging theorem in [106] together with its exponential p-stability result (with $p=2$) for the initial random system using ε sufficiently small.

5.6.5 ▪ Asymptotic Convergence to the Extremum

We define the stopping time [140]

$$\tau_\varepsilon^{\Delta_1(\varepsilon)} := \inf\left\{\forall t \geq 0 : |\varphi(t)| > M|\varphi(\delta)|e^{-\lambda t} + \mathcal{O}(\varepsilon)\right\} \tag{5.85}$$

as the first time when the norm of the error vector does not satisfy the exponential decay property. The error vector norm $|\varphi(t)|$ converges below a residual value $\Delta_1(\varepsilon) = \mathcal{O}(\varepsilon)$ exponentially fast almost surely (5.53) and in probability (5.54). From (5.53) it is clear that $\tau_\varepsilon^{\Delta_1(\varepsilon)}$ most surely approaches infinity as ε goes to zero. Similarly in (5.54), the deterministic function $T(\varepsilon)$ tends to infinity as ε goes to zero. It follows from (5.53) and (5.54) that the exponential convergence is satisfied over an arbitrarily long time interval. Either component of the error vector converges to below $\Delta_1(\varepsilon) = \mathcal{O}(\varepsilon)$, particularly the $\tilde{\theta}_r(t)$ component. Then, we can write $\lim_{\varepsilon\to 0} \mathrm{P}\left\{\limsup_{t\to\infty} |\tilde{\theta}_{\mathrm{r}}(t)|\right\} = \mathcal{O}(\varepsilon)$. From (5.14) and (5.23), one has that $\theta(t) - \theta^* = \tilde{\theta}(t) + a\sin(\eta(t))$. Since the first term in the right-hand side for the reduced system is ultimately of order $\mathcal{O}(\varepsilon)$ and the second term is of order $\mathcal{O}(a)$, then we state (5.55). Hence, the proof of Theorem 5.1 is completed. □

Remark 5.1. In the stochastic ES for dynamical systems with output equilibrium map (5.62), we focus on the stability of the *reduced system* [140], where the closed-loop system has stochastic perturbations and thus generally there is no equilibrium solution or periodic solution so that the singular perturbation (SP) methods [129] are not applicable. Despite the lack of such a SP theorem for FDEs, for the reduced system we could analyze properties of the solution by the developed averaging theory in [106] to obtain the approximation to the extremum of the higher derivatives for the equilibrium map.

5.7 ▪ Simulations

In order to show the effectiveness of the stochastic perturbation technique, we compare it with the periodic (deterministic) method [200] considering the following map:

$$\dot{x} = -\frac{1}{\sqrt{100\pi}}\exp\left(-\frac{(t-40)^2}{100}\right) + \dot{\theta}, \quad x(0) = \theta(0), \tag{5.86}$$

$$y = g(x) \; = \; -\frac{1}{6}x^3 + 3x, \tag{5.87}$$

where we are interested in the maximization of $\nu^{(1)}(\theta)$ with $\nu(\theta)$ in (5.8).

We assume $D = 5s$. For the periodic algorithm, we employ essentially the same predictor (5.50) of the stochastic case, but the signal $z(t)$ is constructed with deterministic perturbations [200], as in the previous chapters. Both scenarios have identical parameters: $c = 20$, $\varepsilon = 1/1000$, $a = 0.1$, $\omega = 35$, $k_I' = 1$, $k_R' = 0.25$. The stochastic perturbation process's standard deviation is $\sigma = 0.1$.

As can be seen in Figure 5.2, all signals converge to the expected values (time-varying for θ^* due to the plant dynamics), despite the delay. The convergence rate of the stochastic method outperforms the periodic one when using identical simulation parameters. This confirms the theoretical train of thought in [142, Chapter 2].

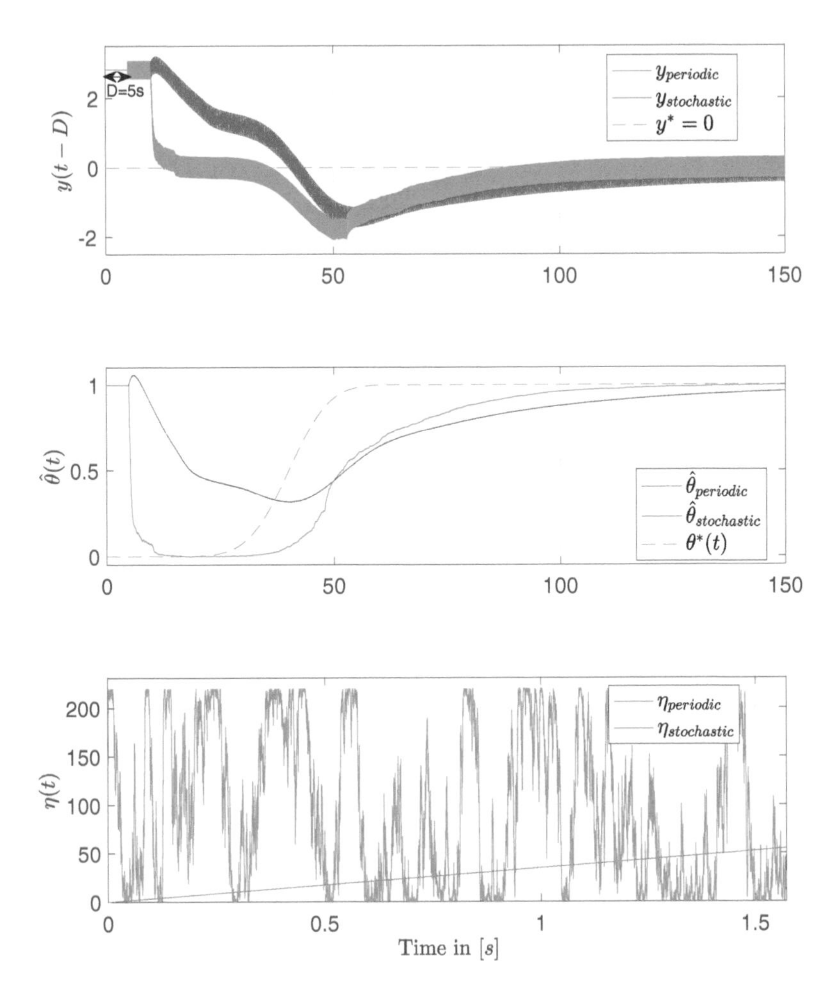

Figure 5.2. *Maximization of* $\nu^{(1)}(\theta)$ *with nonlinear dynamic system* (5.86) *and cubic output function* (5.87) *using stochastic and periodic perturbations:* (a) *delayed output* $y(t-D)$*;* (b) *estimate of maximizing parameter* $\hat{\theta}(t)$*;* (c) *excerpts of the stochastic signal* $\eta(t)$ *and the deterministic* ωt*. The dashed lines indicate the target values* $y^* = 0$ *and* $\theta^*(t) = 0.5 erf((t-40)/10) + 0.5$*, where* $erf(\cdot)$ *denotes the Gauss error function.*

In the presence of plant dynamics, it is prohibited in theory to increase arbitrarily the frequency ω due to the application of the singular perturbation reduction [129] to dynamic maps. The price to be paid in these cases is the decrease of the convergence speed. Additionally, we state that the predictor is vital for the compensation of the delays, since the delays massively degrade the ES performance and lead to instability (curves not shown).

Remark 5.2. In practice, high-frequency switching in the plots of Figure 5.2 may lead to *chattering* or *limit cycles* in actuators. The inability to remove these and achieve equilibrium stabilization in ES may also be associated with actuator constraints, such as magnitude and rate saturation. Here, the best control requirement could be to enforce a stable, "smallest" limit cycle [10, Chapter 5]. However, ES-based controllers whose control efforts vanish as the system approaches equilibrium have been proposed [235]. In [59] and [203], Lie bracket–based ES was introduced to obtain ES feedback with bounded update rates. Thus, limit cycles may not only be reduced, but they can also be completely eliminated.

5.8 ▪ Notes and References

Extremum Seeking (ES) control is a non-model-based adaptive technique for real-time optimization that became very popular in recent years with the Newton-like scheme [3], [166] being attractive due its user-assignable convergence rate [73]. However, there are two noticeable challenges in ES. The first one is not the optimization of the map itself, but rather of the map's higher derivative [18], [88], [230]. Second, and an even more challenging issue, is the effect of time delays in implemented ES, leading to a degradation of the system's behavior or even its instability.

In order to solve the problem of maximizing map sensitivity, the generalization of Newton-based ES [73] was presented in [151] to optimize arbitrary higher derivatives of an unknown map employing sinusoidal perturbations. In addition, by using the mathematical foundation introduced in [179] to handle ES for functional differential equations (FDEs), the authors in [200] expanded the application of the results of [151] to the case of static maps with time delays.

Furthermore, it is a known issue that ES techniques based on periodic perturbations suffer with higher dimensionality [10]. Orthogonality requirements on the elements of the periodic perturbation vector pose an implementation challenge [140]. Limitations of the deterministic ES scheme also include the fact that learning using a periodic excitation signal is rare in probing-based learning and optimization approaches [140], which may lead to slower converge rates. For instance, ES algorithms inspired by biomimicry and others sensitive to deterministic perturbation signals suggest other perturbation techniques rather than periodic ones [142], although periodic excitation signals are intuitive and simple for gradient estimation. Hence, in a more general context, there is a huge demand in developing nonperiodic ES techniques such as stochastic methods [151], where the possibility of achieving faster convergence rate and global maximization/minimization in the presence of local extremum point were made possible [142].

The key contribution of this chapter is to take into account output delays in the higher-derivatives maps, which were not considered in the real-time optimization method proposed by [151]. Moreover, we have also assumed dynamic maps being optimized by stochastic extremum seeking, thus being different from the approach adopted in [200], which was devoted only to static maps. We have proved the local stability of the proposed predictor feedback for delay compensation and have shown convergence to a small neighborhood of the unknown extremum point. As in [179], we employed a semi-model-based approach due to the treatment of the delay. While assuming known delay, our predictor construction follows a model-free approach, where the plant model parameters (Hessian and its inverse) are unknown being estimated using perturbations.

Even though we can apply our approach adaptively to unknown constant delays such as in [32], the result would necessarily be local (the initial estimate of the delay would have to be close to the actual delay) because of the nonlinear parametrization of the delay. This would offer little advantage over the existing robustness findings of the predictor feedback to small perturbations in the delay, discussed in the next chapter.

Chapter 6

Robustness to Delay Mismatch in ES

6.1 ▪ Delay Robustness: Can a Predictor Be as Model-Free as ES?

While extremum seeking is a model-free optimization approach, compensation of delays in extremum seeking is a model-based operation. The need for the Hessian of the unknown map can be circumvented by Hessian's estimation but the delay D is assumed known.

Is there no inconsistency in the formulation of ES under delays—treating the map $h(\theta)$ as unknown and its delay D (at either the input or the output) as completely known? Indeed, this is an inconsistency, practical and theoretical. How might one go about removing or at least ameliorating this inconsistency?

One option is estimating the unknown delay D. But this option, as known from delay-adaptive design in [32], incurs the price of the result being local in the delay estimation error. If one is to be forced to choose the initial delay estimate to be close to the actual delay—which is unknown anyway—why bother with adaptation at all?

This reasoning leads to the question whether predictor-based ES possesses some amount of robustness to a mismatch between the assumed constant delay D_0 which is used in the predictor and the actual delay D which is acting on the map.

This chapter is dedicated to providing an affirmative answer to the question whether ES possesses some amount of delay robustness. The result of the chapter leverages the results of the second author in establishing robustness to delay mismatch of predictor feedback for general plants.

The hallmark of establishing robustness to delay mismatch is that, while the delay is a constant parameter, unlike other constant parameters whose upward and downward deviations from the nominal value can be treated within a single analysis, this is not the case with delay mismatch. Similar to singularly perturbed systems where when a parameter that is nominally zero but assumes a small positive value constitutes a growth in the dynamic order of the system, a delay mismatch changes the dynamic order of a system.

A system with a positive delay is infinite-dimensional to begin with. If a predictor feedback law underestimates the true value of the delay, the feedback is inaccurate but the domain size (the memory length in the case of delay systems) is not altered. In contrast, if a predictor feedback law overestimates the actual delay and begins feeding past values of the input over a window longer than the actual length of the delay, such mismatched predictor feedback is not only inaccurate but it perturbs the system's dimension (it grows the overall delay affecting the system).

Hence, two distinct proofs are entailed in establishing delay robustness of predictor-based ES algorithms.

As we are handling in this book both problems of seeking extrema of unknown maps and of their higher derivatives, in this chapter we study delay robustness of the seekers of extrema of higher derivatives. In contrast to Chapter 5, where we studied stochastic higher-order extremum seeking, in this chapter we study a deterministic version.

In this chapter, we establish sufficient conditions for the robustness to small delay uncertainties of predictor-based ES algorithms. Local stability and exponential convergence to a small neighborhood of the unknown extremum are preserved. A numerical simulation example is presented to illustrate the performance of the predictor-based control scheme for uncertain time-delay compensation.

6.2 ▪ System Description

Scalar ES considers applications in which one wants to maximize (or minimize) the output $y \in \mathbb{R}$ of an *unknown* nonlinear static map $h(\theta)$ by varying the input $\theta \in \mathbb{R}$ in *real time*. But like in many technical applications we have to consider that the output may be time-delayed, and hence we additionally assume that there is a constant and, contrary to our previous chapters, *uncertain* delay $D \geq 0$ such that the output is expressed by

$$y(t) = h(\theta(t-D)). \tag{6.1}$$

As done before, in this chapter we also assume that our map is output-delayed. In Figure 6.1, we illustrate the proposed scalar version of the Newton-based ES for maximization of higher derivatives of the map $h(\cdot)$ based on predictor feedback for delay compensation.

6.2.1 ▪ Problem Formulation

Without loss of generality, let us consider the maximization of nth derivative of the output in the presence of time delay using Newton-based ES, where the maximizing value of θ is denoted by θ^*. We state our optimization problem as follows:

$$\max_{\theta \in \mathbb{R}} h^{(n)}(\theta). \tag{6.2}$$

Assumption 6.1. *Let $h^{(n)}(\cdot)$ be the nth derivative of a smooth function $h(\cdot)$: $\mathbb{R} \to \mathbb{R}$. Now let us define*

$$\theta_{\max} = \{\theta | h^{(n+1)}(\theta) = 0, \quad h^{(n+2)}(\theta) < 0\} \tag{6.3}$$

to be a collection of maxima where $h^{(n)}$ is locally concave. Now assume that $\theta^* \in \theta_{\max}$ and $\theta_{\max} \neq \emptyset$.

In addition, we consider that maps satisfying Assumption 6.1 are also locally approximated by a quadratic function in a neighborhood of θ^*:

$$h^{(n)}(\theta) = Q^* + \frac{H}{2}(\theta - \theta^*)^2 \tag{6.4}$$

for constants $Q^* > 0$ and $H < 0$, where H is the Hessian of the quadratic approximation [10].

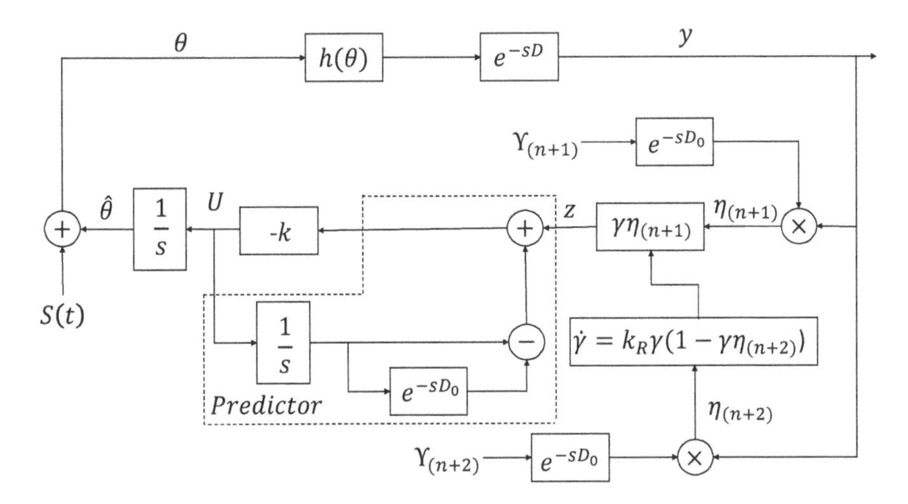

Figure 6.1. *Block diagram of the basic prediction scheme for output-delay compensation in Newton-based ES for maximizing higher derivatives* (6.16) *under delay mismatch, i.e.,* $D = D_0 + \Delta D$. *The predictor feedback with a perturbation-based estimate of the Hessian's inverse obeys* (6.26). *The dither and demodulating signals are given by* $S(t)$, $\Upsilon_{(n+1)}$, *and* $\Upsilon_{(n+2)}$, *which are calculated according to* (6.15) *and* (6.11) *for* $j = 1$ *and* $j = 2$, *respectively.*

The uncertain delay can be represented such that it has a mismatch given by the uncertainty ΔD, which can be either positive or negative, relative to its assumed known nominal value $D_0 > 0$.

Assumption 6.2. *The output time-delay* D *given in Figure* 6.1 *is expressed by*

$$D = D_0 + \Delta D, \tag{6.5}$$

where D_0 *and* ΔD *are constant and* $\Delta D \in (-\delta, \delta)$ *for some small constant* $\delta > 0$.

6.2.2 ▪ Probing and Demodulation Signals

Let $\hat{\theta}$ be the estimate of the maximum point and

$$\tilde{\theta}(t) = \hat{\theta}(t) - \theta^* \tag{6.6}$$

be the estimation error.

From the block diagram in Figure 6.1, the error dynamics can be written as

$$\dot{\tilde{\theta}}(t) = \dot{\hat{\theta}}(t) = U(t). \tag{6.7}$$

Note that the integrator dynamics (6.7) is part of the extremum seeking controller and it is not related to the plant to be optimized, which is indeed a static map defined in (6.1).

Moreover, we have

$$\dot{\gamma} = k_R \gamma (1 - \gamma \eta_{(n+2)}), \tag{6.8}$$

where (6.8) is a Riccati equation [73]. Hence, the Hessian's inverse estimation error is defined by

$$\tilde{\gamma}(t) = \gamma(t) - H^{-1}. \tag{6.9}$$

Now, we rearrange the equations given in [151] by changing them to the following:

$$\theta(t) = \hat{\theta}(t) + S(t), \tag{6.10}$$

$$\Upsilon_j(t) = C_j \sin\left(j\omega t + \frac{\pi}{4}\left(1 + (-1)^j\right)\right), \tag{6.11}$$

$$C_j = \frac{2^j j!}{a^j}(-1)^F, \tag{6.12}$$

$$F = \frac{j - \left|\sin\left(\frac{j\pi}{2}\right)\right|}{2}. \tag{6.13}$$

The demodulated signals

$$\eta_j(t) = y\Upsilon_j(t - D_0) \tag{6.14}$$

are used to estimate the gradient ($j = n+1$) and the Hessian ($j = n+2$) of $h^{(n)}$. Equation (6.8) will be used to generate an estimate of the Hessian's inverse [73].

We define the additive dither signal as

$$S(t) = a\sin(\omega t), \tag{6.15}$$

which is not delayed. However, as shown in Figure 6.1, the demodulation signals $\Upsilon_{(n+1)}$ and $\Upsilon_{(n+2)}$ are delayed by the nominal delay D_0 in order to address the delayed output y [178].

Furthermore, let us define the measurable signal

$$z(t) = \gamma(t)\eta_{(n+1)}(t), \tag{6.16}$$

where γ is updated according to (6.8) and $\eta_{(n+1)}$ is the demodulated signal. We can construct (6.16), because we only use measurable/available signals γ and $\eta_{(n+1)}$.

6.3 ▪ Extremum Seeking for Known Delays

First of all, we will briefly revisit the baseline case where $\Delta D = 0$, i.e., $D(t) = D_0$, in order to review the case of known delays [198], [199], [200].

6.3.1 ▪ Motivation for Predictor Feedback

From (6.7) it follows that

$$\dot{\hat{\theta}}(t - D) = U(t - D). \tag{6.17}$$

Now, using (6.7), the following "shifted" average model[2] can be derived:

$$\dot{\hat{\theta}}_{\mathrm{av}}(t) = U_{\mathrm{av}}(t), \tag{6.18}$$

with $U_{\mathrm{av}} \in \mathbb{R}$ being the resulting average control for $U \in \mathbb{R}$.

[2]Consider a generic nonlinear system $\dot{x} = f(t, x, 1/\omega)$, where $x \in \mathbb{R}^n$ is the state vector, and $f(t, x, 1/\omega)$ is periodic in t with period $T := 2\pi/\omega$, i.e., $f(t+T, x, 1/\omega) = f(t, x, 1/\omega)$. Thus, for $1/\omega > 0$ sufficiently small, we can obtain its average model given by $\dot{x}_{\mathrm{av}} = f_{\mathrm{av}}(x_{\mathrm{av}})$, with $f_{\mathrm{av}}(x_{\mathrm{av}}) = 1/T\int_0^T f(\tau, x_{\mathrm{av}}, 0)d\tau$, where $x_{\mathrm{av}}(t)$ denotes the average version of the state $x(t)$ [108]. According to [83], a similar procedure can be applied to obtain the average model of a delayed system governed by $\dot{x}(t) = f(\omega t, x_t)$ with $x_t(\Theta) = x(t+\Theta)$, and $\Theta \in [-D, 0]$.

One can try to feed back the future state $z_{\rm av}(\sigma+D)$ in the equivalent average system where σ is time. Assuming that $\sigma = t-D$ we get

$$z_{\rm av}(\sigma+D) = z_{\rm av}(t-D+D) = z_{\rm av}(t) \tag{6.19}$$

and thus a time-delay compensated signal which we feed back. The delayed version of (6.18) is $\dot{\hat{\theta}}_{\rm av}(t-D) = U_{\rm av}(t-D)$. From (6.16) and according to the developments carried out in [198], [199], [200], it is possible to obtain the following linearized average system:

$$\dot{z}_{\rm av}(t) = U_{\rm av}(t-D). \tag{6.20}$$

Feeding back the future state $z_{\rm av}(t+D)$ for the delay compensation motivates the use of predictor feedback design.

Applying variation-of-constants formula [119], the future state is being calculated as

$$z_{\rm av}(t+D) = z_{\rm av}(t) + \int_{t-D}^{t} U_{\rm av}(\sigma)d\sigma. \tag{6.21}$$

We derive a controller as follows:

$$U_{\rm av}(t) = \dot{\hat{\theta}}_{\rm av}(t) = -kz_{\rm av}(t+D) \tag{6.22}$$

$\forall t \geq D$ and $k>0$, which yields to an average predictor-based control

$$U_{\rm av}(t) = -k\left[z_{\rm av}(t) + \int_{t-D}^{t} U_{\rm av}(\sigma)d\sigma\right]. \tag{6.23}$$

Furthermore, using (6.21), (6.22) as well as time shifting, we obtain the linearization of the average error dynamics in the following form:

$$\dot{\tilde{\theta}}_{\rm av}(t) = -kz_{\rm av}(t+D) = -k\tilde{\theta}_{\rm av}(t) \tag{6.24}$$

$\forall t \geq D$, with eigenvalues which are determined by $-k$. Thus, the system has an exponentially attractive equilibrium.

6.3.2 ▪ Predictor Feedback for Known Delays

In order to compensate known time delay, in [198], [199], [200], we proposed a predictor-based controller which incorporates low-pass filter [175]:

$$U(t) = \frac{c}{s+c}\left\{-k\left[z(t) + \int_{t-D}^{t} U(\tau)d\tau\right]\right\}. \tag{6.25}$$

Note that here we mix the time- and frequency-domain notation in (6.25) by using the braces $\{\cdot\}$ to denote that the transfer function acts as an operator on a time-domain function. In Chapters 3 and 5 (see also [198], [199], [200]), we proved that the predictor (6.25) compensates the known and constant time delay.

In case of delay mismatch ($\Delta D \neq 0$), the predictor (6.25) would be

$$U(t) = \frac{c}{s+c}\left\{-k\left[z(t) + \int_{t-D_0}^{t} U(\tau)d\tau\right]\right\}, \quad k>0, \tag{6.26}$$

for known nominal delay D_0. However, since the delay mismatch ΔD is unknown, it may affect the convergence or even destabilize the whole system shown in Figure 6.1. Therefore, we have to analyze the behavior of the system for $\Delta D \neq 0$.

6.4 ▪ Robustness to Delay Mismatch

If the output time delay has a mismatch ΔD relative to the assumed plant output delay $D_0 \geq 0$ applied to the demodulation signals $\Upsilon_{(n+1)}$ and $\Upsilon_{(n+2)}$, then the following lemma holds.

Lemma 6.1. *Let Assumptions* 6.1 *and* 6.2 *hold and the output time-delay* $D \geq 0$ *have a mismatch* ΔD. *Then the average versions of signals* $\eta_{(n+1)}$ *and* $\eta_{(n+2)}$ *are given by*

$$\eta^{\rm av}_{(n+1)}(t) = H\tilde{\theta}_{\rm av}(t-D)\cos((n+1)\omega\Delta D), \tag{6.27}$$

$$\eta^{\rm av}_{(n+2)}(t) = H\cos\left((n+2)\omega\Delta D\right). \tag{6.28}$$

Proof. Suppose that Assumptions 6.1 and 6.2 hold. Consider the function $\eta_j := \Upsilon_j y$ which can be written as $\eta_{(n+1)}(t) := \Upsilon_{(n+1)}(t-D_0)y(t)$ for $j = n+1$.

Using (6.11)–(6.13) we get

$$\eta_{(n+1)}(t) = h(\theta(t-D))P\sin\left((n+1)\omega(t-D_0)+\frac{\pi}{4}\left(1+(-1)^{(n+1)}\right)\right), \tag{6.29}$$

where

$$P := C_{(n+1)} = \frac{2^{(n+1)}(n+1)!}{a^{(n+1)}}(-1)^F, \tag{6.30}$$

$$F = \frac{(n+1)-\left|\sin\left(\dfrac{(n+1)\pi}{2}\right)\right|}{2}.$$

Now, recalling (6.4), and assuming w.l.o.g. that $j = n+1$ is odd, we rewrite (6.29) in the neighborhood of θ^* as

$$\eta_{(n+1)} = \left(Q^* + \frac{H}{2}\left[\theta(t-D)-\theta^*\right]^2\right)P\sin\left(j\omega(t-D_0)\right). \tag{6.31}$$

Moreover, one can easily follow that

$$\theta(t-D)-\theta^* = \tilde{\theta} + a\sin\left(\omega(t-D)\right). \tag{6.32}$$

For the sake of simplicity, consider the particular case $n = 0$ such that $j = 1$. Hence, in average we obtain for (6.31)

$$\begin{aligned}\eta_{(n+1)}(t) = {} & Q^* P \underbrace{\sin(\omega(t-D_0))}_{mean=0} \\ & + \frac{H}{2}P\left(\tilde{\theta}(t-D)+a\sin(\omega(t-D))\right)^2 \\ & \times \sin(\omega(t-D_0)),\end{aligned} \tag{6.33}$$

which yields, if we suppress the first term in the RHS of (6.33),

$$\begin{aligned}\eta_{(n+1)}(t) = {} & \frac{H}{2}P\left(\tilde{\theta}(t-D)+a\sin\left(\omega(t-D)\right)\right)^2 \\ & \times \sin(\omega(t-D_0)).\end{aligned} \tag{6.34}$$

Furthermore, we get

$$\begin{aligned}\eta_{(n+1)}(t) = {} & \frac{H}{2}P[\tilde{\theta}^2(t-D)\underbrace{\sin(\omega(t-D_0))}_{mean=0} \\ & + 2a\tilde{\theta}\sin(\omega(t-D))\sin(\omega(t-D_0)) \\ & + a^2\sin^2(\omega(t-D))\sin(\omega(t-D_0))].\end{aligned} \tag{6.35}$$

Now, plugging in $D-\Delta D$ for D_0, from (6.35) we can follow

$$\begin{aligned}\eta_{(n+1)}(t) = &\frac{H}{2}P[2a\tilde{\theta}(t-D)\sin(\omega(t-D))\\ &\times\sin(\omega(t-D+\Delta D))\\ &+a^2\sin^2(\omega(t-D))\sin(\omega(t-D+\Delta D))].\end{aligned} \tag{6.36}$$

Applying trigonometric calculus, it is easily seen that

$$\begin{aligned}\sin(\omega(t-D+\Delta D)) = &\sin(\omega(t-D))\cos(\omega\Delta D)\\ &+\cos(\omega(t-D))\sin(\omega\Delta D).\end{aligned} \tag{6.37}$$

Hence, using (6.37), we get for (6.36) the following expression:

$$\begin{aligned}\eta_{(n+1)}(t) = &\frac{H}{2}P[2a\tilde{\theta}(t-D)\underbrace{\sin^2(\omega(t-D))}_{mean=0.5}\\ &\times\cos(j\omega\Delta D)\\ &+\{2a\tilde{\theta}(t-D)\sin(\omega(t-D))\\ &\times\cos(\omega(t-D))\sin(\omega\Delta D)\}\\ &+a^2\underbrace{\sin^3(\omega(t-D))}_{mean=0}\cos(\omega\Delta D)\\ &+a^2\sin^2(\omega(t-D))\cos(\omega(t-D))\sin(\omega\Delta D)].\end{aligned} \tag{6.38}$$

We rewrite (6.38) to

$$\begin{aligned}\eta_{(n+1)}(t) = &\frac{H}{2}Pa\tilde{\theta}(t-D)\cos(\omega\Delta D)\\ &+[HPa\tilde{\theta}(t-D)\underbrace{\sin(\omega(t-D))\cos(\omega(t-D))}_{=0.5\sin(2\omega(t-D))}\\ &+\frac{H}{2}Pa^2\underbrace{\sin^2(\omega(t-D))\cos(\omega(t-D))]}_{=\sin(\omega(t-D))0.5\sin(2\omega(t-D))}\\ &\times\sin(\omega\Delta D).\end{aligned} \tag{6.39}$$

Using the trigonometric equations

$$\sin x\cos x = \frac{1}{2}\sin 2x, \tag{6.40}$$

$$\sin(2x) = 2\sin x\cos x, \tag{6.41}$$

$$\sin^2 x = 1-\cos^2 x, \tag{6.42}$$

we get

$$\begin{aligned}\eta_{(n+1)}(t) = &\frac{H}{2}Pa\tilde{\theta}(t-D)\cos(\omega\Delta D)\\ &\underbrace{+[HPa\tilde{\theta}(t-D)\frac{1}{2}\underbrace{\sin(2\omega(t-D))}_{mean=0}}_{=0}\\ &\underbrace{+\frac{H}{2}Pa^2(\underbrace{\cos(\omega(t-D))}_{mean=0}-\underbrace{\cos^3(\omega(t-D))}_{mean=0})]}_{=0}\\ &\times\sin(\omega\Delta D),\end{aligned} \tag{6.43}$$

which gives in the average sense

$$\begin{aligned}\eta^{\mathrm{av}}_{(n+1)}(t) &= \frac{H}{2}Pa\tilde{\theta}_{\mathrm{av}}(t-D)\cos(\omega\Delta D)\\ &= H\tilde{\theta}_{\mathrm{av}}(t-D)\cos(\omega\Delta D)\,, \end{aligned}\tag{6.44}$$

since $P=\frac{2}{a}$ for $n=0$. For the general case $j=n+1$ odd or even, a similar sequence of algebraic reductions follows [151, Appendix C] and we arrive at a result identical to (6.27) .

Moreover, we can easily perform an analogous procedure for $\eta_{(n+2)}(t):=\Upsilon_{(n+2)}(t-D_0)y(t)$ in order to obtain the expression in (6.28). This concludes the proof of Lemma 6.1. □

In the case of Lemma 6.1, the estimation error of the Hessian's inverse in (6.9) must be redefined by

$$\tilde{\gamma}(t)=\gamma(t)-[\cos((n+2)\omega\Delta D)H]^{-1}. \tag{6.45}$$

Analogously to [151], the following delayed equations for the average error system can be written as

$$\frac{d\tilde{\theta}_{\mathrm{av}}(t)}{dt} = -k\gamma_{\mathrm{av}}\eta^{\mathrm{av}}_{n+1}(t)\,, \tag{6.46}$$

$$\frac{d\tilde{\gamma}_{\mathrm{av}}(t)}{dt} = k_R\gamma_{\mathrm{av}}\left(1-\gamma_{\mathrm{av}}\eta^{\mathrm{av}}_{n+2}(t)\right), \tag{6.47}$$

with $\gamma\neq 0$ and $\gamma_{\mathrm{av}}\neq 0$, as $c\to\infty$.

Additionally, if the predictor was not applied in the feedback loop, one could also write $\dot{\tilde{\theta}}(t)=-k\gamma(t)\eta_{(n+1)}(t)$. Now using (6.27) and (6.28) we get for the system (6.46)–(6.47) the following expression:

$$\begin{aligned}\frac{d\tilde{\theta}_{\mathrm{av}}(t)}{dt} &= -k\frac{\cos((n+1)\omega\Delta D)}{\cos((n+2)\omega\Delta D))}\tilde{\theta}_{\mathrm{av}}(t-D)\\ &\quad -kH\cos((n+1)\omega\Delta D)\tilde{\gamma}_{\mathrm{av}}(t)\tilde{\theta}_{\mathrm{av}}(t-D)\,,\end{aligned}\tag{6.48}$$

$$\frac{d\tilde{\gamma}_{\mathrm{av}}(t)}{dt} = -k_R\tilde{\gamma}_{\mathrm{av}}(t)-k_RH\cos((n+2)\omega\Delta D)\tilde{\gamma}^2_{\mathrm{av}}(t). \tag{6.49}$$

From (6.48) it is easy to conclude that the best operational mode is $(n+1)\omega\Delta D=0$, i.e., the uncertainty-free case. Moreover, at $\pi/2<|(n+2)\omega\Delta D|<\pi$ the feedback changes its sign in the first term of the RHS in (6.48), and as a result the system would move away from the extremum point even if the corrected prediction action was implemented to counteract the nominal delay D_0. Thus, a primary condition that the uncertainty in the output delay must satisfy is $|(n+2)\omega\Delta D|<\pi/2$, i.e.,

$$|\Delta D|<\frac{\pi}{2(n+2)\omega}\,, \tag{6.50}$$

to guarantee the robustness of the predictor feedback. In addition, from (6.50) it is clear that there is a trade-off between the dither frequency ω, the order of the derivative, and the maximum uncertainty ΔD. For constant dither frequency ω, the tolerated uncertainty decreases with increasing order of the derivative.

However, the inequality (6.50) is not a sufficient condition. In what follows, we discuss the effect of the delay uncertainties ΔD in the predictor design.

In [199], we pointed out the following linearized average expressions:

$$z_{\rm av}(t) = \tilde{\theta}_{\rm av}(t-D_0), \tag{6.51}$$
$$\dot{z}_{\rm av}(t) = U_{\rm av}(t-D_0). \tag{6.52}$$

For the case with mismatch delays, we obtain

$$z_{\rm av}(t) = \frac{\cos((n+1)\omega\Delta D)}{\cos((n+2)\omega\Delta D)}\tilde{\theta}_{\rm av}(t-D), \tag{6.53}$$
$$\dot{z}_{\rm av}(t) = \frac{\cos((n+1)\omega\Delta D)}{\cos((n+2)\omega\Delta D)}U_{\rm av}(t-D). \tag{6.54}$$

Consequently, the resulting average control would be

$$U_{\rm av}(t) = -k\frac{c}{s+c}\left\{z_{\rm av}(t) + G\int_{t-D}^{t} U(\tau)d\tau\right\}, \tag{6.55}$$

with

$$G = \frac{\cos((n+1)\omega\Delta D)}{\cos((n+2)\omega\Delta D)}. \tag{6.56}$$

Thus, the average controller (6.55) would be a good approximation for (6.26) if and only if $(n+1)\omega\Delta D \approx 0$, and $D \approx D_0$, as expected.

Hence, we are able to provide the following theorem.

Theorem 6.1. *Consider the average closed-loop system* (6.54) *with the control law* (6.26). *There exist* $c>0$ *sufficiently large in* (6.55) *and* $\delta>0$ *sufficiently small such that, for*

$$\Delta D \in (-\delta,\delta), \tag{6.57}$$

$z_{\rm av} = 0$ *is exponentially stable in the sense of the state norm*

$$N(t) = \left(|z_{\rm av}(t)|^2 + \int_{t-\bar{D}}^{t} U_{\rm av}(\tau)d\tau\right)^{1/2}, \tag{6.58}$$

where

$$\bar{D} = D_0 + \max\{0,\Delta D\}. \tag{6.59}$$

Proof. We use the same transport PDE formalism[3] as in [119, Chapters 2 and 4]; i.e., the system (6.54) is represented by

$$\dot{z}_{\rm av}(t) = Gu_{\rm av}(0,t), \tag{6.60}$$
$$\partial_t u_{\rm av}(x,t) = \partial_x u_{\rm av}(x,t), \tag{6.61}$$
$$u_{\rm av}(D_0+\Delta D,t) = U_{\rm av}(t), \tag{6.62}$$

where G is given in (6.56) and the spatial domain of the PDE is defined by

$$x \in (\min\{0,\Delta D\}, D_0+\Delta D] \tag{6.63}$$

[3]Recall that we denote the partial derivatives of a function $u(x,t)$ as $\partial_x u(x,t) = \partial u(x,t)/\partial x$ and $\partial_t u(x,t) = \partial u(x,t)/\partial t$.

and

$$u_{\rm av}(x,t) = U_{\rm av}(t+x-D_0-\Delta D), \tag{6.64}$$

from which it follows that the boundary condition

$$u_{\rm av}(0,t) = U_{\rm av}(t-D_0-\Delta D). \tag{6.65}$$

We use the following backstepping transformation and its inverse:

$$w(x,t) = u_{\rm av}(x,t) - \left[\int_0^x KGu_{\rm av}(y,t)dy \ + \ Kz_{\rm av}(t)\right], \tag{6.66}$$

$$u_{\rm av}(x,t) = w(x,t) + \int_0^x Ke^{GK(x-y)}Gw(y,t)dy + Ke^{GKx}z_{\rm av}(t). \tag{6.67}$$

The target system is given by

$$\dot{z}_{\rm av}(t) = Kz_{\rm av}(t) + Gw(0,t), \tag{6.68}$$

$$\partial_t w(x,t) = \partial_x w(x,t) \tag{6.69}$$

and with the boundary condition for $w(D_0+\Delta D,t)$ to be defined later on. First, we notice that the average version of the feedback control law (6.26) approximates (6.55) for δ sufficiently small ($\Delta D \to 0$) and the latter can be written as

$$u_{\rm av}(D_0+\Delta D,t) = K\left[z_{\rm av}(t) + \int_{\Delta D}^{D_0+\Delta D} Gu_{\rm av}(y,t)dy\right] \tag{6.70}$$

as $c \to \infty$ in (6.26) and $G \to 1$ in (6.56), with $K=-k$. Using (6.66) for $x = D_0+\Delta D$ gives us

$$w(D_0+\Delta D,t) = -k\int_0^{\Delta D} Gu_{\rm av}(y,t)dy. \tag{6.71}$$

Then, employing (6.67) under the integral and performing certain calculations, we obtain

$$w(D_0+\Delta D,t) = K\left[(I-e^{GK\Delta D})z_{\rm av}(t) - \int_0^{\Delta D} e^{GK(\Delta D-y)}Gw(y,t)dy\right]. \tag{6.72}$$

One then shows that

$$w(D_0+\Delta D,t)^2 \le 2q_1|z_{\rm av}(t)|^2 + 2q_2\int_{\min\{0,\Delta D\}}^{\max\{0,\Delta D\}} w(x,t)^2dx, \tag{6.73}$$

where the functions $q_1(\Delta D)$ and $q_2(\Delta D)$ are

$$q_1(\Delta D) = \left|K\left(I-e^{GK\Delta D}\right)\right|^2, \tag{6.74}$$

$$q_2(\Delta D) = \int_{\min\{0,\Delta D\}}^{\max\{0,\Delta D\}} \left(Ke^{GK(\Delta D-y)}G\right)^2 dy. \tag{6.75}$$

Notice that

$$q_1(0) = q_2(0) = 0 \tag{6.76}$$

and that q_1 and q_2 are both continuous functions of ΔD.

The cases $\Delta D > 0$ and $\Delta D < 0$ have to be considered separately. The case $\Delta D > 0$ is easier, and the state of the average system is $z_{\rm av}(t), u_{\rm av}(x,t)$, $x \in [0, D_0 + \Delta D]$, i.e., $z_{\rm av}(t), U_{\rm av}(\Theta)$, $\Theta \in [t - D_0 - \Delta D, t]$. The case $\Delta D < 0$ is more complicated since the state of the average system is $z_{\rm av}(t), u_{\rm av}(x,t)$, $x \in [\Delta D, D_0 + \Delta D]$, i.e., $z_{\rm av}(t), U_{\rm av}(\Theta)$, $\Theta \in [t - D_0, t]$.

Caso $\Delta D > 0$: We take the Lyapunov function

$$V(t) = \lambda_P z_{\rm av}^2(t) + \frac{a}{2} \int_0^{D_0 + \Delta D} (1 + x) w(x,t)^2 dx \tag{6.77}$$

with $\lambda_P > 0$ and $a > 0$.

Computing the time derivative of (6.77), one gets

$$\begin{aligned}
\dot{V} = & -\lambda_Q z_{\rm av}^2(t) + 2\lambda_P G z_{\rm av}(t) w(0,t) \\
& + \frac{a}{2}(1 + D) w(D_0 + \Delta D, t)^2 \\
& - \frac{a}{2} w(0,t)^2 - \frac{a}{2} \int_0^{D_0 + \Delta D} w(x,t)^2 dx \\
\leq & -\left(\frac{\lambda_Q}{2} - a(1 + D) q_1(\Delta D)\right) z_{\rm av}^2(t) \\
& - \left(\frac{a}{2} - \frac{2(\lambda_P G)^2}{\lambda_Q}\right) w(0,t)^2 \\
& - a\left(\frac{1}{2} - (1 + D) q_2(\Delta D)\right) \int_0^{D_0 + \Delta D} w(x,t)^2 dx,
\end{aligned} \tag{6.78}$$

where we have denoted $D = D_0 + \Delta D$, such as in (6.5).

This proves the exponential stability of the origin of the transformed system with state $(z_{\rm av}(t), w(x,t), x \in [0, D_0 + \Delta D])$, for sufficiently small ΔD, by choosing

$$a > \frac{4(\lambda_P G)^2}{\lambda_Q} \tag{6.79}$$

and then choosing the sufficiently small $\delta > 0$ as the largest value of $|\Delta D|$ so that

$$\frac{\lambda_Q}{2} > a(1 + D) q_1(\Delta D) \tag{6.80}$$

and

$$\frac{1}{2} > (1 + D) q_2(\Delta D). \tag{6.81}$$

Exponential stability in the norm $N_2(t)$ is obtained using a linear function $N_2^2(t)$ for upper- and lower-bounding $V(t)$, where, for $\Delta D > 0$, one defines

$$N_2(t) = \left(z_{\rm av}^2(t) + \int_{t - D_0 - \Delta D}^{t} U_{\rm av}(\Theta)^2 d\Theta\right)^{1/2}. \tag{6.82}$$

Case $\Delta D < 0$: In this case, we use a different Lyapunov function,

$$\begin{aligned} V(t) =& \lambda_P z_{\text{av}}^2(t) + \frac{a}{2}\int_0^{D_0+\Delta D}(1+x)w(x,t)^2 dx \\ &+ \frac{1}{2}\int_{\Delta D}^0 (D_0+x)w(x,t)^2 dx, \end{aligned} \tag{6.83}$$

and obtain

$$\begin{aligned} \dot{V} \leq& -\left(\frac{\lambda_Q}{2} - a(1+D)q_1(\Delta D)\right)z_{\text{av}}^2(t) \\ &-\left(\frac{a}{2} - \frac{D_0}{2} - \frac{2(\lambda_P G)^2}{\lambda_Q}\right)w(0,t)^2 \\ &-\left(\frac{1}{2} - a(1+D)q_2(\Delta D)\right)\int_{\Delta D}^0 w(x,t)^2 dx \\ &-\frac{D}{2}w(\Delta D,t)^2 - \frac{\max\{a,1\}}{4}\int_{\Delta D}^{D_0+\Delta D} w(x,t)^2 dx. \end{aligned} \tag{6.84}$$

This quantity is made negative definite by first choosing

$$a > D_0 + \frac{4(\lambda_P G)^2}{\lambda_Q} \tag{6.85}$$

and then choosing the sufficiently small $\delta > 0$ as the largest value of $|\Delta D|$ so that

$$\frac{\lambda_Q}{2} > a(1+D)q_1(\Delta D) \tag{6.86}$$

and

$$\frac{1}{2} > a(1+D)q_2(\Delta D). \tag{6.87}$$

Consequently, one can write

$$\begin{aligned} \dot{V} \leq& -\left(\frac{\lambda_Q}{2} - a(1+D)q_1(\Delta D)\right)z_{\text{av}}^2(t) \\ &-\left(\frac{1}{2} - a(1+D)q_2(\Delta D)\right)\int_{\Delta D}^0 w(x,t)^2 dx \\ &-\frac{\max\{a,1\}}{4}\int_{\Delta D}^{D_0+\Delta D} w(x,t)^2 dx \\ \leq& -\left(\frac{\lambda_Q}{2} - a(1+D)q_1(\Delta D)\right)z_{\text{av}}^2(t) \\ &-\left(\frac{1}{2} - a(1+D)q_2(\Delta D)\right)\frac{2}{D_0}\frac{1}{2}\int_{\Delta D}^0 (D_0+x)w(x,t)^2 dx \\ &-\frac{\max\{a,1\}}{4}\frac{2}{a(1+D)}\frac{a}{2}\int_0^{D_0+\Delta D}(1+x)w(x,t)^2 dx, \end{aligned} \tag{6.88}$$

which leads to

$$\dot{V} \leq -\mu V, \tag{6.89}$$

with

$$\begin{aligned}\mu = \min\Bigg\{ & \left(\frac{\lambda_Q}{2} - a(1+D)q_1(\Delta D)\right)\frac{1}{\lambda_P}, \\ & \left(\frac{1}{2} - a(1+D)q_2(\Delta D)\right)\frac{2}{D_0}, \quad \frac{\max\{a,1\}}{2a(1+D)}\Bigg\}.\end{aligned} \tag{6.90}$$

Hence, we get an exponential stability estimate in terms of $|z_{\rm av}(t)|^2 + \int_{\Delta D}^{D_0+\Delta D} w(x,t)^2 dx$. With some further work, we also get an estimate in terms of $|z_{\rm av}(t)|^2 + \int_{\Delta D}^{D_0+\Delta D} u_{\rm av}(x,t)^2 dx$, i.e., in terms of $|z_{\rm av}(t)|^2 + \int_{t-D_0}^{t} U_{\rm av}(\Theta)^2 d\Theta$. We start from

$$\psi_1\left(|z_{\rm av}(t)|^2 + \int_{\Delta D}^{D} w(x,t)^2 dx\right) \leq V(t) \leq \psi_2\left(|z_{\rm av}(t)|^2 + \int_{\Delta D}^{D} w(x,t)^2 dx\right), \tag{6.91}$$

where

$$\psi_1 = \min\left\{\lambda_P, \frac{a}{2}, \frac{D}{2}\right\}, \tag{6.92}$$

$$\psi_2 = \max\left\{\lambda_P, \frac{a(1+D)}{2}, \frac{D_0}{2}\right\}. \tag{6.93}$$

Let us now consider

$$w(x,t) = u_{\rm av}(x,t) - m(x) \star u_{\rm av}(x,t) - K z_{\rm av}(t), \tag{6.94}$$

$$u_{\rm av}(x,t) = w(x,t) + n(x) \star w(x,t) + KN(x) z_{\rm av}(t), \tag{6.95}$$

where $\star$ denotes the convolution operation in x and

$$m = KG, \quad n(s) = KN(s)G, \quad N(x) = e^{(GK)x}. \tag{6.96}$$

It is easy to show, using (6.94) and (6.95), that

$$\int_{\Delta D}^{D} w(x,t)^2 dx \leq \alpha_1 \int_{\Delta D}^{D} u_{\rm av}(x,t)^2 dx + \alpha_2 |z_{\rm av}(t)|^2, \tag{6.97}$$

$$\int_{\Delta D}^{D} u_{\rm av}(x,t)^2 dx \leq \beta_1 \int_{\Delta D}^{D} w(x,t)^2 dx + \beta_2 |z_{\rm av}(t)|^2, \tag{6.98}$$

where

$$\alpha_1 = 3(1 + D_0 m^2), \quad \alpha_2 = 3K^2, \tag{6.99}$$

$$\beta_1 = 3(1 + D_0 ||n||^2), \quad \beta_2 = 3||KN||^2, \tag{6.100}$$

and $||\cdot||$ denotes the $L_2[\Delta D, D]$ norm. Hence, we obtain

$$\phi_1\left(|z_{\rm av}(t)|^2 + \int_{\Delta D}^{D} u_{\rm av}(x,t)^2 dx\right) \leq |z_{\rm av}(t)|^2 + \int_{\Delta D}^{D} w(x,t)^2 dx, \tag{6.101}$$

$$|z_{\rm av}(t)|^2 + \int_{\Delta D}^{D} w(x,t)^2 dx \leq \phi_2\left(|z_{\rm av}(t)|^2 + \int_{\Delta D}^{D} u_{\rm av}(x,t)^2 dx\right), \tag{6.102}$$

where $\phi_1 = \frac{1}{\max\{\beta_1, \beta_2+1\}}$ and $\phi_2 = \max\{\alpha_1, \alpha_2 + 1\}$.

Combining the above inequalities, one gets

$$\phi_1\psi_1 N_2^2(t) \leq V(t) \leq \phi_2\psi_2 N_2^2(t), \tag{6.103}$$

where, for $\Delta D < 0$,

$$N_2(t) = \left(|z_{\rm av}(t)|^2 + \int_{t-D_0}^{t} U_{\rm av}(\Theta)^2 d\Theta\right)^{1/2}. \tag{6.104}$$

Hence, with (6.89), we get

$$N_2^2(t) \leq \frac{\phi_2\psi_2}{\phi_1\psi_1} N_2^2(0) e^{-\mu t}, \tag{6.105}$$

which completes the proof of exponential stability. □

Hence, it is straightforward to conclude that compensating the nominal delay D_0, we are still able to guarantee local exponential stability in the sense of the L_2-norm for maximizing or minimizing the output y in Figure 6.1 or any of its (if existing) higher derivatives, as long as the delay mismatch is bounded as in (6.57).

Finally, since the average closed-loop system is exponentially stable, analogously to Steps 6 and 7 originally presented for the proof of Theorem 2 in [179], we can invoke the averaging theorem in infinite dimensions by [83] (for convenience, see Appendix A) and conclude at least the following local convergence result for delay mismatch[4]:

$$\limsup_{t\to\infty} |\theta(t) - \theta^*| = \mathcal{O}(a + 1/\omega). \tag{6.106}$$

When stating Theorem 6.1, we have not built an upper bound δ^* for δ. Since the expression in (6.50) is only a sufficient condition (not necessary and sufficient), it would be hard to obtain a direct relation with the value of the upper bound $|\Delta D| \leq \delta < \delta^*$ for some $\delta^* > 0$. For this reason, we have conveniently kept the presentation as done before, considering in Theorem 6.1 an existence result for $0 < \delta < \delta^*$ and δ^* sufficiently small.

6.5 ▪ Simulation Results

In Figure 6.2, we consider the maximization of the first derivative of a cubic map rather than simply optimizing an unknown objective function:

$$h(\theta) = 2.5\theta - \frac{(\theta-2)^3}{3}, \tag{6.107}$$

where its first derivative

$$h^{(n)} = 2.5 - (\theta-2)^2, \quad n = 1, \tag{6.108}$$

is to be optimized and is subject to an output delay of $D = 5 + \Delta D$, where the values for ΔD will be given later.

[4]From [108], recall that a vector function $f(t,\epsilon) \in \mathbb{R}^n$ is said to be of order $\mathcal{O}(\epsilon)$ over an interval $[t_1, t_2]$ if $\exists k, \bar{\epsilon} : |f(t,\epsilon)| \leq k\epsilon \, \forall \epsilon \in [0, \bar{\epsilon}]$ and $\forall t \in [t_1, t_2]$. In most cases we use k and $\bar{\epsilon}$ as generic constants, such as in (6.106), and we use $\mathcal{O}(\epsilon)$ to be interpreted as an order-of-magnitude relation for sufficiently small ϵ.

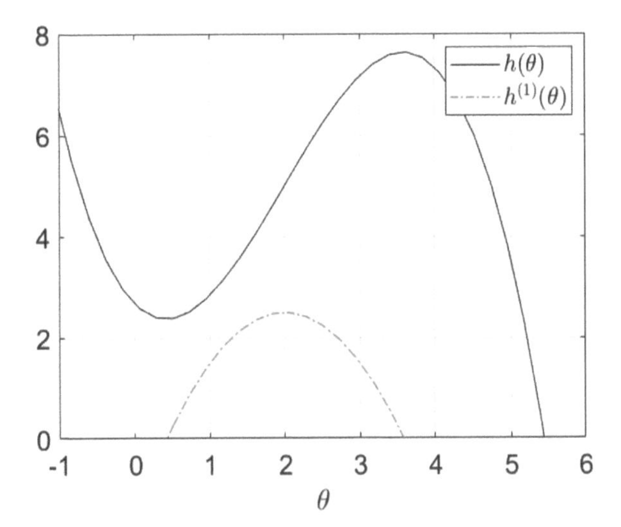

Figure 6.2. *The function* $h(\theta)$ *from* (6.107) *and its first derivative* $h^{(1)}(\theta)$ *with the extremum (maximum) at* $(\theta^*, Q^*) = (2, 2.5)$.

This example better illustrates the main objectives of the paper on real-time optimization of higher-derivatives maps in the presence of uncertain delays. Moreover, this scenario is closer to the applications for refrigeration systems [229, 230], where the optimization is performed for a desired operating point such that the function available for measurement ha sigmoid-like properties and the operating point is the maximum slope.

According to (6.107), the extremum point for $n=1$ is $(\theta^*, Q^*) = (2, 2.5)$ and the Hessian of the map is $H=-2$ with the inverse given by $H^{-1}=-0.5$. In Figure 6.2, we depict (6.107) and its first derivative (6.108). We use washout and low-pass filters with corner frequencies (ω_h and ω_l) to improve the controller performance (see [73, Figure 4]). We use the following parameters: $a=0.2, \omega=10, k=0.2, k_R=0.25, \omega_h=\omega_l=1$, $c=20$, $\hat{\theta}(0)=0$, and $\gamma(0)=-0.1$.

Figure 6.3 shows the output $y(t)$ in 3 situations: (a) free of delays, (b) in the presence of output delay but without any delay compensation and (c) with output-delay and prediction-based compensation. In the latter case, no uncertainties for the delay, i.e., $\Delta D = 0$ and $D = D_0$, were assumed.

Figure 6.4 shows the performance of the Newton-based ES scheme when we take into account a small uncertainty $\Delta D = 0.07$ in the output-time delay. The design parameters were the same used in the previous section supposing a nominal delay of $D_0 = 5s$. Thus, the simulation was performed with $D = 5.07s$.

The maximum delay uncertainty supported by the closed-loop system can be enlarged if ω decreases. As shown in Figure 6.4, we reset ω to $5rad/s$ and an uncertainty of $\Delta D = 0.15$ (about two times larger than the previous one) could be admitted. It reinforces the idea that the predictor feedback would actually have a more positive robustness margin to delay uncertainty.

In Figure 6.5, the robustness of the proposed predictor-based ES to delay mismatch $D_0 = D - \Delta D$ is shown. We assume a relative uncertainty of the delay in the predictor, the additive dither $S(t)$, and demodulation signals $\Upsilon_1(t)$ and $\Upsilon_2(t)$ such that $D = 1$ and $\Delta D = \Delta \in [-0.4, 0.4]$.

For this choice of uncertainty the ES still converges to a neighborhood of the extremum, although the Hessian estimate $\hat{H}_f$ is wrong ($H = -2$). The convergence speed depends on the Hessian of the static map, thus the perturbation-based Hessian estimate. An underestimation of the delay ($D > D_0$) results in a faster convergence to the extremum, but also for higher positive uncertainties it leads to overshooting or instability. Overestimation of the delay ($D < D_0$) results

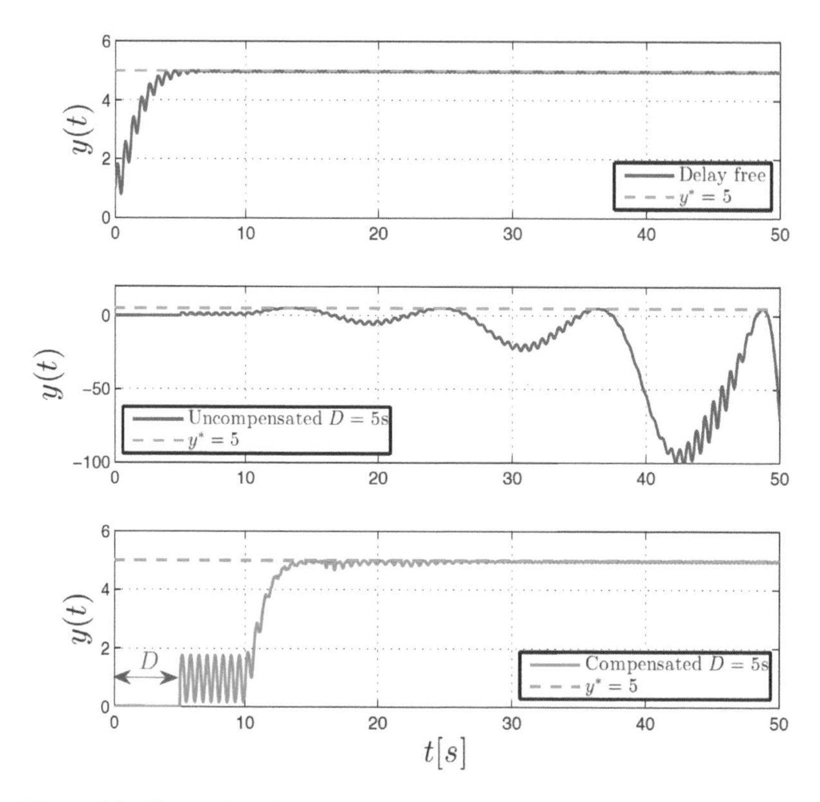

Figure 6.3. *Newton-based ES plus known output delay (time response of y(t)):* (a) *basic ES works well without delays;* (b) *ES goes unstable in the presence of delays;* (c) *predictor fixes this, leading the output* $y(t)$ *to* $y^* = h(\theta^*) = 2.5\theta^* = 5$.

in slower convergence or instability for higher negative uncertainties. Based on this numerical simulation, observations indicate that there exist some lower bound $0 < \underline{D} < D$ and upper bound $0 < D < \bar{D}$ such that for $D_0 \in [\underline{D}, \bar{D}]$ the closed-loop is exponentially stable and converges to a neighborhood of the extremum.

For this particular example, the simulations suggest that the delay mismatches may be greater than those expected by the condition given in (6.50) and that, for this simple strongly convex example, the correspondent uncertainty on the Hessian estimate does not affect the equilibrium of the ES dynamics, but only its speed of convergence.

Remark 6.1. In the ideal scenario $\Delta D = 0$, we would like to obtain

$$\begin{aligned}\dot{\tilde{\theta}}_{\text{av}}(t) &= -k z_{\text{av}}(t) \\ &= -k\gamma_{\text{av}}(t)\Upsilon^{\text{av}}_{n+1}(t - D_0) \\ &= -kH^{-1}H\tilde{\theta}_{\text{av}}(t - D_0) \\ &= -k\tilde{\theta}_{\text{av}}(t - D_0), \quad k > 0,\end{aligned} \tag{6.109}$$

which becomes exponentially stable if the nominal delay D_0 is properly compensated and the convergence rate depends exclusively on the design parameter $k > 0$. When $\Delta D \neq 0$, the signals $\Upsilon^{\text{av}}_{n+1}(t)$ and $\gamma_{\text{av}}(t)$ may assume values such that $z_{\text{av}}(t) \geq \epsilon\tilde{\theta}_{\text{av}}(t - D_0)$, with $\epsilon > H^{-1}H = 1$.

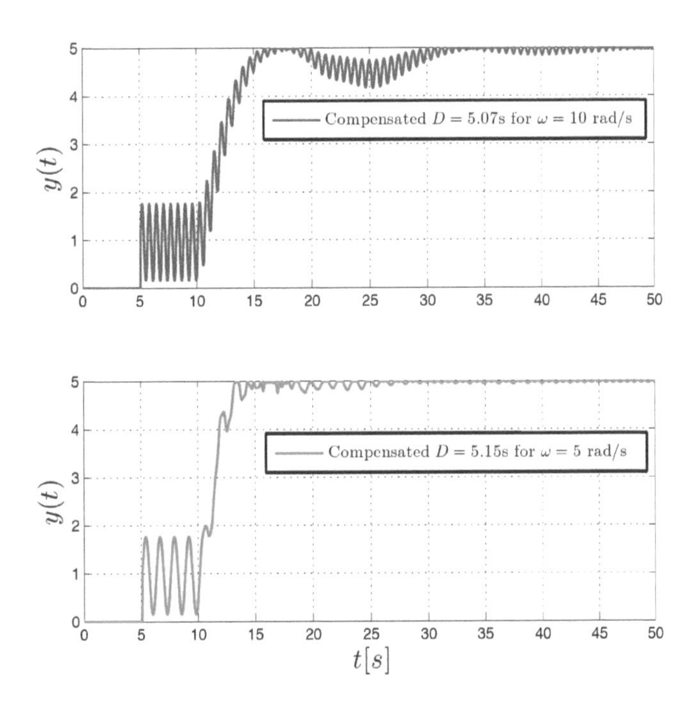

Figure 6.4. *Newton-based ES under uncertain delays:* (a) *time response of the output signal y(t) converging to* $y^* = h(\theta^*) = 2.5\theta^* = 5$ *in the presence of an uncertain delay* $D = 5 + \Delta D$, *with* $\Delta D = 0.07s$; (b) *a lower frequency* $\omega = 5rad/s$ *can increase the margin of delay uncertainty to* $D = 0.15$.

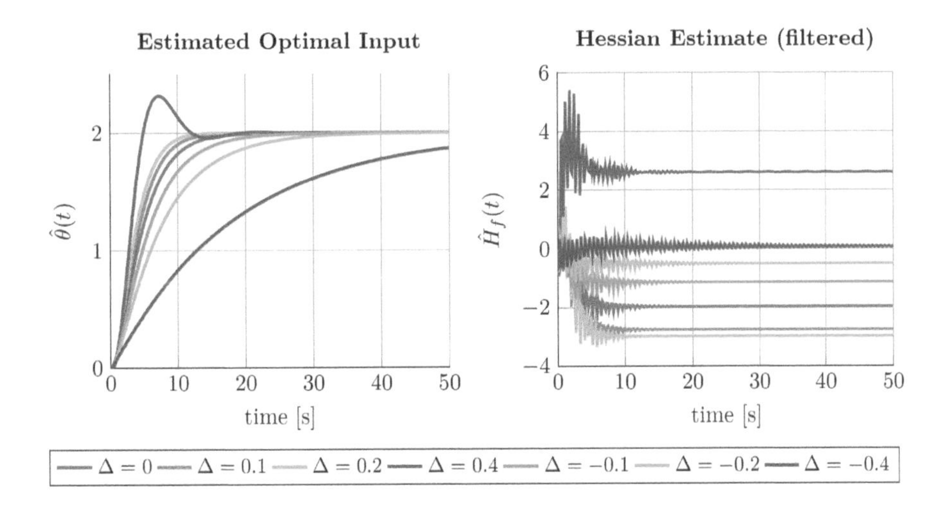

Figure 6.5. *Robustness analysis for different values of* ΔD. *The estimated optimal input still converges to the optimizer* θ^*, *but the Hessian estimate is not correct, which influences the convergence speed.*

In this case, the linearized expression of (6.48) would be

$$\dot{\tilde{\theta}}_{\rm av}(t) = -k \underbrace{\frac{\cos((n+1)\omega\Delta D)}{\cos((n+2)\omega\Delta D)}}_{\epsilon} \tilde{\theta}_{\rm av}(t-D), \tag{6.110}$$

which would result in faster convergence rates since $\epsilon > 1$. However, we could not take it for granted since ϵ may also assume values such that $\epsilon < 1$, depending on how the delay mismatch ΔD affects the estimates of $\Upsilon^{\rm av}_{n+1}(t)$ and $\gamma_{\rm av}(t)$ individually. According to the condition in (6.50), for larger values of ΔD, the parameter ϵ may even become negative and then the closed-loop would be unstable from (6.110) above, independently if the delay $D = D_0 + \Delta D$ was compensated or not.

6.6 ▪ Notes and References

Extremum seeking (ES) has been an extremely active topic in control research in recent years with thousands of applications, among which a few we mention here are [71], [164], [224], [109], [218], [166], [54], [77], [29], [80], [194], [239]. Despite this fact, there was no work which rigorously deals with the problem of ES in the presence of delays.

In [177], [178], [179], we began to tackle this topic and its generalizations, including multiparameter gradient ES. In [198], [199], [200], we considered the Newton-based ES [73] with constant delays.

However, we treated the delay as known, even though our designs were for maps that are completely unknown. Hence, there was a need for a deeper investigation of the delay mismatch and its impact on ES and its task of optimizing a static map or its derivatives of arbitrarily high order, in the presence of delays, using the tools of Newton-based ES.

In the present chapter, we generalized our previous approaches to handle uncertain delays. First of all, we proved that the predictor compensation introduced in our previous chapters is indeed robust to small delay mismatch. The delay mismatch also affects the averaging estimates of the gradient and the unknown second derivative (Hessian) [73] of the nonlinear map to be optimized. Local exponential stability is shown in this chapter for maps whose higher derivatives are locally quadratic, provided the delay mismatch is sufficiently small.

For the sake of clarity and simplicity, we considered only the scalar ES. The multivariable Newton-based ES [73] is discussed in the next chapter.

Chapter 7

Multivariable ES for Distinct Input Delays

7.1 ▪ The Challenges of Distinct Delays Affecting Different Channels of the Unknown Map

Our handling of delays in extremum seeking has been based on results for predictor-based control of linear systems. We have specialized those results for general LTI systems with input delays to the case where the plant is an integrator whose input coefficient is the Hessian of the unknown map. Hence, we have relied on only a very special case from a broad design catalog for delay compensation for LTI systems.

Within the broad class of input-delay compensation problems for LTI systems there exist some that are very difficult. They include plants with multiple inputs, where the individual inputs are subject to distinct delays, and plants with distributed delays on the input. We tackle both of those challenging classes of problems in our book. However, we wish to make it clear that the ES algorithm design never calls for the full generality of the design techniques that have been developed for predictor-based control of general LTI systems. This fact—that ES doesn't call for everything we are capable of doing in the realm of delay compensation—which one might call regrettable and another might call fortunate, is a consequence of the fact that ES systems are all variants of integral control; i.e., the compensator being designed by the ES algorithm designer is always in some general form of an integrator, without a drift term, such as $A\hat{\theta}$. The drift term, if stabilizing, would be resulting in the algorithm undesirably settling at some value $\hat{\theta}$ that is proportional to the gradient, rather than settling, as desired, at the optimizer θ^* when the gradient is zero. Hence, ES design employs only a portion of the predictor feedback toolkit, the portion corresponding to $A = 0$.

Even when $A = 0$, the compensation of delays is challenging in some multivariable situations. Those situations are the ones where the system has multiple inputs, with distinct inputs acting on the delays, and when the input matrix is not diagonal (i.e., in the ES situation, when the Hessian is not diagonal).

With this insight regarding the predictor feedback tools at our disposal, we are in the position to consider various multivariable ES problem statements and the possible approaches that can be applied to these problem.

The problems and the approaches that we consider come in three categories:

- Delay(s) can act on either the map's output or on its inputs.
- If the delays are acting on the map's inputs, the delays can be equal or distinct.
- The algorithm designs can be of the gradient or Newton type.

Hence, we have two possibilities in each of the three categories. This makes for eight combinations of input locations (input or output), delay values (equal or distinct), and algorithm type (gradient or Newton).

Of these eight combinations, four are easy because they deal with a delay on the output and require that a single delay be compensated but in a multi-input ES setting.

This leaves us with four remaining possibilities: equal input delays, gradient; equal input delays, Newton; distinct input delays, gradient; and distinct input delays, Newton. The first two of these four cases are easy as they are essentially no different than the case of the delay on a scalar output.

Hence, it is worth focusing on distinct delays on the inputs, using the gradient and the Newton designs. Surprisingly, while the Newton approach has so far always been the more complex of the two design options, in the presence of input delays it offers a simplification. The Newton approach, which employs an inverse of the estimate of the unknown Hessian to multiply the Hessian of the map in the convergence rate and make the convergence rate matrix diagonal, performs a diagonalization of the system for which the delay compensation needs to be performed. Hence, with the Newton ES approach, the predictor feedback needs to be designed for a vector of scalar and decoupled integrators, with distinct delays. The fact that the delays are distinct no longer matters—one gets a collection of scalar plants for which to compensate for the distinct input delays.

Under gradient design, the problem is more challenging. The input channels are coupled and, while compensating for the input delay in one channel, one needs to account for the input being applied from the other channel. Care must be taken about ordering the channels by delay lengths since the more delayed inputs need to be compensated differently than the less delayed inputs. The integral terms in the predictor feedback laws for each channel contain the inputs from all the other channels, though over different integration intervals. Hence, compensation of distinct input delays in the gradient approach is the most challenging of the multivariable ES problems in the presence of delays.

Of the eight possible problems, in this book we consider four. We leave out the four problems with equal delays on multiple inputs. Or, rather, we convert them into output delay problems and consider them in that form.

For the hardest of the eight problems, the gradient design for distinct input delays, we present two variants. One employs a backstepping-based predictor design, which is the more difficult approach. The other variant takes a shortcut by employing Artstein's reduction transformation, which in the case where $A = 0$ can be seen as a forwarding transformation, and this approach reduces the complexity of the design and the analysis. The case of the highest theoretical interest is the gradient design based on backstepping for distinct input delays.

With a source seeking example, we show the performance of the proposed delay-compensated extremum seeking schemes.

7.2 ▪ Problem Statement

Multiparameter or multivariable ES considers applications in which the goal is to maximize (or minimize) the scalar output $y \in \mathbb{R}$ of an unknown and convex nonlinear static map $y = Q(\theta)$ by varying the input vector $\theta = [\theta_1 \; \theta_2 \; \cdots \; \theta_n]^T$.

7.2.1 ▪ The Basic Idea of ES Algorithms Free of Delays

In this section, we revisit the basic ideas of multivariable gradient ES without delays. In the maximum seeking problem, there exists $\theta^* \in \mathbb{R}^n$ such that

$$\frac{\partial Q(\theta^*)}{\partial \theta} = 0, \tag{7.1}$$

$$\frac{\partial^2 Q(\theta^*)}{\partial \theta^2} = H < 0, \quad H = H^T, \tag{7.2}$$

where θ^* and H are considered unknown. Referring to the Taylor series expansion of the nonlinear map around the peak θ^*, we have

$$y = y^* + \frac{1}{2}(\theta - \theta^*)^T H(\theta - \theta^*) + R(\theta - \theta^*), \tag{7.3}$$

where $R(\theta - \theta^*)$ stands for higher-order terms in $\theta - \theta^*$ and $y^* = Q(\theta^*)$ is the extremum.

The basic Gradient ES algorithm measures the scalar signal $y(t)$ and with the help of the additive dither and demodulation signal

$$S(t) = [a_1 \sin(\omega_1 t) \ \cdots \ a_n \sin(\omega_n t)]^T, \tag{7.4}$$

$$M(t) = \left[\frac{2}{a_1}\sin(\omega_1 t) \ \cdots \ \frac{2}{a_n}\sin(\omega_n t)\right]^T \tag{7.5}$$

constructs $G(t) = M(t)y(t)$ to estimate the unknown gradient $\partial Q(\theta)/\partial \theta$ of the nonlinear map $Q(\theta)$. The actual input $\theta(t) := \hat{\theta}(t) + S(t)$ is based on the real-time estimate $\hat{\theta}(t)$ of θ^* but is perturbed by $S(t)$. The estimate $\hat{\theta}$ is generated with the integrator $\hat{\theta} = (K/s)G$, which locally approximates the gradient update law $\dot{\hat{\theta}}(t) = KH(\hat{\theta}(t) - \theta^*)$, tuning $\hat{\theta}(t)$ to θ^* if $\hat{\theta}(0)$ is close of θ^*. The adaptation gain (diagonal matrix $K > 0$) controls the speed of estimation but cannot be arbitrarily increased a priori due to the limitations on the averaging analysis [129].

To guarantee convergence, the user should choose appropriate frequencies $\omega_i \neq \omega_j$ and nonzero small amplitudes a_i. The former is a key condition that differentiates the multi-input case [73] from the single-input case [129]. The sinusoid feature of (7.4) and (7.5) is only one choice for the dither signals—many other perturbations, from square waves to stochastic noise, can be used in lieu of it, provided they are of zero mean [142, 218].

However, the convergence rate depends on the unknown Hessian H. This weakness of the gradient-based ES algorithm is removed with the Newton-based ES and will be detailed later on in the chapter. Briefly, a multiplicative excitation denoted by $N(t)$ is introduced to generate the estimate of the Hessian H as $\hat{H}(t) = N(t)y(t)$ [166]. According to [73], then a Riccati differential equation inverts this Hessian's estimate and cancels out the term H from the convergence rate, making it user-assignable. A fair overview of gradient- and Newton-based versions of ES free of delays can be found in [123].

7.2.2 ▪ Input-Output Delays

The main contribution of the present chapter is to additionally consider that the nonlinear map to be optimized in real time is subject to delays in the actuator path and/or measurement system. In order to start to formulate the problem, we can initially assume the simplest case of delays in the scalar output. In this case, there exists a constant delay denoted by $D_{\text{out}} \geq 0$ in the measurement system such that the output is given by

$$y(t) = Q(\theta(t - D_{\text{out}})). \tag{7.6}$$

However, it is not difficult to show that input delays could be handled in the same way noting that input delays, denoted by $D_{\text{in}} \geq 0$, can be moved to the output of the static map. The restriction here is that the delay must be the same in each individual input channel. The scenario when input and output delays occur simultaneously could also be tackled assuming that the total delay to be counteracted would be $D_{\text{total}} = D_{\text{in}} + D_{\text{out}}$.

In a more general framework, we can assume the following input-output delay representation:

$$y(t) = Q(\theta(t-D)) = e^{-Ds}[Q(\theta(t))], \tag{7.7}$$

where the constant delay matrix $D = \text{diag}\{D_1, \ldots, D_n\}$ must have the same component delays D_i in each individual input channel of $\theta(t) \in \mathbb{R}^n$, i.e., $D_1 \equiv D_2 \equiv \cdots \equiv D_n \geq 0$ for non-distinct input delays (or output delays). Nevertheless, the main advantage of this representation is the possibility of including multiple and distinct input delays. In this case, without loss of generality, we assume that the inputs have distinct delays which are ordered so that

$$D = \text{diag}\{D_1, \ldots, D_n\}, \quad 0 \leq D_1 \leq D_2 \leq \cdots \leq D_n. \tag{7.8}$$

In addition, we consider that the constants D_i must be *known* for all $i \in \{1, 2, \ldots, n\}$. Note that we mix the time and frequency domains in (7.7) by using the brackets $[\cdot]$ to denote that the transfer function acts as an operator on a time-domain function.

7.2.3 ▪ Locally Quadratic Maps with Delays

Without loss of generality, let us consider the maximum seeking problem such that the maximizing value of θ is denoted by θ^*, satisfying (7.1)–(7.2). For the sake of simplicity, we also assume that the nonlinear multivariable map (7.7) is at least locally quadratic, i.e.,

$$Q(\theta) = y^* + \frac{1}{2}(\theta - \theta^*)^T H(\theta - \theta^*), \tag{7.9}$$

within a neighborhood of the unknown extremum point (θ^*, y^*), where $\theta^* \in \mathbb{R}^n$, $y^* \in \mathbb{R}$, and $H = H^T < 0$ is the $n \times n$ unknown Hessian matrix of this static map.

By plugging (7.9) into (7.7), we obtain the *locally quadratic static map with delay*:

$$y(t) = y^* + \frac{1}{2}(\theta(t-D) - \theta^*)^T H(\theta(t-D) - \theta^*). \tag{7.10}$$

In the general case of multiple and distinct delays in the control channels, the delayed input vector can be represented by

$$\theta(t-D) := \begin{bmatrix} \theta_1(t-D_1) \\ \theta_2(t-D_2) \\ \vdots \\ \theta_n(t-D_n) \end{bmatrix}. \tag{7.11}$$

7.2.4 ▪ Probing and Demodulation Signals

Let $\hat{\theta}$ be the estimate of θ^* and

$$\tilde{\theta}(t) = \hat{\theta}(t) - \theta^* \tag{7.12}$$

be the *estimation error*. Moreover, let us define

$$G(t) = M(t)y(t), \quad \theta(t) = \hat{\theta}(t) + S(t), \tag{7.13}$$

where the vector dither and demodulation signals are given by

$$S(t) = [a_1 \sin(\omega_1(t+D_1)) \cdots a_n \sin(\omega_n(t+D_n))]^T, \tag{7.14}$$

$$M(t) = \left[\frac{2}{a_1}\sin(\omega_1 t) \cdots \frac{2}{a_n}\sin(\omega_n t)\right]^T, \tag{7.15}$$

with nonzero perturbation amplitudes a_i.

The elements of the $n \times n$ demodulating matrix $N(t)$ to construct the signal

$$\hat{H}(t) = N(t)y(t) \tag{7.16}$$

are given by

$$N_{i,i}(t) = \frac{16}{a_i^2}\left(\sin^2(\omega_i t) - \frac{1}{2}\right), \tag{7.17}$$

$$N_{i,j}(t) = \frac{4}{a_i a_j}\sin(\omega_i t)\sin(\omega_j t), \quad i \neq j. \tag{7.18}$$

The probing frequencies ω_i can be selected as

$$\omega_i = \omega_i' \omega = \mathcal{O}(\omega), \quad i \in 1, 2, \ldots, n, \tag{7.19}$$

where ω is a positive constant and ω_i' is a rational number. One possible choice is given in [73] as

$$\omega_i' \notin \left\{\omega_j', \frac{1}{2}(\omega_j' + \omega_k'), \omega_j' + 2\omega_k', \omega_j' + \omega_k' \pm \omega_l'\right\} \tag{7.20}$$

for all distinct i, j, k, and l.

Notice that (7.14) is different from the additive dither signal (7.4) used in standard ES algorithms free of delays. As will be shown, the input/output delays can always be transferred for analysis purposes to the integrator output of the estimation error dynamics (or, equivalently, to its input); thus, the phase shift $+\omega_i D_i$ is applied to compensate the delay effect in the dither signal $S(t)$.

7.2.5 ▪ Basic Averaging Properties

In [73], the following two averaging properties were proved:

$$\frac{1}{\Pi}\int_0^{\Pi} N(\sigma) y d\sigma = H, \tag{7.21}$$

$$\frac{1}{\Pi}\int_0^{\Pi} M(\sigma) y d\sigma = H\tilde{\theta}_{\mathrm{av}} \tag{7.22}$$

if a quadratic map as in (7.9) is considered, which are still valid even in the presence of delays. In other words, we obtain the average signals

$$\hat{H}_{\mathrm{av}} = (Ny)_{\mathrm{av}} = H, \tag{7.23}$$

$$G_{\mathrm{av}}(t) = (My)_{\mathrm{av}} = H\tilde{\theta}_{\mathrm{av}}(t-D) \tag{7.24}$$

for $\hat{H}(t)$ and $G(t)$ with

$$\tilde{\theta}_{\mathrm{av}}(t-D) = [\tilde{\theta}_1^{\mathrm{av}}(t-D_1) \cdots \tilde{\theta}_n^{\mathrm{av}}(t-D_n)]^T \tag{7.25}$$

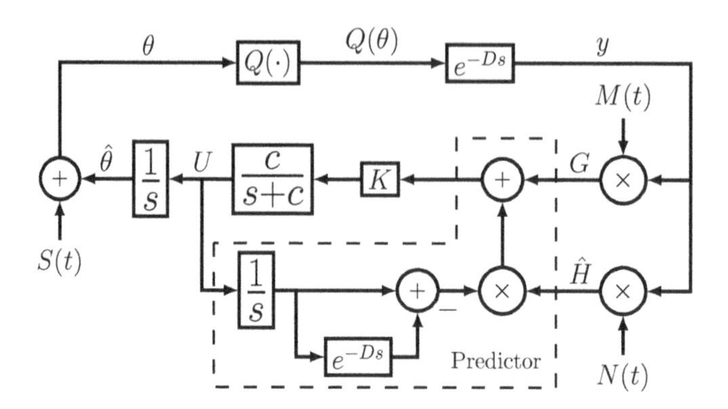

Figure 7.1. *Block diagram of the basic prediction scheme for output-delay compensation in multivariable gradient-based ES, where the delay $D \geq 0$ is a simple scalar. The predictor with a perturbation-based estimate of the Hessian obeys* (7.38), *and the additive dither and demodulation signals are given by* $S(t) = [a_1 \sin(\omega_1(t+D)) \ \cdots \ a_n \sin(\omega_n(t+D))]^T$ *and* $M(t) = \left[\frac{2}{a_1}\sin(\omega_1 t) \ \cdots \ \frac{2}{a_n}\sin(\omega_n t)\right]^T$. *The demodulating matrix* $N(t)$ *is computed by* (7.17)–(7.18).

and Π defined as

$$\Pi = 2\pi \times \text{LCM}\left\{\frac{1}{\omega_i}\right\} \quad \forall i \in \{1,2,\ldots,n\}, \tag{7.26}$$

where LCM stands for the least common multiple.

Basically, we can say that the signals $\hat{H}(t)$ and $G(t)$ provide averaging-based estimates for the Hessian and the delayed Gradient of the nonlinear map (7.9)–(7.10). These properties are fundamental for the predictor design and stability analysis of the Gradient and Newton-based ES schemes developed in the next sections.

7.3 ▪ Gradient-Based ES with Output Delays

For the sake of clarity, we start considering output delays (or same delay in the input channels) for Gradient ES before moving on to the more complex case of multiple and distinct input delays. Therefore, the variables D_i in (7.10) and (7.11) are equal throughout this section and $D > 0$ can be faced as a single *known* constant. The proposed Gradient ES under output delays is shown in the block diagram of Figure 7.1.

7.3.1 ▪ Averaging Analysis without Predictor Compensation

If the filtered predictor based controller was not applied in Figure 7.1, but the standard Gradient ES [129] feedback law $U(t) = KG(t)$, one could write $\dot{\tilde{\theta}}(t) = \dot{\hat{\theta}}(t) = KG(t)$, where $K > 0$ is an $n \times n$ positive diagonal matrix. From (7.13), the closed-loop system equation would be written as

$$\dot{\tilde{\theta}}(t) = KM(t)y(t), \tag{7.27}$$

and using the identity (7.22) to average (7.27), we would obtain the following average model:

$$\frac{d\tilde{\theta}_{\text{av}}(t)}{dt} = KH\tilde{\theta}_{\text{av}}(t-D). \tag{7.28}$$

From (7.28), it is clear that the equilibrium $\tilde{\theta}^e_{\text{av}} = 0$ of the average system is not necessarily stable for arbitrary values of the delay D. This reinforces the necessity of applying the prediction

$U(t) = KG(t+D), \forall t \geq 0$, to stabilize the system. Besides the phase compensation in the additive dither (7.14), the predictor design is the second measure taken in this chapter to deal with the time-delay issue.

7.3.2 ▪ Predictor Feedback with a Perturbation-Based Estimate of the Hessian

From Figure 7.1, the *error dynamics* of (7.12) is written as

$$\dot{\tilde{\theta}}(t-D) = U(t-D). \tag{7.29}$$

The idea of the predictor feedback is to compensate for the delay by feeding back the future state $G(t+D)$, or $G_{\rm av}(t+D)$ in the equivalent average system.

The average version of the vector signal (7.13) is given by (7.24). Hence, from (7.29), the following average models can be obtained:

$$\dot{\tilde{\theta}}_{\rm av}(t-D) = U_{\rm av}(t-D), \tag{7.30}$$

$$\dot{G}_{\rm av}(t) = HU_{\rm av}(t-D), \tag{7.31}$$

where $U_{\rm av} \in \mathbb{R}^n$ denotes the resulting average control for $U \in \mathbb{R}^n$. Given the stabilizing diagonal matrix $K > 0$ for the undelayed system, our wish is to have a controller that achieves

$$U_{\rm av}(t) = KG_{\rm av}(t+D) \quad \forall t \geq 0, \tag{7.32}$$

and it appears to be nonimplementable since it requires future values of the state. However, by applying the variation-of-constants formula to (7.31), we can express the future state as

$$G_{\rm av}(t+D) = G_{\rm av}(t) + H\int_t^{t+D} U_{\rm av}(\tau - D)d\tau, \tag{7.33}$$

where the current state $G_{\rm av}(t)$ is the initial condition. Shifting the time variable under the integral in (7.33), we obtain

$$G_{\rm av}(t+D) = G_{\rm av}(t) + H\int_{t-D}^{t} U_{\rm av}(\tau)d\tau, \tag{7.34}$$

which gives the future state $G_{\rm av}(t+D)$ in terms of the average control signal $U_{\rm av}(\tau)$ from the past window $[t-D,t]$. It yields the following feedback law:

$$U_{\rm av}(t) = K\left[G_{\rm av}(t) + H\int_{t-D}^{t} U_{\rm av}(\tau)d\tau\right]. \tag{7.35}$$

Hence, from (7.34) and (7.35), the average feedback law (7.32) can be obtained indeed as desired. Consequently,

$$\dot{\tilde{\theta}}_{\rm av}(t) = KG_{\rm av}(t+D) \quad \forall t \geq 0. \tag{7.36}$$

Therefore, from (7.24), one has

$$\frac{d\tilde{\theta}_{\rm av}(t)}{dt} = KH\tilde{\theta}_{\rm av}(t) \quad \forall t \geq D, \tag{7.37}$$

with an exponentially attractive equilibrium $\tilde{\theta}^e_{\rm av} = 0$, since $KH < 0$. It means that the delay is perfectly compensated in D units of time, namely that the system evolves as if the delay were absent after D units of time.

The feedback law (7.35) seems to be implicit since $U_{\rm av}$ is present on both sides. However, the input memory $U_{\rm av}(\tau)$, where $\tau \in [t-D,t]$, is part of the state of an infinite-dimensional system, and thus the control law is effectively a complete-state-feedback controller. However, the analysis sketched above in the spirit of "finite spectrum assignment" does not capture the entire system consisting of the ODE in (7.29) and the infinite-dimensional subsystem of the input delay.

Another difficulty arises in the application of the *Averaging Theorem* to infinite dimensions (see Appendix A). For the class of functional differential equations (FDEs) studied here, there is no "off the shelf" averaging theorem result oriented for input-output delays. In general, the theory applies only to state-delay systems. This fact lead us to propose a simple modification of the basic predictor-based controllers, which employs a low-pass filter [115], to achieve our control objectives.

Thus, the averaging-based predictor feedback used in order to compensate output delays is redefined by

$$U(t) = \frac{c}{s+c}\left\{K\left[G(t)+\hat{H}(t)\int_{t-D}^{t} U(\tau)d\tau\right]\right\}, \tag{7.38}$$

where $c > 0$ is sufficiently large. The predictor feedback is of the form of a low-pass filtered of the nonaverage version of (7.35). With some abuse of notation, now we mix the time and frequency domains in (7.38) by using the braces $\{\cdot\}$ to denote that the lag transfer function acts as an operator on a time-domain function.

The predictor (7.38) is infinite-dimensional because the integral involves the control history over the interval $[t-D,t]$. Furthermore, it is averaging-based (perturbation-based) because $\hat{H}$ is updated according to the estimate (7.16) of the unknown Hessian H, satisfying the averaging property (7.21).

In the next section, we derive the predictor equations for the multiple and distinct input delays in gradient ES. After that, we present the complete steps of the stability analysis for all cases of sensor/actuator delays.

7.4 ▪ Gradient-Based ES with Multiple and Distinct Input Delays

Now, let us consider the case where the input delays D_i are distinct and satisfy (7.8). From Figure 7.2 and (7.12), we can easily write

$$\dot{\hat{\theta}}(t-D) = \begin{bmatrix} U_1(t-D_1) \\ U_2(t-D_2) \\ \vdots \\ U_n(t-D_n) \end{bmatrix}, \quad \dot{\hat{\theta}}_i(t-D_i) = U_i(t-D_i), \tag{7.39}$$

and the average model below by using (7.24):

$$\dot{G}_{\rm av}(t) = \sum_{i=1}^{n} H_i U_i^{\rm av}(t-D_i) = H \begin{bmatrix} U_1^{\rm av}(t-D_1) \\ U_2^{\rm av}(t-D_2) \\ \vdots \\ U_n^{\rm av}(t-D_n) \end{bmatrix}, \tag{7.40}$$

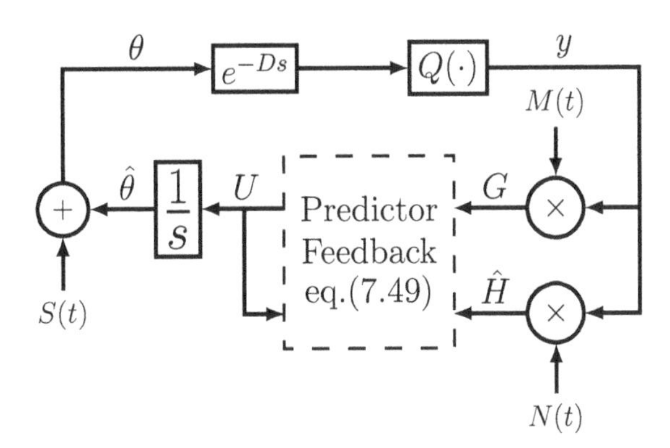

Figure 7.2. *Block diagram of gradient-based ES for multiple-input delay compensation, where* $D = diag\{D_1,\ldots,D_n\}$. *The predictor feedback is implemented according to* (7.49). *The additive dither signal is* $S(t) = a_1\sin(\omega_1(t+D_1)) \cdots a_n\sin(\omega_n(t+D_n))]^T$ *and the demodulating signals* $M(t)$ *and* $N(t)$ *are given by* $M(t) = \left[\frac{2}{a_1}\sin(\omega_1 t) \cdots \frac{2}{a_n}\sin(\omega_n t)\right]^T$ *and* (7.17)–(7.18), *respectively.*

since

$$\dot{\hat{\theta}}_{\text{av}}(t-D) = \begin{bmatrix} U_1^{\text{av}}(t-D_1) \\ U_2^{\text{av}}(t-D_2) \\ \vdots \\ U_n^{\text{av}}(t-D_n) \end{bmatrix}. \tag{7.41}$$

In (7.40), $U_i^{\text{av}}(t) \in \mathbb{R}$ for all $i \in \{1,2,\ldots,n\}$ are the elements of the average control $U_{\text{av}}(t) \in \mathbb{R}^n$ and the Hessian matrix $H = (H_1,H_2,\ldots,H_n) \in \mathbb{R}^{n\times n}$. In this case, there always exists a positive diagonal matrix $K = (K_1,K_2,\ldots,K_n)^T \in \mathbb{R}^{n\times n}$ such that HK is Hurwitz. For the sake of clarity, we say that H_i are column vectors of H and that K_i are row vectors of K $\forall i = 1,\ldots,n$.

Henceforth, the purpose of this section is to find a control feedback which has to perform prediction of the cross-coupling of the channels in (7.40). Applying the variation-of-constants formula to (7.40) gives

$$G_{\text{av}}(t+D_i) = G_{\text{av}}(t) + \sum_{j=1}^{n} \int_{t-D_j}^{t-(D_j-D_i)} H_j U_j^{\text{av}}(\tau)d\tau\,, \tag{7.42}$$

such that the predictor feedback realizes

$$U_i^{\text{av}}(t) = K_i^T G_{\text{av}}(t+D_i)\,. \tag{7.43}$$

From (7.8), each control input U_i^{av} arrives at the plant at a different time. Since the system is causal, values of U_j^{av} with $j > i$ on the interval $(t-(D_j-D_i),t)$ provide no information for prediction of $G_{\text{av}}(t+D_i)$. On the other hand, values of U_j^{av} with $j < i$ on the interval $(t,t+D_i-D_j)$ are necessary to calculate (7.42), but they are unavailable future values. For $i,j \in \{1,\ldots,n\}$ with $i < j$, we have

$$\int_{t-D_j}^{t-(D_j-D_i)} (*)d\tau = \int_{t-D_j}^{t} (*)d\tau + \int_{t}^{t+D_i-D_j} (*)d\tau\,. \tag{7.44}$$

Differently from the results in Section 7.3, the variation-of-constants formula contains an extra term $\int_t^{t+D_i-D_j}(*)d\tau$.

In what follows, we will provide a predictor-based controller that is more consistent with (7.42). To derive it, we have had to extend the backstepping approach [119] introducing a new successive backstepping-like transformation [226]. The explicit equation of this transformation of the delay state is given in the proof of the main theorem.

First of all, let us define the following notation:

$$A_i := \sum_{j=1}^{i} H_j K_j^T, \quad i \in \{0,1,2,\ldots,n\}, \tag{7.45}$$

it being obvious that $A_0 = 0_{n\times n}$ and $A_n = HK$. In addition, the matrix-valued function Φ can be represented as [226]

$$\begin{aligned}\Phi(x,\zeta) &= e^{A_{i-1}(x-D_{i-1})}e^{A_{i-2}(D_{i-1}-D_{i-2})}\cdots e^{A_{j-1}(D_j-\zeta)},\\ &\quad D_{i-1}\le x< D_i, \qquad D_{j-1}\le \zeta < D_j\end{aligned} \tag{7.46}$$

for any $i,j \in \{1,2,\ldots,n\}$ satisfying $i>j$, and

$$\Phi(x,\zeta) = e^{A_{i-1}(x-\zeta)}, \qquad D_{i-1}\le \zeta \le x \le D_i, \tag{7.47}$$

$i \in \{1,2,\ldots,n\}$, where we need to treat D_0 as 0.

In a few words, Φ can be seen as the state-transition matrix of the time-varying system $\dot{X}(t) = A(t)X(t)\ \forall t \ge 0$, where $A(t) \in \mathbb{R}^{n\times n}$ is a piecewise constant function defined by

$$A(t) = \begin{cases} 0_{n\times n}, & t \le D_1,\\ A_i, & D_i < t \le D_{i+1}, \quad i = 1,\ldots,n-1,\\ A_n, & t > D_n. \end{cases} \tag{7.48}$$

As will be shown in the next section, the predictor-based controller

$$\begin{aligned} U_i(t) = \frac{c}{s+c}\Bigg\{ & K_i^T \hat{\Phi}(D_i,0)G(t)\\ & + K_i^T\left(\sum_{j=1}^{i}\int_{t-D_j}^{t}\hat{\Phi}(D_i,\tau-t+D_j)\hat{H}_jU_j(\tau)d\tau\right.\\ & \left.+\sum_{j=i+1}^{n}\int_{t-D_i}^{t}\hat{\Phi}(D_i,\tau-t+D_i)\hat{H}_jU_j(\tau-(D_j-D_i))d\tau\right)\Bigg\} \end{aligned} \tag{7.49}$$

guarantees exponential stability for the closed-loop system, with $\hat{\Phi}$ defined as in (7.46) and (7.47) but replacing A_i in (7.45) by

$$\hat{A}_i(t) := \sum_{j=1}^{i}\hat{H}_j(t)K_j^T, \quad i \in \{0,1,2,\ldots,n\}, \tag{7.50}$$

the vector $G(t)$ given in (7.13) and $\hat{H}_j(t)$ being the columns of the Hessian estimate

$$\hat{H}(t) = (\hat{H}_1(t),\hat{H}_2(t),\ldots,\hat{H}_n(t)) \in \mathbb{R}^{n\times n}$$

given by (7.16). Hence, for ω sufficiently large and a_i sufficiently small, the average version of the predictor feedback form (7.49) can be numerically approximated by

$$U_i^{\rm av}(t) = \frac{c}{s+c}\left\{ K_i^T \Phi(D_i,0) G_{\rm av}(t) \right.$$
$$+ K_i^T \left(\sum_{j=1}^{i} \int_{t-D_j}^{t} \Phi(D_i, \tau - t + D_j) H_j U_j^{\rm av}(\tau) d\tau \right.$$
$$\left.\left. + \sum_{j=i+1}^{n} \int_{t-D_i}^{t} \Phi(D_i, \tau - t + D_i) H_j U_j^{\rm av}(\tau - (D_j - D_i)) d\tau \right) \right\}. \tag{7.51}$$

From (7.48), it is possible to show the term between braces in (7.51) by itself, if applied to (7.40), is enough to conclude that the closed-loop system $\dot{G}_{\rm av}(t) = A_n G_{\rm av}(t)$ is totally delay-compensated $\forall t \geq D_n$, since $A_n = HK$ is Hurwitz. However, due to technical limitations involving averaging results in infinite dimensions, we must include a low-pass filter $c/(s+c)$ in the predictor control loop, as was done in (7.49) and (7.51).

7.5 ▪ Stability Analysis for Gradient-Based ES under Time Delays

The availability of Lyapunov functionals for predictor feedback via backstepping transformation [119] permits the stability analysis of the complete feedback system under delays with a cascade representation of ODE–PDE equations and the infinite-dimensional control law.

The resulting exponential stability estimate in the L_2-norm of the closed-loop infinite-dimensional system is stated in the next theorem for Gradient ES subject to multiple and distinct input delays.

Theorem 7.1 (Multiple and Distinct Input Delays). *Consider the control system in Figure* 7.2 *with multiple and distinct input delays according to* (7.7)–(7.11) *and locally quadratic nonlinear map* (7.9). *There exists* $c^* > 0$ *such that,* $\forall c \geq c^*$, $\exists\, \omega^*(c) > 0$ *such that,* $\forall \omega > \omega^*$, *the closed-loop delayed system* (7.39) *and* (7.49) *with state* $\tilde{\theta}_i(t - D_i)$, $U_i(\tau)$, $\forall \tau \in [t - D_i, t]$ *and* $\forall i \in 1, 2, \ldots, n$, *has a unique locally exponentially stable periodic solution in* t *of period* Π, *denoted by* $\tilde{\theta}_i^{\Pi}(t - D_i)$, $U_i^{\Pi}(\tau)$, $\forall \tau \in [t - D_i, t]$ *and* $\forall i \in \{1, 2, \ldots, n\}$, *satisfying,* $\forall t \geq 0$,

$$\left(\sum_{i=1}^{n} \left[\tilde{\theta}_i^{\Pi}(t - D_i) \right]^2 + \left[U_i^{\Pi}(t) \right]^2 + \int_{t-D_i}^{t} \left[U_i^{\Pi}(\tau) \right]^2 d\tau \right)^{1/2} \leq \mathcal{O}(1/\omega). \tag{7.52}$$

Furthermore,

$$\limsup_{t \to +\infty} |\theta(t) - \theta^*| = \mathcal{O}(|a| + 1/\omega), \tag{7.53}$$
$$\limsup_{t \to +\infty} |y(t) - y^*| = \mathcal{O}(|a|^2 + 1/\omega^2), \tag{7.54}$$

where $a = [a_1\ a_2\ \cdots\ a_n]^T$.

Proof. The proof consists of the **Steps 1** to **7** below.

Step 1: *Transport PDE for Delay Representation*
Each individual delay D_i in (7.39) can be represented using a transport PDE as

$$\dot{\hat{\theta}}_i(t-D_i)=u_i(0,t), \tag{7.55}$$
$$\partial_t u_i(x,t)=\partial_x u_i(x,t), \quad x\in[0,D_i], \tag{7.56}$$
$$u_i(D_i,t)=U_i(t), \quad i=1,2,\ldots,n, \tag{7.57}$$

where the solution of (7.56)–(7.57) is

$$u_i(x,t)=U_i(t+x-D_i) \tag{7.58}$$

and $u(x,t)=[u_1(x,t),\ldots,u_n(x,t)]^T$ is the state of the total delay infinite-dimensional subsystem.

Step 2: *Average Model of the Closed-Loop System*
From (7.55)–(7.57), we can rewrite (7.40) as

$$\dot{G}_{\rm av}(t)=\sum_{i=1}^{n}H_i u_i^{\rm av}(0,t), \tag{7.59}$$
$$\partial_t u_i^{\rm av}(x,t)=\partial_x u_i^{\rm av}(x,t), \quad x\in[0,D_i], \tag{7.60}$$
$$u_i^{\rm av}(D_i,t)=U_i^{\rm av}(t), \quad i=1,2,\ldots,n, \tag{7.61}$$

where the solution of (7.60)–(7.61) is

$$u_i^{\rm av}(x,t)=U_i^{\rm av}(t+x-D_i) \tag{7.62}$$

and the PDE state is $u_{\rm av}(x,t)=[u_1^{\rm av}(x,t),\ldots,u_n^{\rm av}(x,t)]^T$.

By representing the integrand in (7.51) using the transport PDE state, one has the average control law

$$U_i^{\rm av}(t)=\frac{c}{s+c}\left\{K_i^T\left(\Phi(D_i,0)G_{\rm av}(t)\right.\right.$$
$$\left.\left.+\sum_{j=1}^{n}\int_0^{\phi_j(D_i)}\Phi(D_i,\sigma)H_j u_j^{\rm av}(\sigma,t)d\sigma\right)\right\}, \tag{7.63}$$

with $\phi_j:[0,D_n]\to[0,D_j]$, $j\in\{1,2,\ldots,n\}$ being the function defined by

$$\phi_j(x)=\begin{cases}x, & 0\le x\le D_j,\\ D_j, & D_j<x<D_n.\end{cases} \tag{7.64}$$

Finally, substituting (7.63) into (7.61), we have

$$\dot{G}_{\rm av}(t)=\sum_{i=1}^{n}H_i u_i^{\rm av}(0,t), \tag{7.65}$$
$$\partial_t u_i^{\rm av}(x,t)=\partial_x u_i^{\rm av}(x,t), \quad x\in[0,D_i], \tag{7.66}$$
$$\frac{d}{dt}u_i^{\rm av}(D_i,t)=-c\,u_i^{\rm av}(D_i,t)+cK_i^T\left(\Phi(D_i,0)G_{\rm av}(t)\right.$$
$$\left.+\sum_{j=1}^{n}\int_0^{\phi_j(D_i)}\Phi(D_i,\sigma)H_j u_j^{\rm av}(\sigma,t)d\sigma\right). \tag{7.67}$$

Step 3: *Successive Backstepping-Like Transformation, Its Inverse, and the Target System* Consider the infinite-dimensional backstepping-like transformation [226] of the delay state

$$w_i(x,t) = u_i^{\text{av}}(x,t) - K_i^T \left(\Phi(x,0)G_{\text{av}}(t) + \sum_{j=1}^{n} \int_0^{\phi_j(x)} \Phi(x,\sigma)H_j u_j^{\text{av}}(\sigma,t)d\sigma \right), \tag{7.68}$$

which maps the system (7.65)–(7.67) into the target system,

$$\dot{G}_{\text{av}}(t) = A_n G_{\text{av}}(t) + \sum_{i=1}^{n} H_i w_i(0,t), \tag{7.69}$$

$$\partial_t w_i(x,t) = \partial_x w_i(x,t) - \sum_{j=1}^{i-1} \lambda_{ij}(x) w_j(D_j,t), \quad x \in [0,D_i], \tag{7.70}$$

$$w_i(D_i,t) = -\frac{1}{c}\partial_t u_i^{\text{av}}(D_i,t), \quad i = 1,2,\ldots,n, \tag{7.71}$$

where $A_n = HK$ and the coefficients $\lambda_{ij} : [0,D_i] \to \mathbb{R}$ are

$$\lambda_{ij}(x) = \begin{cases} 0, & 0 \le x \le D_j, \\ K_i^T \Phi(x,D_j)H_j, & D_j < x \le D_i. \end{cases} \tag{7.72}$$

Note that the PDE for w_i is not a simple transport equation unless w_i vanishes at the right boundary. Using (7.68) for $x = D_i$ and the fact that $u_i^{\text{av}}(D_i,t) = U_i^{\text{av}}(t)$, we can directly obtain (7.67) and (7.63) from (7.71).

On the other hand, the inverse of (7.68) is given by

$$u_i^{\text{av}}(x,t) = w_i(x,t) + K_i^T \left(e^{A_n x} G_{\text{av}}(t) + \sum_{j=1}^{n} \int_0^{\phi_j(x)} e^{A_n(x-\sigma)} H_j w_j(\sigma,t)d\sigma \right). \tag{7.73}$$

For later use, now we find an expression for $\partial_t w_i(D_i,t)$. Differentiating (7.73) with respect to x on the interval $x \in (D_{i-1},D_i)$ gives

$$\begin{aligned} \partial_x u_i^{\text{av}}(x,t) &= \partial_x w_i(x,t) + \sum_{j=i}^{n} K_i^T H_j w_j(x,t) \\ &\quad + K_i^T A_n \left(e^{A_n x} G_{\text{av}}(t) + \sum_{j=1}^{n} \int_0^{\phi_j(x)} e^{A_n(x-\sigma)} H_j w_j(\sigma,t)dy \right). \end{aligned} \tag{7.74}$$

In light of (7.66)–(7.67) and (7.70)–(7.72), we arrive at

$$\begin{aligned} \partial_t w_i(D_i,t) &= -c w_i(D_i,t) - \sum_{j=1}^{i} K_i^T \Phi(D_i,D_j) H_j w_j(D_j,t) \\ &\quad - \sum_{j=i+1}^{n} K_i^T H_j w_j(D_i,t) - \gamma_i(0)^T G_{\text{av}}(t) \\ &\quad - \sum_{j=1}^{n} \int_0^{\phi_j(D_i)} \gamma_i(\sigma)^T H_j w_j(\sigma,t)d\sigma, \end{aligned} \tag{7.75}$$

where $\gamma_i(x) := e^{A_n^T(D_i - x)} A_n^T K_i$ for each $i \in \{1,2,\ldots,n\}$.

Note that the right-hand side contains $w_j(D_i,t)$ for each j greater than i, which is not a boundary value of w_j. For this reason, a key feature of the Lyapunov functional is the necessity of breaking the domain of integration for the terms $(1+x)w_i(x,t)^2$, as shown in the next step.

Step 4: *Lyapunov–Krasovskii Functional*

Let V be the candidate of Lyapunov function defined by

$$V(t) = G_{\text{av}}(t)^T P G_{\text{av}}(t) + \sum_{i=1}^{n}\sum_{j=1}^{i} \frac{\bar{a}_j}{2} \int_{D_{j-1}}^{D_j} (1+x)w_i(x,t)^2 dx + \frac{1}{2}\sum_{i=1}^{n} w_i(D_i,t)^2, \quad (7.76)$$

where $P = P^T \in \mathbb{R}^{n\times n}$ is the solution of the Lyapunov equation $PA_n + A_n^T P = -Q$ for some $Q = Q^T > 0$. The real constant $\bar{a}_1 > 0$ is determined later. The other constants $\bar{a}_2,\ldots,\bar{a}_n$ are arbitrary real numbers satisfying $\bar{a}_1 < \bar{a}_2 < \cdots < \bar{a}_n$. To shorten notation, we define a function $w : [0,1]\times[0,+\infty) \to \mathbb{R}^n$ by

$$w(\xi,t) = \begin{pmatrix} w_1(D_1\xi,t) & w_2(D_2\xi,t) & \cdots & w_n(D_n\xi,t) \end{pmatrix}^T \quad (7.77)$$

for $0 \le \xi \le 1$. In addition, we omit the dependence on the temporal variable t for simplicity. For instance, we write V and $w_i(x)$ instead of $V(t)$ and $w_i(x,t)$. Differentiating V with respect to t yields

$$\begin{aligned}
\dot{V} &= -G_{\text{av}}^T Q G_{\text{av}} + 2G_{\text{av}}^T P H w(0) \\
&\quad + \sum_{i=1}^{n}\sum_{j=1}^{i} \frac{\bar{a}_j}{2} \int_{D_{j-1}}^{D_j} (1+x)2w_i(x)\partial_t w_i(x)dx + w(1)^T \partial_t w(1) \\
&= -G_{\text{av}}^T Q G_{\text{av}} + 2G_{\text{av}}^T P H w(0) - \frac{\bar{a}_1}{2} w(0)^T w(0) \\
&\quad - \sum_{i=2}^{n}\sum_{j=1}^{i-1} \frac{\alpha_j}{2} w_i(D_j)^2 + w(1)^T \partial_t w(1) \\
&\quad - \sum_{i=1}^{n}\sum_{\ell=1}^{i-1}\sum_{j=\ell+1}^{i} \bar{a}_j w_\ell(D_\ell) \int_{D_{j-1}}^{D_j} (1+x) K_i^T \Phi(x,D_\ell) H_\ell w_i(x) dx \\
&\quad + \frac{1}{2} w(1)^T \Delta w(1) - \sum_{i=1}^{n}\sum_{j=i}^{n} \frac{\bar{a}_i}{2} \int_{D_{i-1}}^{D_i} w_j(x)^2 dx,
\end{aligned} \quad (7.78)$$

where $\alpha_j > 0$ and $\Delta \in \mathbb{R}^{n\times n}$ are defined by

$$\alpha_j = (\bar{a}_{j+1} - \bar{a}_j)(1+D_j), \quad j \in \{1,2,\ldots,n-1\}, \quad (7.79)$$

$$\Delta = \text{diag}\,\{\bar{a}_1(1+D_1), \bar{a}_2(1+D_2), \ldots, \bar{a}_n(1+D_n)\}. \quad (7.80)$$

In what follows we estimate the terms in each line of (7.78).

For the terms in the first line, we have

$$-G_{\text{av}} Q G_{\text{av}} + 2G_{\text{av}}^T P H w(0) - \frac{\bar{a}_1}{2} w(0)^T w(0) \le -G_{\text{av}} \left(Q - \frac{2}{\bar{a}_1} P H H^T P \right) G_{\text{av}}. \quad (7.81)$$

Setting $\bar{a}_1 = 4\lambda_{\max}(PHH^TP)/\lambda_{\min}(Q)$ leads to

$$-G_{\text{av}}QG_{\text{av}} + 2G_{\text{av}}^T PHw(0) - \frac{\bar{a}_1}{2}w(0)^T w(0) \leq -\frac{1}{2}G_{\text{av}}^T QG_{\text{av}}. \tag{7.82}$$

Considering the second line of (7.78), it follows from (7.75) that

$$\begin{aligned}
&-\sum_{i=2}^{n}\sum_{j=1}^{i-1}\frac{\alpha_j}{2}w_i(D_j)^2 + w(1)^T\partial_t w(1) \\
&= -cw(1)^T w(1) - w(1)^T Lw(1) \\
&\quad -\sum_{i=2}^{n}\sum_{j=1}^{i-1}\frac{\alpha_j}{2}\left(w_i(D_j)^2 + K_j^T H_i w_j(D_j) w_i(D_j)\right) \\
&\quad - w(1)^T\Gamma(0)^T G_{\text{av}} - \sum_{i=1}^{n} w_i(D_i)\sum_{j=1}^{n}\int_0^{\phi_j(D_i)}\gamma_i(\sigma)^T H_j w_j(\sigma)d\sigma,
\end{aligned} \tag{7.83}$$

where $L = (L_{ij})$ is the $n \times n$ lower triangular matrix whose (i,j)th entry L_{ij} is given by

$$L_{ij} = \begin{cases} 0, & i < j, \\ K_i^T\Phi(D_i, D_j)H_j. & i \geq j. \end{cases} \tag{7.84}$$

The matrix-valued function $\Gamma : [0, D_n] \to \mathbb{R}^{n\times n}$ is defined to be $\Gamma(\sigma) = (\gamma_1(\sigma), \gamma_2(\sigma), \ldots, \gamma_n(\sigma))$. By completing the square, we see that

$$\begin{aligned}
(7.83) &\leq -cw(1)^T w(1) + \frac{1}{4}G_{\text{av}}^T QG_{\text{av}} \\
&\quad -\frac{1}{2}w(1)\left(L + L^T + B - 2\Gamma(0)^T Q^{-1}\Gamma(0)\right)w(1) \\
&\quad -\sum_{i=1}^{n} w_i(D_i)\sum_{j=1}^{n}\int_0^{\phi_j(D_i)}\gamma_i(\sigma)^T H_j w_j(\sigma)d\sigma,
\end{aligned} \tag{7.85}$$

where $B \in \mathbb{R}^{n\times n}$ is the diagonal matrix whose ith diagonal entry is

$$\frac{1}{\alpha_i}K_i^T\left(\sum_{j=i+1}^{n} H_j H_j^T\right)K_i, \quad i = 1, 2, \ldots, n. \tag{7.86}$$

Substituting (7.82) and (7.85) into (7.78) leads to

$$\begin{aligned}
\dot{V} &\leq -\frac{1}{4}G_{\text{av}}^T QG_{\text{av}} - \frac{\bar{a}_1}{2}\sum_{i=1}^{n}\int_0^{D_i} w_i(x)^2 dx - cw(1)^T w(1) \\
&\quad -\frac{1}{2}w(1)^T\left(L + L^T + B + \Delta - 2\Gamma(0)^T Q^{-1}\Gamma(0)\right)w(1) \\
&\quad -\sum_{i=1}^{n} w_i(D_i)\sum_{j=1}^{n}\int_0^{\phi_j(D_i)}\gamma_i(\sigma)^T H_j w_j(\sigma)d\sigma \\
&\quad -\sum_{i=1}^{n}\sum_{\ell=1}^{i-1}\sum_{j=\ell+1}^{i}\bar{a}_j w_\ell(D_\ell)\int_{D_{j-1}}^{D_j}(1+x)K_i^T\Phi(x, D_\ell)H_\ell w_i(x)dx.
\end{aligned} \tag{7.87}$$

By using the Cauchy–Schwarz and Young's inequalities, we can show there exists a diagonal matrix $\Lambda \in \mathbb{R}^{n\times n}$ such that

$$\begin{aligned}
&-\sum_{i=1}^{n} w_i(D_i)\sum_{j=1}^{n}\int_0^{\phi_j(D_i)} \gamma_i(\sigma)^T H_j w_j(\sigma)d\sigma \\
&-\sum_{i=1}^{n}\sum_{\ell=1}^{i-1}\sum_{j=\ell+1}^{i} \bar{a}_j w_\ell(D_\ell)\int_{D_{j-1}}^{D_j}(1+x)K_i^T\Phi(x,D_\ell)H_\ell w_i(x)dx \\
&\leq \frac{1}{2}w(1)^T\Lambda w(1)+\frac{\bar{a}_1}{4}\sum_{i=1}^{n}\int_0^{D_i} w_i(x)^2dx. \qquad (7.88)
\end{aligned}$$

Then, we have

$$\dot{V} \leq -\frac{1}{4}G_{\rm av}^T Q G_{\rm av} - \frac{\bar{a}_1}{4}\sum_{i=1}^{n}\int_0^{D_i} w_i(x)^2dx - w(1)^T(cI_{n\times n}+R)w(1), \qquad (7.89)$$

where $R := \frac{1}{2}\left(L+L^T+B+\Delta-2\Gamma(0)^TQ^{-1}\Gamma(0)+\Lambda\right)$. Hence, if $c > \lambda_{\min}(R)$, there exists $\mu > 0$ such that

$$\dot{V} \leq -\mu V. \qquad (7.90)$$

Thus, the closed-loop system is exponentially stable in the sense of the full-state norm

$$\left(G_{\rm av}(t)^TG_{\rm av}(t)+\sum_{i=1}^{n}\int_0^{D_i} w_i(x,t)^2dx+w_i(D_i,t)^2\right)^{1/2}, \qquad (7.91)$$

i.e., in the transformed variable $(G_{\rm av}, w)$.

Step 5: *Exponential Stability Estimate (in the L_2-Norm) for the Average System* (7.65)–(7.67) To obtain exponential stability in the sense of the norm

$$\Upsilon(t) \triangleq \left(|G_{\rm av}(t)|^2+\sum_{i=1}^{n}\int_0^{D_i}[u_i^{\rm av}(x,t)]^2dx+[u_i^{\rm av}(D_i,t)]^2\right)^{1/2}, \qquad (7.92)$$

we must show there exist positive α_1 and α_2 such that

$$\alpha_1\Upsilon(t)^2 \leq V(t) \leq \alpha_2\Upsilon(t)^2. \qquad (7.93)$$

This is straightforward to establish by using (7.68), (7.73), (7.76) and employing the Cauchy–Schwarz inequality and other calculations, as in the proof of [119, Theorem 2.1].

Hence, with (7.90), we get

$$\begin{aligned}
&|G_{\rm av}(t)|^2+\sum_{i=1}^{n}\int_0^{D_i}[u_i^{\rm av}(x,t)]^2dx+[u_i^{\rm av}(D_i,t)]^2 \\
&\leq \frac{\alpha_2}{\alpha_1}e^{-\mu t}\left(|G_{\rm av}(0)|^2+\sum_{i=1}^{n}\int_0^{D_i}[u_i^{\rm av}(x,0)]^2dx+[u_i^{\rm av}(D_i,0)]^2\right), \qquad (7.94)
\end{aligned}$$

which completes the proof of exponential stability in the original variable $(G_{\rm av}, u_{\rm av})$.

Step 6: *Invoking the Averaging Theorem*
First, note that the closed-loop system (7.39) and (7.49) can be rewritten as

$$\dot{\tilde{\theta}}_i(t-D_i) = U_i(t-D_i), \quad i=1,\ldots,n, \tag{7.95}$$

$$\begin{aligned} U_i(t) = -cU_i(t) + cK_i^T \Bigg\{ & \hat{\Phi}(D_i,0)G(t) \\ & + \sum_{j=1}^{i} \int_{t-D_j}^{t} \hat{\Phi}(D_i,\tau-t+D_j)\hat{H}_j U_j(\tau)d\tau \\ & + \sum_{j=i+1}^{n} \int_{t-D_i}^{t} \hat{\Phi}(D_i,\tau-t+D_i)\hat{H}_j U_j(\tau-(D_j-D_i))d\tau \Bigg\}, \end{aligned} \tag{7.96}$$

where $\eta(t) = [\tilde{\theta}(t-D), U(t)]^T$ is the state vector. Moreover, from the definitions of $G(t)$ in (7.13) and $\hat{H}(t)$ in (7.16), one has

$$\dot{\eta}(t) = f(\omega t, \eta_t), \tag{7.97}$$

where $\eta_t(\Theta) = \eta(t+\Theta)$ for $-D_n \leq \Theta \leq 0$ and f is an appropriate continuous functional, such that the averaging theorem by [83, 131] in Appendix A can be directly applied considering $\omega = 1/\epsilon$.

From (7.94), the origin of the average closed-loop system (7.65)–(7.67) with transport PDE for delay representation is locally exponentially stable. Then, from (7.24) and (7.25), we can conclude the same results in the norm

$$\left(\sum_{i=1}^{n} \left[\tilde{\theta}_i^{\text{av}}(t-D_i) \right]^2 + \int_0^{D_i} [u_i^{\text{av}}(x,t)]^2 dx + [u_i^{\text{av}}(D_i,t)]^2 \right)^{1/2}$$

since H is nonsingular, i.e., $\left|\tilde{\theta}_i^{\text{av}}(t-D_i)\right| \leq |H^{-1}| \, |G_{\text{av}}(t)|$.

Thus, there exist positive constants α and β such that all solutions satisfy $\Psi(t) \leq \alpha e^{-\beta t}\Psi(0)$ $\forall t \geq 0$, where $\Psi(t) \triangleq \sum_{i=1}^{n} \left[\tilde{\theta}_i^{\text{av}}(t-D_i)\right]^2 + \int_0^{D_i} [u_i^{\text{av}}(x,t)]^2 \, dx + [u_i^{\text{av}}(D_i,t)]^2$ or, equivalently,

$$\Psi(t) \triangleq \sum_{i=1}^{n} \left[\tilde{\theta}_i^{\text{av}}(t-D_i) \right]^2 + \int_{t-D_i}^{t} [U_i^{\text{av}}(\tau)]^2 \, d\tau + [U_i^{\text{av}}(t)]^2, \tag{7.98}$$

using (7.62). Then, according to the averaging theorem by [83, 131] in Appendix A, for ω sufficiently large, (7.55)–(7.57) or (7.39) and (7.49) has a unique locally exponentially stable periodic solution around its equilibrium (origin) satisfying (7.52).

Step 7: *Asymptotic Convergence to the Extremum* (θ^*, y^*)
By using the change of variables $\tilde{\vartheta}_i(t) := \tilde{\theta}_i(t-D_i)$ and then integrating both sides of (7.55) within $[t, \sigma + D_i]$, we have

$$\tilde{\vartheta}_i(\sigma+D_i) = \tilde{\vartheta}_i(t) + \int_t^{\sigma+D_i} u_i(0,s)ds, \qquad i=1,\ldots,n. \tag{7.99}$$

From (7.58), we can rewrite (7.99) in terms of U, namely

$$\tilde{\vartheta}_i(\sigma+D_i) = \tilde{\vartheta}_i(t) + \int_{t-D_i}^{\sigma} U_i(\tau)d\tau. \tag{7.100}$$

Now, note that

$$\tilde{\theta}_i(\sigma) = \tilde{\vartheta}_i(\sigma + D_i) \quad \forall \sigma \in [t - D_i, t]. \tag{7.101}$$

Hence,

$$\tilde{\theta}_i(\sigma) = \tilde{\theta}_i(t - D_i) + \int_{t-D_i}^{\sigma} U_i(\tau)d\tau \quad \forall \sigma \in [t - D_i, t]. \tag{7.102}$$

Applying the supremum norm to both sides of (7.102), we have

$$\begin{aligned}
\sup_{t-D_i\leq\sigma\leq t} |\tilde{\theta}_i(\sigma)| &= \sup_{t-D_i\leq\sigma\leq t} |\tilde{\theta}_i(t-D_i)| + \sup_{t-D_i\leq\sigma\leq t} \left| \int_{t-D_i}^{\sigma} U_i(\tau)d\tau \right| \\
&\leq \sup_{t-D_i\leq\sigma\leq t} |\tilde{\theta}_i(t-D_i)| + \sup_{t-D_i\leq\sigma\leq t} \int_{t-D_i}^{t} |U_i(\tau)|\, d\tau \\
&\leq |\tilde{\theta}_i(t-D_i)| + \int_{t-D_i}^{t} |U_i(\tau)|\, d\tau \quad \text{(by Cauchy–Schwarz)} \\
&\leq |\tilde{\theta}_i(t-D_i)| + \left(\int_{t-D_i}^{t} d\tau \right)^{1/2} \times \left(\int_{t-D_i}^{t} |U_i(\tau)|^2 d\tau \right)^{1/2} \\
&\leq |\tilde{\theta}_i(t-D_i)| + \sqrt{D_i} \left(\int_{t-D_i}^{t} U_i^2(\tau)d\tau \right)^{1/2}.
\end{aligned} \tag{7.103}$$

Now, it is easy to check

$$|\tilde{\theta}_i(t - D_i)| \leq \left(|\tilde{\theta}_i(t - D_i)|^2 + \int_{t-D_i}^{t} U_i^2(\tau)d\tau \right)^{1/2}, \tag{7.104}$$

$$\left(\int_{t-D_i}^{t} U_i^2(\tau)d\tau \right)^{1/2} \leq \left(|\tilde{\theta}_i(t - D_i)|^2 + \int_{t-D_i}^{t} U_i^2(\tau)d\tau \right)^{1/2}. \tag{7.105}$$

By using (7.104) and (7.105), one has

$$|\tilde{\theta}_i(t - D_i)| + \sqrt{D_i} \left(\int_{t-D_i}^{t} U_i^2(\tau)d\tau \right)^{1/2} \leq (1 + \sqrt{D_i}) \left(|\tilde{\theta}_i(t - D_i)|^2 + \int_{t-D_i}^{t} U_i^2(\tau)d\tau \right)^{1/2}. \tag{7.106}$$

From (7.103), it is straightforward to conclude that

$$\sup_{t-D_i\leq\sigma\leq t} |\tilde{\theta}_i(\sigma)| \leq (1+\sqrt{D_i}) \left(|\tilde{\theta}_i(t-D_i)|^2 + \int_{t-D_i}^{t} U_i^2(\tau)d\tau \right)^{1/2} \tag{7.107}$$

and, consequently,

$$|\tilde{\theta}_i(t)| \leq (1+\sqrt{D_i}) \left(|\tilde{\theta}_i(t-D_i)|^2 + \int_{t-D_i}^{t} U_i^2(\tau)d\tau \right)^{1/2}. \tag{7.108}$$

In addition, inequality (7.108) can be given in terms of the periodic solution $\tilde{\theta}_i^{\Pi}(t-D_i)$, $U_i^{\Pi}(\tau)$ $\forall \tau \in [t-D_i, t]$ as follows:

$$|\tilde{\theta}_i(t)| \leq (1+\sqrt{D_i}) \left(\left| \tilde{\theta}_i(t-D_i) - \tilde{\theta}_i^{\Pi}(t-D_i) + \tilde{\theta}_i^{\Pi}(t-D_i) \right|^2 + \int_{t-D_i}^{t} \left[U_i(\tau) - U_i^{\Pi}(\tau) + U_i^{\Pi}(\tau) \right]^2 d\tau \right)^{1/2}. \tag{7.109}$$

By applying Young's inequality and some algebra, the RHS of (7.109) and $|\tilde{\theta}_i(t)|$ can be majorized by

$$|\tilde{\theta}_i(t)| \leq \sqrt{2}\,(1+\sqrt{D_i}) \left(\left| \tilde{\theta}_i(t-D_i) - \tilde{\theta}_i^{\Pi}(t-D_i) \right|^2 + \left| \tilde{\theta}_i^{\Pi}(t-D_i) \right|^2 + \int_{t-D_i}^{t} \left[U_i(\tau) - U_i^{\Pi}(\tau) \right]^2 d\tau + \int_{t-D_i}^{t} \left[U_i^{\Pi}(\tau) \right]^2 d\tau \right)^{1/2}. \tag{7.110}$$

According to the averaging theorem by [83, 131], we can conclude that the actual state converges exponentially to the periodic solution, i.e.,

$$\tilde{\theta}_i(t-D_i) - \tilde{\theta}_i^{\Pi}(t-D_i) \to 0$$

and

$$\int_{t-D_i}^{t} \left[U_i(\tau) - U_i^{\Pi}(\tau) \right]^2 d\tau \to 0.$$

Hence,

$$\limsup_{t\to+\infty} |\tilde{\theta}_i(t)| = \sqrt{2}\left(1+\sqrt{D_i}\right) \times \left(\left| \tilde{\theta}_i^{\Pi}(t-D_i) \right|^2 + \int_{t-D_i}^{t} [U_i^{\Pi}(\tau)]^2 d\tau \right)^{1/2}.$$

Then, from (7.52), we can write $\limsup_{t\to+\infty} |\tilde{\theta}(t)| = \mathcal{O}(1/\omega)$. From (7.12) and recalling that $\theta(t) = \hat{\theta}(t) + S(t)$ with $S(t)$ in (7.14), one has that $\theta(t) - \theta^* = \tilde{\theta}(t) + S(t)$. Since the first term in the right-hand side is ultimately of order $\mathcal{O}(1/\omega)$ and the second term is of order $\mathcal{O}(|a|)$, we state (7.53). From (7.10) and (7.53), we get (7.54). □

Corollary 7.1 (Gradient ES under Output Delays). *It is easy to show that the controller* (7.49) *becomes* (7.38) *in the case of output delays or equal inputs delays. Hence, the local stability/convergence results of the multiparameter Gradient ES in Figure* 7.1 *with delayed output* (7.10) *and* $D \geq 0$ *being simply a scalar can be directly stated for the closed-loop delayed system* (7.29) *and* (7.38) *from Theorem* 7.1.

7.6 ▪ Alternative Solution by Means of the Reduction Approach

In this section, multi-input systems with different time delays in each individual input channel are also dealt with. Unlike the previous sections considering multiparameter extremum seeking and delays, the stability analysis is carried *without* using backstepping transformation, which also eliminates the complexity of the controller. In a nutshell, a simpler implementation scheme and direct analysis without invoking successive backstepping transformation can be assured. As done before, the delays in this modified approach are independent of the dither frequency and system's dimension such that fast convergence rates are still guaranteed. A numerical example illustrates the performance of the new delay-compensated extremum seeking scheme and its simplicity.

7.6.1 ▪ Alternative Predictor Feedback for Gradient Extremum Seeking with Distinct Input Delays

In this section, $e_i \in \mathbb{R}^n$ stands for the ith column of the identity matrix $I_n \in \mathbb{R}^{n\times n}$ for each $i \in \{1,2,\dots,n\}$. Without loss of generality we assume that the inputs have distinct known (constant) delays which are ordered so that

$$D = \mathrm{diag}\{D_1, D_2, \dots, D_n\}, \quad 0 \le D_1 \le \cdots \le D_n. \tag{7.111}$$

Given an $\mathbb{R}^n$-valued signal f, the notation f^D denotes

$$f^D(t) = \begin{bmatrix} f_1(t-D_1) & f_2(t-D_2) & \dots & f_n(t-D_n) \end{bmatrix}^T. \tag{7.112}$$

Let $Q: \mathbb{R}^n \to \mathbb{R}$ be a convex static map with a maximum at $\theta^* \in \mathbb{R}$. We assume that the input-optimal parameter $\theta^* \in \mathbb{R}^n$ is unknown but the output of Q is available for the past input signal. More precisely, the measurable signal is

$$y(t) = Q(\theta^D(t)), \tag{7.113}$$

where $Q(\theta) = y^* + \frac{1}{2}(\theta-\theta^*)^T H(\theta-\theta^*)$, the extremum point is $y^* \in \mathbb{R}$, and $H = H^T < 0$ is the $n \times n$ unknown Hessian matrix of this static map.

The purpose of the extremum seeking is to estimate θ^* from the output y. To this end, define perturbation signals $S(t)$ and $M(t) \in \mathbb{R}^n$ by

$$S(t) = \begin{bmatrix} a_1 \sin(\omega_1(t+D_1)) & \cdots & a_n \sin(\omega_n(t+D_n)) \end{bmatrix}^T, \tag{7.114}$$

$$M(t) = \begin{bmatrix} \frac{2}{a_1}\sin(\omega_1 t) & \cdots & \frac{2}{a_n}\sin(\omega_n t) \end{bmatrix}^T. \tag{7.115}$$

The delayed signal S^D of S is a conventional perturbation signal. We also set the matrix-valued signal $N(t) \in \mathbb{R}^{n\times n}$ as

$$N_{ij}(t) = \begin{cases} \frac{16}{a_i^2}\left(\sin^2(\omega_i t) - \frac{1}{2}\right), & i = j, \\ \frac{4}{a_i a_j}\sin(\omega_i t)\sin(\omega_j t), & i \ne j. \end{cases} \tag{7.116}$$

The probing frequencies ω_i can be selected as

$$\omega_i = \omega_i' \omega = \mathcal{O}(\omega), \quad i \in 1,2,\dots,n, \tag{7.117}$$

where ω is a positive constant and ω_i' is a rational number. One possible choice is given in [73] as

$$\omega_i' \notin \left\{\omega_j',\ \frac{1}{2}(\omega_j' + \omega_k'),\ \omega_j' + 2\omega_k',\ \omega_j' + \omega_k' \pm \omega_l'\right\} \tag{7.118}$$

for all distinct i, j, k, and l.

By using the above signals, we develop an extremum seeking scheme in the presence of input delays. Let the input signal be constructed as

$$\theta(t) = \hat{\theta}(t) + S(t), \tag{7.119}$$

where $\hat{\theta}$ is an estimate of θ^*. Then, the corresponding output signal becomes

$$y(t) = Q(\theta^D(t)) = Q\left(\hat{\theta}^D(t) + S^D(t)\right). \tag{7.120}$$

We introduce the estimation error

$$\tilde{\theta}(t) := \hat{\theta}^D(t) - \theta^*. \tag{7.121}$$

Note that the error is defined with $\hat{\theta}^D$ rather than $\hat{\theta}$. With this error variable, the output signal $y(t)$ can be rewritten as

$$y(t) = Q\left(\theta^* + \tilde{\theta}(t) + S^D(t)\right). \tag{7.122}$$

To compensate the delays, we propose the following predictor-based update law:

$$\dot{\hat{\theta}}(t) = U(t), \tag{7.123}$$

$$\dot{U}(t) = -cU(t) + cK\left(M(t)y(t) + N(t)y(t)\sum_{i=1}^{n} e_i \int_{t-D_i}^{t} U_i(\tau)d\tau\right) \tag{7.124}$$

for some positive constant $c > 0$ and diagonal matrix $K \in \mathbb{R}^{n\times n}$ with positive entries. Since $\dot{\hat{\theta}}^D(t) = U^D(t)$, differentiating the error variable $\tilde{\theta}$ with respect to t yields

$$\dot{\tilde{\theta}}(t) = U^D(t) = \sum_{i=1}^{n} e_i U_i(t - D_i), \tag{7.125}$$

which is in a standard form of a system with input delays. As we will see later, the terms in the parentheses in the RHS of (7.124) correspond to a predicted value of $H\tilde{\theta}$ at some time in the future in the average sense.

The proposed alternative gradient-based ES under multiple input delays is shown in the block diagram of Figure 7.3.

7.6.2 ▪ Stability Analysis

The next theorem summarizes the stability/convergence properties of the closed-loop extremum seeking feedback. In particular, since the dynamic part is a simple integrator in (7.123), or even in (7.125), we will show the predictor feedback (7.124) does not require a backstepping transformation [119] in the stability analysis. Such an analysis for the integrator can be understood as a special case of the analysis in [94, 148], where finite-spectrum assignment is still preserved.

Theorem 7.2. *Consider the control system in Figure 7.3 with multiple and distinct input delays according to (7.111) and locally quadratic nonlinear map (7.113). There exists $c^* > 0$ such that, $\forall c \geq c^*$, $\exists\, \omega^*(c) > 0$ such that, $\forall \omega > \omega^*$, the closed-loop delayed system (7.124) and (7.125) with state $\tilde{\theta}_i(t-D_i)$, $U_i(\tau)$, $\forall \tau \in [t-D_i, t]$ and $\forall i \in 1,2,\ldots,n$, has a unique locally exponentially stable periodic solution in t of period Π, denoted by $\tilde{\theta}_i^{\Pi}(t-D_i)$, $U_i^{\Pi}(\tau)$, $\forall \tau \in [t-D_i,t]$ and $\forall i \in \{1,2,\ldots,n\}$, satisfying, $\forall t \geq 0$,*

$$\left(\sum_{i=1}^{n}\left[\tilde{\theta}_i^{\Pi}(t-D_i)\right]^2 + \left[U_i^{\Pi}(t)\right]^2 + \int_{t-D_i}^{t}\left[U_i^{\Pi}(\tau)\right]^2 d\tau\right)^{1/2} \leq \mathcal{O}(1/\omega). \tag{7.126}$$

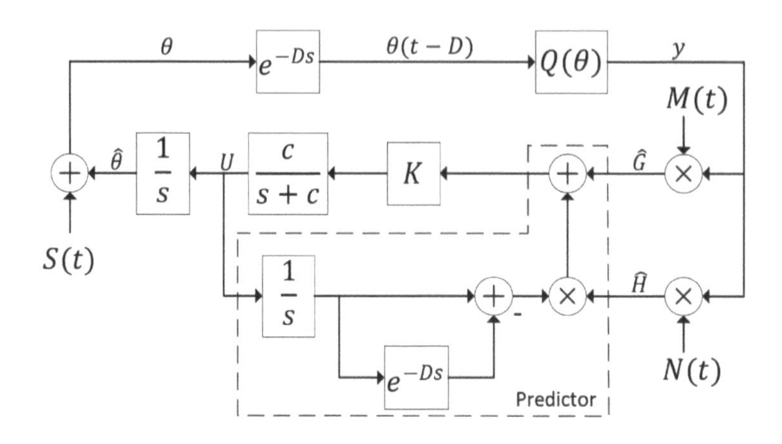

Figure 7.3. *Block diagram of the prediction scheme for multiple input-delay compensation. The (vector) signals $\hat{G}$ and $\hat{H}$ are the gradient and Hessian estimates, respectively. The multiple delays and gains are represented in the compact notation $D = \operatorname{diag}\{D_1, D_2\}$ and $K = \operatorname{diag}\{K_1, K_2\}$.*

Furthermore,

$$\limsup_{t\to+\infty} |\theta(t) - \theta^*| = \mathcal{O}(|a| + 1/\omega), \tag{7.127}$$

$$\limsup_{t\to+\infty} |y(t) - y^*| = \mathcal{O}(|a|^2 + 1/\omega^2), \tag{7.128}$$

where $a = [a_1\ a_2\ \cdots\ a_n]^T$.

Proof. A PDE representation of the closed-loop system (7.124) and (7.125) is given by

$$\dot{\tilde{\theta}}(t) = u(0,t), \tag{7.129}$$

$$u_t(x,t) = D^{-1}u_x(x,t), \quad x \in (0,1), \tag{7.130}$$

$$u(1,t) = U(t), \tag{7.131}$$

$$\dot{U}(t) = -cU(t) + cK\left(M(t)y(t) + N(t)y(t)\int_0^1 Du(x,t)dx\right), \tag{7.132}$$

where $u(x,t) = (u_1(x,t), u_2(x,t), \ldots, u_n(x,t))^T \in \mathbb{R}^n$. It is easy to see that the solution of (7.130) under the condition (7.131) is represented as

$$u_i(x,t) = U_i(D_i x + t - D_i) \tag{7.133}$$

for each $i \in \{1, 2, \ldots, n\}$. Hence, we have

$$\begin{aligned}\int_0^1 D_i u_i(x,t)dx &= \int_{t-D_i}^{t} u_i\left(\frac{\tau - t + D_i}{D_i}, t\right) d\tau \\ &= \int_{t-D_i}^{t} U_i(\tau)d\tau.\end{aligned} \tag{7.134}$$

This means that

$$\int_0^1 Du(x,t)dx = \sum_{i=1}^n e_i \int_0^1 D_i u_i(x,t)dx$$
$$= \sum_{i=1}^n e_i \int_{t-D_i}^t U_i(\tau)d\tau. \tag{7.135}$$

Thus, we can recover (7.125) from (7.130)–(7.132). The average system associated with (7.129)–(7.132) is given by

$$\dot{\tilde{\theta}}_{\mathrm{av}}(t) = u_{\mathrm{av}}(0,t), \tag{7.136}$$
$$u_{\mathrm{av,t}}(x,t) = D^{-1}u_{\mathrm{av,x}}(x,t), \quad x \in (0,1), \tag{7.137}$$
$$u_{\mathrm{av}}(1,t) = U_{\mathrm{av}}(t), \tag{7.138}$$
$$\dot{U}_{\mathrm{av}}(t) = -cU_{\mathrm{av}}(t) + cKH\left(\tilde{\theta}_{\mathrm{av}}(t) + \int_0^1 Du_{\mathrm{av}}(x,t)dx\right), \tag{7.139}$$

where we have used the fact that the averages of $M(t)y(t)$ and $N(t)y(t)$ are calculated as $H\tilde{\theta}_{\mathrm{av}}(t)$ and H, such as in (7.23) and (7.24). For simplicity of notation, let us introduce the following auxiliary variables:

$$\vartheta(t) := H\left(\tilde{\theta}_{\mathrm{av}}(t) + \int_0^1 Du_{\mathrm{av}}(x,t)dx\right), \tag{7.140}$$
$$\tilde{U} = U_{\mathrm{av}} - K\vartheta. \tag{7.141}$$

With this notation, (7.139) can be represented simply as $\dot{U}_{\mathrm{av}} = -c\tilde{U}$. In addition, differentiating (7.140) with respect to t yields

$$\dot{\vartheta}(t) = HU_{\mathrm{av}}(t). \tag{7.142}$$

We prove the exponential stability of the closed-loop system by using the Lyapunov functional defined by

$$V(t) = \vartheta(t)^T K\vartheta(t) + \frac{1}{4}\lambda_{\min}(-H)\int_0^1 \Big((1+x)u_{\mathrm{av}}(x,t)^T$$
$$\times Du_{\mathrm{av}}(x,t)dx\Big) + \frac{1}{2}\tilde{U}(t)^T(-H)\tilde{U}(t). \tag{7.143}$$

Recall that K and D are diagonal matrices with positive entries and that H is a negative-definite matrix. Hence, all of K, D, and $-H$ are positive-definite matrices. For simplicity of notation, we omit explicit dependence of variables on t. The time derivative of V is given by

$$\dot{V} = 2\vartheta^T KHU_{\mathrm{av}} + \frac{1}{2}\lambda_{\min}(-H)U_{\mathrm{av}}^T U_{\mathrm{av}}$$
$$-\frac{1}{4}\lambda_{\min}(-H)u(0)^T u(0) - \frac{1}{4}\lambda_{\min}(-H)$$
$$\times \int_0^1 u_{\mathrm{av}}(x)^T u_{\mathrm{av}}(x)dx + \tilde{U}^T(-H)\left(\dot{U}_{\mathrm{av}} - KHU_{\mathrm{av}}\right)$$
$$\leq 2\vartheta^T KHU_{\mathrm{av}} + \frac{1}{2}U_{\mathrm{av}}^T(-H)U_{\mathrm{av}} - \frac{1}{8D_{\max}}\lambda_{\min}(-H)$$
$$\times \int_0^1 (1+x)u_{\mathrm{av}}(x)^T Du_{\mathrm{av}}(x)dx$$
$$+\tilde{U}^T(-H)\dot{U}_{\mathrm{av}} + \tilde{U}^T(-H)K(-H)U_{\mathrm{av}}. \tag{7.144}$$

Applying Young's inequality to the last term leads to

$$\tilde{U}^T(-H)K(-H)U_{\mathrm{av}} \leq \frac{1}{2}\tilde{U}^T(-HKHKH)\tilde{U} + \frac{1}{2}U_{\mathrm{av}}^T(-H)U_{\mathrm{av}}. \tag{7.145}$$

Then, completing the square yields

$$\begin{aligned}\dot{V} &\leq \tilde{U}^T(-H)\tilde{U} - \vartheta^T K(-H)K\vartheta \\ &\quad - \frac{1}{8D_{\max}}\lambda_{\min}(-H)\int_0^1 (1+x)u_{\mathrm{av}}(x)^T D u_{\mathrm{av}}(x)dx \\ &\quad + \tilde{U}^T(-H)\dot{U}_{\mathrm{av}} + \frac{1}{2}\tilde{U}^T(-HKHKH)\tilde{U} \\ &\leq \tilde{U}^T(-H)\left(\dot{U}_{\mathrm{av}} + c^*\tilde{U}\right) - \vartheta^T K(-H)K\vartheta \\ &\quad - \frac{1}{8D_{\max}}\lambda_{\min}(-H)\int_0^1 (1+x)u_{\mathrm{av}}(x)^T D u_{\mathrm{av}}(x)dx,\end{aligned} \tag{7.146}$$

where $c^* := 1 + \lambda_{\max}(-HKHKH)/\lambda_{\min}(-H)$. Hence, by setting $\dot{U}_{\mathrm{av}} = -c\tilde{U}$ for some $c > c^*$, we see that there exists $\mu > 0$ such that

$$\dot{V} \leq -\mu V. \tag{7.147}$$

Finally, it is not difficult to find positive constants $\alpha, \beta > 0$ such that

$$\begin{aligned}\alpha\left(|\tilde{\theta}_{\mathrm{av}}(t)|^2 + \int_0^1 |u_{\mathrm{av}}(x,t)|^2 dx + |\tilde{U}(t)|^2\right) &\leq V(t) \\ &\leq \beta\left(|\tilde{\theta}_{\mathrm{av}}(t)|^2 + \int_0^1 |u_{\mathrm{av}}(x,t)|^2 dx + |\tilde{U}(t)|^2\right).\end{aligned} \tag{7.148}$$

Therefore, the average system (7.136)–(7.139) is exponentially stable as long as $c > c^*$.

The procedure above eliminates the application of backstepping transformations as in Steps 3 and 4 of the proof of Theorem 7.1, also highlighting the simplicity of the analysis carried out here. Basically, our closed-loop spectrum is finite but our non-backstepping analysis does not exploit a cascade structure of a finite-spectrum system in the target variables after a backstepping transformation, but instead studies a non-cascaded, namely a feedback system. The remaining steps of the proof follow closely the Steps 6 and 7 of Theorem 7.1, as discussed below.

First, noting that $\int_{t-D_i}^t U_i(\tau)d\tau = \int_{-D_i}^0 U_i(t+\tau)d\tau$, the closed-loop system (7.124)–(7.125) can be rewritten as

$$\dot{\eta}(t) = f(\omega t, \eta_t), \tag{7.149}$$

where $\eta(t) = [\tilde{\theta}(t), U(t)]^T$ is the state vector and $\eta_t(\Theta) = \eta(t+\Theta)$ for $-D_n \leq \Theta \leq 0$ and f is an appropriate continuous functional, such that the averaging theorem by [83, 131] in Appendix A can be directly applied considering $\omega = 1/\epsilon$.

From (7.147), the origin of the average closed-loop system (7.136)–(7.139) with transport PDE for delay representation is locally exponentially stable. Then, from (7.140) and (7.141), we can conclude the same results in the norm

$$\left(\sum_{i=1}^n \left[\tilde{\theta}_i^{\mathrm{av}}(t-D_i)\right]^2 + \int_0^{D_i} [u_i^{\mathrm{av}}(x,t)]^2 dx + [u_i^{\mathrm{av}}(D_i,t)]^2\right)^{1/2}$$

since H is nonsingular.

Thus, there exist positive constants α and β such that all solutions satisfy $\Psi(t) \leq \alpha e^{-\beta t}\Psi(0)$ $\forall t \geq 0$, where $\Psi(t) \triangleq \sum_{i=1}^{n}\left[\tilde{\theta}_i^{\rm av}(t-D_i)\right]^2 + \int_0^{D_i}[u_i^{\rm av}(x,t)]^2\,dx + [u_i^{\rm av}(D_i,t)]^2$ or, equivalently,

$$\Psi(t) \triangleq \sum_{i=1}^{n}\left[\tilde{\theta}_i^{\rm av}(t-D_i)\right]^2 + \int_{t-D_i}^{t}[U_i^{\rm av}(\tau)]^2\,d\tau + [U_i^{\rm av}(t)]^2, \tag{7.150}$$

using (7.133). Then, according to the averaging theorem by [83, 131] in Appendix A, for ω sufficiently large, (7.124)–(7.125) or, equivalently, (7.136)–(7.139) has a unique locally exponentially stable periodic solution around its equilibrium (origin) satisfying (7.126).

By using the change of variables $\tilde{\vartheta}_i(t) := \tilde{\theta}_i(t-D_i) = \hat{\theta}_i(t-D_i) - \theta_i^*$ and then integrating both sides of (7.129) within $[t, \sigma + D_i]$, we have

$$\tilde{\vartheta}_i(\sigma + D_i) = \tilde{\vartheta}_i(t) + \int_t^{\sigma + D_i} u_i(0,s)ds, \qquad i = 1, \ldots, n. \tag{7.151}$$

From this point we simply invoke the same arguments provided in **Step 7** to prove Theorem 7.1.

Then, from (7.126), we can write $\limsup_{t\to+\infty}|\tilde{\theta}(t)| = \mathcal{O}(1/\omega)$. From (7.121) and recalling that $\theta(t) = \hat{\theta}(t) + S(t)$ in (7.119) with $S(t)$ in (7.114), one has that $\theta(t) - \theta^* = \tilde{\theta}(t) + S(t)$. Since the first term in the right-hand side is ultimately of order $\mathcal{O}(1/\omega)$ and the second term is of order $\mathcal{O}(|a|)$, then we state (7.127). From (7.113) and (7.127), one can write $y - y^* = \frac{1}{2}(\theta-\theta^*)^T H(\theta-\theta^*)$ and

$$\begin{aligned}\limsup_{t\to+\infty}|y(t) - y^*| &= \mathcal{O}\left(|a| + 1/\omega\right)^2 \\ &= \mathcal{O}\left(|a|^2 + 2|a|/\omega + 1/\omega^2\right).\end{aligned} \tag{7.152}$$

Now, applying Young's inequality, we finally get (7.128). □

7.6.3 ▪ Simulation Results

In order to evaluate the multidimensional version of the delay-compensated extremum seeking, we consider the following static quadratic map:

$$Q(\theta) = 1 + \frac{1}{2}\left(2(\theta_1)^2 + 4(\theta_2 - 1)^2 + 4\theta_1(\theta_2 - 1)\right), \tag{7.153}$$

subject to an output delay of $D = \text{diag}\{35, 40\}$. The extremum point is $\theta^* = (0, 1)$ with $y^* = 1$, and the Hessian's map is

$$H = -\begin{pmatrix} 2 & 2 \\ 2 & 4 \end{pmatrix}. \tag{7.154}$$

For this case ($n = 2$), the predictor controller equation is given by

$$\begin{aligned}\dot{U}(t) = -cU(t) + c\begin{bmatrix} K_1 & 0 \\ 0 & K_2 \end{bmatrix}&\Bigg(M(t)y(t) + N(t)y(t) \\ &\times\left(\begin{bmatrix}1\\0\end{bmatrix}\int_{t-D_1}^{t}U_1(\tau)d\tau + \begin{bmatrix}0\\1\end{bmatrix}\int_{t-D_2}^{t}U_2(\tau)d\tau\right)\Bigg).\end{aligned} \tag{7.155}$$

Since $M(t)$ and $N(t)$ are

$$M(t)=\begin{bmatrix} M_1(t) \\ M_2(t) \end{bmatrix}=\begin{bmatrix} \frac{2}{a_1}\sin(\omega_1 t) \\ \frac{2}{a_2}\sin(\omega_2 t) \end{bmatrix}, \tag{7.156}$$

$$N(t)=\begin{bmatrix} N_{11}(t) & N_{12}(t) \\ N_{21}(t) & N_{22}(t) \end{bmatrix}=\begin{bmatrix} \frac{16}{a_1^2}\left(\sin^2(\omega_1 t)-\frac{1}{2}\right) & \frac{4}{a_1 a_2}\sin(\omega_1 t)\sin(\omega_2 t) \\ \frac{4}{a_2 a_1}\sin(\omega_2 t)\sin(\omega_1 t) & \frac{16}{a_2^2}\left(\sin^2(\omega_2 t)-\frac{1}{2}\right) \end{bmatrix}, \tag{7.157}$$

the predictor equation (7.155) can be written as $\dot{U_1}(t)$ and $\dot{U_2}(t)$ in the following form:

$$\dot{U_1}(t)=-cU_1(t)+cK_1\Big(M_1(t)y(t)+y(t)\Big(N_{11}(t) \times\int_{t-D_1}^{t}U_1(\tau)d\tau+N_{12}(t)\int_{t-D_2}^{t}U_2(\tau)d\tau\Big)\Big), \tag{7.158}$$

$$\dot{U_2}(t)=-cU_2(t)+cK_2\Big(M_2(t)y(t)+y(t)\Big(N_{21}(t) \times\int_{t-D_1}^{t}U_1(\tau)d\tau+N_{22}(t)\int_{t-D_2}^{t}U_2(\tau)d\tau\Big)\Big). \tag{7.159}$$

For this $n=2$ example, $K=\text{diag}\{K_1,K_2\}$. The predictor feedback is implemented according to (7.158) and (7.159), the additive dither is $S(t)=\left[a_1\sin(\omega_1(t+D_1)) \;\; a_2\sin(\omega_2(t+D_2))\right]^T$, and $M(t)$ and $N(t)$ are given by (7.156) and (7.157), respectively. For the sake of comparison, note that the expressions for $U_1(t)$ and $U_2(t)$ in (7.158) and (7.159) are much simpler than the control laws proposed in (7.49)—see (7.223) and (7.224) in Section 7.10.

In our simulations, we use low-pass and washout filters with corner frequencies ω_h and ω_l to improve the controller performance as usual in extremum seeking designs—see [73, Figure 4]. In what follows, we present numerical simulations of the predictor (7.155). We performed our tests with $c=20$, $K_1=\frac{1}{100}$, $K_2=\frac{1}{200}$, $a_1=a_2=0.05$, $\omega=0.5$, $\omega_1=17.5\omega$, $\omega_2=12.5\omega$, $\omega_h=\omega_l=\frac{\omega}{5}$, and $\hat{\theta}(0)=[0,1]^T$.

Figure 7.4 shows the system output $y(t)$ in 3 situations: (a) free of output delays, (b) in the presence of multiple and distinct input delays but without any delay compensation, and (c) with multiple and distinct input delays and predictor-based compensation.

Figure 7.5 presents relevant variables for ES control: (a) the time response of the parameter $\theta(t)$, (b) the control signal $U(t)$, and (c) the Hessian's estimate $\hat{H}=N(t)y(t)$. It is clear the remarkable evolution of the new prediction scheme in searching the maximum and the Hessian's parameters of H^{-1}. This ultimate exact perturbation-based estimation allows the perfect delay compensation.

Remark 7.1. An alternative Gradient extremum seeking controller was developed for multiparameter real-time optimization in the presence of distinct actuator delays. The control scheme introduced here for delay compensation used prediction feedback with perturbation-based estimate of the Hessian associated with an adequate tune of the dither signals. This predictor feedback is much simpler than the previous one presented in Section 7.4; see also [179]. In addition, and unlike other approaches based on sequential predictors [146], there was NO price to be paid for this simplification in terms of restricting the delay duration to increase the convergence rates. On the contrary, our generalization for multi-input-single-output maps performs prediction of the cross-coupling of the channels for arbitrarily long (distinct) input delays without affecting its convergence speed. A further advance in this challenging scenario was that the contributions were achieved without invoking the backstepping methodology usually employed in the publications concerning extremum seeking under delays.

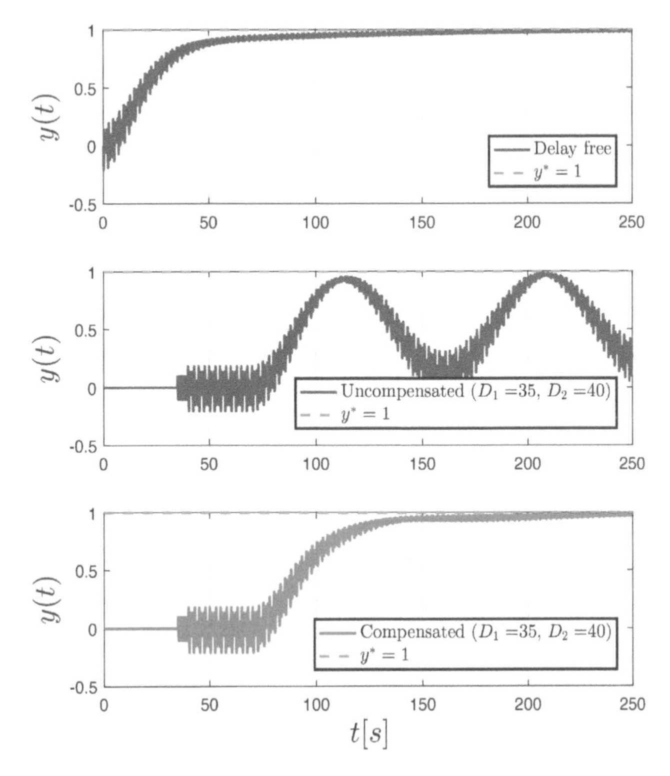

Figure 7.4. *Alternative gradient-based ES under multiple input delays (time response of $y(t)$:* (a) *basic ES works well without delays;* (b) *ES goes unstable in the presence of delays ($D_1 = 35$, $D_2 = 40$);* (c) *predictor fixes this.*

7.7 ▪ Newton-Based ES with Output Delays

The convergence rate of the Gradient algorithms introduced above are severely dependent on the parameters K and H according to (7.37), for instance. Since the elements of the diagonal matrix K are of order $\mathcal{O}(1)$, as otherwise the averaging analysis would fail for K with elements arbitrarily large, then the speed of the response is governed by the unknown Hessian matrix H. Once more, we start to consider the case where $D \geq 0$ is simply a scalar and in Section 7.8 we move on to the more involved case of multiple and distinct input delays.

7.7.1 ▪ Hessian's Inverse Estimation

In [73], the authors presented a multivariable version of the Newton ES algorithm (free of delays), which also ensures that its convergence rate can be user-assignable, rather than being dependent on the Hessian of the static map. In Figure 7.6, we introduce our generalization of such a Newton-based ES in the presence of output delays.

As proved in [73], the Riccati differential equation

$$\dot{\Gamma} = \omega_r \Gamma - \omega_r \Gamma \hat{H} \Gamma, \qquad (7.160)$$

where $\omega_r > 0$ is a design constant, generates an estimate of the Hessian's inverse, avoiding inversions of the Hessian estimates that may be zero during the transient phase. The estimation

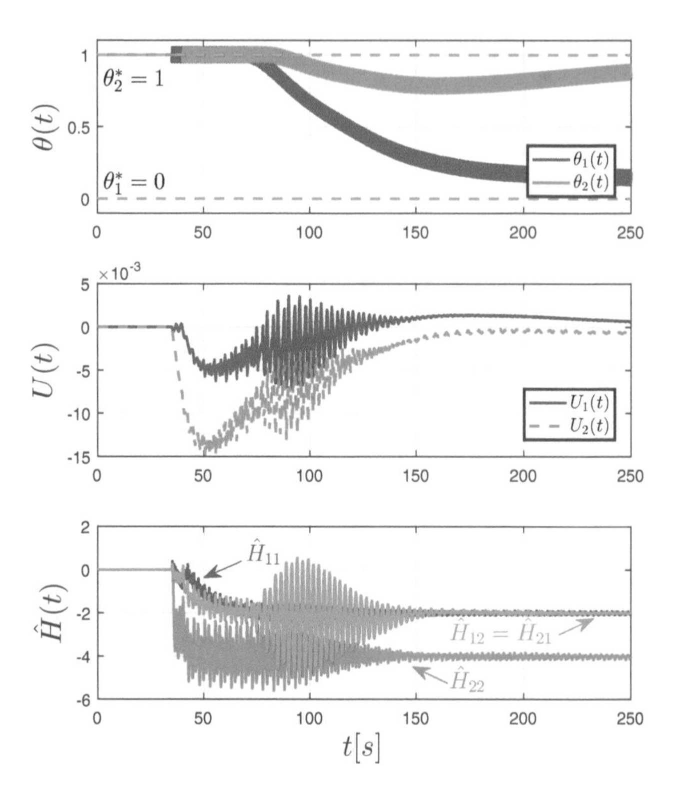

Figure 7.5. *Alternative gradient-based ES under multiple input delays:* (a) *parameter* $\theta(t)$*;* (b) *the control signal* $U(t)$*;* (c) *Hessian's estimate* $\hat{H}(t)$*. The elements of* $\hat{H}(t)$ *converge to the unknown elements of* H*.*

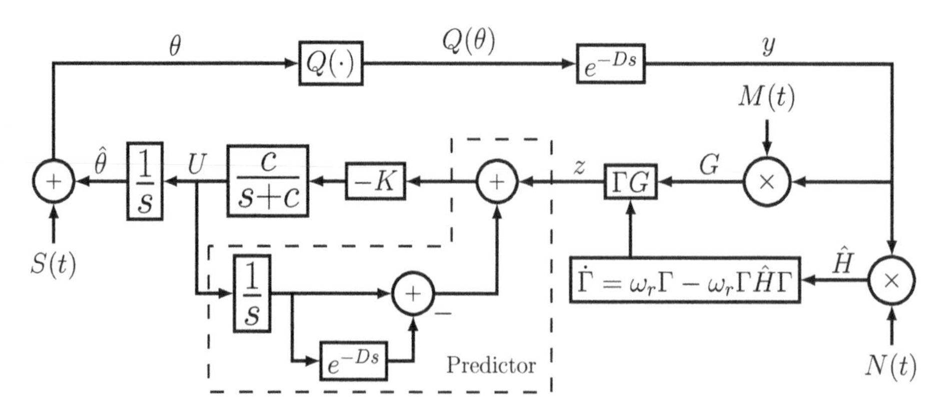

Figure 7.6. *Block diagram of the basic prediction scheme for output-delay compensation in multivariable Newton-based ES, where* $D \geq 0$ *is a simple scalar. The predictor feedback with a perturbation-based estimate of the Hessian's inverse obeys* (7.177)*. The dither and demodulation vector signals are given by* $S(t) = [a_1 \sin(\omega_1(t+D)) \; \cdots \; a_n \sin(\omega_n(t+D))]^T$ *and* $M(t) = \left[\frac{2}{a_1}\sin(\omega_1 t) \; \cdots \; \frac{2}{a_n}\sin(\omega_n t)\right]^T$*. The demodulating signal* $N(t)$ *is computed by* (7.17)–(7.18)*.*

error of the Hessian's inverse can be defined as

$$\tilde{\Gamma}(t) = \Gamma(t) - H^{-1}, \tag{7.161}$$

and its dynamic equation can be written from (7.160)–(7.161) by

$$\dot{\tilde{\Gamma}} = \omega_r[\tilde{\Gamma} + H^{-1}] \times [I_{n\times n} - \hat{H}(\tilde{\Gamma} + H^{-1})]. \tag{7.162}$$

For the quadratic map (7.9), and in the absence of the prediction action, it is easy to conclude that the average model in the error variables $\tilde{\theta}$ and $\tilde{\Gamma}$ would be

$$\frac{d\tilde{\theta}_{\rm av}(t)}{dt} = -K\tilde{\theta}_{\rm av}(t-D) - K\tilde{\Gamma}_{\rm av}(t)H\tilde{\theta}_{\rm av}(t-D), \tag{7.163}$$

$$\frac{d\tilde{\Gamma}_{\rm av}(t)}{dt} = -\omega_r\tilde{\Gamma}_{\rm av}(t) - \omega_r\tilde{\Gamma}_{\rm av}(t)H\tilde{\Gamma}_{\rm av}(t). \tag{7.164}$$

From (7.163), once again it is clear the importance of using a delay compensation strategy. Differently from (7.38) and (7.49), this time we introduce a prediction scheme using (7.160) to estimate H^{-1}.

7.7.2 ▪ Predictor Feedback via Hessian's Inverse Estimation

First of all, let us define the measurable vector signal

$$z(t) = \Gamma(t)G(t). \tag{7.165}$$

By using the averaging analysis, we can verify from (7.13) and (7.165) that

$$z_{\rm av}(t) = \frac{1}{\Pi}\int_0^{\Pi} \Gamma M(\lambda)y d\lambda = \Gamma_{\rm av}(t)H\tilde{\theta}_{\rm av}(t-D). \tag{7.166}$$

From (7.161), equation (7.166) can be written in terms of $\tilde{\Gamma}_{\rm av}(t) = \Gamma_{\rm av}(t) - H^{-1}$ as

$$z_{\rm av}(t) = \tilde{\theta}_{\rm av}(t-D) + \tilde{\Gamma}_{\rm av}(t)H\tilde{\theta}_{\rm av}(t-D). \tag{7.167}$$

The second term in the RHS of (7.167) is quadratic in $(\tilde{\Gamma}_{\rm av},\ \tilde{\theta}_{\rm av})$; thus, the linearization of $\Gamma_{\rm av}(t)$ at H^{-1} and $\tilde{\theta}_{\rm av}(t)$ at zero results in the linearized version of (7.166) given by

$$z_{\rm av}(t) = \tilde{\theta}_{\rm av}(t-D). \tag{7.168}$$

From Figure 7.6 and (7.12), we can repeat the error dynamics (7.29),

$$\dot{\tilde{\theta}}(t-D) = U(t-D), \tag{7.169}$$

and obtain the following average models by using (7.168)–(7.169):

$$\dot{\tilde{\theta}}_{\rm av}(t-D) = U_{\rm av}(t-D), \tag{7.170}$$

$$\dot{z}_{\rm av}(t) = U_{\rm av}(t-D), \tag{7.171}$$

where $U_{\rm av} \in \mathbb{R}^n$ is the resulting average control for $U \in \mathbb{R}^n$.

In order to motivate the predictor feedback design, the idea again is to compensate for the delay by feeding back the future state $z(t+D)$, or $z_{\rm av}(t+D)$ in the equivalent average system. To obtain it with the variation-of-constants formula to (7.171), the future state is written as

$$z_{\rm av}(t+D) = z_{\rm av}(t) + \int_{t-D}^{t} U_{\rm av}(\tau)d\tau \tag{7.172}$$

in terms of the average control signal $U_{\text{av}}(\tau)$ from the past window $[t-D,t]$. Given the same diagonal matrix $K>0$ used before, the average control would be given by

$$U_{\text{av}}(t) = -K\left[z_{\text{av}}(t) + \int_{t-D}^{t} U_{\text{av}}(\tau)d\tau\right], \tag{7.173}$$

resulting in the average control feedback

$$U_{\text{av}}(t) = -Kz_{\text{av}}(t+D) \quad \forall t \geq 0, \tag{7.174}$$

as desired. Hence, the average system would be, $\forall t \geq D$,

$$\frac{d\tilde{\theta}_{\text{av}}(t)}{dt} = -K\tilde{\theta}_{\text{av}}(t) - K\tilde{\Gamma}_{\text{av}}(t+D)H\tilde{\theta}_{\text{av}}(t), \tag{7.175}$$

$$\frac{d\tilde{\Gamma}_{\text{av}}(t)}{dt} = -\omega_r\tilde{\Gamma}_{\text{av}}(t) - \omega_r\tilde{\Gamma}_{\text{av}}(t)H\tilde{\Gamma}_{\text{av}}(t). \tag{7.176}$$

Since $K\tilde{\Gamma}_{\text{av}}H\tilde{\theta}_{\text{av}}$ is quadratic in $(\tilde{\Gamma}_{\text{av}},\tilde{\theta}_{\text{av}})$ and $\omega_r\tilde{\Gamma}_{\text{av}}H\tilde{\Gamma}_{\text{av}}$ is quadratic in $\tilde{\Gamma}_{\text{av}}$, the linearization of the system (7.175)–(7.176) has all its eigenvalues determined by $-K$ and $-\omega_r$. Therefore, the (local) exponential stability of the algorithm could be guaranteed with a convergence rate which is independent of the unknown Hessian H, being user-assignable.

As was done for the gradient case, we propose a predictor feedback in the form of a low-pass filtering [115] of the nonaverage version of (7.173), given by

$$U(t) = \frac{c}{s+c}\left\{-K\left[z(t) + \int_{t-D}^{t} U(\tau)d\tau\right]\right\}, \tag{7.177}$$

where $c>0$ is sufficiently large. Recall that the low-pass filtering is particularly required in the stability analysis when the averaging theorem in infinite dimensions [83, 131] is invoked—see Appendix A.

The predictor feedback control (7.177) is averaging-based because z in (7.165) is updated according to the estimate $\Gamma(t)$ for the unknown Hessian's inverse H^{-1} given by (7.160), with $\hat{H}(t)$ in (7.16) satisfying the averaging property (7.21).

7.8 ▪ Newton-Based ES with Multiple and Distinct Input Delays

From Figure 7.7 and (7.12), we can write

$$\dot{\tilde{\theta}}(t-D) = \begin{bmatrix} U_1(t-D_1) \\ U_2(t-D_2) \\ \vdots \\ U_n(t-D_n) \end{bmatrix}, \quad \dot{\tilde{\theta}}_i(t-D_i) = U_i(t-D_i), \tag{7.178}$$

and its average model

$$\dot{\tilde{\theta}}_i^{\text{av}}(t-D_i) = U_i^{\text{av}}(t-D_i). \tag{7.179}$$

Analogously to the averaging steps performed before to obtain (7.166)–(7.168), we still verify that

$$z_{\text{av}}(t) = \Gamma_{\text{av}}(t)H\tilde{\theta}_{\text{av}}(t-D), \tag{7.180}$$

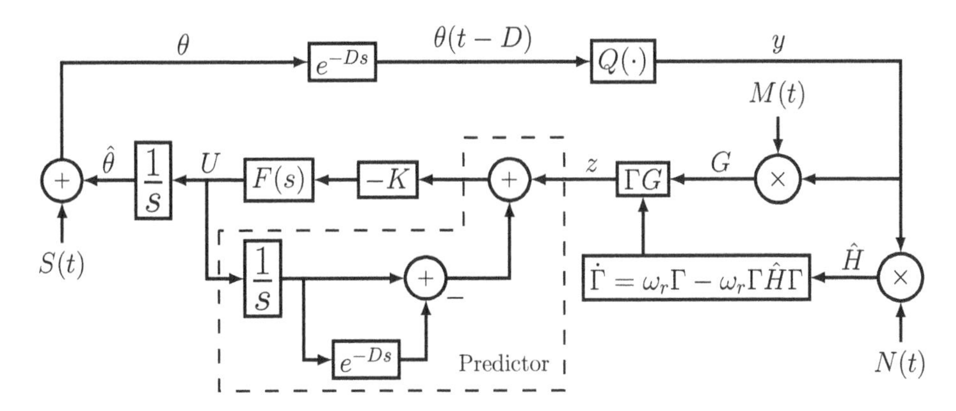

Figure 7.7. *Block diagram of the Newton-based ES control for multiple input-delay compensation, where $D = diag\{D_1,\dots,D_n\}$. The predictor feedback is implemented according to* (7.186) *with $F(s) = diag\{\frac{c_1}{s+c_1},\dots,\frac{c_n}{(s+c_n)}\}$ and $K = diag\{k_1,\dots,k_n\}$. The additive dither signal is modified to be $S(t) = [a_1\sin(\omega_1(t+D_1)) \ \cdots \ a_n\sin(\omega_n(t+D_n))]^T$ and the demodulating signals $M(t)$ and $N(t)$ are defined as in standard Newton-based ES* [73], *i.e., given by $M(t) = \left[\frac{2}{a_1}\sin(\omega_1 t) \ \cdots \ \frac{2}{a_n}\sin(\omega_n t)\right]^T$ and* (7.17)–(7.18), *respectively.*

and its linearized version given by

$$z_{\text{av}}(t) = \tilde{\theta}_{\text{av}}(t-D), \tag{7.181}$$

with

$$\tilde{\theta}_{\text{av}}(t-D) = [\tilde{\theta}_1^{\text{av}}(t-D_1) \ \cdots \ \tilde{\theta}_n^{\text{av}}(t-D_n)]^T. \tag{7.182}$$

Thus, from (7.179) and (7.181), the following average model with state $z_{\text{av}}(t)$ can be obtained:

$$\dot{z}_{\text{av}}(t) = \begin{bmatrix} \dot{z}_1^{\text{av}}(t) \\ \dot{z}_2^{\text{av}}(t) \\ \vdots \\ \dot{z}_n^{\text{av}}(t) \end{bmatrix} = \begin{bmatrix} U_1^{\text{av}}(t-D_1) \\ U_2^{\text{av}}(t-D_2) \\ \vdots \\ U_n^{\text{av}}(t-D_n) \end{bmatrix}. \tag{7.183}$$

Since the input channels in system (7.183) are totally decoupled, we can apply a predictor of the form

$$U_i^{\text{av}}(t) = \frac{c_i}{s+c_i}\left\{-k_i\left[z_i^{\text{av}}(t) + \int_{t-D_i}^{t} U_i^{\text{av}}(\tau)d\tau\right]\right\} \tag{7.184}$$

to each individual subsystem in (7.183) in order to fully stabilize it. Motivated by the control law (7.184) for the average model (7.183), we propose the following predictor feedback to compensate the delays in the nonaverage model (7.178):

$$U_i(t) = \frac{c_i}{s+c_i}\left\{-k_i\left[z_i(t) + \int_{t-D_i}^{t} U_i(\tau)d\tau\right]\right\} \tag{7.185}$$

for all $i = 1,2,\dots,n$, which in the vector form is given by

$$U(t) = F(s)\left\{-Kz(t) - K\begin{bmatrix} \int_{t-D_1}^{t} U_1(\tau)d\tau \\ \int_{t-D_2}^{t} U_2(\tau)d\tau \\ \vdots \\ \int_{t-D_n}^{t} U_n(\tau)d\tau \end{bmatrix}\right\}, \tag{7.186}$$

with $K=\text{diag}\{k_1,k_2,\ldots,k_n\}\in\mathbb{R}^{n\times n}$ being the same positive diagonal matrix ($k_i>0\ \forall i=1,\ldots,n$) assumed before and the filter $F(s)=\text{diag}\left\{\frac{c_1}{s+c_1},\ldots,\frac{c_n}{(s+c_n)}\right\}$.

7.9 ▪ Stability Analysis for Newton-Based ES under Time Delays

Due to its decoupling property, (7.183) can be represented as a set of n cascades of first-order ODE plus PDE equations. Exponential stability of the closed-loop infinite-dimensional system can be guaranteed according to the following theorem for Newton ES under multiple input and distinct delays.

Theorem 7.3 (Multiple and Distinct Input Delays). *Consider the control system in Figure* 7.7 *with multiple and distinct input delays* (7.7)–(7.11) *and locally quadratic nonlinear map* (7.9). *There exists $c^*>0$ such that, $\forall c_i\geq c^*$, $\exists\ \omega^*(c_i)>0$ such that, $\forall\omega>\omega^*$, the closed-loop delayed system* (7.162), (7.178), *and* (7.185) *with state $\tilde{\Gamma}(t)$, $\tilde{\theta}_i(t-D_i)$, $U_i(\tau)$, $\forall\tau\in[t-D_i,t]$ and $\forall i\in\{1,2,\ldots,n\}$, has a unique locally exponentially stable periodic solution in t of period Π, denoted by $\tilde{\Gamma}^{\Pi}(t)$, $\tilde{\theta}_i^{\Pi}(t-D)$, $U_i^{\Pi}(\tau)$, $\forall\tau\in[t-D_i,t]$, satisfying, $\forall t\geq 0$,*

$$\left(\left|\tilde{\Gamma}^{\Pi}(t)\right|^2+\sum_{i=1}^{n}\left[\tilde{\theta}_i^{\Pi}(t-D_i)\right]^2+\left[U_i^{\Pi}(t)\right]^2+\int_{t-D_i}^{t}\left[U_i^{\Pi}(\tau)\right]^2d\tau\right)^{1/2}\leq\mathcal{O}(1/\omega). \quad (7.187)$$

Furthermore,

$$\limsup_{t\to+\infty}|\theta(t)-\theta^*|=\mathcal{O}(|a|+1/\omega), \quad (7.188)$$

$$\limsup_{t\to+\infty}|y(t)-y^*|=\mathcal{O}(|a|^2+1/\omega^2). \quad (7.189)$$

where $a=[a_1\ a_2\ \cdots\ a_n]^T$.

Proof. The proof is organized in order to follow the steps presented in the proof of Theorem 7.1. In the following we highlight the main differences.

Step 1: *PDE Representation for the Closed-Loop System*
After representing the delay in (7.178) using a transport PDE as was done in (7.55)–(7.57) and representing the integrand in (7.185) using the transport PDE state, one has

$$\dot{\tilde{\theta}}_i(t-D_i)=u_i(0,t), \quad (7.190)$$

$$\partial_t u_i(x,t)=\partial_x u_i(x,t),\quad x\in[0,D_i], \quad (7.191)$$

$$u(D_i,t)=\frac{c_i}{s+c_i}\left\{-k_i\left[z_i(t)+\int_0^{D_i}u_i(\sigma,t)d\sigma\right]\right\}. \quad (7.192)$$

Step 2: *Average Model of the Closed-Loop System*
From (7.181) and denoting

$$\tilde{\vartheta}(t)=\tilde{\theta}(t-D), \quad (7.193)$$

we have $\tilde{\vartheta}_{\text{av}}(t)=z_{\text{av}}(t)=\tilde{\theta}_{\text{av}}(t-D)$ and the linearized average version of system (7.190)–

(7.192) can be obtained:

$$\dot{\tilde{\vartheta}}_i^{\rm av}(t) = u_i^{\rm av}(0,t), \tag{7.194}$$

$$\partial_t u_i^{\rm av}(x,t) = \partial_x u_i^{\rm av}(x,t), \quad x \in [0, D_i], \tag{7.195}$$

$$\frac{d}{dt} u_i^{\rm av}(D,t) = -c_i\, u_i^{\rm av}(D_i,t) - c_i\, k_i \left[\tilde{\vartheta}_i^{\rm av}(t) + \int_0^{D_i} u_i^{\rm av}(\sigma,t)d\sigma\right], \tag{7.196}$$

where the solution of the transport PDE (7.195)–(7.196) is

$$u_i^{\rm av}(x,t) = U_i^{\rm av}(t+x-D_i). \tag{7.197}$$

On the other hand, by using (7.21), the average model for the Hessian's inverse estimation error in (7.162) can be deduced as in (7.176). Then, its linearized version at $\tilde{\Gamma}_{\rm av} = 0$ is given by

$$\frac{d\tilde{\Gamma}_{\rm av}(t)}{dt} = -\omega_r \tilde{\Gamma}_{\rm av}(t). \tag{7.198}$$

Step 3: *Standard Backstepping Transformation, Its Inverse, and the Target System*
Consider the infinite-dimensional backstepping transformation of the delay state

$$w_i(x,t) = u_i^{\rm av}(x,t) + k_i \left(\tilde{\vartheta}_i^{\rm av}(t) + \int_0^x u_i^{\rm av}(\sigma,t)d\sigma\right), \tag{7.199}$$

which maps the system (7.194)–(7.196) into the target system:

$$\dot{\tilde{\vartheta}}_i^{\rm av}(t) = -k_i \tilde{\vartheta}_i^{\rm av}(t) + w_i(0,t), \tag{7.200}$$

$$\partial_t w_i(x,t) = \partial_x w_i(x,t), \quad x \in [0, D_i], \tag{7.201}$$

$$w_i(D_i,t) = -\frac{1}{c_i}\partial_t u_i^{\rm av}(D_i,t), \quad i = 1,2,\ldots,n. \tag{7.202}$$

Using (7.199) for $x = D_i$ and the fact that $u_i^{\rm av}(D_i,t) = U_i^{\rm av}(t)$, from (7.202) we get (7.196), i.e.,

$$U_i^{\rm av}(t) = \frac{c_i}{s+c_i}\left\{-k_i\left[\tilde{\vartheta}_i^{\rm av}(t) + \int_0^{D_i} u_i^{\rm av}(\sigma,t)d\sigma\right]\right\}. \tag{7.203}$$

Let us now consider $w_i(D_i,t)$. It is easily seen that

$$\partial_t w_i(D_i,t) = \partial_t u_i^{\rm av}(D_i,t) + k_i u_i^{\rm av}(D_i,t), \tag{7.204}$$

where $\partial_t u_i^{\rm av}(D_i,t) = \dot{U}_i^{\rm av}(t)$. The inverse of (7.199) is

$$u_i^{\rm av}(x,t) = w_i(x,t) - k_i\left[e^{-k_i x}\tilde{\vartheta}_i^{\rm av}(t) + \int_0^x e^{-k_i(x-\sigma)} w_i(\sigma,t)d\sigma\right]. \tag{7.205}$$

Plugging (7.202) and (7.205) into (7.204), we get

$$\begin{aligned}\partial_t w_i(D_i,t) &= -c_i w_i(D_i,t) + k_i w_i(D_i,t)\\ &\quad - k_i^2\left[e^{-k_i D_i}\tilde{\vartheta}_i^{\rm av}(t) + \int_0^{D_i} e^{-k_i(D_i-\sigma)} w_i(\sigma,t)d\sigma\right].\end{aligned} \tag{7.206}$$

Step 4: *Lyapunov–Krasovskii Functional*
The exponential stability of the overall system is established with the Lyapunov functional

$$V(t) = \sum_{i=1}^{n} V_i(t) \quad \forall i = 1,\ldots,n, \tag{7.207}$$

where $V_i(t)$ are functionals

$$V_i(t) = \frac{1}{2}\left[\tilde{\vartheta}_i^{\rm av}(t)\right]^2 + \frac{\bar{a}_i}{2}\int_0^{D_i}(1+x)w_i^2(x,t)dx + \frac{1}{2}w_i^2(D_i,t) \tag{7.208}$$

for each subsystem in (7.200)–(7.202) and $\bar{a}_i > 0$ being appropriate constants to be chosen later. Thus, we have

$$\begin{aligned}
\dot{V}_i(t) &= -k_i\left[\tilde{\vartheta}_i^{\rm av}(t)\right]^2 + \tilde{\vartheta}_i^{\rm av}(t)w_i(0,t)\\
&\quad + \bar{a}_i\int_0^{D_i}(1+x)w_i(x,t)\partial_x w_i(x,t)dx + w_i(D_i,t)\partial_t w_i(D_i,t)\\
&= -k_i\left[\tilde{\vartheta}_i^{\rm av}(t)\right]^2 + \tilde{\vartheta}_i^{\rm av}(t)w_i(0,t)\\
&\quad + \frac{\bar{a}_i(1+D_i)}{2}w_i^2(D_i,t) - \frac{\bar{a}_i}{2}w_i^2(0,t)\\
&\quad - \frac{\bar{a}_i}{2}\int_0^{D_i}w_i^2(x,t)dx + w_i(D_i,t)\partial_t w_i(D_i,t)\\
&\le -k_i\left[\tilde{\vartheta}_i^{\rm av}(t)\right]^2 + \frac{1}{2\bar{a}_i}\left[\tilde{\vartheta}_i^{\rm av}(t)\right]^2 - \frac{\bar{a}_i}{2}\int_0^{D_i}w_i^2(x,t)dx\\
&\quad + w_i(D_i,t)\left[\partial_t w_i(D_i,t) + \frac{\bar{a}_i(1+D_i)}{2}w_i(D_i,t)\right].
\end{aligned}$$

Recalling that $k_i > 0$, let us choose $\bar{a}_i = 1/k_i$. Then

$$\begin{aligned}
\dot{V}_i(t) &\le -\frac{1}{2\bar{a}_i}\left[\tilde{\vartheta}_i^{\rm av}(t)\right]^2 - \frac{\bar{a}_i}{2}\int_0^{D_i}w_i^2(x,t)dx\\
&\quad + w_i(D_i,t)\left[\partial_t w_i(D_i,t) + \frac{\bar{a}_i(1+D_i)}{2}w_i(D_i,t)\right].
\end{aligned} \tag{7.209}$$

Now we consider (7.209) along with (7.206). With a completion of squares, we obtain

$$\begin{aligned}
\dot{V}_i(t) &\le -\frac{1}{4\bar{a}_i}\left[\tilde{\vartheta}_i^{\rm av}(t)\right]^2 - \frac{\bar{a}_i}{4}\int_0^{D_i}w_i^2(x,t)dx\\
&\quad + \bar{a}_i\left|k_i^2 e^{-k_i D_i}\right|^2 w_i^2(D_i,t) + \frac{1}{\bar{a}_i}\left\|k_i^2 e^{-k_i(D_i-\sigma)}\right\|^2 w_i^2(D_i,t)\\
&\quad + \left[\frac{\bar{a}_i(1+D_i)}{2} + k_i\right]w_i^2(D_i,t) - c_i w_i^2(D_i,t),
\end{aligned} \tag{7.210}$$

where the spatial variable $\sigma \in [0,D_i]$ and $\|\cdot\|$ denotes the L_2-norm in σ. Then, from (7.210), we arrive at

$$\dot{V}_i(t) \le -\frac{1}{4\bar{a}_i}\left[\tilde{\vartheta}_i^{\rm av}(t)\right]^2 - \frac{\bar{a}_i}{4}\int_0^{D_i}w_i^2(x,t)dx - (c_i - c_i^*)w^2(D,t), \tag{7.211}$$

where

$$c_i^* = \frac{\bar{a}_i(1+D_i)}{2} + k_i + \bar{a}_i \left| k_i^2 e^{-k_i D_i} \right|^2 + \frac{1}{\bar{a}_i} \left\| k_i^2 e^{-k_i(D_i-\sigma)} \right\|^2 . \tag{7.212}$$

Hence, from (7.211), if c_i is chosen such that $c_i > c_i^*$, we obtain

$$\dot{V}_i(t) \leq -\mu_i V_i(t) \qquad \text{or} \qquad \dot{V}(t) \leq -\mu V(t) \tag{7.213}$$

for some $\mu_i > 0$ and $\mu = \min(\mu_i)$. Thus, the closed-loop system is exponentially stable in the sense of the full-state norm

$$\left(|\tilde{\vartheta}_{\rm av}(t)|^2 + \sum_{i=1}^{n} \int_0^{D_i} w_i^2(x,t)dx + w_i^2(D_i,t) \right)^{1/2}, \tag{7.214}$$

i.e., in the transformed variable $(\tilde{\vartheta}_{\rm av}, w)$.

Step 5: *Exponential Stability Estimate (in the L_2-Norm) for the Average System* (7.194)–(7.196)

To obtain exponential stability in the sense of the norm

$$\Upsilon(t) \triangleq \left(|\tilde{\vartheta}_{\rm av}(t)|^2 + \sum_{i=1}^{n} \int_0^{D_i} [u_i^{\rm av}(x,t)]^2 dx + [u_i^{\rm av}(D_i,t)]^2 \right)^{1/2}, \tag{7.215}$$

we show from (7.199), (7.205), and (7.207)–(7.208) that there exist positive numbers α_1 and α_2 such that $\alpha_1 \Upsilon(t)^2 \leq V(t) \leq \alpha_2 \Upsilon(t)^2$. Hence, with (7.213), we get

$$\Upsilon(t) \leq \frac{\alpha_2}{\alpha_1} e^{-\mu t} \Upsilon(0), \tag{7.216}$$

which completes the proof of exponential stability in the original variable $(\tilde{\vartheta}_{\rm av}, u_{\rm av})$.

Step 6: *Invoking the Averaging Theorem*

First, note that the closed-loop system (7.178), (7.185), and (7.162) can be rewritten as

$$\dot{\tilde{\theta}}_i(t-D_i) = U_i(t-D_i), \quad i = 1, \ldots, n, \tag{7.217}$$

$$\dot{U}_i(t) = -c_i U_i(t) - c_i k_i \left[z_i(t) + \int_{t-D_i}^{t} U_i(\tau) d\tau \right], \tag{7.218}$$

$$\dot{\tilde{\Gamma}}(t) = \omega_r [\tilde{\Gamma}(t) + H^{-1}] \times [I_{n \times n} - \hat{H}(t)(\tilde{\Gamma}(t) + H^{-1})], \tag{7.219}$$

where $\varphi(t) = [\tilde{\Gamma}(t), \tilde{\theta}(t-D), U(t)]^T$ is the state vector. From the definitions of $\hat{H}(t)$ in (7.16) and $z(t)$ in (7.165), one has that the averaging theorem by [83, 131] in Appendix A can be directly applied considering $\omega = 1/\epsilon$ and $\varphi_t(\Theta) = \varphi(t+\Theta)$ for $-D_n \leq \Theta \leq 0$.

From (7.216), the origin of the average closed-loop system (7.194)–(7.196) with transport PDE for delay representation is locally exponentially stable. In addition, we can conclude that the equilibrium $\tilde{\Gamma}_{\rm av}(t) = 0$ of the linearized error system (7.198) is also exponentially stable since $\omega_r > 0$. Thus, using (7.193) and (7.197), there exist constants $\alpha, \beta > 0$ such that all solutions satisfy $\Psi(t) \leq \alpha e^{-\beta t} \Psi(0)$, $\forall t \geq 0$, in the sense of the norm

$$\Psi(t) \triangleq \left|\tilde{\Gamma}_{\rm av}(t)\right|^2 + \sum_{i=1}^{n} \left[\tilde{\theta}_i^{\rm av}(t-D_i) \right]^2 + \int_{t-D_i}^{t} [U_i^{\rm av}(\tau)]^2 d\tau + [U_i^{\rm av}(t)]^2 . \tag{7.220}$$

Then, by applying the averaging theorem by [83, 131] in Appendix A, for ω sufficiently large, we can conclude (7.187).

Step 7: *Asymptotic Convergence to the Extremum* (θ^*, y^*)

Finally, to obtain (7.188) and (7.189), we only have to use the change of variables (7.193) and then integrate both sides of (7.190) within the interval $[t, \sigma + D_i]$ to get

$$\tilde{\vartheta}_i(\sigma + D_i) = \tilde{\vartheta}_i(t) + \int_t^{\sigma + D_i} u_i(0,s)ds\,, \qquad i = 1,\ldots,n. \tag{7.221}$$

After that, we have to reproduce exactly the developments presented in **Step 7** of the proof of Theorem 7.1. □

Corollary 7.2 (Newton ES under Output Delays). *Analogously to Corollary* 7.1, *local exponential stability of the multivariable Newton-based ES in Figure* 7.6 *with delayed output* (7.10) *and* $D \geq 0$ *being a simple scalar can be obtained from Theorem* 7.3 *regarding the closed-loop delayed system* (7.162), (7.169) *and the predictor based control law* (7.177), *but with the advantage of providing a user-assignable convergence rate.*

7.10 ▪ Source Seeking Application

In this example, multivariable ES is used for finding a source of a signal (chemical, acoustic, electromagnetic, etc.) of unknown concentration field as in (7.9) with Hessian and its inverse given by

$$H = \begin{bmatrix} -2 & -2 \\ -2 & -4 \end{bmatrix}, \quad H^{-1} = \begin{bmatrix} -1 & 0.5 \\ 0.5 & -0.5 \end{bmatrix}.$$

The strength of this field decays with the distance and has a local maximum at $y^* = 1$ and unknown maximizer $\theta^* = (\theta_1^*, \theta_2^*) = (0, 1)$. This is achieved without the measurement of the position vector $\theta = (\theta_1, \theta_2)$ and using only the measurement of the output scalar signal y with delay $D_{\text{out}} = 5\,\text{s}$. The two actuator paths of the vehicle are also under distinct delays $D_1^{\text{in}} = 10\,\text{s}$ and $D_2^{\text{in}} = 20\,\text{s}$. Thus, the total delays to be compensated by the predictors are $D_1 = 15\,\text{s}$ and $D_2 = 25\,\text{s}$.

The proposed schemes are slightly modified for the stated task in Figure 7.8 by observing that the integrator, a key adaptation element, is already present in vehicle models where the primary forces or moments acting on the vehicle are those that provide thrust/propulsion [246]. Thus, an application of our result for single and double integrators in control of autonomous vehicles modeled as point mass in the plane is shown to be possible. However, we consider the simplest case of a velocity-actuated point mass only, where the additive dither in (7.14) is changed by $\dot{S}(t)$ since the integrator of the vehicle dynamics can be moved to the ES loop for analysis purposes. For the double integrator case, it would be needed to replace the lag filters used in (7.49) and (7.185)–(7.186) by lead compensators of the form $sc_i/(s + c_i)$, whose role is to recover some of the phase in feedback loop lost due to the addition of the second integrator [246].

We show next that the predictor feedback based ES controllers drive the autonomous vehicle modeled by

$$\dot{\theta}_1 = v_1\,, \qquad \dot{\theta}_2 = v_2 \tag{7.222}$$

to (θ_1^*, θ_2^*), whereas the ES automatically tunes v_1, v_2 to lead the vehicle to the peak of $Q(\theta)$.

In what follows, we initially present numerical simulations for the Newton case with predictor (7.185)–(7.186), where $c_1 = c_2 = 20$ and z is given by (7.165) with G in (7.13) and Γ in (7.160). The control gain matrix was set to $K = \text{diag}\{1,\ 1\}$ and $\omega_r = 0.1$. We perform our tests with the following parameters: $a_1 = a_2 = 0.05$, $\omega = 5$, $\omega_1 = 7\omega$, $\omega_2 = 5\omega$, $\hat{\theta}(0) = (-1, 2)$, and $\Gamma(0) = -1/200\,\text{diag}\{2,\ 1\}$.

Figure 7.9 shows the system output $y(t)$ in 3 situations: (a) free of delays, (b) in the presence of input-output delays but without any delay compensation and (c) with input-output delays and predictor based compensation.

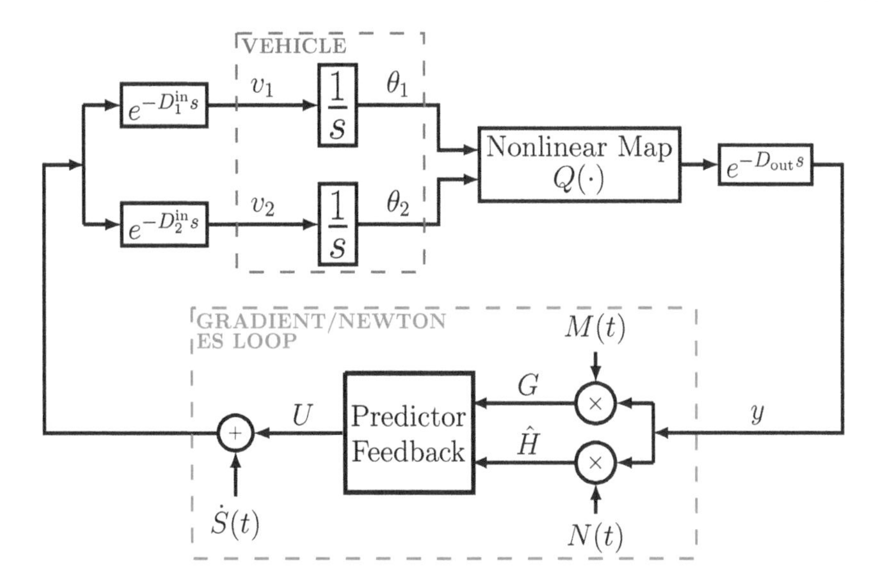

Figure 7.8. *Source seeking under delays for velocity-actuated point mass with additive dither* $\dot{S}(t) = [a_1\omega_1 \cos(\omega_1(t+D_1)) \;\; a_2\omega_2 \cos(\omega_2(t+D_2))]^T$. *The demodulation signals* $M(t)$ *and* $N(t)$ *are chosen according to* (7.15) *and* (7.17)–(7.18). *The predictor-based controllers* (7.49) *or* (7.185)–(7.186) *are used in the ES loop to compensate the total delay* $D_1 = D_1^{\text{in}} + D_{\text{out}}$ *and* $D_2 = D_2^{\text{in}} + D_{\text{out}}$ *in Gradient and Newton schemes, respectively.*

Figure 7.10 presents relevant variables for ES. It is clear the remarkable evolution of the new prediction scheme in searching the maximum and the Hessian's inverse H^{-1}. This exact estimation allow us to cancel the Hessian H and thus guarantee convergence rates that can be arbitrarily assigned by the user.

In order to make a fair comparison with the Gradient ES, all common parameters are chosen the same, except for the gain matrix K. According to [73], we should select $K_G = K_N\Gamma(0)$, where K_G and K_N denote here the gain matrices for gradient-based ES and Newton-based ES, respectively. Hence, we apply the predictor (7.49) for Gradient ES with averaging-based estimate of the Hessian $\hat{H}$ in (7.16), function $\hat{\Phi}$ given by (7.45)–(7.47) for $\hat{A}_i(t)$ rather than A_i, gain matrix $K = 10^{-2}\,\text{diag}\{1,\, 0.5\}$, and $c = 20$.

For the case $n = 2$ (two control inputs), (7.49) is simply

$$\begin{aligned} U_1(t) = \frac{c}{s+c}\Bigg\{ K_1^T \Bigg(G(t) + \int_{t-D_1}^{t} \hat{H}_1 U_1(\tau)d\tau \\ + \int_{t-D_1}^{t} \hat{H}_2 U_2(\tau - (D_2 - D_1))d\tau \Bigg) \Bigg\}, \end{aligned} \tag{7.223}$$

$$\begin{aligned} U_2(t) = \frac{c}{s+c}\Bigg\{ K_2^T \Bigg(e^{\hat{A}_1(D_2-D_1)} G(t) \\ + e^{\hat{A}_1(D_2-D_1)} \int_{t-D_1}^{t} \hat{H}_1 U_1(\tau)d\tau \\ + e^{\hat{A}_1(D_2-D_1)} \int_{t-D_1}^{t} \hat{H}_2 U_2(\tau - (D_2 - D_1))d\tau \\ + \int_{t-(D_2-D_1)}^{t} e^{\hat{A}_1(t-\tau)} \hat{H}_2 U_2(\tau)d\tau \Bigg) \Bigg\}. \end{aligned} \tag{7.224}$$

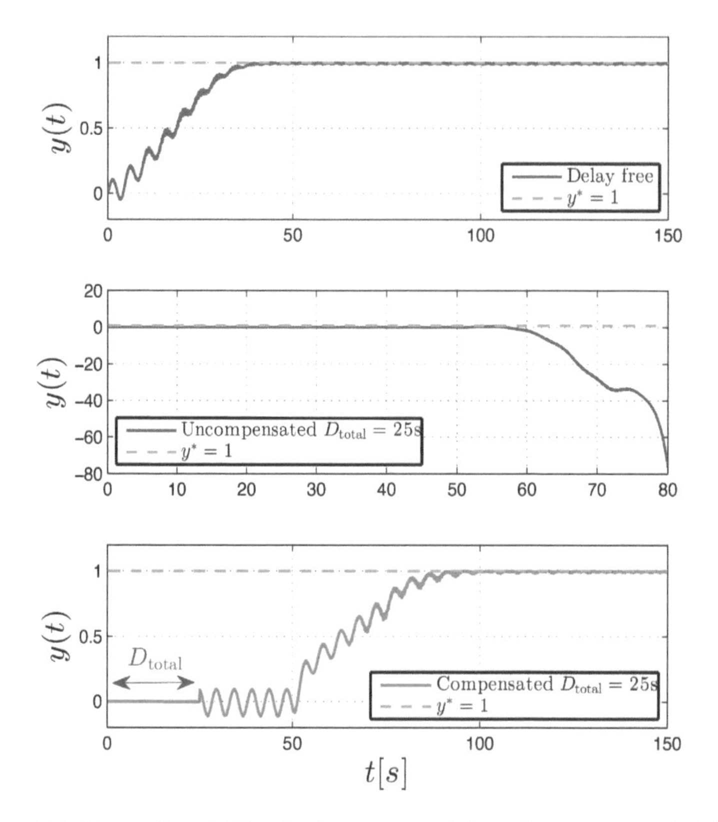

Figure 7.9. ***Newton-based ES under input-output delays (time response of*** $y(t)$***):*** (a) *basic ES works well without delays;* (b) *ES goes unstable in the presence of delays (*$D_{\text{total}} = D_2 = D_2^{\text{in}} + D_{\text{out}}$ *is the longest delay);* (c) *predictor fixes this.*

The issue of robustness to the approximation of the integral terms for prediction was raised by Mondié and Michiels in [156]. It was subsequently shown by Mirkin [152] that it is simply the result of a poor choice of the approximation scheme for the integral. Furthermore, numerical approximation schemes that are robust have been provided recently in [97].

Figure 7.11 shows the estimate of the maximum for both approaches. As expected, the Newton ES converges faster than the Gradient ES, even in the presence of delays.

7.11 ▪ Notes and References

In [178], the authors propose a solution to the problem of multivariable ES algorithms for output- and/or equal input-delay systems via predictor feedback, presenting two approaches to construct a predictor based on perturbation-based estimates of the model. One is based on gradient optimization, where they estimate the Hessian [73, 166] for the purpose of implementing a predictor that compensates the delay. The other approach is based on the Newton optimization, where they estimate the Hessian's inverse for the purpose of making the convergence rate independent of the unknown parameters of the map.

This chapter and the publication [179] extend the result to *multiple* and *distinct* input delays, for both gradient and Newton methods, resulting in a complex predictor design with stability analysis being carried out by using a novel successive backstepping transformation and averaging in infinite dimensions. On the other hand, [146] proposes new designs for multivariable

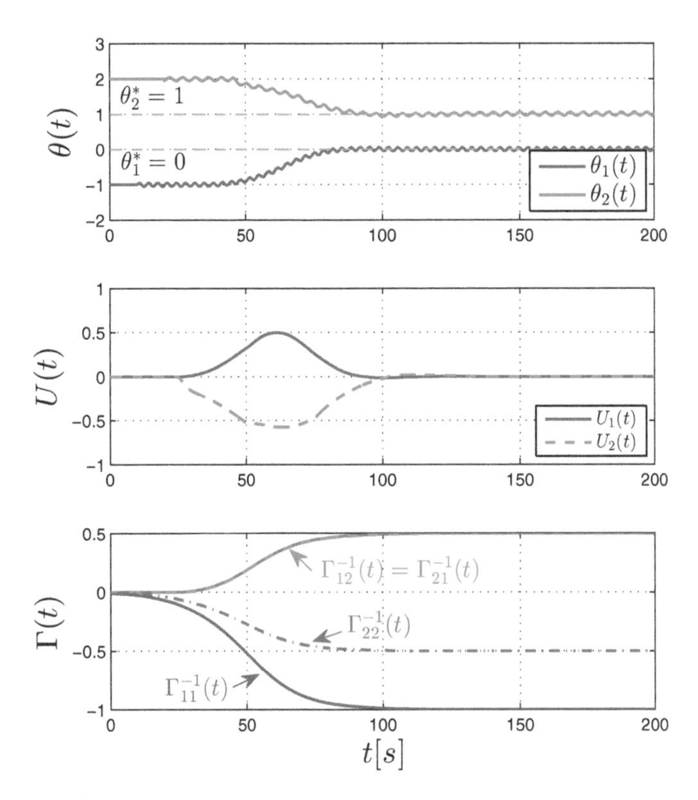

Figure 7.10. ***Newton-based ES under input-output delays:*** (a) *parameter* $\theta(t)$*;* (b) *the control signal* $U(t)$*;* (c) *Hessian's inverse estimate* $\Gamma(t)$*. The elements of* $\Gamma(t)$ *converge to those of* H^{-1}.

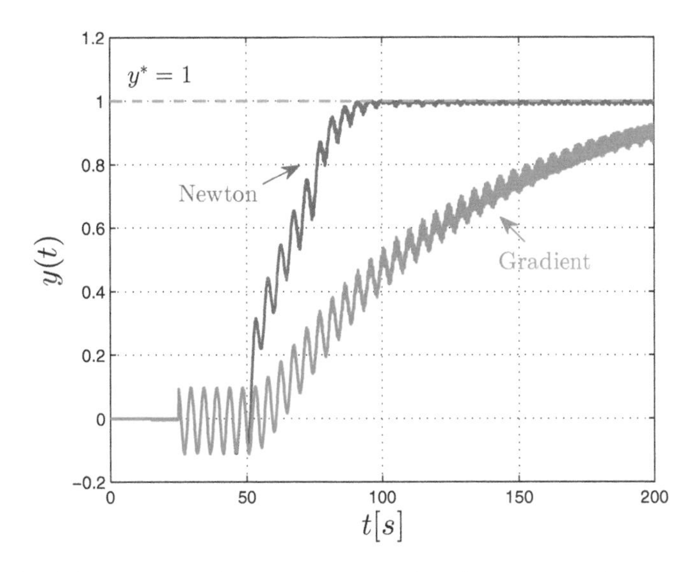

Figure 7.11. ***Newton-based ES versus Gradient ES:*** *Time response of the output* $y(t)$ *subject to input-output delays.*

extremum seeking for static maps with arbitrarily long time delays. Such a sequential predictor based approach eliminates the need for distributed terms. However, in sequential predictors "larger delays or higher-dimension maps require smaller values of the dither excitation frequencies and therefore lead to slower convergence."

An alternative gradient-based extremum seeking scheme is also presented in this chapter that does not require backstepping transformation, which also leads to a much more simple control feedback structure. Indeed, this is the first result for predictor-feedback extremum seeking (with distributed terms) where the backstepping transformation is not employed. Moreover, as in the original reference [179], our advantage over the results in [146] seems to be that the delays in our approach are independent of the dither frequency and the system's dimension, which means that we do not need to consider smaller delays (or lower-order maps) to achieve faster convergence.

Results are given for static plants and the extension to include general nonlinear dynamics seems to be an open problem. We have found many interesting results in the literature of *singular perturbations for FDEs and delayed systems*. However, we have not found any theorem that is directly applicable to our problem and which would allow us to prove stability in the presence of plant dynamics, as done in the absence of delays in [129] and in papers by Tan, Nesić, and coworkers [217, 219, 166]. We need singular perturbation results that apply to distributed delays due to the presence of a (filtered) predictor feedback in our design. In general, the existing works consider a point (fixed) delay, which must be small [67]. There are cases of time-varying or even state-dependent delays but restricted to lower-order (scalar or second-order) systems [188]. In some cases, the delay is distributed, but it is also scaled by the small parameter and occurs in the fast dynamics [13]. There do exist results for linear systems with distributed delays in the slow dynamics, but they are not adequate since they assume that the slow/reduced model has an equilibrium at the origin when the small parameter is set to zero, i.e., that the reduced model is perturbation-free [55]. As a reminder from [129], the averaging step comes second in the stability analysis, which means that, after the singular perturbation approximation, the resulting slow/reduced model must still have the sinusoidal perturbation AND the delayed input, along with the predictor feedback. On the other hand, [223] solves the ES problem of dynamic plants where delays are not directly taken into account, but they may be addressed as a sufficiently large dwell time in the algorithm. However, it would require waiting D time-delay units or longer for each function evaluation, making the transient last for many D's, whereas our transient lasts only D plus whatever the exponential transient of the delay-free ES is.

Even though we can easily compensate known time-varying delays, and unknown time-varying delays that vary sufficiently slowly, the averaging result for FDEs (Appendix A) assumes the delay to be constant. In this sense, the challenges of considering nonconstant delays are discussed in the next chapter.

Chapter 8

ES under Time-Varying and State-Dependent Delays

8.1 ▪ Nonconstant Delays: Physical Examples and Methodological Challenges

The work on predictor feedback has revealed applications in which the input or output delay is not constant. Among the most well-studied of such applications are additive manufacturing, shock waves in traffic flows, cooling systems, and internet congestion.

While there are applications in which the delay varies over time in an open-loop fashion, as a result of drift, aging, and changes in operating conditions, arguably more interesting, and perhaps even prevalent, are situations where delay changes with time because it depends on the state of the system, which itself is time-varying.

Since we have already proved that predictor feedback is capable of compensating delays in extremum seeking problems, it is natural to consider the compensation of delays that are time-varying in ES problems. Predictors for stabilization of systems under delays that are time-varying or state-dependent have already been introduced in the second author's work. Several challenges arise in constructing predictors and in the resulting stability analysis when delays are time-varying or state-dependent. One of the challenges is that, while the delay is known at each time instant, the prediction horizon, namely, the length of time it will take the input signal to reach the plant, is not a priori known at that time instant because that length of time is the inverse function of the difference between the current time and the delay. That time difference is called the "delayed time." When the delay is an open-loop function, the prediction horizon can be found, in principle, as the inverse function of the delayed time. But, when the delay is state-dependent, it is not known, a priori, at what future time the input signal will reach the plant. It, therefore, takes certain transformations of the time variable and a reformulation of the integral equation for the predictor state to determine the prediction horizon and to predict the state value when the current input reaches the plant.

This convoluted dynamic scenario, which bedevils the design process, has its manifestation in the analysis as well. If the delay is growing, and if it is growing too fast, the input signal being generated at the present instant may never reach the plant. It is, therefore, necessary that a rapid growth of the delay be preempted either by a priori assumption, in the case of open-loop time-varying delays, or by design and restriction of the initial condition of the plant, in the case of state-dependent delays. In either case, to be specific, the delay rate (in the direction of growth) must not exceed unity. All of these considerations, which arise in predictor-based stabilization, carry over to extremum seeking. The said challenges have to be faced and dealt with and can be overcome. This is what we do in this chapter.

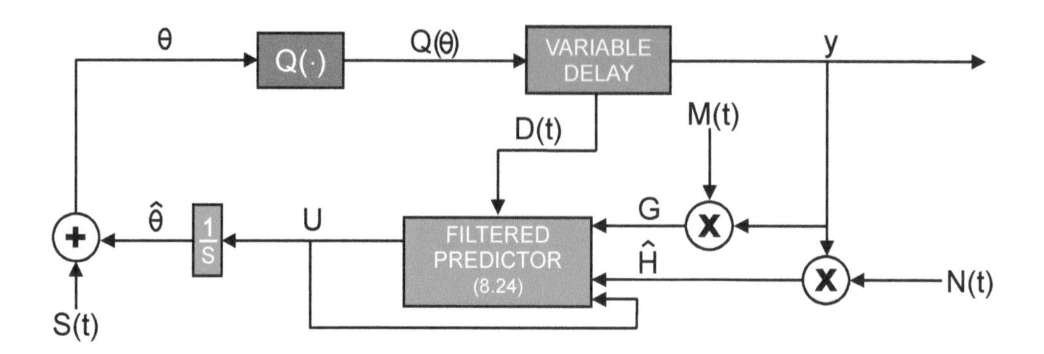

Figure 8.1. *Block diagram of the extremum seeking control system with nonconstant delays.*

We propose a gradient extremum seeking scheme for locally quadratic static maps in the presence of time-varying delays (the Newton-based ES extension is discussed in [201]). A novel predictor design using perturbation-based estimates of the unknown Gradient and Hessian of the map must be introduced to handle this variable nature of the delays, which can arise both in the input and output channels of the nonlinear map to be optimized. The demodulation signals incorporate the time-varying and state-dependent delays.

Local exponential stability and convergence to a small neighborhood of the unknown extremum point are guaranteed. This technical result is assured by using backstepping transformation and averaging theory in infinite dimensions. Implementation aspects of the presented predictor for variable delays as well as the extension to state-dependent delays are also discussed. Near the chapter's end, we introduce the first of this book's ES topics for PDE cascades,[5] in this case of the transport type, interconnected through boundary conditions. Such a transport PDE-PDE cascade is useful to represent simultaneous time- and state-dependent delays. Simulation examples illustrate the effectiveness of the proposed predictor-based extremum seeking approach for time-delay compensation.

8.2 ▪ Problem Statement

Scalar ES addresses the real-time control problem in which the aim is to find the maximum (or minimum) of the output $y \in \mathbb{R}$ of an unknown nonlinear static map $Q(\theta)$ by modifying the input $\theta \in \mathbb{R}$.

In this section, we assume there is an arbitrarily large *time-varying and known* delay $D(t) \geq 0$ in the actuation path or measurement system such that the map's output can be denoted by

$$y(t) = Q(\theta(t - D(t))). \tag{8.1}$$

For the sake of clarity, we assume here that the map is output-delayed, as depicted in the block diagram in Figure 8.1. However, as mentioned in the earlier chapters, the results provided here can be directly generalized to the input-delay case once any input time delay can be moved to the output of the static map. The general setup when input delays $D_{\text{in}}(t)$ and output delays $D_{\text{out}}(t)$ occur concurrently can also be handled by assuming that the total delay to be compensated is $D(t) = D_{\text{in}}(t) + D_{\text{out}}(t)$, with $D_{\text{in}}(t), D_{\text{out}}(t) \geq 0$.

[5]The topic about extremum seeking for cascades of PDEs will be discussed in detail later in Chapter 14.

Remark 8.1. The term "nonconstant delays" in this chapter means variable delays which may depend exclusively on the time variable $D(t)$, exclusively on the state variable $D(\theta(t))$, or simultaneously on both $D(t,\theta(t))$. All of these three cases are handled in the chapter. For the sake of clarity, we start the presentation with the simplest case of time-varying delays $D(t)$ and then we move forward to the more complicated cases involving state-dependent delays $D(\theta(t))$ in Section 8.6 as well as simultaneous time- and state-dependent delays $D(t,\theta(t))$ in Section 8.7.

Without loss of generality, let us consider the maximum seeking problem such that the maximizing value of θ is denoted by θ^*. For the sake of simplicity, we also assume that the nonlinear map is at least locally quadratic:

$$Q(\theta) = y^* + \frac{H}{2}(\theta - \theta^*)^2, \tag{8.2}$$

where besides the constants $\theta^* \in \mathbb{R}$ and $y^* \in \mathbb{R}$ being unknown, the scalar $H < 0$ is the unknown Hessian of the static map. By plugging (8.2) into (8.1), we obtain the *quadratic static map with delay* of interest:

$$y(t) = y^* + \frac{H}{2}(\theta(t - D(t)) - \theta^*)^2. \tag{8.3}$$

8.2.1 ▪ Basic Conditions on Time-Varying Delays

The following are the basic conditions on the arbitrarily large time-varying delay $D(t)$ [119, Chapter 6]:

(a) The delay function $D(t)$ is strictly positive (technical condition ensuring that the state space of the input dynamics can be defined).

(b) The delay function $D(t)$ is uniformly bounded from above, but there is no restriction on the magnitude of its constant upper bound.

(c) The delay-rate function $\dot{D}(t)$ is strictly smaller than 1; i.e., the delay may increase but only at a rate that does not exceed 1.

(d) The delay-rate function $\dot{D}(t)$ is uniformly bounded from below (by an arbitrary negative finite constant); i.e., the delay may decrease at a uniformly bounded rate.

(e) The delay $D(t)$ is not periodic in $T = 2\pi/\omega$, where $\omega > 0$ is the frequency of the dither signal in the ES loop. Any other frequency (different of ω) for a periodic time-varying delay is allowed.

Consider the continuously differentiable functions of the *delay time* $\phi(t)$ and *prediction time* $\phi^{-1}(t)$, defined as

$$\phi(t) := t - D(t), \tag{8.4}$$

where $\phi^{-1}(t)$ is the inverse function of $\phi(t)$. We assume the invertibility of $\phi(t)$ along the chapter. When the delay is constant, $\phi^{-1}(t) = t + D$. The following assumption is also made.

Assumption 8.1. *The output time-varying delay $D(t)$ fulfills*

$$-\infty < \underline{d} < \dot{D}(t) < \overline{d} < 1, \tag{8.5}$$

where $\underline{d}$ and $\overline{d}$ are constants.

This is a natural consequence of (8.4). In addition, from Assumption 8.1 it follows that $\phi(t)$ is strictly increasing, i.e.,

$$\phi(t) \leq t \qquad \forall t \geq 0. \tag{8.6}$$

Inequality (8.6) can be alternatively stated as

$$\phi^{-1}(t) - t > 0. \tag{8.7}$$

8.2.2 ▪ Probing and Demodulation Signals

Let $\hat{\theta}$ be the estimate of θ^* and

$$\tilde{\theta}(t) = \hat{\theta}(t) - \theta^* \tag{8.8}$$

be the *estimation error*. From Figure 8.1 and noting that $\dot{\hat{\theta}}(t) = \dot{\tilde{\theta}}(t)$, the *error dynamics* can be written as

$$\dot{\tilde{\theta}}(t - D(t)) = U(t - D(t)), \tag{8.9}$$

where $U(t)$ is the control signal. Additionally, we have

$$G(t) = M(t)y(t), \quad \theta(t) = \hat{\theta}(t) + S(t), \tag{8.10}$$

with the probing and demodulation signals being given by[6]

$$S(t) = a\sin(\omega t), \quad M(t) = \frac{2}{a}\sin(\omega(t - D(t))), \tag{8.11}$$

with nonzero perturbation amplitude a and frequency ω. The signal

$$\hat{H}(t) = N(t)y(t) \tag{8.12}$$

is applied to obtain an estimate of the unknown Hessian H, where the demodulating signal $N(t)$ is given by

$$N(t) = -\frac{8}{a^2}\cos(2\omega(t - D(t))). \tag{8.13}$$

In [73], it was proved that

$$\frac{1}{T}\int_0^T N(\sigma)y d\sigma = H, \quad T = 2\pi/\omega \tag{8.14}$$

if a quadratic map as in (8.2) is considered. In other words, its average version is $\hat{H}_{\text{av}} = (Ny)_{\text{av}} = H$.

[6] In [182] and [202], the authors have used $S(t) = a\sin(\omega(t + D(t)))$ and $M(t) = \frac{2}{a}\sin(\omega t)$ rather than (8.11). This apparently can be justified since the output delay can be transferred to the integrator output for analysis purposes (or, equivalently, to its input), thus, the phase shift $+\omega D(t)$ would be applied to compensate the delay effects in the additive dither signal $S(t)$. However, this "prediction" property is not valid for time-varying (or nonconstant) delays since $S(t - D(t)) = a\sin(\omega(t - D(t) + D(t - D(t)))) \neq a\sin\omega t$. The correct expression for time-varying delays would be $S(t) = a\sin(\omega(t + \phi^{-1}(t)))$.

8.3 ▪ Predictor Feedback with Hessian's Estimate

By using the averaging analysis and from the basic condition **(e)** in Section 8.2.1, we can verify that the average version of the signal $G(t)$ in (8.10) is given by

$$G_{\mathrm{av}}(t) = H\tilde{\theta}_{\mathrm{av}}(t-D(t)). \tag{8.15}$$

From (8.9), the following average models can be obtained:

$$\dot{\tilde{\theta}}_{\mathrm{av}}(t-D(t)) = U_{\mathrm{av}}(t-D(t)), \tag{8.16}$$

$$\dot{G}_{\mathrm{av}}(t) = HU_{\mathrm{av}}(t-D(t)), \tag{8.17}$$

where $U_{\mathrm{av}} \in \mathbb{R}$ is the resulting average control for $U \in \mathbb{R}$.

In order to motivate the predictor feedback design and recalling (8.4), the idea here is to compensate for the delay by feeding back the future state $G(\phi^{-1}(t))$, or $G_{\mathrm{av}}(\phi^{-1}(t))$ in the equivalent average system. Given any stabilizing gain $k > 0$ for the undelayed system, our wish is to have a control that achieves

$$U_{\mathrm{av}}(t) = kG_{\mathrm{av}}(\phi^{-1}(t)) \quad \forall t \geq 0, \tag{8.18}$$

and it appears to be nonimplementable since it requires future values of the state. However, by applying the variation-of-constants formula to (8.17) we can express the future state [119, Chapter 6] as

$$G_{\mathrm{av}}(\phi^{-1}(t)) = G_{\mathrm{av}}(t) + H\int_{t-D(t)}^{t} \frac{U_{\mathrm{av}}(\sigma)}{\phi'(\phi^{-1}(\sigma))}d\sigma, \tag{8.19}$$

which gives the future state $G_{\mathrm{av}}(\phi^{-1}(t))$ in terms of the average control signal $U_{\mathrm{av}}(\sigma)$ from the past window $[t-D(t), t]$. To obtain (8.19), we have used the basic differentiation rule of the inverse of a function as follows:

$$\frac{d}{d\sigma}\phi^{-1}(\sigma) = \frac{1}{\phi'(\phi^{-1}(\sigma))},$$

where ϕ' denotes the time derivative of $\phi(\cdot)$.

Plugging (8.19) into (8.18) yields the following feedback law:

$$U_{\mathrm{av}}(t) = k\left[G_{\mathrm{av}}(t) + H\int_{t-D(t)}^{t} \frac{U_{\mathrm{av}}(\sigma)}{\phi'(\phi^{-1}(\sigma))}d\sigma\right] \quad \forall t \geq 0. \tag{8.20}$$

Hence, from (8.19) and (8.20), the average feedback law (8.18) can be obtained indeed as desired. Consequently,

$$\dot{\tilde{\theta}}_{\mathrm{av}}(t) = kG_{\mathrm{av}}(\phi^{-1}(t)) \quad \forall t \geq 0. \tag{8.21}$$

Therefore, from (8.15), one has

$$\frac{d\tilde{\theta}_{\mathrm{av}}(t)}{dt} = kH\tilde{\theta}_{\mathrm{av}}(t) \quad \forall t \geq D(t), \tag{8.22}$$

with an exponentially attractive equilibrium $\tilde{\theta}^{e}_{\mathrm{av}} = 0$, since $k > 0$ in the control design and $H < 0$ by assumption.

In the remaining sections of the chapter, we show that the control objectives can still be achieved if a simple modification of the above basic predictor-based controller, which employs a low-pass filter [115], is applied. In this case, we propose the next averaging-based predictor feedback in order to compensate the delay

$$U(t)=\frac{c}{s+c}\left\{k\left[G(t)+\hat{H}(t)\int_{t-D(t)}^{t}\frac{U(\sigma)}{\phi'(\phi^{-1}(\sigma))}d\sigma\right]\right\} \tag{8.23}$$

or, equivalently, in its state-space representation:

$$\dot{U}(t)=-cU(t)+ck\left[G(t)+\hat{H}(t)\int_{t-D(t)}^{t}\frac{U(\sigma)}{\phi'(\phi^{-1}(\sigma))}d\sigma\right] \tag{8.24}$$

$\forall t\geq 0$, where $c>0$ is sufficiently large.

The predictor feedback is of the form of a low-pass filter of the nonaverage version of (8.20). This low-pass filtering was particularly required in the stability analysis when the averaging theorem in infinite dimensions [83] was invoked in the constant delay cases. Basically, this modification allows us to manipulate input-output delays as distributed state delays. In Section 8.8, the possibility of applying such an averaging theorem to nonconstant delays, as well as challenges, are discussed.

8.4 ▪ Approximation and Numerical Implementation of Predictor for Time-Varying Delays

Care is needed to numerically implement the predictor (8.23). Thus, in the next subsections, two simplified methods will be presented for performing this implementation. Alternative implementations and approximations of predictor feedback for delay systems can also be found in [98].

8.4.1 ▪ First Method

Considering (8.4) and (8.17), we can write

$$\dot{G}_{\rm av}(t)=HU_{\rm av}(\phi(t)). \tag{8.25}$$

The prediction of the state at the time when the current control will have an effect on the state is defined as

$$P_{\rm av}(t)=G_{\rm av}(t+D(\tau(t))), \tag{8.26}$$

where the prediction time is given by

$$\tau(t)=t+D(\tau(t)), \tag{8.27}$$

which is derived from the inversion of the time variable $t\to t-D(t)$ in $t\to t+D(\tau(t))$.

In order to utilize (8.27) instead of t, one rewrites (8.25) as

$$\dot{G}_{\rm av}(\tau(t))=HU_{\rm av}(t), \tag{8.28}$$

and rewriting (8.28), we have

$$\frac{dG_{\rm av}(\tau(t))}{dt}=HU_{\rm av}(t)\frac{d\tau(t)}{dt}. \tag{8.29}$$

Moreover, deriving (8.27), we get

$$\frac{d\tau(t)}{dt} = 1 + D'(\tau(t))\frac{d\tau(t)}{dt}, \tag{8.30}$$

$$\frac{d\tau(t)}{dt} = \frac{1}{1-D'(\tau(t))}. \tag{8.31}$$

Thus, (8.26) can be rewritten as

$$\frac{dP_{\text{av}}(t)}{dt} = HU_{\text{av}}(t)\frac{1}{1-D'(\tau(t))}. \tag{8.32}$$

Integrating (8.30) and (8.32) for all $t - D(t) \leq \xi \leq t$, we obtain the following predictor [25]:

$$P_{\text{av}}(t) = G_{\text{av}}(t) + H\int_{t-D(t)}^{t} U_{\text{av}}(\xi)\frac{1}{1-D'(\sigma(\xi))}d\xi, \tag{8.33}$$

$$\tau(t) = t + \int_{t-D(t)}^{t} \frac{1}{1-D'(\tau(\xi))}d\xi, \tag{8.34}$$

Finally, the nonaverage version for the predictor-based control law is given by

$$U(t) = \frac{c}{s+c}\left\{k\left[G(t) + \hat{H}(t)\int_{t-D(t)}^{t} U(\xi)\frac{1}{1-D'(\tau(\xi))}d\xi\right]\right\}. \tag{8.35}$$

Now, for example, using the point-to-point integration rule is easy to calculate (8.35) numerically, as described below.

Computation of the predictor signal: When simulating the predictor feedback controller (8.33)–(8.35), at each time step the ODE for the system (8.25) must be solved (using, for example, a simple Euler scheme) and the length of the delay must be computed (for example, as the integer part of $N(i) = \frac{D(i)}{h}$, say $\overline{N}(i)$, where h is the discretization step). The predictor is then computed by integrating simultaneously the two integral relations (8.33) and (8.34) at each time step, using a numerical integration scheme. For instance, with the left endpoint rule of integration we get

$$P(i) = G(i) + h\hat{H}(i)\sum_{k=i-\overline{N}(i)}^{i-1} U(k)\frac{1}{1-D'(\tau(k))}, \tag{8.36}$$

$$\tau(i) = i + h\sum_{k=i-\overline{N}(i)}^{i-1} \frac{1}{1-D'(\tau(k))}. \tag{8.37}$$

8.4.2 ▪ Second Method

In [36], the authors consider the following predictor:

$$U_{\text{av}}(t) = k\left[G_{\text{av}}(t) + H\int_{t-D(t)}^{t} U_{\text{av}}(s)ds\right] \tag{8.38}$$

for systems of the type $\dot{G}_{\text{av}} = HU_{\text{av}}(t - D(t))$, where H was originally assumed constant. Note that this controller does not exactly match the predicted system state on a time horizon $D(t)$.

Indeed, using the variation-of-constants formula, $\forall t \geq 0$,

$$G_{\mathrm{av}}(\phi^{-1}(t)) = G_{\mathrm{av}}(t) + H \int_{t-D(t)}^{t} U_{\mathrm{av}}(s + D(t) - D(s))ds. \tag{8.39}$$

However, the integral in this prediction may not be implementable as it is not necessarily causal (in detail, this is the case when there exists $s \in [t-D(t), t]$ such that $s - D(s) \geq t - D(t)$, i.e., when the delay $D(t)$ is suddenly high and the signal received at time t is older than those previously received) while the one employed in (8.38) always is [36].

Further, even if one can implement this prediction, the involved integral can be approximated by the one used in (8.38) if $D(t) - D(s) \approx 0$ for *almost every* time instant t, i.e., under the assumption that the variations of the delay $\dot{D}(t)$ are sufficiently small in average. As this assumption is the one which is required in the following Theorem 8.1 to robustly compensate the delay, we rather use the prediction form (8.38), which is always causal and easier to implement.

The implementation of the proposed predictor starts by dividing the integral (8.38) into two parts, according to the following equation:

$$\int_{t-D(t)}^{t} U_{\mathrm{av}}(s)ds = \int_{t-D(t)}^{0} U_{\mathrm{av}}(s)ds + \int_{0}^{t} U_{\mathrm{av}}(s)ds, \tag{8.40}$$

$$\int_{t-D(t)}^{t} U_{\mathrm{av}}(s)ds = -\int_{0}^{t-D(t)} U_{\mathrm{av}}(s)ds + \int_{0}^{t} U_{\mathrm{av}}(s)ds. \tag{8.41}$$

Solving to integral below

$$\int_{0}^{t-D(t)} U_{\mathrm{av}}(s)ds, \tag{8.42}$$

taking into account

$$s = t - D(t), \qquad ds = dt - \dot{D}dt, \tag{8.43}$$

we get $t = D(t)$ for $s = 0$ and $t = t$ for $s = t - D(t)$. Substituting the above conditions at the limits of integration of (8.42), we have

$$\int_{D(t)}^{t} U_{\mathrm{av}}(t - D(t))\left[dt - \dot{D}dt\right] \tag{8.44}$$

$$= \int_{D(t)}^{t} (1 - \dot{D})U_{\mathrm{av}}(t - D(t))dt \tag{8.45}$$

$$= \int_{0}^{t} (1 - \dot{D})U_{\mathrm{av}}(t - D(t))dt, \tag{8.46}$$

being valid if the conditions $\dot{D} \leq 1$ and $1 - \dot{D} \geq 0$ are satisfied and considering that

$$U_{\mathrm{av}}(t - D(t)) = 0, \quad t \in [0, D(t)]. \tag{8.47}$$

Thus, the nonaverage version (8.38) could be rewritten as

$$U(t) = k\left[G(t) + \hat{H}(t)\int_{0}^{t} (1 - \dot{D}(\sigma))U(\sigma - D(\sigma))d\sigma\right]. \tag{8.48}$$

Thus, instead of implementing (8.23), we could simply adopt the filtered version of (8.48).

8.5 ▪ Stability Analysis

Let us consider the following representation of the state of the transport equation:

$$u(x,t) = U(\phi(t + x(\phi^{-1}(t) - t))). \tag{8.49}$$

This rather nonobvious relation—a composition of four functions—was introduced in [119, Chapter 6]. From (8.49), we get boundary values

$$u(0,t) = U(\phi(t)), \tag{8.50}$$
$$u(1,t) = U(t). \tag{8.51}$$

This yields the following system:

$$\dot{\hat{\theta}}(t-D) = u(0,t), \tag{8.52}$$
$$u_t(x,t) = \pi(x,t)\, u_x(x,t), \quad x \in [0,1], \tag{8.53}$$
$$u(1,t) = U(t), \tag{8.54}$$

where the speed of propagation of the transport equation is given by

$$\pi(x,t) = \frac{1 + x\left(\frac{d(\phi^{-1}(t))}{dt} - 1\right)}{\phi^{-1}(t) - t}. \tag{8.55}$$

Equation (8.55) is obtained simply by calculating $\frac{u_t(x,t)}{u_x(x,t)}$ considering $u(x,t)$ in (8.49). Hence, the following condition can be assumed.

Assumption 8.2. *The function* (8.55) *is strictly positive and uniformly bounded from below and from above by finite constants. Additionally, we assume $\pi(x,t)$ to be nonperiodic on $T = 2\pi/\omega$.*

Moreover, we have

$$\pi(0,t) = \frac{1}{\phi^{-1}(t) - t} \geq \pi_0^* > 0, \tag{8.56}$$

where

$$\pi_0^* = \frac{1}{\sup_{\delta \geq \phi^{-1}(0)}(\delta - \phi(\delta))}. \tag{8.57}$$

Furthermore, we can write

$$\phi^{-1}(t) - t > 0, \tag{8.58}$$

which leads to the *delay time* $\phi(t)$ and *prediction time* $\phi^{-1}(t)$ being both positive and uniformly bounded functions.

Now, rewriting the control law (8.23) in terms of $u(x,t)$, we get

$$u(1,t) = \frac{c}{s+c}\left\{k\left[G(t) + \hat{H}(t)\int_0^1 u(\sigma,t)(\phi^{-1}(t) - t)d\sigma\right]\right\}. \tag{8.59}$$

The next initial condition is further assumed:

$$u_0(x) = u(x,0) = U(\phi(\phi^{-1}(0)x)), \quad x \in [0,1], \tag{8.60}$$

with $\tilde{\theta}_0 = \tilde{\theta}(-D(0))$, and considering the change of variables

$$\tilde{\vartheta}(t) = \tilde{\theta}(t - D(t)), \quad \tilde{\vartheta}_{\rm av}(t) = \tilde{\theta}_{\rm av}(t - D(t)), \tag{8.61}$$

the average version of system (8.52)–(8.54) can be directly written as

$$\dot{\tilde{\vartheta}}_{\rm av}(t) = u_{\rm av}(0,t), \tag{8.62}$$
$$\partial_t u_{\rm av}(x,t) = \pi(x,t)\partial_x u_{\rm av}(x,t), \tag{8.63}$$
$$u_{\rm av}(1,t) = U_{\rm av}(t) \tag{8.64}$$

or

$$\dot{\tilde{\vartheta}}_{\rm av}(t) = u_{\rm av}(0,t), \tag{8.65}$$
$$\partial_t u_{\rm av}(x,t) = \pi(x,t)\partial_x u_{\rm av}(x,t), \quad x \in [0,1], \tag{8.66}$$
$$\begin{aligned} \frac{du_{\rm av}(1,t)}{dt} &= -cu_{\rm av}(1,t) \\ &+ckH\left[\tilde{\vartheta}_{\rm av}(t) + \int_0^1 u_{\rm av}(\sigma,t)\left(\phi^{-1}(t) - t\right)d\sigma\right]. \end{aligned} \tag{8.67}$$

Exponential stability of the average system is stated next.

Theorem 8.1. *Consider the closed-loop system consisting of the plant* (8.62)–(8.64) *and the controller* (8.59)*, and let Assumptions* 8.1 *and* 8.2 *hold. There exist positive constants* α_1, α_2, μ *(independent of* ϕ*) such that the closed-loop system* (8.62)–(8.64) *is locally exponentially stable and the solution satisfies,* $\forall t \geq 0$,

$$\begin{aligned} &\left[\left|\tilde{\theta}_{\rm av}(\phi(t))\right|^2 + [U_{\rm av}(t)]^2 + \int_{\phi(t)}^t U_{\rm av}^2(\xi)d\xi\right] \\ &\leq \frac{\alpha_1}{\alpha_2}e^{-\mu t}\left[\left|\tilde{\theta}_{\rm av}(\phi(0))\right|^2 + [U_{\rm av}(0)]^2 + \int_{\phi(0)}^0 U_{\rm av}^2(\xi)d\xi\right]. \end{aligned} \tag{8.68}$$

Proof. Consider the infinite-dimensional backstepping transformation of transport PDE state [121]

$$\begin{aligned} w(x,t) =& u_{\rm av}(x,t) \\ &- kH\left[\tilde{\vartheta}_{\rm av}(t) + \int_0^x u_{\rm av}(\sigma,t)\left(\phi^{-1}(t) - t\right)d\sigma\right], \end{aligned} \tag{8.69}$$

which maps the system (8.65)–(8.68) into the *target system* [119, 121]:

$$\dot{\tilde{\vartheta}}_{\rm av}(t) = kH\tilde{\vartheta}_{\rm av}(t) + w(0,t), \tag{8.70}$$
$$w_t(x,t) = \pi(x,t)w_x(x,t), \quad x \in [0,1], \tag{8.71}$$
$$w(1,t) = -\frac{1}{c}\partial_t u_{\rm av}(1,t). \tag{8.72}$$

In particular, the inclusion of the dynamic boundary condition [115] in (8.72) due to the inclusion of the low-pass filter in (8.23) and (8.24) was not reported in [119, 121] for the PDE representation of time-varying delayed systems.

The system (8.70)–(8.72) is a standard cascade configuration

$$w \to \tilde{\vartheta}_{\rm av}. \tag{8.73}$$

Now, let us consider the term $w(1,t)$. Deriving (8.69) with respect to the spatial variable x and replacing $\partial_t u_{\rm av}(x,t) = \pi(x,t)\partial_x u_{\rm av}(x,t)$ and $w_t(x,t) = \pi(x,t)w_x(x,t)$—see (8.66) and (8.71)—with $x=1$, it is easily seen that

$$w_t(1,t) = \partial_t u_{\rm av}(1,t) - kH\pi(1,t)u_{\rm av}(1,t)[\phi^{-1}(t)-t], \tag{8.74}$$

where $\partial_t u_{\rm av}(1,t) = \dot{U}_{\rm av}(t)$. Equivalently, we can obtain

$$w_t(1,t) = \partial_t u_{\rm av}(1,t) - kH\bar{\pi}^* u_{\rm av}(1,t), \quad \bar{\pi}^* := \frac{d(\phi^{-1}(t))}{dt}, \tag{8.75}$$

by substituting the speed of propagation of the transport equation $\pi(x,t)$ in (8.55) with $x=1$ into (8.74). Equation (8.75) is a more convenient expression, which will be used in the stability analysis since $\bar{\pi}^* = \frac{d(\phi^{-1}(t))}{dt} = 1 + \dot{D}(t)$ is upper bounded such that $|\bar{\pi}^*| \leq 2$, according to Assumption 8.2.

The inverse of (8.69) is given by

$$\begin{aligned} u_{\rm av}(x,t) =& w(x,t) + kHe^{kHx(\phi^{-1}(t)-t)}\tilde{\vartheta}_{\rm av}(t) \\ &+ kH\int_0^x e^{kH(x-\sigma)(\phi^{-1}(t)-t)}w(\sigma,t)(\phi^{-1}(t)-t)d\sigma. \end{aligned} \tag{8.76}$$

Plugging (8.72) and (8.76) into (8.75) with $x=1$, we obtain

$$\begin{aligned} w_t(1,t) =& -cw(1,t) - kH\bar{\pi}^* w(1,t) \\ &- k^2H^2\bar{\pi}^* e^{kH(\phi^{-1}(t)-t)}\tilde{\vartheta}_{\rm av}(t) \\ &- k^2H^2\bar{\pi}^* \int_0^1 e^{kH(1-\sigma)(\phi^{-1}(t)-t)}w(\sigma,t)(\phi^{-1}(t)-t)d\sigma. \end{aligned} \tag{8.77}$$

In [182], the term $\frac{a}{2}\int_0^1 (1+x)w^2(x,t)dx$ was assumed in the Lyapunov–Krasovskii functional. The time derivative of this term lead us to $\int_0^1 (1+x)\pi(x,t)w(x,t)w_x(x,t)dx$. In the proof of our previous conference version [182], we have used the following inequality:

$$\int_0^1 (1+x)\pi(x,t)w(x,t)w_x(x,t)dx \leq \Pi \int_0^1 (1+x)w(x,t)w_x(x,t)dx, \tag{8.78}$$

with $\Pi \geq \pi(x,t) > 0$. However, the inequality (8.78) is not true since both w and w_x can be either positive or negative.

In this context and unlike [182], we now consider the following Lyapunov–Krasovskii functional [121]:

$$V(t) = \frac{\tilde{\vartheta}_{\rm av}^2(t)}{2} + \frac{a}{2}\int_0^1 e^{bx}w^2(x,t)dx + \frac{1}{2}w^2(1,t), \tag{8.79}$$

where the parameters $a>0$ and $b>0$ are to be chosen later. With some abuse of notation, the constant a in (8.79) is not the amplitude of the perturbation signals given in (8.11).

From Assumption 8.2, we have that $\pi(0,t)$ and $\pi(1,t)$ are lower and upper bounded by $0 < \pi_0^* \leq \pi(0,t)$ and $0 < \pi(1,t) \leq \pi_1^*$, with positive constants π_0^* and π_1^*, the former defined in

(8.57). Recalling that $w_t(x,t)=\pi(x,t)w_x(x,t)$ in (8.71) and applying the integration by parts to the integral term, the time derivative of (8.79) along with (8.70)–(8.72) can be written by

$$\begin{aligned}\dot{V}(t) &= kH\tilde{\vartheta}_{\rm av}^2(t)+\tilde{\vartheta}_{\rm av}(t)w(0,t)\\ &\quad +\frac{a}{2}\int_0^1 \overbrace{e^{bx}\pi(x,t)}^{u}\underbrace{2w(x,t)w_x(x,t)dx}_{dv}\\ &\quad +w(1,t)w_t(1,t)\\ &= kH\tilde{\vartheta}_{\rm av}^2(t)+\tilde{\vartheta}_{\rm av}(t)w(0,t)\\ &\quad +\frac{a}{2}e^{b}\pi(1,t)w^2(1,t)-\frac{a}{2}\pi(0,t)w^2(0,t)\\ &\quad -\frac{a}{2}\int_0^1[b\pi(x,t)+\pi_x(x,t)]e^{bx}w^2(x,t)dx\\ &\quad +w(1,t)w_t(1,t),\end{aligned} \tag{8.80}$$

such that

$$\begin{aligned}\dot{V}(t) &\le kH\tilde{\vartheta}_{\rm av}^2(t)+\frac{\tilde{\vartheta}_{\rm av}^2(t)}{2a\pi_0^*}\\ &\quad -\frac{a}{2}\int_0^1[b\pi(x,t)+\pi_x(x,t)]e^{bx}w^2(x,t)dx\\ &\quad +w(1,t)\left[w_t(1,t)+\frac{a}{2}e^{b}\pi_1^*w(1,t)\right],\end{aligned} \tag{8.81}$$

since $-\frac{a}{2}\pi(0,t)w^2(0,t)\le -\frac{a}{2}\pi_0^*w^2(0,t)$ and, from Young's inequality, we have

$$\begin{aligned}\tilde{\vartheta}_{\rm av}(t)w(0,t) &\le \left|\tilde{\vartheta}_{\rm av}(t)\right||w(0,t)|\\ &\le \frac{\tilde{\vartheta}_{\rm av}^2(t)}{2a\pi_0^*}+\frac{a\pi_0^*w^2(0,t)}{2}.\end{aligned} \tag{8.82}$$

Recalling that $k>0$ and $H<0$, let us choose $a=-1/(kH\pi_0^*)$. Then,

$$\begin{aligned}\dot{V}(t) &\le \frac{kH}{2}\tilde{\vartheta}_{\rm av}^2(t)-\frac{a}{2}\int_0^1[b\pi(x,t)+\pi_x(x,t)]e^{bx}w^2(x,t)dx\\ &\quad +w(1,t)\left[w_t(1,t)+\frac{a}{2}e^{b}\pi_1^*w(1,t)\right]\\ &\le -\frac{\tilde{\vartheta}_{\rm av}^2(t)}{2a\pi_0^*}-\frac{a}{2}\int_0^1[b\pi(x,t)+\pi_x(x,t)]e^{bx}w^2(x,t)dx\\ &\quad +w(1,t)\left[w_t(1,t)+\frac{a}{2}e^{b}\pi_1^*w(1,t)\right].\end{aligned} \tag{8.83}$$

Next, we observe that

$$\pi_x(x,t)=\frac{\frac{d(\phi^{-1}(t))}{dt}-1}{\phi^{-1}(t)-t} \tag{8.84}$$

is a function of t only. Hence,

$$\begin{aligned} b\pi(x,t) + \pi_x(x,t) &= \frac{b\left[1+x\left(\frac{d(\phi^{-1}(t))}{dt}-1\right)\right]+\frac{d(\phi^{-1}(t))}{dt}-1}{\phi^{-1}(t)-t} \\ &= \frac{b-1+\frac{d(\phi^{-1}(t))}{dt}+bx\left(\frac{d(\phi^{-1}(t))}{dt}-1\right)}{\phi^{-1}(t)-t}. \end{aligned} \tag{8.85}$$

Since this is a linear function on x, it follows that it has a minimum at either $x=0$ or $x=1$, so we get

$$b\pi(x,t)+\pi_x(x,t) \geq \frac{\min\left\{b-1+\frac{d(\phi^{-1}(t))}{dt},\ (b+1)\frac{d(\phi^{-1}(t))}{dt}-1\right\}}{\phi^{-1}(t)-t}. \tag{8.86}$$

Next, we note that

$$\bar{\pi}^* := \frac{d(\phi^{-1}(t))}{dt} = \frac{1}{\phi'(\phi^{-1}(t))} \geq \frac{1}{\sup_{\delta\geq\phi^{-1}(0)}\phi'(\delta)} = \pi^*, \tag{8.87}$$

which yields

$$b\pi(x,t)+\pi_x(x,t) \geq \frac{\min\{b-1+\pi^*,\ (b+1)\pi^*-1\}}{\phi^{-1}(t)-t}. \tag{8.88}$$

Choosing

$$b \geq (1-\pi^*)\max\left\{1,\ \frac{1}{\pi^*}\right\}, \tag{8.89}$$

and recalling that $\frac{1}{\phi^{-1}(t)-t} \geq \pi_0^*$ from (8.56), we get

$$b\pi(x,t)+\pi_x(x,t) \geq \pi_0^*\beta^*, \tag{8.90}$$

where

$$\beta^* = \min\{b-1+\pi^*,\ (b+1)\pi^*-1\} > 0. \tag{8.91}$$

So, returning to $\dot{V}$, we have that

$$\begin{aligned} \dot{V}(t) \leq & -\frac{1}{2a\pi_0^*}\tilde{\vartheta}_{\mathrm{av}}^2(t) - \frac{a}{2}\pi_0^*\beta^*\int_0^1 e^{bx}w^2(x,t)dx \\ & + w(1,t)\left[w_t(1,t)+\frac{a}{2}e^b\pi_1^*w(1,t)\right]. \end{aligned} \tag{8.92}$$

Now, consider (8.92) along with (8.77). With a completion of squares, we obtain

$$\begin{aligned} \dot{V}(t) \leq & -\frac{1}{4a\pi_0^*}\tilde{\vartheta}_{\mathrm{av}}^2(t) - Y\int_0^1 e^{bx}w^2(x,t)dx \\ & + a\pi_0^*\left|k^2H^2\bar{\pi}^*e^{kH(\phi^{-1}(t)-t)}\right|^2 w^2(1,t) \\ & + X\left\|k^2H^2\bar{\pi}^*\left[\phi^{-1}(t)-t\right]e^{kH(1-\sigma)(\phi^{-1}(t)-t)}\right\|^2 w^2(1,t) \\ & + \left[\frac{a}{2}e^b\pi_1^*+|kH|\bar{\pi}^*\right]w^2(1,t) - cw^2(1,t). \end{aligned} \tag{8.93}$$

To obtain (8.93), we have used

$$\begin{aligned}&-w(1,t)\left\langle k^2H^2\bar{\pi}^*\left[\phi^{-1}(t)-t\right]e^{kH(1-\sigma)(\phi^{-1}(t)-t)},w(\sigma,t)\right\rangle\\&\leq|w(1,t)|\left\|k^2H^2\bar{\pi}^*\left[\phi^{-1}(t)-t\right]e^{kH(1-\sigma)(\phi^{-1}(t)-t)}\right\|\|w(t)\|\\&\leq Y\|w(t)\|^2+X\left\|k^2H^2\bar{\pi}^*\left[\phi^{-1}(t)-t\right]e^{kH(1-\sigma)(\phi^{-1}(t)-t)}\right\|^2w^2(1,t),\end{aligned}\tag{8.94}$$

where the first inequality is the Cauchy–Schwarz and the second is Young's, the notation $\langle\cdot,\cdot\rangle$ denotes the inner product in the spatial variable $\sigma\in[0,1]$, on which both

$$\left[\phi^{-1}(t)-t\right]e^{kH(1-\sigma)(\phi^{-1}(t)-t)}$$

and $w(\sigma,t)$ depend, and $\|\cdot\|$ denotes the L_2-norm in σ. In addition, we have also applied Young's inequality to show that

$$\begin{aligned}&-k^2H^2\bar{\pi}^*e^{kH(\phi^{-1}(t)-t)}\tilde{\vartheta}_{\text{av}}(t)w(1,t)\\&\leq a\pi_0^*\left|k^2H^2\bar{\pi}^*e^{kH(\phi^{-1}(t)-t)}\right|^2w^2(1,t)+\frac{1}{4a\pi_0^*}\left|\tilde{\vartheta}_{\text{av}}(t)\right|^2.\end{aligned}\tag{8.95}$$

Then, from (8.93), we arrive at

$$\dot{V}(t)\leq-\frac{1}{4a\pi_0^*}\tilde{\vartheta}^2_{\text{av}}(t)-Y\int_0^1e^{bx}w^2(x,t)dx-(c-c^*)w^2(1,t),\tag{8.96}$$

where

$$\begin{aligned}c^*=&\,a\pi_0^*\left|k^2H^2\bar{\pi}^*e^{kH(\phi^{-1}(t)-t)}\right|^2\\&+X\left\|k^2H^2\bar{\pi}^*\left[\phi^{-1}(t)-t\right]e^{kH(1-\sigma)(\phi^{-1}(t)-t)}\right\|^2\\&+\left[\frac{a}{2}e^b\pi_1^*+|kH|\bar{\pi}^*\right]\end{aligned}\tag{8.97}$$

for appropriate constants $X>0$ and $Y>0$. Consequently, from (8.96), if c is chosen such that $c>c^*$, we obtain

$$\dot{V}(t)\leq-\mu V(t)\tag{8.98}$$

for some $\mu>0$. Hence, the closed-loop system is exponentially stable in the sense of the full-state norm

$$\left(\left|\tilde{\vartheta}^2_{\text{av}}(t)\right|^2+\int_0^1w^2(x,t)dx+w^2(1,t)\right)^{1/2},\tag{8.99}$$

i.e., in the transformed variable $(\tilde{\vartheta}_{\text{av}},w)$.

From $\left(\left|\tilde{\vartheta}^2_{\text{av}}(t)\right|^2+\int_0^1u^2_{\text{av}}(x,t)dx+u^2_{\text{av}}(1,t)\right)^{1/2}$, we achieve the exponential stability in the sense of the norm L_2 for the original variables $\tilde{\theta}_{\text{av}}$ and u_{av}. In this case, it is necessary to show that there exist positive scalars α_1 and α_2 such that

$$\alpha_1\Psi(t)\leq V(t)\leq\alpha_2\Psi(t),\tag{8.100}$$

with $\Psi(t) = \left|\tilde{\vartheta}_{\rm av}^2(t)\right|^2 + \int_0^1 u_{\rm av}^2(x,t)dx + u_{\rm av}^2(1,t)$, or, equivalently,

$$\Psi(t) = \left|\tilde{\theta}_{\rm av}(t-D(t))\right|^2 + \int_{t-D(t)}^{t} U_{\rm av}^2(\tau)d\tau + U_{\rm av}^2(t). \tag{8.101}$$

This can be directly established by using (8.69), (8.76), (8.79) and applying the Cauchy–Schwarz inequality and other mathematical computations, as in the proof of [119, Theorem 2.1]. Thus, from (8.98), we obtain

$$\Psi(t) \leq \frac{\alpha_2}{\alpha_1} e^{-\mu t}\Psi(0), \tag{8.102}$$

which is equivalent to inequality (8.68). □

8.6 ▪ State-Dependent Delays

In this section, we consider the plant (8.1)–(8.3) subject to a delay function that depends explicitly on the system state [33]:

$$y(t) = Q(\theta(t - D(\theta(t)))). \tag{8.103}$$

This problem is particularly important in distributed extremum seeking control of mobile robots or formation control with multiple agents, where the magnitude of the delay may depend on the distance of the robots/agents from the operator interface or from the virtual leaders [136], [228].

One more interesting feature is that the delay function can even depend on unmeasured signals, but respecting the condition that at least the analytical expression (parametrization) for the delays is still known and the unmeasured signals can be estimated in the average sense. Without loss of generality, let us assume here that the delay function in (8.103) is

$$D(\theta(t)) = Q'(\theta(t)) = H(\theta(t) - \theta^*), \tag{8.104}$$

with $Q(\theta)$ defined in (8.2). In this case, the delay function is exactly the expression of the gradient of the map we want to maximize.

Now, we conveniently redefine the estimation error introduced in (8.8) as

$$\tilde{\theta}(t) = \hat{\theta}(t - D(\theta(t))) - \theta^*, \tag{8.105}$$

and recalling that $\dot{\tilde{\theta}}(t) = \dot{\hat{\theta}}(t - D(\theta(t))) = U(t - D(\theta(t)))$, we can write the associated error dynamics as

$$\dot{\tilde{\theta}}(t) = U(t - D(\theta(t))), \tag{8.106}$$

where $U(t)$ is the control signal.

Similarly to (8.1), but now assuming (8.103) and (8.105), the following average version of the signal $G(t)$ in (8.10) can be derived straightforwardly by

$$G_{\rm av}(t) = H\tilde{\theta}_{\rm av}(t). \tag{8.107}$$

From (8.106) and (8.107), the following average models can also be obtained,

$$\dot{\tilde{\theta}}_{\rm av}(t) = U_{\rm av}(\phi(t)), \tag{8.108}$$
$$\dot{G}_{\rm av}(t) = HU_{\rm av}(\phi(t)), \tag{8.109}$$
$$\phi(t) = t - D(G_{\rm av}(t)), \tag{8.110}$$

since $\theta(t-D(\theta(t))) = \hat{\theta}(t-D(\theta(t))) + S(t-D(\theta(t))) = \tilde{\theta}(t) + \theta^* + a\sin(\omega(t-D(\theta(t))))$ from the definitions in (8.10) and (8.105). Hence, the average signal $[\theta(t-D(\theta(t))) - \theta^*]_{\rm av} = [\tilde{\theta}(t)]_{\rm av}$ since $[a\sin(\omega(t-D(\theta(t))))]_{\rm av} = 0$. By multiplying both sides of the former equation by H, from (8.104) and (8.107), we can write, with some abuse of notation, $[D(\theta(t))]_{\rm av} = G_{\rm av}(t) := D(G_{\rm av}(t))$. The term $U_{\rm av} : [\phi(t_0), \infty) \to \mathbb{R}, t \geq t_0 \geq 0, D \in C^1(\mathbb{R};\, \mathbb{R}_+)$ is the average control of $U \in \mathbb{R}$.

By defining the prediction time as $\sigma(t) = \phi^{-1}(t)$ the inversion of the time variable $t \to t - D(G_{\rm av}(t))$ in $t \to t + D(P_{\rm av}(t))$, the following hold:

$$P_{\rm av}(t) = G_{\rm av}(P_{\rm av}(t)), \tag{8.111}$$

$$\sigma(t) = t + D(P_{\rm av}(t)), \tag{8.112}$$

where $P_{\rm av}(t)$ is the associated predictor state.

Differentiating (8.111) and (8.112) and using (8.109), we arrive at [33]

$$\frac{dG_{\rm av}(\sigma(t))}{dt} = HU_{\rm av}(t)\frac{d\sigma(t)}{dt} \tag{8.113}$$

and

$$\dot{\sigma}(t) = \frac{1}{1 - H\nabla D(P_{\rm av}(t))U_{\rm av}(t)}, \tag{8.114}$$

with $\dot{D} = \nabla D \dot{G}_{\rm av} = \nabla D H U_{\rm av}$, ∇ being the gradient operator.

The following implicit integral relations [33] are derived for the predictor state and the prediction time by integrating (8.113) and (8.114) for all delay interval $\phi(t) \leq \Theta \leq t$:

$$P_{\rm av}(\Theta) = G_{\rm av}(t) + \int_{\phi(t)}^{\Theta} \frac{HU_{\rm av}(s)}{1 - H\nabla D(P_{\rm av}(s))U_{\rm av}(s)} ds, \tag{8.115}$$

$$\sigma(\Theta) = t + \int_{\phi(t)}^{\Theta} \frac{1}{1 - H\nabla D(P_{\rm av}(s))U_{\rm av}(s)} ds. \tag{8.116}$$

Regarding the nominal feedback control gain $k > 0$ for the plant free of delays, the predictor feedback control law can be written as

$$U_{\rm av}(t) = k\left[G_{\rm av}(t) + \int_{\phi(t)}^{t} \frac{HU_{\rm av}(s)}{1 - H\nabla D(P_{\rm av}(s))U_{\rm av}(s)} ds\right]. \tag{8.117}$$

The predictor feedback control law (8.117) is subject to the following feasibility condition [33]:

$$H\nabla D(P_{\rm av}(t))U_{\rm av}(t) < 1. \tag{8.118}$$

This is due to the fact that the prediction signal can be only generated if (8.115) and (8.116) are well-posed. If the condition (8.118) is violated, the control signal is directed toward the opposite direction to the plant.

Thus, the nonaverage version for the filtered predictor-based control law is given by

$$U(t) = \frac{c}{s+c}\{k[G(t) + \Gamma(t)]\}, \quad k > 0, \quad c > 0, \tag{8.119}$$

$$\Gamma(t) = \hat{H}(t)\int_{\phi(t)}^{t} \frac{U(\tau)}{1 - \hat{H}(\tau)\nabla D(G(\tau))U(\tau)} d\tau. \tag{8.120}$$

The stability proof can be established using an analogous average infinite-dimensional representation of the actuator state (8.62)–(8.64) with the propagation speed defined as

$$\pi(x,t) = \frac{1+x(\dot{\sigma}(t)-1)}{\sigma(t)-1}. \tag{8.121}$$

The infinite-dimensional representation for the system (8.108)–(8.110) is

$$\dot{\tilde{\theta}}_{\rm av}(t) = u_{\rm av}(0,t), \tag{8.122}$$

$$\partial_t u_{\rm av}(x,t) = \pi(x,t)\partial_x u_{\rm av}(x,t), \quad x \in [0,1], \tag{8.123}$$

$$\frac{du_{\rm av}(1,t)}{dt} = -cu_{\rm av}(1,t) + ckH\left[\tilde{\theta}_{\rm av} + (\sigma(t)-t)\int_0^1 u_{\rm av}(x,t)dx\right], \tag{8.124}$$

which can be mapped into the *target system* [33]

$$\dot{\tilde{\theta}}_{\rm av}(t) = kH\tilde{\theta}_{\rm av}(t) + w(0,t), \tag{8.125}$$

$$w_t(x,t) = \pi(x,t)w_x(x,t), \quad x \in [0,1], \tag{8.126}$$

$$w(1,t) = -\frac{1}{c}\partial_t u_{\rm av}(1,t), \tag{8.127}$$

with the help of the following invertible backstepping transformation:[7]

$$w(x,t) = u_{\rm av}(x,t) - kH\left[\tilde{\theta}_{\rm av}(t) + (\sigma(t)-t)\int_0^x u_{\rm av}(\sigma,t)d\sigma\right]. \tag{8.128}$$

Considering the next Lyapunov–Krasovskii functional

$$V(t) = \frac{\tilde{\theta}^2_{\rm av}(t)}{2} + \frac{a}{2}\int_0^1 e^{bx}w^2(x,t)dx + \frac{1}{2}w^2(1,t), \tag{8.129}$$

with appropriate constants $a > 0$ and $b > 0$, the proof of stability can be provided following the same procedure for time-varying delays. The main difference is that the delay time $\phi(t)$ depends on the state as well as the prediction time $\phi^{-1}(t)$—here denoted by $\sigma(t)$; see Appendix D.2.

8.7 ▪ Time- and State-Dependent Delays as PDE-PDE Cascades

In this section, we deal with cascades [119, Part V] of hyperbolic transport equations. We concatenate a hyperbolic transport PDE (of variable convection speed) [53] with a constant input delay. This is an option for representing time-varying delays $D(t)$ of Section 8.2, state-dependent delays $D(\theta)$ in Section 8.6, or even simultaneous time- and state-dependent delays $D(t,\theta)$, as discussed in the following.

In this manner, we can represent a time- and state-dependent delay $D(t,\theta) \geq 0$ as [27, 52]:

$$D(t,\theta) = D_0 + \Delta(t,\theta), \tag{8.130}$$

[7] The infinite-dimensional representation of the predictor state is defined following the transformation of the actuator state (8.49), where $\phi(t)$ is now the state-dependent delay given by (8.110).

where the scalar $D_0 > 0$ denotes the constant portion of the delay and the term $\Delta(t,\theta) \in \mathbb{R}$ its variable portion.

Under this sort of delay (8.130) and from Figure 8.2, we can write the equations for the following delay-hyperbolic PDE cascade system [119, Part V]:

$$\alpha(0,t) = \Theta(t), \tag{8.131}$$
$$\partial_t \alpha(x,t) = \pi_\Delta(x,t)\,\partial_x \alpha(x,t), \quad x \in [0,1], \tag{8.132}$$
$$\alpha(1,t) = \theta(t - D_0), \tag{8.133}$$

where $\alpha(x,t)$ is the infinite-dimensional state of the hyperbolic transport PDE with variable convection speed $\pi_\Delta(x,t) > 0$ used to represent $\Delta(t,\theta)$ in (8.130) and spatial domain defined without loss of generality by $x \in [0,1]$. The boundary delay is denoted by D_0 being any arbitrary known constant given in (8.130). In (8.131), we conveniently denote the input of the map $y = Q(\Theta)$ by its inverse function $\Theta = Q^{-1}(y)$.

Consequently, we can rewrite (8.52)–(8.54) into the following delay-hyperbolic PDE dynamic cascade system:

$$\dot{\vartheta}(t) = u(0,t), \tag{8.134}$$
$$\partial_t u(x,t) = \pi_\Delta(x,t)\,\partial_x u(x,t), \quad x \in [0,1], \tag{8.135}$$
$$u(1,t) = v(1,t), \tag{8.136}$$
$$\partial_t v(x,t) = \partial_x v(x,t), \quad x \in [1, 1+D_0], \tag{8.137}$$
$$v(1+D_0,t) = U(t), \tag{8.138}$$

where $U(t)$ is the overall system (control) input and (ϑ, u, v) is the state of the ODE-PDE-PDE cascade. From the transport PDE representation form for the input delay, we know that [119, Chapter 2]

$$v(x,t) = U(t + x - 1 - D_0), \quad x \in [1, 1+D_0]. \tag{8.139}$$

Applying the backstepping transformations (with an appropriate prediction time $\sigma(t)$)

$$\begin{aligned} w_{\mathrm{av}}(x,t) &= u_{\mathrm{av}}(x,t) \\ &\quad - k\left[H\tilde{\vartheta}_{\mathrm{av}}(t) + (\sigma(t)-t)H\int_0^x u_{\mathrm{av}}(\sigma,t)d\sigma\right], \quad x \in [0,1], \end{aligned} \tag{8.140}$$

$$\begin{aligned} \zeta_{\mathrm{av}}(x,t) &= v_{\mathrm{av}}(x,t) - kH\int_1^x v_{\mathrm{av}}(\sigma,t)d\sigma \\ &\quad - kH\int_0^1 u_{\mathrm{av}}(\sigma,t)d\sigma - kH\vartheta_{\mathrm{av}}(t), \quad x \in [1, 1+D_0], \end{aligned} \tag{8.141}$$

to the average version of (8.134)–(8.138), the resulting exponentially stable [119, Part V] average target system is

$$\dot{\vartheta}_{\mathrm{av}}(t) = kH\vartheta_{\mathrm{av}}(t) + w_{\mathrm{av}}(0,t), \tag{8.142}$$
$$\partial_t w_{\mathrm{av}}(x,t) = \pi_\Delta(x,t)\partial_x w_{\mathrm{av}}(x,t), \tag{8.143}$$
$$w_{\mathrm{av}}(1,t) = \zeta_{\mathrm{av}}(1,t), \tag{8.144}$$
$$\partial_t \zeta_{\mathrm{av}}(x,t) = \partial_x \zeta_{\mathrm{av}}(x,t), \quad x \in [1, 1+D_0], \tag{8.145}$$
$$\zeta_{\mathrm{av}}(1+D_0,t) = -\frac{1}{c}\partial_t v_{\mathrm{av}}(1+D_0,t). \tag{8.146}$$

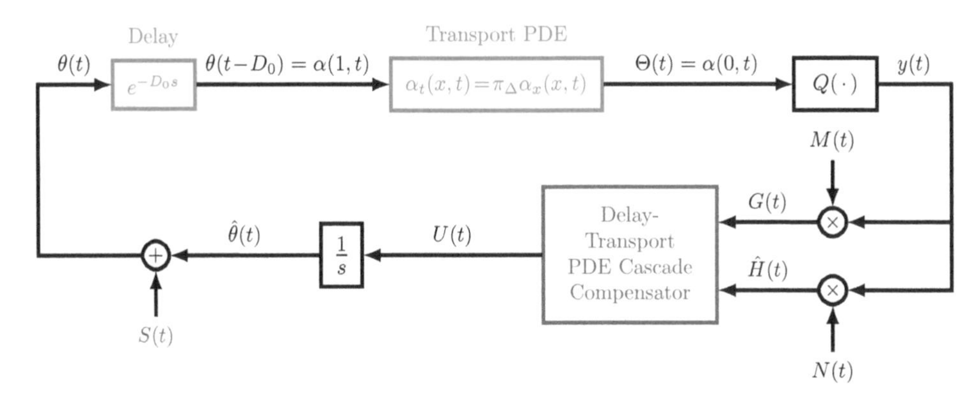

Figure 8.2. *Block diagram for implementation of ES control design for the delay-PDE cascade* (8.131)–(8.133) *at the input of nonlinear convex maps* $y = Q(\Theta)$. *Such a representation encompasses simultaneous time- and state-dependent delays* $D(t,\theta)$ *in* (8.130). *Although the additive dither* $S(t)$ *is the same of the classical ES designs* [129], *the multiplicative demodulation signals* $M(t)$ *and* $N(t)$ *must be redesigned according to* (8.150), *and the application of an adequate boundary control law* (8.148) *for delay-PDE compensation is required.*

From (8.141) and (8.146), one can obtain the expression for $\dot{U}_{\rm av}(t) = -c\zeta_{\rm av}(1+D_0,t)$ such that

$$\begin{aligned}\dot{U}_{\rm av}(t) = -cv_{\rm av}(1+D_0,t)+ckH\int_1^{1+D_0} v_{\rm av}(\sigma,t)d\sigma \\ +ckH\int_0^1 u_{\rm av}(\sigma,t)d\sigma + ckH\vartheta_{\rm av}(t).\end{aligned} \tag{8.147}$$

Plugging (8.139) into (8.147), we obtain the control law

$$U(t) = \frac{c}{s+c}\left\{k\left[G(t)+\hat{H}(t)\int_{t-D_0}^{t} U(\tau)d\tau + \hat{H}(t)\int_0^1 u(x,t)dx\right]\right\} \tag{8.148}$$

for $c > 0$ sufficiently large.

As discussed in [179], the additive dither $S(t)$ is given by

$$S(t) = a\sin(\omega t), \tag{8.149}$$

and the demodulation signals $M(t)$ and $N(t)$ under time- and state-dependent delays would be solved analogously to (8.11) and (8.13) as

$$M(t) = \frac{2}{a}\sin(\omega(t-D(t,\theta))), \quad N(t) = -\frac{8}{a^2}\cos(2\omega(t-D(t,\theta))). \tag{8.150}$$

8.8 ▪ Averaging for Nonconstant Delays

In Theorem 8.1, we proved that the average closed-loop system (8.52)–(8.54) with transport PDE for delay representation is exponentially stable. In [174], [179] we have employed the averaging results for fixed delays. Nevertheless, for time-varying delays (as well as state-dependent delays $D(\theta)$ or simultaneous time- and state-dependent delays $D(t,\theta)$), there is no suitable averaging theorem such as in [83] and [131].

From [83, 131] and under the assumption of constant delays ($\dot{D}(t) \equiv 0$), we would be able to write the closed-loop system (8.9) and (8.24) into the form

$$\dot{\eta}(t) = f(t/\epsilon, \eta_t), \tag{8.151}$$

where $\eta(t) = [\tilde{\theta}(t-D), U(t)]^T$, $\eta_t(r) = \eta(t+r)$ for $-D \leq r \leq 0$, and f is an adequate continuous functional, such that the averaging theorem by [83, 131] (see Appendix A for a compact version) can be straightforwardly applied, assuming $\omega = 1/\epsilon$. The gap here is very subtle: for instance, if the delay is time-varying ($\dot{D}(t) \neq 0$), r would be possible replaced by $r(t)$, which is also time-varying, and we need to prove if the function $r(t)$ would result in $-D_{\max} \leq r(t) \leq 0$, with $D_{\max} > 0$ being a constant upper bound for $D(t)$. This point is not easy to establish, and a formal proof is not available in the literature, which would involve the theory of semigroups for linear operators applied to PDEs [190].

In some particular cases, where the average of the variable term of the delay such as $\Delta(t,\theta)$ in (8.130) has average zero or even constant average ($\Delta_{\mathrm{av}}(t,\theta_{\mathrm{av}}) = 0$, $\Delta_{\mathrm{av}}(t,\theta_{\mathrm{av}}) = \text{constant}$), the cascade average system (8.143)–(8.144) becomes simply a cascade of two transport PDEs with fixed convection speeds or, equivalently, a string of constant delays; for those the representation (8.151) would be valid and the infinite-dimensional averaging theory in [83] and [131] could still be employed. Since the derivation of a completely novel averaging theorem for general variable delays (with distributed terms of the predictor) is not the focus of the current chapter, we leave this challenging generalization for a future research pursuit. Some first steps were already given in [69, 248] for pointwise time-varying delays.

If an appropriate averaging theorem existed, then one would be able to obtain the next theorem for time-varying delays (and an analogous version for the case of state-dependent delays).

Theorem 8.2. *Consider the control system in Figure* 8.1 *with locally quadratic nonlinear map* (8.1)–(8.3) *under time-varying output delays satisfying Assumptions* 8.1 *and* 8.2. *There exists* $c^* > 0$ *such that,* $\forall c \geq c^*$, $\exists w^*(c) > 0$ *such that,* $\forall \omega > \omega^*$, *the delayed closed-loop system* (8.9) *and* (8.24), *with* $G(t)$ *in* (8.10), $\hat{H}(t)$ *calculated as in* (8.12), *and state* $\tilde{\theta}(t-D(t)), U(\sigma)$, $\forall \sigma \in [t-D(t), t]$, *has a unique locally exponentially stable periodic solution in* t *in the period* $T = 2\pi/\omega$, *denoted by* $\tilde{\theta}^T(t-D(t)), U^T(\sigma)$, $\forall \sigma \in [t-D(t), t]$, *satisfying,* $\forall t \geq 0$,

$$\left(\left| \tilde{\theta}^T(t-D(t)) \right|^2 + \left[U^T(t) \right]^2 + \int_{t-D(t)}^{t} \left[U^T(\tau) \right]^2 d\tau \right)^{1/2} \leq \mathcal{O}\left(\frac{1}{\omega} \right). \tag{8.152}$$

Furthermore,

$$\limsup_{t \to +\infty} |\theta(t) - \theta^*| = \mathcal{O}(a + 1/\omega), \tag{8.153}$$

$$\limsup_{t \to +\infty} |y(t) - y^*| = \mathcal{O}(a^2 + 1/\omega^2). \tag{8.154}$$

For the time-varying delays case, the proof of asymptotic convergence to the extremum point (θ^*, y^*), delineated by (8.153) and (8.154) is very close to that constructed in [179] for the case of constant delays, except for some intuitive changes. Please refer to Step 7 of the proofs of Theorem 1 in [179] or Step 8 of the proof for Theorem 2.1 in Chapter 2 for more details.

8.9 ▪ Simulation Results

Unlike most of the results in the literature which consider stabilization, tracking, or disturbance rejection problems [66, 50], as discussed before, we are handling nonconstant time delays in real-time optimization via ES.

Efforts in delay compensation for ES feedback are motivated by the interest to address the lack of delay robustness identified by [174, 179], showing that standard feedback laws for ES have a zero robustness margin to the introduction of a delay in the feedback loop—an arbitrarily

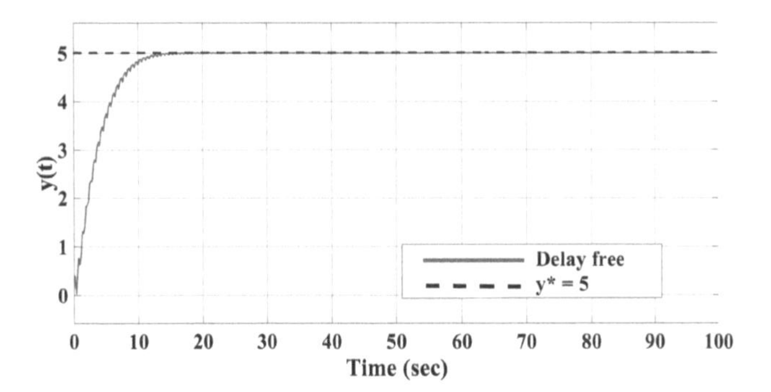

Figure 8.3. *Output $y(t)$ without delays.*

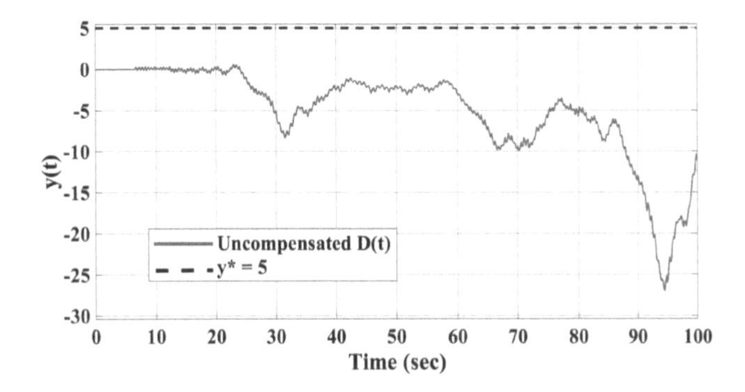

Figure 8.4. *Time evolution of the output $y(t)$ under $D(t)=2\cos(0.1t)+5$ seconds and employing the predictor for constant delays proposed in* [174].

small measurement delay or input delay results in closed-loop instability. In [174, 179], only constant delays were considered.

In order to evaluate the proposed predictor-based ES approach for nonconstant delays compensation, let us assume the following quadratic static map:

$$Q(\theta) = 5 - 0.1(\theta - 2)^2 \tag{8.155}$$

with an output delay $D(t)=2\sin(\omega_0 t)+D_0$ seconds, where $\omega_0=0.1$ rad/s and $D_0=5$. From (8.155), the extremum coordinates are given by $(\theta^*, y^*)=(2,5)$ and the map's Hessian is chosen as $H=-0.2$. In the following, the numerical tests of the predictor (8.23) are carried out with $\hat{H}$ defined by (8.12) and considering $c=20$. The other parameters for simulation were chosen as $a=0.2$, $\omega=10$, $k=0.2$, and initial condition $\theta(0)=-5$.

The Euler method with step-size $h=10^{-4}$ seconds is used for numerical integration. Hence, all signals in the numerical tests are considered continuous since h is chosen sufficiently small. However, it is worth mention that any fast change (of a couple of seconds) within a time range of hundreds of seconds may give a false impression of a discontinuous behavior in the simulation plots.

The output $y(t)$ is shown in three different situations: (a) without delay (Figure 8.3); (b) in the presence of a time-varying output delay, but with predictor-feedback compensation proposed in [174] considering a nominal constant delay of $D_0 = 5$ seconds (Figure 8.4); and (c) with output delay and predictor-based compensator proposed in (8.23)—see Figure 8.5.

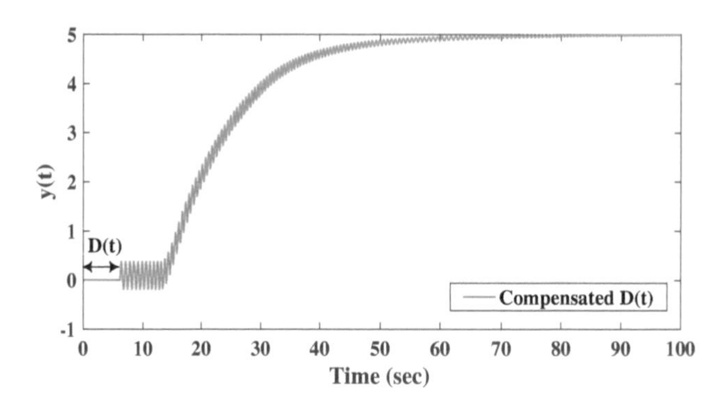

Figure 8.5. *Time evolution of the output $y(t)$ under $D(t) = 2\cos(0.1t) + 5$ seconds and employing the predictor for time-varying delays in* (8.23).

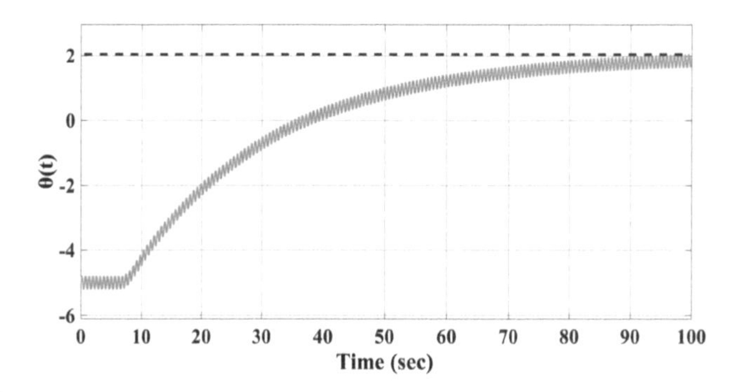

Figure 8.6. *Input parameter $\theta(t)$ converging to the maximizer $\theta^* = 2$.*

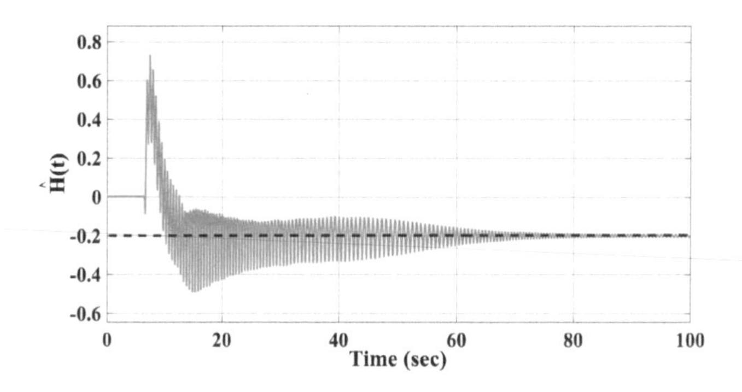

Figure 8.7. *Time evolution of the Hessian estimate $\hat{H}(t)$ converging to $H = -0.2$.*

Although basic ES works well without delays, it goes unstable in the presence of delays if no compensation is applied or if an incorrect prediction scheme for constant delays [174] is implemented instead. The proposed predictor for time-varying delays fixes this.

In Figure 8.6, the input parameter θ is displayed, converging to θ^*. Figure 8.7 shows the estimate of unknown Hessian H, whereas Figure 8.8 presents the plot for the control signal $U(t)$. At last, the variable profile of the considered time-varying delay is displayed in Figure 8.9, whereas its derivative $\dot{D}(t) = -0.2\sin(0.1t)$ is illustrated in Figure 8.10.

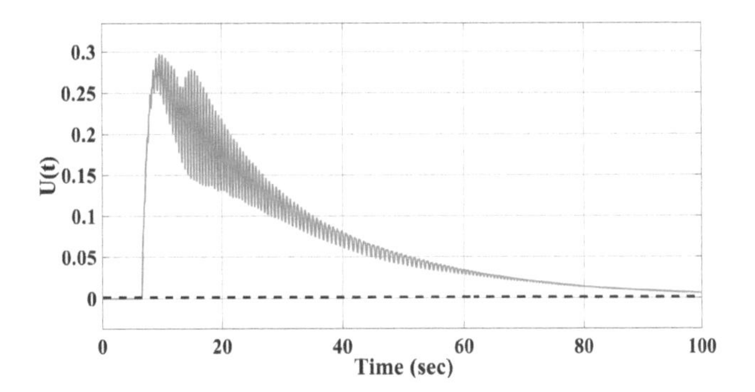

Figure 8.8. *Control signal* $U(t)$.

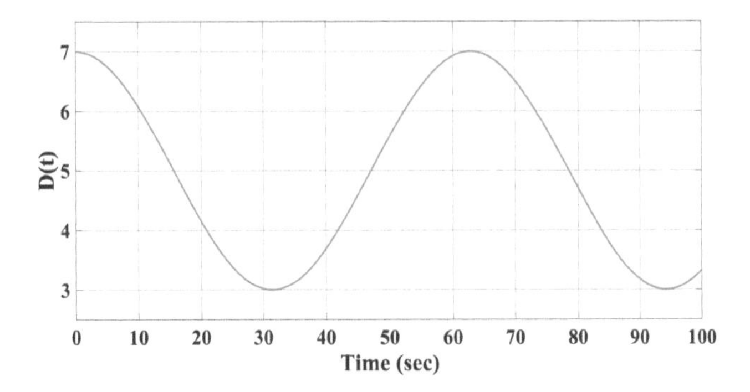

Figure 8.9. *Time-varying nature of the delay* $D(t)=2\cos(0.1t)+5$ *in seconds.*

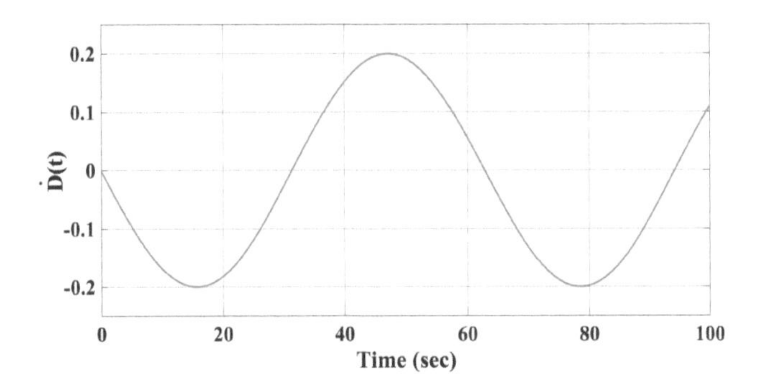

Figure 8.10. *Time-derivative of the delay* $\dot{D}(t) = -0.2\sin(0.1t)$.

Figures 8.11(a) and 8.11(b) and Figures 8.12(a) and 8.12(b) show the robustness and the control performance of the proposed extremum seeking approach in the presence of different types of measurement noise $n(t)$. In Figures 8.11(a) and 8.11(b), a white noise is considered, whereas a colored noise is employed in Figures 8.12(a) and 8.12(b). Under both adverse scenarios, the extremum search is preserved and the ultimate error bound is kept within reasonable limits (convergence to a small residual set around the extremum $y^* = 5$).

Finally, Figures 8.13 and 8.14 present the effect of the time-varying delay function in the convergence of the algorithm for different time-varying delays $D(t) = 5 + 2\epsilon\sin(\omega_0 t)$ with $\omega_0 \in (0,\ 50]$ and $\epsilon = 0.01$. As $\omega_0 \to \omega$, the control system performance is severely affected,

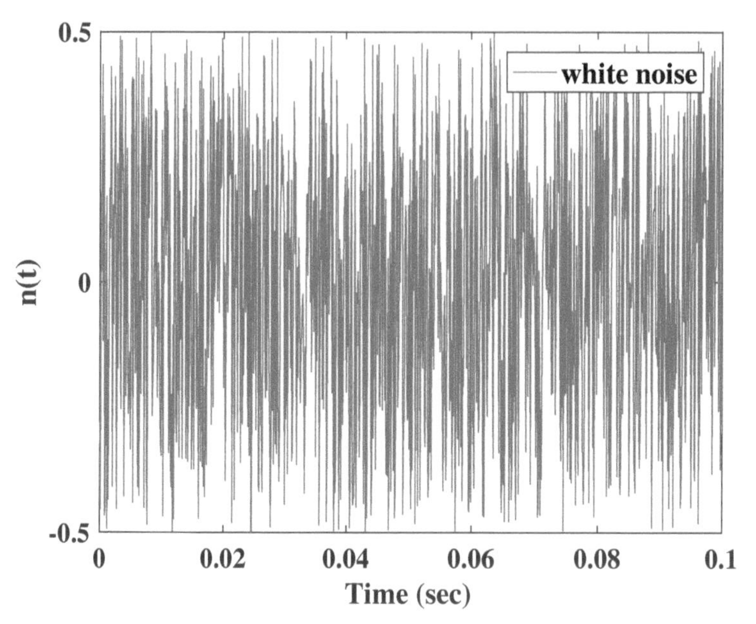

(a) Zoom of the white noise signal $n(t)$ used in the numerical tests.

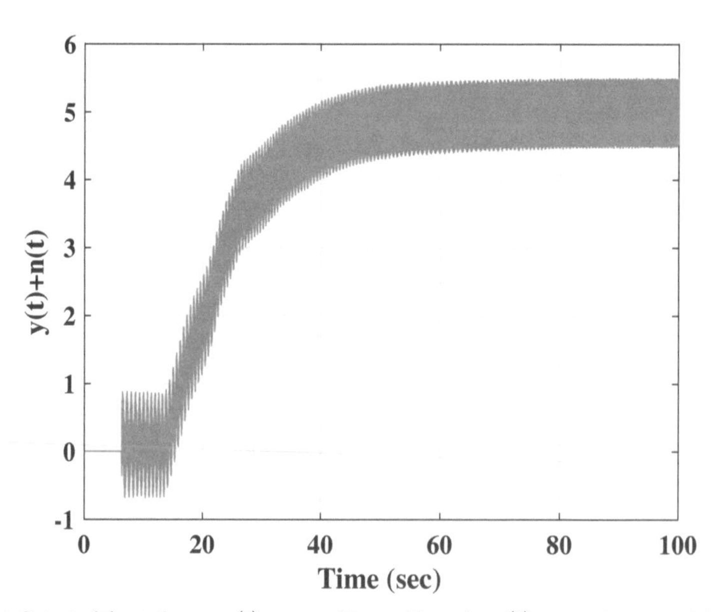

(b) Output of the static map $y(t)$ corrupted by a white noise $n(t)$ converging to a neighborhood of $y^* = 5$.

Figure 8.11. *Measurement noise effects: output signal corrupted by a white noise* $y(t)+n(t)$.

as expected. When $\omega = \omega_0$ the delay function $D(t)$ is periodic in $T = 2\pi/\omega$ such that the basic condition **(e)** in Section 8.2.1 is violated, resulting in a very slow convergence of the ES algorithm to the desired value. In spite of that, the algorithm exhibits some degree of robustness and continues the search for the extremum point.

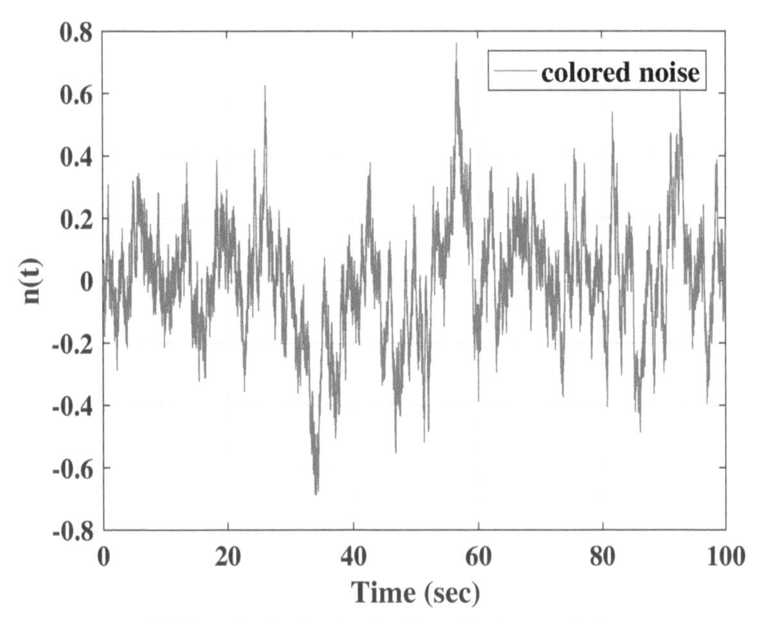

(a) The colored noise signal $n(t)$ used in the numerical tests.

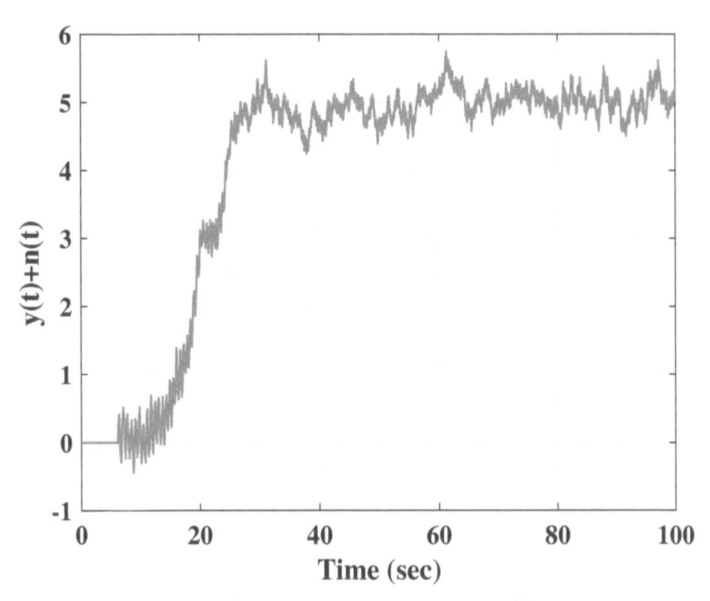

(b) Output of the static map $y(t)$ corrupted by a colored noise $n(t)$ converging to a neighborhood of $y^* = 5$.

Figure 8.12. *Measurement noise effects: output signal corrupted by a colored noise* $y(t)+n(t)$. *The colored noise is generated by filtering out the white noise employed in the plot of Figure* 8.11(a) *through the low-pass filter* $H_0(s) = \frac{k}{s+c}$, *with* $k = 100$ *and* $c = 1$ *rad/s.*

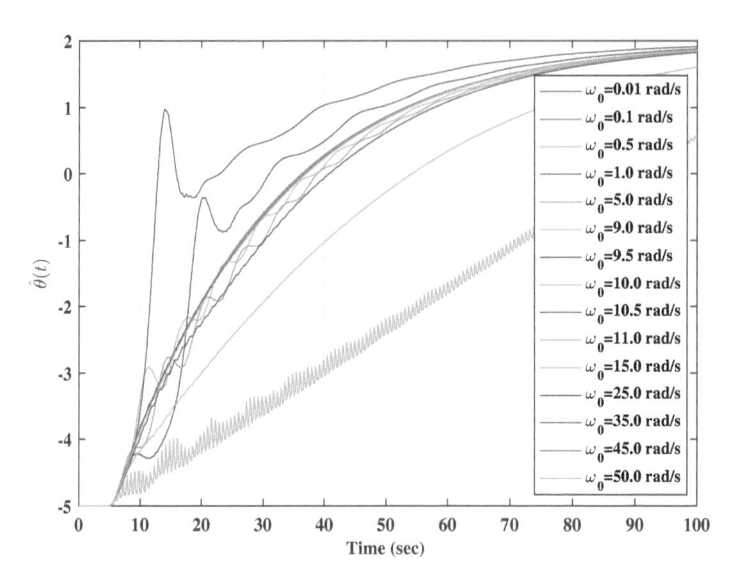

Figure 8.13. *Time evolution of the estimate $\hat{\theta}(t)$ converging to $\theta^* = 2$ under different time-varying delay functions. Effects of the time-varying delay function $D(t) = 5 + 2\epsilon\sin(\omega_0 t)$ in the convergence of the algorithm for different frequencies $\omega_0 \in (0, 50]$ satisfying Assumption 8.1 with $\omega_0 < \frac{1}{2\epsilon}$ and $\epsilon = 0.01$. At the critical ratio of the frequencies $\omega/\omega_0 = 1$, a very slow undesired performance is generated since the delay function $D(t)$ becomes periodic in $T = 2\pi/\omega$ such that the average condition $D_{\mathrm{av}}(t)$=$D(t)$ is violated.*

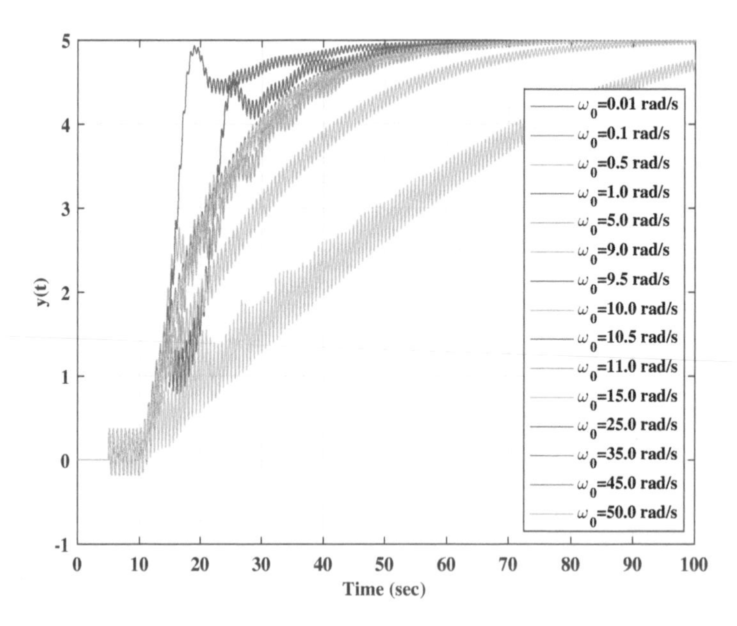

Figure 8.14. *Time evolution of the output signal $y(t)$ converging to a neighborhood of the extremum point $y^* = 5$ under different time-varying delay functions. Effects of the time-varying delay function $D(t) = 5 + 2\epsilon\sin(\omega_0 t)$ in the convergence of the algorithm for different frequencies $\omega_0 \in (0, 50]$ satisfying Assumption 8.1 with $\omega_0 < \frac{1}{2\epsilon}$ and $\epsilon = 0.01$. At the critical ratio of the frequencies $\omega/\omega_0 = 1$, a very slow undesired performance is generated since the delay function $D(t)$ becomes periodic in $T = 2\pi/\omega$ such that the average condition $D_{\mathrm{av}}(t)$=$D(t)$ is violated.*

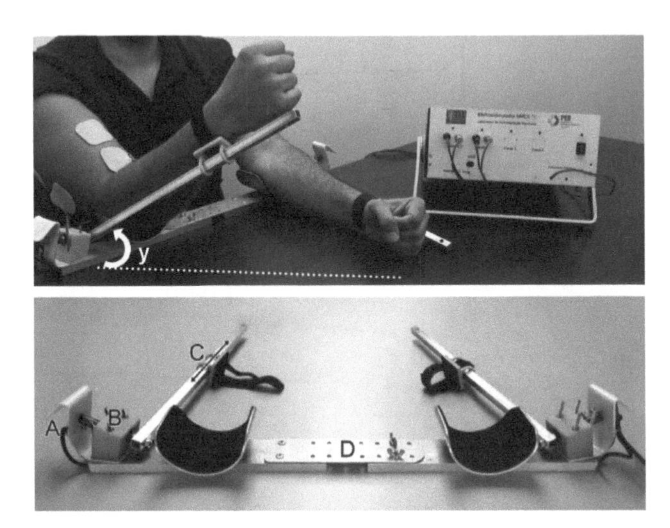

Figure 8.15. *Mechanical apparatus for NMES experimental tests. The point A in the image indicates a goniometer (simple potentiometer) linked to a steel axis B allowing angular displacement readings. Letter C shows that the wrist has an attachment with linear freedom of movement along the aluminum square rod, while D points out that there is an adjustment for the lateral distance of the elbows. In the picture on top, the controlled joint angle, denoted by y, and the NMES equipment are presented.* © 2020 *IEEE. Reprinted, with permission, from P. Paz, T. R. Oliveira, A. V. Pino, and A. P. Fontana, Model-free neuromuscular electrical stimulation by stochastic extremum seeking, IEEE Transactions on Control Systems Technology,* 28 (2020), *pp.* 238–253.

8.10 ▪ Application to Neuromuscular Electrical Stimulation

The recent manuscript [189] on neuromuscular electrical stimulation (NMES) brings a promising application of ES under nonconstant delays. The authors have proposed a stochastic proportional-derivative-integral (PID) automatic tuner via ES and applied it to precise tracking of a flexion-extension reference for NMES and motor relearning for rehabilitation. A picture of the mechanical apparatus as well as the electrical stimulation device for NMES experimental tests is shown in Figure 8.15. The experimental results are innovative since, unlike the referred literature, stroke patients were recruited for the successful tests rather than only healthy subjects. Remarkably, the patients are using such a device at the Public Hospital Universitário Clementino Fraga in Rio de Janeiro, Brazil.

In general, NMES devices applied to clinics work in an open-loop fashion and their parameters must be set at the beginning of the therapy, not facilitating the clinical practice. The levels of electrical stimulation follow pre-calibrated profiles demanding the presence of a practitioner to modify the stimulation parameters. It requires protocols aiming to enhance muscle contraction together with the execution of the intended contractions. The weakness of this procedure is that the device always returns the same portion of electrical assistance to the patient if there is no therapist intervention. In addition, the open-loop devices are not prepared to promote a proper association by means of some feedback error between the subject's intended movement and the artificial activation provided by the NMES system.

In this sense, closed-loop strategies are adequate to generate the NMES electrical current amplitudes based on the angular displacement (or the measurement of some other variable) related to the upper limbs. Although PID controllers have been explored in different engineering applications, the true limitation is that a PID controller is designed for linear systems, but the neuromuscular plant, which is being controlled, is nonlinear, time-varying, and subject to delays.

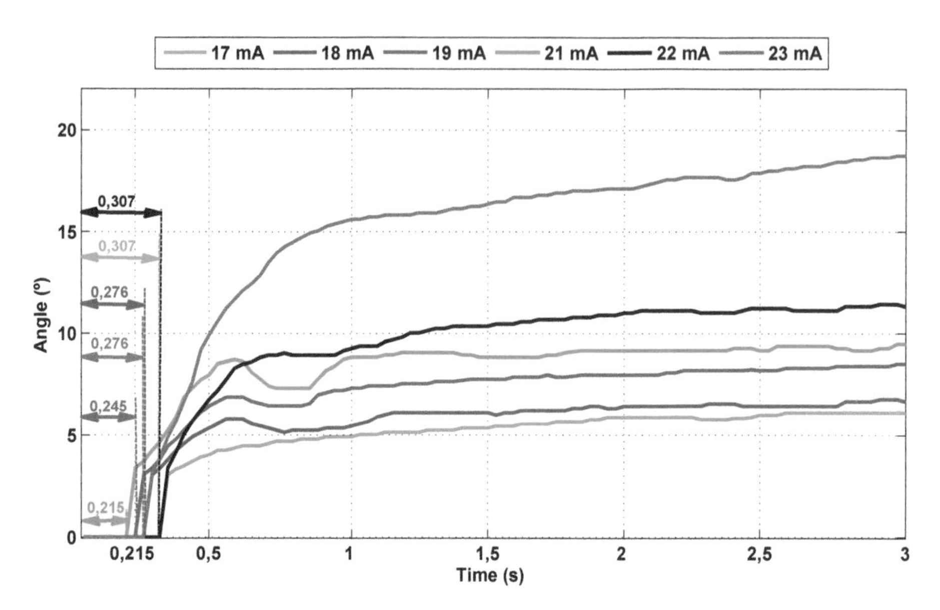

Figure 8.16. *Open-loop responses for different step-current inputs of a stroke patient with delays of order* 300 *ms. Extracted from* [189].

For instance, it is possible to note a time delay in the subject responses shown in Figure 8.16. The electromechanical-neuromuscular delay is in general of a time-varying nature and distinct for each subject. Moreover, it is worth mention that clinicians' knowledge of control systems is limited. Therefore, their expertise in tuning controllers is limited as well. Furthermore, in NMES applications, each patient is unique and requires a particular set of PID parameters. Since it may be difficult to find proper parameters for each patient, better procedures or a more intelligent-adaptive controller are indeed well motivated.

On the other hand, manual tuning is a time-consuming task and analytical methods are based on an exaggerated knowledge of the plant, requiring particular experimental validations to the identification of an acceptable plant model. However, a precise plant model in NMES is not known, and very long identification procedures are not desirable with the patients. This adverse environment of modeling inspires the application of adaptive-robust control methodologies and automatic tuning techniques.

The proposed model-free PID tuner via multiparameter stochastic ES in [189] is shown to be importantly fruitful. The proposed PID tuner eliminates the initial off-line tests with patients since the control gains are automatically computed in order to minimize a cost function according to the tracking error $e(t) := y(t) - r(t)$ between the elbow's angle of the patient's arm $y(t)$ and the reference trajectory $r(t)$. This research is highly successful since the stochastic algorithm gives faster (better transient) responses, which is perfect for a self-tuning, fatigue-resistant control method for neuromuscular-based therapies. Moreover, the parameters of the stochastic ES are simpler to tune since the orthogonality assumption on the dither vector signals of multiparameter deterministic ES imposes additional obstacles in adjusting the frequencies of the sinusoidal perturbations. Finally, deterministic ES may restrict the region of convergence of the algorithm, and the adaptation using a periodic-deterministic perturbation for learning may be rather poor and unusual in some model-free optimization frameworks. Stochastic perturbations overcome those obstacles as well.

Specifically, ES minimizes a cost function which quantifies the performance of the PID controller and iteratively modifies the arguments of the cost function (the PID parameters) so that its

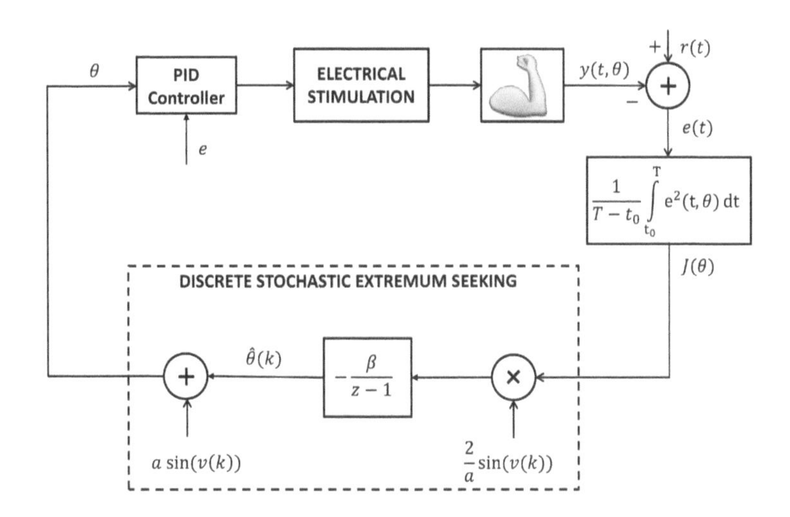

Figure 8.17. *Block diagram of the closed-loop system for NMES using discrete-time stochastic ES, where $\nu(k)$ is the stochastic perturbation vector modeled by Gaussian white noise signals, and the cost function $J(\theta) := \frac{1}{T-t_0}\int_{t_0}^{T} e^2(t,\theta)dt$ is defined over a period between two time instants t_0 and T. The PID control law $u(t) = K_p e(t) + K_i \int_0^t e(\tau)\, d\tau + K_d \frac{de(t)}{dt}$, with $\theta := [K_p, K_i, K_d]^T$ being the PID parameters (gains) to be adapted by means of the ES algorithm.*

output reaches a local minimum. According to the block diagram in Figure 8.17, the time-domain implementation of the discrete-time stochastic ES algorithm is given by

$$\hat{\theta}_i(k+1) = \hat{\theta}_i(k) - \beta\frac{2}{a}\sin(\nu_i(k))J(\hat{\theta}(k) + a\sin(\nu(k))), \tag{8.156}$$

where k is the discrete iteration number, the step size $\beta > 0$ is sufficiently small, the subscript $i = 1,2,3$ indicates the ith entry of a vector, $\nu(k) = [\nu_1(k)\ \nu_2(k)\ \nu_3(k)]^T$, and $\sin(\nu(k)) = [\sin(\nu_1(k))\ \sin(\nu_2(k))\ \sin(\nu_3(k))]^T$. The elements of the stochastic Gaussian perturbation vector $\nu(k)$ are sequentially and mutually independent such that $E\{\nu(k)\} = 0$, $E\{\nu_i^2(k)\} = \sigma_i^2$ and $E\{\nu_i(k)\ \nu_j(k)\} = 0$, $\forall i \neq j$, with $E\{\cdot\}$ denoting the expectation of a signal. In addition, it is assumed the probability density function of the perturbation vector is symmetric about its mean.

Figures 8.18 to 8.20 show the advantages for the closed-loop responses for a stroke patient, originally presented in [189]. In the clinical scenario, Figure 8.18 illustrates that even if the patients know the trajectory of the movement to be performed, they cannot execute it by themselves. Figure 8.19 also highlights that a fixed-gain PID scheme is not able to control adequately the stroke patient as the number of cycles/iterations increase, due to the time-varying nature of the neuromuscular system under delays. Indeed, PID control with fixed gains is not able to bring satisfactory results in long-running tests. Moreover, a unique fixed-gain tuning is not applicable for different individuals. On the other hand, the ES adaptive approach for simple adaptation of PID controller parameters is model-free having the interesting ability of controlling on multiple subjects without tediously tuning the designer or practitioner. On the contrary, the response curves in Figure 8.20 ratify the improved behavior of the adaptive PID control scheme over a fixed-gain PID controller even in this adversary scenario for NMES.

Although the delay discussion was not the focus on [189], it was evidenced there and in previous publications [149, 5, 207] that they may represent a significant challenge in NMES. It motivates the application of predictor feedback developed in the current chapter for delay compensation in extremum seeking algorithms plus PID control or even other techniques [5, 6].

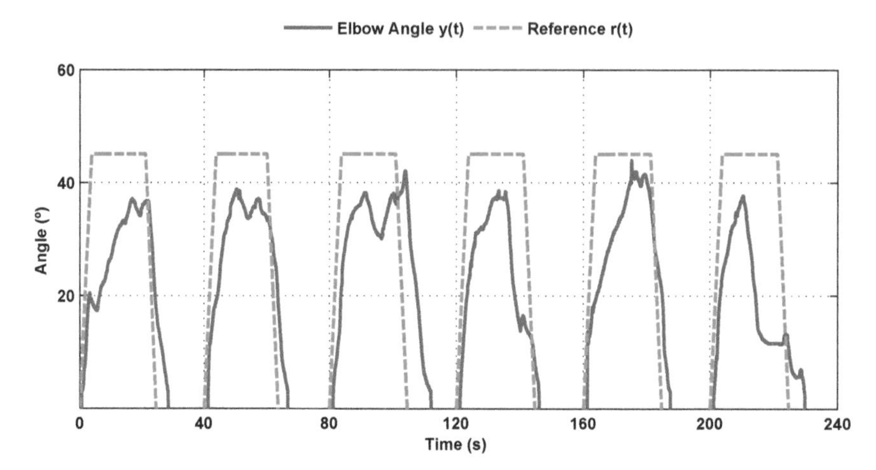

Figure 8.18. *The graphic portrays the angular elbow joint movement performed by the stroke patient without the help of the proposed NMES controller. It can be seen that the stroke subject is not able to actively contract his arm to the final flexion position.*

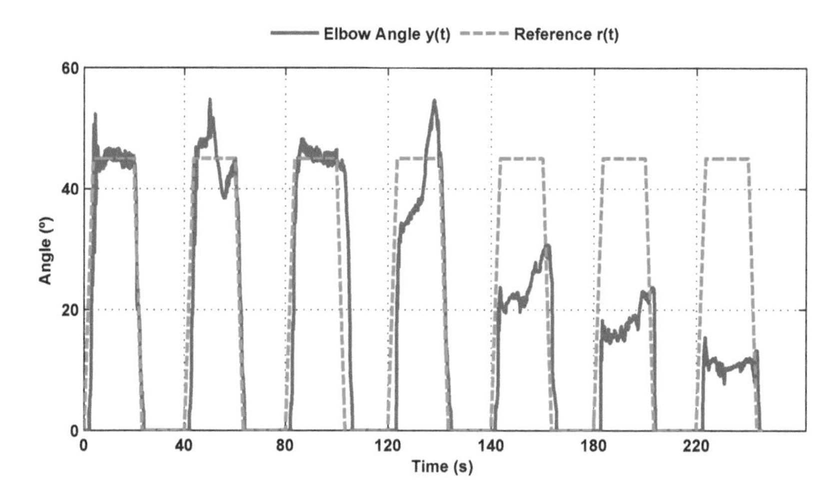

Figure 8.19. *Output responses for a stroke patient: PID with fixed gains* ($K_p = 1, K_i = 1, K_d = 1$) *not guaranteeing an acceptable trajectory tracking after* 120 *seconds.*

8.11 ▪ Notes and References

There is hardly any domain of control engineering practice where the sensors or actuators are not affected by delays. Depending on the specific application, such kind of delays may be known/unknown, small/large, deterministic/stochastic, or constant/variable [119], [25]. This chapter's focus was on variable, long, known, deterministic delays.

Publication [174] has shown that a small amount of (constant) delay is enough to make the extremum-seeking feedback loop unstable when the classic scalar gradient ES is solely considered. In this book's Chapter 2, a solution to the delay problem was proposed by employing two distinct tools: (i) the inclusion of a predictor feedback in the ES loop by incorporating averaging-based estimates for the unknown Gradient and Hessian of the map being optimized, and (ii) the phase compensation of the additive perturbation signal.

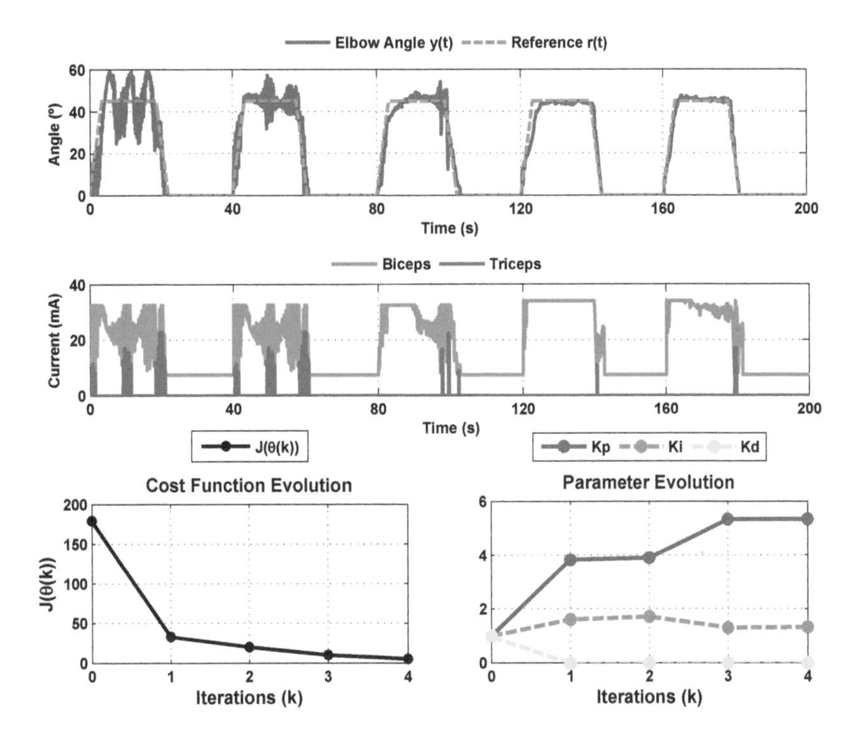

Figure 8.20. *Stochastic extremum-seeking-based PID control for NMES of a stroke patient.*

After this basic delay compensation idea in Chapter 2, extensions have been presented, in the presence of delays, for Newton-based ES [175] with faster responses (Chapter 3), inverse optimality guarantees (Chapter 4), optimization of dynamic maps and its higher derivatives [198] (Chapter 5), robustness to delay mismatches [200] (Chapter 6), expansions for multiparameter ES with output delays (or the same delay in the distinct input channels) [178] and multivariable ES under multiple and different input delays [179] (Chapter 7). In providing all these advances, the delays have been assumed to be constant.

The main goal of this chapter was to generalize our previous results on prediction feedback for ES in order to include a more general class of variable delays, not only time-varying delays but delays depending on the system states and time, simultaneously.

Our preliminary results about ES for PDE-PDE cascades are expanded in Chapter 14 to encompass more general cascades of PDEs from different families rather than first-order hyperbolic PDEs. Future research could investigate experimental applications for mobile robotics or control over networks [33, 136, 228], transportation of materials [53], traffic control [234], and additive manufacturing and 3D printing [52].

Chapter 9
ES for Distributed Delays

9.1 ▪ The Bewildering, Perilous Effect of Distributed Delays

As challenging as discrete delays on the input are for the control designer, such as in the system $\dot{X}(t) = AX(t) + BU(t-D)$, they are benign in comparison with distributed delays. Let us illustrate this.

While distributed delays usually bring to mind a system of the form

$$\dot{X}(t) = AX(t) + \int_{t-D}^{t} B(\tau)U(\tau)d\tau, \tag{9.1}$$

an example as simple as

$$\dot{X}(t) = \beta U(t) + U(t-D), \tag{9.2}$$

where both X and β are scalar, is a system with a distributed delay on the input. Even though this system is little more than an integrator of the input signal, the input signal $U(t)$ reaches $X(t)$ twice—first at time t and again at time $t+D$. Hence, the control designer is not only faced with the task of predicting a future state but also with making up his mind whether he will rely on the presence of the undelayed input term or the delayed input term for control design. If he is designing the controller based on the undelayed $U(t)$ in (9.2), this choice will come back to haunt him at time $t+D$ when the signal reaches the plant through the delayed input $U(t-D)$ as a disturbance. Conversely, if he is designing a feedback law based on the delayed input $U(t-D)$, he will pay the price immediately because such an input acts as a disturbance at the current time t through the undelayed input $U(t)$.

What shall the designer do with such a system in which the input seems to act both as his friend and his enemy?

The answer is provided in Artstein's reduction transformation [12] (see also [24] for a Lyapunov design for distributed delays) and the resulting feedback law

$$U(t) = k\left(X(t) + \int_{t-D}^{t} U(\tau)d\tau\right). \tag{9.3}$$

The spectrum of the closed-loop system (9.2), (9.3) contains only one (finite) eigenvalue, at $k(1+\beta)$. This eigenvalue is arbitrarily assignable with k, except when $\beta = -1$. When $\beta = -1$, system (9.2) is not stabilizable. Such a phenomenon cannot arise in the case with a discrete delay when the pair (A, B) is stabilizable.

As we have illustrated, even the simplest example of distributed delay has the potential to confuse, perplex, and incapacitate the control designer.

In spite of this challenge—or perhaps precisely because of it—we venture into considering ES optimization for maps that have a distributed delay at the input.

Handling a distributed delay necessitates a sophisticated selection of the probing signal, which we provide. We design an update law for an estimate of the unknown argument of the extremum using Artstein's reduction approach. We prove the convergence of the estimation error to a neighborhood of the origin by means of the method of averaging. The effectiveness of the proposed scheme is confirmed by a numerical simulation.

9.2 ▪ Problem Setting

Particular Notation: For an open interval $I \in \mathbb{R}$, the space of (equivalent classes of) Lebesgue square-integrable functions on I is denoted by $L^2(I)$. The first-order Sobolev space as a subset of $L^2(I)$ is denoted by $H^1(I)$. Let $f : [0,\ \infty) \times [0,\ \infty) \to [0\ ,\infty)$ be a nonnegative-valued function. Then, we write $f(t,\epsilon) \leq \mathcal{O}(\epsilon)$ when there exist $k > 0$ and $\bar{\epsilon} > 0$ such that $f(t,\epsilon) \leq k\epsilon$ for any $t \in [0,\ \infty)$, $\epsilon \in [0,\ \bar{\epsilon}]$.

9.2.1 ▪ Review of a Conventional Single-Parameter Extremum Seeking Problem

Consider a static map $f : \mathbb{R} \to \mathbb{R}$. Suppose that f is twice continuously differentiable and that the first and second derivative of f satisfies $f'(\theta^*) = 0$ and $f''(\theta^*) = H \neq 0$ at some point $\theta^* \in \mathbb{R}$. Then, the map f takes an extremum $f^* := f(\theta^*)$ at θ^*. Without loss of generality, we can assume that the Hessian H satisfies $H < 0$. Then, f^* is a local maximum. If f is three-times continuously differentiable, the quadratic map

$$Q(\theta) = f^* + \frac{H}{2}(\theta - \theta^*)^2 \tag{9.4}$$

is an approximation of f around $\theta = \theta^*$. We assume that f^*, H, and θ^* are unknown.

The objective of the extremum seeking is to find f^* without knowledge of H and θ^* by measuring the signal $y(t) = f(\theta(t))$ for an appropriately designed probing signal $\theta(t)$.

A conventional gradient-based extremum seeking scheme is given by

$$\dot{\hat{\theta}}(t) = k\frac{2}{a}\sin(\omega t)y(t) = k\frac{2}{a}\sin(\omega t)f(\theta(t)), \tag{9.5}$$

$$\theta(t) = \hat{\theta}(t) + a\sin(\omega t), \tag{9.6}$$

where $k > 0$ is a feedback gain and the parameters $a > 0$ and $\omega > 0$ are design parameters. The variable $\hat{\theta}$ is an estimate of θ^*. Set the error variable $\tilde{\theta}$ as $\tilde{\theta}(t) = \hat{\theta}(t) - \theta^*$. Since the time derivative of $\tilde{\theta}$ coincides with that of $\hat{\theta}$ and $\hat{\theta}$ is written as $\hat{\theta} = \tilde{\theta} + \theta^*$, the closed-loop system can be expressed in terms of the error variable $\tilde{\theta}$. Indeed, substituting (9.6) into (9.5) yields

$$\dot{\tilde{\theta}}(t) = k\frac{2}{a}\sin(\omega t)f\left(\tilde{\theta} + a\sin(\omega t) + \theta^*\right). \tag{9.7}$$

Normally, the stability analysis of this closed-loop system is conducted with the help of the averaging analysis. Assume that f is approximated by (9.4). Namely, f is replaced with Q in (9.7). Integrating the RHS of (9.7) with fixed $\tilde{\theta}$ with respect to t over $[0,\ 2\pi/\omega]$ and dividing the resulting value by the period $2\pi/\omega$ leads to

$$\dot{\tilde{\theta}}^{\mathrm{av}} = kH\tilde{\theta}^{\mathrm{av}}. \tag{9.8}$$

Since $H < 0$, this system clearly has a unique exponentially stable equilibrium $\tilde{\theta}^{\mathrm{av}} = 0$. Namely, $\tilde{\theta}^{\mathrm{av}}$ converges to 0 as t tends to $+\infty$ for any initial condition. Then, the averaging theorem [108] guarantees some sort of convergence property of the original error variable $\tilde{\theta}$.

9.2.2 ▪ Extremum Seeking under a Distributed Input Delay

In this chapter, we assume that there is a certain kind of delay in the transmission process from the input signal $\theta(t)$ to the measurement output $y(t)$. More precisely, the measurement output y is given by

$$y(t) = f\left(\int_0^D \theta(t-\sigma)d\beta(\sigma)\right) \tag{9.9}$$

for some constant $D > 0$. The integral on the RHS of (9.9) is the Riemann–Stieltjes integral with respect to a function $\beta : [0,\ D] \to \mathbb{R}$ of bounded variation. If β is a continuously differentiable function, the integral is reduced to a common integral of $\theta(t-\sigma)\beta'(\sigma)$ with respect to σ over $[0,\ D]$. As long as it is of bounded variation, β may be a discontinuous function. Indeed, if β is such that

$$\beta(x) = \begin{cases} 0, & 0 \le x < D_0 \\ 1, & D_0 \le x < D, \end{cases} \tag{9.10}$$

for some $D_0 \in (0, D)$, then we have

$$\int_0^D \theta(t-\sigma)d\beta(\sigma) = \theta(t-D_0). \tag{9.11}$$

The extremum seeking schemes developed in our previous chapters about constant delays can handle this case. However, for the case with general β, a new scheme is necessary. We make an assumption on the class of the function β. For each $\omega > 0$, define $\gamma(\omega)$ by

$$\gamma(\omega) = \left(\int_0^D \cos(\omega\sigma)d\beta(\sigma)\right)^2 + \left(\int_0^D \sin(\omega\sigma)d\beta(\sigma)\right)^2. \tag{9.12}$$

Clearly, $\gamma(\omega) \ge 0$ for all $\omega > 0$.

Assumption 9.1. *The function $\beta : [0, D] \to \mathbb{R}$ of bounded variation satisfies $\beta(0) = 0$, and there exists a nondecreasing sequence $\{\omega_i\}_{i=0}^{\infty} \subset (0,\infty) \subset \mathbb{R}$ of positive real numbers such that $\omega_i \to \infty$ as $i \to \infty$ and that γ defined by* (9.12) *satisfies $\gamma(\omega_i) \neq 0$ for any $i \in \mathbb{N}$.*

The first condition does not cause any loss of generality since the Stieltjes integrals with respect to β and $\beta + c_0$ coincide for any constant $c_0 \in \mathbb{R}$. The latter condition is necessary to let the frequency ω be arbitrary large. If we assume that the transmission is lossless, β should take 1 at D. To see this, suppose that $\theta(t) \equiv \theta(0)$ for some $\theta(0) \in \mathbb{R}$. Then we expect that $y(t) = f(\theta_0)$. For β satisfying Assumption 9.1, we have

$$\int_0^D \theta(t-\sigma)d\beta(\sigma) = \theta_0 \int_0^D d\beta(\sigma) = \beta(D)\theta_0. \tag{9.13}$$

Hence, if $\beta(D) = 1$, $y(t) = f(\theta_0)$ as expected. Note that, for β given in (9.10), $\beta(0) = 0$ and, for any $\omega \in (0,\infty)$,

$$\gamma(\omega) = \left(\cos(\omega D_0)\right)^2 + \left(\sin(\omega D_0)\right)^2 = 1 \neq 0. \tag{9.14}$$

Hence, Assumption 9.1 is fulfilled in this case.

The objective of this chapter is to develop an extremum seeking scheme for a static map f that is locally approximated by the quadratic map (9.4) in the presence of distributed delay. As in the previous chapters, the idea employed here is to employ predictor feedback. We construct an estimate of θ^* by using the measurement output (9.9) so that the corresponding update law has the form of predictor feedback control laws. A difficulty in the current problem arises from the fact that the measurement output y only contains an integrated value of θ rather than an actual value of θ at some time t.

9.3 ▪ Extremum Seeking Scheme

In this section, we propose an extremum seeking scheme as a solution to the problem formulated in the previous section.

9.3.1 ▪ Proposed Scheme

Let $\hat{\theta}(t) \in \mathbb{R}$ be an estimate of θ^* and $\overline{\theta}(t) \in \mathbb{R}$ be an intermediate variable. We temporarily write an update law for the intermediate variable $\overline{\theta}$ as

$$\dot{\overline{\theta}} = U. \tag{9.15}$$

The signal $U(t) \in \mathbb{R}$ is to be determined. We define the estimate $\hat{\theta}$ of θ^* by

$$\hat{\theta}(t) = \int_0^D \overline{\theta}(t-\sigma)d\beta(\sigma). \tag{9.16}$$

The use of these two variables $\hat{\theta}$ and $\overline{\theta}$ is a feature of this chapter. Compared with the previous chapters following the notation originally introduced in [179], the definition of the error variable may seem to some extent strange. However, the effectiveness of this novel notation formulation for this particular scenario of distributed delays will be seen later.

In the proposed scheme, we use three perturbation signals $S(t), M(t), N(t) \in \mathbb{R}$ defined by

$$S(t) = \frac{a}{\gamma(\omega)} \int_0^D \sin(\omega(t+\xi))d\beta(\xi), \tag{9.17}$$

$$M(t) = \frac{2}{a}\sin(\omega t), \tag{9.18}$$

$$N(t) = \frac{16}{a^2}\left(\sin^2(\omega t) - \frac{1}{2}\right). \tag{9.19}$$

The first perturbation signal $S(t)$ also has a peculiar form, whereas $M(t)$ and $N(t)$ are relatively simple. The signal $M(t)$ is used in a conventional extremum seeking. The third signal $N(t)$ is introduced in [73] to estimate unknown Hessian H. The signal $N(t)$ will be used for the same purpose. It should be noted that, as reviewed in the previous section, the Hessian H is not necessary in the standard gradient-based extremum seeking in the delay-free case. However, as seen in (9.8), $H\tilde{\theta}^{\text{av}}$ is available in the average analysis but $\tilde{\theta}^{\text{av}}$ itself is not available. Hence, to construct the extremum seeking scheme based on the predictor feedback, an estimate of H will be necessary.

We set the probing signal $\theta(t)$ as

$$\theta(t) = \overline{\theta} + S(t). \tag{9.20}$$

Note that θ includes $\overline{\theta}$ rather than $\hat{\theta}$ unlike conventional extremum seeking schemes. Substituting (9.20) into (9.9) gives an expression of y in terms of the foregoing signals. We claim that, for $\theta(t)$ defined as (9.20), the measurement output y becomes

$$y(t) = f\Big(\hat{\theta}(t) + a\sin(\omega t)\Big). \tag{9.21}$$

Indeed, integrating (9.20) with respect to β yields

$$\begin{aligned}
&\int_0^D \theta(t-\sigma)d\beta(\sigma) \\
&= \hat{\theta}(t) + \frac{a}{\gamma(\omega)}\int_0^D\int_0^D \sin\Big(\omega(t-\sigma+\xi)\Big)d\beta(\sigma)d\beta(\xi) \\
&= \hat{\theta}(t) + \frac{a\sin(\omega t)}{\gamma(\omega)}\int_0^D\int_0^D \cos\Big(\omega(\xi-\sigma)\Big)d\beta(\sigma)d\beta(\xi) \\
&\quad + \frac{a\cos(\omega t)}{\gamma(\omega)}\int_0^D\int_0^D \sin\Big(\omega(\xi-\sigma)\Big)d\beta(\sigma)d\beta(\xi),
\end{aligned} \tag{9.22}$$

where we have used the definition (9.16) of $\hat{\theta}$ and the addition formula of trigonometric functions. The second integral term on the right-hand side is actually equal to 0. This can be seen as follows:

$$\begin{aligned}
&\int_0^D\int_0^D \sin\Big(\omega(\xi-\sigma)\Big)d\beta(\sigma)d\beta(\xi) \\
&= \int_0^D \sin(\omega\xi)d\beta(\xi)\int_0^D \cos(\omega\sigma)d\beta(\sigma) \\
&\quad - \int_0^D \cos(\omega\xi)d\beta(\xi)\int_0^D \sin(\omega\sigma)d\beta(\sigma) = 0.
\end{aligned} \tag{9.23}$$

We next calculate the first integral term in (9.22). It follows from a direct computation that

$$\begin{aligned}
&\int_0^D\int_0^D \cos\Big(\omega(\xi-\sigma)\Big)d\beta(\sigma)d\beta(\xi) \\
&= \int_0^D \cos(\omega\xi)d\beta(\xi)\int_0^D \cos(\omega\sigma)d\beta(\sigma) \\
&\quad + \int_0^D \sin(\omega\xi)d\beta(\xi)\int_0^D \sin(\omega\sigma)d\beta(\sigma) = \gamma(\omega).
\end{aligned} \tag{9.24}$$

Substituting (9.24) into (9.22), we arrive at

$$\begin{aligned}
\int_0^D \theta(t-\sigma)d\beta(\sigma) &= \hat{\theta}(t) + \frac{a\sin(\omega t)}{\gamma(\omega)}\gamma(\omega) \\
&= \hat{\theta}(t) + a\sin(\omega t).
\end{aligned} \tag{9.25}$$

This immediately implies (9.21). Although the estimate $\hat{\theta}$ of θ^* and the signal θ are defined in a somewhat strange manner, the output y is expressed as if the map f received the signal $\hat{\theta} + a\sin(\omega t)$ as in the conventional scheme.

We next derive the dynamic equation that is satisfied by the error variable:

$$\tilde{\theta}(t) := \hat{\theta}(t) - \theta^*. \tag{9.26}$$

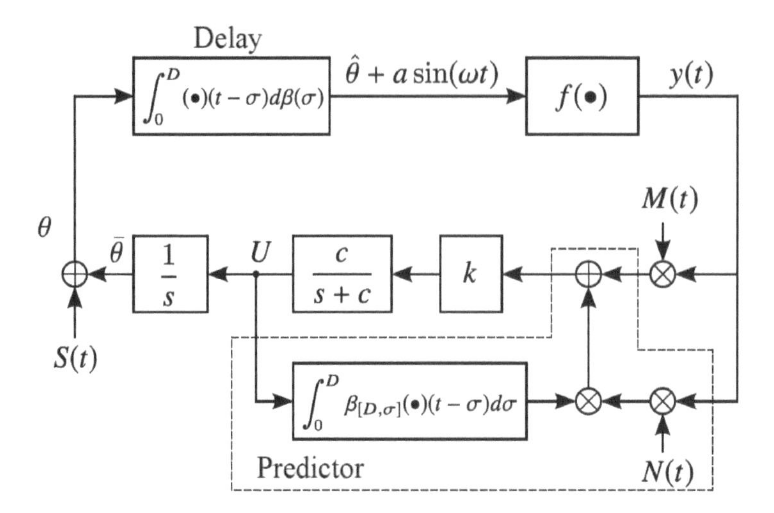

Figure 9.1. *Block diagram of the proposed extremum seeking scheme for compensating distributed delays.*

Differentiating (9.16) with respect to t leads to

$$\dot{\hat{\theta}}(t) = \int_0^D \dot{\bar{\theta}}(t-\sigma)d\beta(\sigma) = \int_0^D U(t-\sigma)d\beta(\sigma). \tag{9.27}$$

Since we have $\dot{\tilde{\theta}}(t) = \dot{\hat{\theta}}$, it follows from (9.27) that

$$\dot{\tilde{\theta}}(t) = \int_0^D U(t-\sigma)d\beta(\sigma). \tag{9.28}$$

The error variable $\tilde{\theta}$ evolves in accordance with this equation. If we regard U as the control input, (9.28) is in a typical form of a driftless linear system with distributed input delay. It is well known that such a system can be stabilized by a predictor feedback control law. In [147, 12], the following control law is proposed:

$$U(t) = -k\left(\tilde{\theta}(t) + \int_t^{t+D}\int_{\tau-t}^{D} U(\tau-\sigma)d\beta(\sigma)d\tau\right), \tag{9.29}$$

where $k > 0$. The terms in the parentheses correspond to a predicted future value of $\tilde{\theta}$. Unfortunately, this control law is not implementable because $\tilde{\theta}$ is unavailable. Our approach aims at realizing (9.29) in the average sense.

We propose to close the loop of our extremum seeking scheme by setting the signal U in the update law (9.15) for $\overline{\theta}$ as the solution to the following time-varying ordinary differential equation:

$$\dot{U}(t) = -cU(t) + ck\left(M(t)y(t) + N(t)y(t)\int_0^D \beta_{[D,\sigma]}U(t-\sigma)d\sigma\right), \tag{9.30}$$

where $\beta_{[D,\sigma]} := \beta(D) - \beta(\sigma)$ and $c > 0$. The corresponding block diagram is given in Figure 9.1.

Equation (9.30) means that U is a filtered version of the signal

$$k\left(M(t)y(t) + N(t)y(t)\int_0^D \beta_{[D,\sigma]}U(t-\sigma)d\sigma\right). \tag{9.31}$$

The filter is represented by the transfer function $c/(s+c)$. As we will see later, the signal (9.31) mimics (9.29) in the average sense. The filter is introduced for a technical reason. As explained in the earlier chapters, U must be a part of the state variable in order that the infinite-dimensional averaging theorem [83] is applicable to the closed-loop system.

9.3.2 ▪ Abstract Formulation of the Closed-Loop System

Recall that the measurement signal y is given by (9.21). This can be rewritten in terms of the error variable $\tilde{\theta}$ as

$$y(t) = f\left(\tilde{\theta}(t) + a\sin(\omega t) + \theta^*\right). \tag{9.32}$$

Hence, the closed-loop system consists of (9.28) and (9.30) with the initial condition $\tilde{\theta}(0) = \tilde{\theta}_0 \in \mathbb{R}$ and $U(\tau - D) = \phi(\tau)$ for each $\tau \in [0,D]$. The initial function ϕ is an element of some function space. To conduct stability analysis, we introduce a partial differential equation (PDE) representation.

Set $u(x,t) = U(x+t-D)$ for each $x \in (0,D]$ and $t \geq 0$ with $x+t \geq 0$. Then, the closed-loop system (9.28) and (9.30) can be expressed as

$$\dot{\tilde{\theta}}(t) = \int_0^D u(D-\sigma,t)d\beta(\sigma), \tag{9.33}$$

$$u_t(x,t) = u_x(x,t), \quad x \in (0,D), \tag{9.34}$$

$$u(D,t) = U(t), \tag{9.35}$$

$$\dot{U}(t) = -cU(t) + ck\left(M(t)y(t) + N(t)y(t)\int_0^D \beta_{[D,\sigma]}u(D-\sigma,t)d\sigma\right) \tag{9.36}$$

with the initial condition $\tilde{\theta}(0) = \tilde{\theta}_0$, $u(x,0) = u_0(x)$, $x \in (0,D)$, and $U(0) = U_0$ In this representation, the state variable is composed of the finite-dimensional components $\tilde{\theta}$, U and the infinite-dimensional one u.

Let the state space be $\mathcal{H} := \mathbb{R} \times L^2(0,D) \times \mathbb{R}$ with the inner product

$$\left([X_1,X_2,X_3]^T,[Y_1,Y_2,Y_3]^T\right)_{\mathcal{H}} := X_1Y_1 + \int_0^D X_2(\xi)Y_2(\xi)d\xi + X_3Y_3 \tag{9.37}$$

for each $[X_1,X_2,X_3]^T,[Y_1,Y_2,Y_3]^T \in \mathcal{H}$. The space $\mathcal{H}$ is a Hilbert space. The norm induced by the inner product (9.37) is denoted by $||\cdot||_{\mathcal{H}}$. Define a linear operator $A : D(A) \subset \mathcal{H} \to \mathcal{H}$ by

$$A\begin{bmatrix} X_1 \\ X_2 \\ X_3 \end{bmatrix} = \begin{bmatrix} \int_0^D X_2(D-\sigma)d\beta(\sigma) \\ \frac{dX_2}{dx} \\ -cX_3 \end{bmatrix} \tag{9.38}$$

with domain

$$D(A) = \{[X_1,X_2,X_3] \in \mathcal{H} \mid X_2 \in \mathcal{H}^1(0,D), X_2(D) = X_3\}. \tag{9.39}$$

The time-varying nonlinear perturbation term is given by

$$F(\omega t, X) = \begin{bmatrix} 0 \\ 0 \\ F_3(\omega t, X) \end{bmatrix}, \tag{9.40}$$

where $F_3 : \mathbb{R} \times \mathcal{H} \to \mathbb{R}$ is defined by

$$F_3(\omega t, X) = ck\left[\frac{2}{a}\sin(\omega t)f(X_1 + a\sin(\omega t) + \theta^*) + \frac{16}{a^2}\left(\sin^2(\omega t) - \frac{1}{2}\right)f(X_1 + a\sin(\omega t) + \theta^*) \times \int_0^D \beta_{[D,\sigma]}X_2(D-\sigma)d\sigma\right]. \tag{9.41}$$

Thus, we arrive at the following abstract evolution equation corresponding to the closed-loop system (9.33)–(9.36):

$$\frac{dX}{dt}(t) = AX + F(\omega t, X). \tag{9.42}$$

Clearly, $F(\omega t + 2\pi, X) = F(\omega t, X)$ for any $X \in \mathcal{H}$. If $1/\omega$ is considered as a small parameter ϵ, we can apply the method of averaging for infinite-dimensional systems developed in [83]; see also Appendix B. Then, the stability of the closed-loop system can be investigated through the corresponding average system. We carry out this investigation in the next section.

9.4 ▪ Stability Analysis

The main goal of this section is to prove our main theorem.

Theorem 9.1. *Let $H < 0$, $D > 0$, and $a > 0$. Consider the system* (9.42) *for f being the quadratic map* (9.4). *Suppose that the function $\beta : [0, D] \to \mathbb{R}$ of bounded variation satisfies Assumption* 9.1 *and the constants $k > 0$ and $c > 0$ are chosen so that $c > -kH$. Then, for each $\rho > 0$, there exist constants $\omega^* > 0$ and $\rho_0 \in (0, \rho)$ such that, for any $\omega > \omega^*$ with $\gamma(\omega) \neq 0$, any solution to* (9.42) *for an initial value $X_0 = [\tilde{\theta}_0, u_0(\cdot), U_0]^T \in D(A)$ with $||X_0||_{\mathcal{H}} \leq \rho_0$ converges to an $\mathcal{O}(1/\omega)$-neighborhood of the origin. In addition, the following estimates hold:*

$$\limsup_{t \to \infty} |y(t) - f^*| = \mathcal{O}(1/\omega^2 + |a|^2). \tag{9.43}$$

To prove the theorem, we consider the averaged version of the system (9.42).

9.4.1 ▪ Average System

Let us obtain the average system associated with the closed-loop system (9.33)–(9.36). The expression (9.32) of the output y has the same form as the one in common delay-free extremum seeking problems. Hence, average computation done in the classical literature (Chapter 1), especially in [73], also works for our problem.

If f is the quadratic map (9.4), the average of (9.41) can be explicitly computed as

$$\frac{\omega}{2\pi}\int_0^{2\pi/\omega} F_3(\omega\tau, X)d\tau = HX_1 + H\int_0^D \beta_{[D,\sigma]}X_2(D-\sigma)d\sigma \tag{9.44}$$

for each $X = [X_1, X_2, X_3]^T \in \mathcal{H}$. From the argument above, the average system associated with the closed-loop system (9.33)–(9.36) is given by

$$\dot{\tilde{\theta}}^{\mathrm{av}}(t) = \int_0^D u^{\mathrm{av}}(D-\sigma,t)d\beta(\sigma), \tag{9.45}$$

$$u_t^{\mathrm{av}}(x,t) = u_x^{\mathrm{av}}(x,t), \quad x \in (0,D), \tag{9.46}$$

$$u^{\mathrm{av}}(D,t) = U^{\mathrm{av}}(t), \tag{9.47}$$

$$\dot{U}^{\mathrm{av}}(t) = -cU^{\mathrm{av}}(t) + ckH\left(\tilde{\theta}^{\mathrm{av}}(t) + \int_0^D \beta_{[D,\sigma]} u^{\mathrm{av}}(D-\sigma,t)d\sigma\right). \tag{9.48}$$

It can be inferred from the relation $u^{\mathrm{av}}(x,t) = U^{\mathrm{av}}(x+t-D)$ that U^{av} is a filtered value of the signal

$$kH\left(\tilde{\theta}^{\mathrm{av}}(t) + \int_0^D \beta_{[D,\sigma]} U^{\mathrm{av}}(t-\sigma)d\sigma\right). \tag{9.49}$$

Let us check the relationship between the predictor feedback (9.29) and (9.49). Observe that integration by parts for the Riemann–Stieltjes integral gives

$$\begin{aligned}
&\int_0^D \int_0^\sigma U^{\mathrm{av}}(t-\xi)d\xi d\beta(\sigma) \\
&= \beta(D)\int_0^D U^{\mathrm{av}}(t-\xi)d\xi - 0 - \int_0^D \beta(\sigma)U^{\mathrm{av}}(t-\sigma)d\sigma \\
&= \int_0^D \beta_{[D,\sigma]} U^{\mathrm{av}}(t-\sigma)d\sigma.
\end{aligned} \tag{9.50}$$

On the other hand, changing the variable of integration from ξ to $\tau = t+\sigma-\xi$ and then reversing the order of integration leads to

$$\begin{aligned}
&\int_0^D \int_0^\sigma U^{\mathrm{av}}(t-\xi)d\xi d\beta(\sigma) \\
&= \int_0^D \int_t^{t+\sigma} U^{\mathrm{av}}(\tau-\sigma)d\tau d\beta(\sigma) \\
&= \int_t^{t+D} \int_{\tau-t}^D U^{\mathrm{av}}(\tau-\sigma)d\beta(\sigma)d\tau.
\end{aligned} \tag{9.51}$$

From (9.50) and (9.51), the expression (9.49) is rewritten as

$$\begin{aligned}
&kH\left(\tilde{\theta}^{\mathrm{av}}(t) + \int_0^D \beta_{[D,\sigma]} U^{\mathrm{av}}(t-\sigma)d\sigma\right) \\
&= kH\left(\tilde{\theta}^{\mathrm{av}} + \int_t^{t+D} \int_{\tau-t}^D U^{\mathrm{av}}(\tau-\sigma)d\beta(\sigma)d\tau\right).
\end{aligned} \tag{9.52}$$

This completely coincides with (9.29) if the gain $-k$ is replaced with kH. Hence, the proposed U given by (9.30) realizes a filtered predictor feedback control law with the gain kH in the average sense.

9.4.2 ▪ Stability of the Average System

The stability of the average system (9.45)–(9.48) is analyzed in this subsection.

Lemma 9.1. *Consider the system* (9.45)–(9.48) *for some* $D > 0$ *and* $H < 0$. *Let* $\beta : [0,D] \to \mathbb{R}$ *be a function of bounded variation satisfying Assumption* 9.1. *Then, for each* $k > 0$, *there exists* $c^* > 0$ *such that, for any* $c > c^*$, *the following assertions hold:*

(i) *The system* (9.45)–(9.48) *admits a unique solution in* $C^1([0,+\infty);\mathcal{H}) \cap C([0,+\infty);D(A))$ *for any given initial data* $[\tilde{\theta}^{\rm av}(0), u^{\rm av}(\cdot,0), U^{\rm av}(0)]^T \in D(A)$, *where the subspace* $D(A) \subset \mathcal{H}$ *is defined by* (9.39).

(ii) *There exist constants* $\lambda > 0$ *and* $M > 0$, *which are independent from the initial data, such that*

$$\begin{aligned} &\left|\left|\tilde{\theta}^{\rm av}(t), u^{\rm av}(\cdot,t), U^{\rm av}(t)\right|\right|_{\mathcal{H}} \\ &\le Me^{-\lambda t}\left|\left|\left[\tilde{\theta}^{\rm av}(0), u^{\rm av}(\cdot,0), U^{\rm av}(0)\right]\right|\right|_{\mathcal{H}}. \end{aligned} \tag{9.53}$$

Proof. The proof starts with transformation of $\tilde{\theta}^{\rm av}$ into another variable. Set $\vartheta(t) \in \mathbb{R}$ as

$$\vartheta(t) := \tilde{\theta}^{\rm av}(t) + \int_0^D \beta_{[D,\sigma]} u^{\rm av}(D-\sigma,t)d\sigma + \frac{1}{c}U^{\rm av}(t). \tag{9.54}$$

Differentiating ϑ with respect to t gives

$$\begin{aligned} \dot{\vartheta}(t) &= \dot{\theta}^{\rm av}(t) + \textstyle\int_0^D \beta_{[D,\sigma]} u_x^{\rm av}(D-\sigma,t)d\sigma + \frac{1}{c}\dot{U}^{\rm av}(t) \\ &= kH\Big(\vartheta(t) - \tfrac{1}{c}U^{\rm av}(t)\Big), \end{aligned} \tag{9.55}$$

where we have used (9.45)–(9.48). Then, the average system (9.45)–(9.48) is considerably simplified as

$$\dot{\vartheta}(t) = kH\Big(\vartheta(t) - \frac{1}{c}U^{\rm av}(t)\Big), \tag{9.56}$$

$$u_t^{\rm av}(x,t) = u_x^{\rm av}(x,t), \quad x \in (0,D), \tag{9.57}$$

$$u^{\rm av}(D,t) = U^{\rm av}(t), \tag{9.58}$$

$$\dot{U}^{\rm av}(t) = -cU^{\rm av}(t) + ckH\Big(\vartheta(t) - \frac{1}{c}U^{\rm av}(t)\Big). \tag{9.59}$$

The PDE-subsystem (9.46), (9.47) does not change. We first investigate the well-posedness and the exponential stability of the transformed system.

Observe that the ordinary differential equations (9.56) and (9.59) are linear and independent from $u^{\rm av}$. Hence, for any $\vartheta(0) \in \mathbb{R}$ and $U^{\rm av}(0) \in \mathbb{R}$ they have the unique infinitely differentiable solutions. Then, it is well-known (see, for example, [44]) that, as long as $u^{\rm av}(\cdot,0) \in \mathcal{H}^1(0,D)$ satisfies $u^{\rm av}(D,0) = U^{\rm av}(0)$, the PDE-subsystem (9.57), (9.58) has a unique solution $u^{\rm av} \in C^1([0,+\infty);L^2(0,D)) \cap C([0,+\infty);\mathcal{H}^1(0,D))$. Since $u^{\rm av}$ and $U^{\rm av}$ are not changed, the transformation (9.54) is trivially invertible. Thus, the first part (i) has been proved.

To prove the second part, let us introduce V defined by

$$V(t) = \frac{c^2}{2}\vartheta(t)^2 + \frac{b}{2}\int_0^D e^{\alpha(x-D)} u^{\rm av}(x,t)^2 dx + \frac{1}{2}U^{\rm av}(t)^2 \tag{9.60}$$

for some $b > 0$ and $\alpha > 0$. We show that V is a Lyapunov functional for the transformed system (9.56)–(9.59). The time derivative of V is given by

$$\begin{aligned}\dot{V}(t) = {} & c^2kH\vartheta(t)\left(\vartheta(t) - \frac{1}{c}U^{\mathrm{av}}(t)\right) + \frac{b}{2}U^{\mathrm{av}}(t)^2 \\ & - \frac{b}{2}e^{-\alpha D}u^{\mathrm{av}}(0,t)^2 - \frac{\alpha b}{2}\int_0^D e^{\alpha(x-D)}u^{\mathrm{av}}(x,t)^2dx \\ & - cU^{\mathrm{av}}(t)^2 + ckHU^{\mathrm{av}}(t)\left(\vartheta(t) - \frac{1}{c}U^{\mathrm{av}}(t)\right)\end{aligned}$$

or, equivalently,

$$\begin{aligned}\dot{V}(t) = {} & c^2kH\left(\vartheta(t) + \frac{1}{c}U^{\mathrm{av}}(t)\right)\left(\vartheta(t) - \frac{1}{c}U^{\mathrm{av}}(t)\right) \\ & - \frac{\alpha b}{2}\int_0^D e^{\alpha(x-D)}u^{\mathrm{av}}(x,t)^2dx - \left(c - \frac{b}{2}\right)U^{\mathrm{av}}(t)^2 \\ & - \frac{b}{2}e^{-\alpha D}u^{\mathrm{av}}(0,t)^2.\end{aligned}$$

Rearranging the terms, we can write

$$\begin{aligned}\dot{V}(t) = {} & -c^2(-kH)\vartheta(t)^2 - \frac{\alpha b}{2}\int_0^D e^{\alpha(x-D)}u^{\mathrm{av}}(x,t)^2dx \\ & - \left(c - \frac{b}{2} + kH\right)U^{\mathrm{av}}(t)^2 - \frac{b}{2}e^{-\alpha D}u^{\mathrm{av}}(0,t)^2.\end{aligned} \tag{9.61}$$

Set $c^* := \frac{b}{2} - kH$. Note that $c^* > 0$ since $H < 0$. Then, it follows from the above equation that

$$\dot{V}(t) \le -2\lambda V(t), \tag{9.62}$$

where $\lambda := \min\{-kH, \alpha/2, c - c^*\} > 0$. The inequality (9.62) implies that

$$V(t) \le e^{-2\lambda t}V(0). \tag{9.63}$$

It is easy to see that there exist constants $c_1, c_2 > 0$ such that

$$c_1\left|\left|\left[\vartheta(t), u^{\mathrm{av}}(\cdot,t), U^{\mathrm{av}}(t)\right]\right|\right|_{\mathcal{H}} \le V(t) \le c_2\left|\left|\left[\vartheta(t), u^{\mathrm{av}}(\cdot,t), U^{\mathrm{av}}(t)\right]\right|\right|_{\mathcal{H}} \tag{9.64}$$

for any $t \ge 0$. Hence, the transformed system (9.56)–(9.59) is exponentially stable. The exponential stability of the original average system follows from the fact that there exist constants $M_1, M_2 > 0$ such that

$$|\vartheta(t)| \le M_1\left|\left|\left[\tilde{\theta}^{\mathrm{av}}(t), u^{\mathrm{av}}(\cdot,t), U^{\mathrm{av}}(t)\right]\right|\right|_{\mathcal{H}}, \tag{9.65}$$

$$|\tilde{\theta}^{\mathrm{av}}(t)| \le M_2\left|\left|\left[\vartheta(t), u^{\mathrm{av}}(\cdot,t), U^{\mathrm{av}}(t)\right]\right|\right|_{\mathcal{H}} \tag{9.66}$$

for each $t \ge 0$. This fact can be inferred from the definition (9.54) of ϑ. The inequalities (9.63)–(9.66) imply (9.53). This completes the proof. □

The exponential stability of the trivial solution to the average system with respect to the norm $||\cdot||_{\mathcal{H}}$ is shown. Thus, we are now in a position to prove Theorem 9.1. Before moving on to the proof, we make a remark. Unlike [179], we do not use a backstepping transformation to construct a Lyapunov functional. Lyapunov functionals similar to (9.60) can be found in [95, 148].

9.4.3 ▪ Proof of Theorem 9.1

The theorem basically follows from the averaging theorem for infinite-dimensional systems [83]. To apply the averaging theorem (Appendix B), we have to check the property of the linear operator A and the perturbation F in (9.42).

In [83], the operator A is assumed to be a generator of a strongly continuous semigroup T_A and the associated semigroup T_A must have the smoothing property **(H)**.

Property 9.1. [83, *Property (H)*] *For any $h:[0,\infty)\to\mathcal{H}$ that is norm continuous, the following relations must hold:*

$$\text{(i)}\quad \int_0^D T_A(t-\tau)h(\tau)d\tau \in D(A), \qquad t\geq 0, \tag{9.67}$$

$$\text{(ii)}\quad \left\| A\int_0^D T_A(t-\tau)h(\tau)d\tau \right\| \leq Me^{\mu t}\max_{0\leq\tau\leq t}||h(\tau)||_{\mathcal{H}}, \quad t\geq 0, \tag{9.68}$$

where $M>0$ and $\mu\in\mathbb{R}$ are independent from h.

It is true that the operator A defined by (9.38) is a generator of a strongly continuous semigroup T_A on $\mathcal{H}$. However, T_A does not fulfill the smoothing property **(H)** in general. Fortunately, the perturbation F defined in (9.40) merely has the finite-dimensional component F_3. Hence, in our problem, T_A only have to satisfy (9.67) and (9.68) for $h:[0,\infty)\to\mathcal{H}$ of the form $h(t)=[0,0,h_3(t)]^T$ for any continuous $h_3:[0,\infty)\to\mathbb{R}$.

For $h(t)=[0,0,h_3(t)]^T$, we can explicitly compute the integral of $T_A(t-\tau)h(\tau)$ with respect to τ from 0 to t as

$$\begin{aligned}\left[\int_0^t T_A(t-\tau)h(\tau)d\tau\right]_1 =& -\frac{1}{c}\int_0^{\max\{0,t-D\}}\int_0^D \left(e^{-c(t-\tau-\sigma)}-1\right)d\beta(\sigma)h_3(\tau)d\tau\\ &-\frac{1}{c}\int_{\max\{0,t-D\}}^t\int_0^{t-\tau}\left(e^{-c(t-\tau-\sigma)}-1\right)d\beta(\sigma)h_3(\tau)d\tau,\end{aligned} \tag{9.69}$$

$$\left[\int_0^t T_A(t-\tau)h(\tau)d\tau\right]_2(x) = \begin{cases} 0, & 0<x\leq D-t,\\ \int_0^{x+t-D} e^{-c(x+t-D-\tau)}h_3(\tau)d\tau, & D-t<0<x\leq D,\end{cases} \tag{9.70}$$

$$\left[\int_0^t T_A(t-\tau)h(\tau)d\tau\right]_3 = \int_0^t e^{-c(t-\tau)}h_3(\tau)d\tau. \tag{9.71}$$

From these expressions, we can see that T_A satisfies the property **(H)** for $h(t)=[0,0,h_3(t)]^T$ with an arbitrary continuous real function h_3.

Take arbitrary $\rho>0$. It follows from (9.41) that F and its Fréchet derivatives are bounded for any $||X||_{\mathcal{H}}\leq\rho$. Hence, we can apply the averaging theorem in [83] in Appendix B to the closed-loop system (9.42). If $c>-kH$, we can find a sufficiently small $b>0$ so that $c>c^*=b/2-kH$. Lemma 9.1 shows the exponential stability of the trivial solution of the average system (9.45)–(9.48). Then, the averaging theorem guarantees the existence of a constant $\omega^*>0$ such that, for each $\omega > \omega^*$ with $\gamma(\omega) \neq 0$, there exists a unique exponentially stable $(2\pi/\omega)$-periodic solution $X^P(t,1/\omega)$ to (9.42). In addition, the periodic solution has the property

$$||X^P(t,1/\omega)||_{\mathcal{H}} \leq \frac{C}{\omega} \qquad \forall t \geq 0 \tag{9.72}$$

for some $C > 0$ independent from ω. We thus prove the exponential convergence of a solution X of (9.42) to an $\mathcal{O}(1/\omega)$-neighborhood of the origin.

It is trivial that $|\tilde{\theta}(t)| \leq ||[\tilde{\theta}(t), u(\cdot,t), U(t)]^T||_{\mathcal{H}}$. Since the right-hand side converges to $||X^P(t,1/\omega)||$ exponentially, we have $\lim_{t\to\infty}\sup|\tilde{\theta}(t)| \leq \mathcal{O}(1/\omega)$. Then the estimate (9.43) is easily deduced from the relation

$$\begin{aligned} |y(t) - f^*| &= \left|\left|\frac{1}{2}H\left(\tilde{\theta}(t) + a\sin(\omega t)\right)^2\right|\right| \\ &\leq |H|\left(\tilde{\theta}(t)^2 + a^2\right), \end{aligned} \tag{9.73}$$

which completes the proof of Theorem 9.1.

9.5 ▪ Numerical Example

In this section, we confirm the effectiveness of the proposed scheme. Let β be given by

$$\beta(x) = \begin{cases} 0, & 0 \leq x \leq \frac{D}{2}, \\ \frac{2x}{D} - 1, & \frac{D}{2} < x \leq D. \end{cases} \tag{9.74}$$

It is clear that this β satisfies the first condition in Assumption 9.1. Note that we have

$$\int_0^D \theta(t-\sigma)d\beta(\sigma) = \frac{2}{D}\int_{D/2}^D \theta(t-\sigma)d\sigma. \tag{9.75}$$

Hence, the map f receives an average of the signal θ over the past interval $[t-D, t-D/2]$ at each t. We next calculate $\gamma(\omega)$ in (9.12). Direct computation shows that

$$\begin{aligned} \gamma(\omega) &= \frac{4}{\omega^2 D^2}\left(\sin(\omega D) - \sin\left(\frac{\omega D}{2}\right)\right)^2 \\ &\quad + \frac{4}{\omega^2 D^2}\left(\cos(\omega D) - \cos\left(\frac{\omega D}{2}\right)\right)^2 \\ &= \frac{16}{\omega^2 D^2}\sin^2\left(\frac{\omega D}{4}\right). \end{aligned} \tag{9.76}$$

We can see that $\gamma(\omega) \neq 0$ as long as $\omega \neq 4m\pi/D$ for each $m \in \mathbb{Z}$. Thus, Assumption 9.1 holds. The perturbation signal S can be obtained explicitly for β in (9.74). Indeed, it follows that

$$\begin{aligned} S(t) &= \frac{2a}{\gamma(\omega)D}\int_{D/2}^D \sin(\omega(t+\xi))d\xi \\ &= \frac{a\omega D}{4\sin(\omega D/4)}\sin\left(\omega\left(t + \frac{3}{4}D\right)\right), \end{aligned} \tag{9.77}$$

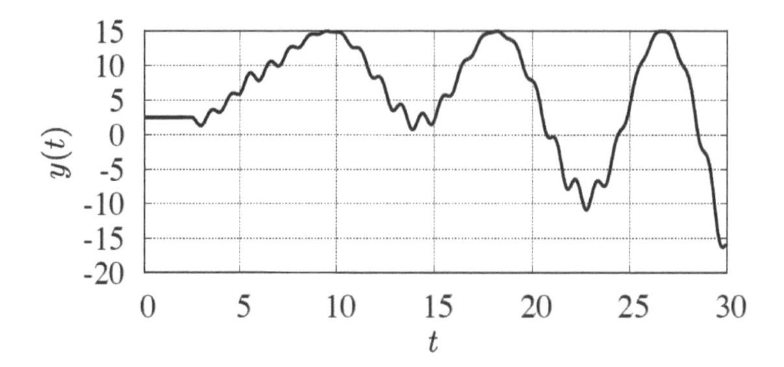

Figure 9.2. *The measured output of the closed-loop system without a prediction term in* (9.30).

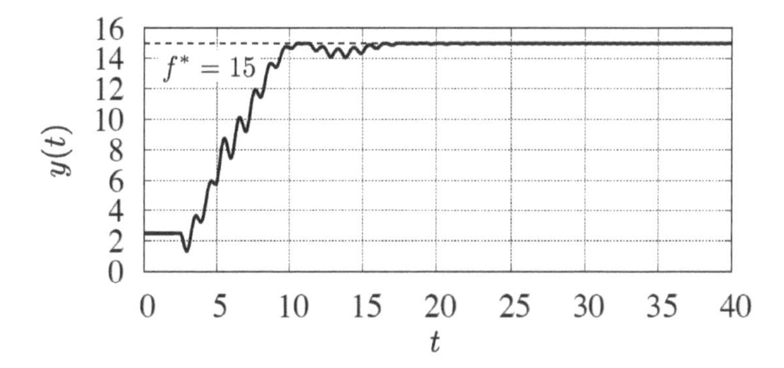

Figure 9.3. *The measurement output* $y(t)$.

where we have used the explicit form (9.76) of $\gamma(\omega)$. Consequently, S is a sinusoidal function, but its amplitude and phase have specific forms.

Now, we conduct a numerical simulation. The unknown parameters of the map are set as $f^* = 15$, $\theta^* = 5$, and $H = -1$. The maximum delay is $D = 5$. The parameters in the proposed extremum seeking scheme are chosen as $a = 0.25$, $\omega = 6$, $k = 0.5$, and $c = 1$. To improve the numerical computation, a high-pass filter $s = s/(s+\omega_h)$ is applied to y. The filtered signal is denoted by z. The signal My in (9.30) is placed with Mz. The signal Ny in (9.30) is also swapped with $\hat{H}(t)$, which is defined as a filtered signal of Nz with the low-pass $\omega_l/(s+\omega_l)$. We set $\omega_h = 1$ and $\omega_l = 0.1$. Initial conditions are such that $\overline{\theta}(0) = 0$ and $U(\sigma) = 0$ for any $\sigma \in [0, D]$. The initial values of the filters' states are also set as 0.

We first show the output y in Figure 9.2 when the predictor is not used. Namely, the second term in the parentheses on the RHS of (9.30) is neglected. We can see that the delay induces the instability.

We next show the simulation results for the proposed scheme in Figures 9.3 to 9.6. The output y is plotted in Figure 9.3 and it is observed that y approaches to a neighborhood of the extremum f^*. Similarly, the estimate $\hat{\theta}$ of θ^* converges to a neighborhood of θ^* as plotted in Figure 9.4. The value of the estimate of the Hessian H is shown in Figure 9.5. Although an oscillatory behavior is observed before $\hat{\theta}(t)$ approaches θ^*, the estimate surely converges to the true value. The control signal $U(t)$ also goes to 0 asymptotically (Figure 9.6). Therefore, the proposed scheme successfully compensates the effect of the distributed delay.

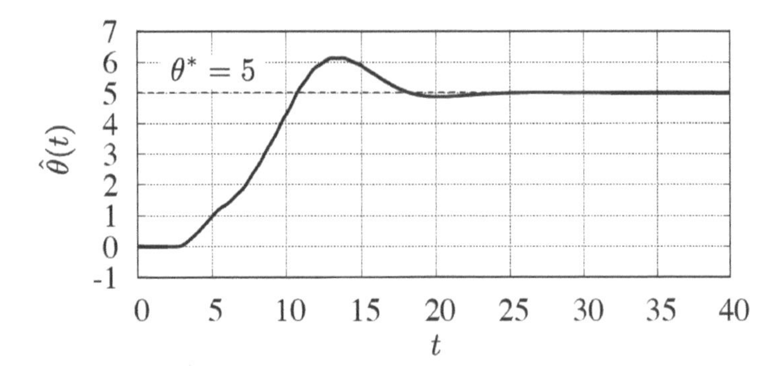

Figure 9.4. *The estimate $\hat{\theta}(t)$ of θ^*.*

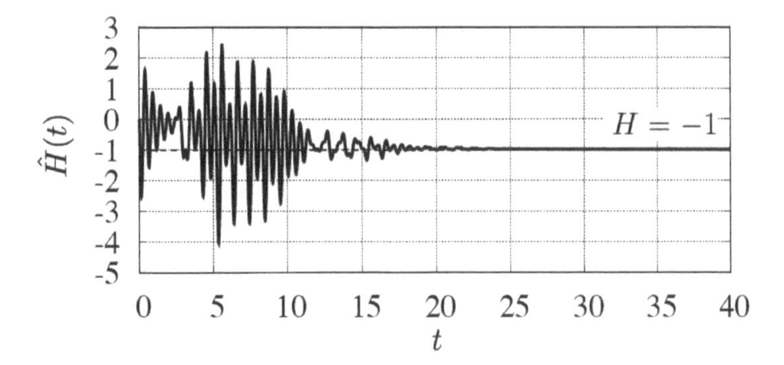

Figure 9.5. *The estimate $\hat{H}(t)$ of H.*

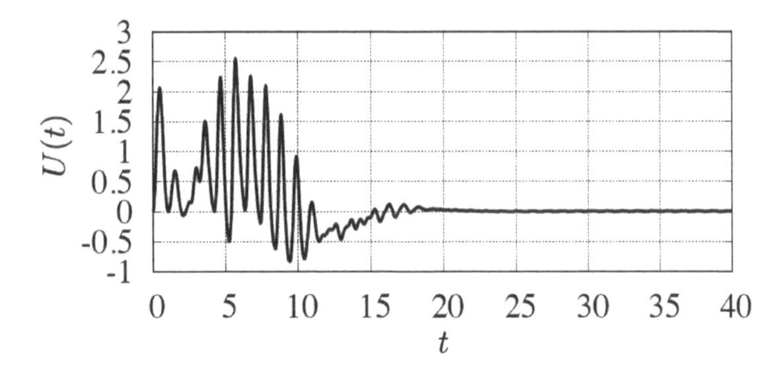

Figure 9.6. *The control signal $U(t)$.*

9.6 ▪ Multivariable Generalization for Distributed Delays

The multiparameter extension is discussed in the following.

9.6.1 ▪ Problem Setting

Consider a quadratic static multivariable map $f : \mathbb{R}^{\mathbb{N}} \to$ defined by

$$f(\theta) = f^* + \frac{1}{2}(\theta - \theta^*)^T H(\theta - \theta^*), \tag{9.78}$$

where f^* is the extremum, $\theta^* \in \mathbb{R}^n$ is the argument of the extremum, and $H = H^T \in \mathbb{R}^{n\times n}$ is a negative-definite Hessian matrix. All these parameters are assumed to be unknown. Suppose that, for an input signal $\theta(t) \in \mathbb{R}^n$, we can measure the signal

$$y(t) = f\left(\int_0^D d\beta(\sigma)\theta(t-\sigma)\right) = f^* + \frac{1}{2}\left(\int_0^D d\beta(\sigma)\theta(t-\sigma) - \theta^*\right)^T H\left(\int_0^D d\beta(\sigma)\theta(t-\sigma) - \theta^*\right), \tag{9.79}$$

where $D > 0$ is the representation of delays and $\beta : \mathbb{R} \to \mathbb{R}^{n\times n}$ is a matrix-valued function whose entries are functions of bounded variation. We also assume that the off-diagonal entries of $\beta(x)$ are 0. Namely, there exist functions $\beta_i : \mathbb{R} \to \mathbb{R}$, $i \in \{1,2,\ldots,n\}$ of bounded variation such that $\beta(x) = \text{diag}(\beta_1(x), \beta_2(x), \ldots, \beta_n(x))$. Then, we have

$$\int_0^D d\beta(\sigma)\theta(t-\sigma) = \left(\int_0^D \theta_1(t-\sigma)d\beta_1(\sigma) \int_0^D \theta_2(t-\sigma)d\beta_2(\sigma) \cdots \int_0^D \theta_n(t-\sigma)d\beta_n(\sigma)\right)^T. \tag{9.80}$$

If a constant signal $\theta(t) \equiv \theta_0 \in \mathbb{R}$ is sent to the map, the corresponding delayed signal is given by

$$\int_0^D d\beta(\sigma)\theta(t-\sigma) = \int_0^D d\beta(\sigma)\theta_0 = (\beta(D) - \beta(0))\theta_0. \tag{9.81}$$

It is natural to expect that this value is θ_0. Hence, β should satisfy $\beta(D) - \beta(0) = I_n$. Since addition of a constant matrix to β does not change the value of the Riemann–Stieltjes integral with respect to β, we can set $\beta(0) = 0$ without loss of generality.

We summarize the assumption on β.

Assumption 9.2. *For each $x \in \mathbb{R}$, the function $\beta(x)$ takes a diagonal matrix value*

$$\text{diag}(\beta_1(x), \beta_2(x), \ldots, \beta_n(x)).$$

Every diagonal entry $\beta_i : \mathbb{R} \to \mathbb{R}$, $i = 1,2,\ldots,n$ is a right-continuous function of bounded variation such that $\beta_i(0) = 0$ and $\beta_i(D) = 1$.

For simplicity of notation, we write the Riemann–Stieltjes integral of θ with respect to β over $[0, D]$ as

$$\theta_\beta(t) = \int_0^D d\beta(\sigma)\theta(t-\sigma). \tag{9.82}$$

With this notation, the measurement output can be rewritten as

$$y(t) = f(\theta_\beta(t)). \tag{9.83}$$

If, for each $i \in \{1,2,\ldots,n\}$, the function β_i is given by

$$\beta_i(x) = \begin{cases} 0, & x < D_i, \\ 1, & D_i \le x, \end{cases} \tag{9.84}$$

for some $D_i \in (0, D)$, then $\theta_\beta(t)$ becomes

$$\int_0^D d\beta(\sigma)\theta(t-\sigma) = \begin{pmatrix}\theta_1(t-D_1) & \theta_2(t-D_2) & \cdots & \theta_n(t-D_n)\end{pmatrix}^T. \tag{9.85}$$

Hence, the above formulation encompasses the case for lumped constant input delays as well.

9.6.2 ▪ Preliminaries

Before developing extremum seeking schemes, we set up signals to be used in the proposed approach.

Probing signal

Let $\hat{\theta}(t) \in \mathbb{R}^n$ be an estimate of θ^*. To determine a suitable update law for $\hat{\theta}$, we also introduce auxiliary variables $\overline{\theta}(t) \in \mathbb{R}^n$ and $U(t) \in \mathbb{R}^n$ satisfying

$$\dot{\overline{\theta}}(t) = U(t). \tag{9.86}$$

Set the estimate $\hat{\theta}$ as

$$\hat{\theta}(t) = \overline{\theta}_\beta(t) = \int_0^D d\beta(\sigma)\overline{\theta}(t-\sigma). \tag{9.87}$$

The probing signal to be applied to the map normally consists of $\hat{\theta}$ and a perturbation signal. We propose the perturbation signal $\overline{S}(t) = \left(\overline{S}_1(t), \overline{S}_2(t) \ldots \overline{S}_n(t)\right) \in \mathbb{R}^n$ given by

$$\overline{S}_i(t) = \frac{a_i}{\gamma_i(\omega_i)} \int_0^D \sin(\omega_i(t+\tau))d\beta_i(\tau), \qquad i \in \{1,2,\ldots,n\}, \tag{9.88}$$

where $a_i > 0$. For each $i \in \{1,2,\ldots,n\}$, the function $\gamma_i : \mathbb{R} \to \mathbb{R}$ is defined by

$$\gamma_i(\omega) := \left(\int_0^D \cos(\omega\sigma)d\beta_i(\sigma)\right)^2 + \left(\int_0^D \sin(\omega\sigma)d\beta_i(\sigma)\right)^2. \tag{9.89}$$

It is clear from (9.88) that $\gamma_i(\omega_i)$ must not vanish.

The signal (9.88) is considerably different from the conventional perturbation signal (see Chapter 1, Equation (1.7)):

$$S(t) = \begin{pmatrix} S_1(t) & S_2(t) & \cdots & S_n(t) \end{pmatrix} := \begin{pmatrix} a_1\sin(\omega_1 t) & a_2\sin(\omega_2 t) & \cdots & a_n\sin(\omega_n t) \end{pmatrix}^T. \tag{9.90}$$

The reason for defining $\overline{S}_i$ by (9.88) is revealed by the next observation. With the angle sum identity for trigonometric functions, it can be shown that

$$\begin{aligned}
\int_0^D \overline{S}_i(t-\sigma)d\beta_i(\sigma) &= \frac{a_i}{\gamma_i(\omega_i)} \int_0^D \int_0^D \sin\left(\omega_i(t-\sigma+\tau)\right)d\beta_i(\tau)d\beta_i(\sigma) \\
&= \frac{a_i}{\gamma_i(\omega_i)}\sin(\omega_i t) \int_0^D \int_0^D \cos\left(\omega_i(\sigma-\tau)\right)d\beta_i(\tau)d\beta_i(\sigma) \\
&\quad - \frac{a_i}{\gamma_i(\omega_i)}\cos(\omega_i t) \int_0^D \int_0^D \sin\left(\omega_i(\sigma-\tau)\right)d\beta_i(\tau)d\beta_i(\sigma) \\
&= \frac{a_i}{\gamma_i(\omega_i)}\sin(\omega_i t)\left(\left(\int_0^D \cos(\omega_i\sigma)d\beta_i(\sigma)\right)^2 + \left(\int_0^D \sin(\omega_i\sigma)d\beta_i(\sigma)\right)^2\right) \\
&\quad - \frac{a_i}{\gamma_i(\omega_i)}\cos(\omega_i t) \\
&\quad \times \left(\int_0^D \cos(\omega_i\sigma)d\beta_i(\sigma)\int_0^D \sin(\omega_i\tau)d\beta_i(\tau)\right. \\
&\quad \left. - \int_0^D \sin(\omega_i\sigma)d\beta_i(\sigma)\int_0^D \cos(\omega_i\tau)d\beta_i(\tau)\right).
\end{aligned} \tag{9.91}$$

Thus,

$$\int_0^D \overline{S}_i(t-\sigma)d\beta_i(\sigma) = a_i \sin(\omega_i t) = S_i(t). \tag{9.92}$$

Hence, setting the probing signal θ as

$$\theta(t) = \overline{\theta}(t) + \overline{S}(t) \tag{9.93}$$

yields the following relation:

$$\begin{aligned}\theta_\beta(t) &= \int_0^D d\beta(\sigma)\big(\overline{\theta}(t-\sigma) + \overline{S}(t-\sigma)\big) \\ &= \int_0^D d\beta(\sigma)\overline{\theta}(t-\sigma) + \int_0^D d\beta(\sigma)\overline{S}(t-\sigma) \\ &= \hat{\theta}(t) + S(t).\end{aligned} \tag{9.94}$$

Note that (9.93) includes $\overline{\theta}$ rather than $\hat{\theta}$. The measurement output is then represented as

$$y(t) = f\big(\theta_\beta(t)\big) = f\big(\hat{\theta}(t) + S(t)\big) = f\big(\tilde{\theta}(t) + S(t) + \theta^*\big), \tag{9.95}$$

where the estimation error is

$$\tilde{\theta}(t) = \hat{\theta}(t) - \theta^*. \tag{9.96}$$

It looks as if delays were absent. Hence, computations in the averaging process developed for the delay-free extremum seeking are still valid for our problem. As seen in (9.88), we need to pay attention to the behavior of γ_i.

Assumption 9.3. *For each $i \in \{1,2,\ldots,n\}$, let $\gamma_i : \mathbb{R} \to \mathbb{R}$ be defined by (9.89) for the ith diagonal entry of $\beta : \mathbb{R} \to \mathbb{R}^{n\times n}$ in Assumption 9.2 satisfying Assumption 9.1. Fix an arbitrary $i \in \{1,2,\ldots,n\}$. Then there exists a non-decreasing sequence $\{\overline{\omega}_{i,j}\}_{j=1}^{\infty} \subset (0,+\infty)$ such that $\gamma_i(\overline{\omega}_{i,j}) \neq 0$ for all $j \in \mathbb{N}$ and that $\overline{\omega}_{i,j}$ tends to ∞ as $j \to \infty$.*

Let us take examples. Suppose that β_i is given by (9.84). In this case, we have

$$\gamma_i(\omega) = \cos^2(\omega t) + \sin^2(\omega t) = 1. \tag{9.97}$$

Hence, the assumption always holds true. As another example, we suppose that β_i is given by

$$\beta_i(x) = \begin{cases} 0, & x < 0, \\ x/D, & 0 \leq x < D, \\ 1, & D \geq x. \end{cases} \tag{9.98}$$

The corresponding γ_i is calculated as

$$\begin{aligned}\gamma_i(\omega) &= \left(\frac{1}{D}\int_0^D \cos(\omega\sigma)d\sigma\right)^2 + \left(\frac{1}{D}\int_0^D \sin(\omega\sigma)d\sigma\right)^2 \\ &= \frac{1}{\omega^2 D^2}\Big(\sin^2(\omega D) + (1-\cos(\omega D))^2\Big) \\ &= \frac{2}{\omega^2 D^2}\big(1-\cos(\omega D)\big).\end{aligned} \tag{9.99}$$

This means that $\gamma_i(\omega) \neq 0$ except at $\omega = 2m\pi/D$ with $m \in \mathbb{N} \cup \{0\}$. Assumption 9.3 is again valid.

Signals to estimate unknown variables

Let us introduce the vector-valued signal $M(t) = (M_1(t), M_2(t), \dots, M_n(t)) \in \mathbb{R}^n$ and the matrix-valued signal $N(t) = (N_{i,j}(t)) \in \mathbb{R}^{n\times n}$ defined for each $i,j \in \{1,2,\dots,n\}$ by

$$M_i(t) = \frac{2}{a_i}\sin(\omega_i t), \tag{9.100}$$

$$N_{i,j} = \begin{cases} \frac{16}{a_i^2}\left(\sin^2(\omega_i t) - \frac{1}{2}\right), & i = j, \\ \frac{4}{a_i a_j}\sin(\omega_i t)\sin(\omega_j t), & i \neq j. \end{cases} \tag{9.101}$$

An assumption is made for the frequencies ω_i, $i = \{1,2,\dots,n\}$.

Assumption 9.4. *There exist real number $\omega > 0$ and rational numbers $\omega_1', \omega_2', \dots, \omega_n' > 0$ such that $\omega_i = \omega\omega_i'$ for each $i = \{1,2,\dots,n\}$ and that*

$$\omega_1' \notin \left\{\omega_j', \frac{1}{2}(\omega_j' + \omega_k'), \omega_j' + 2\omega_k', \omega_j' + \omega_k' \pm \omega' l\right\} \tag{9.102}$$

for all distinct $i,j,k,l \in \{1,2,\dots,n\}$.

Under Assumption 9.4, we can define Π by

$$\Pi = \frac{2\pi}{\omega} \times \text{LCM}\left\{\frac{1}{\omega_1'}, \frac{1}{\omega_2'}, \dots, \frac{1}{\omega_n'}\right\}, \tag{9.103}$$

where LCM stands for the least common multiple. Then, as shown in [73], for any constant $\tilde{\theta} \in \mathbb{R}^n$, we have

$$\frac{1}{\Pi}\int_0^{\Pi} M(\tau) f(\tilde{\theta} + S(\tau) + \theta^*)d\tau = H\tilde{\theta}, \tag{9.104}$$

$$\frac{1}{\Pi}\int_0^{\Pi} N(\tau) f(\tilde{\theta} + S(\tau) + \theta^*)d\tau = H. \tag{9.105}$$

Hence, $H\tilde{\theta}$ and H can be estimated from $M(t)y(t)$ and $N(t)y(t)$, respectively, in the average sense. For later use, we also introduce a signal that estimates the inverse of H. It is also shown in [73] that $\Gamma(t) \in \mathbb{R}^{n\times n}$, satisfying the Riccati filter

$$\dot{\Gamma}(t) = \omega_r\Gamma(t) - \omega_r\Gamma(t)N(t)y(t)\Gamma(t), \tag{9.106}$$

can be an estimate of H^{-1}.

9.6.3 ▪ Gradient- and Newton-Based Multiparameter Extremum Seeking

The remaining task is to determine the update law for the estimate $\hat{\theta}$. Differentiating (9.87) with respect to t leads to

$$\dot{\hat{\theta}}(t) = \int_0^D d\beta(\sigma)\dot{\bar{\theta}}(t-\sigma) = \int_0^D d\beta(\sigma)U(t-\sigma). \tag{9.107}$$

Since $\dot{\tilde{\theta}} = \dot{\hat{\theta}}$, the above relation also implies that

$$\dot{\tilde{\theta}}(t) = \int_0^D d\beta(\sigma)U(t-\sigma). \tag{9.108}$$

If U is the control input, this equation looks like a multi-input linear system with distributed delays. The reduction approach [12] gives the feedback control law

$$U(t) = -K\left(\tilde{\theta}(t) + \int_t^{t+D} d\tau \int_{\tau-t}^{D} d\beta(\sigma)U(t-\sigma)\right) \tag{9.109}$$

for some positive-definite gain $K \in \mathbb{R}^{n\times n}$. Note that the value in the parentheses is not an exact future value of $\tilde{\theta}$ since the interval of integration for the inner integral is not $[0,D]$ but $[\tau - t, D]$. If we use $[0,D]$, future values of U are necessary. Unfortunately, (9.109) cannot be implemented since $\tilde{\theta}$ is not available. The idea is to realize (9.109) in an average sense. We also need to incorporate a filter to make U be a part of the state variable so that the averaging analysis is applicable to the closed-loop system (Appendix B).

We start with simplifying the double integral on the RHS of (9.109). Changing the order of integration gives

$$\begin{aligned}
\int_t^{t+D} d\tau \int_{t-\tau}^{D} d\beta(\sigma)U(\tau-\sigma) &= \int_0^D d\beta(\sigma)\int_t^{t+\sigma} U(\tau-\sigma)d\tau \\
&= \int_0^D d\beta(\sigma)\int_0^{\sigma} U(t-\xi)d\xi \\
&= \int_0^D d\xi \int_{\xi}^{D} d\beta(\sigma)U(t-\xi) \\
&= \int_0^D (I_n - \beta(\sigma))U(t-\sigma)d\sigma.
\end{aligned} \tag{9.110}$$

Based on this expression, we define two extremum seeking schemes: gradient-based and Newton-based schemes. Recall that the estimate $\hat{\theta}$ is given by (9.87) and that the probing signal θ is constructed as (9.93). The intermediate variable $\overline{\theta}$ satisfies (9.86). Hence, we only have to determine U.

As explained, (9.109) cannot be implemented because of the unavailability of $\tilde{\theta}$. However, from (9.104), $H\tilde{\theta}$ is available in the average sense. This fact motivates us to replace $-K$ in (9.109) with KH. The matrix KH must be Hurwitz. Since H is a negative-definite matrix, K is chosen as a diagonal matrix with positive entries. The **gradient-based ES scheme** is given by

$$\dot{U}(t) = -cU(t) + cK\left(M(t)y(t) + N(t)y(t)\int_0^D (I_n - \beta(\sigma))U(t-\sigma)d\sigma\right) \tag{9.111}$$

for some $c > 0$. In the averaging sense, $M(t)y(t)$ and $N(t)y(t)$ correspond to $H\tilde{\theta}$ and H, respectively. Hence, (9.111) imitates the equation

$$\dot{U}(t) = -cU(t) + cKH\left(\tilde{\theta}(t) + \int_0^D (I_n - \beta(\sigma))U(t-\sigma)d\sigma\right). \tag{9.112}$$

In the proposed gradient-based ES scheme, the gain contains the unknown Hessian H, which affects the rate of convergence. In the **Newton-based ES scheme,** $\Gamma(t)$, which is an estimate of

H^{-1}, is used to cancel H. Namely, we proposed the following U:

$$\dot{U}(t) = -cU(t) - cK\left(\Gamma(t)M(t)y(t) + \int_0^D (I_n - \beta(\sigma))U(t-\sigma)d\sigma\right). \tag{9.113}$$

If we regard $M(t)y(t)$ and $\Gamma(t)$ as $H\tilde{\theta}$ and H^{-1}, respectively, then (9.113) can be read as (9.109) in the average sense.

Before proceeding to the next section, we briefly summarize the proposed extremum seeking schemes. The common components are given by

$$\theta(t) = \overline{\theta}(t) + \overline{S}(t), \tag{9.114}$$

$$\dot{\overline{\theta}}(t) = U(t), \tag{9.115}$$

$$\hat{\theta}(t) = \int_0^D d\beta(\sigma)\overline{\theta}(t-\sigma). \tag{9.116}$$

In the **gradient-based ES scheme**, the signal U is determined by (9.111). The closed-loop system is given by

$$\dot{\tilde{\theta}}(t) = \int_0^D d\beta(\sigma)U(t-\sigma), \tag{9.117}$$

$$\dot{U}(t) = -cU(t) + cK\left(M(t)f\left(\tilde{\theta}(t)+S(t)+\theta^*\right) + N(t)f\left(\tilde{\theta}(t)+S(t)+\theta^*\right)\int_0^D (I_n - \beta(\sigma))U(t-\sigma)d\sigma\right). \tag{9.118}$$

On the other hand, U is given by (9.113) together with (9.106) in the **Newton-based ES scheme**. Set $\tilde{\Gamma}(t) := \Gamma(t) - H^{-1}$. Then, the equations of the closed-loop system can be written as

$$\dot{\tilde{\theta}}(t) = \int_0^D d\beta(\sigma)U(t-\sigma), \tag{9.119}$$

$$\begin{aligned}\dot{U}(t) = {} & -cU(t) + cK\left(H^{-1}M(t)f\left(\tilde{\theta}(t)+S(t)+\theta^*\right) + \int_0^D (I_n - \beta(\sigma))U(t-\sigma)d\sigma\right) \\ & + cK\tilde{\Gamma}(t)M(t)f\left(\tilde{\theta}(t)+S(t)+\theta^*\right),\end{aligned} \tag{9.120}$$

$$\dot{\tilde{\Gamma}}(t) = -\omega_r H^{-1}N(t)y(t)\tilde{\Gamma}(t) - \omega_r\tilde{\Gamma}(t)N(t)y(t)\tilde{\Gamma}(t) + \omega_r\left(\tilde{\Gamma}(t)+H^{-1}\right)\left(H - N(t)y(t)\right)H^{-1}. \tag{9.121}$$

9.6.4 ▪ Stability Analysis

Gradient-based scheme

The average equations corresponding to (9.117) and (9.118) are given by

$$\dot{\tilde{\theta}}^{\mathrm{av}}(t) = \int_0^D d\beta(\sigma)U^{\mathrm{av}}(t-\sigma), \tag{9.122}$$

$$\dot{U}^{\mathrm{av}}(t) = -cU^{\mathrm{av}}(t) + cKH\left(\tilde{\theta}^{\mathrm{av}}(t) + \int_0^D (I_n - \beta(\sigma))U^{\mathrm{av}}(t-\sigma)d\sigma\right). \tag{9.123}$$

These equations have the following PDE representation:

$$\dot{\tilde{\theta}}^{\rm av}(t) = \int_0^D d\beta(\sigma) u^{\rm av}(D-\sigma,t), \tag{9.124}$$

$$u_t^{\rm av}(x,t) = u_x^{\rm av}(x,t), \tag{9.125}$$

$$u^{\rm av}(D,t) = U^{\rm av}(t), \tag{9.126}$$

$$\dot{U}^{\rm av}(t) = -cU^{\rm av}(t) + cKH\left(\tilde{\theta}^{\rm av}(t) + \int_0^D (I_n - \beta(\sigma)) u^{\rm av}(D-\sigma,t) d\sigma\right), \tag{9.127}$$

where $u^{\rm av}(x,t) = (u_1^{\rm av}(x,t), u_2^{\rm av}(x,t), \ldots, u_n^{\rm av}(x,t))^T$ is an $\mathbb{R}^n$-valued function defined in $[0,D]\times[0,+\infty)$. It is easy to see that $U^{\rm av}(x+t-D) = u^{\rm av}(x,t)$ for (x,t) satisfying $t \geq D-x$. We use the following reduction transformation:

$$\vartheta(t) := H\left(\tilde{\theta}^{\rm av}(t) + \int_0^D (I_n - \beta(\sigma)) u^{\rm av}(D-\sigma,t) d\sigma\right). \tag{9.128}$$

Differentiating both sides yields

$$\begin{aligned}
\dot{\vartheta}(t) &= H\dot{\tilde{\theta}}^{\rm av}(t) + H\int_0^D (I_n - \beta(\sigma)) u_t^{\rm av}(D-\sigma,t) d\sigma \\
&= H\int_0^D d\beta(\sigma) u^{\rm av}(D-\sigma,t) \\
&\quad + H\int_0^D (I_n - \beta(\sigma)) u_x^{\rm av}(D-\sigma,t) d\sigma \\
&= H\int_0^D d\beta(\sigma) u^{\rm av}(D-\sigma,t) \\
&\quad + H\Big[-(I_n - \beta(\sigma)) u^{\rm av}(D-\sigma,t)\Big]_0^D \\
&\quad - H\int_0^D d\beta(\sigma) u^{\rm av}(D-\sigma,t) \\
&= HU^{\rm av}(t).
\end{aligned} \tag{9.129}$$

This is a delay-free linear system, especially, n integrators. The filter equation (9.127) becomes

$$\dot{U}^{\rm av}(t) = -cU^{\rm av}(t) + cK\vartheta(t) = -c\big(U^{\rm av}(t) - K\vartheta(t)\big). \tag{9.130}$$

Consider a candidate of Lyapunov function

$$\begin{aligned}
V(t) &= \frac{1}{2}\vartheta(t)^T K \vartheta(t) \\
&\quad + \frac{1}{2}\int_0^D e^{-(D-x)} u^{\rm av}(x,t)^T (-H) u^{\rm av}(x,t) dx \\
&\quad + \frac{1}{2}\big(U^{\rm av}(t) - K\vartheta(t)\big)^T \big(U^{\rm av}(t) - K\vartheta(t)\big).
\end{aligned} \tag{9.131}$$

The time derivative of V can be computed as

$$
\begin{aligned}
\dot V(t) &= \vartheta(t)^T KHK\vartheta(t) + \vartheta(t)^T KH\big(U^{\mathrm{av}}(t) - K\vartheta(t)\big) + \frac{1}{2}U^{\mathrm{av}}(t)^T(-H)U^{\mathrm{av}}(t) \\
&\quad - \frac{1}{2}e^{-D}u^{\mathrm{av}}(0,t)^T(-H)u^{\mathrm{av}}(0,t) - \frac{1}{2}\int_0^D e^{-(D-x)}u^{\mathrm{av}}(x,t)^T(-H)u^{\mathrm{av}}(x,t)dx \\
&\quad + \big(U^{\mathrm{av}}(t) - K\vartheta(t)\big)^T\big(-cU^{\mathrm{av}}(t) + cK\vartheta(t) - KHU^{\mathrm{av}}(t)\big) \\
&= \vartheta(t)^T KHK\vartheta(t) + \vartheta(t)^T KH\big(U^{\mathrm{av}}(t) - k\vartheta(t)\big) \\
&\quad + \frac{1}{2}\big(U^{\mathrm{av}}(t) - K\vartheta(t)\big)^T(-H)\big(U^{\mathrm{av}}(t) - K\vartheta(t)\big) \\
&\quad - \big(U^{\mathrm{av}}(t) - K\vartheta(t)\big)^T HK\vartheta(t) - \frac{1}{2}\vartheta(t)^T KHK\vartheta(t) \\
&\quad - \frac{1}{2}e^{-D}u^{\mathrm{av}}(0,t)^T(-H)u^{\mathrm{av}}(0,t) - \frac{1}{2}\int_0^D e^{-(D-x)}u^{\mathrm{av}}(x,t)^T(-H)u^{\mathrm{av}}(x,t)dx \\
&\quad - \big(U^{\mathrm{av}}(t) - KH\vartheta(t)\big)^T\left(cI_n + \frac{1}{2}(KH + HK)\right)\big(U^{\mathrm{av}}(t) - KH\vartheta(t)\big) \\
&\quad + \big(U^{\mathrm{av}}(t) - KH\vartheta(t)\big)^T K(-H)K\vartheta(t) \\
&= -\frac{1}{2}\vartheta(t)^T K(-H)K\vartheta(t) - \frac{1}{2}e^{-D}u^{\mathrm{av}}(0,t)^T(-H)u^{\mathrm{av}}(0,t) \\
&\quad - \frac{1}{2}\int_0^D e^{-(D-x)}u^{\mathrm{av}}(x,t)^T(-H)u^{\mathrm{av}}(x,t)dx \\
&\quad - \big(U^{\mathrm{av}}(t) - KH\vartheta(t)\big)^T\left(cI_n + \frac{1}{2}(KH + HK)\right)\big(U^{\mathrm{av}}(t) - KH\vartheta(t)\big) \\
&\quad + \big(U^{\mathrm{av}}(t) - KH\vartheta(t)\big)^T K(-H)K\vartheta(t). \qquad (9.132)
\end{aligned}
$$

It follows from Young's inequality that

$$
\begin{aligned}
\big(U^{\mathrm{av}}(t) - KH\vartheta(t)\big)^T K(-H)K\vartheta(t) \le \big(U^{\mathrm{av}}(t) - KH\vartheta(t)\big)^T K(-H)K\big(U^{\mathrm{av}}(t) - KH\vartheta(t)\big)^T \\
+ \frac{1}{4}\vartheta(t)^T K(-H)K\vartheta(t). \\
(9.133)
\end{aligned}
$$

We thus have

$$
\begin{aligned}
\dot V(t) &\le \frac{1}{4}\vartheta(t)^T K(-H)K\vartheta(t) + \frac{1}{2}\int_0^D e^{-(D-x)}u^{\mathrm{av}}(x,t)^T(-H)u^{\mathrm{av}}(x,t)dx \\
&\quad - \big(U^{\mathrm{av}}(t) - KH\vartheta(t)\big)^T\left(cI_n + \frac{1}{2}(KH + HK) + KHK\right)\big(U^{\mathrm{av}}(t) - KH\vartheta(t)\big).
\end{aligned}
\qquad (9.134)
$$

Set $c^* := (1/2)\lambda_{\min}(KH + HK + 2KHK)$. Then, $\dot V$ satisfies

$$
\begin{aligned}
\dot V(t) &\le \frac{1}{4}\vartheta(t)^T K(-H)K\vartheta(t) - \frac{1}{2}\int_0^D e^{-(D-x)}u^{\mathrm{av}}(x,t)^T(-H)u^{\mathrm{av}}(x,t)dx \\
&\quad - (c - c^*)\big(U^{\mathrm{av}}(t) - KH\vartheta(t)\big)^T\big(U^{\mathrm{av}}(t) - KH\vartheta(t)\big). \qquad (9.135)
\end{aligned}
$$

Together with other several inequalities, we can infer from (9.135) the exponential stability of the average system (9.124)–(9.127) with respect to the norm

$$||[\theta^{\mathrm{av}}, u^{\mathrm{av}}, U^{\mathrm{av}}]|| = \left(\theta^{\mathrm{av}}(t)^T \theta^{\mathrm{av}}(t) + \int_0^D u^{\mathrm{av}}(x,t)^T u^{\mathrm{av}}(x,t)dx + U^{\mathrm{av}}(t)^T U^{\mathrm{av}}(t) \right)^{1/2}. \tag{9.136}$$

Let $\mathcal{H} = \mathbb{R}^n \times L^2((0,D);\mathbb{R}^n) \times \mathbb{R}^n$ be a Hilbert space equipped with the inner product

$$\Big([X_1, Y_1, Z_1], [X_2, Y_2, Z_2]\Big) = X_1^T X_2 + \int_0^D Y_1(x)^T Y_2(x)dx + Z_1^T Z_2. \tag{9.137}$$

This inner product induces the norm (9.136). The next statement summarizes the results for multiparameter gradient-based ES under distributed delays.

Proposition 9.1. *Consider the system* (9.124)–(9.127), *where* $D > 0$ *is a constant,* $K = K^T \in \mathbb{R}^{n\times n}$ *is a diagonal matrix with positive diagonal entries, and* $H = H^T \in \mathbb{R}^{n\times n}$ *is a negative definite matrix. Assume that* $\beta : \mathbb{R} \to \mathbb{R}^{n\times n}$ *satisfies Assumptions* 9.2 *and* 9.3. *Then there exists* $c^* \in \mathbb{R}$ *depending only on* K *and* H *such that, if* $c > c^*$, *the system* (9.124)–(9.127) *is well-posed in* $\mathcal{H}$ *and the zero-solution is exponentially stable.*

Newton-based scheme

The averaged version of the closed-loop system (9.119)–(9.121) is given by

$$\dot{\tilde{\theta}}^{\mathrm{av}}(t) = \int_0^D d\beta(\sigma) U^{\mathrm{av}}(t-\sigma), \tag{9.138}$$

$$\dot{U}^{\mathrm{av}}(t) = -cU^{\mathrm{av}}(t) - cK\left(\tilde{\theta}^{\mathrm{av}}(t) + \int_0^D (I_n - \beta(\sigma))U^{\mathrm{av}}(t-\sigma)d\sigma\right) + cK\tilde{\Gamma}^{\mathrm{av}}(t)H\tilde{\theta}^{\mathrm{av}}(t), \tag{9.139}$$

$$\dot{\tilde{\Gamma}}^{\mathrm{av}}(t) = -\omega_r \tilde{\Gamma}^{\mathrm{av}}(t) - \omega_r \tilde{\Gamma}^{\mathrm{av}}(t) H \tilde{\Gamma}^{\mathrm{av}}(t). \tag{9.140}$$

Equations (9.139) and (9.140) contain quadratic nonlinear terms. In addition, it is not difficult to see that (9.140) has a locally exponentially stable equilibrium point $\tilde{\Gamma}^{\mathrm{av}}(t) = 0$. Hence, after linearization of the system with respect to $\tilde{\Gamma}^{\mathrm{av}}(t)$ around $\tilde{\Gamma}^{\mathrm{av}}(t) = 0$, the $(\tilde{\theta}^{\mathrm{av}}, U^{\mathrm{av}})$-subsystem and the $\tilde{\Gamma}^{\mathrm{av}}$-subsystem are completely decoupled. In particular, as the linearized equation of (9.139), we have

$$\dot{U}^{\mathrm{av}}(t) = -cU^{\mathrm{av}}(t) - cK\left(\tilde{\theta}^{\mathrm{av}}(t) + \int_0^D (I_n - \beta(\sigma))U^{\mathrm{av}}(t-\sigma)d\sigma\right). \tag{9.141}$$

Similar to the previous subsection, a PDE representation of (9.138), (9.141) is derived as

$$\dot{\tilde{\theta}}^{\mathrm{av}}(t) = \int_0^D d\beta(\sigma) u^{\mathrm{av}}(D-\sigma, t), \tag{9.142}$$

$$u_t^{\mathrm{av}}(x,t) = u_x^{\mathrm{av}}(x,t), \tag{9.143}$$

$$u^{\mathrm{av}}(D,t) = U^{\mathrm{av}}(t), \tag{9.144}$$

$$\dot{U}^{\mathrm{av}}(t) = -cU^{\mathrm{av}}(t) - cK\left(\tilde{\theta}^{\mathrm{av}}(t) + \int_0^D (I_n - \beta(\sigma))u^{\mathrm{av}}(D-\sigma,t)d\sigma\right). \tag{9.145}$$

We show the exponential stability of the average system based on this representation.

We again use the reduction transformation

$$\vartheta(t) = \tilde{\theta}^{\text{av}}(t) + \int_0^D (I_n - \beta(\sigma))u^{\text{av}}(D-\sigma,t)d\sigma. \tag{9.146}$$

Note that the right-hand side does not contain H unlike (9.128) in the gradient-based scheme. Differentiating ϑ defined by (9.146) with respect to t leads to

$$\dot{\vartheta}(t) = \int_0^D d\beta(\sigma)u^{\text{av}}(D-\sigma,t) + u^{\text{av}}(D,t) - \int_0^D d\beta(\sigma)u^{\text{av}}(D-\sigma,t) = U^{\text{av}}(t). \tag{9.147}$$

With this new variable, (9.145) is rewritten as

$$\dot{U}^{\text{av}}(t) = -c\big(U^{\text{av}}(t) + K\vartheta(t)\big). \tag{9.148}$$

Let V be a candidate of the Lyapunov function defined by

$$\begin{aligned} V(t) = &\frac{1}{2}\vartheta(t)^T K^2 \vartheta(t) + \frac{1}{4}\int_0^D e^{-(D-x)}u^{\text{av}}(x,t)^T K u^{\text{av}}(x,t)dx \\ &+ \frac{1}{4}\big(U^{\text{av}}(t) + K\vartheta(t)\big)^T \big(U^{\text{av}}(t) + K\vartheta(t)\big). \end{aligned} \tag{9.149}$$

Differentiating V with respect to t leads to

$$\begin{aligned} \dot{V}(t) = &\vartheta(t)^T K^3 \vartheta(t) + \vartheta(t)^T K^2 \big(U^{\text{av}} + K\vartheta(t)\big) \\ &+ \frac{1}{4}\big(U^{\text{av}}(t) + K\vartheta(t)\big)^T K \big(U^{\text{av}}(t) + K\vartheta(t)\big) \\ &- \frac{1}{2}\big(U^{\text{av}}(t) + K\vartheta(t)\big)^T K^2 \vartheta(t) + \frac{1}{4}\vartheta(t)^T K^3 \vartheta(t) \\ &- \frac{1}{4}e^{-D}u^{\text{av}}(0,t)^T K u^{\text{av}}(0,t) \\ &- \frac{1}{4}\int_0^D e^{-(D-x)}u^{\text{av}}(x,t)^T K u^{\text{av}}(x,t)dx \\ &- \frac{1}{2}\big(U^{\text{av}}(t) + K\vartheta(t)\big)^T \left(cI_n - K\right)\big(U^{\text{av}}(t) + K\vartheta(t)\big) \\ &- \frac{1}{2}\big(U^{\text{av}}(t) + K\vartheta(t)\big)^T K^2 \vartheta(t) \\ = &-\frac{3}{4}\vartheta(t)^T K^3 \vartheta(t) - \frac{1}{4}\int_0^D e^{-(D-x)}u^{\text{av}}(x,t)^T K u^{\text{av}}(x,t)dx \\ &- \frac{1}{4}\big(U^{\text{av}}(t) + K\vartheta(t)\big)^T \left(cI_n - \frac{3}{2}K\right)\big(U^{\text{av}}(t) + K\vartheta(t)\big) \\ &- \frac{1}{4}e^{-D}u^{\text{av}}(0,t)^T K u^{\text{av}}(0,t). \end{aligned} \tag{9.150}$$

Let $c^* > 0$ defined by $c^* = (3/2)\lambda_{\max}(K)$. Then, we have

$$\begin{aligned} \dot{V}(t) \leq &-\frac{3}{4}\vartheta(t)^T K^3 \vartheta(t) - \frac{1}{4}\int_0^D e^{-(D-x)}u^{\text{av}}(x,t)^T K u^{\text{av}}(x,t)dx \\ &- \frac{1}{4}(c - c^*)\big(U^{\text{av}}(t) + K\vartheta(t)\big)^T \big(U^{\text{av}}(t) + K\vartheta(t)\big). \end{aligned} \tag{9.151}$$

Thus, if $c > c^*$, the linearized system (9.142)–(9.145) is exponentially stable with respect to the norm (9.136).

The next statement summarizes the results for multiparameter Newton-based ES under distributed delays.

Proposition 9.2. *Consider the system* (9.142)–(9.145) *and the estimate error system* (9.140) *for the Hessian's inverse* H^{-1} *via Riccati filter* (9.106), *where* $D > 0$ *is a constant,* $K = K^T \in \mathbb{R}^{n\times n}$ *is a diagonal matrix with positive diagonal entries, and* $H = H^T \in \mathbb{R}^{n\times n}$ *is a negative definite matrix. Assume that* $\beta : \mathbb{R} \to \mathbb{R}^{n\times n}$ *satisfies Assumptions* 9.2 *and* 9.3. *Then there exists* $c^* \in \mathbb{R}$ *depending only on* K *such that, if* $c > c^*$, *the closed-loop system* (9.140) *and* (9.142)–(9.145) *is well-posed in* $\mathcal{H}$ *and the zero-solution is exponentially stable.*

9.7 ▪ Notes and References

Extremum seeking is known as a real-time and model-free optimization tool for dynamic nonlinear systems with some tuning parameters as well as nonlinear static maps [220]. In a usual setting, a model of the system or map is unavailable, but we can measure the value of the performance output to be optimized [129, 10]. To obtain information used to search the optimal parameter, we change the value of the parameter in accordance with a sinusoidal perturbation signal. Based on the corresponding output to this oscillatory parameter, an estimate of the optimal parameter is updated. In this way, we seek the optimal parameter.

In those studies, the perturbation signal is assumed to be transmitted to the system immediately. However, sensor and actuator delays or transmission delays are unavoidable in practical situations. Then, the delay might destabilize the extremum seeking loop. Recently, the authors propose an extremum seeking scheme for static maps that can compensate a class of delays [179]. The proposed scheme is constructed based on the idea of predictor feedback control laws [147, 12, 119]. Since the map to be optimized is unknown, a complete predictor cannot be implemented. Hence, we introduce a signal corresponding to a predictor in the average sense.

The class of delays considered in our previous chapters is the point delay. In this chapter, we deal with distributed delays to handle more general situations. As the first step of this study, the single-parameter case is considered and then we move forward to the multivariable setup. In extension of the results in [179] to the distributed delay case, the idea of exploiting predictor feedback laws also works since the original predictor feedback laws proposed in [147, 12] can handle the distributed delays. We have to pay attention to the design of the perturbation and probing signals. Modification of their definition is necessary so that those signals become conventional perturbation and probing signals after delay.

We analyze the stability of a closed-loop system with the aid of the method of averaging. In this approach, the stability of the associated average system is necessary, which will be proved by constructing a Lyapunov functional. The Lyapunov stability of predictor feedback laws for linear systems with distributed input delays is shown in [24]. A Lyapunov functional is constructed by using the backstepping transformation. Although it is possible to follow this approach, we do not employ it. Since our system has a simple structure, we construct a Lyapunov functional without a backstepping transformation.

Part II

EXTREMUM SEEKING FOR PDE SYSTEMS

Chapter 10

ES for Heat PDE

10.1 ▪ Advancing from Compensation of Transport PDE to Heat PDE

In the previous chapters we designed extremum seeking controllers that compensate for delays, i.e., for transport PDEs. These being the simplest of PDEs, how can we stop here?

It is natural to shift our sights next from the simplest (transport) to the second simplest of PDEs—the heat equation. The heat PDE doesn't model only thermal phenomena, including phase change (such as solid-liquid-gas), but numerous other diffusion-driven processes in biological, chemical, and energy storage systems, as well as the dynamics in many social networks, in which diffusion is the dominant mechanism for the transmission of information and opinion.

The backstepping approach to predictor design for delays informs us how to progress with the compensation of diffusion dynamics in ES problems for unknown maps in cascade with heat PDEs. Since the ES update dynamics consist of an integrator, compensating a delay at its input entails using an integral with a constant integration kernel. The approach differs a bit with the heat equation, which is second-order in the spatial variable, rather than being first-order like the transport PDE. Compensating this second spatial derivative at the input to an integrator entails employing not a constant backstepping kernel but a kernel that is linear in the integration variable. This is the biggest difference between compensating transport and heat PDE dynamics in extremum seeking. As with delay compensation, perturbation-based gradient and Hessian estimates of the static map are used in the heat PDE compensation.

The other difference between the delay and the heat PDE is the perturbation signal. While, in the case of a delay, it suffices to advance in time the sinusoidal perturbation, in the case of the heat dynamics one has to employ a solution to the motion planning problem where the output of the heat PDE system is a sinusoid at one of its boundaries and the input is the signal that must be applied on the other boundary in order to generate a sinusoid at the output. This input signal happens to consist of sinusoidal and exponential functions.

The stability analysis of the error-dynamics is based on using Lyapunov's method and applying averaging for infinite-dimensional systems to capture the infinite-dimensional state of the actuator model. Local exponential convergence to a small neighborhood of the optimal point is proven and illustrated by numerical simulations.

At the end of the chapter, a first effort to pursue the extension of the ES approach from the heat PDE (of fixed domain) to the Stefan PDE with moving boundary [112, 113] is also discussed.

In the next section, we introduce the problem statement with setting up the mathematical formulations. Section 10.3 provides the design of the diffusion compensation controller and the derivation of the error-dynamics. Section 10.4 presents the proof of our main theorem on stability and convergence of the error-dynamics for ES with actuation dynamics governed by diffusion PDEs. Numerical simulations are performed in Section 10.5, which validate the stability and convergence property of the closed-loop system. Section 10.6 presents the control design and analysis for the Stefan system with moving boundary. In Section 10.7, we conclude the chapter with some notes and references.

10.2 ▪ Problem Statement

As introduced in Chapter 1, the basic ES for static maps is a real-time optimization control scheme, where the goal is to find and maintain the optimum of an unknown nonlinear static map $Q:\mathbb{R}\to\mathbb{R}$ with optimal unknown output $y^*\in\mathbb{R}$, unknown optimizer $\Theta^*\in\mathbb{R}$, measurable output $y\in\mathbb{R}$, and input $\Theta\in\mathbb{R}$ (see Figure 10.1). Without loss of generality, we consider maximization problems (for minimization use $y=-y$).

10.2.1 ▪ Actuation Dynamics and Output

In addition to the basic ES scheme in Figure 10.1, we consider actuation dynamics which are described by a diffusion process, i.e., a heat equation with the actuator $\theta(t)\in\mathbb{R}$ and the propagated actuator $\Theta(t)\in\mathbb{R}$ given by

$$\Theta(t)=\alpha(0,t), \tag{10.1}$$
$$\partial_t\alpha(x,t)=\partial_{xx}\alpha(x,t), \quad x\in(0,D), \tag{10.2}$$
$$\partial_x\alpha(0,t)=0, \tag{10.3}$$
$$\alpha(D,t)=\theta(t), \tag{10.4}$$

where $\alpha:[0,D]\times\mathbb{R}_+\to\mathbb{R}$ and $D>0$ is the known domain length, i.e., a proportional indicator of the diffusion coefficient. The measurement is defined by the unknown static map with input (10.1) such that

$$y(t)=Q(\Theta(t)). \tag{10.5}$$

For the sake of simplicity, we assume the following.

Assumption 10.1. *The unknown nonlinear static map is quadratic, i.e.,*

$$Q(\Theta)=y^*+\frac{H}{2}(\Theta-\Theta^*)^2, \tag{10.6}$$

where besides the optimal input $\Theta^*\in\mathbb{R}$ and the optimal output $y^*\in\mathbb{R}$ being unknown, the scalar $H<0$ is the unknown Hessian of the static map.

Assumption 10.1 is reasonable since every nonlinear function in $\mathcal{C}^2(\mathbb{R})$ can be approximate as a quadratic function in the neighborhood of its extremum. Therefore, all stability results derived in this chapter hold at least locally.

The output of the static map is then given by

$$y(t)=y^*+\frac{H}{2}(\Theta(t)-\Theta^*)^2. \tag{10.7}$$

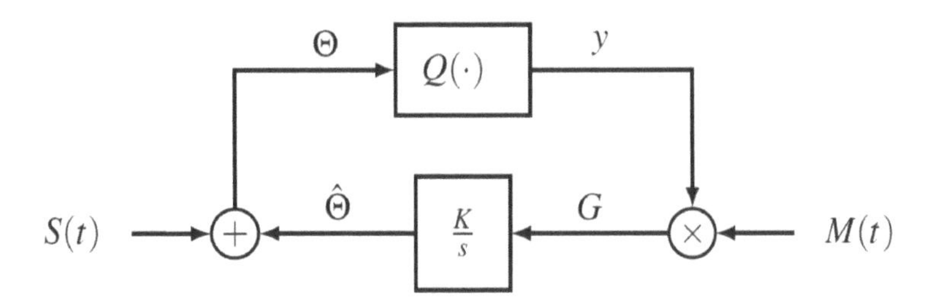

Figure 10.1. *Basic gradient extremum seeking scheme.*

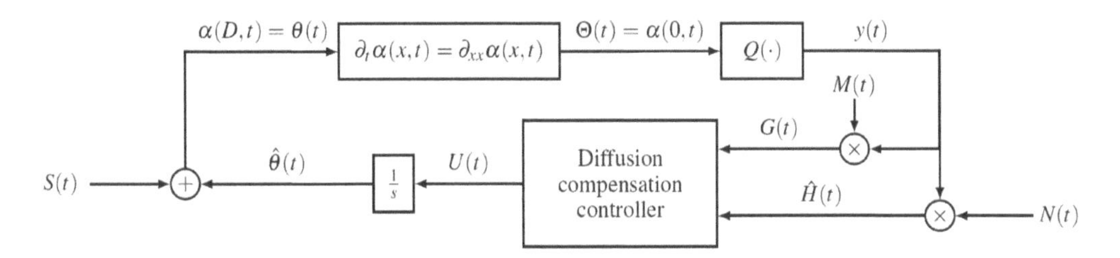

Figure 10.2. *Extremum seeking control loop with actuation dynamics governed by a diffusion PDE.*

Combining the actuation dynamics (10.1)–(10.4) and the basic ES scheme, further adapting the proposed scheme in [179], the closed-loop ES with actuation dynamics governed by a diffusion PDE system under yet unknown diffusion compensation controller is shown in Figure 10.2.

10.2.2 ▪ Probing and Demodulation Signals

We adapt the perturbation signal $S(t)$ from the basic scheme in Figure 10.1, which should add $a\sin(\omega t)$ to $\Theta(t)$ as in the standard extremum seeking scheme; i.e., in the case of actuation dynamics, the diffusion process has to be compensated. Hence, $a\sin(\omega t)$ with perturbation amplitude a and frequency ω is applied as follows:

$$S(t) := \beta(D,t), \tag{10.8}$$

$$\partial_t\beta(x,t) = \partial_{xx}\beta(x,t), \quad x \in (0,D), \tag{10.9}$$

$$\partial_x\beta(0,t) = 0, \tag{10.10}$$

$$\beta(0,t) = a\sin(\omega t), \tag{10.11}$$

where $\beta : [0,D] \times \mathbb{R}_+ \to \mathbb{R}$. Equations (10.8)–(10.11) describe a *trajectory generation problem* for motion planning design as in [128, Chapter 12]. The explicit solution of (10.8) is given by

$$S(t) = \frac{1}{2}ae^{\sqrt{\frac{\omega}{2}}D}\sin\left(\omega t + \sqrt{\frac{\omega}{2}}D\right) + \frac{1}{2}ae^{-\sqrt{\frac{\omega}{2}}D}\sin\left(\omega t - \sqrt{\frac{\omega}{2}}D\right). \tag{10.12}$$

The optimal unknown actuator $\theta(t) \equiv \Theta^*$ drives the static map to its optimal value y^*. Since our objective is to find Θ^*, we introduce the estimators

$$\hat{\theta}(t) = \theta(t) - S(t), \quad \hat{\Theta}(t) = \Theta(t) - a\sin(\omega t). \tag{10.13}$$

Furthermore, we define the estimation errors in the input and propagated input variables as

$$\tilde{\theta}(t) := \hat{\theta}(t) - \Theta^*, \qquad \vartheta(t) := \hat{\Theta}(t) - \Theta^*. \tag{10.14}$$

Let $\bar{\alpha} : [0,D] \times \mathbb{R}_+ \to \mathbb{R}$ be defined as

$$\bar{\alpha}(x,t) := \alpha(x,t) - \beta(x,t) - \Theta^* . \tag{10.15}$$

By (10.1)–(10.4) and (10.8)–(10.11) with the help of (10.13) and (10.14), we have

$$\vartheta(t) = \bar{\alpha}(0,t), \tag{10.16}$$
$$\partial_t \bar{\alpha}(x,t) = \partial_{xx} \bar{\alpha}(x,t), \quad x \in (0,D), \tag{10.17}$$
$$\partial_x \bar{\alpha}(0,t) = 0, \tag{10.18}$$
$$\bar{\alpha}(D,t) = \tilde{\theta}(t). \tag{10.19}$$

Assume that $\bar{\alpha}, \bar{\alpha}_t, \bar{\alpha}_x, \bar{\alpha}_{xx} \in \mathcal{C}^1([0,D] \times \mathbb{R}_+)$. Then, by taking the time derivative of (10.16)–(10.19) and with $\dot{\tilde{\theta}} = U(t)$ (see Figure 10.2), the so-called propagated error-dynamics can be written as

$$\dot{\vartheta}(t) = u(0,t), \tag{10.20}$$
$$\partial_t u(x,t) = \partial_{xx} u(x,t), \quad x \in (0,D), \tag{10.21}$$
$$\partial_x u(0,t) = 0, \tag{10.22}$$
$$u(D,t) = U(t), \tag{10.23}$$

where

$$u(x,t) := \partial_t \bar{\alpha}(x,t) = \partial_{xx} \bar{\alpha}(x,t) . \tag{10.24}$$

The relation among the propagated estimation error $\vartheta(t)$, the propagated input $\Theta(t)$, and the optimizer of the static map Θ^*, derived by combining (10.13) and (10.14), is given by

$$\vartheta(t) + a\sin(\omega t) = \Theta(t) - \Theta^* . \tag{10.25}$$

It remains to define the demodulation signal $N(t)$ which is used to estimate the Hessian of the static map by multiplying it with the output $y(t)$ of the static map. The authors in [73] derived the Hessian estimate as

$$\hat{H}(t) = N(t)y(t) \quad \text{with} \quad N(t) = -\frac{8}{a^2}\cos(2\omega t). \tag{10.26}$$

Note that the dither signal $M(t)$, to estimate the gradient, is the same as in the basic ES approach, such that

$$G(t) = M(t)y(t) \quad \text{with} \quad M(t) = \frac{2}{a}\sin(\omega t). \tag{10.27}$$

10.3 ▪ Controller Design and Error-Dynamics

10.3.1 ▪ Diffusion Compensation Controller

We consider the PDE-ODE cascade (10.20)–(10.23). As in [116], we use the backstepping transformation

$$w(x,t) = u(x,t) - \int_0^x q(x,r)u(r,t)dr - \gamma(x)\vartheta(t), \tag{10.28}$$

with gain kernels $q(x,r) = \bar{K}(x-r)$ and $\gamma(x) = \bar{K}$, which transforms (10.20)–(10.23) into the target system

$$\dot{\vartheta}(t) = \bar{K}\vartheta(t) + w(0,t), \tag{10.29}$$
$$\partial_t w(x,t) = \partial_{xx} w(x,t), \quad x \in (0,D), \tag{10.30}$$
$$\partial_x w(0,t) = 0, \tag{10.31}$$
$$w(D,t) = 0, \tag{10.32}$$

with $\bar{K} < 0$. Since the target system (10.29)–(10.32) is exponentially stable (see [116]), the controller which compensates the diffusion process can be obtained by evaluating the backstepping transformation (10.28) at $x = D$ as

$$U(t) = \bar{K}\vartheta(t) + \bar{K}\int_0^D (D-r)u(r,t)dr. \tag{10.33}$$

However, the proposed control law in (10.33) is not applicable directly because we have no measurement on $\vartheta(t)$. Therefore, we extended an important result of [73] to the parabolic PDE case (proof in Appendix D.3): the average version of the gradient (10.27) and Hessian estimate (10.26) are calculated to

$$G_{\text{av}}(t) = H\vartheta_{\text{av}}(t), \qquad \hat{H}_{\text{av}} = H, \tag{10.34}$$

if a quadratic map as in (10.6) is considered. Accordingly to (10.34), we average (10.33) and choose $\bar{K} = KH$ with $K > 0$ and the introduced unknown Hessian H of the static map, such that

$$U_{\text{av}}(t) = KH\vartheta_{\text{av}}(t) + KH\int_0^D (D-r)u_{\text{av}}(r,t)dr. \tag{10.35}$$

Hence, by plugging the averaged gradient and Hessian estimate (10.34) into (10.35), we obtain

$$U_{\text{av}}(t) = KG_{\text{av}}(t) + K\hat{H}_{\text{av}}\int_0^D (D-r)u_{\text{av}}(r,t)dr. \tag{10.36}$$

Due to technical reasons in the application of the averaging theorem for infinite-dimensional systems (Appendix B) in the following stability proof, we introduce a low-pass filter to the controller in order to write the error-dynamics (10.20)–(10.23) as an infinite-dimensional system including $U(t)$ as a state. Finally, we get the average-based infinite-dimensional control law to compensate the diffusion process by

$$U(t) = \mathcal{T}\left\{K\left[G(t) + \hat{H}(t)\int_0^D (D-r)u(r,t)dr\right]\right\}, \tag{10.37}$$

with the low-pass filter operator

$$\mathcal{T}\{\varphi(t)\} = \mathcal{L}^{-1}\left\{\frac{c}{s+c}\right\} * \varphi(t), \quad \varphi(t) : \mathbb{R}_+ \to \mathbb{R}, \tag{10.38}$$

where the corner frequency $c > 0$ is chosen later, $\mathcal{L}^{-1}\{\cdot\}$ is the inverse Laplace transformation, and $*$ is the convolution operator. Note the feedback law (10.37) depends on the full PDE state information $u(x,t), x \in [0,D]$. Since $u(x,t)$ is an artificial state while it is used for the analysis in the following, the feedback law (10.37) should be described with respect to the plant state $\alpha(x,t)$

and dither signals. Assuming that $\partial_t\alpha(x,t)$ is measured for $x \in [0,D]$, considering (10.15) and (10.24) such that $u(x,t) = \partial_t\bar{\alpha}(x,t) = \partial_t\alpha(x,t) - \partial_t\beta(x,t)$ and applying the integration by parts to the integration of $\partial_t\beta(x,t) = \partial_{xx}\beta(x,t)$, associated with (10.8)–(10.11), the feedback law (10.37) is rewritten as

$$U(t) = \mathcal{T}\left\{K\left[G(t) + \hat{H}(t)\left(\int_0^D (D-r)\partial_t\alpha(r,t)dr - (S(t) - a\sin(\omega t))\right)\right]\right\}. \tag{10.39}$$

In practice, having the measurement of $\Theta(t)$ (input of the static map) is more reasonable than measuring $\partial_t\alpha(x,t)$ which is the time derivative of the spatially distributed plant state. In such a case, the control law (10.39) can be rewritten as

$$U(t) = \mathcal{T}\left\{K\left[G(t) + \hat{H}(t)\left(\hat{\theta}(t) - \Theta(t) + a\sin(\omega t)\right)\right]\right\} \tag{10.40}$$

with the help of the diffusion equation $\partial_t\alpha(x,t) = \partial_{xx}\alpha(x,t)$ and the integration by parts, associated with (10.1)–(10.4) and (10.13)–(10.14).

10.3.2 ▪ Error-Dynamics

With the approximated gradient and Hessian depending on $\vartheta(t)$ and H, calculated by means of (D.59) and (D.60) in Appendix D.3, the infinite-dimensional control law (10.37) is given by

$$\begin{aligned} U(t) = \mathcal{T}\Bigg\{K\Bigg[&y^* + \frac{H}{2}\vartheta^2(t) + Ha\sin(\omega t)\vartheta(t) \\ &+ \frac{Ha^2}{2}\sin^2(\omega t)\Bigg] \times \Bigg[\frac{2}{a}\sin(\omega t) \\ &- \frac{8}{a^2}\cos(2\omega t)\int_0^D (D-r)u(r,t)dr\Bigg]\Bigg\}. \end{aligned} \tag{10.41}$$

Substituting (10.41) into (10.23), we can write the error-dynamics (10.20)–(10.23) as

$$\dot{\vartheta}(t) = u(0,t), \tag{10.42}$$

$$\partial_t u(x,t) = \partial_{xx}u(x,t), \quad x \in (0,\ D), \tag{10.43}$$

$$\partial_x u(0,t) = 0, \tag{10.44}$$

$$\begin{aligned} u(D,t) = \mathcal{T}\{K[&Hf_1(t) + H\vartheta(t) \\ &+ \sin(\omega t)f_2(t) - \cos(2\omega t)f_3(t) \\ &- \sin(3\omega t)f_4(t) + \cos(4\omega t)f_5(t)]\}, \end{aligned} \tag{10.45}$$

where

$$f_1(t) = \int_0^D (D-r)u(r,t)dr, \tag{10.46}$$

$$f_2(t) = (2y^* + H\vartheta^2(t) + 4\vartheta(t)f_1(t) + 3a^2H/4)/a, \tag{10.47}$$

$$f_3(t) = 8y^*f_1(t)/a^2 + H\vartheta(t) + 2H + 4H\vartheta^2(t)f_1(t)/a^2, \tag{10.48}$$

$$f_4(t) = 4H\vartheta(t)f_1(t)/a + Ha/4, \tag{10.49}$$

$$f_5(t) = Hf_1(t). \tag{10.50}$$

10.4 ▪ Stability and Convergence Analysis

The following theorem summarizes the stability and convergence properties of the error-dynamics (10.42)–(10.45) and is proven in this section.

Theorem 10.1. *For a sufficiently large $c > 0$, there exists some $\bar{\omega}(c) > 0$, such that, $\forall \omega > \bar{\omega}$, the error-dynamics* (10.42)–(10.45) *with states $\vartheta(t)$, $u(x,t)$ has an unique exponentially stable periodic solution in t of period $\Pi := 2\pi/\omega$, denoted by $\vartheta^{\Pi}(t), u^{\Pi}(x,t)$, satisfying, $\forall t \geq 0$,*

$$\left(\left|\vartheta^{\Pi}(t)\right|^2 + \left\|u^{\Pi}(t)\right\|^2 + \left\|\partial_x u^{\Pi}(t)\right\|^2 + \left|u^{\Pi}(D,t)\right|^2\right)^{1/2} \leq \mathcal{O}(1/\omega). \tag{10.51}$$

Furthermore,

$$\limsup_{t\to\infty} |\theta(t) - \Theta^*| = \mathcal{O}\left(|a|e^{D\sqrt{\omega/2}} + 1/\omega\right), \tag{10.52}$$

$$\limsup_{t\to\infty} |\Theta(t) - \Theta^*| = \mathcal{O}\left(|a| + 1/\omega\right), \tag{10.53}$$

$$\limsup_{t\to\infty} |y(t) - y^*| = \mathcal{O}\left(|a|^2 + 1/\omega^2\right). \tag{10.54}$$

Proof. Structured into **Steps 1** to **6** below: First, in Steps 1–4, we show the exponential stability of the average error-dynamics of (10.42)–(10.45) via a backstepping transformation. Then, we invoke the averaging theorem for infinite-dimensional systems (Appendix B) in Step 5 to show the exponential stability of the original error-dynamics (10.42)–(10.45). Finally, Step 6 shows the convergence of $(\theta(t), \Theta(t), y(t))$ to a neighborhood of the extremum $(\Theta^*, \Theta^*, y^*)$.

Step 1: *Average error-dynamics*

The average version of the system (10.42)–(10.45) for ω large is

$$\dot{\vartheta}_{\text{av}}(t) = u_{\text{av}}(0,t), \tag{10.55}$$

$$\partial_t u_{\text{av}}(x,t) = \partial_{xx} u_{\text{av}}(x,t), \quad x \in [0,\ D], \tag{10.56}$$

$$\partial_x u_{\text{av}}(0,t) = 0, \tag{10.57}$$

$$\frac{d}{dt} u_{\text{av}}(D,t) = -c u_{\text{av}}(D,t) + cKH\left[\vartheta_{\text{av}}(t) + \int_0^D (D-r) u_{\text{av}}(r,t)dr\right], \tag{10.58}$$

where the low-pass filter is represented in state-space form. To derive (10.55)–(10.58), the terms in (10.45) depending on sine or cosine of argument $k\omega$ for $k = 1, \ldots, 4$ are set to zero, as shown in the proof of (10.34) in Appendix D.3.

Step 2: *Backstepping transformation into the target system*

The backstepping transformation (see [116])

$$w(x,t) = u_{\text{av}}(x,t) - KH\left[\vartheta_{\text{av}}(t) + \int_0^x (x-r) u_{\text{av}}(r,t)dr\right] \tag{10.59}$$

maps the average error-dynamics (10.55)–(10.58) into the exponentially stable target system

(stability shown in Step 3)

$$\dot{\vartheta}_{\rm av}(t) = KH\vartheta_{\rm av}(t) + w(0,t), \tag{10.60}$$

$$\partial_t w(x,t) = \partial_{xx} w(x,t), \quad x \in (0, D), \tag{10.61}$$

$$\partial_x w(0,t) = 0, \tag{10.62}$$

$$\partial_t w(D,t) = -cw(D,t) - KHw(D,t) - (KH)^2 e^{KHD}\left[\int_0^D \left(e^{-KH(D-r)} - 1\right) w(r,t)dr + \vartheta_{\rm av}(t)\right]. \tag{10.63}$$

Equations (10.60)–(10.62) can be derived by plugging in the inverse backstepping transformation (see [116])

$$u_{\rm av}(x,t) = w(x,t) + KHe^{KHx}\vartheta_{\rm av}(t) + KH\int_0^x \left(e^{KH(x-r)} - 1\right) w(r,t)dr \tag{10.64}$$

into the average error-dynamics (10.55)–(10.57). With the boundary equation (10.58) and the backstepping transformation (10.59) we obtain

$$w(D,t) = -\frac{1}{c}\partial_t u_{\rm av}(D,t). \tag{10.65}$$

Calculating the time derivative of the backstepping transformation (10.59) evaluated at $x = D$ yields to

$$\partial_t w(D,t) = \partial_t u_{\rm av}(D,t) - KHu_{\rm av}(D,t), \tag{10.66}$$

and substituting (10.64) and (10.65) into (10.66), we arrive at (10.63).

Step 3: *Exponential stability of the target system*

Consider the Lyapunov–Krasovskii functional

$$\Upsilon(t) = \frac{\vartheta_{\rm av}^2(t)}{2} + \frac{a}{2}\|w(t)\|^2 + \frac{b}{2}\|\partial_x w(t)\|^2 + \frac{d}{2}w^2(D,t), \tag{10.67}$$

where $a,\ b,\ d > 0$ are to be chosen later. We define $\lambda := -KH$ with $\lambda > 0$. Calculating the time derivative of (10.67) associated with the solution of the target system (10.60)–(10.63) and with the help of integration by parts leads to

$$\begin{aligned}\dot{\Upsilon}(t) = & -\lambda\vartheta_{\rm av}^2(t) + \vartheta_{\rm av}(t)w(0,t) \\ & + aw(D,t)\partial_x w(D,t) - a\|\partial_x w(t)\|^2 \\ & + b\partial_x w(D,t)\partial_t w(D,t) - b\|\partial_{xx} w(t)\|^2 \\ & + dw(D,t)\partial_t w(D,t).\end{aligned} \tag{10.68}$$

Applying Young's, Poincaré's, Agmon's, and Cauchy–Schwarz's inequalities on (10.68), whose details are shown in Appendix D.4, and choosing $a = (c-\lambda)/(8D\lambda^3)$, $b = 1/(8D\lambda^3)$, and $d = 1$, we get

$$\dot{\Upsilon}(t) \le -\frac{\lambda}{4}\vartheta_{\rm av}^2(t) + (c_1^* - c)w^2(D,t) + (c_2^* - c)\|\partial_x w(t)\|^2 - \frac{1}{32D\lambda^3}\|\partial_{xx} w(t)\|^2, \tag{10.69}$$

with

$$c_1^* = \frac{3}{2}\lambda^3 + \lambda + \frac{1+2D}{\lambda} + 2D\lambda\zeta(D), \tag{10.70}$$

$$c_2^* = \lambda + 8D\lambda^3 \left[\frac{4D^2+1}{\lambda} + 4D^2\lambda\zeta(D)\right], \tag{10.71}$$

and $\zeta(D) = \int_0^D (e^{-\lambda(D-r)^2} - 1)dr$. Hence, from (10.69), if c is chosen such that $c > \max\{c_1^*, c_2^*\}$, we obtain for some $\mu > 0$

$$\dot{\Upsilon}(t) \leq -\mu\Upsilon(t). \tag{10.72}$$

Thus, the target system (10.60)–(10.63) is exponentially stable in the sense of the $\mathcal{H}_1$-norm

$$\left(\vartheta_{\text{av}}^2(t) + \|w(t)\|^2 + \|\partial_x w(t)\|^2 + w^2(D,t)\right)^{1/2}, \tag{10.73}$$

i.e., in the transformed variable $(\vartheta_{\text{av}}, w)$.

Step 4: *Exponential stability estimate (in the $\mathcal{H}_1$-norm) of the average error-dynamics*

We define

$$\Psi(t) = \vartheta_{\text{av}}^2(t) + \|u_{\text{av}}(t)\|^2 + \|\partial_x u_{\text{av}}(t)\|^2 + u_{\text{av}}^2(D,t). \tag{10.74}$$

Then, there exist upper and lower bounds of the Lyapunov–Krasovskii functional (10.67) with respect to $\Psi(t)$, such that

$$\underline{\rho}\Psi(t) \leq \Upsilon(t) \leq \bar{\rho}\Psi(t), \quad \underline{\rho} = \underline{\tau}\underline{\sigma}, \quad \bar{\rho} = \bar{\tau}\bar{\sigma}, \tag{10.75}$$

with

$$\underline{\sigma} = \min\left\{\frac{1}{2}, \frac{a}{2}, \frac{b}{2}\right\}, \qquad \bar{\sigma} = \max\left\{\frac{1}{2}, \frac{a}{2}, \frac{b}{2}\right\}, \tag{10.76}$$

and $\bar{\tau}, \underline{\tau}$, defined in Appendix D.5. Equation (10.75), along with the exponential stability of the target system (10.72), implies

$$\Psi(t) \leq \frac{\bar{\rho}}{\underline{\rho}} e^{-\mu t}\Psi(0), \tag{10.77}$$

which completes the proof of exponential stability of the average error-dynamics (10.55)–(10.58) in the sense of the $\mathcal{H}_1$-norm $\Psi^{1/2}(t)$ in the variable $(\vartheta_{\text{av}}, u_{\text{av}})$.

Step 5: *Invoking the averaging theorem for infinite-dimensional systems*

To satisfy the conditions of the averaging theorem for infinite-dimensional systems, stated in Appendix B, the error-dynamics (10.42)–(10.45) has to be in the form

$$\dot{z}(t) = \Gamma z(t) + J(\omega t, z(t)), \tag{10.78}$$

where the conditions for Γ and $J(\omega t, z(t))$ are given in the theorem. By the state transformation of (10.42)–(10.45) with $v(x,t) = u(x,t) - U(t)$, we obtain the error-dynamics with homogeneous boundary conditions

$$\dot{\vartheta}(t) = v(0,t) + U(t), \tag{10.79}$$

$$\partial_t v(x,t) = \partial_{xx} v(x,t) - \phi(\vartheta, v, U, t), \quad x \in (0,D), \tag{10.80}$$

$$\partial_x v(0,t) = 0, \tag{10.81}$$

$$v(D,t) = 0, \tag{10.82}$$

$$\dot{U}(t) = \phi(\vartheta, v, U, t), \tag{10.83}$$

with

$$\phi(\vartheta,v,U,t) = -cU(t) + cK\left[G(t) + \hat{H}(t)\int_0^D (D-r)(v(r,t)+U(t))dr\right]. \tag{10.84}$$

Next, we write the PDE system (10.80)–(10.82) as an evolutionary equation (see [145]) in the Banach space $\mathcal{X}$,

$$\dot{V}(t) = \mathcal{A}V(t) - \tilde{\phi}(\vartheta,V,U,t), \quad t>0, \tag{10.85}$$

where $V(t)$ is a function which belongs to the Banach space $\mathcal{X}$. Furthermore, $\mathcal{A}$ is the realization of the second-order derivative with one Dirichlet and Neumann boundary condition in $\mathcal{X}$ with

$$\mathcal{A}\varphi := \frac{\partial^2\varphi}{\partial x^2} \tag{10.86}$$

and the domain

$$D(\mathcal{A}) \quad = \quad \left\{\varphi\in\mathcal{X} : \varphi, \frac{d}{dx}\varphi\in\mathcal{X} \text{ are a.c.}, \frac{d^2}{dx^2}\varphi\in\mathcal{X},\ \frac{d}{dx}\varphi(0)=0, \varphi(D)=0\right\}, \tag{10.87}$$

where a.c. means *absolutely continuous*. To express $v(0,t)$ in the ODE (10.79) in terms of $V(t)$, we introduce the linear boundary operator $\mathcal{B}:\mathcal{X}\to\mathbb{R}$, such that

$$\mathcal{B}V(t) := v(0,t). \tag{10.88}$$

Furthermore, we define the linear operators $\alpha^\top:\mathcal{X}\to\mathbb{R}$ and $\beta:\mathbb{R}\to\mathcal{X}$ as

$$\begin{aligned}
&\alpha^\top V(t) := \int_0^D (D-r)v(r,t)dr,\\
&\beta\zeta := [\beta_1,\beta_2,\ldots]^\top\zeta, \quad \zeta\in\mathbb{R},\\
&\text{with}\quad \beta_k = \int_0^D \psi_k(x)dx\\
&\qquad\quad = -\sqrt{\frac{2}{D}}\frac{2D}{\pi(2k-1)}(-1)^k, \quad k=1,2,\ldots,
\end{aligned} \tag{10.89}$$

where $\psi_k(x)=\sqrt{2/D}\cos(\pi/2(2k-1)x/D), k=1,2,\ldots,$ are the eigenfunctions of $\mathcal{A}$. Finally, the error-dynamics with the infinite-dimensional state vector $z(t)=[\vartheta(t)\ V(t)\ U(t)]^\top$ can be rewritten as (10.78) with

$$\Gamma = \begin{bmatrix} 0 & \mathcal{B} & 1\\ 0 & \mathcal{A} & c\beta\\ 0 & 0 & -c\end{bmatrix}, \ J(\omega t,z) = \begin{bmatrix} 0\\ -c\beta K\left[G(t)+\hat{H}(t)g(z)\right]\\ cK\left[G(t)+\hat{H}(t)g(z)\right]\end{bmatrix}, \tag{10.90}$$

where $g(z)=\frac{1}{2}D^2U(t)+\alpha^\top V(t)$. Since $\mathcal{A}$ is a sectorial operator and therefore generates an analytic semigroup [190], $\mathcal{B}$ is $\mathcal{A}$-bounded ($\|\mathcal{B}V(t)\|\le 4D\sqrt{D}\|\mathcal{A}V(t)\|$) and β, c are bounded, the matrix Γ generates an analytic semigroup by the operator matrix theorem in [161]. Furthermore, $J(\omega t,z)$ in (10.90) is Fréchet differentiable in z, strongly continuous, and almost periodic in t uniformly with respect to z. Hence, all conditions to apply the averaging theorem for infinite-dimensional systems in Appendix B are satisfied and the average error-dynamics is exponentially stable (see (10.75) and (10.77)). Thus, the original error-dynamics (10.42)–(10.45) have an exponentially stable periodic solution $z^{\Pi}(t)$ that satisfies (10.51).

Step 6: *Convergence to a neighborhood of the extremum*

In this step, we prove the convergence statements (10.52)–(10.54). Therefore, applying Agmon's, Poincaré's, and Young's inequalities on the LHS of (10.16), along with (10.16)–(10.19), leads to

$$\tilde{\theta}^2(t) \le 3\vartheta(t)^2 + (4D+1)\|\bar{\alpha}_x\|^2. \tag{10.91}$$

By taking the time derivative of $\bar{\Upsilon}(t) = \frac{\mu}{16D}\|\bar{\alpha}_x(t)\|^2 + \frac{1}{2}\vartheta^2(t) + \frac{1}{2}\|u(t)\|^2 + \frac{1}{2}\|u_x(t)\|^2 + \frac{1}{2}U(t)^2$ along with the exponential stability of the original system, it holds

$$\dot{\bar{\Upsilon}} \le -\mu \min\left\{\frac{1}{4}, \frac{1}{64D^3}\right\}\bar{\Upsilon}$$

with $\mu > 0$. Hence, there exists some $M > 0$, such that

$$\|\bar{\alpha}_x(t)\|^2 \le M e^{-kt} \text{ with } k = \mu \min\left\{\frac{1}{4}, \frac{1}{64D^3}\right\}. \tag{10.92}$$

With (10.91), (10.92) and adding the periodic solution $\vartheta^{\Pi}(t)$, it follows that

$$\limsup_{t\to\infty} |\tilde{\theta}(t)|^2 = \limsup_{t\to\infty} \left\{3|\vartheta(t) + \vartheta^{\Pi}(t) - \vartheta^{\Pi}(t)|^2\right\}. \tag{10.93}$$

By applying Young's inequality on (10.93), it holds that

$$|\vartheta(t) + \vartheta^{\Pi}(t) - \vartheta^{\Pi}(t)|^2 \le \sqrt{2}\left(|\vartheta(t) - \vartheta^{\Pi}(t)|^2 + |\vartheta^{\Pi}(t)|^2\right)$$

and, by the averaging theorem in Appendix B, we have that $\vartheta(t) - \vartheta^{\Pi}(t) \to 0$ exponentially. Hence,

$$\limsup_{t\to\infty} |\tilde{\theta}(t)|^2 = \limsup_{t\to\infty} \left\{3\sqrt{2}|\vartheta^{\Pi}(t)|^2\right\}. \tag{10.94}$$

Along with (10.51) and (10.94), we obtain

$$\limsup_{t\to\infty} |\tilde{\theta}(t)| = \mathcal{O}(1/\omega). \tag{10.95}$$

Since $\theta(t) - \Theta^* = \tilde{\theta}(t) + S(t)$ from (10.14) and Figure 10.2 and $S(t)$ is of order $\mathcal{O}\left(|a|e^{D\sqrt{\omega/2}}\right)$, as shown in (10.12), we finally get with (10.95) the next ultimate bound

$$\limsup_{t\to\infty} |\theta(t) - \Theta^*| = \mathcal{O}\left(|a|e^{D\sqrt{\omega/2}} + 1/\omega\right). \tag{10.96}$$

The convergence of the propagated actuator $\Theta(t)$ to the optimizer Θ^* is proven in the next step. Using (10.25) and taking the absolute value, one has

$$|\Theta(t) - \Theta^*| = |\vartheta(t) + a\sin(\omega(t))|. \tag{10.97}$$

As in the convergence proof of the parameter $\theta(t)$ to the optimal input Θ^* above, we write (10.97) in terms of the periodic solution $\vartheta^{\Pi}(t)$ and follow the same steps by applying Young's

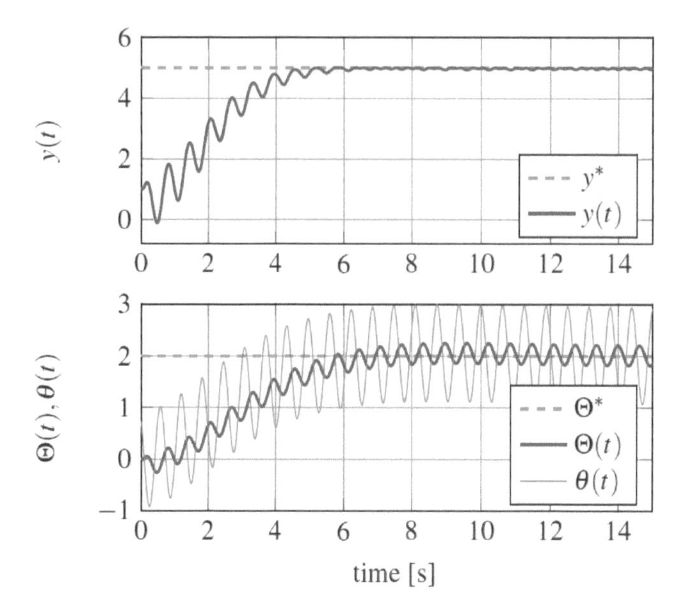

Figure 10.3. *ES with actuation dynamics governed by a diffusion PDE with compensating controller* (10.37) *and perturbation signal* (10.12)*:* (a) *static map* $y(t)$*;* (b) *parameters* $(\theta(t), \Theta(t))$.

inequality and the fact that $\vartheta(t) - \vartheta^{\Pi}(t) \to 0$ exponentially by means of the averaging theorem in Appendix B. Hence, along with (10.51), we finally get

$$\limsup_{t\to\infty} |\Theta(t) - \Theta^*| = \mathcal{O}(|a| + 1/\omega). \tag{10.98}$$

To show the convergence of the output $y(t)$ of the static map to the optimal value y^*, we replace $\Theta(t) - \Theta^*$ in (10.7) by (10.25) and take the absolute value

$$|y(t) - y^*| = \left| \frac{H}{2} [\vartheta(t) + a\sin(\omega(t))]^2 \right|. \tag{10.99}$$

Expanding the quadratic term in (10.99) and applying Young's inequality to the resulting equation, one has $|y(t) - y^*| = \left|H\left[\vartheta(t)^2 + a^2\sin^2(\omega t)\right]\right|$. As before, we add the periodic solution $\vartheta^{\Pi}(t)$, applying Young's inequality, and conclude that $\vartheta(t) - \vartheta^{\Pi}(t) \to 0$ exponentially, according to the averaging theorem [83]. Hence, again with (10.51), we finally get (10.54). □

10.5 ▪ Simulations

Numerical simulations illustrate the stability and convergence properties of the proposed extremum seeking with actuation dynamics governed by diffusion PDEs. Consider a quadratic static map as in (10.6), with Hessian $H = -2$, the optimizer $\Theta^* = 2$, and the optimal value $y^* = 5$ and a domain length $D = 1$ for the actuator dynamics. The parameters of the dither signal and the designed controller are chosen as $\omega = 10$, $a = 0.2$, $c = 10$, and $K = 0.2$. The simulation results of the closed loop are illustrated in Figure 10.3.

One can observe that each variable (θ, Θ, y) converges to a neighborhood of the optimum $(\Theta^*, \Theta^*, y^*)$. Due to the significant roll-off at high frequencies of the actuator dynamics, a larger neighborhood of convergence of $\theta(t)$ is observed, which is consistent with (10.52).

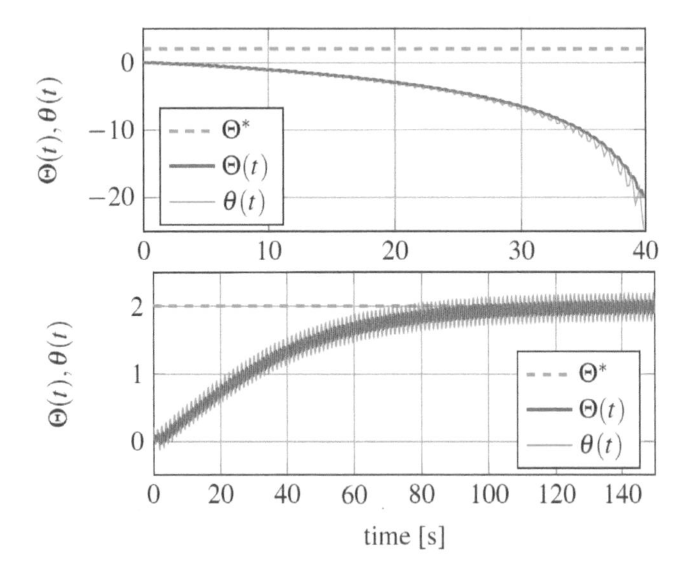

Figure 10.4. *Basic uncompensated ES with dither signals which consider the phase lag introduced by the actuator dynamics:* (a) *instability for large* ω*;* (b) *slow convergence for small* ω*.*

The proposed ES scheme with controller (10.37) and adapted perturbation signal (10.12) is compared to the basic ES scheme with extended dither signals which compensates the known phase lag caused by the actuation dynamics. Hence, the dither and demodulation signals are chosen as $S(t) = a\sin(\omega t - \varphi)$ and $M(t) = \frac{2}{a}\sin(\omega t - \varphi)$, where φ is the phase of the transfer function of (10.1)–(10.4) evaluated at ω. Using the same parameters for ω, a, and K as in the prior simulation results in instability of $\theta(t)$ as illustrated in Figure 10.4(a). On the other hand, Figure 10.4(b) shows the response for a lower frequency $\omega = 5.0$, in which we observe that the convergence speed gets much slower than the result in Figure 10.3. Hence, our proposed ES controller stands as a strong improvement of the basic ES in the presence of actuation dynamics governed with diffusion PDEs, even if the phase lag of the actuator dynamics is compensated in the additive dither signals.

10.6 ▪ Extremum Seeking for Stefan PDE: From Fixed Domain to Moving Boundary

This section presents the design and analysis of the ES for static maps with input governed by a PDE of the diffusion type defined on a time-varying spatial domain described by an ODE. We compensate the average-based actuation dynamics by a controller via backstepping transformation for the moving boundary, which is utilized to transform the original coupled PDE-ODE into a target system whose exponential stability of the average equilibrium of the average system is proved.

10.6.1 ▪ One-Phase Stefan Problem

The physical model which describes the 1-D Stefan problem in a pure one-component material of length L is described in Figure 10.5. The domain $[0, L]$ is divided in two subdomains, $[0, s(t)]$ and $[s(t), L]$, which represent the liquid phase and the solid phase, respectively. The system is controlled by the heat flux $q_c(t)$ at $x = 0$, because we are dealing with a Neumann boundary

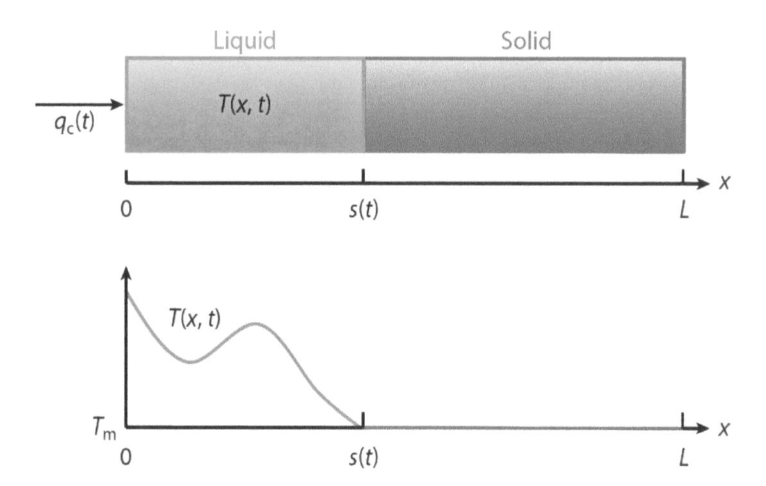

Figure 10.5. *Schematic of one-phase Stefan problem* [112, 113]. *The temperature profile in the solid phase is assumed to be a uniform melting temperature. Used with permission of [Annual Review of Control, Robotics, and Autonomous Systems, Control of the Stefan System and Applications: A Tutorial, by Shumon Koga and Miroslav Krstic, 2022, Vol. 5, p. 550]; permission conveyed through Copyright Clearance Center, Inc.*

actuation as shown below:

$$T_t(x,t) = \alpha T_{xx}(x,t), \quad x \in (0, s(t)), \quad \alpha = \frac{k}{\rho C_p}, \tag{10.100}$$

$$-kT_x(0,t) = q_c(t), \tag{10.101}$$

$$T(s(t),t) = T_m, \tag{10.102}$$

$$\dot{s}(t) = -\beta T_x(s(t),t), \quad \beta = \frac{k}{\rho \Delta H^*}, \tag{10.103}$$

where $T(x,t)$, T_m, $q_c(t)$, k, ρ, C_p, and ΔH^* are the distributed temperature of the liquid phase, melting temperature, manipulated heat flux, liquid heat conductivity, liquid density, liquid heat capacity, and latent heat of fusion, respectively. Equations (10.101) and (10.102) are the boundary conditions of the system, and (10.103) is the Stefan condition, which describes the dynamic of the moving boundary. Figure 10.7 shows the block diagram of the PDE-ODE cascade represented by (10.100)–(10.103).

10.6.2 ▪ Actuation Dynamics and Output Signal

For the sake of simplicity, we consider actuation dynamics which are described by a heat equation with $\alpha, \beta, k = 1$, $\theta(t) \in \mathbb{R}$ and the propagated actuator $\Theta(t) \in \mathbb{R}$ given by

$$\dot{\Theta}(t) = \dot{s}(t) = -\alpha_x(s(t),t)), \quad x \in (0, s(t)), \tag{10.104}$$

$$\partial_t \alpha(x,t) = \partial_{xx}\alpha(x,t), \tag{10.105}$$

$$\alpha(s(t),t) = 0, \tag{10.106}$$

$$-\partial_x \alpha(0,t) = \theta(t), \tag{10.107}$$

where $\alpha : [0, s(t)] \times \mathbb{R}_+ \to \mathbb{R}$ is $\alpha(x,t) = T(x,t) - T_m$ and $s(t) = \Theta(t)$ is the unknown interface represented as the moving boundary. The output is measured by the unknown static map with input (10.104):

$$y(t) = Q(\Theta(t)). \tag{10.108}$$

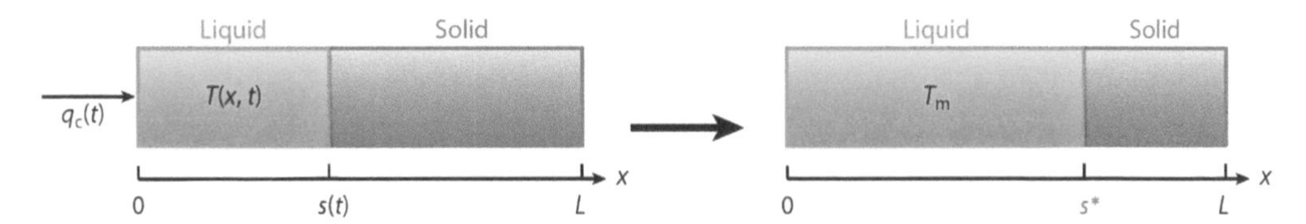

Figure 10.6. *Control objective of the Stefan problem. We aim to design a heat flux input $q_c(t)$ such that the interface position $s(t)$ is driven to the setpoint position s^*. Used with permission of [Annual Review of Control, Robotics, and Autonomous Systems, Control of the Stefan System and Applications: A Tutorial, by Shumon Koga and Miroslav Krstic, 2022, Vol. 5, p. 552]; permission conveyed through Copyright Clearance Center, Inc.*

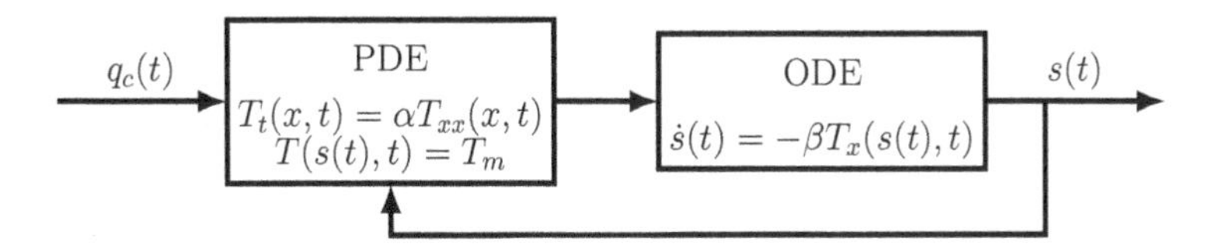

Figure 10.7. *The cascade of the PDE dynamics and the ODE system.*

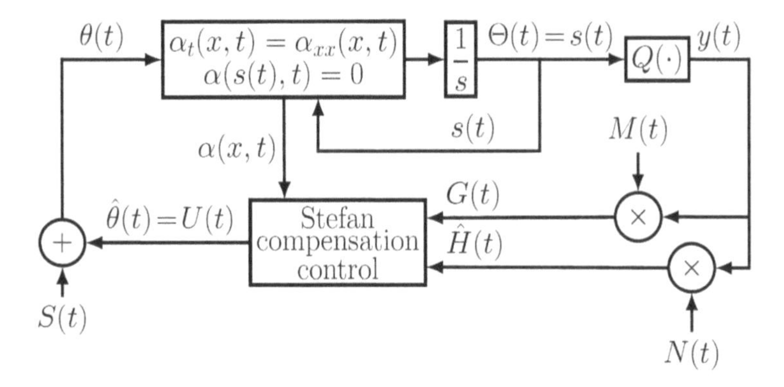

Figure 10.8. *Extremum seeking control loop applied to the one-phase Stefan problem.*

The ES goal is to optimize an unknown static map Q(·) using a real-time optimization control with optimal unknown output y^* and optimizer Θ^* as well as measurable output y and input θ. Consequently, the control objectives of the Stefan problem are achieved, i.e., $\lim_{t\to\infty} s(t) = s^*$ and $\lim_{t\to\infty} T(x,t) = T_m, \forall x \in [0, s^*]$, as illustrated in Figure 10.6.

The unknown nonlinear map is locally quadratic, such that

$$Q(\Theta) = y^* + \frac{H}{2}(\Theta - \Theta^*)^2, \tag{10.109}$$

where $\Theta^*, y^* \in \mathbb{R}$ and $H < 0$ is the Hessian. Hence, the output of the static map is given by

$$y(t) = y^* + \frac{H}{2}(\Theta(t) - \Theta^*)^2. \tag{10.110}$$

Adapting the proposed scheme in [179] and combining (10.104)–(10.107) with the ES approach, the closed-loop ES with actuation dynamics is shown in Figure 10.8.

The demodulation signal $N(t)$, which is used to estimate the Hessian of the static map by multiplying it with the output $y(t)$ of the static map, is defined in [73] as

$$\hat{H}(t) = N(t)y(t) \text{ with } N(t) = -\frac{8}{a^2}\cos(2\omega t), \tag{10.111}$$

whereas the signal $M(t)$ is used to estimate the gradient of the static map as follows:

$$G(t) = M(t)y(t) \text{ with } M(t) = \frac{2}{a}\sin(\omega t). \tag{10.112}$$

10.6.3 ▪ Additive Probing Signal

The perturbation $S(t)$ is adapted from the basic ES to the case of PDE actuation dynamics. The trajectory generation problem as in [128] is described as follows:

$$S(t) := -\partial_x\beta(0,t), \quad x \in (0, s(t)), \tag{10.113}$$

$$\partial_t\beta(x,t) = \partial_{xx}\beta(x,t), \tag{10.114}$$

$$\beta(s(t),t) = 0, \tag{10.115}$$

$$\beta_x(s(t),t) = -a\omega\cos(\omega t), \tag{10.116}$$

where $\beta : [0, s(t)] \times \mathbb{R}_+ \to \mathbb{R}$. The explicit solution of (10.113) is found, respectively, for the reference trajectory and the reference solution postulated by a power series [57]:

$$\xi(t) = a\sin(\omega t), \tag{10.117}$$

$$\beta(x,t) = \sum_{i=0}^{\infty} \frac{a_i(t)}{i!}[x - \xi(t)]^i. \tag{10.118}$$

We can calculate the first coefficients of the power series replacing the boundary conditions (10.115) and (10.116) at (10.118), such that

$$a_0(t) = 0, \quad a_1 = -\dot{\xi}(t). \tag{10.119}$$

The general expression $a_i(t) = \dot{a}_{i-2}(t) - a_{i-1}(t)\dot{\xi}(t)$ is obtained by substituting (10.118) at (10.114). We give the analytic expression of the first four coefficients of the series (10.118) so that one can see how the successive derivatives of $\xi(t)$ appear:

$$a_2(t) = \dot{\xi}(t)^2, \tag{10.120}$$

$$a_3(t) = \ddot{\xi}(t) - \dot{\xi}(t)^3, \tag{10.121}$$

$$a_4(t) = \ddot{\xi}(t)^2 + \dddot{\xi}(t)\dot{\xi}(t) + \dot{\xi}(t)^4. \tag{10.122}$$

The trajectory generation solution which provides all terms of the power series (10.118) is given by [89]

$$\beta(x,t) = \sum_{i=0}^{\infty} \frac{1}{(2i)!}\frac{\partial^i}{\partial t^i}[x - \xi(t)]^{2i}. \tag{10.123}$$

Although (10.123) is not an explicit expression, choosing suitable values for a and ω in (10.117), the series converges with few iterations of the infinite sum, getting the desirable sinusoid signal $\xi(t)$ in the output of the integrator.

According to (10.113), we take the spatial derivative of (10.123) and substitute $x = 0$; therefore, we arrive at the final expression of

$$S(t) = -\sum_{i=0}^{\infty} \frac{1}{(2i-1)!}\frac{\partial^i}{\partial t^i}[-a\sin(\omega t)]^{2i-1}. \tag{10.124}$$

10.6.4 ▪ Estimation Errors and PDE-Error Dynamics

Since our objective is to find Θ^*, which corresponds to the optimal unknown actuator $\theta(t)$, we introduce the estimation errors

$$\hat{\theta}(t) = \theta(t) - S(t), \quad \hat{\Theta}(t) = \Theta(t) - a\sin(\omega t), \tag{10.125}$$

$$\tilde{\theta}(t) := \hat{\theta}(t) - \Theta^*, \quad \vartheta(t) := \hat{\Theta}(t) - \Theta^*, \tag{10.126}$$

reminding one that $\Theta(t) := s(t)$. Combining $\hat{\Theta}(t)$ in (10.125) and (10.126) we get the relation between the propagated estimation error $\vartheta(t)$, the propagated input $\Theta(t)$, and the optimizer of the static map Θ^*:

$$\Theta(t) - \Theta^* = \vartheta(t) + a\sin(\omega t). \tag{10.127}$$

Let us define

$$u(x,t) = \alpha(x,t) - \beta(x,t), \tag{10.128}$$

$$\hat{\theta}(t) = U(t). \tag{10.129}$$

By (10.104)–(10.107) and (10.113)–(10.116) with the help of (10.125) and (10.126), we have our original system:

$$\dot{\vartheta}(t) = -u_x(s(t),t), \quad x \in (0, s(t)), \tag{10.130}$$

$$u_t(x,t) = u_{xx}(x,t), \tag{10.131}$$

$$u(s(t),t) = 0, \tag{10.132}$$

$$-u_x(0,t) = U(t). \tag{10.133}$$

10.6.5 ▪ Controller Design

Stefan compensation

We consider the PDE-ODE cascade (10.130)–(10.133) and use the backstepping transformation

$$\begin{aligned} w(x,t) = u(x,t) - \bar{K}\int_x^{s(t)} (x-\sigma)u(\sigma,t)\,dy \\ - \bar{K}(x - s(t))\vartheta(t) \end{aligned} \tag{10.134}$$

with $\bar{K} > 0$ is an arbitrary controller gain. Equation (10.134) transforms (10.130)–(10.133) into the target system:

$$\dot{\vartheta}(t) = -\bar{K}\vartheta(t) - w_x(s(t),t), \quad x \in (0, s(t)), \tag{10.135}$$

$$w_t(x,t) = w_{xx}(x,t) + \bar{K}\dot{s}(t)\vartheta(t), \tag{10.136}$$

$$w_x(0,t) = 0, \tag{10.137}$$

$$w(s(t),t) = 0. \tag{10.138}$$

The compensation controller can be obtained by taking the derivative of (10.134) with respect to t and x, respectively, along the solution (10.130)–(10.133) and substituting $x = 0$:

$$U(t) = -\bar{K}\left(\vartheta(t) + \int_0^{s(t)} u(x,t)\,dx\right). \tag{10.139}$$

Implementable ES control law

Since we have no measurement on $\vartheta(t)$, (10.139) is not applicable directly. Thus, introducing a result of [73], the average version of the gradient and Hessian estimates are calculated by

$$G_{\rm av}(t) = H\vartheta_{\rm av}(t), \quad \hat{H}_{\rm av}(t) = H. \tag{10.140}$$

Averaging (10.139), choosing $\bar{K} = KH$ with $K < 0$, and plugging in the average gradient and Hessian estimates (10.140), we obtain

$$U_{\rm av}(t) = -KG_{\rm av}(t) - KH\int_0^{s_{\rm av}(t)} u_{\rm av}(x,t)\,dx. \tag{10.141}$$

We introduce a low-pass filter to the controller with the purpose of applying the average theorem for infinite-dimensional systems [83] in the following stability proof, such that

$$U(t) = \frac{c}{s+c}\left\{K\left[G(t) + \hat{H}(t)\int_0^{s(t)} u(x,t)\,dx\right]\right\} \tag{10.142}$$

for $c > 0$ sufficiently large.

10.6.6 ▪ Stability Analysis

The following theorem summarizes the stability properties for the average version of the error dynamics (10.130)–(10.133).

Theorem 10.2. *Assume the model validity conditions $T_{\rm av}(x,t) > T_m$, $\dot{s}_{\rm av}(t) > 0$, $x \in (0, s_{\rm av}(t))$, and $s_0 < s_{\rm av}(t) < s^*$ are satisfied at least in the average sense $\forall t \geq 0$, $s^* = \Theta^*$ and for initial conditions $(T_{\rm av}(x,0), s_0)$ compatible with the control law $U(t)$ in* (10.142)*. Then, for a sufficiently large $c > 0$, the average version of the closed-loop system* (10.130)–(10.133) *is exponentially stable in the sense of the norm $\|u_{\rm av}(t)\|^2_{\mathcal{H}_1} + |\vartheta_{\rm av}(t)|^2$, i.e.,*

$$\|u_{\rm av}(t)\|^2_{\mathcal{H}_1} + |\vartheta_{\rm av}(t)|^2 \leq M(\|u_{\rm av}(0)\|^2_{\mathcal{H}_1} + \vartheta_{\rm av}(0)^2)e^{-nt} \quad \forall t \geq 0, \tag{10.143}$$

and appropriate constants $M, n > 0$.

Proof. The proof is carried out in **Steps 1** to **3** below.

Step 1: *Average closed-loop system*

The average version of the system (10.130)–(10.133) is

$$\dot{\vartheta}_{\rm av}(t) = -(u_{\rm av})_x(s_{\rm av}(t),t), \quad x \in (0, s_{\rm av}(t)), \tag{10.144}$$

$$(u_{\rm av})_t(x,t) = (u_{\rm av})_{xx}(x,t), \tag{10.145}$$

$$u_{\rm av}(s_{\rm av}(t),t) = 0, \tag{10.146}$$

$$\frac{d}{dt}(u_{\rm av})_x(0,t) = -c(u_{\rm av})_x(0,t) - cKH\left[\vartheta_{\rm av}(t) + \int_0^{s_{\rm av}(t)} u_{\rm av}(x,t)\,dx\right], \tag{10.147}$$

where the low-pass filter is represented in the state-space form. To derive (10.147), we plug the relationships $\vartheta(t) + a\sin(\omega t) = \Theta(t) - \Theta^*$, $G(t) = M(t)y(t)$ and (10.111) into (10.142). With the help of the identities $2\sin^2(\omega t) = 1-\cos(2\omega t)$, $2\sin(\omega t)\cos(2\omega t) = \sin(3\omega t) - \sin(\omega t)$, $4\sin^3(\omega t) = 3\sin(\omega t) - \sin(3\omega t)$, and $4\sin^2(\omega t)\cos(2\omega t) = 2\cos(2\omega t) - \cos(4\omega t) - 1$ and applying averaging, we arrive at (10.147).

The backstepping transformation

$$\begin{aligned} w(x,t) = u_{\text{av}}(x,t) - KH\int_x^{s_{\text{av}}(t)} (x-\sigma)u_{\text{av}}(\sigma,t)\,d\sigma \\ - KH(x-s_{\text{av}}(t))\vartheta_{\text{av}}(t) \end{aligned} \tag{10.148}$$

maps the average error-dynamics (10.144)–(10.147) into the exponentially stable target system after assuming $c \to +\infty$ for the sake of simplicity. Consequently,

$$\dot{\vartheta}_{\text{av}}(t) = -KH\vartheta_{\text{av}}(t) - w_x(s_{\text{av}}(t),t), \quad x \in (0, s_{\text{av}}(t)), \tag{10.149}$$

$$w_t(x,t) = w_{xx}(x,t) + KH\dot{s}_{\text{av}}(t)\vartheta_{\text{av}}(t), \tag{10.150}$$

$$w_x(0,t) = 0, \tag{10.151}$$

$$w(s_{\text{av}}(t),t) = 0. \tag{10.152}$$

Step 2: ***Inverse transformation***

To ensure the equivalent stability property between the target system and the original system, the invertibility of the transformation (10.148) needs to be guaranteed. Assume the inverse transformation that maps (10.149)–(10.152) into (10.144)–(10.147),

$$\begin{aligned} u_{\text{av}}(x,t) = w(x,t) + \int_x^{s_{\text{av}}(t)} k(x-\sigma)w(\sigma,t)\,d\sigma \\ + \phi(x-s_{\text{av}}(t))\vartheta_{\text{av}}(t), \end{aligned} \tag{10.153}$$

where $k(x-\sigma)$ and $\phi(x-s_{\text{av}}(t))$ are the kernel functions. Taking the derivatives with respect to t and x, respectively, along the solution of (10.149)–(10.152), the functions $\phi(x)$ and $k(x-\sigma)$ must satisfy

$$\phi''(x) = -KH\phi(x), \ \phi(0) = 0, \ \phi' = KH, \tag{10.154}$$

$$k(x-s_{\text{av}}(t)) = \phi(x-s_{\text{av}}(t)), \tag{10.155}$$

$$\phi'(x-s_{\text{av}}(t)) = KH\left(1 + \int_x^{s_{\text{av}}(t)} k(x-\sigma)\,d\sigma\right). \tag{10.156}$$

The solutions of the gain kernel can be deduced from (10.154)–(10.156), such that

$$\phi(x) = KH\sqrt{\frac{1}{KH}}\sin(\sqrt{KH}x), \tag{10.157}$$

$$k(x-\sigma) = \phi(x-\sigma). \tag{10.158}$$

Hence, replacing (10.157) and (10.158) into (10.153), we have the following inverse transformation:

$$\begin{aligned} u_{\text{av}}(x,t) &= w(x,t) \\ &+ \int_x^{s_{\text{av}}(t)} KH\sqrt{\frac{1}{KH}}\sin(\sqrt{KH}(x-\sigma))w(\sigma,t)\,d\sigma \\ &+ KH\sqrt{\frac{1}{KH}}\sin(\sqrt{KH}(x-s_{\text{av}}(t)))\vartheta_{\text{av}}(t). \end{aligned} \tag{10.159}$$

Step 3: ***Exponential stability***

We prove the exponential stability of the average closed-loop system based on the target system (10.144)–(10.147) using the Lyapunov method. We consider the following Lyapunov functional:

$$V = V_1 + V_2 + V_3, \tag{10.160}$$

$$V_1 = \frac{1}{2}\int_0^{s_{\rm av}(t)} w(x,t)^2\,dx, \tag{10.161}$$

$$V_2 = \frac{1}{2}\int_0^{s_{\rm av}(t)} w_x(x,t)^2\,dx, \tag{10.162}$$

$$V_3 = \rho\frac{1}{2}\vartheta_{\rm av}(t)^2. \tag{10.163}$$

Taking the derivative of (10.161) with respect to t,

$$\begin{aligned}\dot{V}_1 = &-\int_0^{s_{\rm av}(t)} w_x(x,t)^2\,dx \\ &+ KH\dot{s}_{\rm av}(t)\vartheta_{\rm av}(t)\int_0^{s_{\rm av}(t)} w(x,t)\,dx.\end{aligned} \tag{10.164}$$

Taking the derivative of (10.162) with respect to t,

$$\begin{aligned}\dot{V}_2 = {}& w_x(s_{\rm av}(t),t)w_t(x,t) + \frac{1}{2}\dot{s}_{\rm av}(t)w_x(s_{\rm av}(t),t)^2 \\ &- KH\dot{s}_{\rm av}(t)\vartheta_{\rm av}(t)w_x(s_{\rm av}(t),t) - \int_0^{s_{\rm av}(t)} w_{xx}(x,t)^2\,dx.\end{aligned} \tag{10.165}$$

Using the relationship $w_t(s_{\rm av}(t),t) = -\dot{s}_{\rm av}(t)w_x(s_{\rm av}(t),t)$ and replacing it into (10.165), we obtain

$$\begin{aligned}\dot{V}_2 = &-\int_0^{s_{\rm av}(t)} w_{xx}(x,t)^2\,dx - \frac{1}{2}\dot{s}_{\rm av}(t)w_x(s_{\rm av}(t),t)^2 \\ &- KH\dot{s}_{\rm av}(t)\vartheta_{\rm av}(t)w_x(s_{\rm av}(t),t).\end{aligned} \tag{10.166}$$

Taking the derivative of (10.163) with respect to t leads us to

$$\dot{V}_3 = -\rho KH\vartheta_{\rm av}(t)^2 - \rho\vartheta_{\rm av}(t)w_x(s_{\rm av}(t),t). \tag{10.167}$$

Substituting the terms (10.164), (10.166), and (10.167) into the time derivative of (10.160) and using Young's inequality in $-\rho\vartheta_{\rm av}(t)w_x(s_{\rm av}(t),t)$, $KH\dot{s}_{\rm av}(t)\vartheta_{\rm av}(t)\int_0^{s_{\rm av}(t)} w(x,t)\,dx$, and $-KH\dot{s}_{\rm av}(t)\vartheta_{\rm av}(t)w_x(s_{\rm av}(t),t)$, we have

$$\begin{aligned}\dot{V} \le &-\int_0^{s_{\rm av}(t)} w_{xx}(x,t)^2\,dx - \int_0^{s_{\rm av}(t)} w_x(x,t)^2\,dx \\ &- \frac{\rho KH}{2}\vartheta_{\rm av}(t)^2 + \frac{\rho}{2KH}w_x(s_{\rm av}(t),t)^2 \\ &+ \dot{s}_{\rm av}(t)\left(\frac{s^*}{2}\int_0^{s_{\rm av}(t)} w(x,t)^2 + (KH)^2\vartheta_{\rm av}(t)^2\right).\end{aligned} \tag{10.168}$$

By choosing $\rho = \frac{KH}{4s^*}$ and applying Poincaré and Agmon's inequalities at $\int_0^{s_{\rm av}(t)} w(x,t)^2\,dx$ and $w_x(s_{\rm av}(t),t)^2$, respectively, we obtain

$$
\begin{aligned}
\dot{V} \leq & -\frac{1}{8s^{*2}}\int_0^{s_{\rm av}(t)} w_x(x,t)^2 dx - \frac{1}{4s^{*2}}\int_0^{s_{\rm av}(t)} w(x,t)^2 dx \\
& + \dot{s}_{\rm av}(t)\left(\frac{s^*}{2}\int_0^{s_{\rm av}(t)} w(x,t)^2\,dx + (KH)^2\vartheta_{\rm av}(t)^2\right) - \frac{\rho KH}{2}\vartheta_{\rm av}(t)^2 \\
\leq & -mV + n\dot{s}_{\rm av}(t)V,
\end{aligned}
\tag{10.169}
$$

where

$$
n = \max\{1, 8s^*KH\}, \quad m = \min\{1/4s^{*2}, KH\}. \tag{10.170}
$$

The term $n\dot{s}_{\rm av}(t)V$ on the RHS of (10.169) does not let us to directly conclude exponential stability. To deal with it, a new Lyapunov function candidate W is defined according to

$$
W(t) = V(t)e^{-ns_{\rm av}(t)}. \tag{10.171}
$$

The time derivative of (10.171) can be calculated using (10.169) such that

$$
\dot{W}(t) = (\dot{V}(t) - n\dot{s}_{\rm av}(t)V(t))e^{-ns_{\rm av}(t)} \leq -mW(t). \tag{10.172}
$$

Taking into account (10.160), we can establish the following relationship:

$$
\|w(t)\|_{\mathcal{H}_1}^2 + \rho\vartheta_{\rm av}(t)^2 \leq e^{ns^*}(\|w_0\|_{\mathcal{H}_1}^2 + \rho\vartheta_{\rm av}(0)^2)e^{-mt}, \quad w(x,0) = w_0\,. \tag{10.173}
$$

Hence, we can conclude the existence of a positive constant $M > 0$ using the inverse transformation (10.148) combined with Young's and Cauchy–Schwarz's inequalities, such that

$$
\|u_{\rm av}(t)\|_{\mathcal{H}_1}^2 + \vartheta_{\rm av}(t)^2 \leq M(\|u_{\rm av0}\|_{\mathcal{H}_1}^2 + \vartheta_{\rm av}(0)^2)e^{-mt}, \quad u_{\rm av}(x,0) = u_{\rm av0}\,, \tag{10.174}
$$

which completes the proof. □

Asymptotic Convergence to a Neighborhood of the Extremum Point

In Theorem 10.2, we prove that the average closed-loop system (10.144)–(10.147) is exponentially stable. However, there is no suitable averaging theorem for moving-boundary PDE systems and it remains as an open problem in the literature. If this theorem existed such that it was employed for PDEs of fixed domains [83] (see also Appendix B), then we would apply it to (10.130)–(10.133) and conclude for the nonaverage system the existence of a unique exponentially stable periodic solution in t of period $\Pi := 2\pi/\omega$, denoted by $\vartheta^{\Pi}(t)$, $u^{\Pi}(x,t)$, satisfying

$$
\left(\left|\vartheta^{\Pi}(t)\right|^2 + \left\|u^{\Pi}(t)\right\|^2 + \left\|u_x^{\Pi}(t)\right\|^2\right)^{1/2} \leq \mathcal{O}(1/\omega) \quad \forall t \geq 0\,. \tag{10.175}
$$

On the other hand, the asymptotic convergence to a neighborhood of the extremum point would be proved taking the absolute value of (10.127),

$$
|\Theta(t) - \Theta^*| = |\vartheta(t) + a\sin(\omega t)|\,, \tag{10.176}
$$

and writing (10.176) in terms of the periodic solution $\vartheta^{\Pi}(t)$. By applying Young's inequality and using again the appropriate average theorem, one would have

$$\limsup_{t\to\infty} |\Theta(t)-\Theta^*| = \limsup_{t\to\infty} \left|\sqrt{2}\vartheta^{\Pi}(t)+a\sin(\omega t)\right|. \tag{10.177}$$

Finally, with (10.175) we would arrive to

$$\limsup_{t\to\infty} |\Theta(t)-\Theta^*| = \mathcal{O}(|a|+1/\omega). \tag{10.178}$$

In order to show the convergence of the output $y(t)$, we could follow the same steps employed for $\Theta(t)$ by plugging (10.110) into (10.127), such that

$$\limsup_{t\to\infty} |y(t)-y^*| = \limsup_{t\to\infty} \left|H\vartheta^2(t)+Ha^2\sin(\omega t)^2\right|. \tag{10.179}$$

Hence, rewriting (10.179) in terms of $\vartheta^{\Pi}(t)$ and again with the help of (10.175), we finally get

$$\limsup_{t\to\infty} |y(t)-y^*| = \mathcal{O}\left(|a|^2+1/\omega^2\right). \tag{10.180}$$

10.6.7 ▪ Simulations

The numerical simulation employs the quadratic map described in (10.109) and the parameters are chosen as stated in Table 10.1.

Table 10.1. *Simulation parameters.*

	Symbol	Description	Value
	K	controller gain	-0.1
Controller parameters	a	perturbation amplitude	0.1
	c	pole of the low-pass filter [rad/s]	10
	ω	perturbation frequency [rad/s]	10
	L	spatial domain	1
	Θ^*	optimizer static map	0.8
System parameters	y^*	optimal value static map	4
	H	Hessian of the static map	-1
	s_0	initial interface [m]	0.12
	T_0	initial temperature [°C]	110
	T_m	melting temperature [°C]	100

Figure 10.9 corresponds to the numerical plot of the temperature profile for the closed-loop system converging in a three-dimensional space (taking into account the domain L and the time t) to a close neighborhood of T_m.

Figure 10.10 shows the convergence of the moving boundary to the optimizer Θ^*. The sinusoidal movement of $s(t)$ would violate the usual conditions for the Stefan problem that the temperature remains above or below the melting temperature on the whole interval $[0,s(t)]$, forming a periodic chain of liquid and solid. However, since the problem occurs only in the extremum point and its neighborhood, we could redesign the algorithm in order to introduce vanishing probing signals and tapering off the perturbation after the extremum neighborhood is achieved, as studied in [59], [203], and [235].

Finally, Figures 10.11 and 10.12 show the convergence of the output $y(t)$ to y^* and $U(t)$ to 0, respectively.

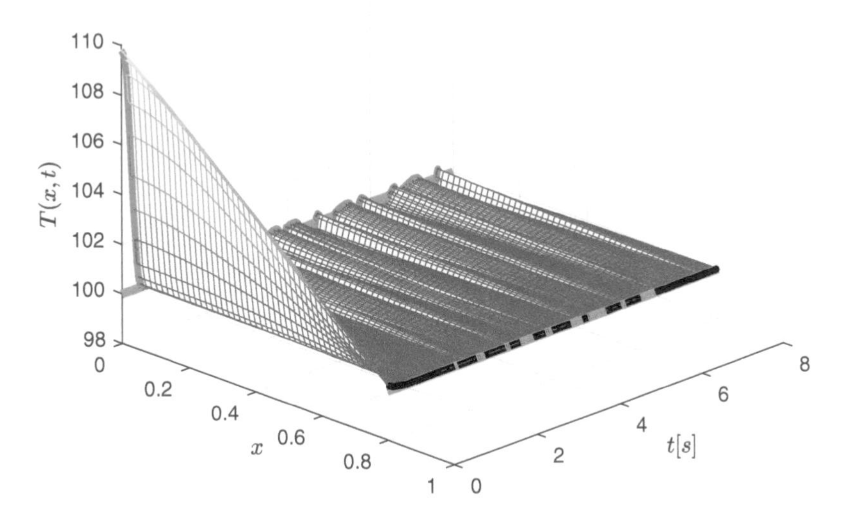

Figure 10.9. *Convergence of $T(0,t)$ (red curve) and $T(s(t),t)$ (black curve) to $T_m = 100°C$ (green curve) in a three-dimensional space for the PDE state $T(x,t)$.*

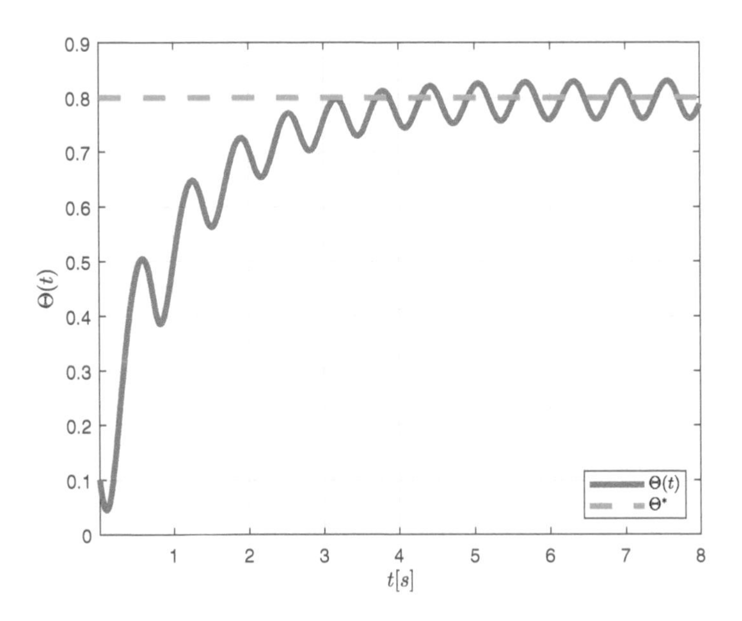

Figure 10.10. *Convergence of $s(t) := \Theta(t)$ to a small neighborhood of Θ^*.*

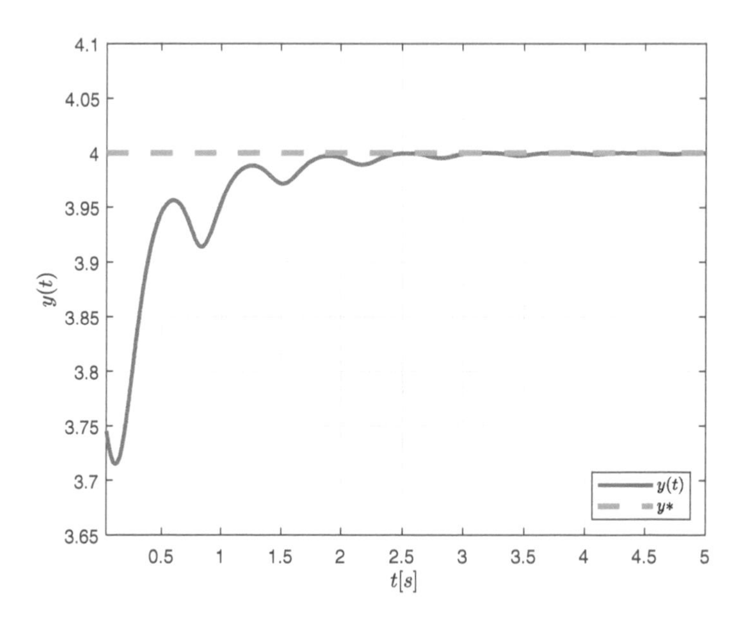

Figure 10.11. *Convergence of the output $y(t)$ to y^*.*

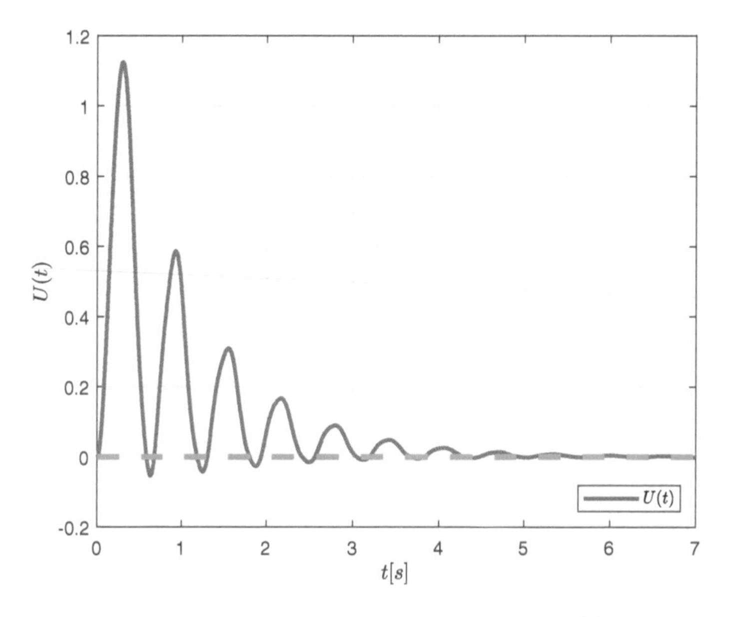

Figure 10.12. *Convergence of the control signal $U(t)$ to 0.*

10.7 ▪ Notes and References

Actuation dynamics described by infinite-dimensional systems has been introduced in several applications. The representative example is a system under actuator delays, in which the delay propagation can be described by a first-order hyperbolic partial differential equation (PDE). Compensation of the actuator and sensor delay has been developed in [127] for linear systems and in [120] for nonlinear systems via the infinite-dimensional backstepping method [128]. Recently, the first contribution of applying ES to infinite-dimensional actuation dynamics was achieved by [179] in the case of actuator delays with known delay time. A stability proof of the error-dynamics and the convergence to the extremum was presented.

Another well-known infinite-dimensional system is a diffusion process which arises in several biological, chemical, and economical systems (see [60], [240]). Compensation of actuator dynamics governed by a diffusion PDE was pioneered in [116] for the stabilization of a linear system. However, there was no work in the literature which concerned ES in the presence of actuation dynamics governed by diffusion PDEs or more general parabolic PDEs.

We tackled this problem with a semi-model-based or so-called partially model-based approach, since the diffusion coefficient is known, but the plant model (map) is unknown and its parameters are estimated using perturbations, as done in [179]. As previously mentioned by learning-based controllers in [29] and [30], such kinds of controllers are partly based on a physics-based model and partly based on a model-free learning algorithm. In this paradigm, model-free learning is used to complement the physics-based model and compensate for the uncertain or the missing part of the model.

In this chapter, we designed ES for static maps with actuation dynamics governed by diffusion PDEs. The controller to compensate the known actuation dynamics is designed via an infinite-dimensional backstepping transformation and is fed with the gradient and Hessian estimate of the static map. The main contribution of this chapter is on stability analysis. First, the transformed target system associated with the low-pass filtered boundary value for ES is shown to be exponentially stable. Second, invoking the averaging theorem for infinite-dimensional systems was successfully applied to the parabolic PDE of the average system via semigroup analysis, while in [179] the system was described as a functional differential equation. Finally, convergence to the neighborhood of the extremum is proven.

On the other hand, a large number of applications in various areas appear as moving boundary or phase change problems, such as [192] and [241]. Usually, these kind of problems arise in heat conduction situations and need to be solved in a time-dependent space domain with a moving boundary condition. For this reason diffusion PDEs with moving boundaries, known as "Stefan problems," have been studied actively for the last few decades.

The dynamics of the position of the moving boundary in the Stefan problem is governed by an ODE which depends on the PDE state, generating a nonlinear coupling of the PDE and ODE dynamics, increasing the complexity of the problem when compared to conventional analysis for PDEs of fixed domains (not depending on time or states) and ODEs.

In this sense, we also developed an ES controller for the Stefan problem. The integrator that is usually employed in the ES scheme can be leveraged as part of the Stefan model, just like the one proposed in [246]. The objective of the ES was to find the maximizer interface s^* of some unknown map $Q(s^*)$ aiming to regulate the phase change interface position to a value that attains the extremum. For this purpose, we designed a compensator of the heat PDE with moving boundary and the probing signal, which is the result of solving the problem of generating a sinusoid at the distal end of a boundary-actuated heat equation.

An important discussion is about the validation of the Stefan model. Although the usual sinusoidal movement provoked by the ES algorithm may violate the phase maintenance when the extremum is achieved or during the transient, we can keep the phase maintenance at least for the average system, thus preserving the convergence analysis.

Chapter 11

ES for Reaction-Advection-Diffusion PDEs

11.1 ▪ From the Pedagogically Inspired Heat PDE to the Practice-Driven Reaction-Advection-Diffusion PDEs

The heat equation, dealt with in Chapter 10, is just the simplest member of the class of parabolic PDEs. What we did in Chapter 10 to compensate for the heat dynamics at the input of an unknown map can be done also in the presence of more general parabolic PDE dynamics, including the Reaction-Advection-Diffusion (RAD) PDEs. Advection or reaction that occurs simultaneously with diffusion is very common in many physical systems and engineering applications, from certain manufacturing systems, to some supply-chain and economic systems, to bioreactors, which we deal with later in this chapter.

We dealt with the heat PDE separately in Chapter 10 for pedagogical reasons. Both of the advancements in the design process relative to the transport PDE, which make the compensation of a heat PDE more complicated than for the transport PDE, become even more complicated—quite a lot more complicated—for the RAD PDE.

First, the probing signal, produced through a trajectory generation or motion planning procedure, given a sinusoid required at one boundary and seeking the input needed on the opposite boundary to produce the sinusoid, is much more complex for the RAD PDE than for the heat PDE.

Second, while the backstepping-based compensation for the heat PDE involves a kernel that is linear in the variable of spatial integration, such a kernel is more complex for RAD PDEs and involves hyperbolic and additional exponential functions.

Our parabolic PDE compensator also employs a perturbation-based (averaging-based) estimate for the Hessian. We prove local stability of the algorithm, real-time maximization of the map, and convergence to a small neighborhood of the desired (unknown) extremum by means of backstepping transformation, Lyapunov functional, and the theory of averaging in infinite dimension. Last, we present the generalization to the scalar Newton-based extremum seeking algorithm, which maximizes the map's higher derivatives in the presence of RAD-like dynamics. By modifying the demodulating signals, the extremum seeking algorithm maximizes the nth derivative only through measurements of the own map. As expected, the Newton-based extremum seeking approach removes the dependence of the convergence rate on the unknown Hessian of the higher derivative, an effort to improve performance and remove limitations of standard gradient-based extremum seeking.

11.2 ▪ Problem Statement

Standard scalar ES considers applications in which one wants to maximize (or minimize) the output $y \in \mathbb{R}$ of an *unknown* nonlinear static map $Q(\Theta)$,

$$y(t) = Q(\Theta(t)), \tag{11.1}$$

by varying the input $\Theta \in \mathbb{R}$ in *real time*.

In this sense and without loss of generality, let us consider the maximization of the output (11.1) using gradient ES, where the maximizing value of Θ is denoted by Θ^*. We state our optimization problem as

$$\max_{\Theta \in \mathbb{R}} Q(\Theta). \tag{11.2}$$

Assumption 11.1. *Let $Q^{(n)}(\cdot)$ be the nth derivative of a smooth function $Q(\cdot)$: $\mathbb{R} \to \mathbb{R}$. Now let us define*

$$\Theta_{\max} = \{\Theta \in \mathbb{R} \quad | \quad Q^{(1)}(\Theta) = 0, \quad Q^{(2)}(\Theta) < 0\} \tag{11.3}$$

to be a collection of maxima where Q is locally concave. Now, assume there exists a unique $\Theta^* \in \Theta_{\max}$ such that $\Theta_{\max} \neq \emptyset$.

In the neighborhood of Θ^*, we can write $Q(\Theta)$ in its quadratic form

$$Q(\Theta) = y^* + \frac{H}{2}(\Theta - \Theta^*)^2, \tag{11.4}$$

where $y^* \in \mathbb{R}$ is the unknown extremum for $Q(\Theta^*)$ and the scalar $H < 0$ is the unknown Hessian of the static map. We can switch from maximization to minimization problem by setting simply $H > 0$ in (11.4).

11.3 ▪ Gradient Extremum Seeking Equations

From Figure 11.1, the actuation dynamics in the input of the map (11.1) is described by a RAD-like parabolic PDE according to

$$\Theta(t) = \alpha(0,t), \tag{11.5}$$

$$\alpha_t(x,t) = \epsilon\alpha_{xx}(x,t) + b\alpha_x(x,t) + \lambda\alpha(x,t), \quad x \in [0,1], \tag{11.6}$$

$$\alpha_x(0,t) = 0, \tag{11.7}$$

$$\alpha(1,t) = \theta(t), \tag{11.8}$$

with known coefficients of diffusion $\epsilon > 0$, advection $b \geq 0$, and reaction $\lambda \geq 0$. Here, we use the term RAD in a generalized sense relative to its common use in chemical engineering [242], [46], [91]. Moreover, we state the following assumption to assure stability of the actuator dynamics.

Assumption 11.2. *We assume the following relationship of the coefficients in the reaction-advection-diffusion PDE* (11.6)*:*

$$\frac{b^2}{4\epsilon} \geq \lambda. \tag{11.9}$$

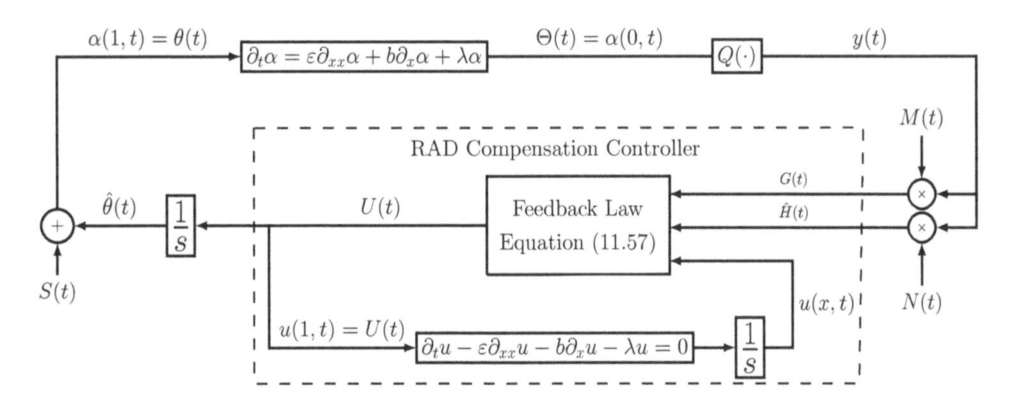

Figure 11.1. *Gradient extremum seeking loop for maximizing unknown scalar maps $y=Q(\Theta)$ with infinite-dimensional input dynamics.*

Furthermore, the propagated error dynamics through the reaction-advection-diffusion process are given by

$$\dot{\vartheta}(t)=u(0,t), \tag{11.10}$$
$$u_t(x,t)=\epsilon u_{xx}(x,t)+bu_x(x,t)+\lambda u(x,t), \quad x\in[0,1], \tag{11.11}$$
$$u_x(0,t)=0, \tag{11.12}$$
$$u(1,t)=U(t), \tag{11.13}$$

where $\vartheta(t)$ is the propagated estimation error

$$\tilde{\theta}(t):=\hat{\theta}(t)-\theta^*, \tag{11.14}$$

as done in Chapter 10—see (10.20) to (10.23)—but now through the reaction-advection-diffusion domain, and $\hat{\theta}(t)$ the estimation of the optimal input θ^*. From the block diagram in Figure 11.1, the error dynamics can be written as

$$\dot{\tilde{\theta}}(t)=\dot{\hat{\theta}}(t)=U(t). \tag{11.15}$$

In contrast to the actuation dynamics governed by diffusion PDEs [65], [169], where $\theta^*=\Theta^*$, now we have $\theta^*\neq\Theta^*$ in general. The relation is given by

$$\theta^*=\Theta^* e^{-\frac{b}{2\epsilon}}\left[\frac{b}{2\epsilon\tau}\sinh(\tau)+\cosh(\tau)\right], \quad \tau=\sqrt{\frac{b^2}{4\epsilon^2}-\frac{\lambda}{\epsilon}} \tag{11.16}$$

and shown in Figure 11.2, where we chose $\Theta^*=1$.

The perturbation signal $S(t)$ is defined in order to solve the following *trajectory generation problem* [128, Chapter 12] with Dirichlet boundary condition.

Definition 11.1. *The perturbation signal $S(t)$ is the boundary value of an inverse reaction-advection-diffusion process defined as follows:*

$$S(t)=\beta(1,t), \tag{11.17}$$
$$\beta_t(x,t)=\epsilon\beta_{xx}(x,t)+b\beta_x(x,t)+\lambda\beta(x,t), \quad x\in[0,1], \tag{11.18}$$
$$\beta_x(0,t)=0, \tag{11.19}$$
$$\beta(0,t)=a\sin(\omega t), \tag{11.20}$$

with nonzero perturbation amplitude a and frequency ω.

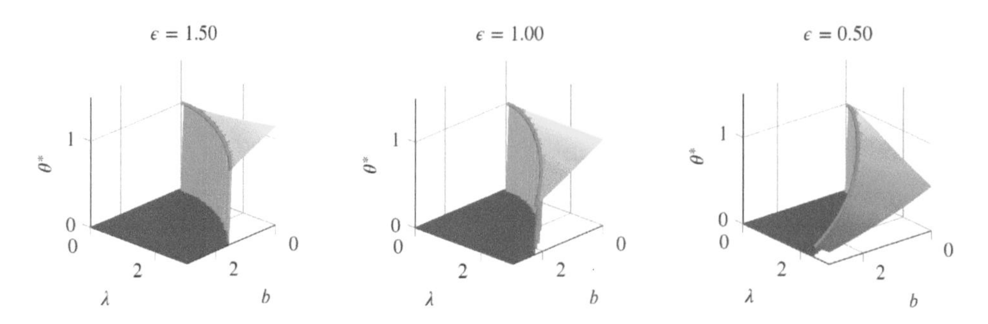

Figure 11.2. *Relation between θ^* and $\Theta^* = 1$ for different values of the diffusion coefficient ϵ. We note that $\theta^* \leq \Theta^*$ if $\epsilon \leq 1$. In the area where $\theta^* = 0$, Assumption 11.2 is not satisfied.*

The explicit solution of (11.17)–(11.20) can be calculated as infinite sums:[8]

$$S(t) = e^{-\frac{b}{2\epsilon}} \sum_{k=0}^{\infty} \frac{a_{2k}(t)}{(2k)!} + \frac{b}{2\epsilon} \frac{a_{2k}(t)}{(2k+1)!}$$

$$\text{with} \quad a_{2k} := \frac{a}{\epsilon^k} \sin(\omega t) \sum_{n=0}^{k} \binom{k}{2n} \xi^{k-2n} \omega^{2n} + \frac{a}{\epsilon^k} \cos(\omega t) \sum_{n=0}^{k} \binom{k}{2n+1} \xi^{k-2n-1} \omega^{2n+1},$$

$$\xi := \frac{b^2}{4\epsilon} - \lambda \quad \text{and} \quad \binom{y}{z} := 0 \text{ for } y < z. \tag{11.21}$$

The detailed derivation of (11.21) can be found in Appendix D.6.

Now, we define the multiplicative demodulation signals $M(t)$ and $N(t)$ as

$$M(t) = \frac{2}{a} \sin(\omega t), \quad N(t) = -\frac{8}{a^2} \cos(2\omega t) \tag{11.22}$$

and, consequently, the gradient and Hessian estimates for $Q(\Theta)$ as

$$G(t) = M(t) y(t), \quad \hat{H}(t) = N(t) y(t). \tag{11.23}$$

11.4 ▪ Backstepping Boundary Control for RAD-PDE Compensation

In this section, we derive the reaction-advection-diffusion compensation controller. Therefore, we consider the PDE-ODE cascade (11.10)–(11.13), as illustrated in Figure 11.3.

First, we transform the system (11.10)–(11.13) with the transformation

$$\bar{z}(x,t) = u(x,t) e^{\frac{b}{2\epsilon} x} \tag{11.24}$$

[8]Another solution can be envisaged with an integral involving the Bessell function. First convert RAD to RD with that ξ coefficient using the exponential scaling (11.28). Then, convert the RD equation into a heat equation using backstepping. Then, solve trajectory generation explicitly for the heat equation, as in [65] and [169]. This will give us the state reference (12.18) and the input reference (12.19) in the book [128, Chapter 12] for the heat equation. Then, go back, using the inverse backstepping transform, to get the state and input references for the RD PDE. This is where the state reference (12.18) gets integrated against the Bessell kernel and one finally applies the inverse of the exponential scaling to get the state and input references for the RAD PDE.

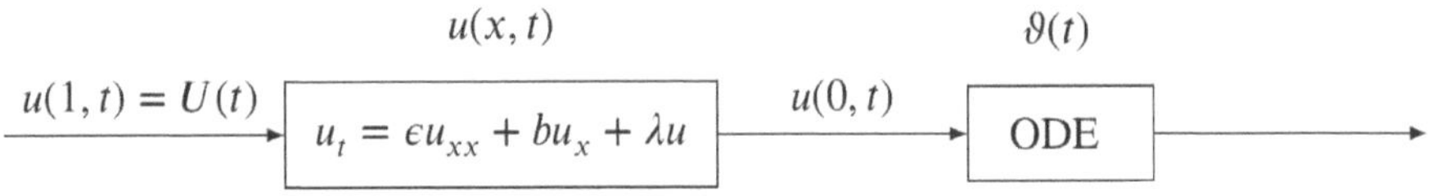

$$u(1,t)=U(t) \rightarrow \boxed{u_t = \epsilon u_{xx} + b u_x + \lambda u} \xrightarrow{u(0,t)} \boxed{\text{ODE}} \rightarrow$$

Figure 11.3. *Cascade of reaction-advection-diffusion PDE with ODE dynamics and input $U(t)$ at $x=1$.*

into the reaction-diffusion system

$$\dot{\vartheta}(t) = \bar{z}(0,t), \tag{11.25}$$

$$\bar{z}_t(x,t) = \epsilon \bar{z}_{xx}(x,t) - \xi \bar{z}(x,t), \quad x \in [0,1], \tag{11.26}$$

$$\bar{z}_x(0,t) = \frac{b}{2\epsilon}\bar{z}(0,t), \tag{11.27}$$

$$\bar{z}(1,t) = U(t)e^{\frac{b}{2\epsilon}}, \tag{11.28}$$

with Robin and Dirichlet boundary conditions and $\xi := b^2/(4\epsilon) - \lambda \geq 0$. As with the usual procedure in backstepping, we propose the transformation of the system (11.25)–(11.28) into an exponentially stable target system.

Proposition 11.1. *The backstepping transformation*

$$w(x,t) = \bar{z}(x,t) - \int_0^x q(x,y)\bar{z}(y,t)dy - \gamma(x)\vartheta(t), \tag{11.29}$$

with the kernels

$$q(x,y) = \bar{K}\frac{1}{\epsilon}\sqrt{\frac{\xi}{\epsilon}}\sinh\left(\sqrt{\frac{\xi}{\epsilon}}(x-y)\right) := \bar{K}\bar{m}(x-y) \tag{11.30}$$

and

$$\gamma(x) = \bar{K}\cosh\left(\sqrt{\frac{\xi}{\epsilon}}x\right) + \frac{\bar{K}b}{2\epsilon}\sqrt{\frac{\epsilon}{\xi}}\sinh\left(\sqrt{\frac{\xi}{\epsilon}}x\right) := \bar{K}\bar{\gamma}(x), \tag{11.31}$$

transforms the PDE-ODE cascade (11.10)–(11.13) *into the exponentially stable target system*

$$\dot{\vartheta}(t) = \bar{K}\vartheta(t) + w(0,t), \quad \bar{K} < 0, \tag{11.32}$$

$$w_t(x,t) = \epsilon w_{xx}(x,t) - \xi w(x,t), \quad x \in [0,1], \tag{11.33}$$

$$w_x(0,t) = \frac{b}{2\epsilon}w(0,t), \tag{11.34}$$

$$w(1,t) = 0. \tag{11.35}$$

Proof. The proof is divided into two steps. First, the exponential stability of the target system (11.32)–(11.35) is proved. Second, the gain kernels $q(x,y)$ and $\gamma(x)$ are derived.

Step 1: Consider the Lyapunov–Krasovskii functional

$$W(t) = \frac{1}{2}\vartheta^2(t) + \frac{1}{2}\int_0^1 w^2(x,t)dx. \tag{11.36}$$

By taking the derivative w.r.t. time of (11.36) and integrating by parts we get

$$\dot{W}(t) = \bar{K}\vartheta^2(t) + \vartheta(t)w(0,t) - \frac{b}{2}w^2(0,t) - \epsilon\|w_x(t)\|^2 - \xi\epsilon\|w(t)\|^2. \tag{11.37}$$

Applying Young's inequality leads to

$$\dot{W}(t) \leq \left(\bar{K} + \frac{\gamma}{2}\right)\vartheta^2(t) + \left(\frac{1}{2\gamma} - \frac{b}{2}\right)w^2(0,t) - \epsilon\|w_x(t)\|^2 - \xi\epsilon\|w(t)\|^2. \tag{11.38}$$

The parameters $\gamma > 0$ and $\bar{K} < 0$ can be chosen such that

$$\exists \mu > 0 : \ \dot{W}(t) \leq -\mu W(t), \tag{11.39}$$

which shows the exponential stability of the target system (11.32)–(11.35).

Step 2: Differentiating the transformation (11.29) with respect to space x twice yields

$$w_x(x,t) = \bar{z}_x(x,t) - q(x,x)\bar{z}(x,t) - \int_0^x q_x(x,y)\bar{z}(y,t)dy - \gamma'(x)\vartheta(t), \tag{11.40}$$

$$\begin{aligned} w_{xx}(x,t) = {} & \bar{z}_{xx}(x,t) - \bar{z}(x,t)\frac{d}{dx}q(x,x) - q(x,x)\bar{z}_x(x,t) \\ & - q_x(x,x)\bar{z}(x,t) - \int_0^x q_{xx}(x,y)\bar{z}(y,t)dy - \gamma''(x)\vartheta(t) \end{aligned} \tag{11.41}$$

and with respect to time t gives

$$\begin{aligned} w_t(x,t) = {} & \epsilon\bar{z}_{xx}(x,t) - \epsilon q(x,x)\bar{z}_x(x,t) + \bar{z}(x,t)\left[\epsilon q_y(x,x) - \xi\right] \\ & + \bar{z}(0,t)\left[\frac{b}{2}q(x,0) - \epsilon q_y(x,0) - \gamma(x)\right] - \int_0^x \bar{z}(y,t)\left[\epsilon q_{yy}(x,y) - \xi q(x,y)\right]dy. \end{aligned} \tag{11.42}$$

By inserting (11.40)–(11.42) into the PDE (11.33), we obtain

$$\begin{aligned} w_t(x,t) - \epsilon w_{xx}(x,t) - \xi w(x,t) = {} & 2\bar{z}(x,t)\frac{d}{dx}q(x,x) + \bar{z}(0,t)\left[\frac{b}{2}q(x,0) - \epsilon q_y(x,0) - \gamma(x)\right] \\ & + \int_0^x \bar{z}(y,t)\left[\epsilon q_{xx}(x,y) - \epsilon q_{yy}(y,t)\right]dy + \vartheta(t)\left[\epsilon\gamma''(x) - \xi\gamma(x)\right]. \end{aligned} \tag{11.43}$$

Considering (11.43) and evaluating (11.40) and (11.41) at $x = 0$, we get the following conditions on the gain kernels $q(x,y)$ and $\gamma(x)$:

$$\gamma''(x) = \frac{\xi}{\epsilon}\gamma(x), \tag{11.44}$$

$$\gamma(0) = \bar{K}, \tag{11.45}$$

$$\gamma'(0) = \bar{K}\frac{b}{2\epsilon}, \tag{11.46}$$

$$q_{xx}(x,y) - q_{yy}(x,y) = 0, \tag{11.47}$$

$$q(x,x) = 0, \tag{11.48}$$

$$q(x,0) = \frac{2\epsilon}{b}q_y(x,0) + \frac{2}{b}\gamma(x). \tag{11.49}$$

Solving the second-order ODE system (11.45)–(11.46) results in

$$\gamma(x) = \bar{K}\bar{\gamma}(x) = \bar{K}\cosh\left(\sqrt{\frac{\xi}{\epsilon}}x\right) + \frac{\bar{K}b}{2\epsilon}\sqrt{\frac{\epsilon}{\xi}}\sinh\left(\sqrt{\frac{\xi}{\epsilon}}x\right). \tag{11.50}$$

Conditions (11.47)–(11.49) comprise a hyperbolic PDE of the Goursat type. Equations (11.47) and (11.48) lead to the convolutional-kernel ansatz

$$q(x,y) = m(x-y), \tag{11.51}$$

where $m(\cdot)$ is a scalar function. With (11.49), we get the ODE

$$m'(x) = -\frac{b}{2\epsilon}m(x) + \frac{1}{\epsilon}\gamma(x) \tag{11.52}$$

to derive the kernel function $q(x,y)$. Hence, using the ansatz $m(x) = \exp(ax)$ and the variation-of-constants method, the gain kernel can be calculated to be

$$q(x,y) = \bar{K}\bar{m}(x-y) = \bar{K}\frac{1}{\epsilon}\sqrt{\frac{\epsilon}{\xi}}\sinh\left(\sqrt{\frac{\xi}{\epsilon}}(x-y)\right). \tag{11.53}$$

This completes the proof. □

Hence, the boundary control law can be stated as

$$\begin{aligned}
U(t) &= \bar{K}e^{-\frac{b}{2\epsilon}}\left[\bar{\gamma}(1)\vartheta(t) + \int_0^1 \bar{m}(1-y)\bar{z}(y,t)dy\right], \qquad (11.54)\\
\bar{\gamma}(1) &= \cosh\left(\sqrt{\frac{\xi}{\epsilon}}\right) + \frac{b}{2\epsilon}\sqrt{\frac{\epsilon}{\xi}}\sinh\left(\sqrt{\frac{\xi}{\epsilon}}\right),\\
\xi &= b^2/(4\epsilon) - \lambda \geq 0,\\
\bar{m}(1-y) &= \frac{1}{\epsilon}\sqrt{\frac{\epsilon}{\xi}}\sinh\left(\sqrt{\frac{\xi}{\epsilon}}(1-y)\right).
\end{aligned}$$

However, the control law in (11.54) is not applicable directly, because we have no measurement on $\vartheta(t)$. Thus, we introduce an important result of [73]: the averaged versions of the gradient and Hessian estimates in (11.23) are

$$G_{\text{av}}(t) = H\vartheta_{\text{av}}(t), \qquad \hat{H}_{\text{av}}(t) = H \tag{11.55}$$

if at least a locally quadratic map as in (11.4) is considered. Regarding (11.55), we average (11.54) and choose $\bar{K} = KH$ with $K > 0$, such that

$$U_{\text{av}}(t) = Ke^{-\frac{b}{2\epsilon}}\left[\bar{\gamma}(1)G_{\text{av}}(t) + H\int_0^1 \bar{m}(1-y)\bar{z}_{\text{av}}(y,t)dy\right]. \tag{11.56}$$

Due to technical reasons in the application of the averaging theorem for infinite-dimensional systems [83] in the following stability proof, we introduce a low-pass filter to the controller. Finally, recalling (11.24), we get the average-based infinite-dimensional control law to compensate the reaction-advection-diffusion process,

$$U(t) = \frac{c}{s+c}\left\{Ke^{-\frac{b}{2\epsilon}}\left[\bar{\gamma}(1)G(t) + \hat{H}(t)\int_0^1 e^{\frac{b}{2\epsilon}y}\bar{m}(1-y)u(y,t)dy\right]\right\}, \quad K>0, \tag{11.57}$$

where constant $c > 0$ is chosen later. For notational convenience, we mix the time and frequency domain in (11.57), where the low-pass filter acts as an operator on the term between braces. Note that the reaction-advection-diffusion process of the actuation dynamics is known, i.e., the coefficients ϵ, b, and λ.

Remark 11.1. Note that the feedback law (11.57) is consistent with the basic cases of delays [179] (Chapter 2) and diffusion PDEs [65] (Chapter 10). In fact, the delay and diffusion PDEs are special cases of this result, with $\epsilon = \lambda = 0$ and $b > 0$ for the former as well as $\lambda = b = 0$ and $\epsilon > 0$ for the latter.

11.5 ▪ Stability and Convergence

The proof of stability and convergence of the closed-loop system (11.10)–(11.13) with controller (11.57) is presented now.

11.5.1 ▪ Average Closed-Loop System and Its Stability

Substituting (11.57) into (11.13) and recalling that $G_{\rm av}(t) = H\vartheta_{\rm av}(t)$ from (11.55), we can write, for ω sufficiently large, the average closed-loop system of (11.10)–(11.13) as

$$\dot{\vartheta}_{\rm av}(t) = u_{\rm av}(0,t), \tag{11.58}$$

$$(u_{\rm av})_t(x,t) = \epsilon(u_{\rm av})_{xx}(x,t) + b(u_{\rm av})_x(x,t) + \lambda u_{\rm av}(x,t), \quad x \in [0,1], \tag{11.59}$$

$$(u_{\rm av})_x(0,t) = 0, \tag{11.60}$$

$$(u_{\rm av})_t(1,t) = -cu_{\rm av}(1,t) + cKHe^{-\frac{b}{2\epsilon}}\left[\bar{\gamma}(1)\vartheta_{\rm av}(t) + \int_0^1 e^{\frac{b}{2\epsilon}y}\bar{m}(1-y)u_{\rm av}(y,t)dy\right]. \tag{11.61}$$

In the next proposition, we show the exponential stability of the closed-loop average system.

Proposition 11.2. *The average closed-loop system* (11.58)–(11.61) *is exponentially stable in the sense of the $\mathcal{H}_1$-norm*

$$\left(|\vartheta_{\rm av}(t)|^2 + \int_0^1 u_{\rm av}^2(x,t)dx + \int_0^1 (u_{\rm av})_x^2(x,t)dx + u_{\rm av}^2(1,t) + u_{\rm av}^2(0,t)\right)^{1/2}. \tag{11.62}$$

Proof. With some abuse of notation, we also use $w(x,t)$ to denote the average transformed state, and from the backstepping transformation (11.29) in the variables $(u_{\rm av}, w)$,

$$w(x,t) = e^{\frac{b}{2\epsilon}x}u_{\rm av}(x,t) - \int_0^x \tilde{q}(x,y)e^{\frac{b}{2\epsilon}y}u_{\rm av}(y,t)dy - \tilde{\gamma}(x)\vartheta_{\rm av}(t), \tag{11.63}$$

we can state its inverse (the complete derivation can be found in Appendix D.7) in the variables $(u_{\rm av}, w)$ as

$$u_{\rm av}(x,t) = e^{-\frac{b}{2\epsilon}x}w(x,t) - e^{-\frac{b}{2\epsilon}x}\int_0^x \tilde{p}(x,y)w(y,t)dy - e^{-\frac{b}{2\epsilon}x}\tilde{\eta}(x)\vartheta_{\rm av}(t), \tag{11.64}$$

with

$$\tilde{q}(x,y) = KH\bar{m}(x-y) = KH\frac{1}{\epsilon}\sqrt{\frac{\epsilon}{\xi}}\sinh\left(\sqrt{\frac{\xi}{\epsilon}}(x-y)\right), \tag{11.65}$$

$$\tilde{\gamma}(x) = KH\bar{\gamma}(x) = KH\cosh\left(\sqrt{\frac{\xi}{\epsilon}}x\right) + \frac{KHb}{2\epsilon}\sqrt{\frac{\epsilon}{\xi}}\sinh\left(\sqrt{\frac{\xi}{\epsilon}}x\right), \tag{11.66}$$

$$\begin{aligned}\tilde{p}(x,y) &= KH\bar{n}(x-y)\\ &= -\frac{KH}{\sqrt{KH+\xi}}\sinh\left(\sqrt{KH+\xi}(x-y)\right),\end{aligned} \tag{11.67}$$

$$\tilde{\eta}(x) = KH\bar{\eta}(x) = -KH\cosh\left(\sqrt{KH+\xi x}\right) - \frac{KHb}{2\epsilon\sqrt{KH+\xi}}\sinh\left(\sqrt{KH+\xi x}\right), \tag{11.68}$$

which transforms the average closed-loop system (11.58)–(11.61) into the target system

$$\dot{\vartheta}_{\rm av}(t) = KH\vartheta(t) + w(0,t), \tag{11.69}$$

$$w_x(x,t) = \epsilon w_{xx}(x,t) - \xi w(x,t), \quad x \in [0,1], \tag{11.70}$$

$$w_x(0,t) = \frac{b}{2\epsilon}w(0,t), \tag{11.71}$$

$$w_t(1,t) = -cw(1,t) - KHw(1,t) - (KH)^2\left[\bar{\eta}(1)\vartheta_{\rm av}(t) + \int_0^1 \bar{n}(1-y)w(y,t)\right]. \tag{11.72}$$

Consider the Lyapunov–Krasovskii functional

$$V(t) = \frac{\vartheta_{\rm av}^2(t)}{2} + \frac{a}{2}\int_0^1 w^2(x,t)dx + \frac{d}{2}\int_0^1 w_x^2(x,t)dx + \frac{e}{2}w^2(1,t) + \frac{f}{2}w^2(0,t) \tag{11.73}$$

and its time derivative along with (11.58)–(11.61) and (11.69)–(11.72):

$$\begin{aligned}\dot{V}(t) = {} & KH\vartheta_{\rm av}^2(t) + \vartheta_{\rm av}(t)w(0,t) - \frac{a}{2}w^2(0,t) - a\epsilon\int_0^1 w_x^2(x,t)dx - a\xi\int_0^1 w^2(x,t)dx\\ & + a\epsilon w(1,t)w_x(1,t) - d\epsilon\int_0^1 w_{xx}^2(x,t)dx - \frac{db}{2\epsilon}w^2(0,t) + dw_x(1,t)w_t(1,t)\\ & + d\xi w_x(1,t)w(1,t) + fw(0,t)w_t(0,t),\end{aligned}$$

where we choose $e = db/(2\epsilon)$ and $a,d,e,f > 0$. Since the nonnegative terms of $\dot{V}$ are the same as of the time derivative of the Lyapunov–Krasovkii functionals in [65] (or Chapter 10), we conclude following similar calculations that there exist $c^* > 0$ and $\mu^* > 0$ such that $\dot{V}(t) \leq -\mu^* V(t)$ and the target system (11.69)–(11.72) is exponentially stable in the $\mathcal{H}_1$-norm

$$\left(|\vartheta_{\rm av}(t)|^2 + \int_0^1 w^2(x,t)dx + \int_0^1 w_x^2(x,t)dx + w^2(1,t) + w^2(0,t)\right)^{1/2} \tag{11.74}$$

for $c > c^*$ sufficiently large in (11.57). Moreover, as in the proof of Theorem 10.1 (Step 4) in Chapter 10, the exponential stability property in the variables $(\vartheta_{\rm av}, w)$ can be transferred to the original variables $(\vartheta_{\rm av}, u_{\rm av})$ by using the inverse transformation (11.64). Hence, the average closed-loop system (11.58)–(11.61) is also exponentially stable in the sense of the $\mathcal{H}_1$-norm (11.62). □

11.5.2 ▪ Invoking the Averaging Theorem

The following proposition shows we can apply the Averaging Theorem for infinite-dimensional systems of [83, Section 2] (in Appendix B) to the closed-loop system so that there exists a unique exponentially stable periodic solution in t of period $\Pi := 2\pi/\omega$, denoted by $\vartheta^{\Pi}(t)$, $u^{\Pi}(t)$, satisfying $\forall t \geq 0$,

$$\left(|\vartheta^{\Pi}(t)|^2 + \int_0^1 [u^{\Pi}(x,t)]^2 dx + \int_0^1 [u_x^{\Pi}(x,t)]^2 dx + [u^{\Pi}(1,t)]^2 + [u^{\Pi}(0,t)]^2\right)^{1/2} \leq \mathcal{O}(1/\omega). \tag{11.75}$$

Proposition 11.3. *The original closed-loop system* (11.10)–(11.13) *with the controller* (11.57) *can be written in the form*

$$\dot{Z}(t) = FZ(t) + J(\omega t, Z(t)), \tag{11.76}$$

where F generates an analytic semigroup, $J(\omega t, Z(t))$ satisfies the smoothness conditions of the Averaging Theorem [83, *Section* 2], *and $Z(t)$ is an infinite-dimensional state vector.*

Proof. The state-transformation $v(x,t) = u(x,t) - U(t)$ transforms the original closed-loop system (11.10)–(11.13) with the controller (11.57) into

$$\dot{\vartheta}(t) = v(0,t) + U(t), \tag{11.77}$$
$$v_t(x,t) = \epsilon v_{xx}(x,t) + bv_x(x,t) + \lambda v(x,t) - \phi(\vartheta, v, U, t), \tag{11.78}$$
$$v_x(0,t) = 0, \tag{11.79}$$
$$v(1,t) = 0, \tag{11.80}$$
$$\dot{U}(t) = \phi(\vartheta, v, U, t), \tag{11.81}$$

with

$$\phi(\vartheta, v, U, t) = -cU(t) + cK\left[\bar{\gamma}(1)G(t) + \hat{H}(t)\int_0^1 e^{\frac{b}{2\epsilon}y}\bar{m}(1-y)(v(y,t)+U(t))dy\right]. \tag{11.82}$$

We write the PDE (11.78) as an evolutionary equation [145] in the Banach space $\mathcal{X} := \mathcal{H}_1([0,D])$,

$$\dot{V}(t) = \mathcal{A}V(t) - \tilde{\phi}(\vartheta, V, U, t) \quad \forall t > 0, \tag{11.83}$$

with

$$\mathcal{A}\varphi := \epsilon\frac{\partial^2\varphi}{\partial x^2} + b\frac{\partial\varphi}{\partial x} + \lambda\varphi, \tag{11.84}$$
$$D(\mathcal{A}) = \left\{\varphi \in \mathcal{X}: \quad \varphi, \frac{d}{dx}\varphi \in \mathcal{X} \text{ are a.c.}, \quad \frac{d^2}{dz^2}\varphi \in \mathcal{X},\ \frac{d}{dz}\varphi(0) = 0,\ \varphi(1) = 0\right\}. \tag{11.85}$$

By introducing the linear operators

$$\boldsymbol{\beta}:\mathbb{R}\to\mathcal{X} \quad \text{s.t.} \quad \boldsymbol{\beta}\zeta := [\beta_1,\beta_2,\ldots]^\top\zeta \quad \text{with}$$

$$\beta_k = \int_0^1 \psi_k(x)dx, \quad \zeta\in\mathbb{R}, \tag{11.86}$$

$$\mathcal{B}:\mathcal{X}\to\mathbb{R} \quad \text{s.t.} \quad \mathcal{B}V(t) := v(0,t), \tag{11.87}$$

$$\boldsymbol{\alpha}^\top:\mathcal{X}\to\mathbb{R} \quad \text{s.t.} \quad \boldsymbol{\alpha}^\top V(t) := \sum_{k=1}^{\infty}\alpha_k v_k^*(t) \quad \text{with}$$

$$\alpha_k = \int_0^1 e^{\frac{b}{2\epsilon}y}\bar{m}(1-y)\tilde{\psi}_k(y)dy, \tag{11.88}$$

and ψ_k being the eigenfunctions and $\tilde{\psi}_k(x)$ being the adjoint eigenfunctions of $\mathcal{A}$, we arrive at the infinite-dimensional system of the form (11.76) with state vector $Z(t) = [\vartheta(t)\ V(t)\ U(t)]^\top$,

$$F = \begin{bmatrix} 0 & \mathcal{B} & 1 \\ 0 & \mathcal{A} & c\beta \\ 0 & 0 & -c \end{bmatrix}, \tag{11.89}$$

$$J(\omega t, Z) = \begin{bmatrix} 0 \\ -c\beta K\left[G(t)+\hat{H}(t)g(Z(t))\right] \\ cK\left[G(t)+\hat{H}(t)g(Z(t))\right] \end{bmatrix}, \tag{11.90}$$

where $g(Z(t)) = \mu U(t) + \alpha^\top V(t)$ and

$$\mu = \int_0^1 e^{\frac{b}{2\epsilon}y}\bar{m}(1-y)dy. \tag{11.91}$$

Since $\mathcal{A}$ is a strongly elliptic operator [75], it is also a generator of an analytic semigroup [190, Theorem 2.7]. The $\mathcal{A}$-boundedness of the boundary operator $\mathcal{B}$ in (11.86) can be easily proven and is satisfied with the parameters $\epsilon_1 = \max\{2,\epsilon,b,\lambda\}$, $\epsilon_2 = 0$ in the operator matrix theorem [161]. Hence, the operator matrix F generates an analytic semigroup. Furthermore, $J(\omega t, Z)$ in (11.89) is Fréchet differentiable in Z, strongly continuous and periodic in t uniformly with respect to Z. □

Hence, along with the exponential stability of the average closed-loop system (11.58)–(11.61), all assumptions to apply the averaging theorem for infinite-dimensional systems in [83, Section 2] (in Appendix B) are satisfied. Thus, the closed-loop system (11.10)–(11.13) with controller (11.57) has an exponentially stable periodic solution $Z^{\Pi}(t)$ with $\|Z^{\Pi}(t)\| \le \mathcal{O}(1/\omega)$ that satisfies (11.75).

11.5.3 ▪ Convergence to a Neighborhood of the Extremum

Employing an analogous procedure carried out in the proof of Theorem 10.1 (Step 6) in Chapter 10, we can state the convergence of $(\theta(t),\Theta(t),y(t))$ to the neighborhood of the optimizers

$(\theta^*, \Theta^*, y^*)$ without further calculations:

$$\limsup_{t\to\infty} |\theta(t) - \theta^*| = \mathcal{O}\left(|a| \exp\left(\sqrt{\frac{\xi + \omega}{\epsilon}}\right) + 1/\omega\right), \quad \xi = \frac{b^2}{4\epsilon} - \lambda, \tag{11.92}$$

$$\limsup_{t\to\infty} |\Theta(t) - \Theta^*| = \mathcal{O}\left(|a| + 1/\omega\right), \tag{11.93}$$

$$\limsup_{t\to\infty} |y(t) - y^*| = \mathcal{O}\left(|a|^2 + 1/\omega^2\right). \tag{11.94}$$

The exponential term in (11.92) results from the order of $S(t)$ in (11.21) by expanding the sums. Finally, (11.92) shows the ultimate convergence of the proposed ES approach with actuation dynamics governed with reaction-advection-diffusion PDEs of Figure 11.1 to a neighborhood of $(\theta^*, \Theta^*, y^*)$ by choosing ω sufficiently large and a sufficiently small.

11.6 ▪ Newton-Based ES for Higher-Derivatives Maps with Parabolic PDEs

As described in Section 11.2, standard scalar gradient ES considers applications in which one wants to maximize (or minimize) the output $y \in \mathbb{R}$ of an *unknown* nonlinear static map $Q(\Theta)$.

In [151] (and Chapter 5 for delayed systems), a new problem formulation was introduced for Newton-based ES: instead of optimizing the output of the map itself, the objective was to optimize any higher derivative of its output (assuming only the measurement of y). Without loss of generality, we consider the maximization of nth derivative of the output (11.1), where the maximizing value of Θ is denoted by Θ^*. In this case, we state our optimization problem as

$$\max_{\Theta \in \mathbb{R}} Q^{(n)}(\Theta). \tag{11.95}$$

Assumption 11.3. *Let $Q^{(n)}(\cdot)$ be the nth derivative of a smooth function $Q(\cdot)$: $\mathbb{R} \to \mathbb{R}$ and*

$$\Theta_{max} = \{\Theta \mid Q^{(n+1)}(\Theta) = 0, \quad Q^{(n+2)}(\Theta) < 0\} \tag{11.96}$$

be a collection of maxima where $Q^{(n)}$ is locally concave. Now, assume that $\Theta^* \in \Theta_{max}$ and $\Theta_{max} \neq \emptyset$.

In the neighborhood of Θ^*, we can write $Q^{(n)}(\Theta)$ in its quadratic form

$$Q^{(n)}(\Theta) = y^* + \frac{H}{2}(\Theta - \Theta^*)^2, \tag{11.97}$$

where $y^* \in \mathbb{R}$ is the unknown extremum for $Q^{(n)}(\Theta)$ and the scalar $H < 0$ is the unknown Hessian of the nth derivative of the static map. According to [151], we switch from maximization to minimization by setting $\text{sgn}(\Gamma_0) = \text{sgn}(Q^{(n+2)}(0))$ with Γ_0 being the initial value of the Riccati filter to be discussed later on.

11.6.1 ▪ Newton-Based Extremum Seeking Equations

Basically, all problem formulation (11.5)–(11.21) constructed before repeats itself for the purpose of obtaining a Newton version of the proposed algorithm. The principal change is the redefinition of the signals $M(t)$, $N(t)$, $\hat{H}(t)$, $G(t)$, and $z(t)$. In this sense, we first recall the following auxiliary signal from [151]:

$$\Upsilon_j(t) = C_j \sin\left(j\omega t + \frac{\pi}{4}(-1)^{n+1}\left(1+(-1)^j\right)\right), \tag{11.98}$$

$$C_j = \frac{2^j j!}{a^j}(-1)^F, \quad F = \frac{j - \left|\sin\left(\frac{j\pi}{2}\right)\right|}{2}. \tag{11.99}$$

In [151] and [200], the signal $\widehat{h^{(j)}}(t) := y(t)\Upsilon_j(t)$ was defined by means of it being possible to obtain an estimate for the gradient of $Q^{(n)}(\Theta)$ if the index j was chosen equal to $j = n+1$ and for its Hessian if $j = n+2$, respectively. Therefore, we define the multiplicative demodulation signals $M(t)$ and $N(t)$ as

$$M(t) = \Upsilon_{n+1}(t), \quad N(t) = \Upsilon_{n+2}(t) \tag{11.100}$$

and, consequently, the gradient and Hessian estimates for $Q^{(n)}(\Theta)$ as

$$G(t) = M(t)y(t), \quad \hat{H}(t) = N(t)y(t). \tag{11.101}$$

Finally, let us define the signal

$$z(t) = \Gamma(t)G(t), \tag{11.102}$$

where $\Gamma(t)$ is updated according to the following Riccati differential equation [73]:

$$\dot{\Gamma} = \omega_r \Gamma - \omega_r \hat{H}\Gamma^2, \tag{11.103}$$

with $\omega_r > 0$ being a design constant. Equation (11.103) generates an estimate of the Hessian's inverse (H^{-1}), avoiding inversions that may cross zero during the transient phase. The estimation error of the Hessian's inverse is defined as

$$\tilde{\Gamma}(t) = \Gamma(t) - H^{-1}, \tag{11.104}$$

and its dynamic equation is written from (11.103) and (11.104) by

$$\dot{\tilde{\Gamma}} = \omega_r(\tilde{\Gamma} + H^{-1})[1 - \hat{H}(\tilde{\Gamma} + H^{-1})]. \tag{11.105}$$

According to Proposition 11.1, we are able to propose a Newton-based ES controller following the same backstepping transformation in (11.29)–(11.31), with $\bar{K} = -K$ and $K > 0$.

Recall that the averaged versions of the gradient and Hessian estimates in (11.101) are (11.55) if at least a locally quadratic map as in (11.97) is considered [151, 200]. Hence, from (11.55) and $z(t)$ in (11.102), we can verify that

$$z_{\rm av}(t) = \frac{1}{\Pi}\int_0^{\Pi} \Gamma M(\lambda) y d\lambda = \Gamma_{\rm av}(t) H \vartheta_{\rm av}(t), \tag{11.106}$$

where $\Pi := 2\pi/\omega$, $\Gamma_{\rm av}(t)$ and $\vartheta_{\rm av}(t)$ denote the average versions of $\Gamma(t)$ and $\vartheta(t)$, respectively. Then, (11.106) can be written in terms of $\tilde{\Gamma}_{\rm av}(t) = \Gamma_{\rm av}(t) - H^{-1}$ as

$$z_{\rm av}(t) = \vartheta_{\rm av}(t) + \tilde{\Gamma}_{\rm av}(t) H \vartheta_{\rm av}(t). \tag{11.107}$$

The second term in the right side of (11.107) is quadratic in $(\tilde{\Gamma}_{\rm av}, \vartheta_{\rm av})$; thus, the linearization of $\Gamma_{\rm av}(t)$ at H^{-1} results in the linearized version of (11.106) or (11.107) given by

$$z_{\rm av}(t) = \vartheta_{\rm av}(t). \tag{11.108}$$

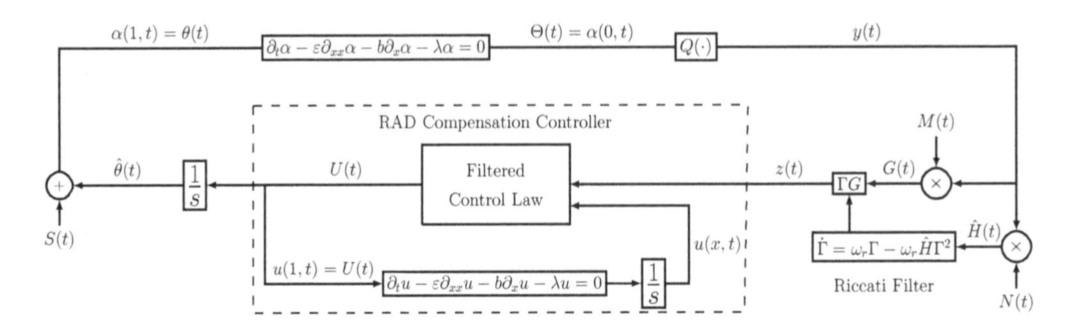

Figure 11.4. *Newton-based ES loop for maximizing higher derivatives $Q^{(n)}(\Theta)$ of unknown maps $y = Q(\Theta)$ with actuation dynamics governed by a RAD-like parabolic PDE.*

Regarding (11.108), we average (11.54) and choose $\bar{K} = -K$ with $K > 0$, such that

$$U_{\rm av}(t) = -Ke^{-\frac{b}{2\epsilon}}\left[\bar{\gamma}(1)z_{\rm av}(t) + \int_0^1 \bar{m}(1-y)\bar{z}_{\rm av}(y,t)dy\right]. \tag{11.109}$$

Analogously to (11.57) and recalling (11.24), we get the filtered average-based infinite-dimensional control law of (11.109) to compensate the reaction-advection-diffusion process:

$$U(t) = \frac{c}{s+c}\left\{-Ke^{-\frac{b}{2\epsilon}}\left[\bar{\gamma}(1)z(t) + \int_0^1 e^{\frac{b}{2\epsilon}y}\bar{m}(1-y)u(y,t)dy\right]\right\}, \quad K > 0, \tag{11.110}$$

where $c > 0$ is an appropriate constant. Figure 11.4 shows an illustrative block diagram of the closed-loop system for the proposed Newton-based ES controller with a RAD-PDE in the input channel.

11.6.2 ▪ Stability Analysis

The closed-loop system consists of the propagated error dynamics (11.10)–(11.13) with the controller (11.110) plus the Hessian's inverse error estimation dynamics (11.105). The following theorem summarizes the stability results of such a feedback system.

Theorem 11.1. *Consider the closed-loop system* (11.10)–(11.13) *with controller* (11.110) *and suppose that Assumption* 11.3 *is satisfied for the map* (11.1). *For a sufficiently large $c > 0$, there exists some $\bar{\omega}(c) > 0$, such that, $\forall \omega > \bar{\omega}$, the closed-loop system with states $\Gamma(t)$, $\vartheta(t)$, $u(x,t)$ has a unique exponentially stable periodic solution in t of period $\Pi := 2\pi/\omega$, denoted by $\Gamma^{\Pi}(t)$, $\vartheta^{\Pi}(t)$, $u^{\Pi}(x,t)$, satisfying, $\forall t \geq 0$,*

$$\left(|\Gamma^{\Pi}(t)|^2 + |\vartheta^{\Pi}(t)|^2 + \int_0^1 [u^{\Pi}(x,t)]^2 dx + \int_0^1 [u_x^{\Pi}(x,t)]^2 dx + [u^{\Pi}(1,t)]^2 + [u^{\Pi}(0,t)]^2\right)^{1/2} \leq \mathcal{O}(1/\omega). \tag{11.111}$$

Furthermore,

$$\limsup_{t\to\infty} |\theta(t) - \theta^*| = \mathcal{O}\left(|a|\exp\left(\sqrt{\frac{\xi+\omega}{\epsilon}}\right) + 1/\omega\right), \tag{11.112}$$

$$\limsup_{t\to\infty} |\Theta(t) - \Theta^*| = \mathcal{O}(|a| + 1/\omega). \tag{11.113}$$

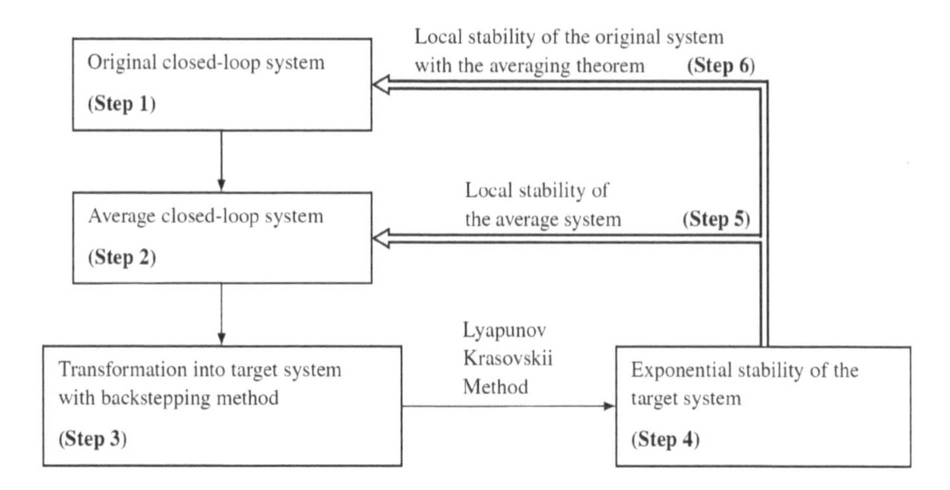

Figure 11.5. *Structure of the stability proof for the closed-loop system.*

Proof. The proof of Theorem 11.1 follows the same ideas carried out to prove Propositions 11.2 and 11.3 in Section 11.5 for the Gradient ES. For the sake of clarity on how we prove the stability and convergence properties, we present the structure of the proof in Figure 11.5, divided in six main steps.

The first step is to write the equations of the closed-loop system in the original variables $\Gamma(t)$, $\vartheta(t)$, $u(x,t)$. Therefore, we expand all the equations to apply the averaging theory in order to obtain the closed-loop average system in the next step. Then, we use a backstepping transformation to first show the stability of the resulting target system with a suitable Lyapunov–Krasovskii functional and further the exponential stability of the average closed-loop system in the original variables, due to the invertibility of the transformation. Invoking the averaging theorem for infinite-dimensional systems [83, Section 2] in Appendix B, we prove the practical exponential stability of the original closed-loop system according to the convergence of the periodic solution. Finally, we can show the convergence of $(\theta(t),\Theta(t))$ to a neighborhood of the extremum (θ^*,Θ^*) for $t\to\infty$.

Step 1: Original Closed-Loop System

Substituting (11.110) into (11.13), we can write the equations of the closed-loop system (11.10)–(11.13) and (11.105) as

$$\dot{\vartheta}(t) = u(0,t), \tag{11.114}$$

$$\partial_t u(x,t) = \epsilon u_{xx}(x,t) + b u_x(x,t) + \lambda u(x,t), \quad x\in[0,1], \tag{11.115}$$

$$u_x(0,t) = 0, \tag{11.116}$$

$$u(1,t) = U(t), \tag{11.117}$$

$$\dot{U}(t) = -cU(t) + c\left\{-Ke^{-\frac{b}{2\epsilon}}\left[\bar{\gamma}(1)z(t) + \int_0^1 e^{\frac{b}{2\epsilon}y}\bar{m}(1-y)u(y,t)dy\right]\right\}, \tag{11.118}$$

$$\dot{\tilde{\Gamma}} = \omega_r(\tilde{\Gamma}+H^{-1})[1-\hat{H}(\tilde{\Gamma}+H^{-1})]. \tag{11.119}$$

Step 2: Average Closed-Loop System

Recalling that $z_{\rm av}(t) = \vartheta_{\rm av}(t)$ from (11.108), we can obtain, for ω sufficiently large, the average closed-loop system of (11.114)–(11.118):

$$\dot{\vartheta}_{\rm av}(t) = u_{\rm av}(0,t), \tag{11.120}$$

$$(u_{\rm av})_t(x,t) = \epsilon (u_{\rm av})_{xx}(x,t) + b(u_{\rm av})_x(x,t) + \lambda u_{\rm av}(x,t), \quad x \in [0,1], \tag{11.121}$$

$$(u_{\rm av})_x(0,t) = 0, \tag{11.122}$$

$$(u_{\rm av})_t(1,t) = -cu_{\rm av}(1,t) - cKe^{-\frac{b}{2\epsilon}}\left[\bar{\gamma}(1)\vartheta_{\rm av}(t) + \int_0^1 e^{\frac{b}{2\epsilon}y}\bar{m}(1-y)u_{\rm av}(y,t)dy\right]. \tag{11.123}$$

On the other hand, the average model for the Hessian's inverse estimation error in (11.105) or, equivalently, (11.119) is

$$\frac{d\tilde{\Gamma}_{\rm av}(t)}{dt} = -\omega_r \tilde{\Gamma}_{\rm av}(t) - \omega_r H \tilde{\Gamma}^2_{\rm av}(t), \tag{11.124}$$

since $\hat{H}_{\rm av} = \frac{1}{\Pi}\int_0^\Pi N(\lambda) y d\lambda = H$.

Step 3: Target System

With some abuse of notation, we also use $w(x,t)$ to denote the average transformed state, and from the backstepping transformation (11.29) in the variables $(u_{\rm av}, w)$,

$$w(x,t) = e^{\frac{b}{2\epsilon}x}u_{\rm av}(x,t) - \int_0^x \tilde{q}(x,y)e^{\frac{b}{2\epsilon}y}u_{\rm av}(y,t)dy - \tilde{\gamma}(x)\vartheta_{\rm av}(t), \tag{11.125}$$

we can state its inverse in the variables $(u_{\rm av}, w)$ as

$$u_{\rm av}(x,t) = e^{-\frac{b}{2\epsilon}x}w(x,t) - e^{-\frac{b}{2\epsilon}x}\int_0^x \tilde{p}(x,y)w(y,t)dy - e^{-\frac{b}{2\epsilon}x}\tilde{\eta}(x)\vartheta_{\rm av}(t), \tag{11.126}$$

with

$$\tilde{q}(x,y) = -K\bar{m}(x-y) = -K\frac{1}{\epsilon}\sqrt{\frac{\epsilon}{\xi}}\sinh\left(\sqrt{\frac{\xi}{\epsilon}}(x-y)\right), \tag{11.127}$$

$$\tilde{\gamma}(x) = -K\bar{\gamma}(x) = -K\cosh\left(\sqrt{\frac{\xi}{\epsilon}}x\right) - \frac{Kb}{2\epsilon}\sqrt{\frac{\epsilon}{\xi}}\sinh\left(\sqrt{\frac{\xi}{\epsilon}}x\right), \tag{11.128}$$

$$\tilde{p}(x,y) = -K\bar{n}(x-y) = \frac{K}{\sqrt{\xi - K}}\sinh\left(\sqrt{\xi - K}(x-y)\right), \tag{11.129}$$

$$\tilde{\eta}(x) = -K\bar{\eta}(x) = K\cosh\left(\sqrt{\xi - K}x\right) + \frac{K}{2\epsilon\sqrt{\xi - K}}\sinh\left(\sqrt{\xi - K}x\right), \tag{11.130}$$

which transforms the average closed-loop system (11.120)–(11.123) into the target system

$$\dot{\vartheta}_{\rm av}(t) = -K\vartheta_{\rm av}(t) + w(0,t), \tag{11.131}$$

$$w_x(x,t) = \epsilon w_{xx}(x,t) - \xi w(x,t), \quad x \in [0,1], \tag{11.132}$$

$$w_x(0,t) = \frac{b}{2\epsilon}w(0,t), \tag{11.133}$$

$$w_t(1,t) = -cw(1,t) + Kw(1,t) - K^2\left[\bar{\eta}(1)\vartheta_{\rm av}(t) + \int_0^1 \bar{n}(1-y)w(y,t)\right]. \tag{11.134}$$

Step 4: Exponential Stability of the Target System in the $\mathcal{H}_1$-Norm

Considering the Lyapunov–Krasovskii functional

$$V(t) = \frac{\vartheta_{\rm av}^2(t)}{2} + \frac{a}{2}\int_0^1 w^2(x,t)dx + \frac{d}{2}\int_0^1 w_x^2(x,t)dx + \frac{e}{2}w^2(1,t) + \frac{f}{2}w^2(0,t) \quad (11.135)$$

and its time derivative along with (11.120)–(11.123) and (11.131)–(11.134), one has

$$\begin{aligned}\dot{V}(t) =& -K\vartheta_{\rm av}^2(t) + \vartheta_{\rm av}(t)w(0,t) - \frac{a}{2}w^2(0,t) - a\epsilon\int_0^1 w_x^2(x,t)dx - a\xi\int_0^1 w^2(x,t)dx \\ &+ a\epsilon w(1,t)w_x(1,t) - d\epsilon\int_0^1 w_{xx}^2(x,t)dx - \frac{db}{2\epsilon}w^2(0,t) + dw_x(1,t)w_t(1,t) \\ &+ d\xi w_x(1,t)w(1,t) + fw(0,t)w_t(0,t),\end{aligned} \quad (11.136)$$

where we choose $e = db/(2\epsilon)$ and $a,d,e,f > 0$. Since the nonnegative terms of $\dot{V}$ are the same as of the time derivative of the Lyapunov–Krasovkii functionals in [65], [169] and Chapter 10, we conclude following similar calculations that there exist $c^* > 0$ and $\mu^* > 0$ such that $\dot{V}(t) \leq -\mu^* V(t)$ and the target system (11.131)–(11.134) is exponentially stable in the $\mathcal{H}_1$-norm

$$\left(|\vartheta_{\rm av}(t)|^2 + \int_0^1 w^2(x,t)dx + \int_0^1 w_x^2(x,t)dx + w^2(1,t) + w^2(0,t)\right)^{1/2} \quad (11.137)$$

for $c > c^*$ sufficiently large in (11.110).

Step 5: Local Exponential Stability of the Average Closed-Loop System

As in the proof of Theorem 10.1 (Step 4) in Chapter 10, the exponential stability property in the variables $(\vartheta_{\rm av}, w)$ can be transferred to the original variables $(\vartheta_{\rm av}, u_{\rm av})$ by using the inverse transformation (11.126). Hence, the average closed-loop system (11.120)–(11.123) is also exponentially stable in the sense of the $\mathcal{H}_1$-norm

$$\left(|\vartheta_{\rm av}(t)|^2 + \int_0^1 u_{\rm av}^2(x,t)dx + \int_0^1 (u_{\rm av})_x^2(x,t)dx + u_{\rm av}^2(1,t) + u_{\rm av}^2(0,t)\right)^{1/2}. \quad (11.138)$$

In addition, the linearized version for the average model of the Hessian's inverse estimation error in (11.124) around $\tilde{\Gamma}_{\rm av}(t) = 0$ is given by

$$\frac{d\tilde{\Gamma}_{\rm av}(t)}{dt} = -\omega_r \tilde{\Gamma}_{\rm av}(t), \quad (11.139)$$

which is also exponentially stable for $\omega_r > 0$.

Hence, the average of the full closed-loop system (11.120)–(11.123) and (11.139) is exponentially stable in the sense of the $\mathcal{H}_1$-norm

$$\left(|\tilde{\Gamma}_{\rm av}(t)|^2 + |\vartheta_{\rm av}(t)|^2 + \int_0^1 u_{\rm av}^2(x,t)dx + \int_0^1 (u_{\rm av})_x^2(x,t)dx + u_{\rm av}^2(1,t) + u_{\rm av}^2(0,t)\right)^{1/2}. \quad (11.140)$$

Step 6: Practical Exponential Stability of the Original Closed-Loop System via Averaging Theorem

The state-transformation $v(x,t) = u(x,t) - U(t)$ transforms the original closed-loop system (11.114)–(11.118) with the controller (11.110) into

$$\dot{\vartheta}(t) = v(0,t) + U(t), \tag{11.141}$$
$$v_t(x,t) = \epsilon v_{xx}(x,t) + b v_x(x,t) + \lambda v(x,t) - \phi(\vartheta, v, U, t), \tag{11.142}$$
$$v_x(0,t) = 0, \tag{11.143}$$
$$v(1,t) = 0, \tag{11.144}$$
$$\dot{U}(t) = \phi(\vartheta, v, U, t), \tag{11.145}$$

with

$$\phi(\vartheta, v, U, t) = -cU(t) - cK\left[\bar{\gamma}(1)z(t) + \int_0^1 e^{\frac{b}{2\epsilon}y}\bar{m}(1-y)(v(y,t)+U(t))dy\right]. \tag{11.146}$$

We write the PDE (11.142) as an evolutionary equation [145] in the Banach space $\mathcal{X} := \mathcal{H}_1([0,1])$

$$\dot{V}(t) = \mathcal{A}V(t) - \tilde{\phi}(\vartheta, V, U, t) \quad \forall t > 0, \tag{11.147}$$

with an appropriate operator $\mathcal{A}$ and function $\tilde{\phi}$ (see Step 5 of Theorem 10.1 in Chapter 10). Thus, we arrive at the infinite-dimensional system of the form

$$\dot{Z}(t) = FZ(t) + J(\omega t, Z(t)), \tag{11.148}$$

with infinite-dimensional state vector $Z(t) = [\vartheta(t)\ V(t)\ U(t)\ \tilde{\Gamma}(t)]^\top$,

$$F = \begin{bmatrix} 0 & \mathcal{B} & 1 & 0 \\ 0 & \mathcal{A} & c\beta & 0 \\ 0 & 0 & -c & 0 \\ 0 & 0 & 0 & \omega_r \end{bmatrix}, \qquad J(\omega t, Z) = \begin{bmatrix} 0 \\ -c\beta K[z(t) + g(Z)] \\ -cK[z(t)+g(Z)] \\ \omega_r H^{-1} - \omega_r \hat{H}(t)\left(\tilde{\Gamma} + H^{-1}\right)^2 \end{bmatrix}, \tag{11.149}$$

where $g(Z) = \mu U(t) + \alpha^\top V(t)$ and $\mu = \int_0^1 e^{\frac{b}{2\epsilon}y}\bar{m}(1-y)dy$ for adequate linear operators $\mathcal{B}$, α, and β. Since $\mathcal{A}$ is a strongly elliptic operator [75], it is also a generator of an analytic semigroup [190, Theorem 2.7]. The $\mathcal{A}$-boundedness of the boundary operator $\mathcal{B}$ can be easily proven and is satisfied with the parameters $\epsilon_1 = \max\{2, \epsilon, b, \lambda\}$, $\epsilon_2 = 0$ in the operator matrix theorem [161] in Appendix B. Hence, the operator matrix F generates an analytic semigroup and $J(\omega t, Z(t))$ satisfies the smoothness conditions of the averaging theorem [83, Section 2], and along with the exponential stability of the average closed-loop system (11.120)–(11.123), all assumptions to apply the averaging theorem for infinite-dimensional systems in [83, Section 2] are satisfied. Thus, the closed-loop system (11.114)–(11.119) with controller (11.110), or simply (11.148), has an exponentially stable periodic solution $Z^\Pi(t)$ with $\|Z^\Pi(t)\| \le \mathcal{O}(1/\omega)$ that satisfies (11.111).

Similar to the procedure carried out in proof of Theorem 10.1 (Step 6) and along with (11.111), it is not difficult to show

$$\limsup_{t\to\infty} |\tilde{\theta}(t)| = \mathcal{O}(1/\omega). \tag{11.150}$$

From (11.14) and Figure 11.4, we can write $\theta(t) - \theta^* = \tilde{\theta}(t) + S(t)$, and recalling $S(t)$ in (11.20) is of order $\mathcal{O}\left(|a|\exp\left(\sqrt{\frac{\xi+\omega}{\epsilon}}\right)\right)$, we finally get with (11.112). The convergence of the propagated actuator $\Theta(t)$ to the optimizer Θ^* is easier to prove. The relation among the propagated estimation error $\vartheta(t)$, the propagated input $\Theta(t)$, and the optimizer of the static map Θ^* is given by

$$\vartheta(t)+a\sin(\omega t)=\Theta(t)-\Theta^*. \tag{11.151}$$

Using (11.151) and taking the absolute value, one has

$$|\Theta(t)-\Theta^*|=|\vartheta(t)+a\sin(\omega(t))|. \tag{11.152}$$

As in the convergence proof of the parameter $\theta(t)$ to the optimal input θ^* above, we write (11.152) in terms of the periodic solution $\vartheta^{\Pi}(t)$ and follow the same steps by applying Young's inequality and $\vartheta(t)-\vartheta^{\Pi}(t)\to 0$ exponentially by the averaging theorem [83, Section 2] in Appendix B. Hence, along with (11.111), we finally get (11.113). □

11.7 ▪ Numerical Simulations

In the next subsections, we present three numerical tests in order to illustrate the applicability of the proposed algorithms for real-time optimization of static maps (11.1) subject to infinite-dimensional input dynamics (11.5)–(11.8). The first test shows a maximization problem (11.2) using a quadratic map (11.4), whereas the second one presents comparison results between gradient and Newton-based ES for flatter maps, highlighting the convergence speed of each one. After that, the third example presents the maximization (11.95) of higher derivatives for an unknown map (11.96) satisfying the quadratic condition in (11.97).

11.7.1 ▪ Gradient ES

The numerical simulation for the gradient ES with actuation dynamics governed by reaction-advection-reaction PDEs is performed with the parameters listed in Table 11.1. The simulation results are shown in Figures 11.6(a) to 11.6(f) for the nonlinear map (11.4).

Table 11.1. *Simulation parameters for ES with actuation dynamics governed by a reaction-advection-diffusion PDE.*

	symbol	description	value
controller parameters	K	controller gain	0.1
	a	perturbation amplitude	0.2
	ω	perturbation frequency [rad/s]	10
	c	controller low-pass corner frequency [rad/s]	20
	$w_{h,l}$	corner frequency high/low-pass filter [rad/s]	1
system parameters	ϵ	diffusion coefficient	1
	b	advection coefficient	1
	λ	reaction coefficient	0.2
	Θ^*	optimizer static map	2
	y^*	optimal value static map	5
	H	Hessian static map	−2

We observe the convergence of $(y(t),\theta(t),\Theta(t))$ to a neighborhood of (y^*,θ^*,Θ^*). The main difference to the pure diffusion case in Chapter 10 is $\theta^*\neq\Theta^*$, shown theoretically in (11.16). The estimated optimal input parameter $\hat{\theta}(t)$ converges to $\theta^*=1.885$, which is the same calculated with (11.16) for the parameters used in the simulation.

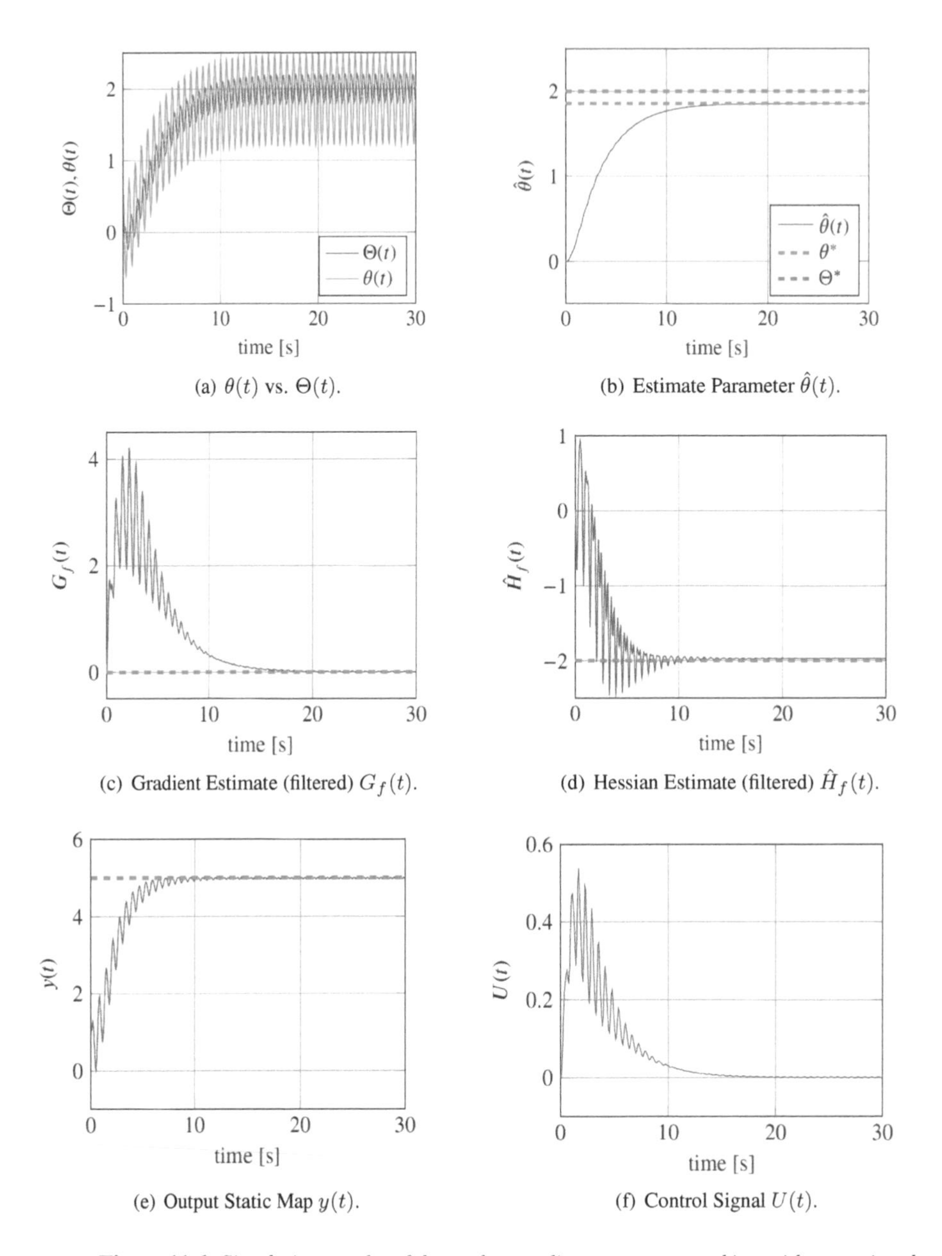

(a) $\theta(t)$ vs. $\Theta(t)$.

(b) Estimate Parameter $\hat{\theta}(t)$.

(c) Gradient Estimate (filtered) $G_f(t)$.

(d) Hessian Estimate (filtered) $\hat{H}_f(t)$.

(e) Output Static Map $y(t)$.

(f) Control Signal $U(t)$.

Figure 11.6. *Simulation results of the scalar gradient extremum seeking with actuation dynamics governed by a reaction-advection-diffusion PDE. The controller parameters are given in Table* 11.1. *In particular, we have used low-pass and washout filters with corner frequencies ω_h and ω_l in order to improve numerical estimates of the gradient $G(t)$ and Hessian $\hat{H}(t)$—denoted by $G_f(t)$ and $\hat{H}_f(t)$, as suggested in* [73, *Figure* 4].

11.7.2 ▪ Gradient ES vs. Newton-Based ES

We still consider a quadratic static map as in the previous section change the Hessian value to $H = -0.2$ in order to obtain a flatter map (11.1)–(11.4). This change allows us to check that faster results can be expected from Newton-based algorithm when compared to classical gradient ES

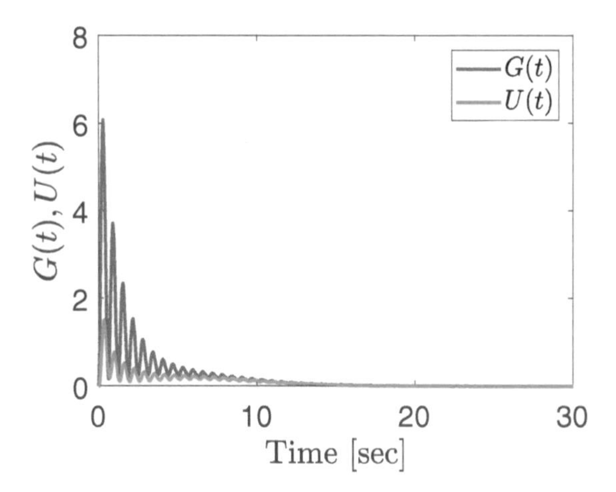

Figure 11.7. *Newton-based ES with a RAD-like PDE in actuation dynamics: Gradient estimate $G(t)$ and control signal $U(t)$.*

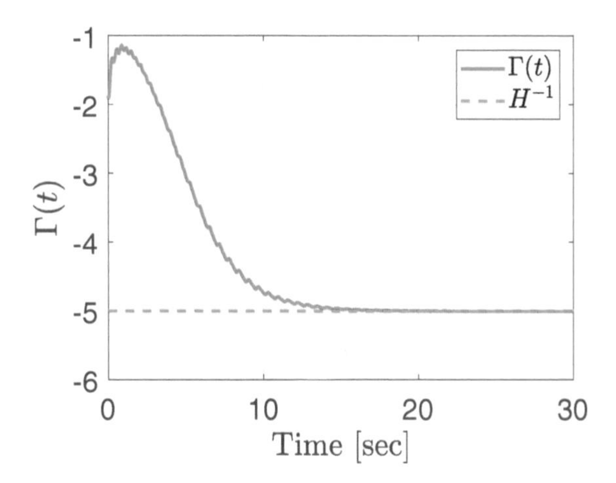

Figure 11.8. *Newton-based ES with a RAD-like PDE in actuation dynamics: Hessian's inverse estimate $\Gamma(t)$.*

design since the latter has convergence rates which are dependent of the Hessian. The remaining parameters are the same shown in Table 11.1. In particular, the estimate of $H^{-1} = -5$ in the proposed Newton-based ES is given by the solution of the Riccati equation (11.103), which was implemented with $\omega_r = 0.5$ and initial condition $\Gamma(0) = -2$.

The simulation results of the closed-loop with Newton-based ES are illustrated in Figures 11.7 to 11.9. The convergence of the gradient and Hessian's inverse estimates are given in Figures 11.7 and 11.8, respectively. In particular, the exact estimation of $\Gamma(t) \to H^{-1}$ in the Newton-based design allows us to cancel the Hessian H and thus guarantee convergence rates that can be arbitrarily assigned by the user. In Figure 11.9, we observe y converging to a neighborhood of the optimum y^*. The improved performance by the Newton-based controller can be seen by comparing with the simulation result of the proposed Gradient ES. As expected, the Newton algorithm converges to the extremum faster than the Gradient scheme, even in the presence of actuation dynamics governed by a RAD-like parabolic PDE.

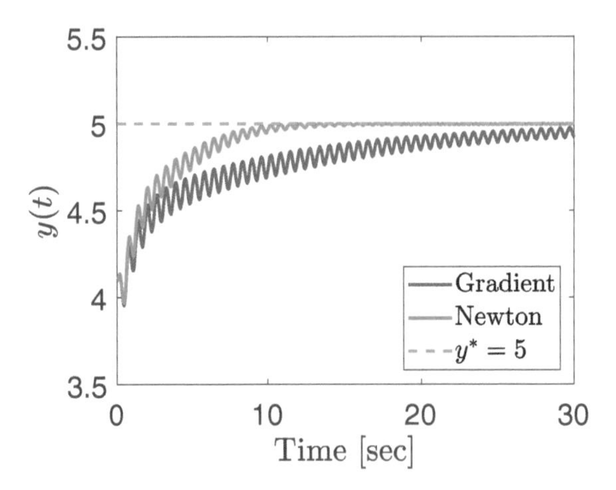

Figure 11.9. *Newton-based ES versus Gradient ES: time response of the output $y(t)$ subject to an actuator RAD-like PDE.*

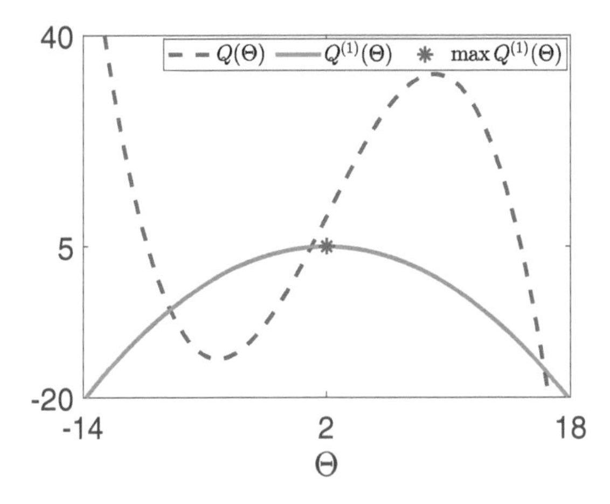

Figure 11.10. *The static map $y = Q(\Theta) = \Theta y^* + (H/6)(\Theta - \Theta^*)^3$ and its first derivative $Q^{(1)}(\Theta) = y^* + (H/2)(\Theta - \Theta^*)^2$, with $H = -0.2$ and maximum at $\Theta^* = 2$.*

11.7.3 ▪ Maximization of Higher Derivatives of Unknown Maps with Newton-Based ES

The numerical test for the Newton-based ES of higher-derivative map (11.4) with actuation dynamics governed by reaction-advection-reaction PDEs is performed with the same parameters listed in Table 11.1, except for the Hessian, which was chosen $H = -0.2$. The original map is chosen as $y = Q(\Theta) = \Theta y^* + (H/6)(\Theta - \Theta^*)^3$, which is shown in Figure 11.10 with a dashed line. The objective is to maximize the first derivative of the map $Q^{(1)}(\Theta)$, which has an extremum at $\Theta^* = 2$—see the solid line in Figure 11.10. The simulation result of the closed-loop with the control law (11.110) is shown in Figure 11.11. As expected, we observe that the variable Θ converges to a neighborhood of the optimum Θ^*.

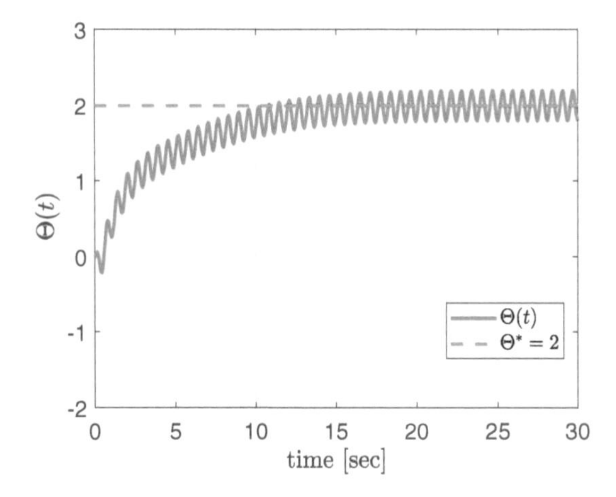

Figure 11.11. *Newton-based ES maximizing the first derivative $Q^{(1)}(\Theta) = y^* + (H/2)(\Theta - \Theta^*)^2$ with a RAD-like PDE in the actuation dynamics of $y = Q(\Theta)$ and $\Theta(t)$ converging to $\Theta^* = 2$.*

11.8 ▪ Application to Bioreactors

So far we have presented the stability and convergence proof of the ES for static maps with actuation dynamics governed by diffusion PDEs—an adaptive semi-model based control concept. We also considered actuation dynamics governed by reaction-advection-diffusion PDEs in the ES loop. In both cases, the actuation dynamics are known. Therefore, we use the term "semi-model based" and not "non-model based" as common for ES. The controllers to compensate the actuation dynamics were derived with the backstepping method and are fed with the perturbation-based gradient and Hessian estimates. The perturbation signal also takes the actuation dynamics into account and compensates that. Stability and convergence to a neighborhood of the extremum have been illustrated with numerical simulations. The convergence to the extremum is only local and depends on the Hessian of the static map; hence, a proper choice of the initial value is important.

In the next two subsections we first present an application idea for this novel control concept: a tubular bioreactor. Second, we address further open problems.

11.8.1 ▪ Potential Application

We consider the optimization problem in Figure 11.12 of a tubular bioreactor, where the goal is to operate the bioreactor at the unknown optimal product rate, e.g., growth of biomass by determining the optimal input, hence the substrate concentration, e.g., glucose concentration. The product concentration x_B is generally not measurable, unlike the product rate.

Since the static map, which determines the product rate depending on the product concentration x_B, is not known or only approximately known, we can apply our introduced control concept. A simple model of a tubular bioreactor is presented by Winkin et al. [242], where the chemical reaction of a reactant R and a product P, given by

$$R \to \lambda P, \tag{11.153}$$

where the stoichiometric coefficient λ of the reaction, is considered. In its simplest form, the model is linear, since the nonlinear reaction term is simplified to a linear kinetic model only depending on the reactant concentration. The control-loop structure to reach and operate the tubular bioreactor at the optimal (highest product rate) is shown in Figure 11.13.

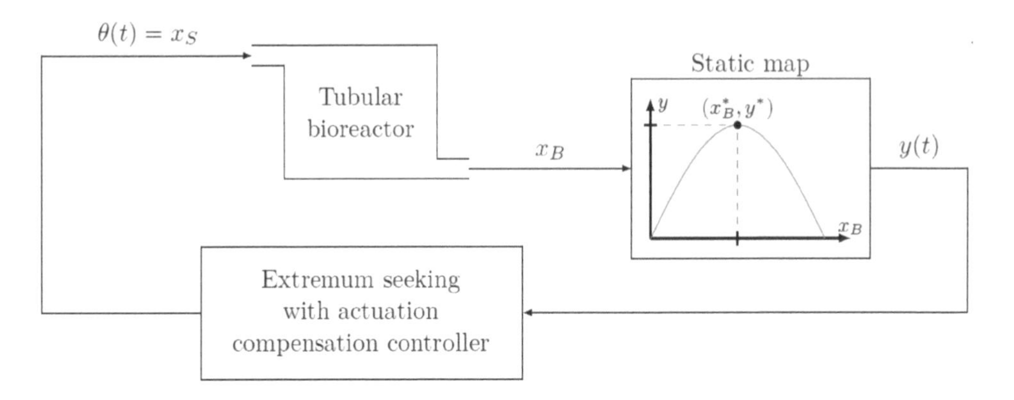

Figure 11.12. *Extremum seeking scheme for tubular bioreactors.*

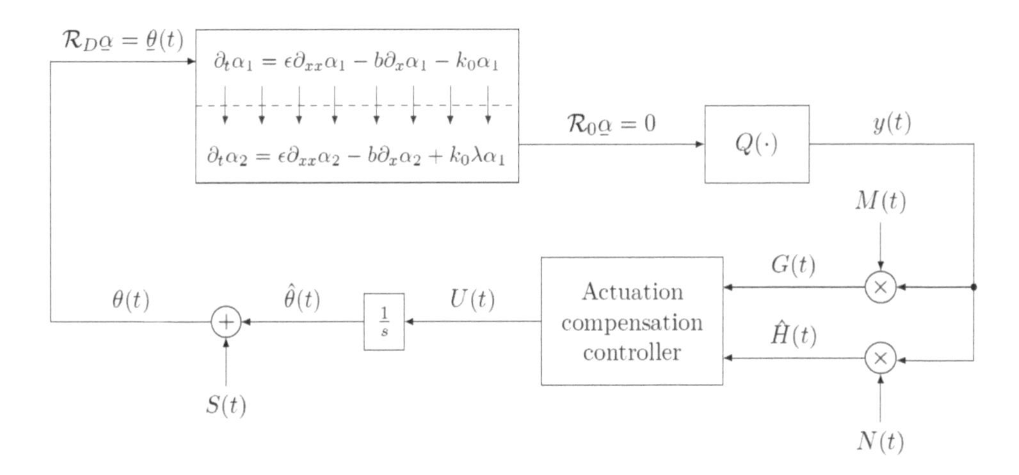

Figure 11.13. *Basic gradient ES scheme for tubular bioreactors modeled by RAD-PDEs.*

The boundary operators of the tubular reactor in Figure 11.13 are defined as

$$
\begin{aligned}
\mathcal{R}_D \underline{\alpha} &= \begin{bmatrix} \epsilon\partial_x \alpha_1(D,t) - b\alpha_1(D,t) \\ \epsilon\partial_x \alpha_2(D,t) - b\alpha_2(D,t) \end{bmatrix} = \begin{bmatrix} -b\theta(t) \\ 0 \end{bmatrix}, \\
\mathcal{R}_0 \underline{\alpha} &= \begin{bmatrix} \epsilon\partial_x \alpha_1(0,t) \\ \epsilon\partial_x \alpha_2(0,t) \end{bmatrix} = \begin{bmatrix} 0 \\ 0 \end{bmatrix}.
\end{aligned}
\tag{11.154}
$$

In a first step, the average-based infinite-dimensional controller to compensate the tubular reactor process, which are the actuator dynamics in this case, has to be derived. Note that the dynamics of the reactor has to be known to apply this control concept. Baccoli et al. [19] considers a similar dynamics, i.e., coupled linear parabolic PDEs, especially reaction-diffusion equations, with boundary input in one variable. But we additionally have an ODE cascade, arising from the integrator in the control loop, which has to be stabilized; see, for example, the propagated error dynamics (11.10)–(11.13). To the best of our knowledge, there is no work that considers a coupled parabolic PDE-ODE cascade with two parabolic PDEs coupled in domain plus boundary input in one variable and boundary measurement in the other variable. The controller derivation and further derivations like the perturbation signal $S(t)$ for this application would go beyond the scope of this book. We emphasize that invoking the averaging theorem for infinite-dimensional systems in Appendix B would work, since the publication [242] showed the analytic semigroup property of the operator, which describes the coupled parabolic PDE system in Figure 11.13. Furthermore, the exponential stability proof of the average system will follow the same steps, but with an extended Lyapunov–Krasovskii functional and more calculation steps. Possibly there will be some restriction on the system parameters ϵ, b, k_0, and λ.

11.8.2 ▪ Open Problems

This chapter and the previous one have considered the dynamics governed by diffusion PDEs and reaction-advection-diffusion PDEs only in the actuation path, and not also in the sensing path, because measurement dynamics described by these phenomena are much harder to deal with in the stability proof. Since we have to look at $\mathcal{D}\{y(t)\}$, where the static map is in general nonlinear, even under the assumption of a quadratic map, the same procedure as in the proof with actuation dynamics does not apply to sensor PDE dynamics.

Our compensators of the parabolic PDE dynamics employ the full state of the PDE. It would be relevant to develop state estimators for such PDEs, under the measurement either at the boundary where the input is being applied or at the boundary where the PDE output feeds into the unknown map. Various Dirichlet and Neumann combinations of actuation, measurement, and connection with the map are possible and worth exploring.

Additionally, we considered only static maps and not dynamical plants as for example in Chapter 5. A drawback of this control concept is the need of the exact knowledge of the actuation dynamics. Therefore, an analytic robustness analysis with respect to uncertain coefficients (such as the diffusion coefficient) is of interest in order to validate the behavior of the numerical simulations in Section 11.7, but now considering parametric uncertainties.

11.9 ▪ Notes and References

In spite of ES being successfully employed in many engineering applications [237], [11], [204], [139], as the authors of [179] point out, the presence of a delay is a major limiting factor in the application of ES in practical situations. Although for ES with distinct input and/or output delays many predictor-based control designs [200], [198] (Chapter 5) have been developed since [179] (Chapter 7), it was not until the results in [179] that a rigorous ES solution has appeared to systems described by Partial Differential Equations (PDEs). In particular, only first-order hyperbolic transport PDEs were originally assumed in [179] to represent pure delays. This key idea has enabled the development of extensions to other classes of PDEs, such as those describing the diffusion phenomenon studied in [65] and [169]. For instance, the former reference considered the Gradient version [129] of the ES algorithm, while the Newton-based method [73] was explored in the latter one.

However, in many chemical and biological processes, the actuation dynamics are not governed by a diffusion process alone, but possibly also by reaction and advection phenomena [211]. For example, tubular reactors [242], fluidized bed and fixed bed reactors [86], or the process of crystal growth from the melt [37] all exhibit this kind of behavior. This motivates us to extend the results in [65], [169] to ES with actuation dynamics governed by RAD-like parabolic PDEs [128].

In this chapter, we first focus on the gradient extremum seeking algorithm [129]. The complete control design employing a compensator for the RAD-like actuation dynamics is developed via backstepping transformation by feeding back the estimates for the gradient and Hessian (first and second derivatives) of the unknown static map to be maximized. Our proofs for local stability of the closed-loop system and the convergence to a small neighborhood of the extremum are based on backstepping methodology for PDE control [128], the construction of a Lyapunov functional, and the use of averaging theorem for infinite-dimensional systems [83]. Second, we consider the Newton-based extremum seeking algorithm [73]. Some of the advantages of the Newton-based approach are its faster responses when compared to the gradient method [73] as well as the possibility of optimizing in real time not only the output of the convex map, but also its arbitrarily higher derivatives, i.e., maximizing map sensitivity [151]. This is a generalization of the results in [200], [198] (Chapter 5) for pure advection PDEs to PDEs that also include diffusion and reaction, besides the advection.

Chapter 12

ES for Wave PDEs

12.1 ▪ From Transport and Heat PDEs to the Wave PDE

After dedicating about a half of this book to compensating PDE dynamics that are first order in both time and space (transport), and two chapters dedicated to PDEs that are first order in time but second order in space (RAD), it would be inexcusable to stop there and not explore ES in the presence of dynamics that are second order in both space and time, namely, the wave PDEs. So, in this chapter we develop compensation of wave dynamics in the actuation path of an unknown map.

As the reader shall observe, compensating a wave PDE is more complex than compensating a transport PDE but less complex than compensating a heat PDE. This may seem counterintuitive—that an increase in the number of derivatives in time reduces the complexity. But the reason for the wave PDE being less difficult to compensate for than the heat PDE is that the wave equation is, in fact, a second-order PDE system that can be represented as a coupled system of two transport (first-order) PDEs and that this coupling can be severed by control, leaving essentially only one transport PDE (i.e., only a delay) to compensate for. The details are a bit more complex than the explanation here implies, but the essence is as stated.

So, in this chapter, we develop extremum seeking algorithms for *wave dynamics* connected in cascade with a static scalar map to be optimized (see Figure 12.1). A distributed-parameter-based control law using the backstepping approach and Neumann actuation is initially proposed. Local exponential stability as well as practical convergence to an arbitrarily small neighborhood of the unknown extremum point is guaranteed by employing Lyapunov–Krasovskii functionals and averaging theory in infinite dimension. Extensions for wave equations with Dirichlet actuation, anti-stable wave PDEs, as well as the design for the delay-wave PDE cascade are also discussed. Numerical simulations illustrate the theoretical results.

12.2 ▪ Example for Drilling Control

The objective of this motivating section is just to bring a potential connection of the proposed ES strategy to a real-world application, while the main focus of the chapter is to pursue designs and a rigorous stability analysis of ES feedback subject to infinite-dimensional actuation dynamics of the wave PDE type.

In this sense, a common type of instability in oil drilling is the friction-induced stick-slip oscillation (see [26] and references therein), which results in torsional vibrations of the drill-string and can severely damage the drilling facilities (see Figure 2 from [1]).

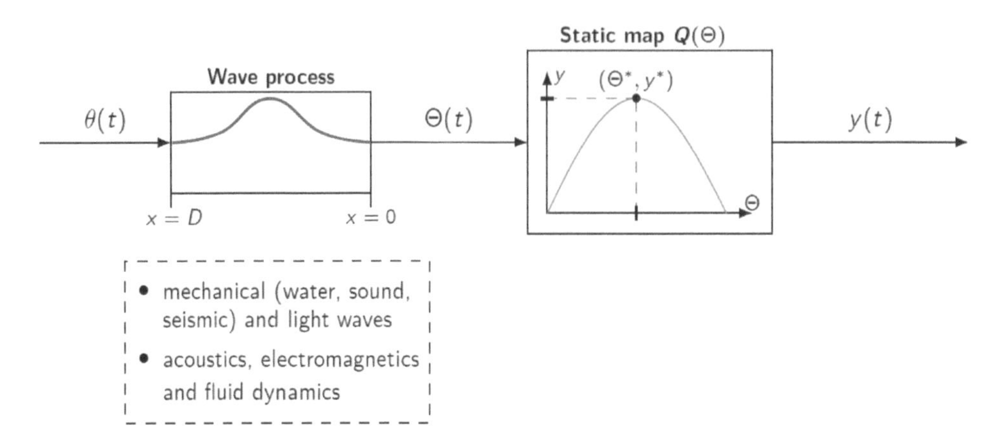

Figure 12.1. *Cascade of wave PDE with a static map $y(\Theta(t)) = Q(\Theta(t))$. The extremum $y(t) = y^*$ is achieved for $\Theta(t) = \Theta^*$. Wave PDEs are used generally to model different sort of processes such as mechanical, acoustics, electromagnetic, and fluid dynamics.*

The picture in Figure 12.2 shows a modern land-based drilling rig. The tower operates like the derrick of a crane: the traveling block is connected by several steel drill lines with one attached to the deadline anchor and the other being spooled on a drum controlled by AC induction motors. Another electric motor, called Top Drive, is connected to the travelling block. The Top Drive is used to rotate the drill-string, a set of hundreds of drill pipes (about 30 feet long each) that conducts the Bore Hole Assembly (BHA). The BHA contains several sensors (pressure,

Figure 12.2. *Picture showing the topside of a drilling rig. Used with permission of the American Society of Mechanical Engineers from T. R. Oliveira and M. Krstić, Extremum seeking feedback with wave partial differential equation compensation, ASME Journal of Dynamic Systems, Measurement, and Control,* 143 (2021), 041002; *permission conveyed through Copyright Clearance Center, Inc.*

temperature, and vibration, among others) and the drill bit itself. There are several different types of drill bit design and materials, adequate for drilling different geological formations.

In analogy, the rig operates similarly to a drill press, but with a drill bit which is several inches wide (4 to 36 inches is a common range) and up to several miles long (an onshore well can be as shallow as 200 yards or as deep as 2 miles). By rotating this drill-string and using its weight to generate an axial force, the BHA mills the rocks, drilling the well. Because of the small diameter when compared to its length, the drill-string is subject to axial and torsional effects, much like a flexible rod. Because of this elasticity, the force and velocity propagation can be modeled by *wave equations*.

In this particular model, the actuation is the velocity of the travelling block, i.e., the axial velocity of the drill-string on the surface. Although not considered here, the rotational velocity also influences the rate of penetration (ROP) in a real scenario. The model output is the Weight On Hook (WOH) which somewhat models the Weight On Bit (WOB). The WOB estimates the contact between the drill bit and the rock formation and it is the downhole boundary condition to be controlled. In [1], the authors have discussed the feasibility of controlling the hook load to optimize ROP while drilling.

The key point that enables such an approach is the concept of bit foundering [1], i.e., the fact that ROP tapers off (and sometimes starts decreasing) with increasing weight on bit past the foundering point. This makes the static mapping between ROP and weight on bit upwards convex in an interval around the foundering point. This transfers to an upwards convex static mapping between the equilibrium hook load set point and feed rate. Consequently, these signals can be used as the plant input and output for the design of a drilling control system. Hence, this physical application motivates our ES scheme for static maps with actuation dynamics described by wave PDEs, as depicted in Figure 12.1.

12.3 ▪ Problem Formulation

We start our presentation in Section 12.3.1 where wave PDEs are considered in the input of the static maps we want to optimize in real-time.

12.3.1 ▪ Wave PDE under Neumann Actuation

We consider actuation dynamics which are described by a *wave PDE process*, where the actuator $\theta(t)$ and the propagated actuator $\Theta(t)$ are given by

$$\Theta(t) = \partial_x \alpha(0,t), \tag{12.1}$$

$$\partial_{tt}\alpha(x,t) = \partial_{xx}\alpha(x,t), \quad x \in [0,D], \tag{12.2}$$

$$\alpha(0,t) = 0, \tag{12.3}$$

$$\partial_x \alpha(D,t) = \theta(t), \tag{12.4}$$

with the domain length D being arbitrary, but known.

The Neumann actuation choice $\partial_x \alpha(D,t) = \theta(t)$ is first pursued because this is a natural physical choice since $\partial_x \alpha(D,t)$ corresponds to a force on the string's boundary. In Section 12.6, we address the case of an alternative actuation choice, Dirichlet actuation via $\alpha(D,t) = \theta(t)$.

The measurement is defined by the unknown static map with input (12.1), such that

$$y(t) = Q(\Theta(t)). \tag{12.5}$$

For the sake of simplicity, we assume the following.

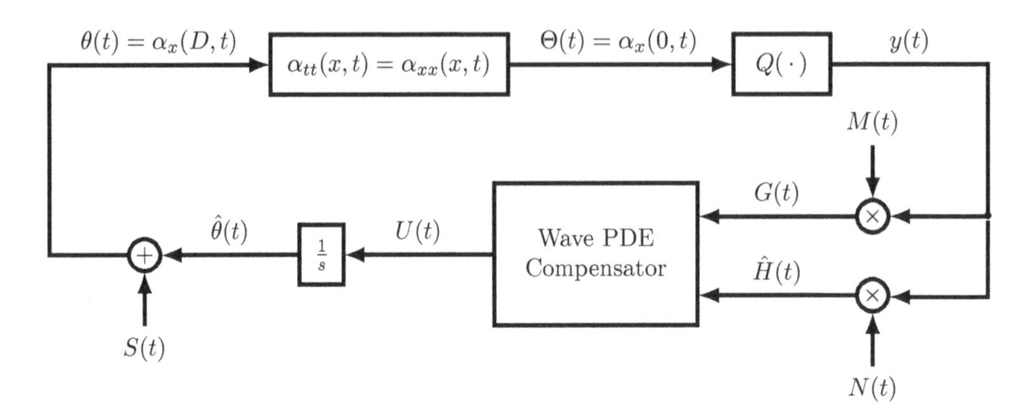

Figure 12.3. *Gradient ES with actuation dynamics governed by a wave PDE with compensating controller* (12.40) *and additive and multiplicative perturbation signals* (12.23), (12.26), *and* (12.27), *respectively.*

Assumption 12.1. *The unknown nonlinear static map is locally quadratic, i.e.,*

$$Q(\Theta) = y^* + \frac{H}{2}(\Theta - \Theta^*)^2, \tag{12.6}$$

in the neighborhood of the extremum, where besides $\Theta^* \in \mathbb{R}$ and $y^* \in \mathbb{R}$ being both unknown, the scalar $H < 0$ is the unknown Hessian of the static map.

Hence, the output of the static map is given by

$$y(t) = y^* + \frac{H}{2}(\Theta(t) - \Theta^*)^2. \tag{12.7}$$

Combining the above actuation dynamics and the basic ES scheme, further adapting the proposed scheme in [179], the closed-loop ES with actuation dynamics governed by a wave PDE system under wave compensation controller is shown in Figure 12.3.

12.3.2 ▪ Probing and Demodulation Signals

As in the basic ES scheme, we define the unknown optimal input θ^* of $\theta(t)$ in (12.4) with respect to the static map (12.7) and the wave process, with the relation $\Theta^* = \theta^*$. Since our goal is to find the unknown optimal input θ^*, we define the estimation error

$$\tilde{\theta}(t) := \hat{\theta}(t) - \theta^*, \tag{12.8}$$

where $\hat{\theta}(t)$ is the estimate of θ^*. To make (12.8) consistent with the optimizer of the static map Θ^*, we introduce the propagated estimation error $\vartheta(t) := \hat{\Theta}(t) - \Theta^*$ through the wave PDE domain

$$\vartheta(t) := \partial_x \bar{\alpha}(0,t), \tag{12.9}$$

$$\partial_{tt}\bar{\alpha}(x,t) = \partial_{xx}\bar{\alpha}(x,t), \quad x \in [0,D], \tag{12.10}$$

$$\bar{\alpha}(0,t) = 0, \tag{12.11}$$

$$\partial_x \bar{\alpha}(D,t) = \tilde{\theta}(t). \tag{12.12}$$

From the control loop in Figure 12.3, we get

$$\dot{\hat{\theta}}(t) = U(t). \tag{12.13}$$

Taking the time derivative of (12.9)–(12.12) and with the help of (12.8) and (12.13), the propagated error dynamics is written as the following cascade of a wave PDE and ODE (integrator) with Neumann interconnection [215]:

$$\dot{\vartheta}(t) = \partial_x u(0,t), \tag{12.14}$$
$$\partial_{tt} u(x,t) = \partial_{xx} u(x,t), \quad x \in [0,D], \tag{12.15}$$
$$u(0,t) = 0, \tag{12.16}$$
$$\partial_x u(D,t) = U(t), \tag{12.17}$$

where $\dot{\tilde{\theta}}(t) = \dot{\hat{\theta}}(t)$, since θ^* is constant, and

$$u(x,t) = \partial_t \bar{\alpha}(x,t). \tag{12.18}$$

As in extremum seeking without actuation through a wave PDE domain, the perturbation signal $S(t)$ should add $a\sin(\omega t)$ to $\Theta(t)$, thus compensating the wave process. Hence, $a\sin(\omega t)$ with perturbation amplitude a and frequency ω is applied as follows:

$$S(t) := \partial_x \beta(D,t), \tag{12.19}$$
$$\partial_{tt} \beta(x,t) = \partial_{xx} \beta(x,t), \quad x \in [0,D], \tag{12.20}$$
$$\beta(0,t) = 0, \tag{12.21}$$
$$\partial_x \beta(0,t) = a\sin(\omega t). \tag{12.22}$$

Equations (12.19)–(12.22) describe a *trajectory generation problem* as in [128, Chapter 12]. The explicit solution of (12.19) is given by

$$S(t) = a\cos(\omega D)\sin(\omega t), \tag{12.23}$$

since $\beta(x,t) = \frac{a}{\omega}\sin(\omega x)\sin(\omega t)$. The relation among the propagated estimation error $\vartheta(t)$, the propagated input $\Theta(t)$, and the optimizer of the static map Θ^* is given by

$$\vartheta(t) + a\sin(\omega t) = \Theta(t) - \Theta^*, \tag{12.24}$$

which can be easily proven since

$$\bar{\alpha}(x,t) = \alpha(x,t) - \beta(x,t) - \Theta^* \tag{12.25}$$

for $x = D$ and considering $\theta(t) = \hat{\theta}(t) + S(t)$ along with the solutions of (12.1)–(12.4), (12.9)–(12.12), and (12.19)–(12.22). It remains to define the dither signal $N(t)$, which is used to estimate the Hessian of the static map by multiplying it with the output $y(t)$ of the static map. In [73], the Hessian estimate is derived as

$$\hat{H}(t) = N(t)y(t) \quad \text{with} \quad N(t) = -\frac{8}{a^2}\cos(2\omega t). \tag{12.26}$$

Note that the dither signal $M(t)$, to estimate the gradient, is the same as in the basic ES such that

$$G(t) = M(t)y(t) \quad \text{with} \quad M(t) = \frac{2}{a}\sin(\omega t). \tag{12.27}$$

12.4 ▪ Controller Design and Closed-Loop System

In this section, we present the proposed filtered boundary control with perturbation-based (averaging-based) estimates of the gradient and Hessian used for wave compensation in the closed-loop extremum seeking feedback of Figure 12.3.

12.4.1 ▪ Wave Compensation by Means of Hessian's Estimation

We consider the PDE-ODE cascade (12.14)–(12.17). As in [215], we use the backstepping transformation

$$w(x,t)=u(x,t)-\int_0^x l(x,\sigma)u_t(\sigma,t)d\sigma-\gamma(x)\vartheta(t), \tag{12.28}$$

with the gain kernels

$$l(x,\sigma)=\gamma(x-\sigma), \tag{12.29}$$

$$\gamma(x)=\bar{K}[0\ \ I]e^{Ax}[I\ \ 0]^T, \quad A=\begin{pmatrix}0 & 0\\ I & 0\end{pmatrix}, \tag{12.30}$$

which transforms (12.14)–(12.17) into the target system

$$\dot{\vartheta}(t)=\bar{K}\vartheta(t)+w_x(0,t), \quad \bar{K}<0, \tag{12.31}$$

$$\partial_{tt}w(x,t)=\partial_{xx}w(x,t), \quad x\in[0,D], \tag{12.32}$$

$$w(0,t)=0, \tag{12.33}$$

$$w_x(D,t)=-\bar{c}w_t(D,t), \quad \bar{c}>0. \tag{12.34}$$

Since the target system (12.31)–(12.34) is exponentially stable (to be shown), the control signal (12.17) which compensates the wave process can be obtained by evaluating the backstepping transformation (12.28) at $x=D$ as

$$U(t)=\bar{c}\left[\bar{K}u(D,t)-\partial_t u(D,t)\right]+\rho(D)\vartheta(t)+\int_0^D \rho(D-\sigma)\partial_t u(\sigma,t)d\sigma, \tag{12.35}$$

where $\rho(s)=\bar{K}[0\ \ I]e^{As}[0\ \ I]^T$. However, the proposed control law in (12.35) is not applicable directly, because we have no measurement on $\vartheta(t)$. Thus, we recall an important result of [73]: the averaged version of the gradient (12.27) and Hessian (12.26) estimates are

$$G_{\mathrm{av}}(t)=H\vartheta_{\mathrm{av}}(t), \qquad \hat{H}_{\mathrm{av}}(t)=H \tag{12.36}$$

if a quadratic map as in (12.6) is considered. For the proof of (12.36), see [73]. Regarding (12.36), we average (12.35) and choose $\bar{K}=KH$ with $K>0$, such that

$$\begin{aligned}U_{\mathrm{av}}(t)=\;&\bar{c}[KHu_{\mathrm{av}}(D,t)-\partial_t u_{\mathrm{av}}(D,t)]+\bar{\rho}(D)KH\vartheta_{\mathrm{av}}(t)\\ &+KH\int_0^D \bar{\rho}(D-\sigma)\partial_t u_{\mathrm{av}}(\sigma,t)d\sigma,\end{aligned} \tag{12.37}$$

with

$$\bar{\rho}(s)=[0\ \ I]e^{As}[0\ \ I]^T, \quad A=\begin{pmatrix}0 & 0\\ I & 0\end{pmatrix}. \tag{12.38}$$

By plugging the averaged estimate (12.36) into (12.37), we obtain

$$U_{\rm av}(t) = \bar{c}[KHu_{\rm av}(D,t) - \partial_t u_{\rm av}(D,t)] + \bar{\rho}(D)KG_{\rm av}(t) + KH\int_0^D \bar{\rho}(D-\sigma)\partial_t u_{\rm av}(\sigma,t)d\sigma . \tag{12.39}$$

Due to technical reasons in the application of the averaging theorem for infinite-dimensional systems [83, Section 5] in the stability proof, we introduce a low-pass filter to the controller so that $U(t)$ can be treated as a state variable. Finally, we get the average-based infinite-dimensional control law to compensate the wave process:

$$U(t) = \frac{c}{s+c}\left\{\bar{c}\left[K\hat{H}(t)u(D,t) - \partial_t u(D,t)\right] + \bar{\rho}(D)KG(t) + K\hat{H}(t)\int_0^D \bar{\rho}(D-\sigma)\partial_t u(\sigma,t)d\sigma\right\}, \tag{12.40}$$

where $c > 0$ is chosen later. For notation convenience, we mix again the time and frequency domain in (12.40), where the low-pass filter acts as an operator on the term between braces.

12.4.2 ▪ Closed-Loop System

Substituting (12.40) into (12.17), we can write the closed-loop system (12.14)–(12.17) as

$$\dot{\vartheta}(t) = \partial_x u(0,t), \tag{12.41}$$

$$\partial_{tt} u(x,t) = \partial_{xx} u(x,t), \quad x \in [0,D], \tag{12.42}$$

$$u(0,t) = 0, \tag{12.43}$$

$$\partial_x u(D,t) = \frac{c}{s+c}\left\{\bar{c}\left[K\hat{H}(t)u(D,t) - \partial_t u(D,t)\right] + \bar{\rho}(D)KG(t) + K\hat{H}(t)\int_0^D \bar{\rho}(D-\sigma)\partial_t u(\sigma,t)d\sigma\right\}. \tag{12.44}$$

The availability of Lyapunov functionals via backstepping transformation [128] permits the stability analysis in the next section of the complete feedback system (12.41)–(12.44) with a representation of the form of cascade ODE-PDE equations and the infinite-dimensional control law.

12.5 ▪ Stability Analysis

The following theorem provides the stability and local convergence properties of the closed-loop system.

Theorem 12.1. *Consider the system in Figure* 12.3 *with the dynamic system being represented by the nonlinear quadratic map* (12.7)*, satisfying Assumption* 12.1*, in cascade with the actuation dynamics governed by the wave PDE in* (12.1)–(12.4)*. For a sufficiently large* $c > 0$*, there exists some* $\bar{\omega}(c) > 0$*, such that,* $\forall \omega > \bar{\omega}$*, the closed-loop system* (12.41)–(12.44) *with states* $\vartheta(t)$*,* $u(x,t)$ *has a unique locally exponentially stable periodic solution in* t *of period* $\Pi := 2\pi/\omega$*, denoted by* $\vartheta^{\Pi}(t), u^{\Pi}(x,t)$*, satisfying,* $\forall t \geq 0$*,*

$$\left(\left|\vartheta^{\Pi}(t)\right|^2 + \left\|\partial_x u^{\Pi}(t)\right\|^2 + \left\|\partial_t u^{\Pi}(t)\right\|^2 + \left|\partial_x u^{\Pi}(D,t)\right|^2\right)^{1/2} \leq \mathcal{O}(1/\omega). \tag{12.45}$$

Furthermore,

$$\limsup_{t\to\infty} |\theta(t)-\theta^*| = \mathcal{O}(a+1/\omega), \tag{12.46}$$

$$\limsup_{t\to\infty} |\Theta(t)-\Theta^*| = \mathcal{O}(a+1/\omega), \tag{12.47}$$

$$\limsup_{t\to\infty} |y(t)-y^*| = \mathcal{O}\left(a^2+1/\omega^2\right). \tag{12.48}$$

Proof. The proof is structured into **Step 1** to **6**, analogously to what has been done in [65] (and Theorem 10.1, Chapter 10).

Step 1: *Average Closed-Loop System*

The average version of the system (12.41)–(12.44) for ω large is

$$\dot{\vartheta}_{\text{av}}(t) = \partial_x u_{\text{av}}(0,t), \tag{12.49}$$

$$\partial_{tt} u_{\text{av}}(x,t) = \partial_{xx} u_{\text{av}}(x,t), \quad x\in[0,\, D], \tag{12.50}$$

$$u_{\text{av}}(0,t) = 0, \tag{12.51}$$

$$\begin{aligned}\frac{d}{dt}\partial_x u_{\text{av}}(D,t) = -c\partial_x u_{\text{av}}(D,t) - c\Bigg[&\bar{c}[KHu_{\text{av}}(D,t)-\partial_t u_{\text{av}}(D,t)] + \bar{\rho}(D)KH\vartheta_{\text{av}}(t)\\ &+KH\int_0^D \bar{\rho}(D-\sigma)\partial_t u_{\text{av}}(\sigma,t)d\sigma\Bigg],\end{aligned} \tag{12.52}$$

where the low-pass filter is represented in state-space form.

Step 2: *Backstepping Transformation into the Target System*

With some abuse of notation, we also use $w(x,t)$ to denote the average transformed state. From (12.28), the backstepping transformation

$$w(x,t) = u_{\text{av}}(x,t) - \int_0^x \gamma(x-\sigma)\partial_t u_{\text{av}}(\sigma,t)d\sigma - \gamma(x)\vartheta_{\text{av}}(t) \tag{12.53}$$

maps the average closed-loop system (12.49)–(12.52) into the exponentially stable target system (shown in Step 3)

$$\dot{\vartheta}_{\text{av}}(t) = KH\vartheta_{\text{av}}(t) + w_x(0,t), \tag{12.54}$$

$$\partial_{tt} w(x,t) = \partial_{xx} w(x,t), \quad x\in[0,\, D], \tag{12.55}$$

$$w(0,t) = 0, \tag{12.56}$$

$$w_t(D,t) = -\frac{1}{\bar{c}}\partial_x w(D,t), \quad \bar{c}>0, \tag{12.57}$$

$$\partial_x w(D,t) = -\frac{1}{c}\partial_t\partial_x u_{\text{av}}(D,t). \tag{12.58}$$

The target system (12.54)–(12.58) can be derived by plugging in the inverse backstepping transformation [215]:

$$u_{\text{av}}(x,t) = w(x,t) + KHn(x)\vartheta_{\text{av}}(t) + KH\int_0^x n(x-\sigma)w_t(\sigma,t)d\sigma, \tag{12.59}$$

with

$$n(x) = [0\ \ I]e^{\bar{A}x}[I\ \ 0]^T, \quad \bar{A} = \begin{pmatrix} 0 & (KH)^2 \\ I & 0 \end{pmatrix}, \tag{12.60}$$

into the average closed-loop system (12.49)–(12.52). Additionally taking the time derivative of the backstepping transformation (12.53) along with (12.52) and its inverse (12.59), we arrive at (12.58) reminding one that $\dot{U}_{\mathrm{av}}(t) = \partial_t \partial_x u_{\mathrm{av}}(D,t)$ or, equivalently,

$$\begin{aligned} \partial_t w_x(D,t) &= -cw_x(D,t) + KHw(D,t) \\ &+ (KH)^2 n(D)\vartheta_{\mathrm{av}}(t) + (KH)^2 \int_0^D n(D-\sigma) w_t(\sigma,t) d\sigma . \end{aligned} \tag{12.61}$$

Step 3: *Exponential Stability of the Target System*

We start by introducing the system norms

$$\Omega(t) = \|\partial_x u_{\mathrm{av}}(t)\|^2 + \|\partial_t u_{\mathrm{av}}(t)\|^2 + |\vartheta_{\mathrm{av}}(t)|^2 + |\partial_x u_{\mathrm{av}}(D,t)|^2, \tag{12.62}$$

$$\Xi(t) = \|w_x(t)\|^2 + \|w_t(t)\|^2 + |\vartheta_{\mathrm{av}}(t)|^2 + |w_x(D,t)|^2. \tag{12.63}$$

To prove the stability of the closed-loop system, we consider the Lyapunov–Krasovskii functional

$$V(t) = \frac{\vartheta_{\mathrm{av}}^2(t)}{2} + \bar{a}E(t) + \frac{b}{2} w_x^2(D,t), \tag{12.64}$$

where the parameters $\bar{a}$, $b > 0$ are to be chosen later and the functional $E(t)$ is defined by [215]:

$$\begin{aligned} E(t) &= \frac{1}{2}\left(||w_x(t)||^2 + ||w_t(t)||^2\right) \\ &+ \delta \int_0^D (1+y)\, w_x(y,t) w_t(y,t) dy , \end{aligned} \tag{12.65}$$

where $\delta > 0$ is also a parameter to be chosen later. We observe that

$$\theta_1 \Xi \leq V \leq \theta_2 \Xi, \tag{12.66}$$

where

$$\theta_1 = \min\left\{\frac{1}{2}, \frac{\bar{a}}{2}[1-\delta(1+D)], \frac{b}{2}\right\}, \tag{12.67}$$

$$\theta_2 = \min\left\{\frac{1}{2}, \frac{\bar{a}}{2}[1+\delta(1+D)], \frac{b}{2}\right\}. \tag{12.68}$$

We choose

$$0 < \delta < \frac{1}{1+D} \tag{12.69}$$

in order to ensure that θ_1 and θ_2 are nonnegative and so the Lyapunov function V in (12.64) is positive definite. Next, we compute the time derivative of $E(t)$:

$$\begin{aligned} \dot{E}(t) &= -\frac{\delta}{2}\left[||w_x(t)||^2 + ||w_t(t)||^2 + w_x(0,t)^2\right] \\ &+ \frac{\delta}{2}(1+D)\left[w_t(D,t)^2 + w_x(D,t)^2\right] \\ &+ w_x(D,t) w_t(D,t) . \end{aligned} \tag{12.70}$$

From (12.57), we substitute the boundary condition $w_x(D,t) = -\bar{c}w_t(D,t)$ and get

$$\dot{E}(t) = -\frac{\delta}{2}\left[||w_x(t)||^2 + ||w_t(t)||^2 + w_x(0,t)^2\right] - \left[\bar{c} - \delta\frac{1+D}{2}\left(1+\bar{c}^2\right)\right] w_t(D,t)^2. \quad (12.71)$$

Choosing now

$$\delta < \frac{2\bar{c}}{(1+D)(1+\bar{c}^2)}, \quad (12.72)$$

we have that the constant between brackets in the second term of (12.71) is positive. Now, computing the complete derivative of $V(t)$, associated with the solution of the target system (12.54)-(12.58), we have

$$\dot{V}(t) = KH\vartheta_{\text{av}}^2(t) + \vartheta_{\text{av}}(t)w_x(0,t) + \bar{a}\dot{E}(t) + bw_x(D,t)\partial_t w_x(D,t). \quad (12.73)$$

By applying Young's inequality to the second term in (12.73), we can write

$$\dot{V}(t) \leq \frac{KH}{2}\vartheta_{\text{av}}^2(t) + \left[\frac{1}{2|KH|} - \bar{a}\frac{\delta}{2}\right] w_x(0,t)^2 - \bar{a}\frac{\delta}{2}\left[||w_x(t)||^2 + ||w_t(t)||^2\right] + bw_x(D,t)\partial_t w_x(D,t). \quad (12.74)$$

By choosing

$$\bar{a} \geq \frac{1}{\delta|KH|}, \quad (12.75)$$

we now obtain

$$\dot{V}(t) \leq \frac{KH}{2}\vartheta_{\text{av}}^2(t) - \bar{a}\frac{\delta}{2}\left[||w_x(t)||^2 + ||w_t(t)||^2\right] + bw_x(D,t)\partial_t w_x(D,t). \quad (12.76)$$

Rigorously, substituting (12.61) into (12.76), the last term in the RHS of (12.76) can be treated analogously to what has been done in [65] and [169] for diffusion processes (Chapters 10 and 11) or even as carried out in [179] with pure delays (Chapter 7), when the parameter $c > 0$ is assumed sufficiently large. The first term $-cw_x(D,t)$ in the RHS of (12.61) when plugged into (12.76) results in $-bcw_x^2(D,t)$. Intuitively, $w_x(D,t) \to 0$ as $c \to +\infty$ according to (12.58) since $w_x(x,t)$ is at least bounded from (12.71) and, consequently, (12.76) becomes negative definite. After lengthy calculations, applying Young's, Poincaré's, Agmon's and Cauchy–Schwarz's inequalities (more than one) and the help of integration by parts, we conclude that there exists $c^* > 0$ (depending on KH and D) such that, for $c > c^*$ sufficiently large in (12.40), one has

$$\dot{V}(t) \leq -\eta V(t), \qquad \eta > 0, \quad (12.77)$$

and the target system (12.54)–(12.58) is exponentially stable in the norm

$$\left(|\vartheta_{\text{av}}(t)|^2 + \|w_x(t)\|^2 + \|w_t(t)\|^2 + |w_x(D,t)|^2\right)^{1/2},$$

i.e., in the transformed variables $(\vartheta_{\text{av}}, w)$.

Step 4: *Exponential Stability Estimate (in the $\mathcal{H}_1$-norm) of the Average Closed-Loop System*

In the previous step, from (12.77), we arrive at the estimate

$$V(t) \le e^{-\eta t} V(0), \quad \eta > 0. \tag{12.78}$$

In order to prove stability of the closed-loop system in its original variables $(\vartheta_{\rm av}, u_{\rm av})$ from (12.78), we provide inequalities relating the variables $u(x,t)$ and $w(x,t)$. From the inverse transformation (12.59), we obtain that

$$\begin{aligned}\partial_x u_{\rm av}(x,t) &= w_x(x,t) + \int_0^x \phi'(x-y)w(y,t)dy \\ &\quad + \int_0^x n'(x-y)w_t(y,t)dy + \psi(x)'\vartheta_{\rm av}(t), \end{aligned} \tag{12.79}$$

$$\begin{aligned}\partial_t u_{\rm av}(x,t) &= w_t(x,t) + \int_0^x \phi(x-y)w_t(y,t)dy \\ &\quad + \int_0^x n'(x-y)w(y,t)dy + \psi(x)KH\vartheta_{\rm av}(t). \end{aligned} \tag{12.80}$$

Applying Poincaré's, Young's, and Cauchy–Schwarz's inequalities, we get

$$\begin{aligned}||\partial_x u_{\rm av}(t)||^2 &\le \alpha_1||w_x(t)||^2 + \alpha_2||w_t(t)||^2 + \alpha_3|\vartheta_{\rm av}(t)|^2, \\ ||\partial_t u_{\rm av}(t)||^2 &\le \beta_1||w_x(t)||^2 + \beta_2||w_t(t)||^2 + \beta_3|\vartheta_{\rm av}(t)|^2, \end{aligned} \tag{12.81}$$

where

$$\alpha_1 = 4(1+4D^3||\phi'||^2), \tag{12.82}$$
$$\alpha_2 = 4D||n'||^2, \tag{12.83}$$
$$\alpha_3 = 4||\psi'||^2, \tag{12.84}$$
$$\beta_1 = 4||n'||^2 \tag{12.85}$$
$$\beta_2 = 4(1+4D^3||\phi||^2), \tag{12.86}$$
$$\beta_3 = 4||\psi KH||^2. \tag{12.87}$$

Applying (12.81), we obtain

$$\Omega(t) \le \theta_4 \Xi(t), \tag{12.88}$$

where

$$\theta_4 = \max\{\alpha_1+\beta_1, \alpha_2+\beta_2, \alpha_3+\beta_3\}. \tag{12.89}$$

With the help of time and space derivatives of (12.28)—see [215] for more details—and applying again Poincaré's, Young's, and Cauchy–Schwarz's inequalities, we obtain the following inequalities:

$$||w_x(t)||^2 \le a_1||\partial_x u_{\rm av}(t)||^2 + a_2||\partial_t u_{\rm av}(t)||^2 + a_3|\vartheta_{\rm av}(t)|^2, \tag{12.90}$$
$$||w_t(t)||^2 \le b_1||\partial_x u_{\rm av}(t)||^2 + b_2||\partial_t u_{\rm av}(t)||^2 + b_3|\vartheta_{\rm av}(t)|^2. \tag{12.91}$$

By means of them, we obtain

$$\theta_3 \Xi \le \Omega(t), \tag{12.92}$$

where

$$\theta_3 = \frac{1}{\max\{a_1+b_1, a_2+b_2, a_3+b_3+1\}}. \tag{12.93}$$

With the help of (12.66), (12.78), (12.88), and (12.92), we get

$$\Omega(t) \leq \frac{\theta_1\theta_3}{\theta_2\theta_4}\Omega(0)e^{-\eta t}, \tag{12.94}$$

which completes the proof of exponential stability of the average closed-loop system in the sense of the norm (12.62) in the variables $(\vartheta_{\text{av}}, u_{\text{av}})$.

Step 5: *Invoking the Averaging Theorem for Infinite-Dimensional Systems*

The main idea is to convert the wave equation in the closed-loop system (12.41)–(12.44) to a cascade of two first-order transport equations which convect in opposite directions. To achieve this, we define the Riemann transformations [26]

$$\bar{\zeta}(x,t) = u_t(x,t) + u_x(x,t), \tag{12.95}$$

$$\bar{\omega}(x,t) = u_t(x,t) - u_x(x,t) \tag{12.96}$$

together with their inverses given by

$$u_t(x,t) = \frac{\bar{\zeta}(x,t) + \bar{\omega}(x,t)}{2}, \tag{12.97}$$

$$u_x(x,t) = \frac{\bar{\zeta}(x,t) - \bar{\omega}(x,t)}{2}. \tag{12.98}$$

Defining

$$\xi(t) = u(0,t) \tag{12.99}$$

and noting that $\bar{\zeta}(0,t) = \dot{\xi}(t) + u_x(0,t)$, the system (12.41)–(12.44) is written as

$$\dot{\vartheta}(t) = \frac{\bar{\zeta}(0,t) - \bar{\omega}(0,t)}{2}, \tag{12.100}$$

$$\dot{\xi}(t) = \frac{\bar{\zeta}(0,t) + \bar{\omega}(0,t)}{2}, \tag{12.101}$$

$$\bar{\omega}_t(x,t) = -\bar{\omega}_x(x,t), \tag{12.102}$$

$$\bar{\omega}(0,t) = \bar{\zeta}(0,t) - 2u_x(0,t), \tag{12.103}$$

$$\bar{\zeta}_t(x,t) = \bar{\zeta}_x(x,t), \tag{12.104}$$

$$\bar{\zeta}(D,t) = U(t) + u_t(D,t), \tag{12.105}$$

$$\dot{U}(t) = -cU(t) + c\left\{\bar{c}\left[K\hat{H}(t)u(D,t) - u_t(D,t)\right]\right.$$
$$\left. + \bar{\rho}(D)KG(t) + K\hat{H}(t)\int_0^D \bar{\rho}(D-\sigma)u_t(\sigma,t)d\sigma\right\}. \tag{12.106}$$

In this new framework, the wave phenomenon is represented as the cascade of two transport PDEs, with two ODE (simple integrators), being driven by the two PDEs. The ODE (12.100) plays a central role since it was made stable by feedback, which is applied through the transport equation (12.104) at the boundary $x = D$ in this new representation form. The second ODE (12.101) is already at the equilibrium $\xi(t) \equiv 0$ by the choice of the boundary condition (12.43) and (12.99). A second transport phenomenon (12.102) is also present, in the opposite direction, accounting for the reflection of the wave at $x = 0$.

The subsystem (12.100) and (12.104)–(12.106) can also be interpreted as an input-delay ordinary differential equation, delayed by D units of time, followed by a stable transport phenomenon (12.102)–(12.103). Hence, the closed-loop system (12.100)–(12.106) is rewritten as

$$\dot{z}(t) = f(\omega t, z_t), \tag{12.107}$$

where $z(t) = [\tilde{\vartheta}(t), \xi(t), U(t)]^T$ is the state vector and the distributed terms are encompassed by $z_t(r) = z(t+r)$ for $-D \leq r \leq 0$, with f being an appropriate continuous functional, such that the averaging theorem by Hale and Lunel [83, Section 5] in Appendix A can be directly applied, considering $\omega = 1/\epsilon$.

Since we have already proved the origin of the average closed-loop system with wave PDE is exponentially stable according to (12.94) and $\xi_{\text{av}}(t) \equiv 0$ from (12.43) and (12.99), by applying the averaging theorem for infinite-dimensional systems developed in [83, Section 5] in Appendix A, for ω sufficiently large, we conclude (12.49)–(12.52) has a unique exponentially stable periodic solution around its equilibrium satisfying (12.45).

Step 6: *Convergence to a Neighborhood of the Extremum (θ^*, Θ^*, y^*)*

Applying Agmon's, Poincaré's, and Young's inequalities on the LHS of (12.9), along with (12.9)–(12.12), we have

$$\tilde{\theta}^2(t) \leq (1+2D)\vartheta(t)^2 + (4D^2+1)\|\bar{\alpha}_x(t)\|^2. \tag{12.108}$$

Using the again the Poincaré inequality

$$\|\bar{\alpha}_x(t)\|^2 \leq 2\bar{\alpha}_x(0,t)^2 + 4D^2\|\bar{\alpha}_{xx}(t)\|^2, \tag{12.109}$$

with $\bar{\alpha}_x(0,t) = \vartheta(t)$ from (12.9) and $\partial_t u(x,t) = \partial_{tt}\bar{\alpha}(x,t) = \partial_{xx}\bar{\alpha}(x,t)$ from (12.10) and (12.18), we can rewrite (12.108) as

$$\tilde{\theta}^2(t) \leq (3+2D+8D^2)\vartheta(t)^2 + (16D^4+4D^2)\|\partial_t u(t)\|^2. \tag{12.110}$$

Inequality (12.110) can be written in terms of the periodic solution $\vartheta^{\Pi}(t)$ and $\partial_t u^{\Pi}(x,t)$ as follows:

$$\begin{aligned}
&\limsup_{t\to\infty} |\tilde{\theta}(t)|^2 \\
&= \limsup_{t\to\infty} \left\{(3+2D+8D^2)|\vartheta(t)+\vartheta^{\Pi}(t)-\vartheta^{\Pi}(t)|^2\right\} \\
&\quad + \limsup_{t\to\infty} \left\{(16D^4+4D^2)\|\partial_t u(t)+\partial_t u^{\Pi}(t)-\partial_t u^{\Pi}(t)\|^2\right\}.
\end{aligned} \tag{12.111}$$

By applying Young's inequality and some algebra, it holds that

$$|\vartheta(t)+\vartheta^{\Pi}(t)-\vartheta^{\Pi}(t)|^2 \leq \sqrt{2}\left(|\vartheta(t)-\vartheta^{\Pi}(t)|^2+|\vartheta^{\Pi}(t)|^2\right).$$

The same procedure can be applied to the second term on the RHS of (12.111). From the averaging theorem [83, Section 5] in Appendix A, we know $\vartheta(t)-\vartheta^{\Pi}(t)\to 0$ and $\partial_t u(t)-\partial_t u^{\Pi}(t)\to 0$, exponentially as $t \to +\infty$. Hence,

$$\limsup_{t\to\infty} |\tilde{\theta}(t)|^2 = \limsup_{t\to\infty} \left\{ \sqrt{2}(3+2D+8D^2)|\vartheta^{\Pi}(t)|^2 \right\}$$
$$+ \limsup_{t\to\infty} \left\{ \sqrt{2}(16D^4+4D^2)\|\partial_t u^{\Pi}(t)\|^2 \right\}. \tag{12.112}$$

Along with (12.45), it is not difficult to show

$$\limsup_{t\to\infty} |\tilde{\theta}(t)| = \mathcal{O}(1/\omega). \tag{12.113}$$

From (12.8) and Figure 12.3, we can write $\theta(t) - \theta^* = \tilde{\theta}(t) + S(t)$, and recalling $S(t)$ in (12.23) is of order $\mathcal{O}(a)$, we finally get with (12.46). The convergence of the propagated actuator $\Theta(t)$ to the optimizer Θ^* is easier to prove. Using (12.24) and taking the absolute value, one has

$$|\Theta(t) - \Theta^*| = |\vartheta(t) + a\sin(\omega(t))|. \tag{12.114}$$

As in the convergence proof of the parameter $\theta(t)$ to the optimal input θ^* above, we write (12.114) in terms of the periodic solution $\vartheta^{\Pi}(t)$ and follow the same steps by applying Young's inequality and recalling that $\vartheta(t) - \vartheta^{\Pi}(t) \to 0$ exponentially according to the averaging theorem [83, Section 5] (Appendix A). Hence, along with (12.45), we finally get (12.47). To show the convergence of the output $y(t)$ of the static map to the optimal value y^*, we replace $\Theta(t) - \Theta^*$ in (12.7) by (12.24) and take the absolute value:

$$|y(t) - y^*| = \left| \frac{H}{2} \left[\vartheta(t) + a\sin(\omega(t))\right]^2 \right|. \tag{12.115}$$

Expanding the quadratic term in (12.115) and applying Young's inequality to the resulting equation, one has $|y(t)-y^*| = \left|H\left[\vartheta(t)^2 + a^2\sin^2(\omega t)\right]\right|$. As before, we add and subtract the periodic solution $\vartheta^{\Pi}(t)$ and use the convergence of $\vartheta(t) - \vartheta^{\Pi}(t) \to 0$ via averaging theorem [83, Section 5] (Appendix A). Hence, again with (12.45), we get (12.48). □

In the next sections, alternative extremum seeking schemes for wave PDE compensation are presented using different actuation topology (Dirichlet rather than Neumann)—Section 12.6—as well as feedback loops and cascades with wave PDEs—Section 12.7.

12.6 ▪ Extremum Seeking for Wave PDEs with Dirichlet Actuation

We now return to the problem formulation as in Section 12.3.1, but with a distinct choice for the actuated variable in the wave PDE.

For the sake of clarity, in what follows we set $c \to +\infty$ so that we can focus our attention on the design of the new feedback controllers and do not distract the readers with technical details of including the low-pass filter in the closed-loop, as done in (12.40). However, as discussed in [179], it is worth mention that the inclusion of the filter $c/(s+c)$ is a fundamental step which allows us to apply the average theorem for infinite-dimensional systems [83] and complete the proof of our theorems.

In this sense, we consider the system

$$\Theta(t) = \alpha(0,t), \tag{12.116}$$
$$\partial_{tt}\alpha(x,t) = \partial_{xx}\alpha(x,t), \quad x \in [0,D], \tag{12.117}$$
$$\partial_x \alpha(0,t) = 0, \tag{12.118}$$
$$\alpha(D,t) = \theta(t), \tag{12.119}$$

where instead of the Neumann actuation choice, $\partial_x \alpha(D,t) = \theta(t)$, we consider Dirichlet actuation, $\alpha(D,t) = \theta(t)$.

In the Dirichlet case, (12.14)–(12.17) can be rewritten as

$$\dot{\vartheta}(t) = u(0,t), \tag{12.120}$$
$$\partial_{tt} u(x,t) = \partial_{xx} u(x,t), \quad x \in [0,D], \tag{12.121}$$
$$\partial_x u(0,t) = 0, \tag{12.122}$$
$$u(D,t) = U(t). \tag{12.123}$$

The control law is obtained as [119, Chapter 16.4]

$$\begin{aligned} U(t) = c_0 K \hat{H}(t) \int_0^D (1 + M(D-y)) u(y,t) dy \\ + K\hat{H}(t) \int_0^D m(D-y) u_t(y,t) dy - c_0 \int_0^D u_t(y,t) dy \\ + KM(D)G(t), \end{aligned} \tag{12.124}$$

where $\hat{H}(t)$ and $G(t)$ are the same signals given by (12.26) and (12.27), respectively. Moreover, the functions $M(\cdot)$ and $m(\cdot)$ are simply

$$M(s) = [I \ \ 0] e^{As} [I \ \ 0]^T, \quad A = \begin{pmatrix} 0 & 0 \\ I & 0 \end{pmatrix}, \tag{12.125}$$

$$m(s) = \int_0^s M(\xi) d\xi. \tag{12.126}$$

On the other hand, the signal $S(t)$ in (12.19)–(12.22) must be redesigned according to the *trajectory generation problem* [128, Chapter 12] for Dirichlet actuation, leading us to

$$S(t) = \frac{a}{2} \left[\sin(\omega(t+D)) + \sin(\omega(t-D)) \right]. \tag{12.127}$$

Notice that the trajectory generation problem for the wave PDE with Dirichlet actuation is a particular case of the wave equation with Kelvin–Voigt damping studied in [208]:

$$\epsilon \partial_{tt} u(x,t) = (1 + d\partial_t) \partial_{xx} u(x,t), \quad \partial_x u(0,t) = 0, \quad \epsilon, d > 0. \tag{12.128}$$

Thus, the resulting average target system for the wave equation with Dirichlet actuation becomes

$$\dot{\vartheta}_{\text{av}}(t) = KH\vartheta_{\text{av}}(t) + w(0,t), \tag{12.129}$$
$$\partial_{tt} w(x,t) = \partial_{xx} w(x,t), \quad x \in [0, D], \tag{12.130}$$
$$\partial_x w(0,t) = c_0 \partial_t w(0,t), \quad c_0 > 0, \tag{12.131}$$
$$w(D,t) = 0, \tag{12.132}$$

which is exponentially stable with the following spectrum (decay rate):

$$\text{eig}\{KH\} \bigcup \left\{ -\frac{1}{2} \ln \left| \frac{1+c_0}{1-c_0} \right| + j\frac{\pi}{D} \begin{cases} n + \frac{1}{2}, & 0 \leq c_0 < 1 \\ n, & c_0 > 1 \end{cases} \right\}, \quad n \in \mathbb{Z}. \tag{12.133}$$

The fact that we employ Dirichlet actuation at $x = D$ prevents us from applying damping at this end; hence, we induce boundary damping at the opposite end.

Although we do not detail the analysis here, a similar stability theorem can be proved as for the case of Neumann actuation in Theorem 12.1.

Remark 12.1. The control law (12.124) can be expressed in terms of $U(t)$ rather than $u(x,t)$, according to [119, Chapter 16.5]:

$$\begin{aligned} U(t) = & \frac{1}{1+c_0\tanh(Ds)}[KM(D)G(t)] \\ & + \frac{1}{1+e^{-2s}+c_0(1-e^{-2Ds})}[K\hat{H}(t)\mathcal{D}(t)], \end{aligned} \tag{12.134}$$

where

$$\mathcal{D}(t) = \int_{t-D}^{t}\rho(t-\tau)U(\tau)d\tau - \int_{t-2D}^{t-D}\rho(\tau-t+2D)U(\tau)d\tau, \tag{12.135}$$

with $\rho(\tau) = c_0 + (1+c_0)M(\tau)$.

12.7 ▪ Extremum Seeking for PDE-PDE Cascades and Feedback Loops

In this section, we deal with distinct feedback loops and cascades of hyperbolic PDEs with wave equations. We tackle with (a) an anti-stable wave PDE described as two coupled transport (delay) equations and (b) an anti-stable wave PDE with input delay (hyperbolic transport equation). A deeper study will be carried out in Chapter 14 when cascades involving PDEs of different nature (hyperbolic AND parabolic) will be considered as well.

In this section, our presentation is restricted to derive the extremum seeking feedback laws $U(t)$, obtain the additive dither $S(t)$ which solves the trajectory generation problem [128, Chapter 12], and briefly make statements of the closed-loop properties. Unlike the developments of Section 12.5, we forgo a detailed Lyapunov stability analysis and the associated estimates for the transformations between the plant and the target systems. Again, a complete and detailed analysis for PDE-PDE cascades is conducted in Chapter 14.

12.7.1 ▪ Anti-stable Wave PDE as a Feedback Loop of Two Transport (Delay) PDEs

In this section, we consider wave PDEs with boundary anti-damping:

$$\Theta(t) = \alpha(0,t), \tag{12.136}$$

$$\partial_{tt}\alpha(x,t) = \partial_{xx}\alpha(x,t), \quad x\in[0,D], \tag{12.137}$$

$$\partial_x\alpha(0,t) = -q\partial_t\alpha(0,t), \quad |q|\neq 1, \tag{12.138}$$

$$\partial_x\alpha(D,t) = \theta(t), \tag{12.139}$$

where θ is the input appearing in the form of Neumann actuation and the output Θ as a Dirichlet sensor. The damping coefficient $q\geq 0$ is considered to be known. As discussed in [119, Chapter 19], when $q=1$, the real part of the plant (infinite) eigenvalues is $+\infty$ while, for $q\neq 1$ and $q\geq 0$, the real part is positive but finite. Consequently, we can rewrite (12.9)–(12.12) into

$$\vartheta(t) := \bar{\alpha}(0,t), \tag{12.140}$$

$$\partial_{tt}\bar{\alpha}(x,t) = \partial_{xx}\bar{\alpha}(x,t), \quad x\in[0,D], \tag{12.141}$$

$$\partial_x\bar{\alpha}(0,t) = -q\partial_t\bar{\alpha}(0,t), \quad |q|\neq 1, \tag{12.142}$$

$$\partial_x\bar{\alpha}(D,t) = \tilde{\theta}(t) \tag{12.143}$$

and, recalling that $u(x,t) = \partial_t \bar{\alpha}(x,t)$, (12.14)–(12.17) such that

$$\dot{\vartheta}(t) = u(0,t), \tag{12.144}$$
$$\partial_{tt} u(x,t) = \partial_{xx} u(x,t), \quad x \in [0,D], \tag{12.145}$$
$$\partial_x u(0,t) = -q \partial_t u(0,t), \quad |q| \neq 1, \tag{12.146}$$
$$\partial_x u(D,t) = U(t). \tag{12.147}$$

Analogous to the developments carried out in Step 5 for the proof of Theorem 12.1, we use again the Riemann-like transformations [26] but now for the system (12.136)–(12.139) such that we can redefine

$$\bar{\zeta}(x,t) = \frac{1}{1-q}[u_t(x,t) + u_x(x,t)], \tag{12.148}$$
$$\bar{\omega}(x,t) = \frac{1}{1+q}[u_t(x,t) - u_x(x,t)], \tag{12.149}$$

with their corresponding inverses given by

$$u_t(x,t) = \frac{(1-q)\bar{\zeta}(x,t) + (1+q)\bar{\omega}(x,t)}{2}, \tag{12.150}$$
$$u_x(x,t) = \frac{(1-q)\bar{\zeta}(x,t) - (1+q)\bar{\omega}(x,t)}{2}. \tag{12.151}$$

Hence, (12.144)–(12.147) is reshaped as

$$\textbf{ODE:} \quad \dot{\vartheta}(t) = \bar{\zeta}(0,t), \tag{12.152}$$
$$\textbf{PDE 1:} \begin{cases} \bar{\zeta}_t(x,t) = \bar{\zeta}_x(x,t), \quad x \in [0,D], \\ \bar{\zeta}(D,t) = \frac{1}{1-q}[U(t) + u_t(D,t)], \end{cases} \tag{12.153}$$
$$\textbf{PDE 2:} \begin{cases} \bar{\omega}_t(x,t) = -\bar{\omega}_x(x,t), \quad x \in [0,D], \\ \bar{\omega}(0,t) = \bar{\zeta}(0,t). \end{cases} \tag{12.154}$$

Applying the backstepping transformations [26]

$$w(x,t) = \bar{\zeta}(x,t) - KH \int_0^x \bar{\zeta}(\sigma,t) d\sigma - KH\vartheta(t), \tag{12.155}$$
$$\varpi(x,t) = \bar{\omega}(x,t) + KH \int_0^x \bar{\omega}(\sigma,t) d\sigma - KH\vartheta(t), \tag{12.156}$$

the resulting average target system for (12.152)–(12.154) is

$$\dot{\vartheta}_{\text{av}}(t) = KH\vartheta_{\text{av}}(t) + w_{\text{av}}(0,t), \tag{12.157}$$
$$\partial_t w_{\text{av}}(x,t) = \partial_x w_{\text{av}}(x,t), \quad w_{\text{av}}(0,t) = \varpi_{\text{av}}(0,t), \tag{12.158}$$
$$\partial_t \varpi_{\text{av}}(x,t) = -\partial_x \varpi_{\text{av}}(x,t), \tag{12.159}$$
$$w_{\text{av}}(D,t) = -\frac{1}{c}\partial_t \partial_x u_{\text{av}}(D,t), \tag{12.160}$$

which is exponentially stable for $c > 0$ sufficiently large. This result is not difficult to prove since (12.157) is exponentially Input-to-State Stable (ISS) [99] with respect to $w_{\text{av}}(0,t)$. On the other hand, $w_{\text{av}}(x,t)$ is finite-time stable with respect to $\varpi_{\text{av}}(0,t)$ and $\varpi_{\text{av}}(x,t)$ is asymptotically stable for $c \to +\infty$ (or $w_{\text{av}}(D,t) \to 0$) [26].

Plugging (12.155) with $x = D$ into (12.160), we can write $\dot{U}_{\rm av}(t) = -cw_{\rm av}(D,t)$ according to (12.147) as

$$\dot{U}_{\rm av}(t) = -c\bar{\zeta}_{\rm av}(D,t) + cK\left[H\vartheta_{\rm av}(t) + H\int_0^D \bar{\zeta}_{\rm av}(x,t)dx\right]. \tag{12.161}$$

Plugging the average version of (12.153) and (12.148) into (12.161), one has

$$\begin{aligned}\dot{U}_{\rm av}(t) = &-\frac{c}{1-q}[U_{\rm av}(t) + \partial_t u_{\rm av}(D,t)] \\ &+ cK\left[H\vartheta_{\rm av}(t) + H\int_0^D \frac{1}{1-q}[\partial_t u_{\rm av}(x,t) + \partial_x u_{\rm av}(x,t)]dx\right],\end{aligned} \tag{12.162}$$

which can be rewritten as

$$\begin{aligned}\dot{U}_{\rm av}(t) = &-\bar{c}U_{\rm av}(t) - \bar{c}\partial_t u_{\rm av}(D,t) \\ &+ \bar{c}K\left[(1-q)H\vartheta_{\rm av}(t) + H\int_0^D [\partial_t u_{\rm av}(x,t) + \partial_x u_{\rm av}(x,t)]dx\right]\end{aligned} \tag{12.163}$$

or, equivalently, as

$$\begin{aligned}U_{\rm av}(t) = \frac{\bar{c}}{s+\bar{c}}\Bigg\{&-\partial_t u_{\rm av}(D,t) + K\Bigg[(1-q)H\vartheta_{\rm av}(t) \\ &+ H\int_0^D [\partial_t u_{\rm av}(x,t) + \partial_x u_{\rm av}(x,t)]dx\Bigg]\Bigg\},\end{aligned} \tag{12.164}$$

where $\bar{c} = \frac{c}{1-q}$. Recalling that $G_{\rm av}(t) = H\vartheta_{\rm av}(t)$ and $\hat{H}_{\rm av}(t) = H$ from (12.36), one can finally obtain the implementable control law

$$\begin{aligned}U(t) = \frac{\bar{c}}{s+\bar{c}}\Bigg\{&-u_t(D,t) + K\Bigg[(1-q)G(t) \\ &+ \hat{H}(t)\int_0^D [u_t(x,t) + u_x(x,t)]dx\Bigg]\Bigg\},\end{aligned} \tag{12.165}$$

with $\hat{H}(t)$ and $G(t)$ defined by (12.26) and (12.27), respectively.

The last stage is to obtain the additive dither $S(t)$ solving the following trajectory generation problem:

$$S(t) := \partial_x\beta(D,t), \tag{12.166}$$
$$\partial_{tt}\beta(x,t) = \partial_{xx}\beta(x,t), \quad x \in [0,D], \tag{12.167}$$
$$\partial_x\beta(0,t) = -q\partial_t\beta(0,t), \quad |q| \neq 1, \tag{12.168}$$
$$\beta(0,t) = a\sin(\omega t). \tag{12.169}$$

The explicit solution of (12.166) is given according to [128, Chapter 12]

$$S(t) = -a\omega\sin(\omega D)\cos(\omega t) - aq\omega\cos(\omega D)\cos(\omega t), \tag{12.170}$$

since $\beta(x,t) = a\cos(\omega x)\sin(\omega t) - aq\sin(\omega x)\cos(\omega t)$.

Remark 12.2. A similar result can be provided for the Dirichlet actuation $\alpha(D,t)=\theta(t)$, $\bar{\alpha}(D,t)=\tilde{\theta}(t)$, and $u(D,t)=U(t)$ by redefining the Riemann transformations (12.148) and (12.149) to

$$\bar{\zeta}(x,t)=\frac{1}{1-q}\left[\bar{\alpha}_t(x,t)+\bar{\alpha}_x(x,t)\right], \tag{12.171}$$

$$\bar{\omega}(x,t)=\frac{1}{1+q}\left[\bar{\alpha}_t(x,t)-\bar{\alpha}_x(x,t)\right]. \tag{12.172}$$

In this case, the control law can be expressed in terms of $\bar{\alpha}(x,t)$ rather than $u(x,t)$, according to

$$U(t)=\frac{\bar{c}}{s+\bar{c}}\left\{-\partial_x\bar{\alpha}(D,t)+K\left[(1-q)G(t)\right.\right.$$
$$\left.\left.+\hat{H}(t)\int_0^D[u(x,t)+\partial_x\bar{\alpha}(x,t)]dx\right]\right\}, \tag{12.173}$$

with $\hat{H}(t)$ and $G(t)$ defined by (12.27) and (12.26), respectively. Recalling (12.25), the control law (12.173) is indeed implementable since $\partial_x\bar{\alpha}(x,t)$ can be written in terms of measurable signals

$$\partial_x\bar{\alpha}(x,t)=\partial_x\alpha(x,t)-\partial_x\beta(x,t), \tag{12.174}$$

and the integral in (12.173) is given by

$$\begin{aligned}\int_0^D &\partial_x\bar{\alpha}(x,t)dx=\bar{\alpha}(D,t)-\bar{\alpha}(0,t)\\ &=\alpha(D,t)-\beta(D,t)-\Theta^*-\alpha(0,t)+\beta(0,t)+\Theta^*\\ &=\theta(t)-\beta(D,t)-\Theta(t)+a\sin(\omega t).\end{aligned} \tag{12.175}$$

The term $\beta(x,t)$ is defined like the trajectory generation problem (12.166)–(12.169) but replacing (12.166) by $S(t):=\beta(D,t)$, which leads to [128, Chapter 12]

$$S(t)=a\cos(\omega D)\sin(\omega t)-aq\sin(\omega D)\cos(\omega t), \tag{12.176}$$

since $\beta(x,t)=a\cos(\omega x)\sin(\omega t)-aq\sin(\omega x)\cos(\omega t)$. Notice that (12.176) exactly matches (after some manipulations) to (12.127) when $q=0$ is set in (12.176).

12.7.2 ▪ Anti-stable Wave PDE with Input Delay

Inspired by [122], we discuss the extension of our ES approach for the same anti-stable wave PDE of the previous section but now with a delayed input. This is particularly an important problem since Datko et al. [47] showed that standard feedback laws for wave equations have a zero robustness margin to the introduction of a delay in the feedback loop.

Hence, we consider the delay-wave cascade system

$$\Theta(t)=\alpha(0,t), \tag{12.177}$$

$$\partial_{tt}\alpha(x,t)=\partial_{xx}\alpha(x,t),\quad x\in[0,1], \tag{12.178}$$

$$\partial_x\alpha(0,t)=-q\partial_t\alpha(0,t),\quad |q|\neq 1, \tag{12.179}$$

$$\alpha(1,t)=\theta(t-D), \tag{12.180}$$

where $\alpha(x,t)$ is the infinite-dimensional state of the anti-stable wave PDE with spatial domain defined without loss of generality by $x\in[0,1]$. The boundary delay is denoted by $D>0$ being any arbitrary known constant.

The delay-wave system is alternatively written as [122]

$$\dot{\vartheta}(t) = u(0,t), \tag{12.181}$$
$$\partial_{tt}u(x,t) = \partial_{xx}u(x,t), \quad x \in [0,1], \tag{12.182}$$
$$\partial_x u(0,t) = -q\partial_t u(0,t), \quad |q| \neq 1, \tag{12.183}$$
$$u(1,t) = v(1,t), \tag{12.184}$$
$$\partial_t v(x,t) = \partial_x v(x,t), \quad x \in [1,1+D], \tag{12.185}$$
$$v(1+D,t) = U(t), \tag{12.186}$$

where $U(t)$ is the overall system (control) input and (ϑ,u,v) is the state of the ODE-PDE-PDE cascade. From the transport PDE representation form for the input delay, we know that [119, Chapter 2]

$$v(x,t) = U(t+x-1-D), \quad x \in [1,1+D]. \tag{12.187}$$

Recalling that (12.182)–(12.184) can be represented into (12.153)–(12.154), applying the transformations

$$w(x,t)=\bar{\zeta}(x,t)-KH\int_0^x \bar{\zeta}(\sigma,t)d\sigma-KH\vartheta(t), \quad x\in[0,1], \tag{12.188}$$
$$\varpi(x,t)=\bar{\omega}(x,t)+KH\int_0^x \bar{\omega}(\sigma,t)d\sigma-KH\vartheta(t), \quad x\in[0,1], \tag{12.189}$$
$$\begin{aligned}\zeta(x,t)=&\,v(x,t)-KH\int_1^x v(\sigma,t)d\sigma\\ &+\frac{1}{1-q}\bar{\alpha}_x(1,t)-KH\int_0^1 \bar{\zeta}(\sigma,t)d\sigma-KH\vartheta(t), \quad x\in[1,1+D],\end{aligned} \tag{12.190}$$

the resulting average target system for (12.181)–(12.186) is

$$\dot{\vartheta}_{\rm av}(t)=KH\vartheta_{\rm av}(t)+w_{\rm av}(0,t), \tag{12.191}$$
$$\partial_t w_{\rm av}(x,t)=\partial_x w_{\rm av}(x,t), \quad w_{\rm av}(0,t)=\varpi_{\rm av}(0,t), \tag{12.192}$$
$$\partial_t \varpi_{\rm av}(x,t)=-\partial_x \varpi_{\rm av}(x,t), \tag{12.193}$$
$$w_{\rm av}(1,t)=\zeta_{\rm av}(1,t), \tag{12.194}$$
$$\partial_t\zeta_{\rm av}(x,t)=\partial_x\zeta_{\rm av}(x,t), \quad x \in [1,1+D], \tag{12.195}$$
$$\zeta_{\rm av}(1+D,t)=-\frac{1}{c}\partial_t v_{\rm av}(1+D,t). \tag{12.196}$$

From (12.190) and (12.196), one can obtain the expression for $\dot{U}_{\rm av}(t) = -c\zeta_{\rm av}(1+D,t)$ such that

$$\begin{aligned}\dot{U}_{\rm av}(t) = &-cv_{\rm av}(1+D,t)+cKH\int_1^{1+D} v_{\rm av}(\sigma,t)d\sigma\\ &-\frac{c}{1-q}\bar{\alpha}_x(1,t)+cKH\int_0^1 \bar{\zeta}_{\rm av}(\sigma,t)d\sigma+cKH\vartheta_{\rm av}(t).\end{aligned} \tag{12.197}$$

Plugging (12.187) into (12.197), we can obtain the control law:

$$U(t) = \frac{c}{s+c}\left\{\frac{-1}{1-q}\bar{\alpha}_x(1,t)+K\left[G(t)+\hat{H}(t)\int_{t-D}^{t}U(\tau)d\tau\right.\right.$$
$$\left.\left.+\hat{H}(t)\int_0^1\frac{1}{1-q}\left[u(x,t)+\bar{\alpha}_x(x,t)\right]dx\right]\right\} \tag{12.198}$$

for $c > 0$ sufficiently large, $u(x,t) = \bar{\alpha}_t(x,t)$, and $\bar{\alpha}_x(x,t)$ satisfying (12.174)–(12.175).

As discussed in [179], the explicit solution of the *trajectory generation problem* [128, Chapter 12] under delays would be solved from (12.176) as

$$S(t-D) = a\cos(\omega D)\sin(\omega t) - aq\sin(\omega D)\cos(\omega t)\,,$$

leading us to

$$S(t){=}a\cos(\omega D)\sin(\omega(t{+}D)){-}aq\sin(\omega D)\cos(\omega(t{+}D))\,. \tag{12.199}$$

12.8 ▪ Numerical Simulations

We explore an academic example here, which can be regarded as a very simplified version of the drilling control problem described in Section 12.2. No particular modeling of specific nonlinear phenomena for the oil drilling problem is taken into account in this section. However, pairing this real-time optimal drilling problem and the proposed PDE-based extremum seeking methods shows promise. It appears that we can improve the drilling performance by compensating the PDE-drill dynamics within the extremum-seeking algorithm. We can assume the drill-string model as simple as delay dynamics or as a wave PDE, as discussed in [1].

In oil drilling engineering it is customary to model the downhole boundary as an ODE, coupled with the PDE of the drill-string. We do not think this specific type of modeling is required in order to produce an ES algorithm that operates effectively but it would be of benefit if we were able to handle such a dynamic system structure. We have not studied the compensation of wave PDEs in the presence of ODE dynamics but we do not see why such a structure would not be tractable. We just need to assume that the ODE is "fast" enough, namely, we have to use the frequency ω (of the signals $S(t)$, $M(t)$ and $N(t)$), which is slower than the ODE's time constants. With such a relationship among the time scales of the ODE and the probing signal, we would estimate the gradient/derivative of the map in the quasi-steady state (as a static map of Assumption 12.1). This would mirror what is done with the singular perturbation approach in the paper [129] for ES design for general nonlinear ODE dynamics.

There may be an obstacle to generating a convergence guarantee theorem because there may not exist a singular perturbation theorem covering the exact system in our application, which incorporates both a nonlinear ODE and delays (distributed in the compensation portion of the ES feedback law) and/or wave PDEs. But, in light of this attractive application, it would be worth considering the study of this problem in future research. And, insofar as simply using the algorithm is concerned, which may be the practitioner's primary concern, this should be simple—as said above, ω and the "dither" signal $\sin(\omega t)$ just have to be chosen slow compared to the ODE dynamics.

Here, for the sake of simplicity, we just consider the wave PDE dynamics in cascade with the static map as described in (12.1)–(12.6) and illustrated in Figure 12.1. No other unmodeled dynamics are addressed such that the dither signals can still be chosen with a sufficiently large frequency. We focus on showing the numerical simulations only for the extremum seek-

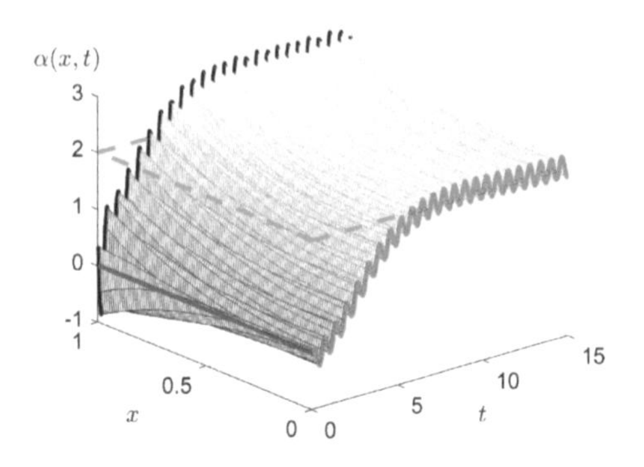

(a) Parameter $\Theta(t)$ (red) converging to Θ^* (dashed-green).

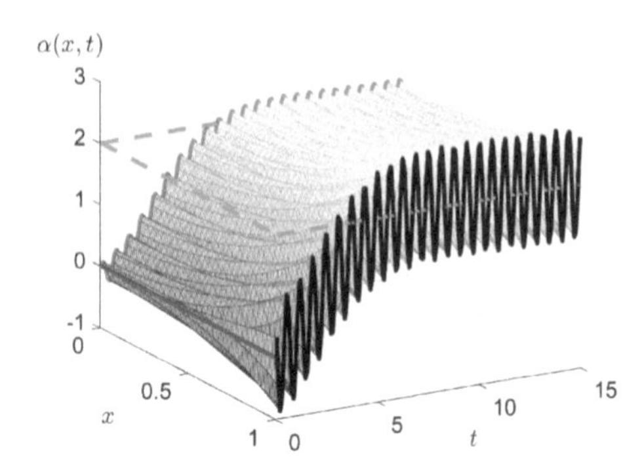

(b) Parameter $\theta(t)$ (black) converging to θ^* (dashed-green).

Figure 12.4. *Compensation of anti-stable wave PDEs in actuator dynamics for extremum seeking feedback with boundary Dirichlet actuation:* $\alpha(0,t)=\Theta(t)$ *and* $\alpha(D,t)=\theta(t)$, *with* $D=1$, $q=5$, *and* $\Theta^*=\theta^*=2$.

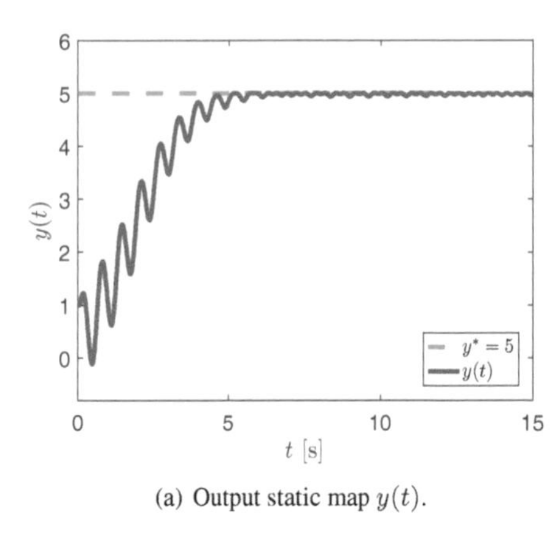

(a) Output static map $y(t)$.

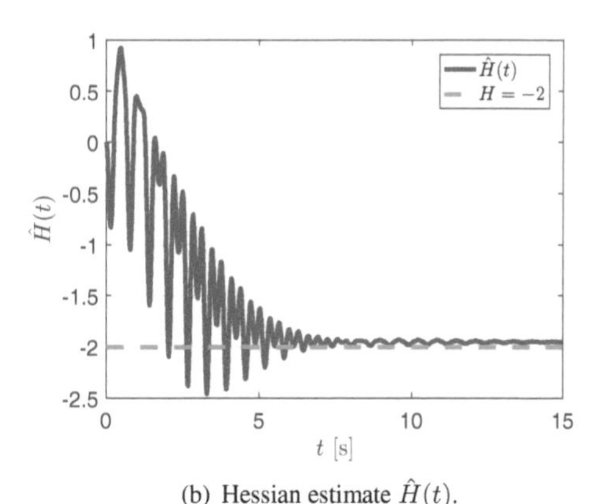

(b) Hessian estimate $\hat{H}(t)$.

Figure 12.5. *Compensation of anti-stable wave PDEs in actuator dynamics for extremum seeking feedback with boundary Dirichlet actuation: time response of* $y(t)$ *converging to the extremum* $y^*=5$ *and the Hessian's estimate* $\hat{H}(t)$ *converging to* $H=-2$.

ing feedback for anti-stable wave PDEs with Dirichlet boundary actuation of Section 12.7.1 (Remark 12.2).

We consider a quadratic static map as in (12.6), with Hessian $H=-2$, optimizer $\Theta^*=\theta^*=2$, and optimal value $y^*=5$. The domain length of the wave PDE and the damping coefficient are set to $D=1$ and $q=5$, respectively. The parameters of the proposed ES are chosen as $\omega=10$, $a=0.2$, $c=10$, $\bar{c}=0.5$, and $K=0.2$.

The corresponding numerical plots of the closed-loop system are given in Figures 12.4(a) and 12.4(b). We note that the signals Θ and θ converge to a close neighborhood of the optimizer $\Theta^*=\theta^*$.

Figures 12.5(a) and 12.5(b) present the relevant variables for ES. To be noted is the remarkable evolution of the output signal $y(t)$ and the Hessian's estimate $\hat{H}(t)$, ultimately achieving the extremum $y^*=5$ and the correct Hessian value $H=-2$, even in the presence of the wave PDE.

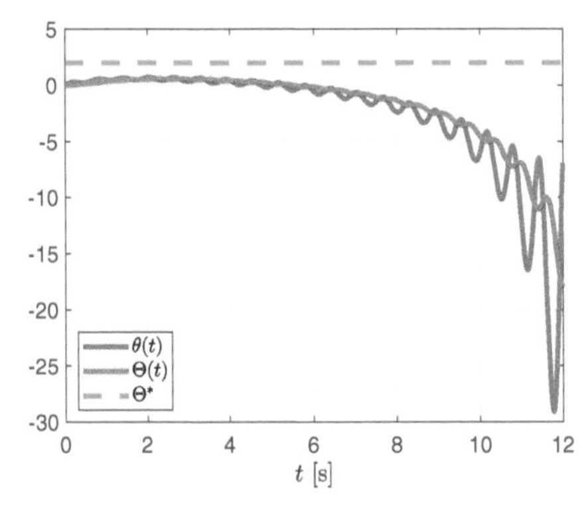

(a) Evolution of the signals $\theta(t)$ and $\Theta(t)$ for large ω.

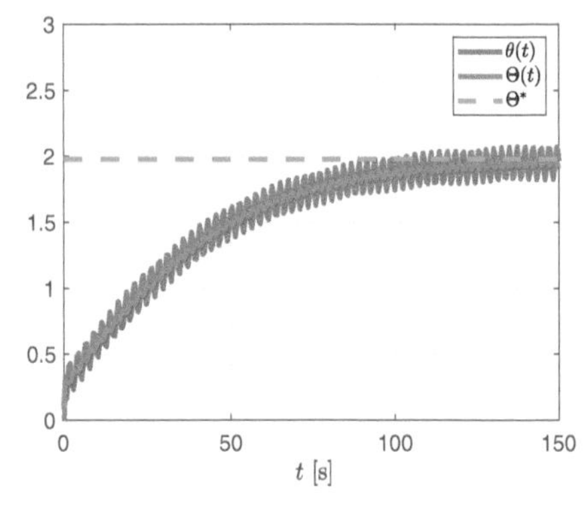

(b) Evolution of the signals $\theta(t)$ and $\Theta(t)$ for small ω.

Figure 12.6. *Classical ES with uncompensated wave PDE actuator dynamics:* (a) *instability for large* ω*;* (b) *very slow convergence for small* ω*.*

Remark 12.3. Persistent perturbations are a necessity if the algorithm is to remain alert to changes in the system [140]. The persistence of oscillation allows one to re-optimize the operation as equipment ages and the drill bit wears and passes from one rock type to another. In practice, high-frequency oscillations in the plots of Figures 12.4(a) and 12.4(b), due to the persistent excitation signal $S(t)$, may also lead to chattering or limit cycles in actuators. The inability to remove the oscillation and achieve equilibrium stabilization in ES may also be associated with actuator constraints, such as magnitude and rate saturation. Here, the best control requirement could be to enforce a stable, "smallest" limit cycle [10, Chapter 5]. However, ES-based controllers whose control efforts vanish as the system approaches equilibrium have been proposed in [235] and [158]. In [203], [204], [59], and [77], Lie bracket–based ES was also introduced as a tool for designing feedback controllers with bounded update rates. Hence, limit cycles may not only be reduced, but they can also be completely eliminated in these cases.

In Figures 12.6(a) and 12.6(b), the proposed ES scheme with PDE compensation is compared to the classical ES design [129], which does not incorporate the wave PDE compensation. Using the same parameters for ω, a, and K of the prior simulation results in instability of $\theta(t)$ and $\Theta(t)$, as illustrated in Figure 12.6(a). On the other hand, Figure 12.6(b) shows the response for a lower frequency $\omega = 0.4$ rad/s, in which we observe that the stability can be recovered but the convergence speed gets much slower (around 30 times!) than the results in Figures 12.4(a) and 12.4(b). Hence, our proposed ES controller stands as a strong improvement of the classical ES in the presence of actuation dynamics governed by wave PDEs.

Figures 12.7(a) and 12.7(b) present the effect of the dither frequency ω in the convergence of the algorithm for different values of $\omega \in [0.006,\ 30]$ rad/s. As $\omega \to 0.006$, the control system performance is severely affected, as expected, resulting in a slower convergence of the ES algorithm to the desired values. For instance, with $\omega = 0.5$ rad/s, the convergence is indeed settled after 60 seconds (not shown). In spite of that, the algorithm continues the search for the extremum point.

Similar tests can be performed for the design parameters K and a (curves not shown). By increasing the adaptation gain K we can accelerate the closed-loop responses, whereas by decreasing the constant a we can reduce the amplitude of the residual oscillations, according to (12.46)–(12.48). Recall that the ultimate residual set for the error $\theta(t) - \theta^*$ in (12.46) is of the order $\mathcal{O}(a + aq + 1/\omega)$ for the anti-stable wave PDE, also depending on the damping coefficient q due to the amplitude of the dither $S(t)$ defined in (12.176).

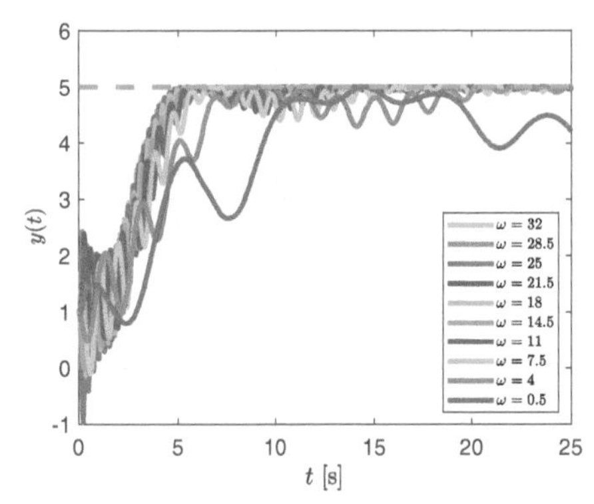

(a) Evolution of the signal $y(t)$ for different values of ω.

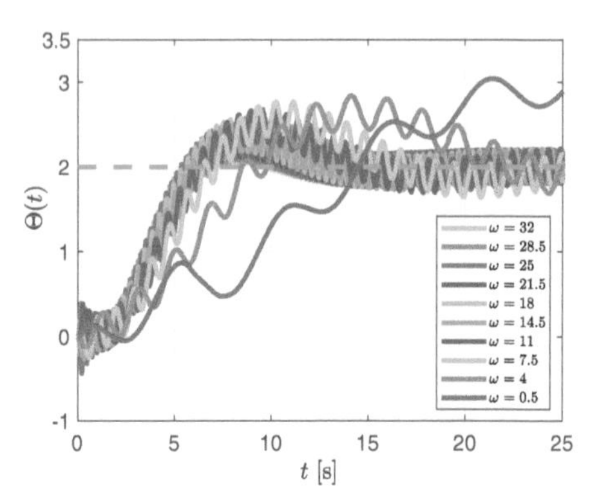

(b) Evolution of the signal $\Theta(t)$ for different values of ω.

Figure 12.7. *Time evolution of the signals $y(t)$ and $\Theta(t)$ converging to a neighborhood of the extremum point $y^* = 5$ and the maximizer $\Theta^* = 2$ under different values of the dither frequency $\omega \in [0.006, 30]$ rad/s.*

12.9 ▪ Notes and References

In [179, 198], we have rigorously analyzed the effects of the delays in ES feedback and proposed predictor-based solutions for delay compensation of delayed static and dynamic nonlinear maps. Both gradient and Newton-based designs were studied. Particularly, only pure delays were considered in [179, 198], which were represented by a first-order hyperbolic transport Partial Differential Equation (PDE). This infinite-dimensional representation was crucial to allow the generalization of the ES method to different families of PDEs, such as diffusion (heat) processes [65, 169] and wider classes of parabolic reaction-advection-diffusion PDEs in [172, 171] as well. Although there are many publications, mostly on the theory but some also applicable to the practice of stabilization of PDEs [122, 159, 205, 74, 236, 208] and time-delay systems [78, 63, 168, 167], none of the references has considered the ES approach.

The author in [122] has developed a boundary stabilizing control that compensates an arbitrarily long delay at the input of an anti-stable wave equation system. Reference [159] develops an adaptive PDE observer for battery state-of-charge and state-of-health estimation. On the other hand, the authors of [205] have designed a controller for flow-induced vibrations of an infinite-band membrane. The model of the flow-induced vibration is given by a wave PDE with an anti-damping term throughout the 1D domain. The paper [74] presents a deterministic hybrid PDE model which accounts for thermostatically controlled loads populations and which facilitates the aggregate synthesis of power control in power networks. Reference [236] deals with the axial vibrations of the cable lifting up a cage with miners via a mining cable. These vibration dynamics can be described by a coupled wave PDE and an ODE system with a Neumann interconnection on a time-varying spatial domain. Explicit motion-planning reference solutions are presented for flexible beams with Kelvin–Voigt damping and wave equations in [208]. The goal is to generate periodic reference signals for the displacement and deflection angle at the free end of the beam using only actuation at the base. Publication [78] gives a broad overview of the stability and control of time-delay systems, and [63] addresses the challenging adaptive posicast control problem for uncertain systems in the presence of time delays. The stability of a general class of linear time-invariant-neutral time-delay systems was studied in [168], whereas a new analytic approach to obtain the complete solution for systems of delay differential equations based on the concept of Lambert functions was presented in [167].

In this chapter, we extend the class of PDEs for which ES feedback can be employed, by considering *wave dynamics* connected in cascade with the static scalar map to be optimized [176]. The problem tackled in the chapter is inspired by a specific off-shore drilling application [26] as well as by its optimal control [1], where the real-time optimization approach is affected by a wave PDE in the actuation dynamics. The wave process is challenging due to the fact that all of its (infinitely many) eigenvalues are on the imaginary axis, and a limited (finite) speed of propagation (large control amplitudes do not help) [117]. While at the beginning of this chapter we argued that the heat PDE introduces more complexity in the generation of the probing signal and the compensation of its dynamics in the ES algorithm, one can also argue, conversely, that the wave problem, studied in this chapter, is more challenging than the diffusion case in [65, 169] (Chapter 10) due to the fact that the wave PDE is second order in time, which means that the state is "doubly infinite dimensional" (distributed displacement and velocity). This is not so much of a problem dimensionally, as it is a challenge in constructing the state transformations for compensating the PDEs [117]. One has to deal with the coupling of two infinite-dimensional states.

The complete control design employing a compensator for the wave actuation dynamics is developed via backstepping transformation by feeding back the estimates for the gradient and Hessian (first and second derivatives) of the static map to be maximized. Our proofs for local stability of the closed-loop system and the convergence to a small neighborhood of the extremum are also based on the backstepping methodology for PDEs [128], the construction of a Lyapunov functional, and the use of the averaging theorem for infinite-dimensional systems [83]. ES control design for wave PDEs with both Neumann and Dirichlet actuation is considered. ES design for an anti-stable wave PDE with boundary anti-damping is presented as well, as is also the design for an anti-stable wave PDE with input delay [122].

Although we have only pursued the gradient-based method, the proposed strategy can be extended to the Newton-based extremum seeking [73] following the procedure presented in [179]. The multivariable extension of the proposed algorithm can also be achieved according to [170], studied in the next chapter. As for open problems, the authors expect continued efforts in data validation and experiments to demonstrate the capability of the proposed extremum seeking methodology in resolving some specific real-time optimization problems in oil drilling scenarios [1].

Chapter 13

Multivariable ES for Distinct Families of PDEs

The methods introduced in Chapters 10, 11, and 12 have opened up the possibility of handling PDEs of types more general than the simple pure delay (transport PDE). However, so far the designs presented have been only for single-input maps.

The extension of the ES framework to multi-input maps under distinct PDEs in the actuation paths is not easy since the Hessian matrix (the second derivative) of the nonlinear map to be optimized is not diagonal and appears in the PDE formulation by coupling the different actuation channels. This is the key challenge in multivariable ES when compared to ES for single-input maps, originally proposed in [65] and [169]. The seeming obstacles to the control design and stability analysis are removed in this chapter through the Newton-based framework [73]—see also Chapter 7.

Although we focus in this chapter's technical analysis on the case where each actuator is governed by a distinct diffusion process, our approach possesses the generality that allows to simultaneously handle different types of PDEs in each individual input channel, including parabolic reaction-advection-diffusion (RAD) PDEs, second-order hyperbolic PDEs (wave equations), first-order hyperbolic transport PDEs (pure delays, constant or time-varying), and distributed delays.

13.1 ▪ Problem Formulation

Our goal is to maximize in real time the output $y(t) \in \mathbb{R}$ of an nonlinear static map $Q(\cdot)$ with optimizer $\Theta^* \in \mathbb{R}^n$ and the optimal value $y^* = Q(\Theta^*) \in \mathbb{R}$. However, unlike the classical problem [129], we consider the challenging multiparameter ES of multi-input maps with *actuation dynamics* described by *multiple diffusion processes.*

13.1.1 ▪ Multivariable Maps with PDE Actuation Dynamics

As shown in Figure 13.1, distinct heat equations with Dirichlet actuation are assumed in the actuator vector $\theta(t) \in \mathbb{R}^n$. Thus, the propagated actuator vector $\Theta(t) \in \mathbb{R}^n$ is given by

$$\Theta_i(t) = \alpha_i(0,t) \quad \forall i \in \{1,2,\ldots,n\}, \tag{13.1}$$

$$\partial_t \alpha_i(x,t) = \partial_{xx} \alpha_i(x,t), \quad x \in (0,D_i), \tag{13.2}$$

$$\partial_x \alpha_i(0,t) = 0, \tag{13.3}$$

$$\alpha_i(D_i,t) = \theta_i(t), \tag{13.4}$$

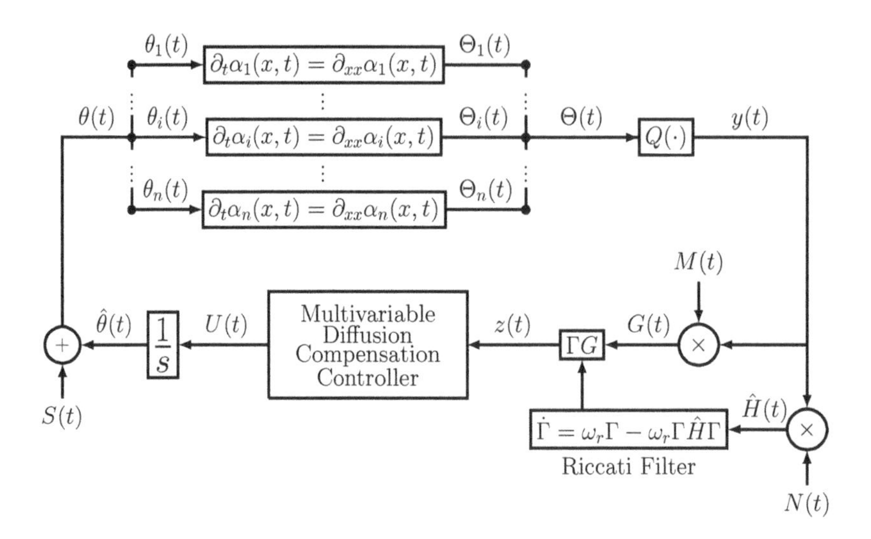

Figure 13.1. *Multivariable Newton-based extremum seeking under distinct diffusion PDEs in the actuation dynamics, with boundary conditions $\theta_i(t)=\alpha_i(D_i,t)$, $\Theta_i(t)=\alpha_i(0,t)$, and $i\in\{1,2,\ldots,n\}$.*

where $\alpha_i:[0,D_i]\times\mathbb{R}_+\to\mathbb{R}$ and each different domain length D_i is known. The solution of (13.1)–(13.4) is given by

$$\alpha_i(x,t)=\mathcal{L}^{-1}\left[\frac{\cosh(x\sqrt{s})}{\cosh(D_i\sqrt{s})}\right]*\theta_i(t), \tag{13.5}$$

where $\mathcal{L}^{-1}(\cdot)$ denotes the inverse Laplace transformation and $*$ is the convolution operator. Given this relation, we define the *diffusion operator* for the PDE (13.1)–(13.4) with boundary input and measurement given by $\mathcal{D}=\text{diag}\{\mathcal{D}_1,\ldots,\mathcal{D}_n\}$ with

$$\mathcal{D}_i[\varphi(t)]=\mathcal{L}^{-1}\left[\frac{1}{\cosh(D_i\sqrt{s})}\right]*\varphi(t), \quad \text{s.t.} \quad \Theta(t)=\mathcal{D}[\theta(t)]. \tag{13.6}$$

The scalar measurement $y\in\mathbb{R}$ is defined by the unknown static map with propagated input (13.1), such that

$$y(t)=Q(\mathcal{D}[\theta(t)])=Q(\Theta(t)), \quad \Theta(t)=[\Theta_1(t),\ldots,\Theta_n(t)]^T. \tag{13.7}$$

For the sake of simplicity, we assume the unknown nonlinear static map is locally quadratic in the neighborhood of the extremum point y^*, i.e.,

$$y(t)=Q(\Theta(t)):=y^*+\frac{1}{2}(\Theta(t)-\Theta^*)^TH(\Theta(t)-\Theta^*), \tag{13.8}$$

where besides $\Theta^*\in\mathbb{R}^n$ and $y^*\in\mathbb{R}$ being unknown, the Hessian matrix $H^T=H<0$ of the static multivariable map is also unknown. We can also define the optimal input θ^* of $\theta(t)$ with respect to the static map and the diffusion process with the relation $\mathcal{D}[\theta^*]:=\Theta^*$.

13.1.2 ▪ Estimation Error and Error Dynamics

Since our goal is to find the unknown optimal inputs θ^* and Θ^*, we define the estimation errors

$$\tilde{\theta}(t)=\hat{\theta}(t)-\theta^*, \qquad \vartheta(t)=\hat{\Theta}(t)-\Theta^*, \tag{13.9}$$

where the vectors $\hat{\theta}(t)$ and $\hat{\Theta}(t)$ are the estimates of θ^* and Θ^*. In order to make (13.9) coherent with the optimizer of the static map Θ^*, we apply the diffusion operator (13.6) to $\tilde{\theta}_i$ in (13.9) and get

$$\vartheta_i(t) := \bar{\alpha}_i(0,t), \tag{13.10}$$

$$\partial_t \bar{\alpha}_i(x,t) = \partial_{xx} \bar{\alpha}_i(x,t), \quad x \in (0, D_i), \tag{13.11}$$

$$\partial_x \bar{\alpha}_i(0,t) = 0, \tag{13.12}$$

$$\bar{\alpha}_i(D_i,t) = \tilde{\theta}_i(t), \tag{13.13}$$

where $\bar{\alpha}_i : [0, D_i] \times \mathbb{R}_+ \to \mathbb{R}$ and $\vartheta(t) := \mathcal{D}[\tilde{\theta}(t)] = \hat{\Theta}(t) - \Theta^*$ is the propagated estimation error $\tilde{\theta}(t)$ through the diffusion domain. For $\lim_{t\to\infty} \theta(t) = \theta_c$, we have $\lim_{t\to\infty} \Theta(t) = \Theta_c = \theta_c$, where the index c indicates a constant signal. Indeed, from (13.5) and (13.6), for a constant input $\theta = \theta_c$, one has $\mathcal{L}\{\theta_c\} = \theta_c/s$, and applying the Laplace limit theorem we get $\lim_{t\to\infty} \mathcal{D}_i[\theta_{ci}] = \lim_{s\to 0} \left\{ \frac{\theta_{ci}}{\cosh(\sqrt{s} D_i)} \right\} = \theta_{ci}$. In the particular case $\theta = \theta_c = \theta^*$, one has $\Theta^* = \theta^*$.

From the block diagram in Figure 13.1, one has

$$\dot{\hat{\theta}}(t) = U(t) \tag{13.14}$$

and, consequently,

$$\dot{\tilde{\theta}}_i(t) = U_i(t) \qquad \forall i \in \{1, 2, \ldots, n\}, \tag{13.15}$$

since $\dot{\tilde{\theta}}(t) = \dot{\hat{\theta}}(t)$, once θ^* is constant.

Taking the time derivative of (13.10)–(13.13) and with the help of (13.9) and (13.15), the *propagated error dynamics* is written as

$$\dot{\vartheta}_i(t) = u_i(0,t), \tag{13.16}$$

$$\partial_t u_i(x,t) = \partial_{xx} u_i(x,t), \quad x \in (0, D_i), \tag{13.17}$$

$$\partial_x u_i(0,t) = 0, \tag{13.18}$$

$$u_i(D_i,t) = U_i(t), \tag{13.19}$$

where $u_i : [0, D_i] \times \mathbb{R}_+ \to \mathbb{R}$ and $u_i(x,t) := \partial_t \bar{\alpha}_i(x,t)$.

13.1.3 ▪ Additive Dither Signal: $S(t) := [S_1(t), \ldots, S_n(t)]^T$ and the Trajectory Generation Problem

Analogously to the classical ES [129] without actuation through a diffusion domain, the perturbation signals $S_i(t)$ should add $a_i \sin(\omega_i t)$ to the vector $\Theta(t)$ in order to compensate the diffusion process. Hence, $a_i \sin(\omega_i t)$ with perturbation amplitude a_i and frequency ω_i is applied as follows:

$$S_i(t) := \beta_i(D_i,t), \tag{13.20}$$

$$\partial_t \beta_i(x,t) = \partial_{xx} \beta_i(x,t), \quad x \in (0, D_i), \tag{13.21}$$

$$\partial_x \beta_i(0,t) = 0, \tag{13.22}$$

$$\beta_i(0,t) = a_i \sin(\omega_i t), \tag{13.23}$$

with $\beta_i : [0, D_i] \times \mathbb{R}_+ \to \mathbb{R}$. Equations (13.20)–(13.23) describe a *trajectory generation problem* [196, 128, 150] with Dirichlet boundary condition so that the effects of the PDE dynamics in the additive dither signals are properly annihilated. The explicit solution of (13.20) for all $i \in \{1,2,\ldots,n\}$ is given by

$$S_i(t) = \frac{1}{2} a_i e^{\sqrt{\frac{\omega_i}{2}} D_i} \sin\left(\omega_i t + \sqrt{\frac{\omega_i}{2}} D_i\right) + \frac{1}{2} a_i e^{-\sqrt{\frac{\omega_i}{2}} D_i} \sin\left(\omega_i t - \sqrt{\frac{\omega_i}{2}} D_i\right). \quad (13.24)$$

The modified perturbation signal $S_i(t)$ is the boundary value of an inverse diffusion process described in (13.20)–(13.23). We can define this inverse diffusion operator for (13.20)–(13.23) in the same fashion as the diffusion operator (13.6), such that $\mathcal{D}_i^{-1}[\varphi(t)] = \mathcal{L}^{-1}[\cosh(D_i\sqrt{s})] * \varphi(t)$. Hence,

$$S_i(t) = \mathcal{D}_i^{-1}[a_i \sin(\omega_i t)]. \quad (13.25)$$

The PDE state $\bar{\alpha}_i(x,t)$ in (13.10)–(13.13) can be defined as

$$\bar{\alpha}_i(x,t) := \alpha_i(x,t) - \beta_i(x,t) - \Theta_i^*. \quad (13.26)$$

From its particular boundary value $\alpha_i(D_i,t)$ and recalling $\Theta_i^* = \theta_i^*$, we have

$$\theta_i(t) - \theta_i^* = \tilde{\theta}_i(t) + S_i(t). \quad (13.27)$$

By plugging (13.25) into (13.27), we get

$$\theta_i(t) - \theta_i^* = \tilde{\theta}_i(t) + \mathcal{D}_i^{-1}[a_i \sin(\omega_i t)]. \quad (13.28)$$

Finally, we apply the diffusion operator (13.6) on both sides of (13.28). With $\mathcal{D}[\theta(t)] = \Theta(t)$, $\theta^* = \Theta^*$, and $\mathcal{D}[\tilde{\theta}(t)] = \vartheta(t)$, we arrive at

$$\vartheta_i(t) + a_i \sin(\omega_i t) = \Theta_i(t) - \Theta_i^*. \quad (13.29)$$

13.1.4 ▪ Estimation of the Gradient, Hessian, and Its Inverse

It remains to define the multiplicative demodulation vector signals $M(t)$ and $N(t)$ which are used to estimate the gradient and Hessian of the static map, respectively, by multiplying them with the output $y(t)$ of the static map.

In [73], the gradient estimate is given as

$$G(t)=M(t)y(t), \quad M(t)=\left[\frac{2}{a_1}\sin(\omega_1 t) \cdots \frac{2}{a_n}\sin(\omega_n t)\right]^T. \quad (13.30)$$

On the other hand, the Hessian estimate is given by

$$\hat{H}(t) = N(t)y(t). \quad (13.31)$$

The elements of the $n \times n$ demodulating matrix $N(t)$ are

$$N_{i,i}(t) = \frac{16}{a_i^2}\left(\sin^2(\omega_i t) - \frac{1}{2}\right), \quad (13.32)$$

$$N_{i,j}(t) = \frac{4}{a_i a_j}\sin(\omega_i t)\sin(\omega_j t), \quad i \neq j, \quad (13.33)$$

with nonzero perturbation amplitudes a_i and frequencies $\omega_i \neq \omega_j$. The probing frequencies ω_i's can be selected as

$$\omega_i = \omega_i' \omega = \mathcal{O}(\omega) \quad \forall i \in \{1, 2, \ldots, n\}, \tag{13.34}$$

where ω is a positive constant and ω_i' is a rational number.

Remark 13.1. An additional advantage of the Newton approach is the possibility of optimizing in real time not only the output of the (strongly) convex map $y = Q(\Theta)$ in (13.8), but also maximizing the nth derivative $Q^{(n)}(\Theta)$ of the output (13.8), as shown in Chapter 5. Maximizing map sensitivity [151] can be achieved by redesigning the demodulating signals $M(t)$ in (13.30) and $N(t)$ (13.32)–(13.33) according to [151] or [172], independently of the presence of actuation dynamics described by PDEs.

As discussed in [73] (see also Chapter 7), generating an estimate of the Hessian matrix in non-model-based optimization is not the only challenge. The other challenge is that the Newton algorithm requires an inverse of the Hessian matrix H^{-1}. The estimate of this matrix, as it evolves in continuous time, does not necessarily remain invertible. We tackle this challenge as in [73] by employing a dynamic system for generating the inverse asymptotically. This dynamic system is a filter in the form of a Riccati differential equation [73]:

$$\dot{\Gamma} = \omega_r \Gamma - \omega_r \Gamma \hat{H} \Gamma, \tag{13.35}$$

with $\omega_r > 0$ being a design constant.

The estimation error of the Hessian's inverse can be defined as

$$\tilde{\Gamma}(t) = \Gamma(t) - H^{-1}, \tag{13.36}$$

and its dynamic equation is written from (13.35) and (13.36) by

$$\dot{\tilde{\Gamma}} = \omega_r (\tilde{\Gamma} + H^{-1})[I_{n \times n} - \hat{H}(\tilde{\Gamma} + H^{-1})]. \tag{13.37}$$

In particular, the convergence rates of both estimators of the gradient $G(t)$ in (13.30) and the Hessian inverse $\Gamma(t)$ in (13.35) are independent of the unknown Hessian and can be assigned arbitrarily by the user [73]. Due to this property, using (13.35), we are able to define the following measurable signal:

$$z(t) = \Gamma(t) G(t), \tag{13.38}$$

which will be employed in the next section to derive the proposed control law.

13.2 ▪ Multivariable Newton-Based ES Control with Diffusion Compensation

13.2.1 ▪ Control Design

First of all, consider the PDE-ODE cascade (13.16)–(13.19). Now, we use the backstepping transformation [116]:

$$w_i(x,t) = u_i(x,t) - \int_0^x q_i(x,\sigma) u_i(\sigma,t) d\sigma - \gamma_i(x) \vartheta_i(t), \tag{13.39}$$

with the gain kernels $q_i(x,\sigma) = \bar{K}_i(x-\sigma)$ and $\gamma_i(x) = \bar{K}_i$, which transforms (13.16)–(13.19) into the target system

$$\dot{\vartheta}_i(t) = \bar{K}_i \vartheta_i(t) + w_i(0,t), \tag{13.40}$$

$$\partial_t w_i(x,t) = \partial_{xx} w_i(x,t), \quad x \in (0, D_i), \tag{13.41}$$

$$\partial_x w_i(0,t) = 0, \tag{13.42}$$

$$w_i(D_i,t) = 0, \tag{13.43}$$

with $\bar{K}_i < 0$. Since the target system (13.40)–(13.43) is exponentially stable (see [116]), the controller which compensates the diffusion process can be obtained by evaluating the backstepping transformation (13.39) at $x = D_i$ as

$$U_i(t) = \bar{K}_i \vartheta_i(t) + \bar{K}_i \int_0^{D_i} (D_i - \sigma) u_i(\sigma,t) d\sigma. \tag{13.44}$$

However, the proposed control law in (13.44) is not applicable directly, because we have no measurement on $\vartheta_i(t)$. However, from [73], it is known that the average versions of the gradient (13.30) and Hessian (13.31) estimates are

$$G_{\text{av}}(t) = H\vartheta_{\text{av}}(t), \qquad \hat{H}_{\text{av}}(t) = H \tag{13.45}$$

if a quadratic map as in (13.8) is considered. In addition, from (13.45) and $z(t) = [z_1(t), \ldots, z_n(t)]^T$ in (13.38), we can verify [73]

$$z_{\text{av}}(t) = \frac{1}{\Pi} \int_0^{\Pi} \Gamma M(\lambda) y d\lambda = \Gamma_{\text{av}}(t) H \vartheta_{\text{av}}(t), \tag{13.46}$$

where $\Pi = 2\pi \times \text{LCM}\{1/\omega_i\}$ with the acronym LCM standing for the least common multiple and $\Gamma_{\text{av}}(t)$ and $\vartheta_{\text{av}}(t)$ denoting the average versions of $\Gamma(t)$ and $\vartheta(t)$, respectively.

Then, (13.46) can be written in terms of $\tilde{\Gamma}_{\text{av}}(t) = \Gamma_{\text{av}}(t) - H^{-1}$ as

$$z_{\text{av}}(t) = \vartheta_{\text{av}}(t) + \tilde{\Gamma}_{\text{av}}(t) H \vartheta_{\text{av}}(t). \tag{13.47}$$

The second term in the RHS of (13.47) is bilinear in $(\tilde{\Gamma}_{\text{av}}, \vartheta_{\text{av}})$; thus, the linearization of $\Gamma_{\text{av}}(t)$ at H^{-1} results in the linearized version of (13.46) given by

$$z_{\text{av}}(t) = \vartheta_{\text{av}}(t). \tag{13.48}$$

Regarding (13.48), we average (13.44) and choose $\bar{K}_i = -k_i$ with $k_i > 0$ to obtain

$$U_i^{\text{av}}(t) = -k_i z_i^{\text{av}}(t) - k_i \int_0^{D_i} (D_i - \sigma) u_i^{\text{av}}(\sigma,t) d\sigma. \tag{13.49}$$

Due to technical reasons in the application of the averaging theorem for infinite-dimensional systems [83] (Appendix B), we introduce a low-pass filter to the controller with an average-based diffusion compensation feedback via Hessian's inverse estimation of the form

$$U_i(t) = \frac{c_i}{s + c_i} \left\{ -k_i \left[z_i(t) + \int_0^{D_i} (D_i - \sigma) u_i(\sigma,t) d\sigma \right] \right\} \tag{13.50}$$

to each individual subsystem in (13.19) in order to fully stabilize it, where $c_i > 0$ and $k_i > 0$ are design constants.

As discussed in [65, Remark 2] and analogously to (10.40) in Chapter 10, the control law (13.50) can be rewritten as

$$U_i(t) = \frac{c_i}{s+c_i}\left\{-k_i\left[z_i(t)+\hat{\theta}_i(t)-\Theta_i(t)+a_i\sin(\omega_i t)\right]\right\}, \tag{13.51}$$

using the diffusion equations $\partial_t\alpha_i(x,t)=\partial_{xx}\alpha_i(x,t)$ and the integration by parts, associated with (13.1)–(13.4), (13.9), and (13.29).

13.2.2 ▪ Closed-Loop System

Substituting the control law (13.50) into (13.19), we can write the closed-loop system (13.16)–(13.19) as

$$\dot{\vartheta}_i(t) = u_i(0,t), \tag{13.52}$$

$$\partial_t u_i(x,t) = \partial_{xx}u_i(x,t), \quad x\in(0,D_i), \tag{13.53}$$

$$\partial_x u_i(0,t) = 0, \tag{13.54}$$

$$u_i(D_i,t) = \frac{c_i}{s+c_i}\left\{-k_i\left[z_i(t)+\int_0^{D_i}(D_i-\sigma)u_i(\sigma,t)d\sigma\right]\right\}. \tag{13.55}$$

In the next section, the availability of Lyapunov functionals via backstepping transformation [116, 128] permits the stability analysis of the complete ES feedback system.

13.3 ▪ Stability Analysis

The Sobolev space $\mathcal{H}_1$ is to be regarded as the state space of the PDE in the closed-loop system [119], as this is the norm in which stability is studied in the next theorem. However, the regularity of all the derivatives involved in the Lyapunov calculation is not proved but only assumed in this chapter.

Theorem 13.1. *Consider the system in Figure 13.1 with the plant being represented by the nonlinear map in (13.7)–(13.8) and multiple inputs channels with actuation dynamics governed by distinct diffusion PDEs according to (13.1)–(13.4). There exists $c^*>0$ such that, $\forall c_i \geq c^*$, $\exists\ \omega^*(c_1,\dots,c_n)>0$ such that, $\forall \omega>\omega^*$, the closed-loop system (13.16)–(13.19) with control law U_i in (13.50) and state variables defined by $\tilde{\Gamma}(t)=\Gamma(t)-H^{-1}$, $\vartheta_i(t)$, $u_i(x,t)$, and $i\in 1,2,\dots,n$ has a unique exponentially stable periodic solution in t of period $\Pi=2\pi\times LCM\{1/\omega_i\}$, denoted by $\tilde{\Gamma}^{\Pi}(t)$, $\vartheta_i^{\Pi}(t)$, $u_i^{\Pi}(x,t)$, satisfying, $\forall t\geq 0$,*

$$\left(\left|\tilde{\Gamma}^{\Pi}(t)\right|^2+\left|\vartheta^{\Pi}(t)\right|^2+\sum_{i=1}^{n}\left\|u_i^{\Pi}(t)\right\|^2+\left\|\partial_x u_i^{\Pi}(t)\right\|^2+\left|u_i^{\Pi}(D_i,t)\right|^2\right)^{1/2}\leq \mathcal{O}(1/\omega). \tag{13.56}$$

Furthermore,

$$\limsup_{t\to\infty}|\theta(t)-\theta^*| = \mathcal{O}\left(|a|e^{D\sqrt{\omega/2}}+1/\omega\right), \tag{13.57}$$

$$\limsup_{t\to\infty}|\Theta(t)-\Theta^*| = \mathcal{O}\left(|a|+1/\omega\right), \tag{13.58}$$

$$\limsup_{t\to\infty}|y(t)-y^*| = \mathcal{O}\left(|a|^2+1/\omega^2\right), \tag{13.59}$$

where $a=[a_1\ a_2\ \cdots\ a_n]^T$ and $D=\max(D_i)$.

Proof. The proof is structured in Sections 13.3.1 to 13.3.6 below.

13.3.1 ▪ Average Model of the Closed-Loop System

From (13.48), the *uncoupled-linearized* average version of the system (13.52)–(13.55) for ω large is

$$\dot{\vartheta}_i^{\mathrm{av}}(t) = u_i^{\mathrm{av}}(0,t), \tag{13.60}$$

$$\partial_t u_i^{\mathrm{av}}(x,t) = \partial_{xx} u_i^{\mathrm{av}}(x,t), \quad x \in (0,D_i), \tag{13.61}$$

$$\partial_x u_i^{\mathrm{av}}(0,t) = 0, \tag{13.62}$$

$$\frac{d}{dt} u_i^{\mathrm{av}}(D_i,t) = -c_i\, u_i^{\mathrm{av}}(D_i,t) - c_i\, k_i \left[\vartheta_i^{\mathrm{av}}(t) + \int_0^{D_i} (D_i - \sigma) u_i^{\mathrm{av}}(\sigma,t) d\sigma \right], \tag{13.63}$$

where the low-pass filter in (13.55) is represented in state-space form.

From the second equation in (13.45), the average model for the Hessian's inverse estimation error in (13.37) is

$$\frac{d\tilde{\Gamma}_{\mathrm{av}}(t)}{dt} = -\omega_r \tilde{\Gamma}_{\mathrm{av}}(t) - \omega_r \tilde{\Gamma}_{\mathrm{av}}(t) H \tilde{\Gamma}_{\mathrm{av}}(t). \tag{13.64}$$

13.3.2 ▪ Backstepping Transformation and Target System

Since in (13.60)–(13.63) we have n-independent ODE-PDE cascades, for each of them we can apply the "purely diagonal" (channel-by-channel) infinite-dimensional backstepping transformation (13.39) of the PDE state with $\bar{K}_i = -k_i$ such that

$$w_i(x,t) = u_i^{\mathrm{av}}(x,t) + k_i \left(\vartheta_i^{\mathrm{av}}(t) + \int_0^x (x-\sigma) u_i^{\mathrm{av}}(\sigma,t) d\sigma \right), \tag{13.65}$$

which maps the system (13.60)–(13.63) into the target system:

$$\dot{\vartheta}_i^{\mathrm{av}}(t) = -k_i \vartheta_i^{\mathrm{av}}(t) + w_i(0,t), \tag{13.66}$$

$$\partial_t w_i(x,t) = \partial_{xx} w_i(x,t), \quad x \in (0,D_i), \tag{13.67}$$

$$\partial_x w_i(0,t) = 0, \tag{13.68}$$

$$w_i(D_i,t) = -\frac{1}{c_i} \partial_t u_i^{\mathrm{av}}(D_i,t), \quad i = 1,2,\ldots,n. \tag{13.69}$$

Using (13.65) for $x = D_i$ and the fact that $u_i^{\mathrm{av}}(D_i,t) = U_i^{\mathrm{av}}(t)$, from (13.69) we get (13.63). Calculating the time derivative of (13.65) evaluated at $x = D_i$, it is easily seen that

$$\partial_t w_i(D_i,t) = \partial_t u_i^{\mathrm{av}}(D_i,t) + k_i u_i^{\mathrm{av}}(D_i,t), \tag{13.70}$$

where $\partial_t u_i^{\mathrm{av}}(D_i,t) = \dot{U}_i^{\mathrm{av}}(t)$. The inverse of (13.65) is [116, Theorem 1]

$$u_i^{\mathrm{av}}(x,t) = w_i(x,t) - k_i \left[e^{-k_i x} \vartheta_i^{\mathrm{av}}(t) + \int_0^x (e^{-k_i(x-\sigma)} - 1) w_i(\sigma,t) d\sigma \right]. \tag{13.71}$$

Plugging (13.69) and (13.71) with $x = D_i$ into (13.70), we get

$$\partial_t w_i(D_i,t) = -c_i w_i(D_i,t) + k_i w_i(D_i,t) - k_i^2 e^{-k_i D_i} \vartheta_i^{\mathrm{av}}(t) - k_i^2 \int_0^{D_i} (e^{-k_i(D_i-\sigma)} - 1) w_i(\sigma,t) d\sigma. \tag{13.72}$$

13.3.3 ▪ Lyapunov–Krasovskii Functional

Recalling that the linearized version of (13.64) is given by

$$\frac{d\tilde{\Gamma}_{\mathrm{av}}(t)}{dt} = -\omega_r \tilde{\Gamma}_{\mathrm{av}}(t), \tag{13.73}$$

which is locally exponentially stable since $\omega_r > 0$, and the $(\vartheta_{\mathrm{av}}, w)$-subsystem and the $\tilde{\Gamma}_{\mathrm{av}}$-subsystem are completely decoupled, then, the exponential stability of the overall system can established simply with the Lyapunov functional

$$V(t) = \sum_{i=1}^{n} V_i(t), \tag{13.74}$$

where $V_i(t)$ are functionals of the form

$$V_i(t) = \frac{1}{2}[\vartheta_i^{\mathrm{av}}(t)]^2 + \frac{a_i}{2}\|w_i(t)\|^2 + \frac{b_i}{2}\|\partial_x w_i(t)\|^2 + \frac{d_i}{2} w_i^2(D_i,t) \tag{13.75}$$

for each subsystem in (13.66)–(13.69) and $a_i, b_i, d_i > 0$ being appropriate constants to be chosen later $\forall i = 1, \ldots, n$.

We define $\lambda_i := k_i$ with $\lambda_i > 0$ by construction. Calculating the time derivative of (13.75), associated with the solution of the target system (13.66)–(13.69),

$$\dot{V}_i(t) = \vartheta_i^{\mathrm{av}}(t)\dot{\vartheta}_i^{\mathrm{av}}(t) + a_i \int_0^{D_i} \overbrace{w_i(x,t)}^{u} \underbrace{\overbrace{\partial_{xx} w_i(x,t)}^{dv}}_{=\partial_t w_i(x,t)} dx \tag{13.76}$$

$$+ b_i \int_0^{D_i} \overbrace{\partial_x w_i(x,t)}^{u} \underbrace{\overbrace{\partial_{xxx} w_i(x,t)}^{dv}}_{=\partial_x \partial_t w_i(x,t)} dx + d_i w_i(D_i,t)\partial_t w_i(D_i,t),$$

and with the help of integration by parts (recalling that $\partial_x w_i(0,t) = 0$), we obtain

$$\begin{aligned} \dot{V}_i(t) = & -\lambda_i [\vartheta_i^{\mathrm{av}}(t)]^2 + \vartheta_i^{\mathrm{av}}(t) w_i(0,t) \\ & + a_i w_i(D_i,t)\partial_x w_i(D_i,t) - a_i \|\partial_x w_i(t)\|^2 \\ & + b_i \partial_x w_i(D_i,t)\partial_t w_i(D_i,t) - b_i \|\partial_{xx} w_i(t)\|^2 \\ & + d_i w_i(D_i,t)\partial_t w_i(D_i,t). \end{aligned} \tag{13.77}$$

We apply Young's inequality on $\vartheta_i^{\mathrm{av}}(t) w_i(0,t)$ so that we can write $\vartheta_i^{\mathrm{av}}(t) w_i(0,t) \le \frac{1}{2\gamma_1}[\vartheta_i^{\mathrm{av}}(t)]^2 + \frac{\gamma_1}{2} w_i^2(0,t), \gamma_1 = 2/\lambda_i$. Using Agmon's, Poincaré's, and Young's inequalities, one has

$$\begin{aligned} w_i(0,t)^2 &\le \max_{x\in[0,D_i]} |w_i(x,t)|^2 \\ &\le w_i(D_i,t)^2 + 2\|w_i(t)\|\|\partial_x w_i(t)\|, \end{aligned} \tag{13.78}$$

$$\|w_i(t)\|^2 \le 2D_i w_i(D_i,t)^2 + 4D_i^2 \|\partial_x w_i(t)\|^2, \tag{13.79}$$

$$2\|w_i(t)\|\|\partial_x w_i(t)\| \le \|w_i(t)\|^2 + \|\partial_x w_i(t)\|^2. \tag{13.80}$$

Consequently,

$$w_i(0,t)^2 \le (1+2D_i) w_i(D_i,t)^2 + (4D_i^2+1)\|\partial_x w_i(t)\|^2. \tag{13.81}$$

Following analogous derivations detailed in the proof of Theorem 10.1 (Step 3) in Chapter 10, considering the relation (13.72), and applying Young's, Poincaré's, Agmon's, and Cauchy–Schwarz's inequalities [128] (more than once) to the cross terms in (13.77) in a similar fashion as done in (13.78)–(13.81) with $a_i = (c_i - \lambda_i)/(8D_i\lambda_i^3)$, $b_i = 1/(8D_i\lambda_i^3)$, $d_i = 1$, we get

$$\dot{V}_i(t) \leq -\frac{\lambda_i}{4}[\vartheta_i^{\text{av}}(t)]^2 + (c_1^* - c_i)w_i^2(D_i,t) + (c_2^* - c_i)\|\partial_x w_i(t)\|^2 - \frac{1}{32D_i\lambda_i^3}\|\partial_{xx}w_i(t)\|^2, \tag{13.82}$$

where

$$c_1^* = \frac{3}{2}\lambda_i^3 + \lambda_i + \frac{1+2D_i}{\lambda_i} + 2D_i\lambda_i\xi(D_i), \tag{13.83}$$

$$c_2^* = \lambda_i + 8D_i\lambda_i^3\left[\frac{4D_i^2+1}{\lambda_i} + 4D_i^2\lambda_i\xi(D_i)\right], \tag{13.84}$$

and $\xi(D_i) := \int_0^{D_i}(e^{-\lambda_i(D_i-\sigma)} - 1)^2 d\sigma$. By applying Poincaré's inequality [128]

$$\|\partial_x w_i(t)\|^2 \leq 2D_i\partial_x w_i(0,t)^2 + 4D_i^2\|\partial_{xx}w_i(t)\|^2 \tag{13.85}$$

to the last term in (13.82), and recalling $\partial_x w_i(0,t) = 0$, we get

$$\dot{V}_i(t) \leq -\frac{\lambda_i}{4}[\vartheta_i^{\text{av}}(t)]^2 - (c_i - c_1^*)w_i^2(D_i,t) - (c_i - c_2^*)\|\partial_x w_i(t)\|^2 - \frac{1}{128D_i^3\lambda_i^3}\|\partial_x w_i(t)\|^2. \tag{13.86}$$

By applying again Poincaré's inequality (second case) [128]

$$\|w_i(t)\|^2 \leq 2w_i(D,t)^2 + 4D_i^2\|\partial_x w_i(t)\|^2 \tag{13.87}$$

to the term $-\frac{1}{128D_i^3\lambda_i^3}\|\partial_x w_i(t)\|^2$ in (13.86), we obtain

$$\begin{aligned}\dot{V}_i(t) \leq & -\frac{\lambda_i}{4}[\vartheta_i^{\text{av}}(t)]^2 - (c_i - c_1^*)w_i^2(D_i,t) - (c_i - c_2^*)\|\partial_x w_i(t)\|^2 \\ & - \frac{1}{512D_i^5\lambda_i^3}\|w_i(t)\|^2 + \frac{1}{256D_i^5\lambda_i^3}w_i^2(D_i,t).\end{aligned} \tag{13.88}$$

Finally, by defining $c_3^* = c_1^* + \frac{1}{256D_i^5\lambda_i^3}$, and rearranging the terms, we can write

$$\dot{V}_i(t) \leq -\frac{\lambda_i}{4}[\vartheta_i^{\text{av}}(t)]^2 - \frac{1}{512D_i^5\lambda_i^3}\|w_i(t)\|^2 - (c_i - c_2^*)\|\partial_x w_i(t)\|^2 - (c_i - c_3^*)w_i^2(D_i,t). \tag{13.89}$$

Hence, from (13.86), if c_i is chosen such as $c_i > \max\{c_2^*, c_3^*\}$, we obtain

$$\dot{V}_i(t) \leq -\mu_i V_i(t) \qquad \text{or} \qquad \dot{V}(t) \leq -\mu V(t) \tag{13.90}$$

for some $\mu_i > 0$ and $\mu = \min(\mu_i)$.

Since the diffusion constants $D_i > 0$ are considered to be known, there is no adverse impact on the results to satisfy the bound estimates in the technical developments. More importantly, our estimates do not depend on the Hessian H.

Thus, the target system is exponentially stable in the sense of the $\mathcal{H}_1$-norm

$$\left(\sum_{i=1}^{n}[\vartheta_i^{\mathrm{av}}(t)]^2+\int_0^{D_i} w_i^2(x,t)dx+\int_0^{D_i}\partial_x w_i^2(x,t)dx+w_i^2(D_i,t)\right)^{1/2} \tag{13.91}$$

or in the compact form

$$\left(\vartheta_{\mathrm{av}}^2(t)+\|w(t)\|^2+\|\partial_x w(t)\|^2+w^2(D,t)\right)^{1/2}, \tag{13.92}$$

i.e., in the transformed variable $(\vartheta_{\mathrm{av}}, w)$.

13.3.4 ▪ Exponential Stability Estimate (in the $\mathcal{H}_1$-Norm) for the Average Closed-Loop System (13.60)–(13.63)

To obtain exponential stability in the sense of the norm

$$\Upsilon^2(t) \triangleq \left(|\vartheta_{\mathrm{av}}(t)|^2+\sum_{i=1}^{n}\|u_i^{\mathrm{av}}(x,t)\|^2+\|\partial_x u_i^{\mathrm{av}}(x,t)\|^2+|u_i^{\mathrm{av}}(D_i,t)|^2\right) \tag{13.93}$$

we need to show there exist $\alpha_1, \alpha_2 > 0$ such that

$$\alpha_1\Upsilon^2(t) \leq V(t) \leq \alpha_2\Upsilon^2(t). \tag{13.94}$$

This is straightforward to establish by using (13.65), (13.71), (13.74), (13.75) and employing the Cauchy–Schwarz inequality and other calculations, as in the proof of [119, Theorem 2.1].

Hence, with (13.90), we get

$$\Upsilon(t) \leq \frac{\alpha_2}{\alpha_1}e^{-\mu t}\Upsilon(0), \tag{13.95}$$

which completes the proof of exponential stability in the original variable $(\vartheta_{\mathrm{av}}, u_{\mathrm{av}})$.

13.3.5 ▪ Invoking the Averaging Theorem

The averaging theorem for infinite-dimensional systems [83, Section 2] (Appendix B) applies to systems in the form

$$\dot{Z}(t) = AZ(t)+F(\omega t, Z(t)), \tag{13.96}$$

where $Z(t) \in \mathcal{X} := \mathcal{H}_1([0,D])$ is the infinite-dimensional state vector. By the state-transformation of (13.52)–(13.55) with $v_i(x,t) = u_i(x,t)-U_i(t)$, we obtain the closed-loop system with homogeneous boundary conditions

$$\dot{\vartheta}_i(t) = v_i(0,t)+U_i(t), \tag{13.97}$$

$$\partial_t v_i(x,t) = \partial_{xx}v_i(x,t)-\phi_i(\vartheta_i, v_i, U_i, t), \quad x \in (0, D_i), \tag{13.98}$$

$$\partial_x v_i(0,t) = 0, \tag{13.99}$$

$$v_i(D_i,t) = 0, \tag{13.100}$$

$$\dot{U}_i(t) = \phi_i(\vartheta_i, v_i, U_i, t), \tag{13.101}$$

with

$$\phi_i(\vartheta_i, v_i, U_i, t) = -c_iU_i(t) - c_ik_i\left[z_i(t)+\int_0^{D_i}(D_i-\sigma)(v_i(\sigma,t)+U_i(t))d\sigma\right]. \tag{13.102}$$

Next, we write the PDE system (13.98)–(13.100) as an evolutionary equation in the Banach space $\mathcal{X}$,

$$\dot{\mathcal{V}}_i(t) = \mathcal{A}_i \mathcal{V}_i(t) - \tilde{\phi}_i(\vartheta_i, \mathcal{V}_i, U_i, t) \quad \forall t > 0, \tag{13.103}$$

where $\mathcal{V}_i(t)$ is an infinite-dimensional vector function which belongs to the Banach space $\mathcal{X}$. Furthermore, $\mathcal{A}_i$ is the realization of the second-order derivative with one Dirichlet and Neumann boundary condition in $\mathcal{X}$ with

$$\mathcal{A}_i \varphi_i := \frac{\partial^2 \varphi_i}{\partial x^2}, \tag{13.104}$$

and the domain

$$\bar{\mathcal{D}}(\mathcal{A}_i) = \left\{ \varphi_i \in \mathcal{X} : \varphi_i,\ \frac{d}{dx}\varphi_i \in \mathcal{X} \text{ are a.c.}, \right.$$
$$\left. \frac{d^2}{dx^2}\varphi_i \in \mathcal{X},\ \frac{d}{dx}\varphi_i(0) = 0,\ \varphi_i(D_i) = 0 \right\}, \tag{13.105}$$

where a.c. means *absolutely continuous*. To express $v_i(0,t)$ in the ODE (13.97) in terms of $\mathcal{V}_i(t)$, we introduce the linear boundary operator $\mathcal{B}_i : \mathcal{X} \to \mathbb{R}$ such that $\mathcal{B}_i \mathcal{V}_i(t) := v_i(0,t)$. Furthermore, we define the linear operators $\alpha^\top : \mathcal{X} \to \mathbb{R}$ and $\beta_i : \mathbb{R} \to \mathcal{X}$ as

$$\begin{aligned}
&\alpha^\top \mathcal{V}_i(t) := \int_0^{D_i} (D_i - \sigma) v_i(\sigma, t) d\sigma, \\
&\beta_i \zeta := [(\beta_i)_1, (\beta_i)_2, \ldots]^\top \zeta, \quad \zeta \in \mathbb{R}, \\
&\text{with} \quad (\beta_i)_k = \int_0^{D_i} \psi_k^{[i]}(x) dx \\
&\qquad\qquad = -\sqrt{\frac{2}{D_i}} \frac{2D_i}{\pi(2k-1)} (-1)^k, \quad k = 1, 2, \ldots, \infty,
\end{aligned} \tag{13.106}$$

where $\psi_k^{[i]}(x) = \sqrt{2/D_i} \cos(\pi/2(2k-1)x/D_i)$ are the eigenfunctions of $\mathcal{A}_i$ and $\tilde{\phi}_i(\vartheta_i, \mathcal{V}_i, U_i, t)$ in (13.103) can be rewritten as

$$\tilde{\phi}_i(\vartheta_i, \mathcal{V}_i, U_i, t) = -c_i \beta_i U_i(t) - c_i \beta_i k_i \left[z_i(t) + \left(\frac{1}{2} D_i^2 U_i(t) + \alpha^\top \mathcal{V}_i(t) \right) \right]. \tag{13.107}$$

Finally, taking into account the $\tilde{\Gamma}$-dynamics given in (13.37), the closed-loop system with infinite-dimensional state vector $Z(t) = [Z_1(t) \ \ldots \ Z_n(t) \ \tilde{\Gamma}(t)]^\top$ can be rewritten as (13.96)

$$\dot{Z} = \underbrace{\begin{bmatrix} A_1 & \ldots & 0 & 0 \\ \vdots & \ddots & \vdots & 0 \\ 0 & \ldots & A_n & 0 \\ 0 & \ldots & 0 & \omega_r \end{bmatrix}}_{A} Z + \underbrace{\begin{bmatrix} F_1(\omega t, Z_1) \\ \vdots \\ F_n(\omega t, Z_n) \\ \omega_r H^{-1} - \omega_r \left[\tilde{\Gamma} + H^{-1}\right] \hat{H}(t) \left[\tilde{\Gamma} + H^{-1}\right] \end{bmatrix}}_{F(\omega t, Z(t))} \tag{13.108}$$

with $Z_i(t) = [\vartheta_i(t) \ \mathcal{V}_i(t) \ U_i(t)]^\top$, $i = 1, \ldots, n$,

$$A_i = \begin{bmatrix} 0 & \mathcal{B}_i & 1 \\ 0 & \mathcal{A}_i & -c_i \beta_i \\ 0 & 0 & -c_i \end{bmatrix}, \quad F_i(\omega t, Z_i) = \begin{bmatrix} 0 \\ -c_i \beta_i k_i [z_i(t) + g_i(Z_i)] \\ -c_i k_i [z_i(t) + g_i(Z_i)] \end{bmatrix}, \tag{13.109}$$

and $g_i(Z_i) = \frac{1}{2}D_i^2 U_i(t) + \alpha^\top \mathcal{V}_i(t)$. Note that the signals $z_i(t)$ and $\hat{H}(t)$ are explicit functions of ωt by construction since $\omega_i = \mathcal{O}(\omega)$ according to (13.34). Since the operator A in (13.108) generates an analytic semigroup by the operator matrix theorem in [161] and $F(\omega t, Z)$ is Fréchet differentiable in Z, strongly continuous and periodic in t, uniformly with respect to Z, hence, all assumptions to apply the averaging theorem for infinite-dimensional systems [83, Section 2] in Appendix B are satisfied.

The solution of the original closed-loop system (13.52)–(13.55) behaves as the averaging system $\dot{Z}_{\rm av} = AZ_{\rm av} + F_0(Z_{\rm av})$ with $F_0(Z_{\rm av}) = \lim_{\Pi\to 0} \frac{1}{\Pi}\int_0^{\Pi} F(\tau, Z_{\rm av})d\tau$. Along with the local exponential stability of the average system we showed in (13.95) and for the equilibrium $\tilde{\Gamma}_{\rm av}(t) = 0$ of (13.73), there exists an exponentially stable periodic solution $Z^{\Pi}(t)$ with $\|Z^{\Pi}(t)\| \le \mathcal{O}(1/\omega)$ of the original system (13.52)–(13.55) and (13.37). Thus, the original closed-loop system has an exponentially stable periodic solution $Z^{\Pi}(t)$ that satisfies (13.56) for ω sufficiently large.

13.3.6 ▪ Asymptotic Convergence to a Neighborhood of the Extremum $(\theta^*, \Theta^*, y^*)$

First, consider the PDE system (13.10)–(13.13). From Agmon's, Poincaré's, and Young's inequalities [128], one has

$$\begin{aligned} \bar{\alpha}_i(D_i,t)^2 &\le \max_{x\in[0,D_i]} |\bar{\alpha}_i(x,t)|^2 \\ &\le \bar{\alpha}_i(0,t)^2 + 2\|\bar{\alpha}_i(t)\|\|\partial_x\bar{\alpha}_i(t)\|, \qquad (13.110) \\ \|\bar{\alpha}_i(t)\|^2 &\le 2D_i\bar{\alpha}_i(0,t)^2 + 4D_i^2\|\partial_x\bar{\alpha}_i(t)\|^2, \qquad (13.111) \\ 2\|\bar{\alpha}_i(t)\|\|\partial_x\bar{\alpha}_i(t)\| &\le \|\bar{\alpha}_i(t)\|^2 + \|\partial_x\bar{\alpha}_i(t)\|^2 \qquad (13.112) \end{aligned}$$

and, finally,

$$\bar{\alpha}_i(D_i,t)^2 \le (2D_i+1)\bar{\alpha}_i(0,t)^2 + (4D_i^2+1)\|\partial_x\bar{\alpha}_i(t)\|. \qquad (13.113)$$

Plugging the boundary values (13.10) and (13.13) into (13.113), one has

$$\tilde{\theta}_i^2(t) \le (2D_i+1)\vartheta_i(t)^2 + (4D_i^2+1)\|\partial_x\bar{\alpha}_i\|^2. \qquad (13.114)$$

Now, we introduce the Lyapunov–Krasovskii functional

$$\Lambda_1(t) = \frac{1}{2}\|\partial_x\bar{\alpha}_i(t)\|^2 \qquad (13.115)$$

and show the exponential stability of $\partial_x\bar{\alpha}_i(x,t)$. By taking the time derivative of (13.115) along with (13.11)–(13.12), one has

$$\dot{\Lambda}_1(t) = \partial_x\bar{\alpha}_i(D_i,t)\partial_t\bar{\alpha}_i(D_i,t) - \|\partial_{xx}\bar{\alpha}_i(t)\|^2. \qquad (13.116)$$

We use Young's, Agmon's, and Poincaré's inequalities [128] again and the identities $\partial_t\bar{\alpha}_i(D_i,t) = \dot{\tilde{\theta}}_i(t) = U_i(t)$, $\partial_x\bar{\alpha}_i(0,t) = 0$, such that from (13.116) we get

$$\dot{\Lambda}_1(t) \le -\frac{1}{8D_i^2}\|\partial_x\bar{\alpha}_i(t)\|^2 + 2D_iU_i(t)^2. \qquad (13.117)$$

Hence, the Λ_1-dynamics is Input-to-State (exponentially) Stable [99] with respect to $U_i(t)$. With (13.114), some appropriate $k > 0$, and defining $\bar{k} := 2D_i + 1$, we have

$$\limsup_{t\to\infty} |\tilde{\theta}_i(t)|^2 = \limsup_{t\to\infty} \{\bar{k}|\vartheta_i(t)|^2 + k|U_i(t)|^2\}. \qquad (13.118)$$

Equation (13.118) can be written in terms of the periodic solutions $\vartheta_i^{\Pi}(t)$ and $U_i^{\Pi}(t)=u_i^{\Pi}(D_i,t)$ as

$$\limsup_{t\to\infty}|\tilde{\theta}_i(t)|^2=\limsup_{t\to\infty}\left\{\bar{k}\sqrt{2}\left(|\vartheta_i(t)-\vartheta_i^{\Pi}(t)|^2+|\vartheta_i^{\Pi}(t)|^2\right)\right\}$$
$$+\limsup_{t\to\infty}\left\{k\sqrt{2}\left(|U_i(t)-U_i^{\Pi}(t)|^2+|U_i^{\Pi}(t)|^2\right)\right\}. \tag{13.119}$$

From the averaging theorem [83] in Appendix B, we know $\vartheta_i(t)-\vartheta_i^{\Pi}(t)\to 0$ as well as $U_i(t)-U_i^{\Pi}(t)\to 0$ exponentially as $t\to\infty$. Hence,

$$\limsup_{t\to\infty}|\tilde{\theta}_i(t)|^2=\limsup_{t\to\infty}\left\{\bar{k}\sqrt{2}|\vartheta_i^{\Pi}(t)|^2+k\sqrt{2}|U_i^{\Pi}(t)|^2\right\}. \tag{13.120}$$

With (13.56) and (13.120), we obtain $\limsup_{t\to\infty}|\tilde{\theta}(t)|=\mathcal{O}(1/\omega)$. Using the definition of $\tilde{\theta}(t)$ in (13.9) and the relation $\theta(t)=\hat{\theta}(t)+S(t)$, we get $\theta(t)-\theta^*=\tilde{\theta}(t)+S(t)$. Since $S(t)$ is of order $\mathcal{O}\left(|a|e^{D\sqrt{\omega/2}}\right)$ with $D=\max(D_i)$, as shown in (13.24), we finally obtain (13.57). The convergence of the propagated actuator $\Theta(t)$ to the optimizer Θ^* is even easier to prove. Taking the 2-norm of (13.29), we have

$$|\Theta_i(t)-\Theta_i^*|=|\vartheta_i(t)+a_i\sin(\omega_i(t))|. \tag{13.121}$$

As in the convergence proof of the parameter $\theta(t)$ to the optimal parameter θ^* above, we write (13.121) in terms of the periodic solution $\vartheta_i^{\Pi}(t)$ and follow the same steps by applying Young's inequality [128] and the exponential convergence of $\vartheta_i(t)-\vartheta_i^{\Pi}(t)\to 0$ as $t\to\infty$, according to the averaging theorem [83, Section 2] (Appendix B). Hence,

$$\limsup_{t\to\infty}|\Theta_i(t)-\Theta_i^*|=\limsup_{t\to\infty}\left|\sqrt{2}\vartheta_i^{\Pi}(t)+a_i\sin(\omega_i t)\right| \tag{13.122}$$

and finally, with (13.56), we can write (13.58).

In order to show the convergence of the output $y(t)$ of the static map to the optimal value y^*, we rewrite (13.8) as

$$y(t)-y^*=\frac{1}{2}(\Theta(t)-\Theta^*)^T H(\Theta(t)-\Theta^*). \tag{13.123}$$

By applying the Rayleigh inequality to the quadratic function in (13.123), we obtain $|y(t)-y^*|\le\lambda_{\max}(H)|\Theta(t)-\Theta^*)|^2$, where $\lambda_{\max}(H)$ denotes the largest eigenvalue of H. Hence,

$$\limsup_{t\to\infty}|y(t)-y^*|=\lambda_{\max}(H)\limsup_{t\to\infty}|\Theta(t)-\Theta^*)|^2. \tag{13.124}$$

Finally, by replacing (13.58) into (13.124) and applying Young's inequality [128] to the result, we have (13.59).

13.4 ▪ Application to Different Classes of PDEs

In order to show that the proposed ES approach in Figure 13.1 is general and applicable to a wider class of PDEs, we have formulated in Table 13.1 the stabilizing boundary control (BC) law $U_i(t)$ and given the explicit solutions to the trajectory generation problem of $S_i(t)$ in (13.20)–(13.23) for five other classes of distributed parameter systems [128]: (a) reaction-advection-diffusion (RAD) PDEs [172] (Chapter 11), (b) wave equations [176] (Chapter 12), (c) hyperbolic transport

Table 13.1. *Multivariable ES for Distinct Classes of PDE Systems.*

RAD Equation	PDE : $\partial_t\alpha_i(x,t)=\epsilon_i\partial_{xx}\alpha_i(x,t)+b_i\partial_x\alpha_i(x,t)+\lambda_i\alpha_i(x,t)$, $x\in[0,1]$ BC (Dirichlet): $U_i(t)=\frac{c_i}{s+c_i}\left\{-k_ie^{-\frac{b_i}{2\epsilon_i}}\left[\gamma(1)z_i(t)+\int_0^1 e^{\frac{b_i}{2\epsilon_i}\sigma}m(1-\sigma)u(\sigma,t)d\sigma\right]\right\}$, $\gamma(x)=\cosh\left(\sqrt{\frac{\xi}{\epsilon_i}}x\right)+\frac{b_i}{2\epsilon_i}\sqrt{\frac{\epsilon_i}{\xi}}\sinh\left(\sqrt{\frac{\xi}{\epsilon_i}}x\right)$, $\xi:=b_i^2/(4\epsilon_i)-\lambda_i\geq 0$, $m(x-\sigma)=\frac{1}{\epsilon_i}\sqrt{\frac{\epsilon_i}{\xi}}\sinh\left(\sqrt{\frac{\xi}{\epsilon_i}}(x-\sigma)\right)$, $\epsilon_i>0, b_i\geq 0, \lambda_i\geq 0$ Trajectory Generation : $S_i(t)=e^{-\frac{b_i}{2\epsilon_i}}\sum_{k=0}^{\infty}\frac{a_{2k}(t)}{(2k)!}+\frac{b_i}{2\epsilon_i}\frac{a_{2k}(t)}{(2k+1)!}$, $a_{2k}:=\frac{a_i}{\epsilon_i^k}\sin(\omega_i t)\sum_{n=0}^{k}\binom{k}{2n}\xi^{k-2n}\omega_i^{2n}+\frac{a_i}{\epsilon_i^k}\cos(\omega_i t)\sum_{n=0}^{k}\binom{k}{2n+1}\xi^{k-2n-1}\omega_i^{2n+1}$
Wave Dynamics	PDE : $\partial_{tt}\alpha_i(x,t)=\partial_{xx}\alpha_i(x,t)$, $x\in[0,D_i]$ BC (Neumann): $U_i(t)=\frac{c_i}{s+c_i}\{c[-k_iu_i(D_i,t)-\partial_t u_i(D_i,t)]+\rho(D_i)z_i+$ $\int_0^{D_i}\rho(D_i-\sigma)\partial_t u_i(\sigma,t)d\sigma\}$, $\rho(s)=-k_i[0\ \ I]e^{As}[0\ \ I]^T$, $A=\begin{pmatrix}0 & 0\\ I & 0\end{pmatrix}$ Trajectory Generation : $S_i(t)=a_i\cos(\omega_iD_i)\sin(\omega_i t)$
Constant Delay	PDE : $\partial_t\alpha_i(x,t)=\partial_x\alpha_i(x,t)$, $x\in[0,D_i]$ BC (Dirichlet) : $U_i(t)=\frac{c_i}{s+c_i}\left\{-k_i\left[z_i(t)+\int_0^{D_i}u_i(\sigma,t)d\sigma\right]\right\}$ Trajectory Generation : $S_i(t)=a_i\sin(\omega_i(t+D_i))$
Variable Delay	PDE : $\partial_t\alpha_i(x,t)=\pi_i(x,t)\partial_x\alpha_i(x,t)$, $x\in[0,1]$, $\pi_i(x,t)=\frac{1+x[\frac{d(\phi^{-1}(t))}{dt}-1]}{\phi^{-1}(t)-t}$ BC (Dirichlet): $U_i(t)=\frac{c_i}{s+c_i}\{-k_i[z_i(t)+\int_0^1 u_i(\sigma,t)(\phi^{-1}(t)-t)\,d\sigma]\}$ Trajectory Generation : $S_i(t)=a_i\sin(\omega_i t)$, $\phi(t):=t-D_i(t)$ Demodulation : $M_i(t)=\frac{2}{a_i}\sin(\omega_i(t-D_i(t)))$, $N_i(t)=-\frac{8}{a_i^2}\cos(2\omega_i(t-D_i(t)))$
Distributed Delay	PDE : $\partial_t\alpha_i(x,t)=\partial_x\alpha_i(x,t)$, $x\in[0,D_i]$, $y=Q\left(\int_0^{D_i}\Theta(t-\sigma)d\beta(\sigma)\right)$ BC (Dirichlet): $U_i(t)=\frac{c_i}{s+c_i}\left\{-k_i\left[z_i(t)+\int_0^{D_i}(1-\beta(\sigma))u_i(D_i-\sigma,t)d\sigma\right]\right\}$ Trajectory Generation : $S_i(t)=\frac{a_i}{\gamma(\omega_i)}\int_0^{D_i}\sin(\omega_i(t+\xi))d\beta(\xi)$

PDEs —for constant delays [179] (Chapters 2 and 7), (d) time-varying delays [182] (Chapter 8), and (e) distributed delays [227] (Chapter 9).

As for the diffusion case in (13.16)–(13.19), the term $u_i(x,t)$ which appears in $U_i(t)$ of Table 13.1 is the state of the infinite-dimensional system corresponding to a copy of the PDE model of the actuator dynamics.

For the distinct cases involving ES plus PDEs in Table 13.1, we point out the following applications and its corresponding references: (1) *Traffic Control* for linearized Lighthill–Whitham–Richards (LWR) macroscopic PDE models transformed into constant delays [232, 179]; (2) *Neuromuscular Electrical Stimulation* (NMES) problem for ES under time-varying delays [189, 182]; (3) *Optimal Oil Drilling Control* with ES for wave models [1]; (4) *Bioreactors* considering ES for models described by parabolic PDEs (reaction-diffusion equations) (see [46, 91]); and (5) additive manufacturing modeled by the *Stefan PDE* [110].

13.5 ▪ Simulation Example

Consider a quadratic static map as in (13.8), with Hessian $H=-\begin{pmatrix}2 & 2\\ 2 & 4\end{pmatrix}$, optimizer $\Theta^*=(1,0)$ and optimal value $y^*=5$. The domains length for the diffusion processes are set to $D_1=3$ and $D_2=4$. In what follows, we present numerical simulations of the proposed multivariable ES based on the filtered diffusion compensator (13.51). We performed our tests with

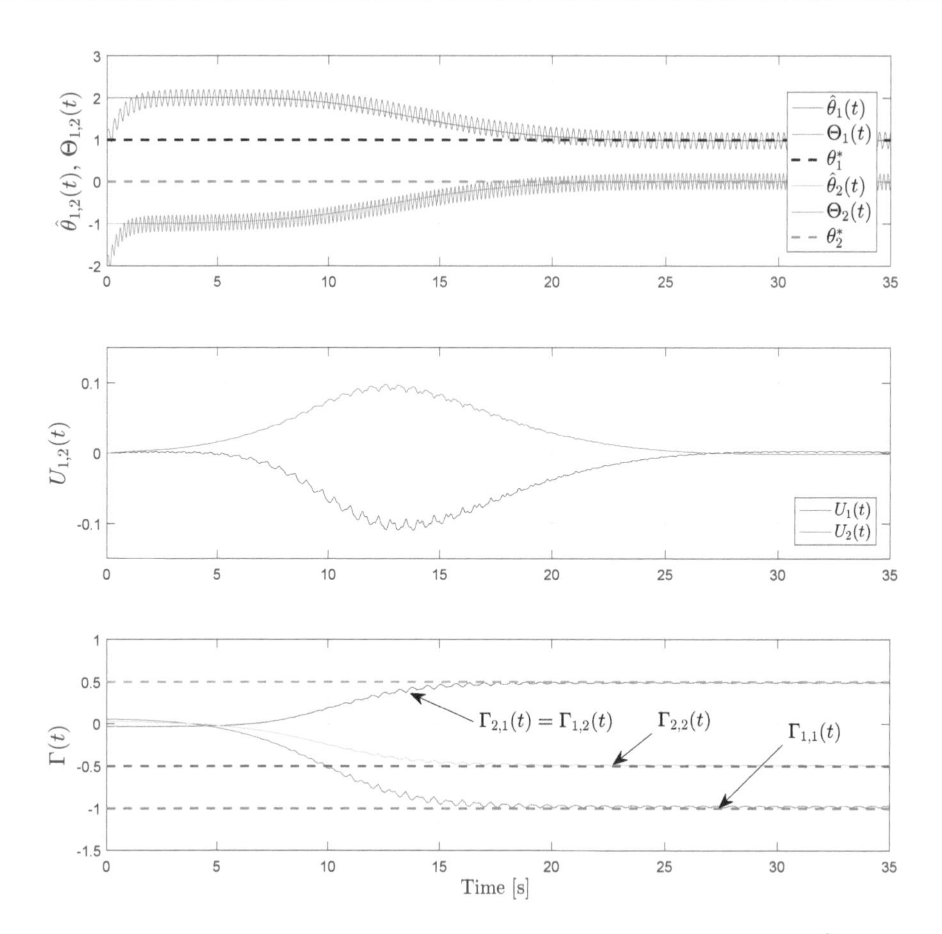

Figure 13.2. *Newton-based ES under multiple diffusion PDEs:* (a) *parameters* $\hat{\theta}_i(t)$ *and* $\Theta_i(t)$*;* (b) *control signals* $U_1(t)$ *and* $U_2(t)$*;* (c) *Hessian's inverse estimate* $\Gamma(t)$*. The elements of the matrix* $\Gamma(t)$ *converge to the unknown elements of* $H^{-1} = [-1 \ \ 0.5; 0.5 \ \ -0.5]$.

$c_1 = c_2 = 10$, $k_1 = k_2 = 0.17$, $a_1 = a_2 = 0.2$, $\omega = 5$, $\omega_1 = 5\omega$, $\omega_2 = 7\omega$, $\omega_r = 0.5$, $\Gamma(0) = -[1/100 \ \ 0; 0 \ \ 1/200]$, and $\hat{\theta}(0) = (2, -1)$.

Figure 13.2 presents relevant variables for the Newton-based ES: (a) each variable $(\hat{\theta}(t), \Theta(t))$ converging to a neighborhood of the optimum (θ^*, Θ^*), (b) the control signals $U_1(t)$ and $U_2(t)$, and (c) the Hessian's inverse estimate given by $\Gamma(t)$. This exact perturbation-based estimation of the inverse of the Hessian allows the perfect diffusion compensation.

In Figure 13.3, we note the remarkable evolution of new ES scheme in searching the maximum in spite of the presence of diffusion in the input channels and despite the difference in the length of the input channels.

13.6 ▪ Notes and References

With the preceding chapters on heat, RAD, and wave PDEs dealing with single-input problems, in this chapter we dealt with an extension to multi-input maps, with PDEs in their input channels, and proposed a Newton-based multivariable ES algorithm for a map with distinct PDE systems in its distinct actuation paths. This contribution generalizes our multivariable results for time-

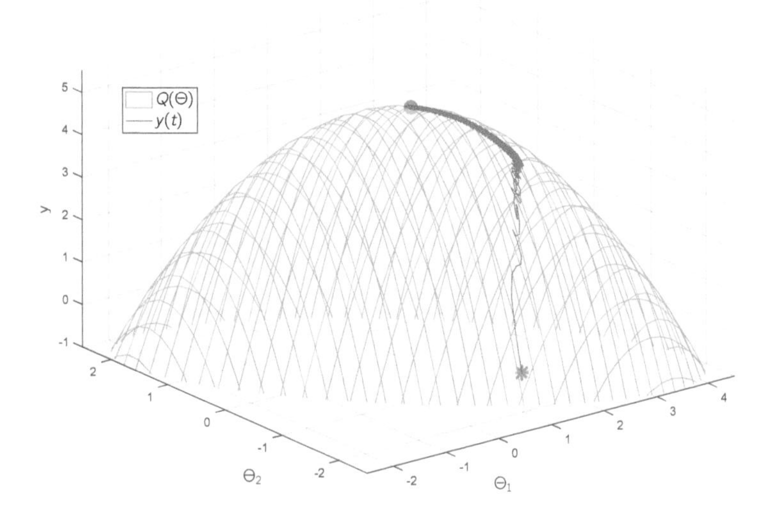

Figure 13.3. *Extremum seeking in the 3D-space: parametrized quadratic map* $Q(\Theta)$, *the initial point* $(\Theta_1(0), \Theta_2(0)) = (1.25, -1.75)$ *denoted by* $*$, *and the output trajectory* $y(t) \to y^* \pm \mathcal{O}\left(|a|^2 + 1/\omega^2\right)$ *in the state-space, with* $y^* = 5$ *(denoted by* $\bullet$*) corresponding to the optimizer* $(\Theta_1^*, \Theta_2^*) = (1, 0)$.

delay systems to a wider range of PDEs, encompassing multiple diffusion processes (each with a distinct spatial domain), and amenable to extension to reaction-advection-diffusion equations and wave equations.

Unlike Gradient, the Newton method estimates the Hessian's inverse and effectively diagonalizes the multivariable map by canceling the coupling of its multiple inputs via Hessian matrix, which allows for decentralized PDE-compensators for each input channel of the map. Thanks to this decentralization feature, the main contribution with respect to our previous papers [65] and [169] is that now we consider the *multivariable* ES problem for a quite wide class of PDE-based systems, while only the simplest single-input maps were treated in [65] and [169] for a single PDE in its actuation channel.

Our result is not a simple combination of ES and PDE compensation. In general, the design of a feedback controller to compensate for the PDE dynamics in the actuation path requires a known model of the ODE. In our problem, while we assume known diffusion coefficients, the parameters of the nonlinear static map (and, most notably, its Hessian) are unknown. So we present one semi-model-based approach to construct a PDE boundary control based on perturbation-based estimates of the model. Our approach is based on the Newton optimization where we estimate the Hessian's inverse by a Riccati filter [73] for the purpose of making the design of the PDE compensator simple. With the inverse of the Hessian estimated, the average plant for which the compensator is designed becomes, essentially, an integrator with a known gain, the simplest special case of the existing backstepping literature [128] for stabilization of PDE-ODE cascades with full modeling knowledge. The scalar PDE-integrator cascade compensator design is then generalized to the multi-input case and finally employed in this "partially model-free" ES setting. Moreover, each element of the vector with periodic dither signals is chosen individually, following the trajectory generation approach [196, 128, 150], with distinct amplitude and frequency parameters.

The analysis builds on the same mathematical foundation originally established in [179], but now accounting for the fact that we are dealing with a number of distinct (and uncoupled) PDE systems. We extend this analysis methodology to collections of PDEs more general than delays on inputs, indicating in our exposition the precise sequence of analytical steps for averaging in infinite dimension [83], the backstepping transformations, and for the overall infinite-dimensional system with nonlinearities and periodic perturbations. Local exponential stability for the overall closed-loop system is guaranteed as well as the convergence to a small neighborhood of the extremum point.

The next step is the extension of the proposed approach to a scenario where the PDE coefficients can be considered unknown [119, Section 15.3] following the same guidelines explored in Chapter 6 for the time-delay case. Another open problem would be the study of ES designs for other infinite-dimensional systems.

Chapter 14

ES for PDE-PDE Cascades

In this final chapter of the book's Part II, which has dealt with PDEs at the input into an unknown map, and in which we have already advanced from transport PDEs to reaction-advection-diffusion PDEs to wave PDEs, we now consider one last configuration in which the input pathway to the map contains a cascade of PDEs from distinct classes, as depicted in the general block diagram in Figure 14.1. First, we deal with PDEs with input delays such as, for example, the notorious problem of a wave PDE with an input delay where, if the delay is left uncompensated, an arbitrarily short delay destroys the closed-loop stability. Then, we move forward to an even more challenging class of problems for parabolic-hyperbolic cascades of PDEs, focusing on a heat equation at the input of a wave PDE. The treatment of such systems with PDE-PDE cascades is performed by means of boundary control. Local exponential stability and convergence to a small neighborhood of the unknown extremum point are guaranteed by using a backstepping transformation and averaging in infinite dimension. Although our presentation focuses mostly on the classical Gradient extremum seeking, a generalization to the Newton-based approach is also provided.

PDE-PDE cascades have a great deal in common with PDE-ODE cascades. For instance, a cascade of a delay into a PDE is just a much generalized version of an integrator with an input delay, which has been the focus of the book's Part I. However, while in a delay-integrator cascade the design can be pursued within the predictor feedback framework, with backstepping just employed for an interpretation and for analysis, in delay-PDE cascades a predictor for a PDE is too complicated of a mathematical object to be of value. Instead, the design is pursued entirely by the backstepping approach. Similarly, while the heat-integrator cascade became familiar in Chapter 10, and was dealt with through the backstepping design, a more general backstepping design is applied to a heat-wave PDE cascade in a part of this chapter.

14.1 ▪ Static Maps with Actuation Dynamics Governed by Wave PDEs with Input Delays

In the simplest case of ES for static maps, the goal is to vary in real-time the input $\Theta(t)$ of an unknown nonlinear static map

$$y(t) = Q(\Theta(t)) \tag{14.1}$$

in order to find and maintain the output $y(t)$ at the unknown extremum y^*, where Θ^* is the unknown optimizer.

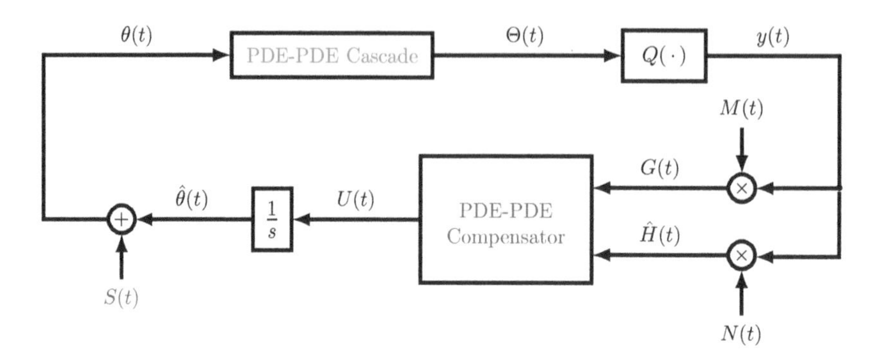

Figure 14.1. *Block diagram for implementation of ES control design for PDE-PDE cascades (in red) at the input of nonlinear convex maps $Q(\cdot)$. Although the multiplicative perturbation signals $M(t)$ and $N(t)$ are the same of the classical ES designs* [129], *the additive dither $S(t)$ (in blue) must be redesigned using the trajectory generation paradigm* [128, *Chapter* 12] *and the application of an adequate boundary control law (in blue) for PDE-PDE compensation is mandatory.*

14.1.1 ▪ Plant and Basic Assumptions

Without loss of generality, we consider maximization problems. We further assume the map (14.1) satisfies the following assumption.

Assumption 14.1. *The unknown nonlinear static map* (14.1) *is locally quadratic, i.e.,*

$$Q(\Theta) = y^* + \frac{H}{2}(\Theta - \Theta^*)^2, \tag{14.2}$$

in the neighborhood of the extremum, where besides $\Theta^* \in \mathbb{R}$ and $y^* \in \mathbb{R}$ being unknown, the scalar $H < 0$ is the unknown Hessian of the static map.

Assumption 14.1 states that if maps are at least locally quadratic, our methodology can be applied, and provide guarantees, in some neighborhood of the extremum. For maps that are not locally quadratic, and may not yield exponential stability of the average system, an approach by [217] might lead to *asymptotic* practical stability, not exponential, but the form of averaging theory that they use is not available for systems on Banach spaces (PDEs, delays).

Hence, by plugging (14.2) into (14.1), the output of the static map can be written as

$$y(t) = y^* + \frac{H}{2}(\Theta(t) - \Theta^*)^2. \tag{14.3}$$

Inspired by [122], we discuss the extension of our ES approach for wave PDEs with a delayed input as a PDE-PDE cascade. As discussed in Section 12.7.2, this is particularly an important problem since Datko et al. [47] showed that standard feedback laws for wave equations have a zero robustness margin to the introduction of a delay in the feedback loop.

In this sense, we consider the following actuation dynamics where the actuator $\theta(t)$ and the propagated actuator $\Theta(t)$ are related according to the following delay-wave PDE cascade system:

$$\Theta(t) = \alpha(0,t), \tag{14.4}$$

$$\partial_{tt}\alpha(x,t) = \partial_{xx}\alpha(x,t), \quad x \in [0,1], \tag{14.5}$$

$$\partial_x\alpha(0,t) = 0, \tag{14.6}$$

$$\alpha(1,t) = \theta(t - D), \tag{14.7}$$

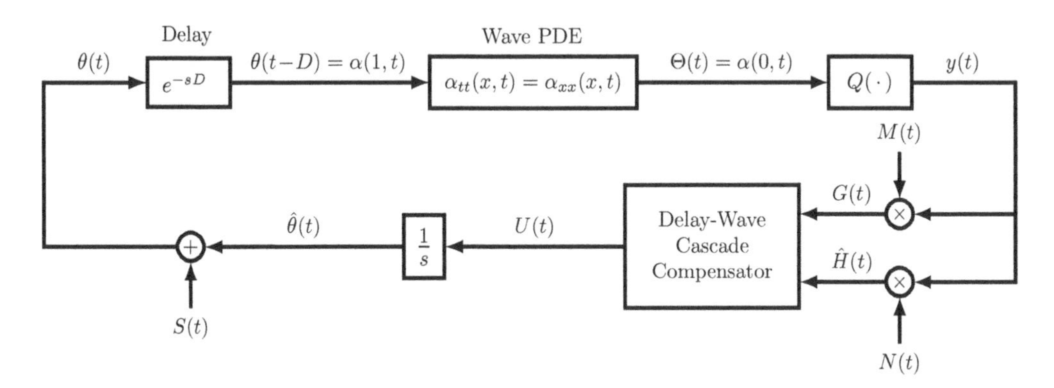

Figure 14.2. *Basic block diagram of the extremum seeking control system with a delay-wave PDE cascade system.*

where $\alpha(x,t)$ is the infinite-dimensional state of the wave PDE with spatial domain defined without loss of generality by $x\in[0,1]$. The boundary delay is denoted by $D>0$ being any arbitrary known constant.

Combining the actuation dynamics (14.4)–(14.7) and the static map (14.3), the closed-loop ES with actuation dynamics governed by a delay-wave PDE system under boundary control is illustrated in Figure 14.2.

14.1.2 ▪ Estimation Errors

As in the basic ES scheme, we define the unknown optimal input θ^* of $\theta(t)$ with respect to the static map and the wave process, with the relation $\Theta^*=\theta^*$. Since our goal is to find the unknown optimal input θ^*, we define the *estimation error*

$$\tilde{\theta}(t):=\hat{\theta}(t)-\theta^*, \tag{14.8}$$

where $\hat{\theta}(t)$ is the estimate of θ^*. To make (14.8) consistent with the optimizer of the static map Θ^*, we introduce the *propagated estimation error*

$$\vartheta(t):=\hat{\Theta}(t)-\Theta^* \tag{14.9}$$

through the delay-wave PDE cascade

$$\vartheta(t):=\bar{\alpha}(0,t), \tag{14.10}$$
$$\partial_{tt}\bar{\alpha}(x,t)=\partial_{xx}\bar{\alpha}(x,t), \quad x\in[0,1], \tag{14.11}$$
$$\bar{\alpha}_x(0,t)=0, \tag{14.12}$$
$$\bar{\alpha}(1,t)=\tilde{\theta}(t-D). \tag{14.13}$$

From the control loop in Figure 14.2, we get

$$\dot{\hat{\theta}}(t)=U(t). \tag{14.14}$$

Taking the time derivative of (14.10)–(14.13) and with the help of (14.8) and (14.14), the propagated error dynamics is written as the following cascade of a delay-wave PDE and ODE (integrator) with Dirichlet interconnection:

$$\dot{\vartheta}(t) = u(0,t), \tag{14.15}$$
$$\partial_{tt}u(x,t) = \partial_{xx}u(x,t), \quad x \in [0,1], \tag{14.16}$$
$$u_x(0,t) = 0, \tag{14.17}$$
$$u(1,t) = U(t-D), \tag{14.18}$$

where $\dot{\tilde{\theta}}(t-D) = \dot{\hat{\theta}}(t-D)$, since θ^* is constant, and

$$u(x,t) = \partial_t \bar{\alpha}(x,t). \tag{14.19}$$

14.1.3 ▪ Probing and Demodulation Signals

As in ES approach without actuation through a delay-wave PDE cascade, the perturbation signal $S(t)$ should add $a\sin(\omega t)$ to the input $\Theta(t)$ of the map (14.1), thus compensating the effects of the delay-wave process in the map's input.

Hence, $a\sin(\omega t)$ with perturbation amplitude a and frequency ω is applied as follows:

$$\beta(0,t) = a\sin(\omega t), \tag{14.20}$$
$$\partial_{tt}\beta(x,t) = \partial_{xx}\beta(x,t), \quad x \in [0,1], \tag{14.21}$$
$$\beta_x(0,t) = 0, \tag{14.22}$$
$$\beta(1,t) = \bar{\beta}(1,t) = S(t-D), \tag{14.23}$$
$$\bar{\beta}_t(x,t) = \bar{\beta}_x(x,t), \quad x \in [1, 1+D], \tag{14.24}$$
$$\bar{\beta}(1+D,t) = S(t), \tag{14.25}$$

where the delay term was represented by the transport PDE (14.24)–(14.25). Equations (14.20)–(14.25) describe a *trajectory generation problem* as in [128, Chapter 12]. The explicit solution of (14.25) is given by

$$S(t) = a\cos(\omega D)\sin(\omega(t+D)), \tag{14.26}$$

since $\beta(x,t) = a\cos(\omega x)\sin(\omega t)$ and $\bar{\beta}(x,t) = S(t+x-1-D)$.

The relation among the propagated estimation error $\vartheta(t)$, the propagated input $\Theta(t)$, and the optimizer of the static map Θ^* is given by

$$\vartheta(t) + a\sin(\omega t) = \Theta(t) - \Theta^*, \tag{14.27}$$

which can be easily proven for $x = 1$ since

$$\bar{\alpha}(x,t) = \alpha(x,t) - \beta(x,t) - \Theta^*, \tag{14.28}$$

and by considering $\theta(t) = \hat{\theta}(t) + S(t)$ along with the solutions of (14.4)–(14.7), (14.10)–(14.13), and (14.20)–(14.25). It remains to define the demodulation signal $N(t)$, which is used to estimate the Hessian of the static map by multiplying it with the output $y(t)$ of the static map. In [73], the

Hessian estimate for locally quadratic maps (14.3) is derived as

$$\hat{H}(t) = N(t)y(t) \quad \text{with} \quad N(t) = -\frac{8}{a^2}\cos(2\omega t). \tag{14.29}$$

Note that the demodulation signal $M(t)$, to estimate the gradient, is the same as in the basic ES (see Chapter 1), such that

$$G(t) = M(t)y(t) \quad \text{with} \quad M(t) = \frac{2}{a}\sin(\omega t). \tag{14.30}$$

Now, we introduce an important result of [73], where the averaged version of the gradient (14.30) and Hessian (14.29) estimates are

$$G_{\text{av}}(t) = H\vartheta_{\text{av}}(t), \qquad \hat{H}_{\text{av}}(t) = H \tag{14.31}$$

if a quadratic map as in (14.3) is considered. For the proof of (14.31), see [73].

14.2 ▪ Baseline Case ($D = 0$): Wave PDE as a Feedback Loop of Two Transport PDEs

Before we continue with the analysis of the delay-compensating design for the formulation in the previous section, we revisit the feedback law (in Section 12.7.1 with $q = 0$) that is obtained in the absence of delay ($D = 0$). This special case is of interest in its own right since it achieves stabilization of the wave equation, as performed in [180].

The main idea is to convert the wave equation in (14.15)–(14.18) into a feedback loop of two first-order transport equations which convect in opposite directions. To achieve this, we define the following Riemann transformations [26]:

$$\bar{\zeta}(x,t) = \bar{\alpha}_t(x,t) + \bar{\alpha}_x(x,t), \tag{14.32}$$

$$\bar{\omega}(x,t) = \bar{\alpha}_t(x,t) - \bar{\alpha}_x(x,t), \tag{14.33}$$

with their corresponding inverses given by

$$\bar{\alpha}_t(x,t) = \frac{\bar{\zeta}(x,t) + \bar{\omega}(x,t)}{2}, \tag{14.34}$$

$$\bar{\alpha}_x(x,t) = \frac{\bar{\zeta}(x,t) - \bar{\omega}(x,t)}{2}. \tag{14.35}$$

Hence, (14.15)–(14.18) is reshaped as

$$\textbf{ODE:} \quad \dot{\vartheta}(t) = \bar{\zeta}(0,t), \tag{14.36}$$

$$\textbf{PDE 1:} \begin{cases} \bar{\zeta}_t(x,t) = \bar{\zeta}_x(x,t), & x \in [0,1], \\ \bar{\zeta}(1,t) = U(t) + \bar{\alpha}_x(1,t), \end{cases} \tag{14.37}$$

$$\textbf{PDE 2:} \begin{cases} \bar{\omega}_t(x,t) = -\bar{\omega}_x(x,t), & x \in [0,1], \\ \bar{\omega}(0,t) = \bar{\zeta}(0,t). \end{cases} \tag{14.38}$$

Applying the backstepping transformations [26]

$$w(x,t) = \bar{\zeta}(x,t) - KH\int_0^x \bar{\zeta}(\sigma,t)d\sigma - KH\vartheta(t), \tag{14.39}$$

$$\varpi(x,t) = \bar{\omega}(x,t) + KH\int_0^x \bar{\omega}(\sigma,t)d\sigma - KH\vartheta(t), \tag{14.40}$$

the resulting average target system for (14.36)–(14.38) is

$$\dot{\vartheta}_{\rm av}(t) = KH\vartheta_{\rm av}(t) + w_{\rm av}(0,t), \tag{14.41}$$

$$\partial_t w_{\rm av}(x,t) = \partial_x w_{\rm av}(x,t), \quad w_{\rm av}(0,t) = \varpi_{\rm av}(0,t), \tag{14.42}$$

$$\partial_t \varpi_{\rm av}(x,t) = -\partial_x \varpi_{\rm av}(x,t), \tag{14.43}$$

$$w_{\rm av}(1,t) = -\frac{1}{c}\partial_t u_{\rm av}(1,t), \tag{14.44}$$

which is exponentially stable for $c > 0$ sufficiently large. This result is not difficult to prove since (14.41) is exponentially Input-to-State Stable (ISS) [99] with respect to $w_{\rm av}(0,t)$. On the other hand, $w_{\rm av}(x,t)$ is finite-time stable with respect to $\varpi_{\rm av}(0,t)$ and $\varpi_{\rm av}(x,t)$ is asymptotically stable for $c \to +\infty$ (or $w_{\rm av}(1,t) \to 0$) [26].

Since $u(1,t) = U(t)$ for $D = 0$, plugging (14.39) with $x = 1$ into (14.44), we can write $\dot{U}_{\rm av}(t) = -cw_{\rm av}(1,t)$ as

$$\dot{U}_{\rm av}(t) = -c\bar{\zeta}_{\rm av}(1,t) + cK\left[H\vartheta_{\rm av}(t) + H\int_0^1 \bar{\zeta}_{\rm av}(x,t)dx\right]. \tag{14.45}$$

Plugging the average version of (14.32) and (14.37) into (14.45), one has

$$\begin{aligned}\dot{U}_{\rm av}(t) &= -c[U_{\rm av}(t) + \partial_x\bar{\alpha}_{\rm av}(1,t)] \\ &\quad + cK\left[H\vartheta_{\rm av}(t) + H\int_0^1 [\partial_t\bar{\alpha}_{\rm av}(x,t) + \partial_x\bar{\alpha}_{\rm av}(x,t)]dx\right],\end{aligned} \tag{14.46}$$

which can be rewritten as

$$\begin{aligned}\dot{U}_{\rm av}(t) &= -cU_{\rm av}(t) + c\Bigg\{-\partial_x\bar{\alpha}_{\rm av}(1,t) + K\Bigg[H\vartheta_{\rm av}(t) \\ &\quad + H\int_0^1 [\partial_t\bar{\alpha}_{\rm av}(x,t) + \partial_x\bar{\alpha}_{\rm av}(x,t)]dx\Bigg]\Bigg\}.\end{aligned} \tag{14.47}$$

Recalling that $G_{\rm av}(t) = H\vartheta_{\rm av}(t)$ and $\hat{H}_{\rm av}(t) = H$ from (14.31) and (14.19), one can finally obtain the implementable control law

$$\dot{U}(t) = -cU(t) + c\left\{-\bar{\alpha}_x(1,t) + K\left[G(t) + \hat{H}(t)\int_0^1 [u(x,t) + \bar{\alpha}_x(x,t)]dx\right]\right\}, \tag{14.48}$$

with $\hat{H}(t)$ and $G(t)$ defined by (14.29) and (14.30), respectively.

Recalling (14.28), the control law (14.48) is indeed implementable since $\bar{\alpha}_x(x,t)$ can be written in terms of measurable signals

$$\bar{\alpha}_x(x,t) = \alpha_x(x,t) - \beta_x(x,t) \tag{14.49}$$

and the integral in (14.48) is given by

$$\begin{aligned}\int_0^1 \bar{\alpha}_x(x,t)dx &= \bar{\alpha}(1,t) - \bar{\alpha}(0,t)\\ &= \alpha(1,t) - \beta(1,t) - \Theta^* - \alpha(0,t) + \beta(0,t) + \Theta^*\\ &= \theta(t) - \beta(1,t) - \Theta(t) + a\sin(\omega t)\,. \end{aligned} \tag{14.50}$$

14.3 ▪ ES Boundary Control Law and Closed-Loop System for Wave PDE with Input Delay

Taking the time derivative of (14.10)–(14.13), the delay-wave PDE system is alternatively written as [122]

$$\dot{\vartheta}(t) = u(0,t), \tag{14.51}$$
$$\partial_{tt} u(x,t) = \partial_{xx} u(x,t), \quad x \in [0,1], \tag{14.52}$$
$$\partial_x u(0,t) = 0, \tag{14.53}$$
$$u(1,t) = v(1,t)\,, \tag{14.54}$$
$$\partial_t v(x,t) = \partial_x v(x,t), \quad x \in [1,1+D], \tag{14.55}$$
$$v(1+D,t) = U(t)\,, \tag{14.56}$$

where $U(t)$ is the overall system (control) input and (ϑ, u, v) is the state of the ODE-PDE-PDE cascade. From the transport PDE representation form for the input delay, we know that [119, Chapter 2]

$$v(x,t) = U(t+x-1-D), \quad x \in [1,1+D]\,. \tag{14.57}$$

Recalling that (14.52)–(14.54) can be represented into (14.37)–(14.38) with the help of (14.32) and (14.33), we can write

$$\textbf{ODE:} \quad \dot{\vartheta}(t) = \bar{\zeta}(0,t)\,, \tag{14.58}$$

$$\textbf{PDE 1:}\begin{cases} \partial_t \bar{\zeta}(x,t) = \partial_x \bar{\zeta}(x,t), & x \in [0,1]\,,\\ \bar{\zeta}(1,t) = v(1,t) + \bar{\alpha}_x(1,t)\,, \end{cases} \tag{14.59}$$

$$\textbf{PDE 2:}\begin{cases} \bar{\omega}_t(x,t) = -\bar{\omega}_x(x,t), & x \in [0,1]\,,\\ \bar{\omega}(0,t) = \bar{\zeta}(0,t)\,, \end{cases} \tag{14.60}$$

$$\textbf{PDE 3:}\begin{cases} v_t(x,t) = v_x(x,t), & x \in [1,1+D]\,,\\ v(1+D,t) = U(t)\,. \end{cases} \tag{14.61}$$

Applying the backstepping transformations

$$w(x,t) = \bar{\zeta}(x,t) - KH\int_0^x \bar{\zeta}(\sigma,t)d\sigma - KH\vartheta(t), \quad x \in [0,1], \tag{14.62}$$

$$\varpi(x,t) = \bar{\omega}(x,t) + KH\int_0^x \bar{\omega}(\sigma,t)d\sigma - KH\vartheta(t), \quad x \in [0,1], \tag{14.63}$$

$$\begin{aligned}\zeta(x,t) = {}& v(x,t) - KH\int_1^x v(\sigma,t)d\sigma\\ &+ \bar{\alpha}_x(1,t) - KH\int_0^1 \bar{\zeta}(\sigma,t)d\sigma - KH\vartheta(t), \quad x \in [1,1+D], \end{aligned} \tag{14.64}$$

into (14.58)–(14.61), the resulting average target system for (14.51)–(14.56) is given by

$$\textbf{ODE:} \quad \dot{\vartheta}_{\rm av}(t) = KH\vartheta_{\rm av}(t) + w_{\rm av}(0,t), \tag{14.65}$$

$$\textbf{PDE 1:} \begin{cases} \partial_t w_{\rm av}(x,t) = \partial_x w_{\rm av}(x,t), & x \in [0,1], \\ w_{\rm av}(1,t) = \zeta_{\rm av}(1,t), \end{cases} \tag{14.66}$$

$$\textbf{PDE 2:} \begin{cases} \partial_t \varpi_{\rm av}(x,t) = -\partial_x \varpi_{\rm av}(x,t), & x \in [0,1], \\ \varpi_{\rm av}(0,t) = w_{\rm av}(0,t), \end{cases} \tag{14.67}$$

$$\textbf{PDE 3:} \begin{cases} \partial_t \zeta_{\rm av}(x,t) = \partial_x \zeta_{\rm av}(x,t), & x \in [1,1+D], \\ \zeta_{\rm av}(1+D,t) = -\frac{1}{c}\partial_t v_{\rm av}(1+D,t). \end{cases} \tag{14.68}$$

From the target system (14.65)–(14.68), we can realize that $\vartheta_{\rm av}(t)$ is ISS [99] with respect to $w_{\rm av}(0,t)$; $w_{\rm av}(x,t)$ is ISS with respect to $\zeta_{\rm av}(1,t)$; and $\zeta_{\rm av}(x,t)$ is stabilized by the boundary control $U_{\rm av}(t) = v_{\rm av}(1+D,t)$, with $\dot{U}_{\rm av}(t) = \partial_t v_{\rm av}(1+D,t)$.

14.3.1 ▪ ES Boundary Control

From (14.64) and (14.68), one can obtain the expression for $\dot{U}_{\rm av}(t) = -c\zeta_{\rm av}(1+D,t)$ such that

$$\begin{aligned} \dot{U}_{\rm av}(t) = &-cv_{\rm av}(1+D,t) + cKH\int_1^{1+D} v_{\rm av}(\sigma,t)d\sigma \\ &- c\bar{\alpha}_x(1,t) + cKH\int_0^1 \bar{\zeta}_{\rm av}(\sigma,t)d\sigma + cKH\vartheta_{\rm av}(t). \end{aligned} \tag{14.69}$$

Recalling (14.32) and plugging (14.57) into (14.69), we can obtain the control law:

$$\begin{aligned} \dot{U}(t) = -cU(t) + c\Bigg\{ &-\bar{\alpha}_x(1,t) + K\Bigg[G(t) + \hat{H}(t)\int_{t-D}^{t} U(\tau)d\tau \\ &+ \hat{H}(t)\int_0^1 [u(x,t) + \bar{\alpha}_x(x,t)]\,dx\Bigg]\Bigg\} \end{aligned} \tag{14.70}$$

for $c > 0$ sufficiently large, $u(x,t) = \bar{\alpha}_t(x,t)$, and $\bar{\alpha}_x(x,t)$ satisfying (14.49)–(14.50). As in (14.48), the signals $\hat{H}(t)$ and $G(t)$ are still calculated by (14.29) and (14.30), respectively.

14.3.2 ▪ Closed-Loop System

Substituting the control law (14.70) into (14.56), we can write the closed-loop system (14.51)–(14.56) as

$$\dot{\vartheta}(t) = u(0,t), \tag{14.71}$$

$$\partial_{tt} u(x,t) = \partial_{xx} u(x,t), \quad x \in [0,1], \tag{14.72}$$

$$\partial_x u(0,t) = 0, \tag{14.73}$$

$$u(1,t) = v(1,t), \tag{14.74}$$

$$\partial_t v(x,t) = \partial_x v(x,t), \quad x \in [1,1+D], \tag{14.75}$$

$$\begin{aligned} v_t(1+D,t) = &-cv(1+D,t) - c\bar{\alpha}_x(1,t) \\ &+ cK\left[G(t) + \hat{H}(t)\int_1^{1+D} v(x,t)dx + \hat{H}(t)\int_0^1 [u(x,t) + \bar{\alpha}_x(x,t)]\,dx\right] \end{aligned} \tag{14.76}$$

or, equivalently, from (14.32)–(14.35) and (14.58)–(14.61) as

$$\dot{\vartheta}(t) = \bar{\zeta}(0,t), \tag{14.77}$$

$$\partial_t \bar{\zeta}(x,t) = \partial_x \bar{\zeta}(x,t), \quad x \in [0,1], \tag{14.78}$$

$$\bar{\zeta}(1,t) = 2v(1,t) - \bar{\omega}(1,t), \tag{14.79}$$

$$\bar{\omega}_t(x,t) = -\bar{\omega}_x(x,t), \quad x \in [0,1], \tag{14.80}$$

$$\bar{\omega}(0,t) = \bar{\zeta}(0,t), \tag{14.81}$$

$$v_t(x,t) = v_x(x,t), \qquad x \in [1,1+D], \tag{14.82}$$

$$v_t(1+D,t) = -cv(1+D,t) - \frac{c}{2}\left[\bar{\zeta}(1,t) - \bar{\omega}(1,t)\right]$$
$$+cK\left[G(t) + \hat{H}(t)\int_1^{1+D} v(x,t)dx + \hat{H}(t)\int_0^1 \bar{\zeta}(x,t)dx\right]. \tag{14.83}$$

In the next section, the availability of a small parameter in the hyperbolic PDEs-ODE loop of the target system (14.65)–(14.68), obtained via backstepping transformation [128, 119] for (14.77)–(14.83), permits the stability analysis of the complete ES feedback system through PDE small-gain results for ISS cascades [99].

14.4 ▪ Stability Analysis

The L_∞ space is to be regarded as the state space of the PDE in the closed-loop system, as this is the norm in which stability is studied in the next theorem.

Theorem 14.1. *(Delay-Wave Cascade) Consider the system in Figure* 14.2 *with the dynamic system being represented by the nonlinear quadratic map* (14.3), *satisfying Assumption* 14.1, *in cascade with the actuation dynamics governed by the delay-wave PDE in* (14.4)–(14.7). *For a sufficiently large $c > 0$, there exists some $\omega^*(c) > 0$, such that, $\forall \omega > \omega^*$, the closed-loop system* (14.77)–(14.83) *with states $\vartheta(t)$, $\bar{\zeta}(x,t)$, $\bar{\omega}(x,t)$, $v(x,t)$ has a unique locally exponentially stable periodic solution in t of period $\Pi := 2\pi/\omega$, denoted by $\vartheta^\Pi(t), \bar{\zeta}^\Pi(x,t), \bar{\omega}^\Pi(x,t), v^\Pi(x,t)$, satisfying, $\forall t \geq 0$,*

$$\left|\vartheta^\Pi(t)\right| + \left\|\bar{\zeta}^\Pi(t)\right\|_\infty + \left\|\bar{\omega}^\Pi(t)\right\|_\infty + \left\|v^\Pi(t)\right\|_\infty \leq \mathcal{O}(1/\omega). \tag{14.84}$$

Furthermore,

$$\limsup_{t\to\infty} |\theta(t) - \theta^*| = \mathcal{O}(a + 1/\omega), \tag{14.85}$$

$$\limsup_{t\to\infty} |\Theta(t) - \Theta^*| = \mathcal{O}(a + 1/\omega), \tag{14.86}$$

$$\limsup_{t\to\infty} |y(t) - y^*| = \mathcal{O}\left(a^2 + 1/\omega^2\right). \tag{14.87}$$

Proof. The proof of Theorem 14.1 follows analogous steps employed to prove [179, Theorem 1] (Theorem 7.1, Chapter 7) and [65, Theorem 1] (Theorem 10.1, Chapter 10). In this sense, we will focus our attention in order to show the main different parts of the proof.

The first step of the proof is to write the average version of the closed-loop system (14.77)–(14.83):

$$\dot{\vartheta}_{\rm av}(t) = \bar{\zeta}_{\rm av}(0,t)\,, \tag{14.88}$$

$$\partial_t\bar{\zeta}_{\rm av}(x,t) = \partial_x\bar{\zeta}_{\rm av}(x,t), \qquad x\in[0,1]\,, \tag{14.89}$$

$$\bar{\zeta}_{\rm av}(1,t) = 2v_{\rm av}(1,t) - \bar{\omega}_{\rm av}(1,t)\,, \tag{14.90}$$

$$\partial_t\bar{\omega}_{\rm av}(x,t) = -\partial_x\bar{\omega}_{\rm av}(x,t), \quad x\in[0,1]\,, \tag{14.91}$$

$$\bar{\omega}_{\rm av}(0,t) = \bar{\zeta}_{\rm av}(0,t)\,, \tag{14.92}$$

$$\partial_t v_{\rm av}(x,t) = \partial_x v_{\rm av}(x,t), \qquad x\in[1,1+D]\,, \tag{14.93}$$

$$\begin{aligned}\partial_t v_{\rm av}(1+D,t) &= -cv_{\rm av}(1+D,t) - \frac{c}{2}\left[\bar{\zeta}_{\rm av}(1,t) - \bar{\omega}_{\rm av}(1,t)\right] \\ &+cKH\left[\vartheta_{\rm av}(t) + \int_1^{1+D} v(x,t)dx + \int_0^1 \bar{\zeta}(x,t)dx\right]\,.\end{aligned} \tag{14.94}$$

The main idea is to apply the backstepping transformation (14.62)–(14.64) into (14.88)–(14.94) in order to obtain the average target system (14.65)–(14.68).

While in [179, Theorem 1] and [65, Theorem 1] it was possible to prove the local exponential stability of the average closed-loop system using a Lyapunov functional [119], a different approach is adopted here. We will show that it is possible to guarantee the local exponential stability for the average closed-loop system (14.65)–(14.68) by a small-gain analysis.

Note that each subsystem (**ODE**, **PDE1**, **PDE2**, **PDE3**) in (14.65)–(14.68) is ISS with respect to its corresponding input signal or boundary input [99]. Moreover, the last ISS loop in **PDE3** has a small parameter

$$\mu = 1/c \tag{14.95}$$

in the boundary condition (14.68) if c is chosen sufficiently large.

In this case, by applying successively the Small-Gain Theorem [99, Theorem 8.1, p. 198] for hyperbolic PDE-ODE loops, we can conclude the exponential stability of the average target system since the small-gain condition (8.2.20) in [99, Theorem 8.1, p. 198] holds provided $0<\mu<1$ is sufficiently small. Therefore, if such a small-gain condition holds, then [99, Theorem 8.1, p. 198] allows us to conclude that there exist constants $\delta,\Delta>0$ such that for every $\vartheta_0^{\rm av}\in\mathbb{R}$, $w_0^{\rm av}\in C^0([0,1])$, $\varpi_0^{\rm av}\in C^0([0,1])$, $\zeta_0^{\rm av}\in C^0([1,1+D])$, the unique generalized solution of this initial-boundary value problem, with $\vartheta_{\rm av}(0)=\vartheta_0^{\rm av}$, $w_{\rm av}(x,0)=w_0^{\rm av}$, $\varpi_{\rm av}(x,0)=\varpi_0^{\rm av}$, and $\zeta_{\rm av}(x,0)=\zeta_0^{\rm av}$, satisfies the following estimate:

$$\begin{aligned}&|\vartheta_{\rm av}(t)| + \|w_{\rm av}(t)\|_\infty + \|\varpi_{\rm av}(t)\|_\infty + \|\zeta_{\rm av}(t)\|_\infty \\ &\quad\le \Delta(|\vartheta_0^{\rm av}| + \|w_0^{\rm av}\|_\infty + \|\varpi_0^{\rm av}\|_\infty + \|\zeta_0^{\rm av}\|_\infty)\exp(-\delta t) \quad \forall t\ge 0\,.\end{aligned} \tag{14.96}$$

Hence, we conclude the origin of the average target system (14.65)–(14.68) is exponentially stable under the assumption of $0<\mu<1$ in (14.68) and (14.95) being sufficiently small. Then, from the inverse of the backstepping transformations (14.62)–(14.64), we can conclude the same result for the closed-loop average system (14.88)–(14.94) for the variables

$$(\vartheta_{\rm av}(t), \bar{\zeta}_{\rm av}(x,t), \bar{\omega}_{\rm av}(x,t), v_{\rm av}(x,t))$$

such that $\Omega(t)\le M\Omega(0)e^{-\lambda t}$, for $M,\lambda>0$, in the norm

$$\Omega(t) := |\vartheta_{\rm av}(t)| + \|\bar{\zeta}_{\rm av}(t)\|_\infty + \|\bar{\omega}_{\rm av}(t)\|_\infty + \|v_{\rm av}(t)\|_\infty\,. \tag{14.97}$$

As developed in the proofs of [179, Theorem 1] and [65, Theorem 1], the next step is the application of the local averaging theory for infinite-dimensional systems in [83] (Appendix A) showing the periodic solutions indeed satisfy inequality (14.84).

When we use the Riemann transformation [26] given in (14.32)–(14.33) to convert the wave equation (14.72)–(14.74) in the closed-loop system (14.71)–(14.76) to a feedback loop of two first-order transport equations which convect in opposite directions, as described in **PDE 1** and **PDE 2** subsystems of the representation (14.58)–(14.61), the complete **ODE-PDE 1-PDE 2-PDE 3** cascade can also be interpreted as an input-state delayed[9] ordinary differential equation [26], delayed by $1+D$ units of time. Hence, the complete closed-loop system (14.58)–(14.61) with $U(t)$ in (14.70), represented in the form (14.77)–(14.83), can be rewritten as

$$\dot{z}(t) = f(\omega t, z_t), \tag{14.98}$$

where $z(t)=[\tilde{\vartheta}(t), \xi(t), U(t)]^T$ is the state vector, with $\xi(t)$ representing the equivalent delayed state [26] of the **PDE 1** in (14.59) and **PDE 2** in (14.60) of the functional differential equation. Defining $\xi(t) = \bar{\alpha}(0,t)$ and noting that $\bar{\zeta}(0,t) = \dot{\xi}(t) + \bar{\alpha}_x(0,t)$ from (14.32), one can write $\dot{\xi}(t) = \frac{\bar{\zeta}(0,t)+\bar{\omega}(0,t)}{2}$, according to (14.35). The distributed terms of $\dot{U}(t) = v_t(1+D,t)$ in (14.83) are encompassed by $z_t(r)=z(t+r)$ for $-(1+D) \leq r \leq 0$, with f being an appropriate continuous functional, such that the averaging theorem by Hale and Lunel [83, Section 5] (see Appendix A for a compact version) can be directly applied, considering $\omega = 1/\epsilon$.

Since we have already proved the origin of the average closed-loop system (14.88)–(14.94) is exponentially stable in the first part of the proof, by applying the averaging theorem for infinite-dimensional systems developed in [83, Section 5] (Appendix A), for ω sufficiently large, we conclude (14.77)–(14.83) has a unique exponentially stable periodic solution around its equilibrium satisfying inequality (14.84).

The final step to complete the proof is the conclusion of the attractiveness of the extremum point $(\theta^*, \Theta^*, y^*)$ according to (14.85)–(14.87). Applying Agmon's, Poincaré's, and Young's inequalities on the LHS of (14.10), along with (14.10)–(14.13), we have

$$\tilde{\theta}^2(t-D) \leq (1+2D)\vartheta(t)^2 + (4D^2+1)\|\bar{\alpha}_x(t)\|^2. \tag{14.99}$$

From (14.35), we can rewrite (14.99) as

$$\tilde{\theta}^2(t-D) \leq (1+2D)\vartheta(t)^2 + (4D^2+1)\left\|\frac{1}{2}\left[\bar{\zeta}(t) - \bar{\omega}(t)\right]\right\|^2. \tag{14.100}$$

Inequality (14.100) can be written in terms of the periodic solution $\vartheta^{\Pi}(t)$, $\bar{\zeta}^{\Pi}(x,t)$ and $\bar{\omega}^{\Pi}(x,t)$ as follows:

$$\begin{aligned}
&\limsup_{t\to\infty} |\tilde{\theta}(t-D)|^2 \\
&= \limsup_{t\to\infty} \left\{(1+2D)|\vartheta(t) + \vartheta^{\Pi}(t) - \vartheta^{\Pi}(t)|^2\right\} \\
&\quad + \limsup_{t\to\infty} \left\{(4D^2+1)\left\|\frac{1}{2}\left[\bar{\zeta}(t) + \bar{\zeta}^{\Pi}(t) - \bar{\zeta}^{\Pi}(t) - \left(\bar{\omega}(t) + \bar{\omega}^{\Pi}(t) - \bar{\omega}^{\Pi}(t)\right)\right]\right\|^2\right\}.
\end{aligned} \tag{14.101}$$

[9] From (14.61), recall that any delayed signal $U(t-D)$ with constant delay $D>0$ can be modeled by the following first-order hyperbolic PDE, also referred to "transport PDE": $v_t(x,t) = v_x(x,t)$, $x \in [1, 1+D]$, with $v(x,t) = U(t+x-1-D)$ such that $v(1,t) = U(t-D)$ and $v(1+D,t) = U(t)$. On the other hand, from the transport equations (14.59) and (14.60), we have that $\bar{\zeta}(x,t) = \bar{\zeta}(\sigma, t+x-\sigma)$ and $\bar{\omega}(x,t) = \bar{\omega}(\sigma, t-x+\sigma) = \bar{\zeta}(x, t-2x)$ for any $0 \leq \sigma \leq x \leq 1$ and any $t \geq 0$. In particular, $\bar{\zeta}(x,t) = W(t-1+x)$ and $\bar{\omega}(x,t) = W(t-1-x)$ for any $x \in [0,1]$ and $t \geq 0$, where $W(t) = \bar{\zeta}(1,t)$. For more details, see [34, Remark 1].

By applying Young's inequality and some algebra it holds

$$|\vartheta(t)+\vartheta^{\Pi}(t)-\vartheta^{\Pi}(t)|^2 \leq \sqrt{2}\left(|\vartheta(t)-\vartheta^{\Pi}(t)|^2+|\vartheta^{\Pi}(t)|^2\right).$$

The same procedure can be applied to the second term on the RHS of (14.101). From the averaging theorem [83, Section 5] (Appendix A), we know that $\vartheta(t)-\vartheta^{\Pi}(t)\to 0$, $\bar{\zeta}(t)-\bar{\zeta}^{\Pi}(t)\to 0$ and $\bar{\omega}(t)-\bar{\omega}^{\Pi}(t)\to 0$, exponentially as $t\to+\infty$. Hence,

$$\begin{aligned}\limsup_{t\to\infty}|\tilde{\theta}(t-D)|^2 &= \limsup_{t\to\infty}\left\{\sqrt{2}(1+2D)|\vartheta^{\Pi}(t)|^2\right\}\\ &+\limsup_{t\to\infty}\left\{\sqrt{2}(4D^2+1)\left\|\frac{1}{2}\left[\bar{\zeta}^{\Pi}(t)-\bar{\omega}^{\Pi}(t)\right]\right\|^2\right\}.\end{aligned} \tag{14.102}$$

Along with (14.84), we can conclude

$$\limsup_{t\to\infty}|\tilde{\theta}(t-D)| = \limsup_{t\to\infty}|\tilde{\theta}(t)| = \mathcal{O}(1/\omega). \tag{14.103}$$

From (14.8) and Figure 14.2, we can write $\theta(t)-\theta^* = \tilde{\theta}(t)+S(t)$, and recalling $S(t)$ in (14.26) is of order $\mathcal{O}(a)$, we finally get with (14.85). The convergence of the propagated actuator $\Theta(t)$ to the optimizer Θ^* is now discussed. Using (14.27) and taking its absolute value, one has

$$|\Theta(t)-\Theta^*| = |\vartheta(t)+a\sin(\omega(t))|. \tag{14.104}$$

As in the convergence proof of the parameter $\theta(t)$ to the optimal input θ^* above, we write (14.104) in terms of the periodic solution $\vartheta^{\Pi}(t)$ and follow the same steps by applying Young's inequality and recalling that $\vartheta(t)-\vartheta^{\Pi}(t)\to 0$ exponentially according to the averaging theorem [83, Section 5] (Appendix A). Hence, along with (14.84), we finally get (14.86). To show the convergence of the output $y(t)$ of the static map to the optimal value y^*, we replace $\Theta(t)-\Theta^*$ in (14.3) by (14.27) and take the absolute value:

$$|y(t)-y^*| = \left|\frac{H}{2}\left[\vartheta(t)+a\sin(\omega(t))\right]^2\right|. \tag{14.105}$$

Expanding the quadratic term in (14.105) and applying Young's inequality to the resulting equation, one has $|y(t)-y^*|=\left|H\left[\vartheta(t)^2+a^2\sin^2(\omega t)\right]\right|$. As before, we add and subtract the periodic solution $\vartheta^{\Pi}(t)$ and use the convergence of $\vartheta(t)-\vartheta^{\Pi}(t)\to 0$ via the averaging theorem [83, Section 5] (Appendix A). Hence, again with (14.84), we get (14.87). □

14.5 ▪ Other PDE-PDE Cascades: Wave Equation with Heat Equation at Its Input

In this section, we deal with a distinct cascade of parabolic/hyperbolic PDEs represented by a diffusion equation at the input of a wave PDE. This combination is representative and distinct from that considered in Section 14.3, leading to a potentially challenging problem. For instance, the heat-wave cascade was considered intractable in [119, Section 20.1] since the resulting target system obtained there was not even solvable.

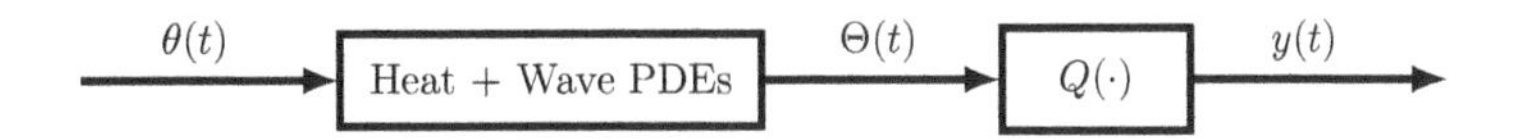

Figure 14.3. *Cascade of heat-wave PDEs followed by the nonlinear map* $y = Q(\Theta)$.

Additionally, in contrast to [247], where the heat and wave equations are *coupled* through two boundary conditions, our two PDEs are in a cascade connection—one feeding into the other.

14.5.1 ▪ Plant with Infinite-Dimensional Actuator

We consider the same locally quadratic map studied in (14.3),

$$y(t) = Q(\Theta(t)) = y^* + \frac{H}{2}(\Theta(t) - \Theta^*)^2, \tag{14.106}$$

but with a different actuation dynamics in the form of a heat-wave PDE cascade, as depicted in Figure 14.3. The infinite-dimensional actuator dynamics can be written as

$$\Theta(t) := \alpha(0,t), \tag{14.107}$$
$$\partial_{tt}\alpha(x,t) = \partial_{xx}\alpha(x,t), \quad x \in [0,1], \tag{14.108}$$
$$\alpha_x(0,t) = 0, \tag{14.109}$$
$$\alpha(1,t) = \rho(1,t), \tag{14.110}$$
$$\rho_t(x,t) = \rho_{xx}(x,t), \quad x \in [1,1+D], \tag{14.111}$$
$$\rho_x(1,t) = 0, \tag{14.112}$$
$$\rho(1+D,t) = \theta(t), \tag{14.113}$$

where $\alpha(x,t)$ is the state of the wave PDE with spatial domain defined without loss of generality by $x \in [0,1]$ and $\rho(x,t)$ the corresponding state of the heat equation within the domain $x \in [1,1+D]$, with $D > 0$.

14.5.2 ▪ Estimation Error Dynamics

Analogously to (14.10)–(14.13), the propagated estimation error $\vartheta(t)$ defined in (14.9) through the heat-wave PDE cascade is simply

$$\vartheta(t) := \bar{\alpha}(0,t), \tag{14.114}$$
$$\partial_{tt}\bar{\alpha}(x,t) = \partial_{xx}\bar{\alpha}(x,t), \quad x \in [0,1], \tag{14.115}$$
$$\bar{\alpha}_x(0,t) = 0, \tag{14.116}$$
$$\bar{\alpha}(1,t) = \tilde{\alpha}(1,t), \tag{14.117}$$
$$\tilde{\alpha}_t(x,t) = \tilde{\alpha}_{xx}(x,t), \quad x \in [1,1+D], \tag{14.118}$$
$$\tilde{\alpha}_x(1,t) = 0, \tag{14.119}$$
$$\tilde{\alpha}(1+D,t) = \tilde{\theta}(t) := \hat{\theta}(t) - \theta^*, \tag{14.120}$$

where $\tilde{\theta}$ in (14.120) is the estimation error such as defined in (14.8).

Taking the time derivative of (14.114)–(14.120), we obtain

$$\dot{\vartheta}(t) = u(0,t), \tag{14.121}$$
$$u_{tt}(x,t) = u_{xx}(x,t), \quad x \in [0,1], \tag{14.122}$$
$$u_x(0,t) = 0, \tag{14.123}$$
$$u(1,t) = v(1,t), \tag{14.124}$$
$$v_t(x,t) = v_{xx}(x,t), \quad x \in [1,1+D], \tag{14.125}$$
$$v_x(1,t) = 0, \tag{14.126}$$
$$v(1+D,t) = U(t), \tag{14.127}$$

where $U(t)$ is the control input signal, (u,u_t,v) is the infinite-dimensional state,

$$v(x,t) = \tilde{\alpha}_t(x,t) \quad \text{and} \quad v(1,t) = \tilde{\alpha}_t(1,t) = \bar{\alpha}_t(1,t). \tag{14.128}$$

14.5.3 ▪ Auxiliary Target System

Again, with the help of the Riemann transformations [26] (14.32)–(14.33), we consider the following backstepping transformations analogously to those used in (14.62)–(14.64):

$$w(x,t) = \bar{\zeta}(x,t) - KH\int_0^x \bar{\zeta}(\sigma,t)d\sigma - KH\vartheta(t), \quad x\in[0,1], \tag{14.129}$$

$$\varpi(x,t) = \bar{\omega}(x,t) + KH\int_0^x \bar{\omega}(\sigma,t)d\sigma - KH\vartheta(t), \quad x\in[0,1], \tag{14.130}$$

$$\begin{aligned} z(x,t) = {} & v(x,t) - KH\int_1^x (x-\sigma)v(\sigma,t)d\sigma \\ & + \bar{\alpha}_x(1,t) - KH\int_0^1 \bar{\zeta}(\sigma,t)d\sigma - KH\vartheta(t), \quad x\in[1,1+D], \end{aligned} \tag{14.131}$$

which transform the cascade PDE system (14.121)–(14.127) into the average target system

$$\dot{\vartheta}_{\rm av}(t) = KH\vartheta_{\rm av} + w_{\rm av}(0,t), \tag{14.132}$$
$$\partial_t w_{\rm av}(x,t) = \partial_x w_{\rm av}(x,t), \quad x \in [0,1], \tag{14.133}$$
$$w_{\rm av}(1,t) = z_{\rm av}(1,t), \tag{14.134}$$
$$\partial_t \varpi_{\rm av}(x,t) = -\partial_x \varpi_{\rm av}(x,t), \quad x \in [0,1], \tag{14.135}$$
$$\varpi_{\rm av}(0,t) = w_{\rm av}(0,t), \tag{14.136}$$
$$\partial_t z_{\rm av}(x,t) = \partial_{xx} z_{\rm av}(x,t), \quad x \in [1,1+D], \tag{14.137}$$
$$\partial_x z_{\rm av}(1,t) = 0, \tag{14.138}$$
$$z_{\rm av}(1+D,t) = -\frac{1}{c}\partial_t v_{\rm av}(1+D,t). \tag{14.139}$$

14.5.4 ▪ ES Boundary Control and Additive Dither

The target system (14.132)–(14.139) is exponentially stable with the average version $\dot{U}_{\rm av}(t) = -cz_{\rm av}(1+D,t)$ of the control law obtained from (14.131) and (14.139) with $U_{\rm av} = v_{\rm av}(1+D,t)$

in (14.127), which results in

$$\dot{U}(t) = -cU(t) + c\left\{-\bar{\alpha}_x(1,t) + K\left[G(t) + \hat{H}(t)\int_1^{1+D}(1+D-\sigma)v(\sigma,t)d\sigma\right.\right.$$
$$\left.\left.+\hat{H}(t)\int_0^1 \left[u(x,t) + \bar{\alpha}_x(x,t)\right]dx\right]\right\} \tag{14.140}$$

for $c > 0$ and $K > 0$, recalling the averaging-based estimates in (14.31). The signals $\hat{H}(t)$ and $G(t)$ are defined according to (14.29) and (14.30), respectively.

The term $\bar{\alpha}_x(x,t)$ which appears in (14.140) can be written in terms of measurable signals as

$$\bar{\alpha}_x(x,t) = \alpha_x(x,t) - \beta_x(x,t), \tag{14.141}$$

similarly to (14.49). On the other hand, the term $\beta_x(x,t)$ comes from the *trajectory generation problem* of the heat-wave PDE cascade system [128, Chapter 12],

$$\beta(0,t) = a\sin(\omega t), \tag{14.142}$$
$$\beta_{tt}(x,t) = \beta_{xx}(x,t), \quad x \in [0,1], \tag{14.143}$$
$$\beta_x(0,t) = 0, \tag{14.144}$$
$$\beta(1,t) = \bar{\beta}(1,t) = a\cos(\omega)\sin(\omega t), \tag{14.145}$$
$$\bar{\beta}_t(x,t) = \bar{\beta}_{xx}(x,t), \quad x \in [1,1+D], \tag{14.146}$$
$$\bar{\beta}_x(1,t) = 0, \tag{14.147}$$
$$\bar{\beta}(1+D,t) = S(t), \tag{14.148}$$

where the explicit solution $S(t)$ is given by [65]

$$S(t) = \frac{1}{2}a\cos(\omega)e^{\sqrt{\frac{\omega}{2}}D}\sin\left(\omega t + \sqrt{\frac{\omega}{2}}D\right)$$
$$+\frac{1}{2}a\cos(\omega)e^{-\sqrt{\frac{\omega}{2}}D}\sin\left(\omega t - \sqrt{\frac{\omega}{2}}D\right). \tag{14.149}$$

14.5.5 ▪ Stability Analysis

In order to conduct the stability analysis with the help of (14.32)–(14.35), we first provide the following equivalent PDE representation of the closed-loop system (14.121)–(14.127) under the control law (14.140):

$$\dot{\vartheta}(t) = \bar{\zeta}(0,t), \tag{14.150}$$
$$\partial_t\bar{\zeta}(x,t) = \partial_x\bar{\zeta}(x,t), \quad x \in [0,1], \tag{14.151}$$
$$\bar{\zeta}(1,t) = 2v(1,t) - \bar{\omega}(1,t), \tag{14.152}$$
$$\bar{\omega}_t(x,t) = -\bar{\omega}_x(x,t), \quad x \in [0,1], \tag{14.153}$$
$$\bar{\omega}(0,t) = \bar{\zeta}(0,t), \tag{14.154}$$
$$v_t(x,t) = v_{xx}(x,t), \quad x \in [1,1+D], \tag{14.155}$$
$$v_x(1,t) = 0, \tag{14.156}$$

$$v_t(1+D,t) = -cv(1+D,t) - \frac{c}{2}\left[\bar{\zeta}(1,t) - \bar{\omega}(1,t)\right]$$
$$+cK\left[G(t) + \hat{H}(t)\int_1^{1+D}(1+D-x)v(x,t)dx + \hat{H}(t)\int_0^1 \bar{\zeta}(x,t)dx\right]. \tag{14.157}$$

In this representation, the state vector is composed of the finite-dimensional components $\vartheta(t)$, $U(t) := v(1+D,t)$ and the infinite-dimensional ones $\bar{\zeta}(x,t), \bar{\omega}(x,t), v(x,t)$.

Let the state space be $\mathcal{H} := \mathbb{R} \times L_{\infty[0,1]} \times L_{\infty[0,1]} \times L_{\infty[1,1+D]} \times \mathbb{R}$. The induced norm is denoted by $\|\cdot\|_{\mathcal{H}}$. Define a linear operator $A : \mathcal{D}(A) \subset \mathcal{H} \to \mathcal{H}$ by

$$A \begin{bmatrix} X_1 \\ X_2 \\ X_3 \\ X_4 \\ X_5 \end{bmatrix} = \begin{bmatrix} X_2(0) & \dfrac{dX_2}{dx} & -\dfrac{dX_3}{dx} & -\dfrac{d^2X_4}{dx^2} & -cX_5 \end{bmatrix}^T, \tag{14.158}$$

with the domain

$$\mathcal{D}(A) = \Bigg\{ [X_1, X_2, X_3, X_4, X_5] \in \mathcal{H} \mid$$
$$X_2(1) = 2X_5(1) - X_3(1), X_3(0) = X_2(0), \frac{dX_5(1)}{dx} = 0 \Bigg\}. \tag{14.159}$$

The time-varying nonlinear perturbation term is given by

$$F(\omega t, X) = [0 \;\; 0 \;\; 0 \;\; 0 \;\; F_5(\omega t, X)]^T, \tag{14.160}$$

where $F_5 : \mathbb{R} \times \mathcal{H} \to \mathbb{R}$ is defined by

$$\begin{aligned} F_5(\omega t, X) = & -\frac{c}{2}[X_2(1) - X_3(1)] + cK\frac{2}{a}\sin(\omega t)y \\ & - cK\frac{8}{a^2}\cos(\omega t)y \left[\int_1^{1+D} (1+D-\sigma)X_5(\sigma)d\sigma + \int_0^1 X_2(\sigma)d\sigma \right] \end{aligned} \tag{14.161}$$

since $\hat{H}(t)$ and $G(t)$ are given by (14.29) and (14.30). The output y in (14.106) can be written as

$$y = Q(X_1 + a\sin(\omega t) + \Theta^*), \tag{14.162}$$

according to (14.27). Hence, we arrive at the next evolution equation corresponding to the closed-loop system (14.150)–(14.157):

$$\frac{dX}{dt} = AX + F(\omega t, X), \tag{14.163}$$

where $X(t) = [\vartheta(t), \bar{\zeta}(\cdot,t), \bar{\omega}(\cdot,t), v(\cdot,t), U(t)]^T$.

Clearly, $F(\omega(t+\Pi), X) = F(\omega t, X)$, for $\Pi = 2\pi/\omega$, for any $X \in \mathcal{H}$. In this case, if $1/\omega$ is considered as a small parameter ϵ, we can apply the method of averaging for infinite-dimensional systems developed in [83, Section 2] (Appendix B) to conclude the results of the next theorem.

Theorem 14.2. *(Heat-Wave Cascade) Consider the dynamic system being represented by the nonlinear quadratic map* (14.106) *with actuation dynamics governed by the heat-wave cascade of PDEs in* (14.107)–(14.113). *For $c > 0$ and for each $\varepsilon > 0$, there exist constants $\omega^* > 0$ and $\varepsilon_0 \in (0,\varepsilon)$ such that, for any $\omega > \omega^*$, any solution to* (14.163) *for an initial value $X_0 = [\vartheta(0), \bar{\zeta}(\cdot,0), \bar{\omega}(\cdot,0), v(\cdot,0), U(0)]^T \in \mathcal{D}(A)$ with $\|X_0\|_{\mathcal{H}} \leq \varepsilon_0$ converges to an $\mathcal{O}(1/\omega)$-neighborhood of the origin. In addition, the following estimates hold:*

$$\limsup_{t\to\infty} |\theta(t) - \theta^*| = \mathcal{O}\left(ae^{\sqrt{\frac{\omega}{2}}D} + 1/\omega\right), \tag{14.164}$$

$$\limsup_{t\to\infty} |\Theta(t) - \Theta^*| = \mathcal{O}(a + 1/\omega), \tag{14.165}$$

$$\limsup_{t\to\infty} |y(t) - y^*| = \mathcal{O}(a^2 + 1/\omega^2). \tag{14.166}$$

Proof. Let us obtain the average system associated with the closed-loop system (14.150)–(14.157). The expression (14.162) of the output y has the same form as the one in common PDE-free ES problems. Hence, average computation done in the literature, especially in [73], also works for our problem.

If $Q(\cdot)$ is the quadratic map (14.106), the average of (14.161) can be explicitly computed as

$$\frac{\omega}{2\pi}\int_0^{2\pi/\omega} F_5(\omega\tau, X^{\mathrm{av}})d\tau = -\frac{c}{2}[X_2^{\mathrm{av}}(1) - X_3^{\mathrm{av}}(1)] + cKHX_1^{\mathrm{av}}$$
$$-cKH\left[\int_1^{1+D}(1+D-\sigma)X_5^{\mathrm{av}}(\sigma)d\sigma + \int_0^1 X_2^{\mathrm{av}}(\sigma)d\sigma\right] \tag{14.167}$$

for each $X^{\mathrm{av}} = [X_1^{\mathrm{av}}, X_2^{\mathrm{av}}, X_3^{\mathrm{av}}, X_4^{\mathrm{av}}, X_5^{\mathrm{av}}]^T \in \mathcal{H}$. From the argument above, the average system associated with the closed-loop system (14.150)–(14.157) is given by

$$\dot{\vartheta}_{\mathrm{av}}(t) = \bar{\zeta}_{\mathrm{av}}(0,t), \tag{14.168}$$
$$\partial_t \bar{\zeta}_{\mathrm{av}}(x,t) = \partial_x \bar{\zeta}_{\mathrm{av}}(x,t), \quad x \in [0,1], \tag{14.169}$$
$$\bar{\zeta}_{\mathrm{av}}(1,t) = 2v_{\mathrm{av}}(1,t) - \bar{\omega}_{\mathrm{av}}(1,t), \tag{14.170}$$
$$\partial_t \bar{\omega}_{\mathrm{av}}(x,t) = -\partial_x \bar{\omega}_{\mathrm{av}}(x,t), \quad x \in [0,1], \tag{14.171}$$
$$\bar{\omega}_{\mathrm{av}}(0,t) = \bar{\zeta}_{\mathrm{av}}(0,t), \tag{14.172}$$
$$\partial_t v_{\mathrm{av}}(x,t) = \partial_{xx} v_{\mathrm{av}}(x,t), \qquad x \in [1,1+D], \tag{14.173}$$
$$\partial_x v_{\mathrm{av}}(1,t) = 0, \tag{14.174}$$
$$\partial_t v_{\mathrm{av}}(1+D,t) = -cv_{\mathrm{av}}(1+D,t) - \frac{c}{2}\left[\bar{\zeta}_{\mathrm{av}}(1,t) - \bar{\omega}_{\mathrm{av}}(1,t)\right]$$
$$+cKH\left[\vartheta_{\mathrm{av}}(t) + \int_1^{1+D}(1+D-x)v_{\mathrm{av}}(x,t)dx + \int_0^1 \bar{\zeta}_{\mathrm{av}}(x,t)dx\right]. \tag{14.175}$$

Applying the backstepping transformation (14.129)–(14.131) into (14.168)–(14.175), we obtain the average target system (14.132)–(14.139).

Now, from the ISS properties [99] of each subsystem of (14.132)–(14.139), we can write the following ISS inequalities:

$$|\vartheta_{\mathrm{av}}(t)| \le M_1|\vartheta_{\mathrm{av}}(0)|e^{-\lambda_1 t} + \gamma_1 \max_{0\le s\le t}(|w_{\mathrm{av}}(0,s)|), \tag{14.176}$$
$$\|\varpi_{\mathrm{av}}(t)\|_\infty \le M_2\|\varpi_{\mathrm{av}}(0)\|_\infty e^{-\lambda_2 t} + \gamma_2 \max_{0\le s\le t}(|w_{\mathrm{av}}(0,s)|), \tag{14.177}$$
$$\|w_{\mathrm{av}}(t)\|_\infty \le M_3\|w_{\mathrm{av}}(0)\|_\infty e^{-\lambda_3 t} + \gamma_3 \max_{0\le s\le t}(|z_{\mathrm{av}}(1,s)|), \tag{14.178}$$
$$\|z_{\mathrm{av}}(t)\|_\infty \le M_4\|z_{\mathrm{av}}(0)\|_\infty e^{-\lambda_4 t} + \gamma_4 \max_{0\le s\le t}(|v_{\mathrm{av}}(1+D,s)|) \tag{14.179}$$

for $\lambda_1 = -KH > 0$ in (14.176) and the remaining constants M_i, λ_i, and γ_i being appropriate positive constants. To obtain (14.179), we first need to compute $\partial_t z_{\mathrm{av}}(1+D,t)$. Recalling (14.128) and $\bar{\zeta}_{\mathrm{av}}(1,t) = v(1,t) + \bar{\alpha}_x(1,t)$, by means of the integration by parts we can write

$$\partial_t z_{\mathrm{av}}(1+D,t) = -cz_{\mathrm{av}}(1+D,t) - KHv_{\mathrm{av}}(1+D,t)$$
$$+KHv_{\mathrm{av}}(1,t) - \partial_t v_{\mathrm{av}}(1,t) - KH\bar{\zeta}_{\mathrm{av}}(1,t) + \partial_t \bar{\zeta}_{\mathrm{av}}(1,t). \tag{14.180}$$

On the other hand, (14.129) and (14.131) can be used to obtain $\partial_t \bar{\zeta}_{\rm av}(1,t)$ and $\partial_t v_{\rm av}(1,t)$, such that

$$KHv_{\rm av}(1,t)-\partial_t v_{\rm av}(1,t)-KH\bar{\zeta}_{\rm av}(1,t)+\partial_t\bar{\zeta}_{\rm av}(1,t) \\ =KHz_{\rm av}(1,t)-\partial_t z_{\rm av}(1,t)-KHw_{\rm av}(1,t)+\partial_t w_{\rm av}(1,t)=0, \tag{14.181}$$

according to (14.134). Then, in the time scale $\tau=\mu t$ (with $\mu=1/c$), the system (14.139) and (14.180) is rewritten as

$$Z_1^{'}=-Z_2, \quad Z_2^{'}=-Z_2-\mu KHZ_1, \tag{14.182}$$

where $Z_1(t):=v_{\rm av}(1+D,t)$, $Z_2(t):=z_{\rm av}(1+D,t)$, $Z_1^{'}:=\frac{dZ_1}{d\tau}$, and $Z_2^{'}:=\frac{dZ_2}{d\tau}$. Hence, for any $\mu=1/c>0$, we can conclude

$$v_{\rm av}(1+D,t)\to 0, \quad z_{\rm av}(1+D,t)\to 0, \tag{14.183}$$

exponentially as $t\to+\infty$. In particular, in this new analysis the filter pole "c" in (14.140) does not need to be made sufficiently large to guarantee stability. Finally, from (14.183) and (14.176)–(14.179), we can show (14.132)–(14.139) is exponentially stable by means of [99, Theorem 11.5, p. 277]. Hence, there exist constants $M>0$ and $\lambda>0$, which are independent from the initial data, such that, $\forall t\geq 0$,

$$\left\|[\vartheta(t),\bar{\zeta}(\cdot,t),\bar{\omega}(\cdot,t),v(\cdot,t),U(t)]\right\|_{\mathcal{H}}\leq M\left\|[\vartheta(0),\bar{\zeta}(\cdot,0),\bar{\omega}(\cdot,0),v(\cdot,0),U(0)]\right\|_{\mathcal{H}}e^{-\lambda t}. \tag{14.184}$$

It is worth mention that a similar result (for a slightly different norm) could be obtained through the following Lyapunov–Krasovskii functional:

$$V(t)=\frac{\bar{a}}{2}\vartheta_{\rm av}^2(t)+\bar{b}\int_0^1 e^{\bar{g}(1+x)}w_{\rm av}^2(x,t)dx+\bar{c}\int_0^1 e^{\bar{h}(1-x)}\varpi_{\rm av}^2(x,t)dx \\ +\frac{\bar{d}}{2}\int_1^{1+D} z_{\rm av}^2(x,t)dx+\frac{\bar{e}}{2}\int_1^{1+D}\partial_x z_{\rm av}^2(x,t)dx+\frac{\bar{f}}{2}z_{\rm av}^2(1+D,t) \tag{14.185}$$

for some $\bar{a},\bar{b},\bar{c},\bar{d},\bar{e},\bar{f},\bar{g},\bar{h}>0$. Following analogous derivations detailed in the proof of [65, Theorem 1, Step 3] (Theorem 10.1, Chapter 10), we get $\dot{V}(t)\leq-\lambda V(t)$, for $\lambda>0$, if $\bar{b}=2\bar{c}$, $\bar{g}=\bar{h}$, and recalling $KH<0$.

Since the closed-loop average system is exponentially stable, it is time to invoke the averaging theorem for infinite dimensions [83, Section 2] (see Appendix B for a compact version) to conclude the existence of a periodic solution of order $\mathcal{O}(1/\omega)$ for which the state vector ultimately converges. To apply the averaging theorem [83, Section 2], we need to check that the operator A in (14.158) and (14.163) is a generator of strongly continuous semigroup T_A on $\mathcal{H}$ and the generated semigroup T_A has a smoothing property. The required property is as follows: for any $h:[0,\infty)\to\mathcal{H}$ being norm continuous, the following relations hold:

$$\textbf{(i)} \quad \int_0^t T_A(t-\tau)h(\tau)d\tau\in\mathcal{D}(A) \quad \forall t\geq 0, \tag{14.186}$$

$$\textbf{(ii)} \quad \left\|A\int_0^t T_A(t-\tau)h(\tau)d\tau\right\|\leq Me^{-\bar{\lambda}t}\max_{0\leq\tau\leq t}\|h(\tau)\|_{\mathcal{H}}, \tag{14.187}$$

where $M>0$ and $\bar{\lambda}\in\mathbb{R}$ are independent from h. This is called in [83] and Section 1.10.1 of **Property (H)**.

On the other hand, the operator A in (14.163) and defined by (14.158) is surely the generator of strongly continuous semigroup T_A on $\mathcal{H}$ [190]. However, T_A does not fulfill the smoothing **Property (H)** in general. Fortunately, the perturbation $F(\omega t, X)$ defined in (14.160) merely has the finite-dimensional component $F_5(\omega t, X)$. Hence, in our problem, T_A only has to satisfy (14.186) and (14.187) for $h:[0,\infty)\to\mathcal{H}$ of the form $h(t)=[0,0,0,0,h_5(t)]^T$ for any continuous $h_5:[0,\infty)\to\mathbb{R}$. We can explicitly compute the integral of $T_A(t-\tau)h(\tau)$ with respect to τ from 0 to t for given $h(t)=[0,0,0,0,h_5(t)]^T$. This can be inferred from the resulting expression that T_A satisfies **Property (H)**. Then, the theorem follows from the exponential stability of the closed-loop average system and the averaging theorem [83, Section 2].

The remaining steps of the proof are quite similar to those employed in Theorem 14.1, recalling that the final residual set for the error $\theta(t)-\theta^*$ in (14.164) depends on $ae^{\sqrt{\frac{\omega}{2}}D}$ due to the amplitude of the additive dither $S(t)$ in (14.149). □

14.6 ▪ Newton-Based Approach for PDE-PDE Cascades

The advantages of Newton-based over Gradient ES in the absence of PDEs were deeply studied in [73] and the discussion can be summarized in the fact that the former removes the dependence of the convergence rate on the unknown second derivative (Hessian) of the nonlinear map to be optimized. The guarantee of this property even in the presence of a class of parabolic PDEs (not cascade of PDEs) was recently proved in two companion papers [170, 171] (see also Chapters 11 and 13), where we estimate the Hessian's inverse rather than only the Hessian. In this boundary control design, the PDE compensation can be achieved with an arbitrarily assigned convergence rate, thus improving the controller performance.

As discussed in [73], generating an estimate of the Hessian matrix in non-model-based optimization is not the only challenge. The other challenge is that the Newton algorithm requires an inverse of the Hessian matrix H^{-1}. The estimate of this matrix, as it evolves in continuous time, need not necessarily remain invertible. We tackle this challenge as in [73] by employing a dynamic system for generating the inverse asymptotically. This dynamic system is a filter in the form of a Riccati differential equation [73]:

$$\dot{\Gamma}=\omega_r\Gamma-\omega_r\hat{H}\Gamma^2, \tag{14.188}$$

with $\omega_r>0$ being a design constant and $\hat{H}(t)$ given in (14.29).

The estimation error of the Hessian's inverse is defined as

$$\tilde{\Gamma}(t)=\Gamma(t)-H^{-1}, \tag{14.189}$$

and its dynamic equation is written from (14.188) and (14.189) by

$$\dot{\tilde{\Gamma}}=\omega_r(\tilde{\Gamma}+H^{-1})[1-\hat{H}(\tilde{\Gamma}+H^{-1})]. \tag{14.190}$$

Using (14.188), we are able to define the following signal:

$$z(t)=\Gamma(t)G(t), \tag{14.191}$$

with $G(t)$ given in (14.30), which will be employed in what follows to derive the proposed control laws.

Analogously to the developments of Chapters 11 and 13, we are able to propose Newton-based ES controllers for the delay-wave and heat-wave cascades following the same backstepping transformation in (14.62)–(14.64) and (14.129)–(14.131), respectively, but replacing KH by $-K$, with $K>0$.

Recall that the averaged versions of the Hessian and gradient estimates in (14.29) and (14.30) are (14.31) if at least a locally quadratic map as in (14.2) is considered [73]. Hence, from (14.31) and $z(t)$ in (14.191), we can verify that

$$z_{\mathrm{av}}(t) = \frac{1}{\Pi}\int_0^{\Pi} \Gamma M(\lambda) y d\lambda = \Gamma_{\mathrm{av}}(t) H \vartheta_{\mathrm{av}}(t), \tag{14.192}$$

where $\Pi := 2\pi/\omega$, with $\Gamma_{\mathrm{av}}(t)$ and $\vartheta_{\mathrm{av}}(t)$ denoting the average versions of $\Gamma(t)$ and $\vartheta(t)$, respectively. Then, (14.192) can be written in terms of $\tilde{\Gamma}_{\mathrm{av}}(t) = \Gamma_{\mathrm{av}}(t) - H^{-1}$ as

$$z_{\mathrm{av}}(t) = \vartheta_{\mathrm{av}}(t) + \tilde{\Gamma}_{\mathrm{av}}(t) H \vartheta_{\mathrm{av}}(t). \tag{14.193}$$

The second term in the right side of (14.193) is bilinear in $(\tilde{\Gamma}_{\mathrm{av}}, \vartheta_{\mathrm{av}})$; thus, the linearization of $\Gamma_{\mathrm{av}}(t)$ at H^{-1} results in the linearized version of (14.192) given by

$$z_{\mathrm{av}}(t) = \vartheta_{\mathrm{av}}(t). \tag{14.194}$$

Regarding (14.194) and recalling again (14.31), we average (14.70) and (14.140), replacing KH by $-K$ with $K > 0$, such that the following average control laws for the *delay-wave cascade*

$$\begin{aligned}\dot{U}_{\mathrm{av}}(t) = -cU_{\mathrm{av}}(t) + c\Bigg\{ &-\partial_x \bar{\alpha}_{\mathrm{av}}(1,t) - K\Bigg[z_{\mathrm{av}}(t) + \int_{t-D}^{t} U_{\mathrm{av}}(\tau) d\tau \\ &+ \int_0^1 \left[u_{\mathrm{av}}(x,t) + \partial_x \bar{\alpha}_{\mathrm{av}}(x,t) \right] dx \Bigg] \Bigg\}\end{aligned} \tag{14.195}$$

and for the *heat-wave cascade*

$$\begin{aligned}\dot{U}_{\mathrm{av}}(t) = -cU_{\mathrm{av}}(t) + c\Bigg\{ &-\partial_x \bar{\alpha}_{\mathrm{av}}(1,t) - K\Bigg[z_{\mathrm{av}}(t) + \int_1^{1+D} (1+D-\sigma) v_{\mathrm{av}}(\sigma,t) d\sigma \\ &+ \int_0^1 \left[u_{\mathrm{av}}(x,t) + \partial_x \bar{\alpha}_{\mathrm{av}}(x,t) \right] dx \Bigg] \Bigg\}\end{aligned} \tag{14.196}$$

can be obtained for an appropriate constant $c > 0$.

Table 14.1 shows the stabilizing boundary control laws $U(t)$ and gives explicit solutions to the equivalent trajectory generation problem of $S(t)$ (in (14.20)–(14.25) and (14.142)–(14.148)) for the two cases of delay-wave and heat-wave cascades of PDEs using the Newton-based ES approach. Figure 14.4 shows an illustrative block diagram of the closed-loop system for the proposed Newton-based ES controllers with PDE-PDE cascades.

In the next theorem, we present the stability/convergence results for the delay-wave (heat-wave) compensation by means of Newton-based extremum seeking.

Theorem 14.3. *(Delay-Wave and Heat-Wave Cascades) Consider the generic Newton-based diagram of Figure* 14.4, *where* $y(t) = Q(\Theta(t)) = y^* + \frac{H}{2}(\Theta(t) - \Theta^*)^2$, $H < 0$, *and the block "PDE-PDE cascade" can represent Delay-Wave or Heat-Wave cascades of PDEs. Hence, for* ω *sufficiently large, we can conclude the following:*

(a) *For the Delay-Wave Cascade* (14.4)–(14.7), *the state variables* $\tilde{\Gamma}(t)$, $\vartheta(t)$, $\bar{\zeta}(x,t)$, $\bar{\omega}(x,t)$, $v(x,t)$ *have a unique locally exponentially stable periodic solution in* t *of period* $\Pi := 2\pi/\omega$, *denoted by* $\tilde{\Gamma}^{\Pi}(t), \vartheta^{\Pi}(t), \bar{\zeta}^{\Pi}(x,t), \bar{\omega}^{\Pi}(x,t), v^{\Pi}(x,t)$, *satisfying,* $\forall t \geq 0$,

$$\left|\tilde{\Gamma}^{\Pi}(t)\right| + \left|\vartheta^{\Pi}(t)\right| + \left\|\bar{\zeta}^{\Pi}(t)\right\|_{\infty} + \left\|\bar{\omega}^{\Pi}(t)\right\|_{\infty} + \left\|v^{\Pi}(t)\right\|_{\infty} \leq \mathcal{O}(1/\omega). \tag{14.197}$$

Table 14.1. *Newton-based ES Boundary Control for PDE-PDE Cascades.*

Delay-Wave Cascade	PDE-PDE Cascade : Equations (14.4) to (14.7) Boundary Control : $\dot{U}(t) = -cU(t) + c\left\{-\bar{\alpha}_x(1,t) - K\left[z(t) + \int_{t-D}^{t} U(\tau)d\tau + \int_0^1 [u(x,t) + \bar{\alpha}_x(x,t)]\,dx\right]\right\}$, with $c>0$, $u(x,t)$, $\bar{\alpha}(x,t)$ and $z(t)$ given in (14.19), (14.28) and (14.191), respectively. Trajectory Generation : $S(t) = a\cos(\omega D)\sin(\omega(t+D))$ Multiplicative Perturbations: $M(t) = \frac{2}{a}\sin(\omega t)$, $N(t) = -\frac{8}{a^2}\cos(2\omega t)$ Riccati Filter: $\dot{\Gamma}(t) = \omega_r\Gamma(t) - \omega_r\hat{H}(t)\Gamma^2(t)$, with $\hat{H}(t) = N(t)y(t)$
Heat-Wave Cascade	PDE-PDE Cascade : Equations (14.107) to (14.113) Boundary Control : $\dot{U}(t) = -cU(t) + c\left\{-\bar{\alpha}_x(1,t) - K\left[z(t) + \int_1^{1+D}(1+D-\sigma)v(\sigma,t)d\sigma + \int_0^1 [u(x,t) + \bar{\alpha}_x(x,t)]\,dx\right]\right\}$, with $c>0$, $u(x,t)$, $v(x,t)$, $\bar{\alpha}_x(x,t)$ and $z(t)$ given in (14.121)–(14.124), (14.128), (14.141) and (14.191), respectively. Trajectory Generation : $S(t) = \frac{1}{2}a\cos(\omega)e^{\sqrt{\frac{\omega}{2}}D}\sin\left(\omega t + \sqrt{\frac{\omega}{2}}D\right) + \frac{1}{2}a\cos(\omega)e^{-\sqrt{\frac{\omega}{2}}D}\sin\left(\omega t - \sqrt{\frac{\omega}{2}}D\right)$ Multiplicative Perturbations: $M(t) = \frac{2}{a}\sin(\omega t)$, $N(t) = -\frac{8}{a^2}\cos(2\omega t)$ Riccati Filter: $\dot{\Gamma}(t) = \omega_r\Gamma(t) - \omega_r\hat{H}(t)\Gamma^2(t)$, with $\hat{H}(t) = N(t)y(t)$

(b) For the Heat-Wave Cascade (14.107)–(14.113), *for each $\varepsilon > 0$, there exists a constant $\varepsilon_0 \in (0,\varepsilon)$ such that any solution for an initial value $X_0 = [\vartheta(0), \bar{\zeta}(\cdot,0), \bar{\omega}(\cdot,0), v(\cdot,0), U(0), \tilde{\Gamma}(0)]^T$ with $\|X_0\|_{\mathcal{H}} \leq \varepsilon_0$ converges to an $\mathcal{O}(1/\omega)$-neighborhood of the origin in the state-space $\mathcal{H} := \mathbb{R} \times L_{\infty[0,1]} \times L_{\infty[0,1]} \times L_{\infty[1,1+D]} \times \mathbb{R} \times \mathbb{R}$.*

Moreover, the same set of ultimate error bounds given in (14.85)–(14.87) *and* (14.164)–(14.166) *can still be guaranteed for the Newton-based ES control algorithm of Table* 14.1 *according to the results of Theorem* 14.1 *and Theorem* 14.2, *respectively. For both cases, we can also write*

$$\limsup_{t\to\infty} |\Gamma(t) - H^{-1}| = \mathcal{O}(1/\omega)\,. \quad (14.198)$$

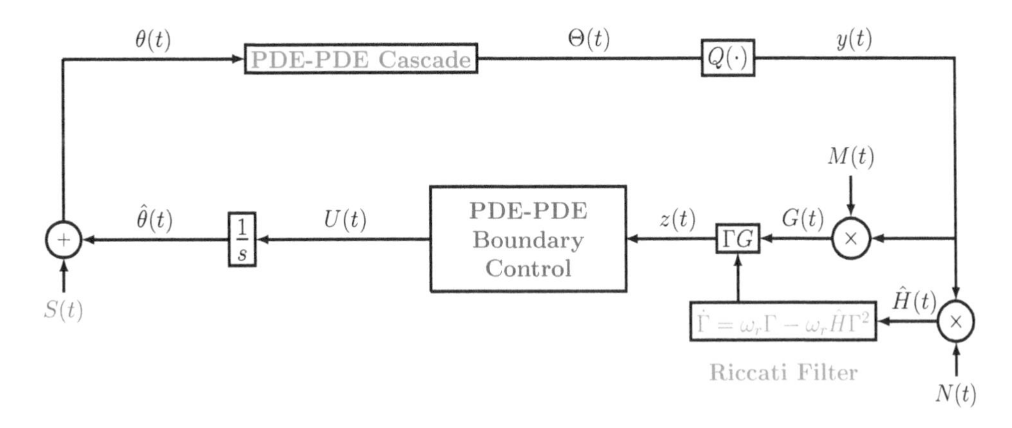

Figure 14.4. *Basic block diagram of the Newton-based extremum seeking control system with a generic PDE-PDE cascade system. In green we highlight the inclusion of a Riccati filter* (14.188) *responsible to estimate Hessian's inverse and to make the convergence rate of the algorithm user-assignable.*

Proof. From the second equation in (14.31), the average model for the Hessian's inverse estimation error in (14.190) is

$$\frac{d\tilde{\Gamma}_{\rm av}(t)}{dt} = -\omega_r \tilde{\Gamma}_{\rm av}(t) - \omega_r H \tilde{\Gamma}_{\rm av}^2(t). \tag{14.199}$$

On the other hand, the linearized version of (14.199) at $\tilde{\Gamma}_{\rm av}(t)=0$ is

$$\frac{d\tilde{\Gamma}_{\rm av}(t)}{dt} = -\omega_r \tilde{\Gamma}_{\rm av}(t), \tag{14.200}$$

which is locally exponentially stable since $\omega_r > 0$. Once the $\tilde{\Gamma}_{\rm av}$-subsystem (14.200) is completely decoupled from the other variables of the closed-loop feedback, the exponential stability of the overall systems (in some appropriate norm) can be orchestrated by following the same analysis carried out in Theorem 14.1 for the Delay-Wave cascade and Theorem 14.2 for the Heat-Wave cascade. The main differences are discussed in the following for the two cases.

• **Delay-Wave Cascade:** By recalling (14.194) and applying the following backstepping transformations

$$w(x,t) = \bar{\zeta}(x,t) + K\int_0^x \bar{\zeta}(\sigma,t)d\sigma + Kz(t), \quad x\in[0,1], \tag{14.201}$$

$$\varpi(x,t) = \bar{\omega}(x,t) - K\int_0^x \bar{\omega}(\sigma,t)d\sigma + Kz(t), \quad x\in[0,1], \tag{14.202}$$

$$\begin{aligned}\zeta(x,t) &= v(x,t) + K\int_1^x v(\sigma,t)d\sigma \\ &\quad + \bar{\alpha}_x(1,t) + K\int_0^1 \bar{\zeta}(\sigma,t)d\sigma + Kz(t), \quad x\in[1,1+D],\end{aligned} \tag{14.203}$$

into (14.58)–(14.61), the resulting average target system for (14.51)–(14.56) is given by

$$\textbf{ODE:} \qquad \dot{\vartheta}_{\rm av}(t) = -K\vartheta_{\rm av}(t) + w_{\rm av}(0,t), \tag{14.204}$$

$$\textbf{PDE 1:} \begin{cases} \partial_t w_{\rm av}(x,t) = \partial_x w_{\rm av}(x,t), & x\in[0,1], \\ w_{\rm av}(1,t) = \zeta_{\rm av}(1,t), \end{cases} \tag{14.205}$$

$$\textbf{PDE 2:} \begin{cases} \partial_t \varpi_{\rm av}(x,t) = -\partial_x \varpi_{\rm av}(x,t), & x\in[0,1], \\ \varpi_{\rm av}(0,t) = w_{\rm av}(0,t), \end{cases} \tag{14.206}$$

$$\textbf{PDE 3:} \begin{cases} \partial_t \zeta_{\rm av}(x,t) = \partial_x \zeta_{\rm av}(x,t), & x\in[1,1+D], \\ \zeta_{\rm av}(1+D,t) = -\frac{1}{c}\partial_t v_{\rm av}(1+D,t). \end{cases} \tag{14.207}$$

• **Heat-Wave Cascade:** Analogously, by employing the next backstepping transformations along with (14.194)

$$w(x,t) = \bar{\zeta}(x,t) + K\int_0^x \bar{\zeta}(\sigma,t)d\sigma + Kz(t), \quad x\in[0,1], \tag{14.208}$$

$$\varpi(x,t) = \bar{\omega}(x,t) - K\int_0^x \bar{\omega}(\sigma,t)d\sigma + Kz(t), \quad x\in[0,1], \tag{14.209}$$

$$\begin{aligned}z(x,t) &= v(x,t) + K\int_1^x (x-\sigma)v(\sigma,t)d\sigma \\ &\quad + \bar{\alpha}_x(1,t) + K\int_0^1 \bar{\zeta}(\sigma,t)d\sigma + Kz(t), \quad x\in[1,1+D],\end{aligned} \tag{14.210}$$

into (14.121)–(14.127), the resulting average target system is

$$\dot{\vartheta}_{\rm av}(t) = -K\vartheta_{\rm av} + w_{\rm av}(0,t), \tag{14.211}$$

$$\partial_t w_{\rm av}(x,t) = \partial_x w_{\rm av}(x,t), \qquad x \in [0,1], \tag{14.212}$$

$$w_{\rm av}(1,t) = z_{\rm av}(1,t), \tag{14.213}$$

$$\partial_t \varpi_{\rm av}(x,t) = -\partial_x \varpi_{\rm av}(x,t), \quad x \in [0,1], \tag{14.214}$$

$$\varpi_{\rm av}(0,t) = w_{\rm av}(0,t), \tag{14.215}$$

$$\partial_t z_{\rm av}(x,t) = \partial_{xx} z_{\rm av}(x,t), \quad x \in [1,1+D], \tag{14.216}$$

$$\partial_x z_{\rm av}(1,t) = 0, \tag{14.217}$$

$$z_{\rm av}(1+D,t) = -\frac{1}{c}\partial_t v_{\rm av}(1+D,t). \tag{14.218}$$

Notice that (14.205)–(14.207) (and (14.212)–(14.218)) are exactly the same as (14.66)–(14.68) (and (14.133)–(14.139)). In addition, (14.204) (and (14.211)) differs from (14.65) (and (14.132)) only by the convergence rate in the ISS relation ($-K$ rather than KH).

After that, we need to incorporate the dynamics of $\tilde{\Gamma}(t)$ given in (14.190) to (14.98) such that the augmented state is redefined by $z(t) = [\tilde{\vartheta}(t), \xi(t), U(t), \tilde{\Gamma}(t)]^T$ for the Delay-Wave case. Analogously, for the Heat-Wave cascade, we need to rewrite the evolution equation (14.163) with $X(t) = [\vartheta(t), \bar{\zeta}(\cdot,t), \bar{\omega}(\cdot,t), v(\cdot,t), U(t), \tilde{\Gamma}(t)]^T$, and the time-varying nonlinear perturbation term is given by

$$F(\omega t, X) = [0 \;\; 0 \;\; 0 \;\; 0 \;\; F_5(\omega t, X) \;\; F_6(\omega t, X)]^T, \tag{14.219}$$

where $F_5(\omega t, X)$ is the same as (14.161) and

$$F_6(\omega t, X) = \omega_r \left(X_6 + H^{-1}\right)\left[1 + \frac{8}{a^2}\cos(2\omega t) y \left(X_6 + H^{-1}\right)\right]. \tag{14.220}$$

The remaining steps of the proof are quite similar to those employed in Theorem 14.1 for the Delay-Wave PDEs and Theorem 14.2 for the Heat-Wave cascades, recalling that the final residual set for the error $\tilde{\Gamma}(t)$ in (14.198) is a direct consequence of the exponential stability of the averaging system and the application of the averaging theorem [83] for the closed-loop systems (14.200) and (14.204)–(14.207) (or (14.211)–(14.218)). In this sense, $\tilde{\Gamma}(t) - \tilde{\Gamma}^{\Pi}(t) \to 0$, exponentially as $t \to +\infty$, and $\tilde{\Gamma}^{\Pi}(t)$ is of order $\mathcal{O}(1/\omega)$.

In particular, the convergence rate of both estimators of the gradient $\vartheta_{\rm av}(t)$ in (14.204) (and (14.211)) as well as the Hessian inverse $\Gamma_{\rm av}(t)$ in (14.200) are independent of the unknown Hessian and can be assigned arbitrarily by the user [73] by choosing K and ω_r. □

14.7 ▪ Simulation Example

Here, we will focus on showing the numerical simulations only for the extremum seeking feedback for wave PDEs with input delay of Section 14.3 (Figure 14.2).

In this context, we consider a quadratic static map as in (14.2), with Hessian $H = -2$, optimizer $\Theta^* = \theta^* = 2$, and optimal value $y^* = 5$. The input delay is set to $D = 1$. As discussed in [47], this small delay is enough to produce instability of the open-loop system composed by the cascade of delay-wave PDEs. The parameters of the proposed ES are chosen as $\omega = 10$, $a = 0.2$, $c = 10$, and $K = 0.4$. In particular, we have used low-pass and washout filters with corner frequencies $\omega_h = 1$ and $\omega_l = 1$ in order to improve the numerical estimates of the gradient $G(t)$ and Hessian $\hat{H}(t)$, as suggested in Figures 1.4 and 1.5 of Section 1.5.

The simulation results of the closed-loop system are illustrated in Figures 14.5(a) and 14.5(b), Figures 14.6(a) and 14.6(b), and Figures 14.7(a) and 14.7(b).

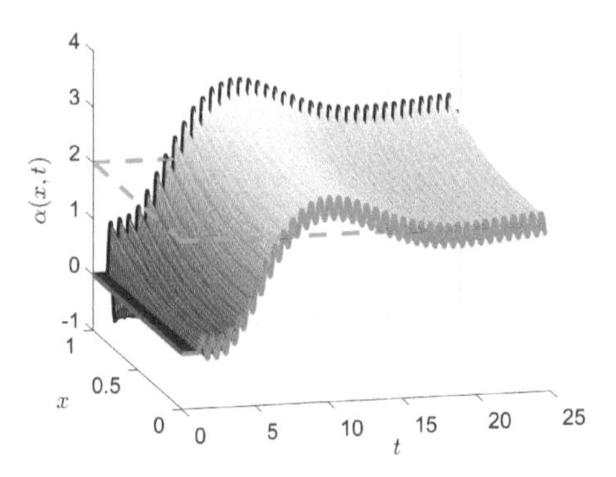

(a) Parameter $\Theta(t)$ (red) converging to Θ^* (dashed-green).

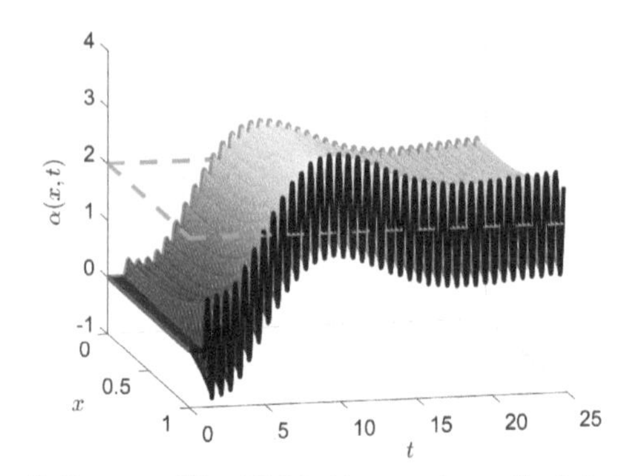

(b) Parameter $\theta(t-D)$ (black) converging to θ^* (dashed-green).

Figure 14.5. *Evolution of the infinite-dimensional state $\alpha(x,t)$ of the wave PDE model with delayed input ($D=1$) from $\alpha(1,t)=\theta(t-D)$ to $\alpha(0,t)=\Theta(t)$, with $\Theta^*=\theta^*=2$.*

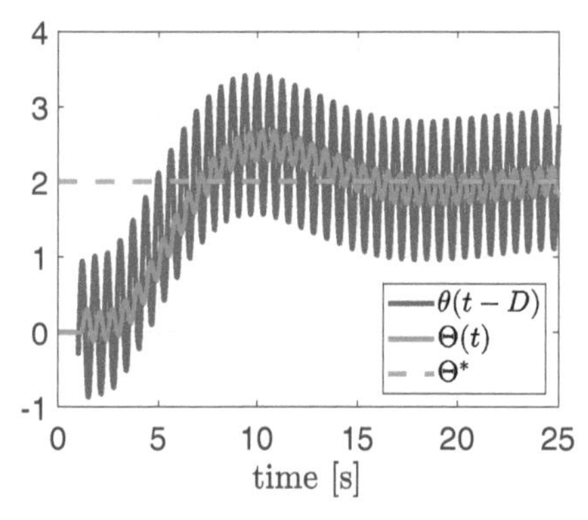

(a) Input of the wave PDE $\theta(t-D)$ and input of the static map $\Theta(t)$.

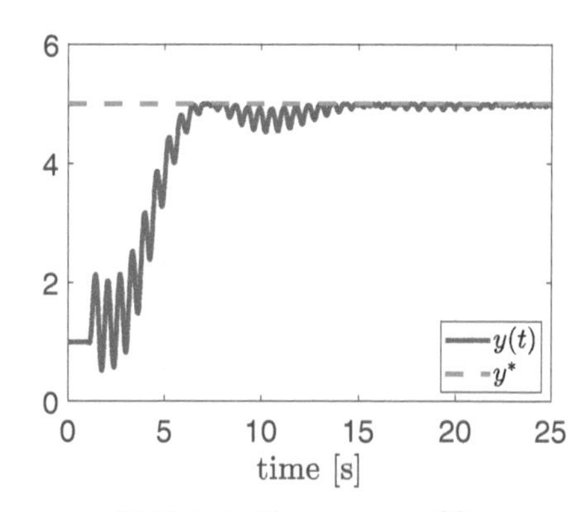

(b) Output of the static map $y(t)$.

Figure 14.6. *Compensation of a wave PDE with input delays in ES using boundary Dirichlet actuation: time responses of $\theta(t-D)$, $\Theta(t)$ converging to the maximizers $\theta^*=\Theta^*=2$ and $y(t)$ converging to the extremum $y^*=5$.*

The evolution of the PDE state $\alpha(x,t)$ modeled by the wave PDE model (14.4)–(14.7) with input delay is shown in Figures 14.5(a) and 14.5(b). The values of the boundary input $\theta(t-D)$ and the boundary output $\Theta(t)$ are highlighted with colors black/red and the initial condition is highlighted with color blue. One can observe in Figure 14.6(a) that after a delay of $D=1$ second, the variables Θ and θ converge to a neighborhood of the optimum θ^* and Θ^*, respectively.

Figure 14.6(b) and Figures 14.7(a) and 14.7(b) also present relevant variables for ES. It is clear the remarkable evolution of the output signal $y(t)$ as well as the Hessian's estimates $\hat{H}(t)$ ultimately achieving the extremum $y^*=5$ and the correct Hessian value $H=-2$, even in the presence of the delay-wave cascade of PDEs. As expected, the gradient's estimate is driven to zero as the output of the map $y(t)$ approximates the extremum.

When the nonlinear map $y=Q(\Theta)$ in (14.1) and (14.2) is flat with the Hessian satisfying the inequality $|H|<1$, the convergence rate with the Gradient ES in the neighborhood of the extremum is expected to become slow. While the convergence rate of the Gradient method is dictated by the unknown Hessian, the Newton-based scheme is independent of that and thus the

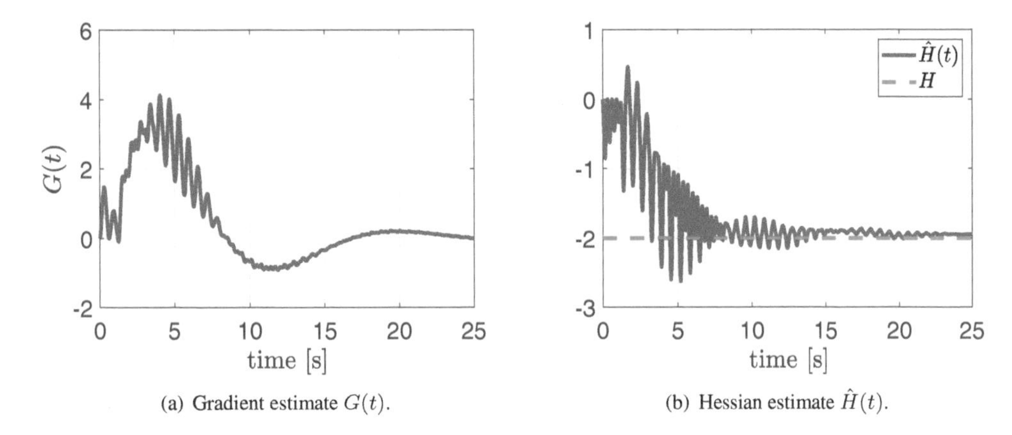

(a) Gradient estimate $G(t)$.

(b) Hessian estimate $\hat{H}(t)$.

Figure 14.7. *Compensation of a wave PDE with input delay in ES using boundary Dirichlet actuation: time response of the gradient estimate $G(t)$ converging to zero and the Hessian's estimate $\hat{H}(t)$ converging to $H = -2$.*

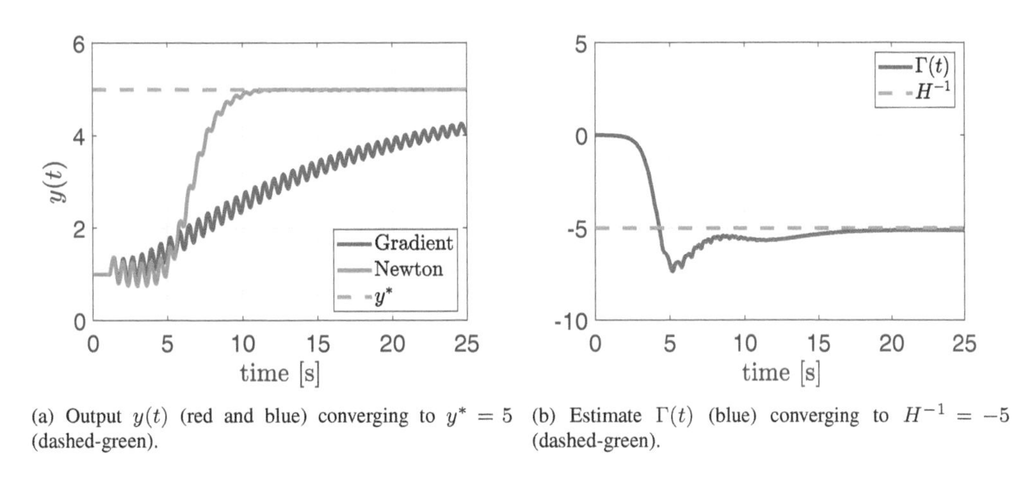

(a) Output $y(t)$ (red and blue) converging to $y^* = 5$ (dashed-green).

(b) Estimate $\Gamma(t)$ (blue) converging to $H^{-1} = -5$ (dashed-green).

Figure 14.8. *Newton-based ES versus Gradient ES: time responses of the output $y(t)$ and the Hessian's inverse estimate $\Gamma(t)$, both subject to a delay-wave cascade.*

PDE-PDE compensation can be achieved with an arbitrarily assigned convergence rate, improving the controller performance. Such a property can be verified in Figure 14.8(a), where we have performed the simulation test with a different value for the Hessian, given by $H = -0.2$. This is possible due to the exact estimation of the Hessian's inverse H^{-1} provided by the Riccati filter (14.188) with $\omega_r = 2$ (see Figure 14.8(b)), which cancels the effects of the Hessian in the convergence rate of the Newton-based ES.

14.8 ▪ Notes and References

Papers [179, 200, 198, 65, 169, 172, 176, 180] contain our efforts on expanding extremum seeking (ES) approach to a more general class of dynamic systems, described not only by finite-dimensional Ordinary Differential Equations (ODEs) [129, 114, 224, 244, 218, 73, 109, 59, 204, 135, 14, 15, 79, 193, 77, 243, 137]. In this direction, we've studied ES for time-delay systems [179, 200, 198], diffusion processes [65, 169], general classes of parabolic reaction-advection-diffusion plants [172], and wave equations [176, 180]. All of them [179, 200, 198, 65, 169, 172, 176, 180] could be described by infinite-dimensional Partial Differential Equations (PDEs).

If the infinite-dimensional system is not properly compensated in the ES feedback, the closed-loop system goes unstable [179] or, in some particular cases, the probing signals must be made sufficiently slow to perform the correct search of the extremum, but with the slow search resulting in quite slow convergence [65].

Our general design process for real-time optimization in the presence of a PDE at the input of a nonlinear convex map consists of two steps: a design of an additive probing signal through a trajectory generation approach for the PDE [128, Chapter 12] and the application of PDE backstepping [128, 119] to compensate for the effect of the PDE on the ES algorithm.

Unlike any of our previous chapters, where a single PDE was considered at the input of the nonlinear map to be optimized, in this chapter we face a more general setup with a cascade of two PDEs. The topic of stabilization of PDE cascades was introduced in the second author's book [119, Part V] and the companion papers. An example of an effort on a PDE-PDE cascade which precedes stabilization is the work on stability (in the absence of feedback design) and on controllability of a heat-wave system by [247]. The effort in PDE cascades is motivated partly by problems in fluid-structure interaction and other interactive physical processes [119, Chapter 20].

We started with a presentation of the control design for the wave PDE with input delay [122]. For clarity and historical context, we first specialized this design to the baseline problem without input delay. Efforts in delay-PDE cascades are motivated by the interest to address the lack of delay robustness identified by [47], where it is shown that standard feedback laws for wave equations have a zero robustness margin to the introduction of a delay in the feedback loop—an arbitrarily small measurement delay or input delay results in closed-loop instability. Such a result does not arise with finite-dimensional plants or with parabolic PDEs [118].

Next, we dealt with cascades of parabolic and hyperbolic PDEs. The parabolic-hyperbolic cascade was represented by a heat equation at the input of a wave equation. Both versions of the extremum seeking algorithm (gradient- and Newton-based) were considered [73].

Stability analysis for cascades of stable PDEs from different classes, when interconnected through a boundary, virtually explodes in complexity despite the seemingly simple structure where one PDE is autonomous and exponentially stable and feeds into the other PDE. The difficulty arises for two reasons. One is that the connectivity through the boundary gives rise to an unbounded input operator in the interconnection. The second reason is that the two subsystems are from different PDE classes, with different numbers of derivatives in space or time (or both). This requires delicate combinations of norms in the Lyapunov functions for the overall systems [119]; the use of PDE small-gain results in various state norms or in the supremum norm [99].

In this sense, we rigorously conducted the stability analysis for the average closed-loop system starting with a development of the backstepping transformation and a proof of stability of the target system [128, 119] by means of the Input-to-State Stability (ISS) analysis for cascades of dynamical ODE and PDE systems [99] (with and without small-gain loops), followed by the corresponding equivalence in the original variables of the system, and a derivation of the estimates on the transformation kernels, which complete the overall stability proof for the average closed-loop system. Finally, the averaging theorem in infinite dimension [83] was invoked to conclude the local exponential stability of the original nonaverage system. We derived explicit formulas for the control gains, and we also found the explicit formulas for the solutions of the closed-loop system showing that a small neighborhood of the extremum point is ultimately achieved.

14.9 ▪ Open Problems

While in Section 14.5 we hint at the opportunities that exist in extending the results from Section 14.2 and 14.3 to other classes of PDE-PDE cascades, much still remains to be done in terms of the implementability of the control laws and the analysis of the closed-loop systems for cascaded PDEs from different families (parabolic-hyperbolic in either order) which are interconnected through boundary conditions.

We can apply our approach adaptively to unknown constant wave coefficients or heat parameters that vary sufficiently slowly, as in [210], but we have to be careful with this. On the one hand, all the parameters of a reaction-advection-diffusion equation can be permitted to be unknown if the full PDE state is *measured*. If, on the other hand, the PDE state is not measured, we cannot adaptively handle an unknown diffusion, for instance. However, because the wave/diffusion parameter enters nonlinearly into the additive dithers, the result would necessarily be local (the initial estimate of the wave/diffusion parameter would have to be close to the actual wave/diffusion coefficient). This is similar to the delay-adaptive work in early publications by [31, 32, 35], where one gets only a local result, with no advantage relative to relying on delay robustness (Chapter 6). In the same sense, adaptation would offer little advantage over the reliance on the robustness of the diffusion compensator feedback to small perturbations in the diffusion parameter, which is proved in [119, Section 15.3].

It is likely possible to develop results, for extremum seeking, on (a) delay robustness, (b) diffusion robustness, and (c) perhaps Young modulus robustness—this is the least predictable due to the Datko counterexamples for wave PDEs [47] but the approach by Auriol and Di Meglio [16] is encouraging.

As for adaptation for unknown delay, diffusion, or Young modulus, even if we can do it outside of the extremum seeking (ES) context, doing it under ES is problematic. Adaptive control systems are never exponentially stable, unless there is persistency of excitation. In ES we have persistency of excitation to estimate the input parameter θ, but we generally do not have a persistency of excitation for estimating unknown delays or diffusion. So, we would get an average system that has neutral stability plus the regulation of $\hat{\theta}$ to the true θ. Then, we would not be able to apply an averaging theorem to establish the regulation of $\hat{\theta}$ to θ in the nonaverage system. Guay, Dochain, and co-authors [80, 2, 81, 82] have made important advances in the general direction of combining model-free ES with model-based adaptive control, for systems governed by ordinary differential equations (ODEs), but PDEs and nonlinear parametrizations bring a whole additional set of challenges.

Part III

NONCOOPERATIVE GAMES THROUGH DELAYS AND PDEs

Chapter 15

Nash Equilibrium Seeking with Arbitrarily Delayed Player Actions

15.1 ▪ A Game with Noncooperative Players, Each Employing Extremum Seeking

In Parts I and II of the book we dealt with model-free online optimization problems in which, whether there is only a single input, or multiple inputs, such as in Chapters 7 and 13, only a single payoff is being maximized (or cost minimized).

Optimization problems become far more interesting when there are multiple payoffs, being maximized by the respective multiple inputs/actors/players. When the different payoffs are distinct functions of the inputs by the different players, it is clear that conflicting objectives may arise among the players. This scenario is called a noncooperative game, a suitable term given that the players are involved in a competition, though not necessary with diametrically opposed objectives, but nevertheless distinct objectives.

The realization that extremum seeking is not only applicable to single-player optimization (possibly with a vector of inputs maximizing a single payoff), in an online model-free setting, but also by multiple competing players in the noncooperative game setting, came in the 2012 paper [70], twelve years after the question of stability in single-player optimization was settled in [129].

But why would multiple players all be employing extremum seeking algorithms? The answer is that it is mathematically provably wise to do so. Each player employing an ES algorithm, whether in an isolated optimization scenario or in a noncooperative game, is maximizing his payoff, irrespective of what the other players' actions are. It is known that, if all the players play rationally or strategically, whether this is the result of the functional forms of the payoffs being known to all the players and the strategies being determined from such models or the strategic actions are determined in some other manner (possibly even by a lucky guess), no player can improve his payoff over the payoff that results from applying a certain optimal action. The collection of such optimal actions by the players is called a Nash equilibrium.

When all the players play their Nash-optimal actions, the game settles in a stalemate. It is easy to miss the meaning of a Nash equilibrium and be disappointed by a winnerless outcome—that one's favorite player doesn't "win" by playing his Nash strategy. The point of playing a Nash strategy is not to win but to *not lose*. With the Nash strategy, one succeeds by not failing.

It is proven in [70] that if all the players are employing ES algorithms, they collectively converge to a Nash equilibrium. In other words, each of the players finds his optimal strategy. And they achieve this in a model-free online fashion. They don't know the analytical forms of the payoff functions. They don't even have access to the actions applied by the other players or the payoffs achieved by the other players. They are merely "minding their own business" by performing the seeking of their own extremum, and yet ES finds the strategy that ensures they achieve the best possible outcomes for themselves in the noncooperative game scenario. That such a model-free finding of a Nash equilibrium is possible was quite unexpected before [70], both at the collective level, that the vector of optimal strategies can be found by the players pursuing their own (decentralized) ES algorithms, and at the individual level, that a player who is possibly even unaware of the presence of the other players, and is merely employing an ES algorithm for the maximization of his payoff, will find what is actually his Nash-optimal action.

For a reader with no prior exposure to noncooperative games, it is useful to think of the following scenario in which real-time model-free seeking of maximum payoffs is reasonable to pursue by each player applying his own ES algorithm. Imagine a good being sold in real time by different companies and also priced in real time (daily, hourly, or in other relatively short intervals). An example of a good that is re-priced daily is gasoline at different gas stations, at least in the United States. Clearly, the companies that price the good more attractively will end up selling more of the good. But, by pricing lower, the company will earn less profit per unit of the good sold. To make the problem more realistic, and less mathematically trivial, suppose that the good being sold by different companies is either of different quality, or not equally desirable, or the convenience of the retail locations is not equal among the companies. Clearly, some companies will be able to set a higher price than others. But how much higher? Nobody can create an accurately predictive mathematical model of the customer's behavior in response to pricing. It is, for this reason, in a model-free real-time iterative pricing scenario, reasonable for companies to employ an ES (profit or revenue maximizing) algorithm of some sort. In fact, many do—the pricing adjustments by gas stations is little other than manual ES.

The basics of Nash equilibrium seeking were introduced in Section 1.6. In Part III of the book we consider noncooperative games where players act through PDE dynamics, starting in this chapter where the players act through arbitrarily long delays. The delays may be distinct and, in general, each player knows only the length of his own delay.

Specifically, in this chapter we consider quadratic noncooperative games where acquisition of information (of two different types) incurs delays: (a) one, which we call cooperative scenario, where each player employs the knowledge of the functional form of his payoff and knowledge of other players' actions, but with delays; and (b) the second one, which we term the noncooperative scenario, where the players have access only to their own payoff values, again with delay.

In order to compensate distinct delays at the inputs of the players, we employ the predictor feedback. We apply a small-gain analysis as well as averaging theory in infinite dimension, due to the infinite-dimensional state of the time delays, in order to obtain local convergence results for the unknown quadratic payoffs to a small neighborhood of the Nash equilibrium. We quantify the size of these residual sets and corroborate the theoretical results numerically on an example of a two-player game with delays.

15.2 ▪ N-Players Game with Quadratic Payoffs and Delays: General Formulation

Game theory provides an important framework for mathematically modeling and analysis of scenarios involving different agents (players) where there is coupling in their actions, in the sense that their respective outcomes (outputs) $y_i(t) \in \mathbb{R}$ do not depend exclusively on their own actions/strategies (input signals) $\theta_i(t) \in \mathbb{R}$, with $i = 1, \ldots, N$, but at least on a subset of others'. Moreover, defining $\theta := [\theta_1, \ldots, \theta_N]^T$, each player's payoff function $J_i(\theta) : \mathbb{R}^N \to \mathbb{R}$ depends on the action θ_j of at least one other Player j, $j \neq i$. An N-tuple of actions, θ^* is said to be in Nash equilibrium if no Player i can improve his payoff by unilaterally deviating from θ_i^*, this

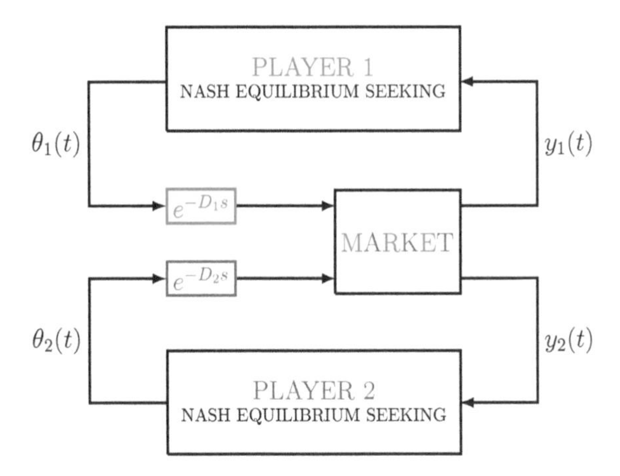

Figure 15.1. *Nash equilibrium seeking schemes applied by two players ($N = 2$) in a duopoly market structure with delayed players' actions.*

being so for all i [21]. Despite the vast number of publications on Nash equilibrium seeking [70], its study under time delays is still an open problem.

For instance, the applications in economic analysis can be used to motivate the problem. Consider the policies of two gas stations, where each station is supplied by a different oil refinery. Basically, the price at the pumps is adjusted based on the current price a barrel of oil and stocks bought at previous values. Thus, the stations take time to pass on to consumers variations in the price of a barrel of oil in refineries. In this context of game theory, each gas station could be viewed as a player and the aforementioned phenomenon described can be interpreted as distinct delays D_i applied to their strategies (price at the pumps). An increase in the pumps' price of the ith gas station results in lower, but not zero, sales at the ith station, with a profit $y_i(t)$, and, consequently, increased sales for the other gas stations; see Figure 15.1.

Hence, we consider games where the payoff function of each player is quadratic, expressed as a strictly concave combination of their delayed actions

$$J_i(\theta(t-D)) = \frac{1}{2}\sum_{j=1}^{N}\sum_{k=1}^{N}\epsilon^i_{jk}H^i_{jk}\theta_j(t-D_j)\theta_k(t-D_k) + \sum_{j=1}^{N}h^i_j\theta_j(t-D_j) + c_i\,, \qquad (15.1)$$

where $\theta_j(t-D_j)\in\mathbb{R}$ is the decision variable of Player j delayed by $D_j\in\mathbb{R}^+$ units of time, H^i_{jk}, h^i_j, $c_i\in\mathbb{R}$ are constants, $H^i_{ii}<0$, $H^i_{jk}=H^i_{kj}$, and $\epsilon^i_{jk}=\epsilon^i_{kj}>0\ \forall i,j,k$.

Without loss of generality, we assume that the inputs have distinct known (constant) delays which are ordered so that

$$D = \mathrm{diag}\{D_1, D_2, \ldots, D_N\}, \quad 0\le D_1\le\cdots\le D_N\,. \qquad (15.2)$$

Moreover, given any $\mathbb{R}^N$-valued signal f, the notation f^D denotes

$$f^D(t) := f(t-D) = \begin{bmatrix} f_1(t-D_1) & f_2(t-D_2) & \ldots & f_N(t-D_N)\end{bmatrix}^T. \qquad (15.3)$$

Quadratic payoff functions are of particular interest in game theory, first because they constitute second-order approximations to other types of non-quadratic payoff functions, and second because they are analytically tractable, leading in general to closed-form equilibrium solutions which provide insight into the properties and features of the equilibrium solution concept under consideration [21].

For the sake of completeness, we provide here in mathematical terms the definition of a Nash equilibrium $\theta^* = [\theta_1^*, \ldots, \theta_N^*]^T$ in an N-player game:

$$J_i(\theta_i^*, \theta_{-i}^*) \geq J_i(\theta_i, \theta_{-i}^*) \quad \forall \theta_i \in \Theta_i, \quad i \in \{1, \ldots, N\}, \tag{15.4}$$

where J_i is the payoff function of Player i, the term θ_i corresponds to its action, while Θ_i is its action set and θ_{-i} denotes the actions of the other players. Hence, no player has an incentive to unilaterally deviate its action from θ^*. In the duopoly example just mentioned, $\Theta_1 = \Theta_2 = \mathbb{R}$, where $\mathbb{R}$ denotes the set of real numbers.

In order to determine the Nash equilibrium solution in strictly concave[10] quadratic games with N players, where each action set is the entire real line, one should differentiate J_i with respect to $\theta_i(t-D_i)$ $\forall i = 1, \ldots, N$, setting the resulting expressions equal to zero and solving the set of equations thus obtained. This set of equations, which also provides a sufficient condition due to the strict concavity, is

$$\sum_{j=1}^{N} \epsilon_{ij}^i H_{ij}^i \theta_j^* + h_i^i = 0, \quad i = 1, \ldots, N, \tag{15.5}$$

which can be written in the form of matrices as

$$\begin{bmatrix} \epsilon_{11}^1 H_{11}^1 & \epsilon_{12}^1 H_{12}^1 & \cdots & \epsilon_{1N}^1 H_{1N}^1 \\ \epsilon_{21}^2 H_{21}^2 & \epsilon_{22}^2 H_{22}^2 & \cdots & \epsilon_{2N}^2 H_{2N}^2 \\ \vdots & \vdots & & \vdots \\ \epsilon_{N1}^N H_{N1}^N & \epsilon_{N2}^N H_{N2}^N & \cdots & \epsilon_{NN}^N H_{NN}^N \end{bmatrix} \begin{bmatrix} \theta_1^* \\ \theta_2^* \\ \vdots \\ \theta_N^* \end{bmatrix} = - \begin{bmatrix} h_1^1 \\ h_2^2 \\ \vdots \\ h_N^N \end{bmatrix}. \tag{15.6}$$

Defining the Hessian matrix H and vectors θ^* and h by

$$H := \begin{bmatrix} \epsilon_{11}^1 H_{11}^1 & \epsilon_{12}^1 H_{12}^1 & \cdots & \epsilon_{1N}^1 H_{1N}^1 \\ \epsilon_{21}^2 H_{21}^2 & \epsilon_{22}^2 H_{22}^2 & \cdots & \epsilon_{2N}^2 H_{2N}^2 \\ \vdots & \vdots & & \vdots \\ \epsilon_{N1}^N H_{N1}^N & \epsilon_{N2}^N H_{N2}^N & \cdots & \epsilon_{NN}^N H_{NN}^N \end{bmatrix}, \quad \theta^* := \begin{bmatrix} \theta_1^* \\ \theta_2^* \\ \vdots \\ \theta_N^* \end{bmatrix}, \quad h := \begin{bmatrix} h_1^1 \\ h_2^2 \\ \vdots \\ h_N^N \end{bmatrix}, \tag{15.7}$$

there exists only one Nash equilibrium at $\theta^* = -H^{-1}h$ if H is invertible:

$$\begin{bmatrix} \theta_1^* \\ \theta_2^* \\ \vdots \\ \theta_N^* \end{bmatrix} = - \begin{bmatrix} \epsilon_{11}^1 H_{11}^1 & \epsilon_{12}^1 H_{12}^1 & \cdots & \epsilon_{1N}^1 H_{1N}^1 \\ \epsilon_{21}^2 H_{21}^2 & \epsilon_{22}^2 H_{22}^2 & \cdots & \epsilon_{2N}^2 H_{2N}^2 \\ \vdots & \vdots & & \vdots \\ \epsilon_{N1}^N H_{N1}^N & \epsilon_{N2}^N H_{N2}^N & \cdots & \epsilon_{NN}^N H_{NN}^N \end{bmatrix}^{-1} \begin{bmatrix} h_1^1 \\ h_2^2 \\ \vdots \\ h_N^N \end{bmatrix}. \tag{15.8}$$

For more details, see [21, Chapter 4].

The *control objective* is to design a novel extremum seeking-based strategy to reach the Nash equilibrium in (non)cooperative games subjected to distinct delays in the decision variables of the players (input signals).

Figure 15.2 contains a schematic diagram that summarizes the proposed Nash equilibrium policy for each ith player where its output is given by

$$y_i(t) = J_i(\theta(t-D)), \tag{15.9}$$

[10] By strict concavity, we mean $J_i(\theta)$ is strictly concave in θ_i for all θ_{-i}, this being so for each $i = 1, \ldots, N$.

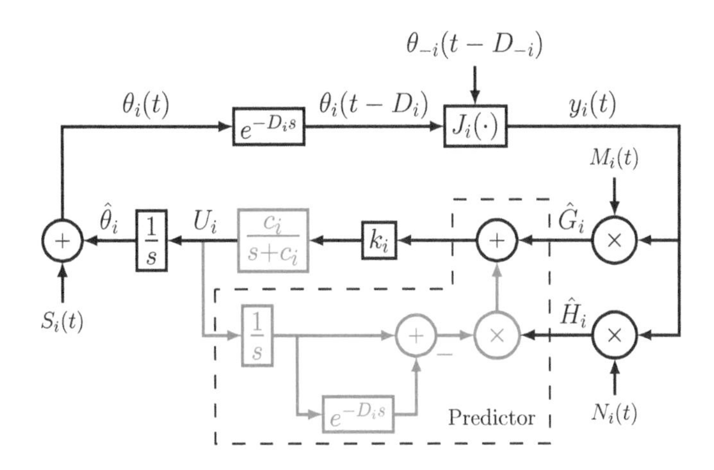

Figure 15.2. *Block diagram illustrating the Nash equilibrium seeking strategy performed for each player. In red color, the predictor feedback is used to compensate the individual delay D_i for the noncooperative case. With some abuse of notation, the constants c_i were chosen to denote the parameters of the filters $c_i/(s+c_i)$, but they are not necessarily the same constants which appear in the payoff functions given in* (15.1).

where the vector $\theta_{-i}(t-D_{-i})$ in Figure 15.2 represents the delayed actions of all other players. The additive-multiplicative dither signals $S_i(t)$ and $M_i(t)$ are

$$S_i(t) = a_i \sin(\omega_i(t+D_i)), \tag{15.10}$$

$$M_i(t) = \frac{2}{a_i}\sin(\omega_i t), \tag{15.11}$$

with nonzero constant amplitudes $a_i > 0$ at frequencies $\omega_i \neq \omega_j$. Such probing frequencies ω_i can be selected as

$$\omega_i = \omega_i' \omega = \mathcal{O}(\omega), \quad i \in 1, 2, \ldots, N, \tag{15.12}$$

where ω is a positive constant and ω_i' is a rational number. One possible choice is given in [73] as

$$\omega_i' \notin \left\{ \omega_j', \ \frac{1}{2}(\omega_j' + \omega_k'), \ \omega_j' + 2\omega_k', \ \omega_j' + \omega_k' \pm \omega_l' \right\} \tag{15.13}$$

for all distinct i, j, k, and l. The remaining signals are defined throughout the chapter depending on the type of game in question: cooperative games (Section 15.3) or noncooperative games (Section 15.4).

By considering $\hat{\theta}_i(t)$ as an estimate of θ_i^*, one can define the *estimation error*:

$$\tilde{\theta}_i(t) = \hat{\theta}_i(t) - \theta_i^*. \tag{15.14}$$

Then, from (15.10), (15.14), and Figure 15.2, it is easy to get

$$\begin{aligned} \theta_i(t) &= S_i(t) + \hat{\theta}_i(t) && (15.15) \\ &= a_i \sin(\omega_i t + \omega_i D_i) + \tilde{\theta}_i(t) + \theta_i^*, && (15.16) \end{aligned}$$

with the following time-delay version:

$$\theta_i(t-D_i) = a_i \sin(\omega_i t) + \tilde{\theta}_i(t-D_i) + \theta_i^*. \tag{15.17}$$

Therefore, from (15.1) and (15.17), the ith output signal in (15.9) can be rewritten as

$$\begin{aligned} y_i(t) =& \frac{1}{2}\sum_{j=1}^{N}\sum_{k=1}^{N} \epsilon^i_{jk} H^i_{jk} \left[a_j \sin(\omega_j t)\tilde{\theta}_k(t-D_k) + a_k \sin(\omega_k t)\tilde{\theta}_j(t-D_j)\right] \\ &+ \frac{1}{2}\sum_{j=1}^{N}\sum_{k=1}^{N} \epsilon^i_{jk} H^i_{jk} \left[\theta^*_k a_j \sin(\omega_j t) + \theta^*_j a_k \sin(\omega_k t)\right] \\ &+ \frac{1}{2}\sum_{j=1}^{N}\sum_{k=1}^{N} \epsilon^i_{jk} H^i_{jk} \left[\theta^*_k \tilde{\theta}_j(t-D_j) + \theta^*_j \tilde{\theta}_k(t-D_k)\right] \\ &+ \frac{1}{2}\sum_{j=1}^{N}\sum_{k=1}^{N} \epsilon^i_{jk} H^i_{jk} a_j a_k \sin(\omega_j t)\sin(\omega_k t) + \frac{1}{2}\sum_{j=1}^{N}\sum_{k=1}^{N} \epsilon^i_{jk} H^i_{jk} \theta^*_j \theta^*_k \\ &+ \frac{1}{2}\sum_{j=1}^{N}\sum_{k=1}^{N} \epsilon^i_{jk} H^i_{jk} \tilde{\theta}_j(t-D_j)\tilde{\theta}_k(t-D_k) \\ &+ \sum_{j=1}^{N} h^i_j a_j \sin(\omega_j t) + \sum_{j=1}^{N} h^i_j \tilde{\theta}_j(t-D_j) + \sum_{j=1}^{N} h^i_j \theta^*_j + c_i . \end{aligned} \tag{15.18}$$

The estimate $\hat{G}_i$ of the unknown gradient of each payoff J_i is given by

$$\hat{G}_i(t) = M_i(t) y_i(t) . \tag{15.19}$$

Plugging (15.11) and (15.18) into (15.19) and computing the average of the resulting signal lead us to

$$\begin{aligned} \hat{G}^{\mathrm{av}}_i(t) &= \frac{1}{\Pi}\int_0^{\Pi} M_i(\tau) y_i d\tau = \sum_{j=1}^{N} \epsilon^i_{ij} H^i_{ij} \tilde{\theta}^{\mathrm{av}}_j(t-D_j) + \underbrace{\sum_{j=1}^{N} \epsilon^i_{ij} H^i_{ij} \theta^*_j + h^i_i}_{=\,0,\ \text{from (15.5)}} , \\ &= \sum_{j=1}^{N} \epsilon^i_{ij} H^i_{ij} \tilde{\theta}^{\mathrm{av}}_j(t-D_j) , \end{aligned} \tag{15.20}$$

with Π defined as

$$\Pi := 2\pi \times \mathrm{LCM}\left\{\frac{1}{\omega_i}\right\} \tag{15.21}$$

and LCM standing for the least common multiple.

At this point, if we neglect the prediction loop and the low-pass filter (both indicated in red color) in Figure 15.2, the control law $U_i(t) = k_i \hat{G}_i(t)$ could be obtained as in the classical ES approach. In this case, from (15.14) and (15.20), we could write the average version of

$$\dot{\hat{\theta}}_i(t) = U_i(t) \tag{15.22}$$

such as

$$\begin{aligned} \dot{\hat{\theta}}^{\mathrm{av}}_i(t) &= k_i \hat{G}^{\mathrm{av}}_i(t) \\ &= k_i \sum_{j=1}^{N} \epsilon^i_{ij} H^i_{ij} \tilde{\theta}^{\mathrm{av}}_j(t-D_j) . \end{aligned} \tag{15.23}$$

Therefore, by defining $\tilde{\theta}^{\rm av}(t) := [\tilde{\theta}_1^{\rm av}(t), \tilde{\theta}_2^{\rm av}(t), \dots, \tilde{\theta}_N^{\rm av}(t)]^T \in \mathbb{R}^N$ in order to take into account all players, one has

$$\dot{\tilde{\theta}}^{\rm av}(t) = KH\tilde{\theta}^{\rm av}(t-D), \tag{15.24}$$

with $K := \text{diag}\{k_1, \dots, k_N\}$ and H given by (15.7). Equation (15.24) means that, even if KH was a Hurwitz matrix, the equilibrium $\tilde{\theta}_{\rm e}^{\rm av} = 0$ of the closed-loop average system would not necessarily be stable for arbitrary values of the time delays D_i. This reinforces the demand of employing the prediction feedback $U_i(t) = k_i\hat{G}_i(t+D_i)$—or even its filtered version—for each player to stabilize collectively the closed-loop system, as illustrated with red color in Figure 15.2.

On the other hand, the derivative of (15.20) is

$$\dot{\hat{G}}_i^{\rm av}(t) = \sum_{j=1}^{N} \epsilon_{ij}^i H_{ij}^i \dot{\tilde{\theta}}_j^{\rm av}(t-D_j). \tag{15.25}$$

By delaying by D_i units the time argument of both sides of the average version of (15.22), we obtain

$$\dot{\tilde{\theta}}_i^{\rm av}(t-D_i) = U_i^{\rm av}(t-D_i). \tag{15.26}$$

Thus, (15.25) can be rewritten as

$$\dot{\hat{G}}_i^{\rm av}(t) = \sum_{j=1}^{N} \epsilon_{ij}^i H_{ij}^i U_j^{\rm av}(t-D_j). \tag{15.27}$$

Taking into account all players, from (15.20) and (15.27), it is possible to find a compact form for the overall average estimated gradient $\hat{G}^{\rm av}(t) := [\hat{G}_1^{\rm av}(t), \dots, \hat{G}_N^{\rm av}(t)]^T \in \mathbb{R}^N$ according to

$$\hat{G}^{\rm av}(t) = H\tilde{\theta}^{\rm av}(t-D), \tag{15.28}$$

$$\dot{\hat{G}}^{\rm av}(t) = HU^{\rm av}(t-D), \tag{15.29}$$

where H is given in (15.7) and $U^{\rm av}(t) := [U_1^{\rm av}(t), U_2^{\rm av}(t), \dots, U_N^{\rm av}(t)]^T \in \mathbb{R}^N$.

Throughout the chapter the key idea is to design control laws (policies) for each player in order to achieve a small neighborhood of the Nash equilibrium. To this end, we use an extremum seeking strategy based on prediction feedback to compensate multiple and distinct delays in the players' actions. Basically, the control laws are able to ensure exponential stabilization of $\hat{G}^{\rm av}(t)$ and, consequently, of $\tilde{\theta}^{\rm av}(t)$. From (15.28), it is clear that, if H is invertible, $\tilde{\theta}^{\rm av}(t) \to 0$ as $\hat{G}^{\rm av}(t) \to 0$. Hence, the convergence of $\tilde{\theta}^{\rm av}(t)$ to the origin results in the convergence of $\theta(t)$ to a small neighborhood of θ^* in (15.4) via averaging theory [83].

15.3 ▪ Cooperative Scenario with Delays

In the cooperative scenario, the purpose of the extremum seeking is to estimate the Nash equilibrium vector θ^* by sharing among the players their outputs (payoffs)

$$y_i(t) = J_i(\theta(t-D)) \tag{15.30}$$

in (15.1), as well as their own actions $\theta_i(t)$ and control laws $U_i(t)$. In this sense, we are able to formulate the closed-loop system in a *centralized* fashion with a multivariable framework in which each state variable corresponds to its corresponding player.

In this section, $e_i \in \mathbb{R}^N$ stands for the ith column of the identity matrix $I_N \in \mathbb{R}^{N\times N}$ for each $i \in \{1, 2, \dots, N\}$.

15.3.1 ▪ Centralized Predictor with Shared Hessian Information among Players

To this end, we redefine perturbation signals in (15.10) and (15.11) through a vector form $S(t)$ and $M(t) \in \mathbb{R}^N$ by

$$S(t) = \begin{bmatrix} a_1 \sin(\omega_1(t+D_1)) & \cdots & a_N \sin(\omega_N(t+D_N)) \end{bmatrix}^T, \tag{15.31}$$

$$M(t) = \begin{bmatrix} \frac{2}{a_1}\sin(\omega_1 t) & \cdots & \frac{2}{a_N}\sin(\omega_N t) \end{bmatrix}^T. \tag{15.32}$$

Notice that the delayed signal S^D of S is a conventional perturbation signal used in [73]. We also set the matrix-valued signal $N(t) \in \mathbb{R}^{N\times N}$ as

$$N_{ij}(t) = \begin{cases} \frac{16}{a_i^2}\left(\sin^2(\omega_i t) - \frac{1}{2}\right), & i = j, \\ \frac{4}{a_i a_j}\sin(\omega_i t)\sin(\omega_j t), & i \neq j. \end{cases} \tag{15.33}$$

By using the above signals, we develop a multivariable extremum seeking scheme in order to compensate collectively the presence of multiple-input delays. Let the input signals (15.14) be constructed in the vector form as

$$\theta(t) = \hat{\theta}(t) + S(t), \tag{15.34}$$

where $\hat{\theta}$ is an estimate of θ^*. We also introduce the estimation error

$$\tilde{\theta}(t) := \hat{\theta}^D(t) - \theta^*. \tag{15.35}$$

Note that the error is defined here with $\hat{\theta}^D$ rather than $\hat{\theta}$. With this error variable, the individual output signals or payoffs $y_i(t)$ can be rewritten as

$$y_i(t) = J_i\left(\theta^* + \tilde{\theta}(t) + S^D(t)\right). \tag{15.36}$$

To compensate the delays, we propose the following predictor-based update law:

$$\dot{\hat{\theta}}(t) = U(t), \tag{15.37}$$

$$\dot{U}_i(t) = -cU_i(t) + ck_i\left(M_i(t)y_i(t) + N_i(t)y_i(t)\sum_{j=1}^{N} e_j \int_{t-D_j}^{t} U_j(\tau)d\tau\right) \tag{15.38}$$

for some positive constants $c, k_i > 0$ with k_i being the elements of the diagonal matrix $K \in \mathbb{R}^{N\times N}$. Without loss of generality, we consider $c_i = c$ in Figure 15.2. Moreover, in the particular case of the cooperative scenario, the signal $N_i(t)$ used in (15.38) is simply defined by

$$N_i(t) := [N_{i1}(t) \ldots\ N_{iN}(t)], \tag{15.39}$$

representing the vector with the elements of each row of the matrix $N(t)$ with N_{ij} given in (15.33).

Since $\dot{\hat{\theta}}^D(t) = U^D(t)$, differentiating the error variable $\tilde{\theta}$ with respect to t yields

$$\dot{\tilde{\theta}}(t) = U^D(t) = \sum_{i=1}^{N} e_i U_i(t-D_i), \tag{15.40}$$

which is in a standard form of a system with input delays. As we will see later, the terms in the parentheses on the RHS of (15.38) correspond to a predicted value of $H\tilde{\theta}$ at some time in the future in the average sense, i.e.,

$$\hat{G}^{\mathrm{av}}(t+D) = \hat{G}^{\mathrm{av}}(t) + H\sum_{i=1}^{N} e_i \int_{t-D_i}^{t} U_i^{\mathrm{av}}(\tau)d\tau. \tag{15.41}$$

We obtain the future state (15.41) by simply applying the variation-of-constants formula to (15.29).

15.3.2 ▪ Reduction Approach: Exponential Stability Deduced from the Explicit Solutions

For the sake of simplicity, we assume $c \to +\infty$ in (15.38) resulting in the following expression:

$$U_i(t) = k_i \left(M_i(t)y_i(t) + N_i(t)y_i(t)\sum_{j=1}^{N} e_j \int_{t-D_j}^{t} U_j(\tau)d\tau \right), \tag{15.42}$$

such that the delayed closed-loop system (15.40) and (15.42) can be written in the corresponding PDE representation form, given by [179]

$$\dot{\tilde{\theta}}(t) = \sum_{i=1}^{N} e_i u_i(0,t), \tag{15.43}$$

$$\partial_t u_i(x,t) = \partial_x u_i(x,t), \quad x \in \left]0, D_i\right[, \quad i = 1,2,\ldots,N, \tag{15.44}$$

$$u_i(D_i,t) = U_i(t). \tag{15.45}$$

The relation between u_i and U_i is given by $u_i(x,t) = U_i(x+t-D_i)$.

In the *reduction approach* [12] (or finite-spectrum assignment), we use the transformation

$$Z(t) = \tilde{\theta}(t) + \sum_{i=1}^{N}\int_{t}^{t+D_i} e_i U_i(\tau - D_i)d\tau$$
$$= \tilde{\theta}(t) + \sum_{i=1}^{N}\int_{0}^{D_i} e_i u_i(\xi,t)d\xi. \tag{15.46}$$

It is not difficult to see that Z satisfies

$$\dot{Z}(t) = \sum_{i=1}^{N} e_i U_i(t). \tag{15.47}$$

This is the key fact in the reduction approach since (15.47) can be written in the simple form

$$\dot{Z}(t) = U(t). \tag{15.48}$$

By employing the feedback law $U(t) = -\bar{K}Z(t)$ into (15.48), which replaces (15.43), with $\bar{K} > 0$ being a diagonal matrix $\bar{K} = \mathrm{diag}(\bar{k}_1 \ \cdots \ \bar{k}_N)$, $\bar{k}_i > 0$, then the closed-loop system (15.43)–(15.45) becomes

$$\dot{Z}(t) = -\bar{K}Z(t), \tag{15.49}$$

$$\partial_t u_i(x,t) = \partial_x u_i(x,t), \quad x \in \left]0, D_i\right[, \quad i = 1,2,\ldots,N, \tag{15.50}$$

$$u_i(D_i,t) = -\bar{k}_i Z_i(t). \tag{15.51}$$

Exponential stability of the closed-loop system can be shown directly from (15.49)–(15.51) since the solution of the Z-subsystem is easily calculated as

$$Z(t) = \exp(-\bar{K}t)Z(0)\,. \tag{15.52}$$

Hence, $Z(t) \to 0$ exponentially as $t \to +\infty$. Then, for $t > D_i$, the solution to the u_i-subsystem in (15.51) is obtained as

$$u_i(x,t) = -\bar{k}_i e_i^T \exp(-\bar{K}(x+t-D_i))Z(0)\,, \quad x \in \,]0, D_i[\,, \quad t > D_i\,. \tag{15.53}$$

Clearly, for each $x \in \,]0, D_i[$, the state variables $u_i(x,t)$, and consequently $u(x,t)$, converge to 0 as $t \to +\infty$. The rate of convergence is exponential.

However, the control law $U(t) = -\bar{K}Z(t)$ cannot be implemented since $Z(t)$ in (15.46) was constructed with the unmeasured signal $\tilde{\theta}(t)$ in (15.35). However, in the average sense, the average version of (15.42) returns

$$U^{\text{av}}(t) = \underbrace{KH}_{-\bar{K}} \left(\underbrace{\tilde{\theta}^{\text{av}}(t) + \sum_{j=1}^{N} e_j \int_{t-D_j}^{t} U_j^{\text{av}}(\tau)d\tau}_{Z^{\text{av}}(t)} \right), \tag{15.54}$$

since

$$\hat{G}_i^{\text{av}}(t) = \frac{1}{\Pi}\int_0^{\Pi} M_i(\tau)y_i d\tau = \sum_{j=1}^{N} \epsilon_{ij}^i H_{ij}^i \tilde{\theta}_j^{\text{av}}(t-D_j)\,, \tag{15.55}$$

$$\hat{H}_i^{\text{av}}(t) = \frac{1}{\Pi}\int_0^{\Pi} N_i(\tau)y_i d\tau = \sum_{j=1}^{N} \epsilon_{ij}^i H_{ij}^i e_j^T\,, \tag{15.56}$$

or, equivalently, $\hat{G}^{\text{av}}(t) = H\tilde{\theta}^{\text{av}}(t-D)$ and $\hat{H}^{\text{av}}(t) = H$, where $\hat{H}_i^{\text{av}}$ is the ith row vector of the Hessian H.

In the next subsection, we show the average control law (15.54)—or its equivalent filtered version in (15.38)—is indeed able to stabilize in the average sense the closed-loop system (15.43)–(15.45).

15.3.3 ▪ Stability Analysis

In what follows, we make the same assumption as in [70] concerning the Hessian matrix H, which describes the interactions among the players in our cooperative game.

Assumption 15.1. *The Hessian matrix H given by* (15.7) *is strictly diagonal dominant, i.e.,*

$$\sum_{j \neq i}^{N} |\epsilon_{ij}^i H_{ij}^i| < |\epsilon_{ii}^i H_{ii}^i|\,, \quad i \in \{1, \dots, N\}\,. \tag{15.57}$$

By Assumption 15.1, the Nash equilibrium θ^* exists and is unique since strictly diagonally dominant matrices are nonsingular by the Levy–Desplanques theorem [90]. To attain θ^* stably in real time, without any model information (except for the delays D_i), each Player i employs the cooperative extremum seeking strategy (15.38) via predictor feedback with shared information. The next theorem provides the stability/convergence properties of the closed-loop extremum seeking feedback for the quadratic N-player cooperative game with delays under the cooperative scenario.

Theorem 15.1. *Consider the closed-loop system* (15.38) *and* (15.40) *under Assumption* 15.1 *and multiple and distinct input delays D_i for an N-player game with payoff functions given in* (15.1) *and under the cooperative scenario. There exists a constant $c^* > 0$ such that, $\forall c \geq c^*$, $\exists\, \omega^*(c) > 0$ such that, $\forall \omega > \omega^*$, the closed-loop delayed system* (15.38) *and* (15.40) *with state $\tilde{\theta}_i(t-D_i)$, $U_i(\tau)$, $\forall \tau \in [t-D_i, t]$ and $\forall i \in 1,2,\ldots,N$, has a unique locally exponentially stable periodic solution in t of period Π, denoted by $\tilde{\theta}_i^{\Pi}(t-D_i)$, $U_i^{\Pi}(\tau)$ $\forall \tau \in [t-D_i,t]$ satisfying, $\forall t \geq 0$,*

$$\left(\sum_{i=1}^{N}\left[\tilde{\theta}_i^{\Pi}(t-D_i)\right]^2+\left[U_i^{\Pi}(t)\right]^2+\int_{t-D_i}^{t}\left[U_i^{\Pi}(\tau)\right]^2 d\tau\right)^{1/2} \leq \mathcal{O}(1/\omega). \qquad (15.58)$$

Furthermore,

$$\limsup_{t\to+\infty} |\theta(t)-\theta^*| = \mathcal{O}(|a|+1/\omega), \qquad (15.59)$$

where $a = [a_1\ a_2\ \cdots\ a_N]^T$ and θ^* is the unique Nash equilibrium given by (15.8).

Proof. The proof is carried out in **5 steps** as follows.

First, in Steps 1 and 2, we write the equations of the closed-loop system as well as its corresponding average version by employing a PDE representation for the transport delays. In Step 3, we show the exponential stability of the average closed-loop system using a Lyapunov–Krasovskii functional. Then, we invoke the averaging theorem for infinite-dimensional systems [83] (Appendix A) in Step 4 to show the exponential (practical) stability of the original closed-loop system. Finally, Step 5 shows the convergence of $\theta(t)$ to a small neighborhood of the Nash equilibrium θ^*.

Step 1: *Closed-Loop System*

A PDE representation of the closed-loop system (15.38) and (15.40) is given by

$$\dot{\hat{\theta}}(t) = u(0,t), \qquad (15.60)$$

$$u_t(x,t) = D^{-1}u_x(x,t), \quad x \in \,]0,1[, \qquad (15.61)$$

$$u(1,t) = U(t), \qquad (15.62)$$

$$\dot{U}(t) = -cU(t)+cK\left(\hat{G}(t)+\hat{H}(t)\int_0^1 Du(x,t)dx\right), \qquad (15.63)$$

where $u(x,t) = (u_1(x,t), u_2(x,t), \ldots, u_N(x,t))^T \in \mathbb{R}^N$ and

$$\hat{G}(t) = [M_1(t)y_1(t)\ \ldots\ M_N(t)y_N(t)]^T \in \mathbb{R}^{N\times N}, \qquad (15.64)$$

$$\hat{H}(t) = [N_1(t)y_1(t)\ \ldots\ N_N(t)y_N(t)]^T \in \mathbb{R}^{N\times N}. \qquad (15.65)$$

It is easy to see that the solution of (15.61) under the condition (15.62) is represented as

$$u_i(x,t) = U_i(D_ix+t-D_i) \qquad (15.66)$$

for each $i \in \{1,2,\ldots,N\}$. Hence, we have

$$\int_0^1 D_i u_i(x,t)dx = \int_{t-D_i}^t u_i\left(\frac{\tau - t + D_i}{D_i}, t\right) d\tau = \int_{t-D_i}^t U_i(\tau)d\tau. \tag{15.67}$$

This means that

$$\int_0^1 Du(x,t)dx = \sum_{i=1}^N e_i \int_0^1 D_i u_i(x,t)dx$$
$$= \sum_{i=1}^N e_i \int_{t-D_i}^t U_i(\tau)d\tau. \tag{15.68}$$

Thus, we can recover (15.40) from (15.61)–(15.63).

Step 2: *Average Closed-Loop System*

The average system associated with (15.60)–(15.63) is given by

$$\dot{\tilde{\theta}}_{\rm av}(t) = u_{\rm av}(0,t), \tag{15.69}$$

$$u_{\rm av,t}(x,t) = D^{-1} u_{\rm av,x}(x,t), \quad x \in \left]0,1\right[, \tag{15.70}$$

$$u_{\rm av}(1,t) = U_{\rm av}(t), \tag{15.71}$$

$$\dot{U}_{\rm av}(t) = -cU_{\rm av}(t) + cKH\left(\tilde{\theta}_{\rm av}(t) + \int_0^1 Du_{\rm av}(x,t)dx\right), \tag{15.72}$$

where we have used the fact that the averages of $\hat{G}(t)$ and $\hat{H}(t)$ are calculated as $H\tilde{\theta}_{\rm av}(t)$ and H.

Step 3: *Exponential Stability via a Lyapunov–Krasovskii Functional*

For simplicity of notation, let us introduce the following auxiliary variables:

$$\vartheta(t) := H\left(\tilde{\theta}_{\rm av}(t) + \int_0^1 Du_{\rm av}(x,t)dx\right), \tag{15.73}$$

$$\tilde{U} = U_{\rm av} - K\vartheta. \tag{15.74}$$

With this notation, (15.72) can be represented simply as $\dot{U}_{\rm av} = -c\tilde{U}$. In addition, differentiating (15.73) with respect to t yields

$$\dot{\vartheta} = HU_{\rm av}(t). \tag{15.75}$$

We prove the exponential stability of the closed-loop system by using the Lyapunov functional defined by

$$V(t) = \vartheta(t)^T K\vartheta(t) + \frac{1}{4}\lambda_{\min}(-H)\int_0^1 \Big((1+x)u_{\rm av}(x,t)^T$$
$$\times Du_{\rm av}(x,t)dx\Big) + \frac{1}{2}\tilde{U}(t)^T(-H)\tilde{U}(t). \tag{15.76}$$

Recall that K and D are diagonal matrices with positive entries and that H is a negative-definite matrix. Hence, all of K, D, and $-H$ are positive-definite matrices. In particular, we use the Gershgorin Circle Theorem [90, Theorem 6.1.1] to guarantee $\lambda(H) \subseteq \bigcup_{i=1}^N \rho_i$, where $\lambda(H)$

denotes the spectrum of H and ρ_i is a Gershgorin disc:

$$\rho_i = \left\{ z \in \mathbb{C} : |z - \epsilon_{ii}^i H_{ii}^i| < \sum_{j \neq i} |\epsilon_{ij}^i H_{ij}^i| \right\}. \tag{15.77}$$

Since $\epsilon_{ii}^i H_{ii}^i < 0$ and H is strictly diagonally dominant, the union of the Gershgorin discs lies strictly in the left half of the complex plane, and we conclude that $\mathrm{Re}\{\lambda\} < 0$ for all $\lambda \in \lambda(H)$.

For simplicity of notation, we also suppress explicit dependence of the variables on t. The time derivative of V is given by

$$\begin{aligned}
\dot{V} &= 2\vartheta^T KHU_{\mathrm{av}} + \frac{1}{2}\lambda_{\min}(-H)U_{\mathrm{av}}^T U_{\mathrm{av}} \\
&\quad - \frac{1}{4}\lambda_{\min}(-H)u(0)^T u(0) - \frac{1}{4}\lambda_{\min}(-H) \\
&\quad \times \int_0^1 u_{\mathrm{av}}(x)^T u_{\mathrm{av}}(x)dx + \tilde{U}^T(-H)\left(\dot{U}_{\mathrm{av}} - KHU_{\mathrm{av}}\right) \\
&\le 2\vartheta^T KHU_{\mathrm{av}} + \frac{1}{2}U_{\mathrm{av}}^T(-H)U_{\mathrm{av}} - \frac{1}{8D_{\max}}\lambda_{\min}(-H) \\
&\quad \times \int_0^1 (1+x)u_{\mathrm{av}}(x)^T D u_{\mathrm{av}}(x)dx \\
&\quad + \tilde{U}^T(-H)\dot{U}_{\mathrm{av}} + \tilde{U}^T(-H)K(-H)U_{\mathrm{av}}.
\end{aligned} \tag{15.78}$$

Applying Young's inequality to the last term leads to

$$\tilde{U}^T(-H)K(-H)U_{\mathrm{av}} \le \frac{1}{2}\tilde{U}^T(-HKHKH)\tilde{U} + \frac{1}{2}U_{\mathrm{av}}^T(-H)U_{\mathrm{av}}. \tag{15.79}$$

Then, completing the square yields

$$\begin{aligned}
\dot{V} &\le \tilde{U}^T(-H)\tilde{U} - \vartheta^T K(-H)K\vartheta \\
&\quad - \frac{1}{8D_{\max}}\lambda_{\min}(-H)\int_0^1 (1+x)u_{\mathrm{av}}(x)^T D u_{\mathrm{av}}(x)dx \\
&\quad + \tilde{U}^T(-H)\dot{U}_{\mathrm{av}} + \frac{1}{2}\tilde{U}^T(-HKHKH)\tilde{U} \\
&\le \tilde{U}^T(-H)\left(\dot{U}_{\mathrm{av}} + c^*\tilde{U}\right) - \vartheta^T K(-H)K\vartheta \\
&\quad - \frac{1}{8D_{\max}}\lambda_{\min}(-H)\int_0^1 (1+x)u_{\mathrm{av}}(x)^T D u_{\mathrm{av}}(x)dx,
\end{aligned} \tag{15.80}$$

where $c^* := 1 + \lambda_{\max}(-HKHKH)/\lambda_{\min}(-H)$. Hence, by setting $\dot{U}_{\mathrm{av}} = -c\tilde{U}$ for some $c > c^*$, we see that there exists $\mu > 0$ such that

$$\dot{V} \le -\mu V. \tag{15.81}$$

Finally, it is not difficult to find positive constants $\alpha, \beta > 0$ such that

$$\begin{aligned}
\alpha\left(|\tilde{\theta}_{\mathrm{av}}(t)|^2 + \int_0^1 |u_{\mathrm{av}}(x,t)|^2 dx + |\tilde{U}(t)|^2\right) &\le V(t) \\
&\le \beta\left(|\tilde{\theta}_{\mathrm{av}}(t)|^2 + \int_0^1 |u_{\mathrm{av}}(x,t)|^2 dx + |\tilde{U}(t)|^2\right).
\end{aligned} \tag{15.82}$$

Therefore, the average system (15.69)–(15.72) is exponentially stable as long as $c > c^*$.

Step 4: *Invoking the Averaging Theorem for Infinite-Dimensional Systems*

Indeed, the closed-loop system (15.38)–(15.40) can be rewritten as

$$\dot{\eta}(t) = f(\omega t, \eta_t), \tag{15.83}$$

where $\eta(t) = [\eta_1(t), \eta_2(t)]^T := [\tilde{\theta}(t), U(t)]^T$ is the state vector and the distributed delayed term is generically represented by $\eta_t(\Theta) = \eta(t+\Theta)$ for $-D_N \leq \Theta \leq 0$. The vector field f is given in (15.84) below, such that $\dot{\eta}(t) = \left[\dot{\tilde{\theta}}(t), \dot{U}(t)\right]^T = f(\omega t, \eta_t)$. At this point, it is worth to note the variable $\eta_t \in C([-D_N, 0]; \mathbb{R}^{2N})$ does not include only terms that suffer a discrete delay action (for instance, $\tilde{\theta}_i(t-D_i)$ and $U_i(t-D_i)$), but this representation also includes operations with terms of distributed delays such as $\int_{-D_i}^{0} U_i(t+\tau)d\tau$. For the sake of clarity, let us express the discrete terms by

$$\eta_{t1} = \left[\tilde{\theta}_1(t-D_1), \ldots, \tilde{\theta}_N(t-D_N)\right]^T$$

and

$$\eta_{t2} = [U_1(t-D_1), \ldots, U_N(t-D_N)]^T$$

for $\Theta = -D_i$ in each element of the vector, whereas

$$\eta_{t3} = [U_1(t+\Theta_1), \ldots, U_N(t+\Theta_N)]^T$$

denotes the distributed terms for $\Theta_i \in \Theta = [-D_N, 0]$. The variable $v = g(t, \eta_t)$ with $g: \mathbb{R}_+ \times \Omega \to \mathbb{R}^N$ and

$$v = \left[\int_{-D_1}^{0} \eta_{t3}^{[1]}(\tau)d\tau, \ldots, \int_{-D_N}^{0} \eta_{t3}^{[N]}(\tau)d\tau\right]^T,$$

where $\eta_{t3}^{[i]}$ represents the ith element of the vector η_{t3}, while $f: \mathbb{R}_+ \times \Omega \to \mathbb{R}^{2N}$ is a continuous functional from a neighborhood Ω of 0 of the supremum-normed Banach space $X = C([-D_N, 0]; \mathbb{R}^{2N})$ of continuous functions from $[-D_N, 0]$ to $\mathbb{R}^{2N}$.

From (15.40), one has

$$\dot{\tilde{\theta}}(t) = U(t-D),$$

with $U(t-D) = \left[U_1(t-D_1), \quad U_2(t-D_2), \quad \ldots, \quad U_N(t-D_N)\right]^T$. Noting that the term $\int_{t-D_i}^{t} U_i(\tau)d\tau = \int_{-D_i}^{0} U_i(t+\tau)d\tau$ and plugging (15.12), (15.32), and (15.33) into (15.38), one can write

$$\begin{aligned}\dot{U}_i(t) = -cU_i(t) + ck_i y_i(t) &\left\{\frac{2}{a_i}\sin(\omega_i' \omega t) + \frac{16}{a_i^2}\left[\sin^2(\omega_i' \omega t) - \frac{1}{2}\right]\right. \\ &\left.\times \int_{-D_i}^{0} U_i(t+\tau)d\tau + \sum_{j \neq i} \frac{4}{a_i a_j}\sin(\omega_i' \omega t)\sin(\omega_j' \omega t)\int_{-D_j}^{0} U_j(t+\tau)d\tau\right\}.\end{aligned}$$

Then, the dynamics of η is simply

$$
\begin{aligned}
\dot{\eta}_1^{[1]} &= f_1 = \eta_{t2}^{[1]}, \\
&\vdots \quad \vdots \\
\dot{\eta}_1^{[i]} &= f_i = \eta_{t2}^{[i]}, \\
&\vdots \quad \vdots \\
\dot{\eta}_1^{[N]} &= f_N = \eta_{t2}^{[N]}, \\
\dot{\eta}_2^{[1]} &= f_{N+1} = -c\eta_2^{[1]} + ck_1 y_1 \left\{ \frac{2}{a_1}\sin(\omega_1'\omega t) + \frac{16}{a_1^2}\left[\sin^2(\omega_1'\omega t) - \frac{1}{2}\right] v_1 \right. \\
&\quad \left. + \sum_{j=2}^{N} \frac{4}{a_1 a_j}\sin(\omega_1'\omega t)\sin(\omega_j'\omega t) v_j \right\}, \\
&\vdots \quad \vdots \\
\dot{\eta}_2^{[i]} &= f_{N+i} = -c\eta_2^{[i]} + ck_i y_i \left\{ \frac{2}{a_i}\sin(\omega_i'\omega t) + \frac{16}{a_i^2}\left[\sin^2(\omega_i'\omega t) - \frac{1}{2}\right] v_i \right. \\
&\quad \left. + \sum_{j\neq i} \frac{4}{a_i a_j}\sin(\omega_i'\omega t)\sin(\omega_j'\omega t) v_j \right\}, \\
&\vdots \quad \vdots \\
\dot{\eta}_2^{[N]} &= f_{2N} = -c\eta_2^{[N]} + ck_N y_N \left\{ \frac{2}{a_N}\sin(\omega_N'\omega t) + \frac{16}{a_N^2}\left[\sin^2(\omega_N'\omega t) - \frac{1}{2}\right] v_N \right. \\
&\quad \left. + \sum_{j=1}^{N-1} \frac{4}{a_N a_j}\sin(\omega_N'\omega t)\sin(\omega_j'\omega t) v_j \right\},
\end{aligned}
\tag{15.84}
$$

where $y_i = y_i(t) = y_i(\omega t, \eta_{t1})$ according to (15.18) for all $i \in \{1,\ldots,N\}$, satisfying

$$
\begin{aligned}
y_i &= \frac{1}{2}\sum_{j=1}^{N}\sum_{k=1}^{N} \epsilon_{jk}^i H_{jk}^i \left[a_j \sin(\omega_j'\omega t)\eta_{t1}^{[k]} + a_k \sin(\omega_k'\omega t)\eta_{t1}^{[j]}\right] \\
&\quad + \frac{1}{2}\sum_{j=1}^{N}\sum_{k=1}^{N} \epsilon_{jk}^i H_{jk}^i \left[\theta_k^* a_j \sin(\omega_j'\omega t) + \theta_j^* a_k \sin(\omega_k'\omega t)\right] \\
&\quad + \frac{1}{2}\sum_{j=1}^{N}\sum_{k=1}^{N} \epsilon_{jk}^i H_{jk}^i a_j a_k \sin(\omega_j'\omega t)\sin(\omega_k'\omega t) \\
&\quad + \frac{1}{2}\sum_{j=1}^{N}\sum_{k=1}^{N} \epsilon_{jk}^i H_{jk}^i \left[\theta_k^*\eta_{t1}^{[j]} + \theta_j^*\eta_{t1}^{[k]}\right] + \frac{1}{2}\sum_{j=1}^{N}\sum_{k=1}^{N} \epsilon_{jk}^i H_{jk}^i \eta_{t1}^{[j]}\eta_{t1}^{[k]} \\
&\quad + \frac{1}{2}\sum_{j=1}^{N}\sum_{k=1}^{N} \epsilon_{jk}^i H_{jk}^i \theta_j^*\theta_k^* + \sum_{j=1}^{N} h_j^i a_j \sin(\omega_j'\omega t) + \sum_{j=1}^{N} h_j^i \eta_{t1}^{[j]} + \sum_{j=1}^{N} h_j^i \theta_j^* + c_i .
\end{aligned}
$$

Therefore, $f(\omega t,\eta_t) := [f_1,\ldots,f_{2N}]^T$ is simply given by the RHS of (15.84) such that the averaging theorem by [83] (see also Appendix A) can be directly applied, considering $\omega = 1/\epsilon$.

From (15.81), the origin of the average closed-loop system (15.69)–(15.72) with transport PDE for delay representation is locally exponentially stable. Then, from (15.73) and (15.74), we can conclude the same results in the norm

$$\left(\sum_{i=1}^{N}\left[\tilde{\theta}_i^{\text{av}}(t-D_i)\right]^2+\int_0^{D_i}[u_i^{\text{av}}(x,t)]^2dx+[u_i^{\text{av}}(D_i,t)]^2\right)^{1/2},$$

since H is nonsingular.

Thus, there exist positive constants α and β such that all solutions satisfy

$$\Psi(t)\le \alpha e^{-\beta t}\Psi(0)\quad \forall t\ge 0,$$

where $\Psi(t)\triangleq\sum_{i=1}^{N}\left[\tilde{\theta}_i^{\text{av}}(t-D_i)\right]^2+\int_0^{D_i}\left[u_i^{\text{av}}(x,t)\right]^2dx+\left[u_i^{\text{av}}(D_i,t)\right]^2$ or, equivalently,

$$\Psi(t)\triangleq\sum_{i=1}^{N}\left[\tilde{\theta}_i^{\text{av}}(t-D_i)\right]^2+\int_{t-D_i}^{t}\left[U_i^{\text{av}}(\tau)\right]^2d\tau+\left[U_i^{\text{av}}(t)\right]^2, \tag{15.85}$$

using (15.66). Then, according to the averaging theorem by [83] (Appendix A), for ω sufficiently large, (15.38)–(15.40), or equivalently (15.69)–(15.72), has a unique locally exponentially stable periodic solution around its equilibrium (origin) satisfying (15.58).

Step 5: *Asymptotic Convergence to a Neighborhood of the Nash Equilibrium*

By first using the change of variables $\tilde{\vartheta}_i(t) := \tilde{\theta}_i(t-D_i)=\hat{\theta}_i(t-D_i)-\theta_i^*$, and then integrating both sides of (15.60) over $[t,\sigma+D_i]$, we have

$$\tilde{\vartheta}_i(\sigma+D_i)=\tilde{\vartheta}_i(t)+\int_t^{\sigma+D_i}u_i(0,s)ds,\qquad i=1,\ldots,N. \tag{15.86}$$

From (15.66), we can rewrite (15.86) in terms of U, namely

$$\tilde{\vartheta}_i(\sigma+D_i)=\tilde{\vartheta}_i(t)+\int_{t-D_i}^{\sigma}U_i(\tau)d\tau. \tag{15.87}$$

Now, note that

$$\tilde{\theta}_i(\sigma)=\tilde{\vartheta}_i(\sigma+D_i)\quad \forall\sigma\in[t-D_i,t]. \tag{15.88}$$

Hence,

$$\tilde{\theta}_i(\sigma)=\tilde{\theta}_i(t-D_i)+\int_{t-D_i}^{\sigma}U_i(\tau)d\tau\quad \forall\sigma\in[t-D_i,t]. \tag{15.89}$$

Applying the supremum norm to both sides of (15.89), we have

$$
\begin{aligned}
&\sup_{t-D_i\le\sigma\le t}\left|\tilde{\theta}_i(\sigma)\right| \\
&= \sup_{t-D_i\le\sigma\le t}\left|\tilde{\theta}_i(t-D_i)\right| + \sup_{t-D_i\le\sigma\le t}\left|\int_{t-D_i}^{\sigma} U_i(\tau)d\tau\right| \\
&\le \sup_{t-D_i\le\sigma\le t}\left|\tilde{\theta}_i(t-D_i)\right| + \sup_{t-D_i\le\sigma\le t}\int_{t-D_i}^{t} \left|U_i(\tau)\right| d\tau \\
&\le \left|\tilde{\theta}_i(t-D_i)\right| + \int_{t-D_i}^{t} \left|U_i(\tau)\right| d\tau \quad \text{(by Cauchy–Schwarz)} \\
&\le \left|\tilde{\theta}_i(t-D_i)\right| + \left(\int_{t-D_i}^{t} d\tau\right)^{1/2} \times \left(\int_{t-D_i}^{t} \left|U_i(\tau)\right|^2 d\tau\right)^{1/2} \\
&\le \left|\tilde{\theta}_i(t-D_i)\right| + \sqrt{D_i}\left(\int_{t-D_i}^{t} U_i^2(\tau)d\tau\right)^{1/2}.
\end{aligned}
\tag{15.90}
$$

Now, it is easy to check

$$
\left|\tilde{\theta}_i(t-D_i)\right| \le \left(\left|\tilde{\theta}_i(t-D_i)\right|^2 + \int_{t-D_i}^{t} U_i^2(\tau)d\tau\right)^{1/2}, \tag{15.91}
$$

$$
\left(\int_{t-D_i}^{t} U_i^2(\tau)d\tau\right)^{1/2} \le \left(\left|\tilde{\theta}_i(t-D_i)\right|^2 + \int_{t-D_i}^{t} U_i^2(\tau)d\tau\right)^{1/2}. \tag{15.92}
$$

By using (15.91) and (15.92), one has

$$
\left|\tilde{\theta}_i(t-D_i)\right| + \sqrt{D_i}\left(\int_{t-D_i}^{t} U_i^2(\tau)d\tau\right)^{1/2} \le (1+\sqrt{D_i})\left(\left|\tilde{\theta}_i(t-D_i)\right|^2 + \int_{t-D_i}^{t} U_i^2(\tau)d\tau\right)^{1/2}. \tag{15.93}
$$

From (15.90), it is straightforward to conclude that

$$
\sup_{t-D_i\le\sigma\le t}\left|\tilde{\theta}_i(\sigma)\right| \le (1+\sqrt{D_i})\left(\left|\tilde{\theta}_i(t-D_i)\right|^2 + \int_{t-D_i}^{t} U_i^2(\tau)d\tau\right)^{1/2} \tag{15.94}
$$

and, consequently,

$$
\left|\tilde{\theta}_i(t)\right| \le (1+\sqrt{D_i})\left(\left|\tilde{\theta}_i(t-D_i)\right|^2 + \int_{t-D_i}^{t} U_i^2(\tau)d\tau\right)^{1/2}. \tag{15.95}
$$

Inequality (15.95) can be given in terms of the periodic solution $\tilde{\theta}_i^{\Pi}(t-D_i)$, $U_i^{\Pi}(\tau)$ $\forall\tau \in [t-D_i, t]$ as follows:

$$
\begin{aligned}
\left|\tilde{\theta}_i(t)\right| \le (1+\sqrt{D_i})\Bigg(&\left|\tilde{\theta}_i(t-D_i) - \tilde{\theta}_i^{\Pi}(t-D_i) + \tilde{\theta}_i^{\Pi}(t-D_i)\right|^2 \\
&+ \int_{t-D_i}^{t} \left[U_i(\tau) - U_i^{\Pi}(\tau) + U_i^{\Pi}(\tau)\right]^2 d\tau\Bigg)^{1/2}.
\end{aligned}
\tag{15.96}
$$

By applying Young's inequality and some algebra, the RHS of (15.96) and $\left|\tilde{\theta}_i(t)\right|$ can be majorized by

$$|\tilde{\theta}_i(t)| \leq \sqrt{2}\,(1+\sqrt{D_i})\left(\left|\tilde{\theta}_i(t-D_i)-\tilde{\theta}_i^{\Pi}(t-D_i)\right|^2+\left|\tilde{\theta}_i^{\Pi}(t-D_i)\right|^2 + \int_{t-D_i}^{t}\left[U_i(\tau)-U_i^{\Pi}(\tau)\right]^2 d\tau+\int_{t-D_i}^{t}\left[U_i^{\Pi}(\tau)\right]^2 d\tau\right)^{1/2}. \tag{15.97}$$

According to the averaging theorem by [83] (Appendix A), we can conclude that the actual state converges exponentially to the periodic solution, i.e.,

$$\tilde{\theta}_i(t-D_i)-\tilde{\theta}_i^{\Pi}(t-D_i)\to 0$$

and

$$\int_{t-D_i}^{t}\left[U_i(\tau)-U_i^{\Pi}(\tau)\right]^2 d\tau \to 0,$$

exponentially. Hence,

$$\limsup_{t\to+\infty}|\tilde{\theta}_i(t)| = \sqrt{2}\left(1+\sqrt{D_i}\right) \times\left(\left|\tilde{\theta}_i^{\Pi}(t-D_i)\right|^2+\int_{t-D_i}^{t}[U_i^{\Pi}(\tau)]^2 d\tau\right)^{1/2}.$$

Then, from (15.58), we can write $\limsup_{t\to+\infty}|\tilde{\theta}(t)|=\mathcal{O}(1/\omega)$. From (15.35) and recalling that $\theta(t)=\hat{\theta}(t)+S(t)$ in (15.34) with $S(t)$ in (15.31), one has that $\theta(t)-\theta^*=\tilde{\theta}(t)+S(t)$. Since the first term on the right-hand side is ultimately of order $\mathcal{O}(1/\omega)$ and the second term is of order $\mathcal{O}(|a|)$, we arrive at (15.59). □

15.4 ▪ Noncooperative Scenario with Delays

In the game under the noncooperative scenario, again subject to multiple and distinct delays, the purpose of the extremum seeking is still to estimate the Nash equilibrium vector θ^*, but without allowing any sharing of information among the players. Each player only needs to measure the value of its own payoff function

$$y_i(t) = J_i(\theta(t-D)), \tag{15.98}$$

with J_i given by (15.1). In this sense, we are able to formulate the closed-loop system in a *decentralized* fashion, where no knowledge about the payoffs y_{-i} or actions θ_{-i} of the other players is required.

15.4.1 ▪ Decentralized Predictor Using Only the Known Diagonal Terms of the Hessian

In such a decentralized scenario, the dither frequencies ω_{-i}, the excitation amplitudes a_{-i}, and consequently, the individual control laws $U_{-i}(t)$ are not available to Player i. Recalling that the model of the payoffs (15.1) is also assumed to be unknown, it becomes impossible to reconstruct individually or estimate completely the Hessian matrix H given in (15.7) by using demodulating signals such as in (15.33). Hence, the predictor design presented in Section 15.3 must be reformulated for noncooperative games.

However, it is still possible to design Nash equilibrium seeking control laws for a class of weakly coupled noncooperative games, as shown in the following. For such a class of games, we consider $\epsilon_{ii}^i = 1$ and $\epsilon_{jk}^i = \epsilon_{kj}^i = \epsilon$, $\forall j \neq k$, such that $0 < \epsilon < 1$ in (15.1) and (15.7).

Following the non-sharing information paradigm, the ith player is only able to estimate the element H_{ii}^i of the H matrix (15.7) by itself, and this being so for all players. Therefore, only the diagonal of H can be properly recovered in the average sense. In this way, the signal $N_i(t)$ is now simply defined by

$$N_i(t) := N_{ii}(t) = \frac{16}{a_i^2}\left(\sin^2(\omega_i t) - \frac{1}{2}\right), \tag{15.99}$$

according to (15.33). Then, the average version of

$$\hat{H}_i(t) = N_i(t) y_i(t) \tag{15.100}$$

is given by

$$\hat{H}_i^{\text{av}}(t) = [N_i(t) y_i(t)]_{\text{av}} = H_{ii}^i. \tag{15.101}$$

In order to compensate the time delays, we propose the following predictor-based update law:

$$\dot{\hat{\theta}}_i(t) = U_i(t), \tag{15.102}$$

$$\dot{U}_i(t) = -c_i U_i(t) + c_i k_i \left(\hat{G}_i(t) + \hat{H}_i(t) \int_{t-D_i}^{t} U_i(\tau) d\tau\right) \tag{15.103}$$

for positive constants k_i and c_i.

15.4.2 ▪ ISS-Like Properties for PDE Representation

For the sake of simplicity, we assume $c_i \to +\infty$ in (15.103), resulting in the following expression:

$$U_i(t) = k_i\left(\hat{G}_i(t) + \hat{H}_i(t) \int_{t-D_i}^{t} U_i(\tau) d\tau\right), \tag{15.104}$$

such that the delayed closed-loop system (15.27) and (15.104) can be written in the corresponding PDE representation form, given by

$$\dot{\hat{G}}_i^{\text{av}}(t) = \sum_{j=1}^{N} \epsilon_{ij}^i H_{ij}^i u_j^{\text{av}}(0,t), \tag{15.105}$$

$$\partial_t u_i^{\text{av}}(x,t) = \partial_x u_i^{\text{av}}(x,t), \quad x \in \,]0, D_i[, \quad i = 1, 2, \ldots, N, \tag{15.106}$$

$$u_i^{\text{av}}(D_i, t) = U_i^{\text{av}}(t). \tag{15.107}$$

The relation between u_i and U_i is given by $u_i^{\text{av}}(x,t) = U_i^{\text{av}}(x + t - D_i)$.

Analogous to the developments carried out in Section 15.3.2, we use the transformation [12]

$$\begin{aligned} \bar{G}_i^{\text{av}}(t) &= \hat{G}_i^{\text{av}}(t) + \sum_{j=1}^{N} \epsilon_{ij}^i H_{ij}^i \int_{t-D_j}^{t} U_j^{\text{av}}(\tau) d\tau \\ &= \hat{G}_i^{\text{av}}(t) + \sum_{j=1}^{N} \epsilon_{ij}^i H_{ij}^i \int_{0}^{D_j} u_j^{\text{av}}(\xi, t) d\xi. \end{aligned} \tag{15.108}$$

Taking the time derivative of (15.108), we obtain

$$\dot{\bar{G}}_i^{\mathrm{av}}(t) = \dot{\hat{G}}_i^{\mathrm{av}}(t) + \sum_{j=1}^{N} \epsilon_{ij}^i H_{ij}^i \int_0^{D_j} \partial_t u_j^{\mathrm{av}}(\xi,t)d\xi \,. \tag{15.109}$$

Then, by plugging (15.105) and (15.106) into (15.109), one has

$$\begin{aligned}\dot{\bar{G}}_i^{\mathrm{av}}(t) &= \sum_{j=1}^{N} \epsilon_{ij}^i H_{ij}^i u_j^{\mathrm{av}}(0,t) + \sum_{j=1}^{N} \epsilon_{ij}^i H_{ij}^i \int_0^{D_j} \partial_x u_j^{\mathrm{av}}(\xi,t)d\xi \\ &= \sum_{j=1}^{N} \epsilon_{ij}^i H_{ij}^i u_j^{\mathrm{av}}(0,t) + \sum_{j=1}^{N} \epsilon_{ij}^i H_{ij}^i \left[u_j^{\mathrm{av}}(D_j,t) - u_j^{\mathrm{av}}(0,t)\right] \\ &= \sum_{j=1}^{N} \epsilon_{ij}^i H_{ij}^i u_j^{\mathrm{av}}(D_j,t) \,. \end{aligned} \tag{15.110}$$

Substituting (15.107) in (15.110), it is not difficult to see that $\bar{G}_i$ satisfies

$$\dot{\bar{G}}_i^{\mathrm{av}}(t) = \sum_{j=1}^{N} \epsilon_{ij}^i H_{ij}^i U_j^{\mathrm{av}}(t) \,. \tag{15.111}$$

Now, after adding and subtracting the next terms in blue and red into (15.104), $U_i(t)$ can be rewritten as

$$\begin{aligned} U_i(t) = k_i &\left(\hat{G}_i(t) + \hat{H}_i(t) \int_{t-D_i}^{t} U_i(\tau)d\tau + \sum_{j\neq i} \epsilon_{ij}^i H_{ij}^i \int_{t-D_j}^{t} U_j(\tau)d\tau \right) \\ &- k_i \sum_{j\neq i} \epsilon_{ij}^i H_{ij}^i \int_{t-D_j}^{t} U_j(\tau)d\tau \,, \end{aligned} \tag{15.112}$$

whose average is

$$\begin{aligned} U_i^{\mathrm{av}}(t) &= k_i \left(\hat{G}_i^{\mathrm{av}}(t) + \hat{H}_i^{\mathrm{av}}(t) \int_{t-D_i}^{t} U_i^{\mathrm{av}}(\tau)d\tau + \sum_{j\neq i} \epsilon_{ij}^i H_{ij}^i \int_{t-D_j}^{t} U_j^{\mathrm{av}}(\tau)d\tau \right) \\ &\quad - k_i \sum_{j\neq i} \epsilon_{ij}^i H_{ij}^i \int_{t-D_j}^{t} U_j^{\mathrm{av}}(\tau)d\tau \\ &= k_i \left(\hat{G}_i^{\mathrm{av}}(t) + \epsilon_{ii}^i H_{ii}^i \int_{t-D_i}^{t} U_i^{\mathrm{av}}(\tau)d\tau + \sum_{j\neq i} \epsilon_{ij}^i H_{ij}^i \int_{t-D_j}^{t} U_j^{\mathrm{av}}(\tau)d\tau \right) \\ &\quad - k_i \sum_{j\neq i} \epsilon_{ij}^i H_{ij}^i \int_{t-D_j}^{t} U_j^{\mathrm{av}}(\tau)d\tau \\ &= k_i \left(\hat{G}_i^{\mathrm{av}}(t) + \sum_{j=1}^{N} \epsilon_{ij}^i H_{ij}^i \int_{t-D_j}^{t} U_j^{\mathrm{av}}(\tau)d\tau \right) \\ &\quad - k_i \sum_{j\neq i} \epsilon_{ij}^i H_{ij}^i \int_{t-D_j}^{t} U_j^{\mathrm{av}}(\tau)d\tau \,. \end{aligned} \tag{15.113}$$

By defining the auxiliary signals

$$\phi_i(D,t) := -\sum_{j\neq i} H^i_{ij} \int_{t-D_j}^{t} U_j(\tau)d\tau\,,$$
$$\phi_i(1,t) := -\sum_{j\neq i} H^i_{ij} \int_0^1 D_j u_j(\xi,t)d\xi, \tag{15.114}$$

and recalling that $\epsilon^i_{jk} = \epsilon \; \forall j \neq k$, (15.113) can be rewritten from (15.108) as

$$U_i^{\mathrm{av}}(t) = k_i \bar{G}_i^{\mathrm{av}}(t) + \epsilon k_i \phi_i^{\mathrm{av}}(D,t)\,. \tag{15.115}$$

Taking into account all players and defining

$$\bar{G}(t) := [\bar{G}_1(t),\ldots,\bar{G}_N(t)]^T \in \mathbb{R}^N,$$
$$U(t) := [U_1(t),\ldots,U_N(t)]^T \in \mathbb{R}^N,$$

and $\phi(D,t) := [\phi_1(D,t),\ldots,\phi_N(D,t)]^T \in \mathbb{R}^N$, it is possible to find a compact form for the overall average game from (15.111) and (15.115) such as

$$\dot{\bar{G}}^{\mathrm{av}}(t) = HK\bar{G}^{\mathrm{av}}(t) + \epsilon HK\phi^{\mathrm{av}}(1,t)\,, \tag{15.116}$$
$$\partial_t u^{\mathrm{av}}(x,t) = D^{-1}\partial_x u^{\mathrm{av}}(x,t)\,, \quad x \in \;]0,1[\,, \tag{15.117}$$
$$u^{\mathrm{av}}(1,t) = K\bar{G}^{\mathrm{av}}(t) + \epsilon K\phi^{\mathrm{av}}(1,t)\,, \tag{15.118}$$

$K = \mathrm{diag}\{k_1,\ldots,k_N\}$ being a positive-definite diagonal matrix, with entries $k_i > 0$.

From (15.116), it is clear that the dynamics of the ODE state variable $\bar{G}^{\mathrm{av}}(t)$ is exponentially Input-to-State Stable (ISS) [99] with respect to the PDE state $u^{\mathrm{av}}(x,t)$ by means of the function $\phi^{\mathrm{av}}(1,t)$. Moreover, the PDE subsystem (15.117) is ISS (finite-time stable) [99] with respect to $\bar{G}^{\mathrm{av}}(t)$ in the boundary condition $u^{\mathrm{av}}(1,t)$.

15.4.3 ▪ Stability Analysis

In this subsection, we will show that this hyperbolic ODE-PDE loop (15.116)–(15.118) contains a small-parameter ϵ which can lead to closed-loop stability if it is chosen sufficiently small. To this end, in addition to Assumption 15.1 formulated in Section 15.3.3, we further assume/formalize the following condition for noncooperative games.

Assumption 15.2. *The parameters ϵ^i_{jk} and ϵ^i_{kj} which appear in the Hessian matrix H given by (15.7) satisfy the conditions below:*

$$\epsilon^i_{ii} = 1\,, \quad \epsilon^i_{jk} = \epsilon^i_{kj} = \epsilon \quad \forall j \neq k\,, \tag{15.119}$$

with $0 < \epsilon < 1$ in the payoff functions (15.1).

Assumption 15.2 could be relaxed to consider different values of the coupling parameters ϵ_i for each Player i. However, without loss of generality, we have assumed the same weights for the interconnection channels among the players in order to facilitate the proof of the next theorem, but also to guarantee that the considered noncooperative game is not favoring any specific player. To attain θ^* stably in real time, without any model information (except for the delays D_i), each Player i employs the noncooperative extremum seeking strategy (15.103) via predictor feedback.

The next theorem provides the stability/convergence properties of the closed-loop extremum seeking feedback for the N-player noncooperative game with delays and non-sharing of information.

Theorem 15.2. *Consider the closed-loop system* (15.105)–(15.107) *under Assumptions* 15.1 *and* 15.2 *and multiple and distinct input delays D_i for the N-player quadratic noncooperative game with no information sharing, with payoff functions given in* (15.1) *and control laws $U_i(t)$ defined in* (15.103). *There exist $c_i > 0$ and $\omega > 0$ sufficiently large as well as $\epsilon > 0$ sufficiently small such that the closed-loop system with state $\tilde{\theta}_i(t-D_i)$, $U_i(\tau)$, $\forall \tau \in [t-D_i, t]$ and $\forall i \in 1,2,\ldots,N$, has a unique locally exponentially stable periodic solution in t of period Π in* (15.21), *denoted by $\tilde{\theta}_i^{\Pi}(t-D_i)$, $U_i^{\Pi}(\tau)$ $\forall \tau \in [t-D_i, t]$, satisfying, $\forall t \geq 0$,*

$$\left(\sum_{i=1}^{N} \left[\tilde{\theta}_i^{\Pi}(t-D_i) \right]^2 + \int_{t-D_i}^{t} \left[U_i^{\Pi}(\tau) \right]^2 d\tau \right)^{1/2} \leq \mathcal{O}(1/\omega). \tag{15.120}$$

Furthermore,

$$\limsup_{t\to+\infty} |\theta(t) - \theta^*| = \mathcal{O}(|a| + 1/\omega), \tag{15.121}$$

where $a = [a_1\ a_2\ \cdots\ a_N]^T$ and θ^* is the unique Nash equilibrium given by (15.8).

Proof. The proof of Theorem 15.2 follows steps similar to those employed to prove Theorem 15.1. In this sense, we will simply point out the main differences instead of giving the full independent proof.

While in Theorem 15.1 it was possible to prove the local exponential stability of the average closed-loop system using a Lyapunov functional (15.76), a different approach is adopted here for the noncooperative scenario. We will show that it is possible to guarantee the local exponential stability for the average closed-loop system (15.116)–(15.118) by means of a small-gain analysis.

First, consider the equivalent hyperbolic PDE-ODE representation (15.116)–(15.118) rewritten for each Player i:

$$\dot{\bar{G}}_i^{\text{av}}(t) = H_{ii}^i k_i \bar{G}_i^{\text{av}}(t) + \epsilon H_{ii}^i k_i \phi_i^{\text{av}}(1,t), \tag{15.122}$$

$$\partial_t u_i^{\text{av}}(x,t) = D_i^{-1} \partial_x u_i^{\text{av}}(x,t), \quad x \in \,]0,1[, \tag{15.123}$$

$$u_i^{\text{av}}(1,t) = k_i \bar{G}_i^{\text{av}}(t) + \epsilon k_i \phi_i^{\text{av}}(1,t), \tag{15.124}$$

where $H_{ii}^i < 0$, $k_i > 0$, $0 < \epsilon < 1$, and $D_i^{-1} > 0$. The average closed-loop system (15.122)–(15.124) satisfies both assumptions **(H1)** and **(H2)** of the Small-Gain Theorem [99, Theorem 8.1, p. 198] for the hyperbolic PDE-ODE loops (see also Appendix C) with $n = N$, $c = D_i^{-1}$, $F(\bar{G}_i^{\text{av}}, u^{\text{av}}, 0) = H_{ii}^i k_i \bar{G}_i^{\text{av}} + \epsilon H_{ii}^i k_i \phi_i^{\text{av}}(1)$, $a(x) = f(x,t) = g(x, \bar{G}_i^{\text{av}}, u^{\text{av}}) = 0$, $\varphi(0, u^{\text{av}}, \bar{G}_i^{\text{av}}) = k_i \bar{G}_i^{\text{av}} + \epsilon k_i \phi_i^{\text{av}}(1)$, $\bar{N} = \max(k_i, \epsilon k_i k_H D_N)$, $L = \max(|H_{ii}^i| k_i, \epsilon |H_{ii}^i| k_i k_H D_N)$, $\gamma_2 = k_i$, $A = \gamma_1 = 0$, $B = \epsilon k_i k_H D_N$, and $b_2 = 0$. Assumption **(H1)** holds with $M = 1$, $\gamma_3 = \epsilon |H_{ii}^i| k_i k_H D_N$, $\sigma = |H_{ii}^i| k_i$ as it can be readily verified by means of the variation-of-constants formula

$$\bar{G}_i^{\text{av}}(t) = \exp(-|H_{ii}^i| k_i t) \bar{G}_i^{\text{av}}(0) + \int_0^1 \exp(-|H_{ii}^i| k_i (t+s)) \epsilon H_{ii}^i k_i \phi_i^{\text{av}}(1,s) ds$$

and from the application of the Cauchy–Schwarz inequality to the term $\phi^{\text{av}}(1,t)$ in (15.114),

$$\phi_i^{\text{av}}(1,t) \le \sum_{j\neq i} |H_{ij}^i| \left(\int_0^1 D_j^2\right)^{\frac{1}{2}} \times \left(\int_0^1 [u_j^{\text{av}}(\xi,t)]^2 d\xi\right)^{\frac{1}{2}}$$

$$\le k_H D_N \left(\int_0^1 u^{\text{av}}(\xi,t)^T u^{\text{av}}(\xi,t) d\xi\right)^{\frac{1}{2}}, \tag{15.125}$$

since $\sum_{j\neq i}^N |H_{ij}^i| < k_H < \frac{1}{\epsilon}|H_{ii}^i|$, where k_H is a positive constant of order $\mathcal{O}(1)$, according to Assumptions 15.1 and 15.2.

It follows that the small-gain condition in Theorem C.1 in Appendix C holds provided $0 < \epsilon < 1$ is sufficiently small. Therefore, if such a small-gain condition holds, then Theorem C.1 allows us to conclude that there exist constants $\delta, \Delta > 0$ such that for every $u_0^{\text{av}} \in C^0([0,1])$, $\bar{G}_{i,0}^{\text{av}} \in \mathbb{R}$, the unique generalized solution of this initial-boundary value problem, with $u^{\text{av}}(x,0) = u_0^{\text{av}}$ and $\bar{G}_i^{\text{av}}(0) = \bar{G}_{i,0}^{\text{av}}$, satisfies the following estimate:

$$|\bar{G}_i^{\text{av}}(t)| + \|u^{\text{av}}(t)\|_\infty \le \Delta(|\bar{G}_{i,0}^{\text{av}}| + \|u_0^{\text{av}}\|_\infty)\exp(-\delta t) \quad \forall t \ge 0. \tag{15.126}$$

Therefore, we conclude that the origin of the average closed-loop system (15.116)–(15.118) is exponentially stable under the assumption of $0 < \epsilon < 1$ being sufficiently small. Then, from (15.28) and (15.108), we can conclude the same results in the norm

$$\left(\sum_{i=1}^N [\tilde{\theta}_i^{\text{av}}(t-D_i)]^2 + \int_0^{D_i} [u_i^{\text{av}}(\tau)]^2 d\tau\right)^{1/2} \tag{15.127}$$

since H is nonsingular, i.e., $|\tilde{\theta}_i^{\text{av}}(t-D_i)| \le |H^{-1}||\hat{G}^{\text{av}}(t)|$.

As developed in the proof of Theorem 15.1, the next steps to complete the proof would be the application of the local averaging theory for infinite-dimensional systems in [83] (Appendix A) showing that the periodic solutions indeed satisfy inequality (15.120) and then the conclusion of the attractiveness of the Nash equilibrium θ^* according to (15.121). □

15.5 ▪ Trade-Off between Measurement Requirements and System Restrictions

In the cooperative approach, there is a kind of information sharing that collectively facilitates the elaboration of the control law implemented by each player. In the context of extremum seeking, such information are the frequencies $\{\omega_1,\ldots,\omega_N\}$ and the amplitudes $\{a_1,\ldots,a_N\}$ of the dither signals as well as the players' actions $\{U_1,\ldots,U_N\}$ and the delays $\{D_1,\ldots,D_N\}$, which are known to all players in the prediction process. From this perspective, through appropriate signals $S(t)$, $M(t)$, and $N(t)$, each ith player is able to individually estimate (on average) the portion of the advanced gradient of his payoff function, i.e., $\hat{G}_i^{\text{av}}(t+D_i)$, and the Hessian matrix H in (15.7) by employing (15.55) and (15.56). Note that (15.38) can be interpreted as a filtered version of the advanced estimate $\hat{G}_i^{\text{av}}(t+D_i)$. In other words, by assuming c to be sufficiently large ($c \to +\infty$) in (15.38) and using (15.41), it is possible to rewrite (15.42) in the following vector form:

$$U(t) = K\hat{G}(t+D), \quad K = \text{diag}(k_1 \cdots k_N), \quad k_i > 0,$$

$$\hat{G}(t+D) = \hat{G}(t) + \hat{H}(t)\sum_{j=1}^N e_j \int_{t-D_j}^t U_j(\tau)d\tau, \tag{15.128}$$

where $e_i \in \mathbb{R}^N$ stands again for the ith column of the identity matrix $I_N \in \mathbb{R}^{N\times N}$ for each $i \in \{1,2,\ldots,N\}$, and $\hat{G}(t)$, $\hat{H}(t)$ are given in (15.64) and (15.65), respectively. Therefore, in the cooperative approach, the control law is able to predict totally the gradient such that the average closed-loop equation in (15.29) is rewritten by

$$\dot{\hat{G}}^{\text{av}}(t) = HU^{\text{av}}(t-D) = HK\hat{G}^{\text{av}}(t),$$

with $U^{\text{av}}(t-D) = \left[U_1^{\text{av}}(t-D_1), \quad U_2^{\text{av}}(t-D_2), \quad \ldots \quad , U_N^{\text{av}}(t-D_N)\right]^T$. Hence, for HK Hurwitz, we can conclude $\hat{G}^{\text{av}}(t) \to 0$. Moreover, from (15.28) and the assumption of H being invertible, one has $\tilde{\theta}^{\text{av}}(t) \to 0$. Consequently, $\hat{\theta}^{\text{av}}(t) \to \theta^*$; i.e., the Nash equilibrium is reached at least asymptotically, as shown in Theorem 15.1.

In the noncooperative scenario, there is no sharing of information among the players and the construction of an estimate of the advanced gradient $\hat{G}(t+D)$ as in (15.41) cannot be conceived as described before for the cooperative case. On the other hand, the control law (15.103) is designed using only information from the ith player. In this case, the closed-loop system is rigorously analyzed under a new perspective, by exploring the small gains in the interconnections of the PDE-ODE cascades. In practice, the control law (15.103) differs from (15.38) by estimating only the diagonal terms of the matrix $\hat{H}(t)$ in (15.65), such that (15.104) can be written as

$$U(t) = K\underbrace{\left(\hat{G}(t) + \text{diag}\left\{\hat{H}(t)\right\}\sum_{j=1}^{N} e_j \int_{t-D_j}^{t} U_j(\tau)d\tau\right)}_{\neq\, \hat{G}(t+D) \text{ in } (15.128)}.$$

Thus, unlike (15.41), there is no classical prediction a priori of $\hat{G}(t+D)$ in the noncooperative scheme. However, after some mathematical manipulations carried out in Section 15.4.2, it is possible to re-write (15.112) in the vector form

$$U(t) = K\bar{G}(t) + \epsilon K\phi(D,t),$$
$$\bar{G}(t) = \hat{G}(t) + \left[\text{diag}\left\{\hat{H}(t)\right\} + H - \text{diag}\{H\}\right]\sum_{j=1}^{N} e_j \int_{t-D_j}^{t} U_j(\tau)d\tau,$$

where $\bar{G}(t) := [\bar{G}_1(t),\ldots,\bar{G}_N(t)]^T \in \mathbb{R}^N$ and $\phi(D,t) := [\phi_1(D,t),\ldots,\phi_N(D,t)]^T \in \mathbb{R}^N$, with $\bar{G}_i(t)$ and $\phi_i(D,t)$ defined in (15.108) and (15.114), respectively. Now, it is easy to verify that, even with $\bar{G}(t) \neq \hat{G}(t+D)$, the average estimate is

$$\bar{G}^{\text{av}}(t) = \hat{G}^{\text{av}}(t+D) = \hat{G}^{\text{av}}(t) + H\sum_{j=1}^{N} e_j \int_{t-D_j}^{t} U_j^{\text{av}}(\tau)d\tau,$$

since $\left[\text{diag}\left\{\hat{H}(t)\right\}\right]_{\text{av}} = \text{diag}\{H\}$ according to (15.7) and (15.101). Hence,

$$\begin{aligned} U^{\text{av}}(t) &= K\bar{G}^{\text{av}}(t) + \epsilon K\phi^{\text{av}}(D,t) \\ &= K\hat{G}^{\text{av}}(t+D) + \epsilon K\phi^{\text{av}}(D,t). \end{aligned} \tag{15.129}$$

Finally, by plugging (15.129) into (15.29), we obtain

$$\dot{\hat{G}}^{\text{av}}(t) = HU^{\text{av}}(t-D) = HK\hat{G}^{\text{av}}(t) + \epsilon HK\phi^{\text{av}}(D,t-D).$$

Under the condition of HK being Hurwitz (with H being invertible), the small-gain analysis performed in the proof of Theorem 15.2 guarantees that the Nash equilibrium can be reached if $\epsilon > 0$ is sufficiently small, even in this more restrictive scenario of measurement requirements and system restrictions. Additionally, in the limiting case where $\epsilon = 0$, both cooperative and noncooperative Nash seeking strategies become equivalent.

15.6 ▪ Simulations with a Two-Player Game under Constant Delays

For an example of a noncooperative game with two players that employ the proposed extremum seeking strategy for delay compensation, we consider the payoff functions (15.1) subject to distinct delays $D_1 = 20$ and $D_2 = 15$ in the players' decisions, $i \in \{1,2\}$,

$$\begin{aligned} J_1(\theta(t-D)) &= -5\,\theta_1^2(t-D_1) + 5\,\epsilon\theta_1(t-D_1)\theta_2(t-D_2) \\ &\quad + 250\,\theta_1(t-D_1) - 150\,\theta_2(t-D_2) - 3000, \end{aligned} \tag{15.130}$$

$$\begin{aligned} J_2(\theta(t-D)) &= -5\,\theta_2^2(t-D_2) + 5\,\epsilon\theta_1(t-D_1)\theta_2(t-D_2) \\ &\quad - 150\,\theta_1(t-D_1) + 150\,\theta_2(t-D_2) + 6000, \end{aligned} \tag{15.131}$$

which, according to (15.8), yield the unique Nash equilibrium

$$\theta_1^* = \frac{100+30\epsilon}{4-\epsilon^2}, \tag{15.132}$$

$$\theta_2^* = \frac{60+50\epsilon}{4-\epsilon^2}. \tag{15.133}$$

In order to attain the Nash equilibrium (15.132) and (15.133) without any knowledge of modeling information, the players implement a non-model-based real-time optimization strategy, e.g., the proposed deterministic extremum seeking with sinusoidal perturbations based on prediction of Section 15.4, to set their best actions despite the delays (see Figure 15.3). Specifically, the players P_1 and P_2 set their actions, θ_1 and θ_2, according to (15.15) with adaptation laws $\dot{\hat{\theta}}_i(t)$ in (15.102) and (15.103), respectively. The gradient estimates $\hat{G}_i$ for each player and the diagonal estimate $\hat{H}_i$ of the Hessian are given, respectively, in (15.19) and (15.100), where the additive dither and demodulation signals $S_i(t)$, $M_i(t)$, and $N_i(t)$ are presented in (15.10), (15.11), and (15.99).

For comparison purposes, except for the delays, the plant and controller parameters were chosen as in [70] in all simulation tests: $\epsilon = 1$, $a_1 = 0.075$, $a_2 = 0.05$, $k_1 = 2$, $k_2 = 5$, $\omega_1 = 26.75$ rad/s, $\omega_2 = 22$ rad/s, $\theta_1(0) = \hat{\theta}_1(0) = 50$, and $\theta_2(0) = \hat{\theta}_2(0) = \theta_2^* = 110/3$. In addition, the time constants of the predictor filters were set to $c_1 = c_2 = 100$.

Unlike other classical strategies for noncooperative games [21] (free of delays), when using the proposed extremum seeking algorithm, the players only need to measure the value of their own payoff functions, J_1 and J_2.

In Figures 15.4(a) and 15.4(b), we can see that the extremum seeking approach proposed in [70] is effective when delays in the decision variables are not taken into account. The players find the Nash equilibrium in about 100 seconds. On the other hand, in the presence of delays D_1 and D_2 in the input signals θ_1 and θ_2, but without considering any kind of delay compensation, Figures 15.5(a) and 15.5(b) show that the game collapses with the explosion of its variables. On the other hand, Figures 15.6(a) and 15.6(b) show that the proposed predictor scheme fixes this with a remarkable evolution in searching the Nash equilibrium and simultaneously compensating the effect of the delays in our noncooperative game.

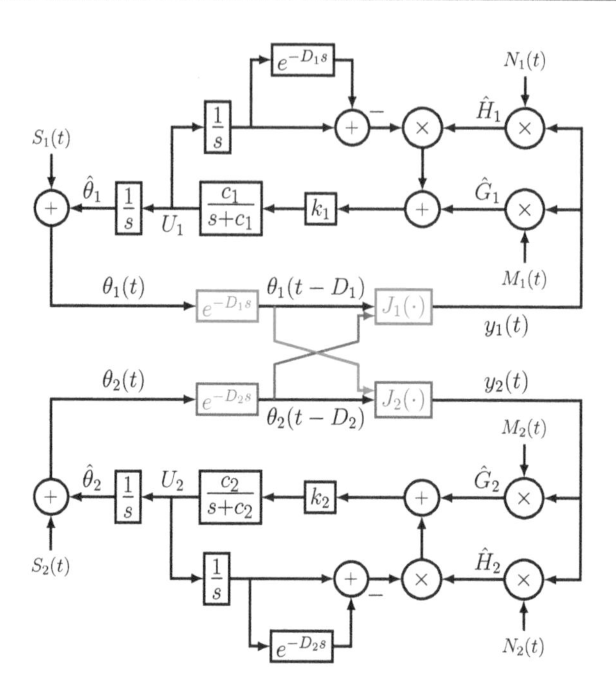

Figure 15.3. *Nash equilibrium seeking schemes with delay compensation for a two-player noncooperative game.*

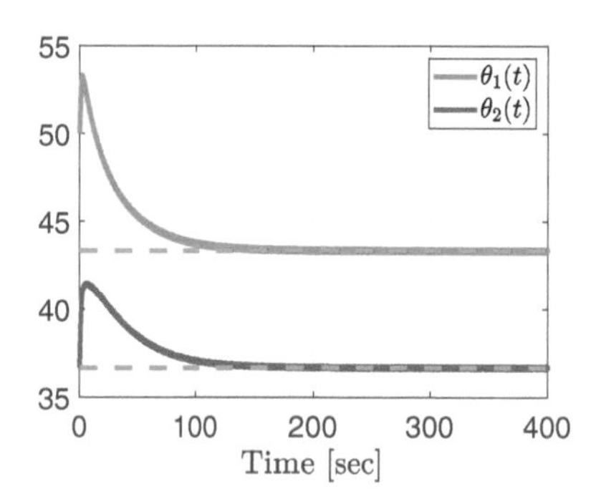

(a) Actions time histories for P_1 and P_2 when implementing the Nash equilibrium seeker of [70] for $\epsilon = 1$. The dashed lines denote the values at the Nash equilibrium, $\theta_1^* = 43.33$ and $\theta_2^* = 36.67$.

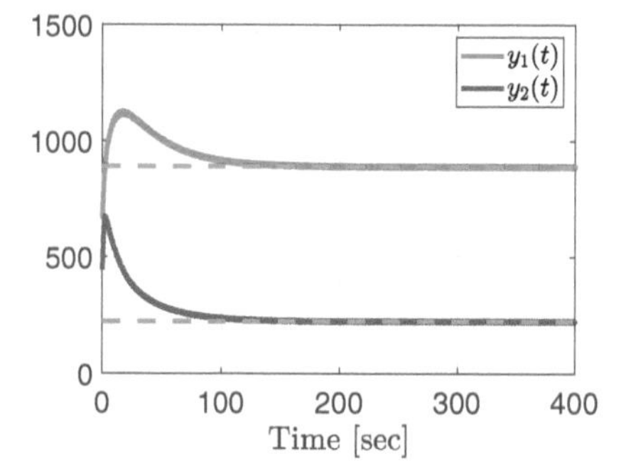

(b) Payoffs time histories for P_1 and P_2 when implementing the Nash equilibrium seeker of [70] for $\epsilon = 1$. The dashed lines denote the corresponding values J_1^* and J_2^* at the Nash equilibrium.

Figure 15.4. *Delay Free.*

This first set of simulations indicates that even under an adversarial scenario of strong coupling between the players with $\epsilon = 1$, the proposed approach has behaved successfully. This suggests our stability analysis may be conservative and the theoretical assumption $0 < \epsilon < 1$ may be relaxed given the performance of the closed-loop control system.

In Figures 15.6(c) and 15.6(d), different values of $\epsilon = 0.5$ and $\epsilon = 0.1$ are considered in order to evaluate the robustness of the proposed scheme under different levels of coupling between the two players and the corresponding impact on the transient responses (the smaller coupling ϵ is, the faster is the convergence rate).

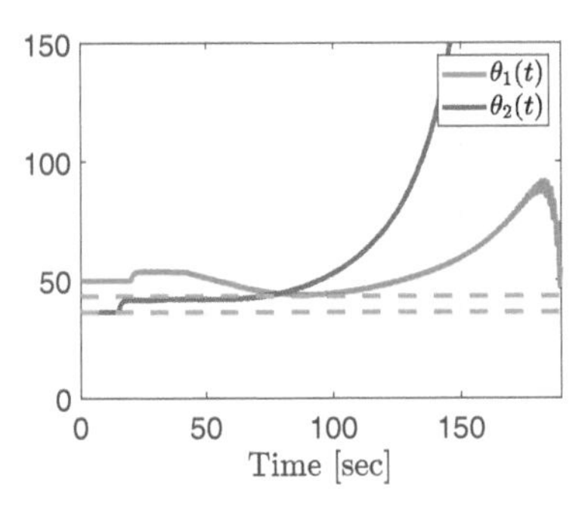

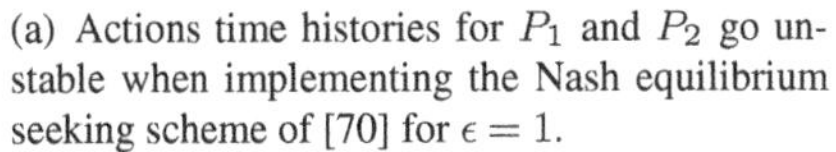
(a) Actions time histories for P_1 and P_2 go unstable when implementing the Nash equilibrium seeking scheme of [70] for $\epsilon = 1$.

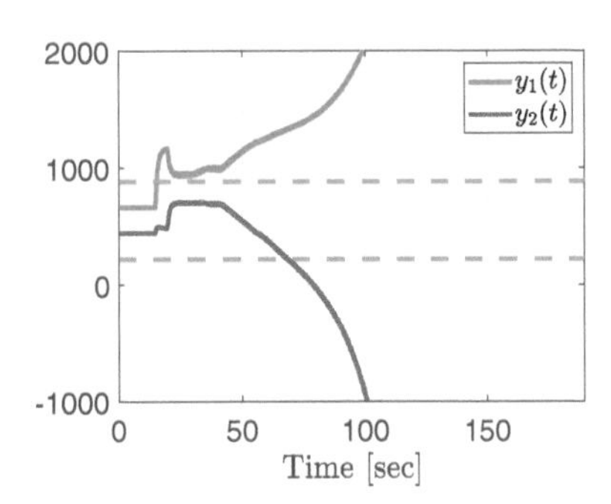

(b) Payoffs time histories for P_1 and P_2 go unstable when implementing the Nash equilibrium seeking scheme proposed in [70] for $\epsilon = 1$.

Figure 15.5. *Uncompensated Delays.*

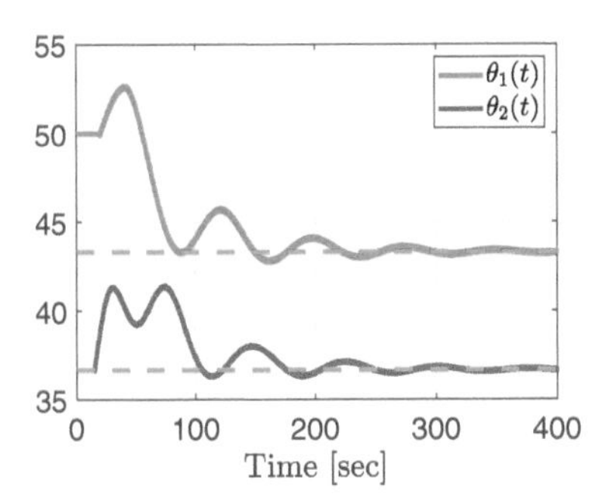

(a) Actions time histories for P_1 and P_2 when implementing the proposed Nash equilibrium seeker for $\epsilon = 1$. The dashed lines denote the values $\theta_1^* = 43.33$ and $\theta_2^* = 36.67$.

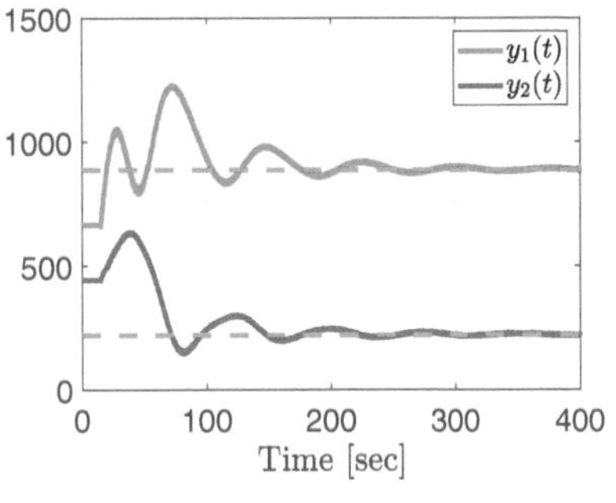

(b) Payoffs time histories for P_1 and P_2 when implementing the proposed Nash equilibrium seeker for $\epsilon = 1$. The dashed lines denote the values of J_1^* and J_2^* at the Nash equilibrium.

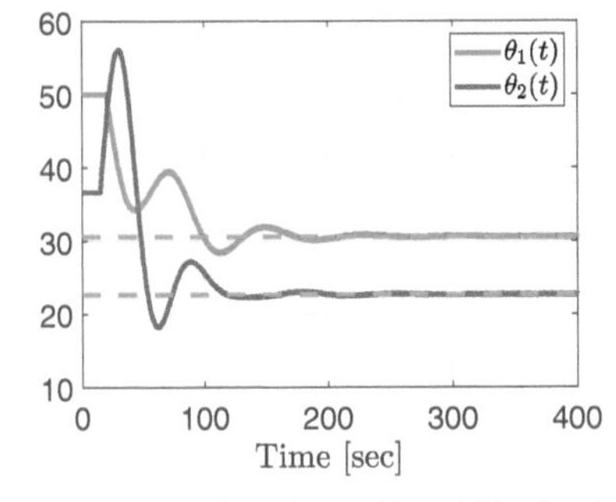

(c) Actions time histories for P_1 and P_2 when implementing the proposed Nash equilibrium seeker for $\epsilon = 0.5$. The dashed lines denote the values $\theta_1^* = 30.67$ and $\theta_2^* = 22.67$.

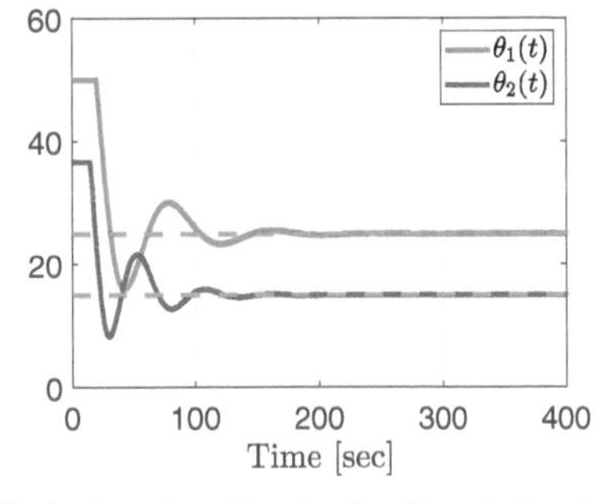

(d) Actions time histories for P_1 and P_2 when implementing the proposed Nash equilibrium seeker for $\epsilon = 0.1$. The dashed lines denote the values $\theta_1^* = 25.81$ and $\theta_2^* = 16.29$.

Figure 15.6. *Compensated Delays.*

In a nutshell, independently of the values used for ϵ, the simulation tests indicate that, even with multiple and distinct delays in the player actions, both of them were able to optimize their payoff functions in the noncooperative game by seeking the desired Nash equilibrium.

15.7 ▪ Notes and References

Time delays are some of the most common phenomena that arise in engineering practice and industry, involving networking problems in areas such as network virtualization, software defined networks, cloud computing, the Internet of Things, context-aware networks, green communications, and security [85, 8, 7]. Hence, the motivation for employing ES to optimize such engineering processes commonly modeled by a game-theoretic framework is clear and justified. We can even find publications on differential games with delays [42, 157, 105, 61, 76, 187, 41], but the literature has not addressed the model-free online and delay-compensating context, as is done here with ES.

In spite of the large number of publications on delay compensation via predictor feedback, there has been no work (other than our own) that develops ES or NES in the presence of time delays. The reason is that delay compensation (such as predictor feedback [119]) is inherently model-based, whereas ES is inherently non-model-based. In addition, since the key purpose of ES is convergence, with a good convergence rate, whereas a delay, when it is simply ignored, severely restricts the convergence rate or destabilizes the closed-loop system, delay-compensated Nash equilibrium seeking is an important practical problem. In our papers [179, 183] (and in Part I of this book), we solve the problems of designing multivariable ES algorithms for delayed systems via predictors based on perturbation-based (averaging-based) estimates of the model. In this chapter we advance such designs from the multi-input-single-output ES scenario to the multi-input-multi-output Nash equilibrium seeking scenario under delays. This is challenging since in the literature for multi-input delay compensation, most authors consider a centralized approach for the predictor design (all vectors multiplying the control inputs needs to be known), while games are decentralized.

In this chapter we pursue two game designs: one with full sharing of information and another, more challenging, where there are restrictions on sharing of information among the players.

For the former problem (cooperative game ES), the predictor-based delay compensator for the average system needs to be multivariable, which means that each player would need to know at least about the existence of other players, including how many of them there are, and possibly also know something about their payoff functions. The key is the square matrix of second derivatives (Hessian) in [70] due to the players' payoffs, which must be estimated using the perturbation signals of each player. Such a perturbation-based estimate of the matrix needs to be shared by all the players. In this case, the game exhibits some kind of "cooperation" among the players for a collective delay compensation. Basically, the players are forced to cooperate minimally so that the Nash equilibrium can be achieved in a scenario with delays.

For the latter problem (noncooperative game NES), we develop a result for N-player games with discrete (point) delays, where the players estimate only the diagonal entries of the Hessian matrix. Hence, we were also able to dominate sufficiently small off-diagonal terms using a small-gain argument [99] for the average system.

Our analysis employed properly sequences steps of averaging in infinite dimension [83], a Lyapunov functional construction [183] for the cooperative result, and, in the noncooperative case, an application of a Small-Gain Theorem for Input-to-State Stable (ISS) cascades of dynamical ODE and PDE systems [99], to prove closed-loop stability. For both scenarios of games, a small neighborhood of the Nash equilibrium is achieved, even in the presence of arbitrarily long (but fixed) delays.

In addition to allowing for the players to have both their inputs delayed and their measured payoffs delayed simultaneously, extensions are of interest to general non-quadratic payoff functions. For games, we only explored the gradient extremum seeking approach and it is of interest whether the Newton-based extremum seeking would also be viable for games.

In the next two chapters, the proposed extremum seeking design facing games with infinite-dimensional PDE dynamics [65, 170] other than pure delays is addressed. A more general combination of PDE-based systems involving, e.g., PDE-PDE cascades (Chapter 14) can be considered as well.

Chapter 16

Nash Equilibrium Seeking with Players Acting through Heat PDE Dynamics

Similar to the transition from the ES design through delays in Part I to the ES design through PDEs in Part II, in this chapter we progress from a Nash equilibrium seeking design in Chapter 15 to developing a non-model-based strategy for locally stable convergence to Nash equilibria in quadratic noncooperative (duopoly) games with player actions subject to diffusion (heat) PDE dynamics with distinct diffusion coefficients and each player having access only to his own payoff value.

The proposed approach employs extremum seeking, with sinusoidal perturbation signals employed to estimate the Gradient (first derivative) and Hessian (second derivative) of unknown quadratic functions. The material in the present chapter is the first instance of noncooperative games being tackled in a model-free fashion in the presence of heat PDE dynamics. In order to compensate distinct diffusion processes in the inputs of the two players, we employ boundary control with averaging-based estimates. We apply a small-gain analysis for the resulting Input-to-State Stable (ISS) parabolic PDE-ODE loop as well as averaging theory in infinite dimension, due to the infinite-dimensional state of the heat PDEs, in order to obtain local convergence to a small neighborhood of the Nash equilibrium. We quantify the size of these residual sets and illustrate the theoretical results numerically on an example of a two-player game under heat PDEs.

16.1 ▪ Duopoly Game with Quadratic Payoffs and PDEs: General Formulation

In a duopoly game, the optimality of the respective outcomes (outputs) of Players P1 and P2, respectively $y_1(t) \in \mathbb{R}$ and $y_2(t) \in \mathbb{R}$, do not depend exclusively on their own strategies (input signals) $\Theta_1(t) \in \mathbb{R}$ and $\Theta_2(t) \in \mathbb{R}$. Moreover, defining $\Theta := [\Theta_1, \Theta_2]^T$, each player's payoff function J_i depends also on Θ_j of the other Player j, $j \neq i$. A 2-tuple of $\Theta^* = [\Theta_1^*, \Theta_2^*]^T$ is said to be in Nash equilibrium if no Player i can improve his payoff by unilaterally deviating from Θ_i^*, this being so for all $i \in \{1, 2\}$ [72].

As shown in Figures 16.1 and 16.2, distinct heat equations (with Dirichlet actuation) are assumed in the vector of player actions $\theta(t) \in \mathbb{R}^2$. Thus, the propagated actuator vector

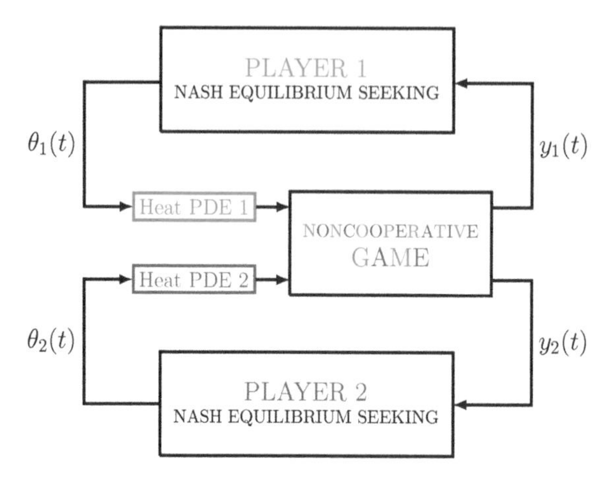

Figure 16.1. *Nash equilibrium seeking in a duopoly noncooperative game with players acting through heat PDE dynamics.*

$\Theta(t) \in \mathbb{R}^2$ is given by

$$\Theta_i(t) = \alpha_i(0,t) \quad \forall i \in \{1,2\}, \tag{16.1}$$
$$\partial_t \alpha_i(x,t) = \partial_{xx}\alpha_i(x,t), \quad x \in (0,D_i), \tag{16.2}$$
$$\partial_x \alpha_i(0,t) = 0, \tag{16.3}$$
$$\alpha_i(D_i,t) = \theta_i(t), \tag{16.4}$$

where $\alpha_i : [0,D_i] \times \mathbb{R}_+ \to \mathbb{R}$ and each domain length D_i is known. The solution of (16.1)–(16.4) is given by

$$\alpha_i(x,t) = \mathcal{L}^{-1}\left[\frac{\cosh(x\sqrt{s})}{\cosh(D_i\sqrt{s})}\right] * \theta_i(t), \tag{16.5}$$

where $\mathcal{L}^{-1}(\cdot)$ denotes the inverse Laplace transformation and $*$ is the convolution operator. Given this relation, we define the *diffusion operator* for the PDE (16.1)–(16.4) with boundary input and measurement given by $\mathcal{D}=\text{diag}\{\mathcal{D}_1,\mathcal{D}_2\}$ with

$$\mathcal{D}_i[\varphi(t)]=\mathcal{L}^{-1}\left[\frac{1}{\cosh(D_i\sqrt{s})}\right] * \varphi(t), \quad \text{s.t.} \quad \Theta(t)=\mathcal{D}[\theta(t)]. \tag{16.6}$$

We consider games where the payoff function $y_i(t)=J_i(\mathcal{D}[\theta(t)])=J_i(\Theta(t))$ of each player is quadratic [21], expressed as a strictly concave combination of their actions propagated through distinct heat PDE dynamics

$$\begin{aligned} J_1(\Theta(t)) = {} & \frac{H_{11}^1}{2}\Theta_1^2(t) + \frac{H_{22}^1}{2}\Theta_2^2(t) + \epsilon H_{12}^1\Theta_1(t)\Theta_2(t) \\ & + h_1^1\Theta_1(t) + h_2^1\Theta_2(t) + c_1, \end{aligned} \tag{16.7}$$
$$\begin{aligned} J_2(\Theta(t)) = {} & \frac{H_{11}^2}{2}\Theta_1^2(t) + \frac{H_{22}^2}{2}\Theta_2^2(t) + \epsilon H_{21}^2\Theta_1(t)\Theta_2(t) \\ & + h_1^2\Theta_1(t) + h_2^2\Theta_2(t) + c_2, \end{aligned} \tag{16.8}$$

where $J_1(\Theta), J_2(\Theta) : \mathbb{R}^2 \to \mathbb{R}$, H_{jk}^i, h_j^i, $c_i \in \mathbb{R}$ are constants, $H_{ii}^i < 0 \ \forall i,j,k \in \{1,2\}$, and $\epsilon > 0$ without loss of generality.

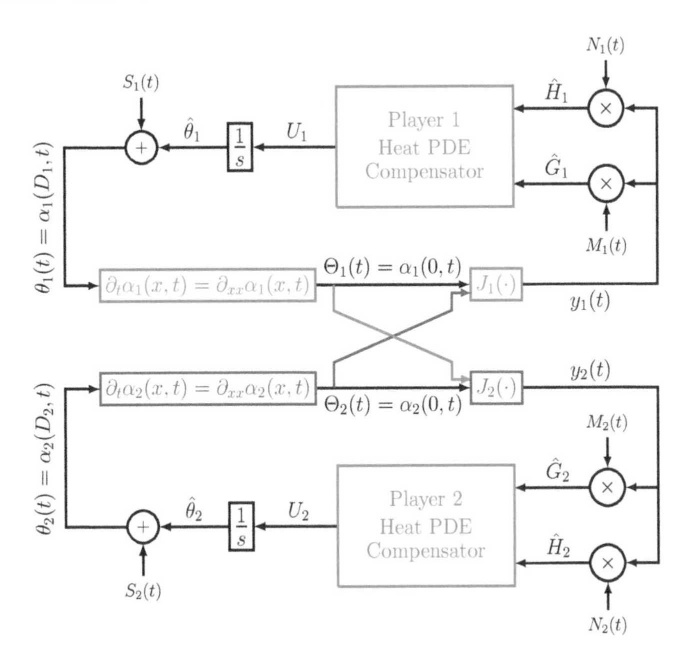

Figure 16.2. *Block diagram illustrating the Nash equilibrium seeking strategy performed for each player. In magenta color are the boundary controllers used to compensate the individual heat PDEs for the noncooperative game.*

As discussed in Chapter 15, quadratic payoff functions are of particular interest in game theory, first because they constitute second-order approximations to other types of non-quadratic payoff functions, and second because they are analytically tractable, leading to, in general, closed-form equilibrium solutions which provide insight into the properties and features of the equilibrium solution concept under consideration [21].

For the sake of completeness, we provide here in mathematical terms the definition of a Nash equilibrium $\Theta^* = [\Theta_1^*, \Theta_2^*]^T$ in a 2-player game:

$$J_1(\Theta_1^*, \Theta_2^*) \geq J_1(\Theta_1, \Theta_2^*) \text{ and } J_2(\Theta_1^*, \Theta_2^*) \geq J_2(\Theta_1^*, \Theta_2). \tag{16.9}$$

Hence, no player has an incentive to unilaterally deviate its action from Θ^*.

In order to determine the Nash equilibrium solution in strictly concave quadratic games with 2 players, where each action set is the entire real line, one should differentiate J_1 and J_2, respectively, with respect to $\Theta_1(t)$ and $\Theta_2(t)$, setting the resulting expressions equal to zero and solving the set of equations thus obtained. This set of equations, which also provides a sufficient condition due to the strict concavity, is

$$\begin{cases} H_{11}^1 \Theta_1^* + \epsilon H_{12}^1 \Theta_2^* + h_1^1 = 0, \\ \epsilon H_{21}^2 \Theta_1^* + H_{22}^2 \Theta_2^* + h_2^2 = 0, \end{cases} \tag{16.10}$$

which can be written in the matrix form as

$$\begin{bmatrix} H_{11}^1 & \epsilon H_{12}^1 \\ \epsilon H_{21}^2 & H_{22}^2 \end{bmatrix} \begin{bmatrix} \Theta_1^* \\ \Theta_2^* \end{bmatrix} = - \begin{bmatrix} h_1^1 \\ h_2^2 \end{bmatrix}. \tag{16.11}$$

Defining the Hessian matrix H and vectors Θ^* and h by

$$H := \begin{bmatrix} H_{11}^1 & \epsilon H_{12}^1 \\ \epsilon H_{21}^2 & H_{22}^2 \end{bmatrix}, \quad \Theta^* := \begin{bmatrix} \Theta_1^* \\ \Theta_2^* \end{bmatrix}, \quad h := \begin{bmatrix} h_1^1 \\ h_2^2 \end{bmatrix}, \tag{16.12}$$

there exists a unique Nash equilibrium at $\Theta^* = -H^{-1}h$ if H is invertible:

$$\begin{bmatrix} \Theta_1^* \\ \Theta_2^* \end{bmatrix} = - \begin{bmatrix} H_{11}^1 & \epsilon H_{12}^1 \\ \epsilon H_{21}^2 & H_{22}^2 \end{bmatrix}^{-1} \begin{bmatrix} h_1^1 \\ h_2^2 \end{bmatrix} . \tag{16.13}$$

For more details, see [21, Chapter 4].

The *control objective* is to design a novel extremum seeking-based strategy to reach the Nash equilibrium in noncooperative games subjected to heat PDEs in the decision variables of the players (input signals).

Since our goal is to find the unknown optimal inputs Θ^* (and θ^*), we define the estimation errors

$$\tilde{\theta}(t) = \hat{\theta}(t) - \theta^* , \qquad \vartheta(t) = \hat{\Theta}(t) - \Theta^* , \tag{16.14}$$

where the vectors $\hat{\theta}(t)$ and $\hat{\Theta}(t)$ are the estimates of θ^* and Θ^*. In order to make (16.14) coherent with the optimizer of the static map Θ^*, we apply the diffusion operator (16.6) to $\tilde{\theta}_i$ in (16.14) and get

$$\vartheta_i(t) := \bar{\alpha}_i(0,t) \quad \forall i \in \{1,2\} , \tag{16.15}$$

$$\partial_t \bar{\alpha}_i(x,t) = \partial_{xx} \bar{\alpha}_i(x,t), \quad x \in (0, D_i) , \tag{16.16}$$

$$\partial_x \bar{\alpha}_i(0,t) = 0 , \tag{16.17}$$

$$\bar{\alpha}_i(D_i,t) = \tilde{\theta}_i(t) , \tag{16.18}$$

where $\bar{\alpha}_i : [0, D_i] \times \mathbb{R}_+ \to \mathbb{R}$ and $\vartheta(t) := \mathcal{D}[\tilde{\theta}(t)] = \hat{\Theta}(t) - \Theta^*$ is the propagated estimation error $\tilde{\theta}(t)$ through the diffusion domain. For $\lim_{t\to\infty} \theta(t) = \theta_c$, we have $\lim_{t\to\infty} \Theta(t) = \Theta_c = \theta_c$, where the index c indicates a constant signal. Indeed, from (16.5), for a constant input $\theta = \theta_c$, one has $\mathcal{L}\{\theta_c\} = \theta_c/s$, and applying the Laplace limit theorem we get $\lim_{t\to\infty} \mathcal{D}_i[\theta_{ci}] = \lim_{s\to 0} \left\{ \frac{\theta_{ci}}{\cosh(\sqrt{s D_i})} \right\} = \theta_{ci}$. In the particular case $\theta = \theta_c = \theta^*$, one has

$$\Theta^* = \theta^* . \tag{16.19}$$

Figure 16.2 depicts a schematic diagram that summarizes the proposed Nash equilibrium policy for each player, where their outputs are given by

$$\begin{cases} y_1(t) & = J_1(\Theta(t)) , \\ y_2(t) & = J_2(\Theta(t)) . \end{cases} \tag{16.20}$$

The additive dither signals in the presence of heat PDE dynamics [65] are defined according to

$$\begin{cases} S_1(t) & = \frac{1}{2} a_1 e^{\sqrt{\frac{\omega_1}{2}} D_1} \sin\left(\omega_1 t + \sqrt{\frac{\omega_1}{2}} D_1\right) + \frac{1}{2} a_1 e^{-\sqrt{\frac{\omega_1}{2}} D_1} \sin\left(\omega_1 t - \sqrt{\frac{\omega_1}{2}} D_1\right) , \\ S_2(t) & = \frac{1}{2} a_2 e^{\sqrt{\frac{\omega_2}{2}} D_2} \sin\left(\omega_2 t + \sqrt{\frac{\omega_2}{2}} D_2\right) + \frac{1}{2} a_2 e^{-\sqrt{\frac{\omega_2}{2}} D_2} \sin\left(\omega_2 t - \sqrt{\frac{\omega_2}{2}} D_2\right) , \end{cases} \tag{16.21}$$

and the multiplicative demodulation signals are given by

$$\begin{cases} M_1(t) & = \frac{2}{a_1} \sin(\omega_1 t) , \\ M_2(t) & = \frac{2}{a_2} \sin(\omega_2 t) , \end{cases} \tag{16.22}$$

with nonzero constant amplitudes a_1, $a_2 > 0$ at frequencies $\omega_1 \neq \omega_2$. Such probing frequencies ω_i can be selected as

$$\omega_i = \omega_i' \omega = \mathcal{O}(\omega) , \quad i = 1 \text{ or } 2 , \tag{16.23}$$

where ω is a positive constant and ω_i' is a rational number—one possible choice is given in [73].

Following the non-sharing information paradigm, only the diagonal elements of H can be properly recovered in the average sense by Player P1 or P2. In this sense, the signals $N_1(t)$ and $N_2(t)$ are simply defined by [73]

$$\begin{cases} N_1(t) = \frac{16}{a_1^2}\left(\sin^2(\omega_1 t) - \frac{1}{2}\right), \\ N_2(t) = \frac{16}{a_2^2}\left(\sin^2(\omega_2 t) - \frac{1}{2}\right). \end{cases} \tag{16.24}$$

Then, the average version of

$$\begin{cases} \hat{H}_1(t) = N_1(t) y_1(t), \\ \hat{H}_2(t) = N_2(t) y_2(t) \end{cases} \tag{16.25}$$

is given by

$$\begin{cases} \hat{H}_1^{\mathrm{av}}(t) = [N_1(t) y_1(t)]_{\mathrm{av}} = H_{11}^1, \\ \hat{H}_2^{\mathrm{av}}(t) = [N_2(t) y_2(t)]_{\mathrm{av}} = H_{22}^2. \end{cases} \tag{16.26}$$

Considering $\hat{\theta}_1(t)$ and $\hat{\theta}_2(t)$ as the estimates of θ_1^* and θ_2^*, one can define from (16.14) the individual estimation errors:

$$\begin{cases} \tilde{\theta}_1(t) = \hat{\theta}_1(t) - \theta_1^*, \quad \vartheta_1(t) = \hat{\Theta}_1(t) - \Theta_1^*, \\ \tilde{\theta}_2(t) = \hat{\theta}_2(t) - \theta_2^*, \quad \vartheta_2(t) = \hat{\Theta}_2(t) - \Theta_2^*. \end{cases} \tag{16.27}$$

The estimate of the unknown gradients of the payoff functions are given by

$$\begin{cases} \hat{G}_1(t) \quad = M_1(t) y_1(t), \\ \hat{G}_2(t) \quad = M_2(t) y_2(t), \end{cases} \tag{16.28}$$

and computing the average of the resulting signal leads us to

$$\begin{cases} \hat{G}_1^{\mathrm{av}}(t) = H_{11}^1 \vartheta_1^{\mathrm{av}}(t) + \epsilon H_{12}^1 \vartheta_2^{\mathrm{av}}(t), \\ \hat{G}_2^{\mathrm{av}}(t) = \epsilon H_{21}^2 \vartheta_1^{\mathrm{av}}(t) + H_{22}^2 \vartheta_2^{\mathrm{av}}(t). \end{cases} \tag{16.29}$$

Additionally, from the block diagram in Figure 16.2, one has

$$\dot{\hat{\theta}}_i(t) = U_i(t) \quad \forall i \in \{1,2\} \tag{16.30}$$

and, consequently,

$$\dot{\tilde{\theta}}_i(t) = U_i(t) \quad \forall i \in \{1,2\}, \tag{16.31}$$

since $\dot{\tilde{\theta}}(t) = \dot{\hat{\theta}}(t)$ once θ^* is constant. Taking the time derivative of (16.15)–(16.18) and with the help of (16.14) and (16.31), the *propagated error dynamics* is written as

$$\dot{\vartheta}_i(t) = u_i(0,t) \quad \forall i \in \{1,2\}, \tag{16.32}$$

$$\partial_t u_i(x,t) = \partial_{xx} u_i(x,t), \quad x \in (0, D_i), \tag{16.33}$$

$$\partial_x u_i(0,t) = 0, \tag{16.34}$$

$$u_i(D_i,t) = U_i(t), \tag{16.35}$$

where $u_i : [0, D_i] \times \mathbb{R}_+ \to \mathbb{R}$ and $u_i(x,t) := \partial_t \bar{\alpha}_i(x,t)$.

Hence, from (16.29) and (16.32)–(16.35), it is possible to find a compact form for the overall average estimated gradient according to

$$\hat{G}^{\text{av}}(t) = H\vartheta^{\text{av}}(t), \tag{16.36}$$

$$\dot{\hat{G}}^{\text{av}}(t) = H\dot{\vartheta}^{\text{av}}(t) = H\mathcal{D}[U^{\text{av}}(t)], \tag{16.37}$$

where the Hessian H is given in (16.12) and $\vartheta^{\text{av}}(t) := [\vartheta_1^{\text{av}}(t), \vartheta_2^{\text{av}}(t)]^T \in \mathbb{R}^2$, $\hat{G}^{\text{av}}(t) := [\hat{G}_1^{\text{av}}(t), \hat{G}_2^{\text{av}}(t)]^T \in \mathbb{R}^2$, and $U^{\text{av}}(t) := [U_1^{\text{av}}(t), U_2^{\text{av}}(t)]^T \in \mathbb{R}^2$ are the average versions of $U(t) := [U_1(t), U_2(t)]^T$, $\vartheta(t) := [\vartheta_1(t), \vartheta_2(t)]^T$, and $\hat{G}(t) := [\hat{G}_1(t), \hat{G}_2(t)]^T$, respectively.

Throughout this chapter the key idea is to design control laws (policies) for each player in order to achieve a small neighborhood of the Nash equilibrium point. To this end, we use an extremum seeking strategy based on boundary control to compensate the diffusion operator $\mathcal{D}[\cdot]$ in (16.37) due to the multiple and distinct heat PDEs in the players' actions. Basically, the control laws must be able to ensure exponential stabilization of $\hat{G}^{\text{av}}(t)$ and, consequently, of $\vartheta^{\text{av}}(t) = \hat{\Theta}^{\text{av}}(t) - \Theta^*$. From (16.36), it is clear that, if H is invertible, $\vartheta^{\text{av}}(t) \to 0$ as $\hat{G}^{\text{av}}(t) \to 0$. Hence, the convergence of $\vartheta^{\text{av}}(t)$ to the origin results in the convergence of $\Theta(t)$ to a small neighborhood of Θ^* in (16.9) via averaging theory [83].

16.2 ▪ Noncooperative Duopoly Games with Heat PDE Dynamics

As mentioned earlier, in our noncooperative game subject to heat PDE dynamics, the purpose of the ES is still to estimate the Nash equilibrium $\Theta^* = \theta^*$ (see (16.19)), but without sharing any information between the players. Each player only needs to measure the value of his own payoff function (16.20).

16.2.1 ▪ Decentralized PDE Boundary Control Using Only the Known Diagonal Terms of the Hessian

In this sense, we are able to formulate the closed-loop system in a *decentralized* fashion, where no knowledge about the payoff or action of the other player is required.

Inspired by [184], where ES was considered for PDE compensation but not in the context of games, we propose the following boundary-based update laws $\dot{\hat{\theta}}_i(t) = U_i(t)$, $i \in \{1,2\}$:

$$\begin{cases} \dot{U}_1(t) &= -c_1 U_1(t) + c_1 k_1 \left(\hat{G}_1(t) + \hat{H}_1(t) \int_0^{D_1} (D_1 - \tau) u_1(\tau,t) d\tau\right), \\ \dot{U}_2(t) &= -c_2 U_2(t) + c_2 k_2 \left(\hat{G}_2(t) + \hat{H}_2(t) \int_0^{D_2} (D_2 - \tau) u_2(\tau,t) d\tau\right) \end{cases} \tag{16.38}$$

for positive constants $k_1 > 0$, $k_2 > 0$, $c_1 > 0$, and $c_2 > 0$, in order to compensate the heat PDEs in (16.32)–(16.35). Again, with some abuse of notation, constants c_1 and c_2 were chosen to denote the parameters of the control laws, but they are not related to those which appear in the payoff functions given in (16.7) and (16.8).

As discussed in [65, Remark 2], the boundary control law (16.38) could be rewritten as

$$\dot{U}_i(t) = -c_i U_i(t) + c_i k_i \left[\hat{G}_i(t) + \hat{H}_i(t) \left(\hat{\theta}_i(t) - \Theta_i(t) + a_i \sin(\omega_i t)\right)\right], \tag{16.39}$$

using the diffusion equations $\partial_t \alpha_i(x,t) = \partial_{xx} \alpha_i(x,t)$, $\forall i \in \{1,2\}$, and the integration by parts associated with (16.1)–(16.4), (16.14), and recalling that $\vartheta_i + a_i \sin(\omega_i t) = \Theta_i(t) - \Theta_i^*$, analogously to [65, equation (25)].

16.2.2 ▪ ISS-Like Properties for PDE Representation

For the sake of simplicity, we assume $c_1, c_2 \to +\infty$ in (16.38), resulting in the following general expression:

$$U_i(t) = k_i\left(\hat{G}_i(t) + \hat{H}_i(t)\int_0^{D_i}(D_i - \tau)u_i(\tau,t)d\tau\right). \tag{16.40}$$

Recalling (16.32)–(16.35), the infinite-dimensional closed-loop system (16.37) and (16.40) in its average version can be written in the corresponding PDE representation form, given by

$$\dot{\hat{G}}^{\text{av}}(t) = Hu^{\text{av}}(0,t), \tag{16.41}$$

$$\partial_t u^{\text{av}}(x,t) = D^{-2}\partial_{xx}u^{\text{av}}(x,t), \quad x \in (0,1), \tag{16.42}$$

$$\partial_x u^{\text{av}}(0,t) = 0, \tag{16.43}$$

$$u^{\text{av}}(1,t) = U^{\text{av}}(t), \tag{16.44}$$

with $D = \text{diag}\{D_1, D_2\}$.

In the *reduction-like approach* [12] (or finite-spectrum assignment), we use the transformation (for $i \in \{1,2\}$)

$$\begin{aligned}\bar{G}_i^{\text{av}}(t) &= \hat{G}_i^{\text{av}}(t) + \sum_{j=1}^{2}\epsilon_{ij}^i H_{ij}^i \int_0^{D_j}(D_j - \tau)u_j^{\text{av}}(\tau,t)d\tau \\ &= \hat{G}_i^{\text{av}}(t) + \sum_{j=1}^{2}\epsilon_{ij}^i H_{ij}^i \int_0^{1} D_j^2(1-\xi)u_j^{\text{av}}(\xi,t)d\xi,\end{aligned} \tag{16.45}$$

where $\int_0^{D_j}(D_j - \tau)u_j^{\text{av}}(\tau,t)d\tau = \int_0^1 D_j^2(1-\xi)u_j^{\text{av}}(\xi,t)d\xi$, whereas $\epsilon_{11}^1 = \epsilon_{22}^2 = 1$ and $\epsilon_{12}^1 = \epsilon_{21}^2 = \epsilon$. With some mathematical manipulations (see Chapter 15), it is not difficult to see that $\bar{G}^{\text{av}}$ satisfies

$$\dot{\bar{G}}^{\text{av}}(t) = HU^{\text{av}}(t). \tag{16.46}$$

Now, after adding and subtracting the next terms in blue and red into (16.40), it is rewritten as

$$\begin{aligned}U_1(t) = k_1\Bigg(&\hat{G}_1(t) + \hat{H}_1(t)\int_0^{D_1}(D_1 - \tau)u_1(\tau,t)d\tau \\ &+ \epsilon H_{12}^1\int_0^{D_2}(D_2 - \tau)u_2(\tau,t)d\tau\Bigg) \\ &- k_1\epsilon H_{12}^1\int_0^{D_2}(D_2 - \tau)u_2(\tau,t)d\tau,\end{aligned} \tag{16.47}$$

$$\begin{aligned}U_2(t) = k_2\Bigg(&\hat{G}_2(t) + \hat{H}_2(t)\int_0^{D_2}(D_2 - \tau)u_2(\tau,t)d\tau \\ &+ \epsilon H_{21}^2\int_0^{D_1}(D_1 - \tau)u_1(\tau,t)d\tau\Bigg) \\ &- k_2\epsilon H_{21}^2\int_0^{D_1}(D_1 - \tau)u_1(\tau,t)d\tau,\end{aligned} \tag{16.48}$$

whose average compact form is

$$U^{\mathrm{av}}(t) = K\bar{G}^{\mathrm{av}}(t) + \epsilon K\phi^{\mathrm{av}}(D,t), \tag{16.49}$$

where the matrix $K := \mathrm{diag}\{k_1, k_2\}$ with entries $k_1 > 0$, $k_2 > 0$, and the auxiliary variable $\phi(D,t)$ is defined as

$$\phi(D,t) := -\begin{bmatrix} H_{12}^1 \int_0^{D_2}(D_2-\tau)u_2(\tau,t)d\tau \\ H_{21}^2 \int_0^{D_1}(D_1-\tau)u_1(\tau,t)d\tau \end{bmatrix},$$
$$\phi(1,t) := -\begin{bmatrix} H_{12}^1 \int_0^1 D_2^2(1-\xi)u_2(\xi,t)d\xi \\ H_{21}^2 \int_0^1 D_1^2(1-\xi)u_1(\xi,t)d\xi \end{bmatrix}. \tag{16.50}$$

Then, it is possible to find a compact form for the overall average game from (16.46) and (16.49), such as

$$\dot{\bar{G}}^{\mathrm{av}}(t) = HK\bar{G}^{\mathrm{av}}(t) + \epsilon HK\phi^{\mathrm{av}}(1,t), \tag{16.51}$$
$$\partial_t u^{\mathrm{av}}(x,t) = D^{-2}\partial_{xx}u^{\mathrm{av}}(x,t), \quad x \in (0,1), \tag{16.52}$$
$$\partial_x u^{\mathrm{av}}(0,t) = 0, \tag{16.53}$$
$$u^{\mathrm{av}}(1,t) = K\bar{G}^{\mathrm{av}}(t) + \epsilon K\phi^{\mathrm{av}}(1,t). \tag{16.54}$$

From (16.51), if HK is Hurwitz, it is clear that the dynamics of the ODE state variable $\bar{G}^{\mathrm{av}}(t)$ is exponentially Input-to-State Stable (ISS) [99] with respect to the PDE state $u(x,t)$ by means of the function $\phi^{\mathrm{av}}(1,t)$. Moreover, the PDE subsystem (16.52) is ISS [99] with respect to $\bar{G}^{\mathrm{av}}(t)$ in the boundary condition $u^{\mathrm{av}}(1,t)$.

16.2.3 ▪ Stability Analysis

In this section, we will show that this parabolic PDE-ODE loop (16.51)–(16.54) contains a small-parameter ϵ which can lead to closed-loop stability if it is chosen sufficiently small. To this end, we assume the following condition for noncooperative games [70].

Assumption 16.1. *The Hessian matrix H given by* (16.12) *is strictly diagonal dominant, i.e.,*

$$\begin{cases} |\epsilon H_{12}^1| < |H_{11}^1|, \\ |\epsilon H_{21}^2| < |H_{22}^2|. \end{cases} \tag{16.55}$$

By Assumption 16.1, the Nash equilibrium Θ^* exists and is unique since strictly diagonally dominant matrices are nonsingular by the Levy–Desplanques theorem [90].

The next theorem provides the stability/convergence properties of the closed-loop extremum seeking feedback for the 2-player noncooperative game with heat PDEs.

Theorem 16.1. *Consider the closed-loop system* (16.41)–(16.44) *under Assumption* 16.1 *and multiple heat PDEs* (16.1)–(16.4) *with distinct diffusion coefficients D_1 and D_2 for a duopoly quadratic game with payoff functions given in* (16.7) *and* (16.8) *and control laws $U_i(t)$ defined in* (16.38). *There exist $c_1 > 0$, $c_2 > 0$, and $\omega > 0$ sufficiently large as well as $\epsilon > 0$ sufficiently small such that the closed-loop system with state $\vartheta_i(t)$, $u_i(x,t)$, $\forall i \in \{1,2\}$, has a unique locally exponentially stable periodic solution in t of period $\Pi := 2\pi \times LCM\{1/\omega_i\}$, with ω_i in* (16.23) *of order $\mathcal{O}(\omega)$, denoted by $\vartheta_i^{\Pi}(t)$, $u_i^{\Pi}(x,t)$ and satisfying, $\forall t \geq 0$,*

$$\left(\sum_{i=1}^{2}\left[\vartheta_i^{\Pi}(t)\right]^2 + \int_0^{D_i}\left[u_i^{\Pi}(x,t)\right]^2 dx\right)^{1/2} \leq \mathcal{O}(1/\omega). \tag{16.56}$$

Furthermore,

$$\limsup_{t\to+\infty} |\Theta(t)-\Theta^*| = \mathcal{O}(|a|+1/\omega), \tag{16.57}$$

$$\limsup_{t\to+\infty} |\theta(t)-\theta^*| = \mathcal{O}\left(|a|e^{\max(D_i)\sqrt{\omega/2}}+1/\omega\right), \tag{16.58}$$

where $a=[a_1\ a_2]^T$ *and* $\theta^*=\Theta^*$ *is the unique (unknown) Nash equilibrium given by* (16.13).

Proof. The proof of Theorem 16.1 follows steps similar to those employed to prove the results about extremum seeking under diffusion PDEs in [65]. In this sense, we will simply point out the main differences for the case of games (not classical extremum seeking), instead of giving a full independent proof.

First, consider the equivalent parabolic PDE-ODE representation (16.51)–(16.54) rewritten for each Player i, $i\in\{1,2\}$:

$$\dot{\bar{G}}_i^{\mathrm{av}}(t) = H_{ii}^i k_i \bar{G}_i^{\mathrm{av}}(t) + \epsilon H_{ii}^i k_i \phi_i^{\mathrm{av}}(1,t), \tag{16.59}$$

$$\partial_t u_i^{\mathrm{av}}(x,t) = D_i^{-2}\partial_{xx} u_i^{\mathrm{av}}(x,t), \quad x\in(0,1), \tag{16.60}$$

$$\partial_x u_i^{\mathrm{av}}(0,t) = 0, \tag{16.61}$$

$$u_i^{\mathrm{av}}(1,t) = k_i \bar{G}_i^{\mathrm{av}}(t) + \epsilon k_i \phi_i^{\mathrm{av}}(1,t). \tag{16.62}$$

Hence, the average closed-loop system (16.59)–(16.62) satisfies all the assumptions **(A1)** to **(A7)** of the Small-Gain Theorem [99, Theorem 8.2, p. 205] for the parabolic PDE-ODE loops with $p(z)=1$, $r(z)=D_i^2$, $q(z)=0$, $F(\bar{G}_i^{\mathrm{av}},u^{\mathrm{av}},0)=H_{ii}^i k_i \bar{G}_i^{\mathrm{av}}+\epsilon H_{ii}^i k_i \phi_i^{\mathrm{av}}(1)$, $g(x,\bar{G}_i^{\mathrm{av}},u^{\mathrm{av}})=0$, $f(x,t)=0$, $\varphi_0(0,u_i^{\mathrm{av}},\bar{G}_i^{\mathrm{av}})=b_1 u_i^{\mathrm{av}}(0,t)$, $b_1<0$, $b_2=1$, $\varphi_1(0,u^{\mathrm{av}},\bar{G}_i^{\mathrm{av}})=k_i\bar{G}_i^{\mathrm{av}}+\epsilon k_i\phi_i^{\mathrm{av}}(1)$, $a_1=1$, $a_2=0$, $L=\max(|H_{ii}^i|k_i, \frac{1}{\sqrt{3}}\epsilon|H_{ii}^i|k_i k_H D_j^2)$, $K_0=1$, $B_0=C_0=0$, γ_0 is of order $\mathcal{O}(\epsilon)$, $K_1=\frac{1}{\sqrt{3}}\epsilon k_i k_H D_j^2$, $B_1=k_i$, $C_1=0$, γ_1 is of order $\mathcal{O}(1)$, $K_2=B_2=0$, and $i\neq j$. The constant $k_H>0$ is defined in the following just after inequality (16.63). Assumption **(A6)** of [99, Theorem 8.2, p. 205] holds with $M=1$, $\gamma_3=\frac{1}{\sqrt{3}}\epsilon|H_{ii}^i|k_i k_H D_j^2$, $\sigma=|H_{ii}^i|k_i$, as it can be readily verified by means of the variation-of-constants formula,

$$\bar{G}_i^{\mathrm{av}}(t) = \exp(-|H_{ii}^i|k_i t)\bar{G}_i^{\mathrm{av}}(0) + \int_0^1 \exp(-|H_{ii}^i|k_i(t+s))\epsilon H_{ii}^i k_i \phi_i^{\mathrm{av}}(1,s)ds,$$

and from the application of the Cauchy–Schwarz inequality to the term $\phi^{\mathrm{av}}(1,t)$ in (16.50),

$$\begin{aligned}\phi_i^{\mathrm{av}}(1,t) &\le |H_{ij}^i| D_j^2 \left(\int_0^1 (1-\tau)^2 d\tau\right)^{\frac{1}{2}} \times \left(\int_0^1 [u_j^{\mathrm{av}}(\xi,t)]^2 d\xi\right)^{\frac{1}{2}} \\ &\le \frac{1}{\sqrt{3}} k_H D_j^2 \left(\int_0^1 [u_j^{\mathrm{av}}(\xi,t)]^2 d\xi\right)^{\frac{1}{2}},\end{aligned} \tag{16.63}$$

since $|H_{ij}^i|<k_H<\frac{1}{\epsilon}|H_{ii}^i|$ according to Assumption 16.1, where k_H is a positive constant of order $\mathcal{O}(1)$. It follows that the small-gain condition [99, inequality (8.3.24)]

$$\begin{aligned}&\max(\gamma_0 K_0, \gamma_1 K_1) + \sigma^{-1} K_2 < 1, \\ &\gamma_3 \max(\gamma_0 B_0, \gamma_1 B_1) + \gamma_3 \sigma^{-1} B_2 < 1\end{aligned} \tag{16.64}$$

holds provided $0<\epsilon<1$ is sufficiently small. Therefore, if such a small-gain condition holds, then [99, Theorem 8.2, p. 205] allows us to conclude that there exist constants $\delta,\Delta>0$ such that for every $u_0^{\rm av} \in C^0([0,1])$, $\bar{G}_0^{\rm av} \in \mathbb{R}^n$, the unique generalized solution of this initial-boundary value problem, with $u^{\rm av}(x,0) = u_0^{\rm av}$ and $\bar{G}^{\rm av}(0) = \bar{G}_0^{\rm av}$, satisfies the following estimate:

$$|\bar{G}^{\rm av}(t)| + \|u^{\rm av}(t)\|_\infty \leq \Delta(|\bar{G}_0^{\rm av}| + \|u_0^{\rm av}\|_\infty)\exp(-\delta t). \tag{16.65}$$

Therefore, we conclude that the origin of the average closed-loop system (16.51)–(16.54) is exponentially stable under the assumption of $0<\epsilon<1$ being sufficiently small. Then, from (16.36) and (16.45), we conclude the same results in the norm

$$\left(\sum_{i=1}^{2} [\vartheta_i^{\rm av}(t)]^2 + \int_0^{D_i} [u_i^{\rm av}(x,t)]^2\, dx\right)^{1/2} \tag{16.66}$$

since H is nonsingular, i.e., $|\vartheta_i^{\rm av}(t)| \leq |H^{-1}||\hat{G}^{\rm av}(t)|$. As developed in [65] (Theorem 10.1, Chapter 10), the next steps of the proof would be the application of the local averaging theory for infinite-dimensional systems in [83, Section 2] (Appendix B), showing that the periodic solutions satisfy (16.56) for ω sufficiently large, and then the conclusion of the attractiveness of the Nash equilibrium Θ^* according to (16.57). The final residual set for the error $\theta(t)-\theta^*$ in (16.58) depends on $|a|e^{\max(D_i)\sqrt{\frac{\omega}{2}}}$ due to the amplitude of the additive dithers $S_i(t)$ in (16.21). □

16.3 ▪ Simulations with a Duopoly Game under Heat PDEs

For an example of a noncooperative game with 2 players that employ the proposed ES strategy for PDE compensation, we consider the following payoff functions (16.7) and (16.8) subject to heat PDEs (16.1)–(16.4) with distinct diffusion coefficients $D_1 = 1$ and $D_2 = 3$ in the players' decisions, $i \in \{1,2\}$:

$$J_1(\Theta(t)) = -5\,\Theta_1^2(t) + 5\,\epsilon\Theta_1(t)\Theta_2(t) + 250\,\Theta_1(t) - 150\,\Theta_2(t) - 3000, \tag{16.67}$$

$$J_2(\Theta(t)) = -5\,\Theta_2^2(t) + 5\,\epsilon\Theta_1(t)\Theta_2(t) - 150\,\Theta_1(t) + 150\,\Theta_2(t) + 2500, \tag{16.68}$$

which, according to (16.13), yield the unique Nash equilibrium

$$\Theta_1^* = \theta_1^* = \frac{100+30\epsilon}{4-\epsilon^2}, \quad \Theta_2^* = \theta_2^* = \frac{60+50\epsilon}{4-\epsilon^2}. \tag{16.69}$$

In order to attain the Nash equilibrium (16.69) without any knowledge of the payoffs parameters, the players implement the non-model-based real-time optimization strategy of Section 16.2 acting through heat PDE dynamics (see Figure 16.2).

For comparison purposes, except for the heat PDEs in the player's input signals, the plant and the controller parameters were chosen similarly to [70] in the simulation tests: $\epsilon=1$, $a_1=0.075$, $a_2=0.05$, $k_1=2$, $k_2=5$, $\omega_1=26.75$ rad/s, $\omega_2=22$ rad/s and $\theta_1(0)=\hat{\theta}_1(0)=50$, $\theta_2(0)=\hat{\theta}_2(0)=\theta_2^*=110/3$. In addition, the time constants of the predictor filters were set to $c_1=c_2=100$.

We can check that the ES approach proposed in [70] is effective when heat PDEs are not present in the decision variables. The players find the Nash equilibrium within 300 seconds (curves not shown). However, in the presence of heat PDEs in the input signals θ_1 and θ_2, but without considering any kind of PDE compensation, Figures 16.3(a) and 16.3(b) show that the game collapses with the explosion of its variables. On the other hand, Figures 16.4(a) and 16.4(b) show that the proposed boundary control–based scheme fixes this with a remarkable evolution in searching the Nash equilibrium and simultaneously compensates the effect of the heat PDEs in our noncooperative game.

The evolution of the infinite-dimensional states $\alpha_1(x,t)$ and $\alpha_2(x,t)$ modeled by the heat PDEs (16.1)–(16.4) is shown in Figures 16.5(a) to 16.5(d). The values of the boundary inputs

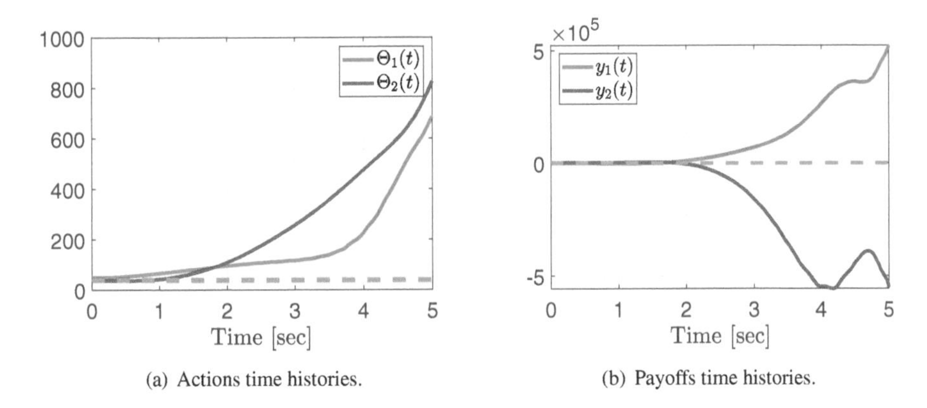

(a) Actions time histories. (b) Payoffs time histories.

Figure 16.3. *Uncompensated heat PDEs:* (a) *actions time histories and* (b) *payoffs time histories for P_1 and P_2 go unstable when implementing the Nash equilibrium seeking scheme of* [70] *for* $\epsilon = 1$.

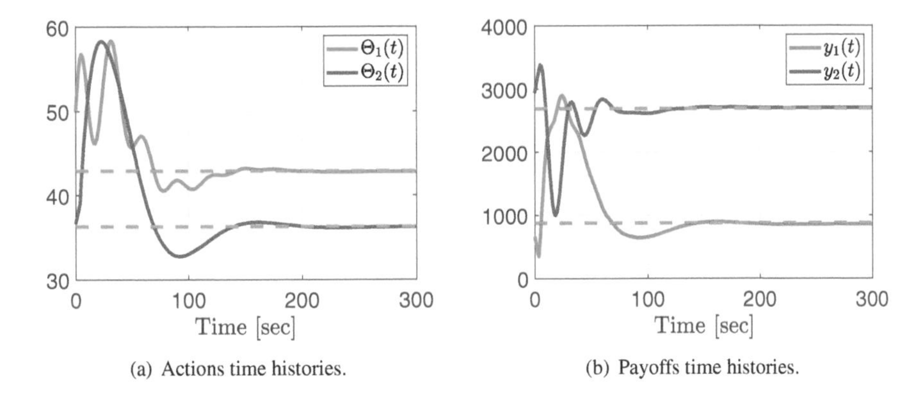

(a) Actions time histories. (b) Payoffs time histories.

Figure 16.4. *Compensated heat PDEs:* (a) *actions time histories and* (b) *payoffs time histories for P_1 and P_2 when implementing the proposed Nash equilibrium seeker for $\epsilon = 1$. The dashed lines denote the values at the Nash equilibrium, $\Theta_1^* = 43.33$ and $\Theta_2^* = 36.67$ (with $J_1(\Theta^*) = 889$ and $J_2(\Theta^*) = 2722$).*

$\theta_1(t)$ and $\theta_2(t)$ as well as the boundary outputs $\Theta_1(t)$ and $\Theta_2(t)$ are highlighted with colors black and red, respectively. The initial condition is highlighted with color blue.

This first set of simulations indicates that even under an adversarial scenario of strong coupling between the players with $\epsilon = 1$, the proposed approach has behaved successfully. This suggests that our stability analysis may be conservative and the theoretical assumption $0 < \epsilon < 1$ may be relaxed given the performance of the control system. In Figures 16.6(a) and 16.6(b), different values of $\epsilon = 0.5$ and $\epsilon = 0.1$ are considered to evaluate the robustness of the proposed scheme under different levels of coupling between the two players and the corresponding impact on the transient responses.

16.4 ▪ Notes and References

This chapter generalized the ES results obtained in [184] (Chapter 15) for pure delays, as well as the classical results for Ordinary Differential Equations (ODEs) [21] (revisited in Section 1.6, Chapter 1), to a wider class of infinite-dimensional systems governed by Partial Differential Equations (PDEs). Such a generalization is challenging since in the literature for multi-input

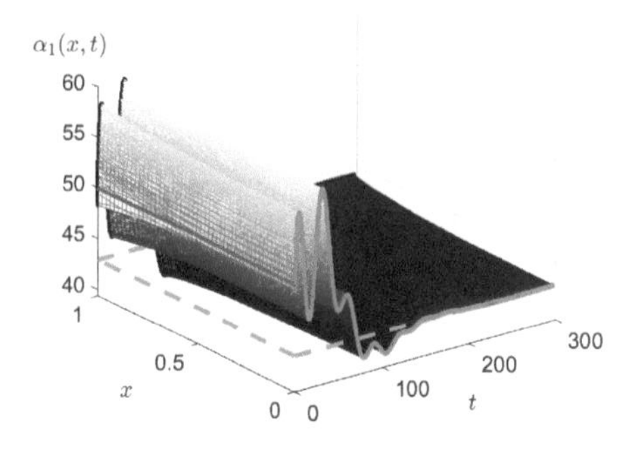

(a) Parameter $\Theta_1(t)$ (red) converging to a $\mathcal{O}(|a| + 1/\omega)$-neighborhood of Θ_1^* (dashed-green) according to (16.58).

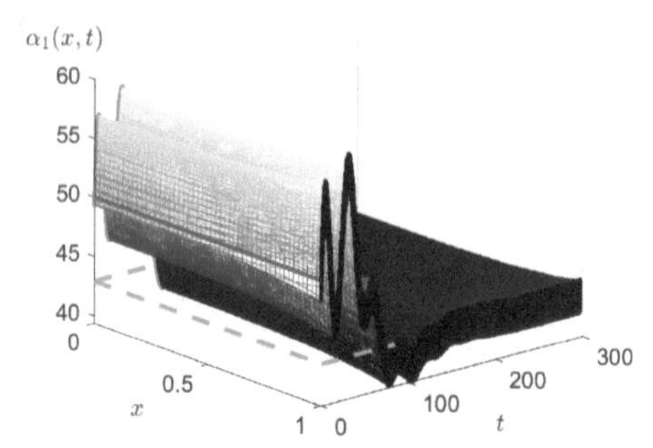

(b) Parameter $\theta_1(t)$ (black) converging to a $\mathcal{O}(|a|e^{\max(D_i)\sqrt{\omega/2}} + 1/\omega)$-neighborhood of θ_1^* (dashed-green) according to (16.57).

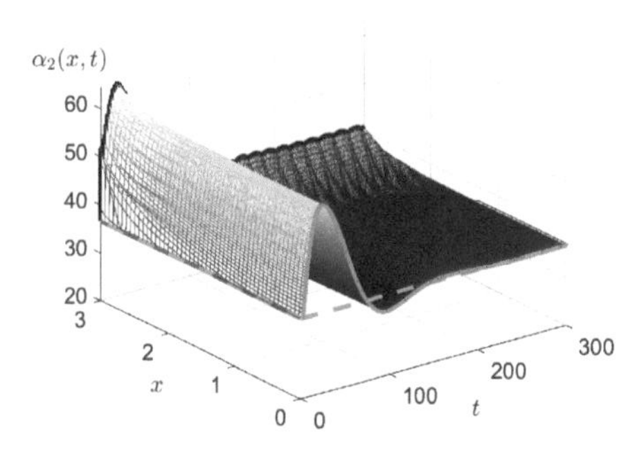

(c) Parameter $\Theta_2(t)$ (red) converging to a $\mathcal{O}(|a| + 1/\omega)$-neighborhood of Θ_2^* (dashed-green) according to (16.58).

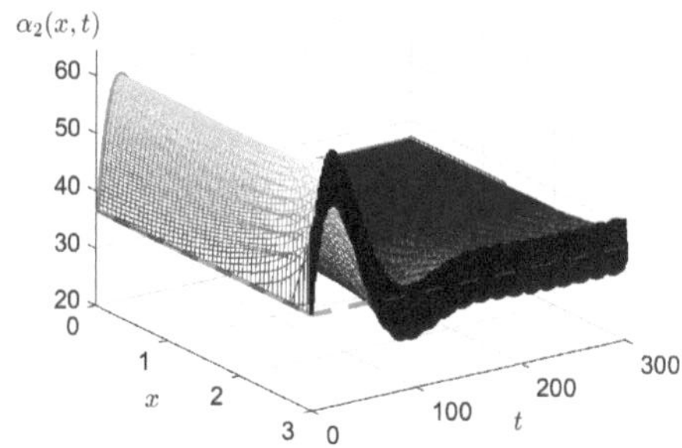

(d) Parameter $\theta_2(t)$ (black) converging to a $\mathcal{O}(|a|e^{\max(D_i)\sqrt{\omega/2}} + 1/\omega)$-neighborhood of θ_2^* (dashed-green) according to (16.57).

Figure 16.5. *Evolution of the infinite-dimensional states $\alpha_1(x,t)$ and $\alpha_2(x,t)$ of the heat PDEs in a duopoly game with boundary Dirichlet actuation: from $\alpha_1(D_1,t) = \theta_1(t)$ to $\alpha_1(0,t) = \Theta_1(t)$, with $D_1 = 1$ for Player P_1, and from $\alpha_2(D_2,t) = \theta_2(t)$ to $\alpha_2(0,t) = \Theta_2(t)$, with $D_2 = 3$ for Player P_2.*

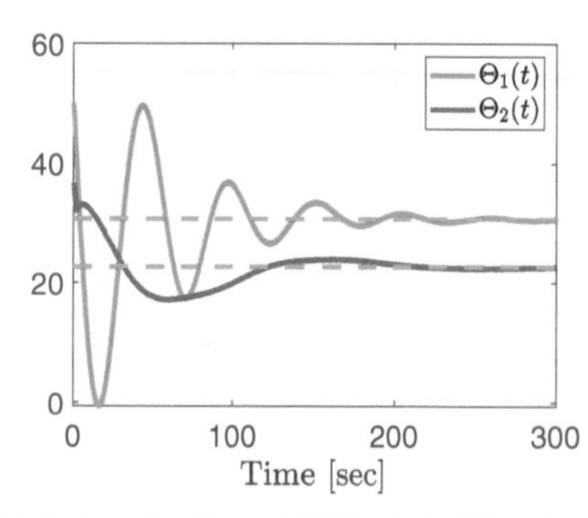

(a) Actions plots for $\epsilon = 0.5$. The dashed lines denote $\Theta_1^* = 30.67$ and $\Theta_2^* = 22.67$.

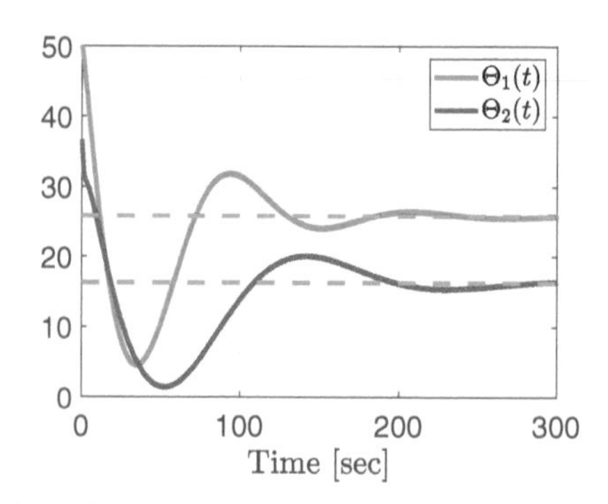

(b) Actions plots for $\epsilon = 0.1$. The dashed lines denote $\Theta_1^* = 25.81$ and $\Theta_2^* = 16.29$.

Figure 16.6. *Compensated heat PDEs: actions time histories for P_1 and P_2 when implementing the proposed Nash equilibrium seeker for distinct values of the coupling coefficient ($\epsilon = 0.5$ and $\epsilon = 0.1$).*

PDE compensation [128], only a centralized approach for the PDE boundary control design is considered (with all the vectors multiplying the control inputs needing to be known), while games are decentralized. We presented a noncooperative game ES design for two players acting through heat PDE dynamics, which may represent behaviors as in the Black–Scholes model of financial markets [197, 132]. In this scenario, we were able to develop a result for games with distinct diffusion coefficients for the heat PDEs in the action paths of each player, where the players estimate only the diagonal entries of the Hessian matrix due to the players' payoffs in the corresponding real-time Nash seeking problem. Hence, we were also able to dominate sufficiently small off-diagonal terms using a small-gain argument [99] for the average system.

Our analysis presented a properly designed sequence of steps of averaging in infinite dimension [83] and Small-Gain Theorem for Input-to-State Stable (ISS) cascades of dynamical ODE-PDE systems [99], resulting in a guarantee of closed-loop convergence, in the presence of nonlinearities, periodic perturbations, and diffusion (heat) PDEs, to a small neighborhood of the Nash equilibrium.

The introduction to boundary control for noncooperative games was presented in this chapter for parabolic PDEs. There is no strong pedagogical reason why the exposition provided could not have been conducted on some of the other classes of PDEs. However, parabolic PDEs are particularly convenient because they are at the same time sufficiently simple and sufficiently general to serve as a design template using which the reader can extend the Nash equilibrium seeking design to other classes of PDEs [128, 170].

Chapter 17

Heterogeneous Duopoly Games with Delays and Heat PDEs

17.1 ▪ Competing Players Acting through PDEs from Distinct Classes

When players are competing in a noncooperative game, there is no reason to assume that players are pursuing even remotely related objectives. One could be maximizing a profit while another maximizes some social good.

Likewise, there is no reason to assume that different players are subject to the same types of physics. One might be propagating an input through some social opinion dynamics while another may be propagating it through some epidemiological dynamics.

This means that, while in Chapters 15 and 16 we provided useful designs and guarantees of Nash equilibrium attainment in the presence of delays or heat dynamics, i.e., in the presence of hyperbolic and parabolic PDE dynamics, it is of interest to see whether, when different players, unaware of the competing optimization pursuits in different physical domains, are employing ES algorithms with compensation of their own specific PDEs, will be able to achieve the Nash-optimality under the interference of other players who are compensating their own PDE dynamics, which are from different classes.

One can formulate a problem of a noncooperative game with dynamics from different PDE classes (hyperbolic, parabolic, Korteweg–de–Vries, Schroedinger, Kuramoto–Sivashinsky, etc.) in great generality. We are not willing to invest the level of effort demanded for achieving such generality. Instead, we demonstrate the achievement of convergence to a Nash equilibrium in a game with just two players (a duopoly) where one player is compensating a transport PDE and the other a heat PDE. This is the simplest example of a heterogeneous PDE system. While not coupled directly, the coupling of the players in the payoff functions, and the coupling that results from the ES algorithms acting at the boundary conditions of the PDEs, results in a coupled heterogeneous pair of PDEs in the closed-loop system, which happens also to be time-varying and nonlinear.

In this chapter, we seek Nash equilibria in a quadratic noncooperative (duopoly) game with player actions subject to heterogeneous PDE dynamics—one player compensating for a delay (transport PDE) and the other for a heat (diffusion) PDE—and each player having access only to his own payoff value. The proposed approach employs extremum seeking, with sinusoidal perturbation signals applied to estimate the Gradient (first derivative) and Hessian (second derivative) of unknown quadratic functions. These perturbation signals, as well as the PDE compensators, have already been designed in Chapters 15 and 16. In this chapter we just analyze the effects of their coupling.

This is the first instance of noncooperative games being tackled in a model-free fashion in the presence of both heat PDE dynamics and delays. We apply a small-gain analysis for the resulting Input-to-State Stable (ISS) coupled hyperbolic-parabolic PDE system as well as averaging theory in infinite dimension, due to the infinite-dimensional state of the heat PDE and the delay, in order to obtain local convergence results to a small neighborhood of the Nash equilibrium. We quantify the size of these residual sets and illustrate the theoretical results numerically on an example combining hyperbolic and parabolic dynamics in a 2-player setting.

17.2 ▪ Duopoly with Quadratic Payoffs and Heterogeneous PDEs: General Formulation

In a duopoly game, the optimality of the respective outcomes (outputs) of Players P_1 and P_2, respectively $y_1(t) \in \mathbb{R}$ and $y_2(t) \in \mathbb{R}$, do not depend exclusively on their own actions/strategies (input signals) $\Theta_1(t) \in \mathbb{R}$ and $\Theta_2(t) \in \mathbb{R}$. Moreover, defining $\Theta(t) := [\Theta_1(t), \Theta_2(t)]^T$, each player's payoff function J_i depends also on the action Θ_j of the other Player j, $j \neq i$. A 2-tuple of actions, $\Theta^* = [\Theta_1^*, \Theta_2^*]^T$, is said to be in Nash equilibrium if no Player i can improve his payoff by unilaterally deviating from Θ_i^*, this being so for all $i \in \{1,2\}$ [72].

As shown in Figures 17.1 and 17.2, distinct (transport and heat) PDEs (with Dirichlet actuation) are assumed in the vector of player actions $\theta(t) \in \mathbb{R}^2$. Thus, the propagated actuator vector $\Theta(t) \in \mathbb{R}^2$ is given by the following transport PDE for Player P_1,

$$\Theta_1(t) = \theta_1(t - D_1) = \alpha_1(0,t), \tag{17.1}$$

$$\partial_t \alpha_1(x,t) = \partial_x \alpha_1(x,t), \quad x \in (0, D_1), \tag{17.2}$$

$$\partial_x \alpha_1(0,t) = 0, \tag{17.3}$$

$$\alpha_1(D_1,t) = \theta_1(t), \tag{17.4}$$

and the next heat PDE for Player P_2,

$$\Theta_2(t) = \alpha_2(0,t), \tag{17.5}$$

$$\partial_t \alpha_2(x,t) = \partial_{xx} \alpha_2(x,t), \quad x \in (0, D_2), \tag{17.6}$$

$$\partial_x \alpha_2(0,t) = 0, \tag{17.7}$$

$$\alpha_2(D_2,t) = \theta_2(t), \tag{17.8}$$

where $\alpha_i : [0, D_i] \times \mathbb{R}_+ \to \mathbb{R}\ \forall i \in \{1,2\}$, and each domain length D_i is known.

The solution of (17.1)–(17.4) is

$$\alpha_1(x,t) = \theta_1(t + x - D_1), \tag{17.9}$$

which represents a delayed action for Player P_1 at $x = 0$.

On the other hand, the solution of (17.5)–(17.8) is given by

$$\alpha_2(x,t) = \mathcal{L}^{-1}\left[\frac{\cosh(x\sqrt{s})}{\cosh(D_2\sqrt{s})}\right] * \theta_2(t), \tag{17.10}$$

where $\mathcal{L}^{-1}(\cdot)$ denotes the inverse Laplace transformation and $*$ is the convolution operator. Given these relations, we define the *heterogeneous transport-diffusion operator* $\mathcal{D} = \operatorname{diag}\{\mathcal{D}_1, \mathcal{D}_2\}$ for the PDEs (17.1)–(17.4) and (17.5)–(17.8) with boundary inputs and measurements given by

$$\begin{aligned} \mathcal{D}_1[\varphi(t)] &= \varphi(t + x - D_1), \quad \text{s.t.} \quad \Theta_1(t) = \mathcal{D}_1[\theta_1(t)], \\ \mathcal{D}_2[\varphi(t)] &= \mathcal{L}^{-1}\left[\frac{1}{\cosh(D_2\sqrt{s})}\right] * \varphi(t), \quad \text{s.t.} \quad \Theta_2(t) = \mathcal{D}_2[\theta_2(t)]. \end{aligned} \tag{17.11}$$

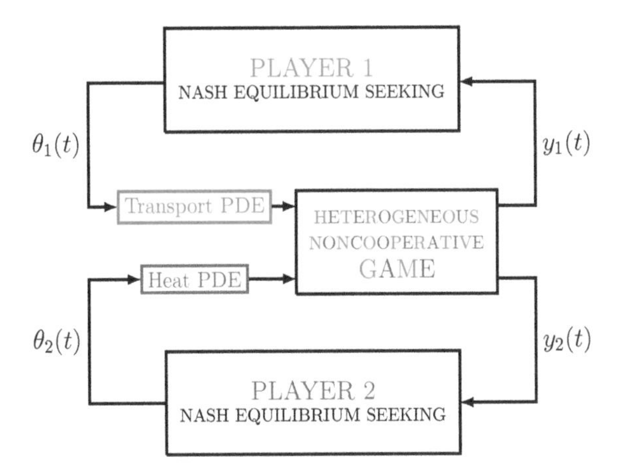

Figure 17.1. *Nash equilibrium seeking in a heterogeneous noncooperative game with players acting through transport-heat PDE dynamics.*

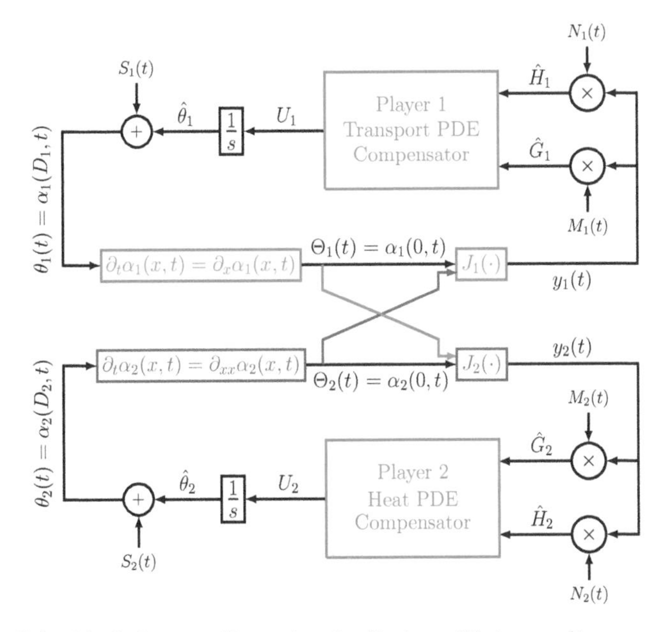

Figure 17.2. *Block diagram illustrating the Nash equilibrium seeking strategy performed for each player. In magenta color are the boundary controllers used to compensate the individual transport-heat PDEs for the heterogeneous noncooperative game.*

We consider games where the payoff function $y_i(t) = J_i(\Theta(t))$, $\forall i \in \{1,2\}$, of each player is quadratic [21], expressed as a strictly concave combination of their actions propagated through distinct transport-heat PDE dynamics

$$J_1(\Theta(t)) = \frac{H_{11}^1}{2}\Theta_1^2(t) + \frac{H_{22}^1}{2}\Theta_2^2(t) + \epsilon H_{12}^1\Theta_1(t)\Theta_2(t) + h_1^1\Theta_1(t) + h_2^1\Theta_2(t) + c_1, \tag{17.12}$$

$$J_2(\Theta(t)) = \frac{H_{11}^2}{2}\Theta_1^2(t) + \frac{H_{22}^2}{2}\Theta_2^2(t) + \epsilon H_{21}^2\Theta_1(t)\Theta_2(t) + h_1^2\Theta_1(t) + h_2^2\Theta_2(t) + c_2, \tag{17.13}$$

where $J_1(\Theta), J_2(\Theta): \mathbb{R}^2 \to \mathbb{R}$, H^i_{jk}, h^i_j, $c_i \in \mathbb{R}$ are constants, $H^i_{ii} < 0$ $\forall i,j,k \in \{1,2\}$, and $\epsilon > 0$ without loss of generality.

For the sake of completeness, we provide here in mathematical terms the definition of a Nash equilibrium $\Theta^* = [\Theta_1^*, \Theta_2^*]^T$ in a 2-player game:

$$J_1(\Theta_1^*, \Theta_2^*) \geq J_1(\Theta_1, \Theta_2^*) \text{ and } J_2(\Theta_1^*, \Theta_2^*) \geq J_2(\Theta_1^*, \Theta_2). \tag{17.14}$$

Hence, no player has an incentive to unilaterally deviate its action from Θ^*.

As done in Chapter 16, to determine the Nash equilibrium solution in strictly concave quadratic games with 2 players, where each action set is the entire real line, one should differentiate J_1 and J_2, respectively, with respect to $\Theta_1(t)$ and $\Theta_2(t)$, setting the resulting expressions equal to zero and solving the set of equations thus obtained. This set of equations, which also provides a sufficient condition due to the strict concavity, is

$$\begin{cases} H^1_{11}\Theta_1^* + \epsilon H^1_{12}\Theta_2^* + h^1_1 = 0, \\ \epsilon H^2_{21}\Theta_1^* + H^2_{22}\Theta_2^* + h^2_2 = 0, \end{cases} \tag{17.15}$$

which can be written in the matrix form as

$$\begin{bmatrix} H^1_{11} & \epsilon H^1_{12} \\ \epsilon H^2_{21} & H^2_{22} \end{bmatrix} \begin{bmatrix} \Theta_1^* \\ \Theta_2^* \end{bmatrix} = - \begin{bmatrix} h^1_1 \\ h^2_2 \end{bmatrix}. \tag{17.16}$$

Defining the Hessian matrix H and vectors Θ^* and h by

$$H := \begin{bmatrix} H^1_{11} & \epsilon H^1_{12} \\ \epsilon H^2_{21} & H^2_{22} \end{bmatrix}, \quad \Theta^* := \begin{bmatrix} \Theta_1^* \\ \Theta_2^* \end{bmatrix}, \quad h := \begin{bmatrix} h^1_1 \\ h^2_2 \end{bmatrix}, \tag{17.17}$$

there exists a unique Nash equilibrium at $\Theta^* = -H^{-1}h$ if H is invertible:

$$\begin{bmatrix} \Theta_1^* \\ \Theta_2^* \end{bmatrix} = - \begin{bmatrix} H^1_{11} & \epsilon H^1_{12} \\ \epsilon H^2_{21} & H^2_{22} \end{bmatrix}^{-1} \begin{bmatrix} h^1_1 \\ h^2_2 \end{bmatrix}. \tag{17.18}$$

For more details, see [21, Chapter 4].

The *control objective* is to design a novel extremum seeking-based strategy to reach the Nash equilibrium in heterogeneous noncooperative games subjected to transport-heat PDEs in the decision variables of the players (input signals).

Since our goal is to find the unknown optimal inputs Θ^* (and θ^*), we define the estimation errors

$$\tilde{\theta}(t) = \hat{\theta}(t) - \theta^*, \qquad \vartheta(t) = \hat{\Theta}(t) - \Theta^*, \tag{17.19}$$

where the vectors $\hat{\theta}(t)$ and $\hat{\Theta}(t)$ are the estimates of θ^* and Θ^*. In order to make (17.19) coherent with the optimizer of the static map Θ^*, we apply the heterogeneous transport-diffusion operator (17.11) to $\tilde{\theta}_i$ in (17.19) getting

$$\vartheta_1(t) = \tilde{\theta}_1(t - D_1) := \bar{\alpha}_1(0,t), \tag{17.20}$$

$$\partial_t \bar{\alpha}_1(x,t) = \partial_x \bar{\alpha}_1(x,t), \quad x \in (0, D_1), \tag{17.21}$$

$$\partial_x \bar{\alpha}_1(0,t) = 0, \tag{17.22}$$

$$\bar{\alpha}_1(D_1,t) = \tilde{\theta}_1(t) \tag{17.23}$$

and

$$\vartheta_2(t) := \bar{\alpha}_2(0,t), \tag{17.24}$$
$$\partial_t \bar{\alpha}_2(x,t) = \partial_{xx}\bar{\alpha}_2(x,t), \quad x \in (0,D_2), \tag{17.25}$$
$$\partial_x \bar{\alpha}_2(0,t) = 0, \tag{17.26}$$
$$\bar{\alpha}_2(D_2,t) = \tilde{\theta}_2(t), \tag{17.27}$$

where $\bar{\alpha}_i : [0,D_i] \times \mathbb{R}_+ \to \mathbb{R}$, $\forall i \in \{1,2\}$, and $\vartheta(t) := \mathcal{D}[\tilde{\theta}(t)] = \hat{\Theta}(t) - \Theta^*$ is the propagated estimation error $\tilde{\theta}(t)$ through the transport-diffusion domain. For $\lim_{t\to\infty}\theta(t) = \theta_c$, we have $\lim_{t\to\infty}\Theta(t) = \Theta_c = \theta_c$, where the index c indicates a constant signal. Indeed, from (17.11), for a constant input $\theta = \theta_c$, one has $\lim_{t\to\infty}\mathcal{D}_1[\theta_{c1}] = \theta_{c1}$ for Player P_1. For Player P_2, on has $\mathcal{L}\{\theta_{c2}\} = \theta_{c2}/s$, and applying the Laplace limit theorem, we get $\lim_{t\to\infty}\mathcal{D}_2[\theta_{c2}] = \lim_{s\to 0}\left\{\frac{\theta_{c2}}{\cosh(\sqrt{s}D_2)}\right\} = \theta_{c2}$. Thus, in the particular case $\theta = \theta_c = \theta^*$, one has

$$\Theta^* = \theta^*. \tag{17.28}$$

Figure 17.2 depicts a schematic diagram that summarizes the proposed Nash equilibrium policy for each player, where their outputs are given by

$$\begin{cases} y_1(t) = J_1(\Theta(t)), \\ y_2(t) = J_2(\Theta(t)). \end{cases} \tag{17.29}$$

The additive dither signals in the presence of transport-heat PDE dynamics [179, 65] are defined according to

$$\begin{cases} S_1(t) = a_1 \sin(\omega_1(t+D_1)), \\ S_2(t) = \frac{1}{2}a_2 e^{\sqrt{\frac{\omega_2}{2}}D_2}\sin\left(\omega_2 t + \sqrt{\frac{\omega_2}{2}}D_2\right) + \frac{1}{2}a_2 e^{-\sqrt{\frac{\omega_2}{2}}D_2}\sin\left(\omega_2 t - \sqrt{\frac{\omega_2}{2}}D_2\right), \end{cases} \tag{17.30}$$

and the multiplicative demodulation signals are given by

$$\begin{cases} M_1(t) = \frac{2}{a_1}\sin(\omega_1 t), \\ M_2(t) = \frac{2}{a_2}\sin(\omega_2 t), \end{cases} \tag{17.31}$$

with nonzero constant amplitudes a_1, $a_2 > 0$ at frequencies $\omega_1 \neq \omega_2$. Such probing frequencies ω_i can be selected as

$$\omega_i = \omega_i'\omega = \mathcal{O}(\omega), \quad i = 1 \text{ or } 2, \tag{17.32}$$

where ω is a positive constant and ω_i' is a rational number—one possible choice is given in [73].

Following the non-sharing information paradigm, only the diagonal elements of H can be properly recovered in the average sense by Player P_1 or P_2. In this sense, the signals $N_1(t)$ and $N_2(t)$ are simply defined by [73]

$$\begin{cases} N_1(t) = \frac{16}{a_1^2}\left(\sin^2(\omega_1 t) - \frac{1}{2}\right), \\ N_2(t) = \frac{16}{a_2^2}\left(\sin^2(\omega_2 t) - \frac{1}{2}\right). \end{cases} \tag{17.33}$$

Then, the average version of

$$\begin{cases} \hat{H}_1(t) = N_1(t)y_1(t), \\ \hat{H}_2(t) = N_2(t)y_2(t) \end{cases} \tag{17.34}$$

is given by

$$\begin{cases} \hat{H}_1^{\mathrm{av}}(t) = [N_1(t)y_1(t)]_{\mathrm{av}} = H_{11}^1, \\ \hat{H}_2^{\mathrm{av}}(t) = [N_2(t)y_2(t)]_{\mathrm{av}} = H_{22}^2. \end{cases} \tag{17.35}$$

Considering $\hat{\theta}_1(t)$ and $\hat{\theta}_2(t)$ as the estimates of θ_1^* and θ_2^*, one can define from (17.19) the individual estimation errors:

$$\begin{cases} \tilde{\theta}_1(t) = \hat{\theta}_1(t) - \theta_1^*, \quad \vartheta_1(t) = \hat{\Theta}_1(t) - \Theta_1^*, \\ \tilde{\theta}_2(t) = \hat{\theta}_2(t) - \theta_2^*, \quad \vartheta_2(t) = \hat{\Theta}_2(t) - \Theta_2^*. \end{cases} \tag{17.36}$$

The estimate of the unknown gradients of the payoff functions are given by

$$\begin{cases} \hat{G}_1(t) \quad = M_1(t)y_1(t), \\ \hat{G}_2(t) \quad = M_2(t)y_2(t), \end{cases} \tag{17.37}$$

and computing the average of the resulting signal leads us to

$$\begin{cases} \hat{G}_1^{\mathrm{av}}(t) = H_{11}^1\vartheta_1^{\mathrm{av}}(t) + \epsilon H_{12}^1\vartheta_2^{\mathrm{av}}(t), \\ \hat{G}_2^{\mathrm{av}}(t) = \epsilon H_{21}^2\vartheta_1^{\mathrm{av}}(t) + H_{22}^2\vartheta_2^{\mathrm{av}}(t). \end{cases} \tag{17.38}$$

Additionally, from the block diagram in Figure 17.2, one has

$$\dot{\hat{\theta}}_i(t) = U_i(t), \quad \dot{\tilde{\theta}}_i(t) = U_i(t) \quad \forall i \in \{1,2\}, \tag{17.39}$$

since $\dot{\tilde{\theta}}(t) = \dot{\hat{\theta}}(t)$, once θ^* is constant. Taking the time derivative of (17.20)–(17.23) and (17.24)–(17.27), with the help of (17.19) and (17.39), the *propagated error dynamics* is written as

$$\dot{\vartheta}_1(t) = U(t-D_1) = u_1(0,t), \tag{17.40}$$
$$\partial_t u_1(x,t) = \partial_x u_1(x,t), \quad x \in (0,D_1), \tag{17.41}$$
$$\partial_x u_1(0,t) = 0, \tag{17.42}$$
$$u_1(D_1,t) = U_1(t) \tag{17.43}$$

and

$$\dot{\vartheta}_2(t) = u_2(0,t), \tag{17.44}$$
$$\partial_t u_2(x,t) = \partial_{xx} u_2(x,t), \quad x \in (0,D_2), \tag{17.45}$$
$$\partial_x u_2(0,t) = 0, \tag{17.46}$$
$$u_2(D_2,t) = U_2(t), \tag{17.47}$$

where $u_i : [0,D_i] \times \mathbb{R}_+ \to \mathbb{R}$, $u_i(x,t) := \partial_t \bar{\alpha}_i(x,t)$, $\forall i \in \{1,2\}$, and $\bar{\alpha}_i(x,t) = \alpha_i(x,t) - \beta_i(x,t) - \Theta_i^*$. The term $\beta_i(x,t)$ is the PDE state of the *trajectory generation problem*

[128, Chapter 12] solved to obtain $S_1(t) = \beta_1(D_1,t)$ and $S_2(t) = \beta_2(D_2,t)$ in (17.30)—for more details, see [170, equations (19)–(22)].

Hence, from (17.38), (17.40)–(17.43), and (17.44)–(17.47), it is possible to find a compact form for the overall average estimated gradient according to

$$\hat{G}^{\mathrm{av}}(t) = H\vartheta^{\mathrm{av}}(t), \tag{17.48}$$

$$\dot{\hat{G}}^{\mathrm{av}}(t) = H\dot{\vartheta}^{\mathrm{av}}(t) = H\mathcal{D}[U^{\mathrm{av}}(t)], \tag{17.49}$$

where the Hessian H is given in (17.17), $\vartheta^{\mathrm{av}}(t) := [\vartheta_1^{\mathrm{av}}(t), \vartheta_2^{\mathrm{av}}(t)]^T \in \mathbb{R}^2$, $\hat{G}^{\mathrm{av}}(t) := [\hat{G}_1^{\mathrm{av}}(t), \hat{G}_2^{\mathrm{av}}(t)]^T \in \mathbb{R}^2$, and $U^{\mathrm{av}}(t) := [U_1^{\mathrm{av}}(t), U_2^{\mathrm{av}}(t)]^T \in \mathbb{R}^2$ are the average versions of $U(t) := [U_1(t), U_2(t)]^T$, $\vartheta(t) := [\vartheta_1(t), \vartheta_2(t)]^T$, and $\hat{G}(t) := [\hat{G}_1(t), \hat{G}_2(t)]^T$, respectively.

Throughout the chapter the key idea is to design control laws (policies) for each player in order to achieve a small neighborhood of the Nash equilibrium point. To this end, we use an extremum seeking strategy based on boundary control to compensate the heterogeneous transport-diffusion operator $\mathcal{D}[\cdot]$ in (17.49) due to the distinct transport-heat PDEs in the players' actions. Basically, the control laws must be able to ensure exponential stabilization of $\dot{\hat{G}}^{\mathrm{av}}(t)$ and, consequently, of $\vartheta^{\mathrm{av}}(t) = \hat{\Theta}^{\mathrm{av}}(t) - \Theta^*$. From (17.48), it is clear that, if H is invertible, $\vartheta^{\mathrm{av}}(t) \to 0$ as $\hat{G}^{\mathrm{av}}(t) \to 0$. Hence, the convergence of $\vartheta^{\mathrm{av}}(t)$ to the origin results in the convergence of $\Theta(t)$ to a small neighborhood of Θ^* in (17.14) via averaging theory [83].

17.3 ▪ Noncooperative Duopoly with Heterogeneous Transport-Heat PDE Dynamics

As mentioned earlier, in our heterogeneous noncooperative game subject to transport-heat PDE dynamics, the purpose of the ES is still to estimate the Nash equilibrium $\Theta^* = \theta^*$ (see 17.28)), but without sharing any information between the players. Each player only needs to measure the value of its own payoff function (17.29).

17.3.1 ▪ Decentralized PDE Boundary Control Using Only the Known Diagonal Terms of the Hessian

In this sense, we are able to formulate the closed-loop system in a *decentralized* fashion, where no knowledge about the payoff or action of the other player is required.

Inspired by [179, 65], where ES was considered for PDE compensation but not in the context of games, we propose the following boundary-based update laws $\dot{\hat{\theta}}_i(t) = U_i(t)$, $i \in \{1,2\}$:

$$\begin{cases} \dot{U}_1(t) &= -c_1U_1(t) + c_1k_1\left(\hat{G}_1(t) + \hat{H}_1(t)\int_0^{D_1} u_1(\tau,t)d\tau\right), \\ \dot{U}_2(t) &= -c_2U_2(t) + c_2k_2\left(\hat{G}_2(t) + \hat{H}_2(t)\int_0^{D_2} (D_2-\tau)u_2(\tau,t)d\tau\right) \end{cases} \tag{17.50}$$

for positive constants $k_1 > 0$, $k_2 > 0$, $c_1 > 0$, and $c_2 > 0$, in order to compensate the transport-heat PDEs in (17.40)–(17.43) and (17.44)–(17.47). As in the previous two chapters and with some abuse of notation, constants c_1 and c_2 were chosen to denote the parameters of the control laws, but they have no relation to those which appear in the payoffs given by (17.12) and (17.13).

The boundary control law (17.50) could be rewritten as

$$\dot{U}_1(t) = -c_1U_1(t) + c_1k_1\left(\hat{G}_1(t) + \hat{H}_1(t)\int_{t-D_1}^{t} U_1(\tau,t)d\tau\right),$$

$$\dot{U}_2(t) = -c_2U_2(t) + c_2k_2\left[\hat{G}_2(t) + \hat{H}_2(t)\left(\hat{\theta}_2(t) - \Theta_2(t) + a_2\sin(\omega_2 t)\right)\right], \tag{17.51}$$

using the relation $u_1(x,t) = U_1(t+x-D_1)$ for the transport PDE and the diffusion equation $\partial_t \alpha_2(x,t) = \partial_{xx}\alpha_2(x,t)$ as well as the integration by parts associated with (17.5)–(17.8), (17.19), and recalling that $\vartheta_2 + a_2\sin(\omega_2 t) = \Theta_2(t) - \Theta_2^*$, analogously to [65, equation (25)].

17.3.2 ▪ ISS-Like Properties for PDE Representation

For the sake of simplicity, we assume $c_1, c_2 \to +\infty$ in (17.50), resulting in the following general expression:

$$U_1(t) = k_1\left(\hat{G}_1(t) + \hat{H}_1(t)\int_0^{D_1} u_1(\tau,t)d\tau\right),$$

$$U_2(t) = k_2\left(\hat{G}_2(t) + \hat{H}_2(t)\int_0^{D_2} (D_2-\tau)u_2(\tau,t)d\tau\right). \tag{17.52}$$

Recalling (17.40)–(17.43) and (17.44)–(17.47), the infinite-dimensional closed-loop system (17.49) and (17.52) in its average version can be written in the corresponding PDE representation form, given by

$$\dot{\hat{G}}_1^{\rm av}(t) = H_{11}^1 u_1^{\rm av}(0,t) + \epsilon H_{12}^1 u_2^{\rm av}(0,t), \tag{17.53}$$

$$\partial_t u_1^{\rm av}(x,t) = D_1^{-1}\partial_x u_1^{\rm av}(x,t), \quad x \in (0,1), \tag{17.54}$$

$$\partial_x u_1^{\rm av}(0,t) = 0, \tag{17.55}$$

$$u_1^{\rm av}(1,t) = U_1^{\rm av}(t) \tag{17.56}$$

and

$$\dot{\hat{G}}_2^{\rm av}(t) = \epsilon H_{21}^2 u_1^{\rm av}(0,t) + H_{22}^2 u_2^{\rm av}(0,t), \tag{17.57}$$

$$\partial_t u_2^{\rm av}(x,t) = D_2^{-2}\partial_{xx} u_2^{\rm av}(x,t), \quad x \in (0,1), \tag{17.58}$$

$$\partial_x u_2^{\rm av}(0,t) = 0, \tag{17.59}$$

$$u_2^{\rm av}(1,t) = U_2^{\rm av}(t). \tag{17.60}$$

In the *reduction-like approach* [12] (or finite-spectrum assignment), we use the following transformations:

$$\bar{G}_1^{\rm av}(t) = \hat{G}_1^{\rm av}(t) + \epsilon_{11}^1 H_{11}^1 \int_0^{D_1} u_1^{\rm av}(\tau,t)d\tau + \epsilon_{12}^1 H_{12}^1 \int_0^{D_2} (D_2-\tau)u_2^{\rm av}(\tau,t)d\tau \tag{17.61}$$

and

$$\bar{G}_2^{\rm av}(t) = \hat{G}_2^{\rm av}(t) + \epsilon_{22}^2 H_{22}^2 \int_0^{D_2} (D_2-\tau)u_2^{\rm av}(\tau,t)d\tau + \epsilon_{21}^2 H_{21}^2 \int_0^{D_1} u_1^{\rm av}(\tau,t)d\tau, \tag{17.62}$$

where $\epsilon_{11}^1 = \epsilon_{22}^2 = 1$ and $\epsilon_{12}^1 = \epsilon_{21}^2 = \epsilon$.

With some mathematical manipulations (see Chapters 15 and 16), it is not difficult to see that $\bar{G}^{\rm av}$ satisfies

$$\dot{\bar{G}}^{\rm av}(t) = HU^{\rm av}(t). \tag{17.63}$$

After adding and subtracting the next terms in blue and red into (17.52), it can be rewritten as

$$U_1(t) = k_1\left(\hat{G}_1(t) + \hat{H}_1(t)\int_0^{D_1} u_1(\tau,t)d\tau + \epsilon H_{12}^1\int_0^{D_2}(D_2-\tau)u_2(\tau,t)d\tau\right) - k_1\epsilon H_{12}^1\int_0^{D_2}(D_2-\tau)u_2(\tau,t)d\tau, \tag{17.64}$$

$$U_2(t) = k_2\left(\hat{G}_2(t) + \hat{H}_2(t)\int_0^{D_2}(D_2-\tau)u_2(\tau,t)d\tau + \epsilon H_{21}^2\int_0^{D_1} u_1(\tau,t)d\tau - \epsilon H_{21}^2\int_0^{D_1} u_1(\tau,t)d\tau\right), \tag{17.65}$$

whose average compact form is

$$U^{\mathrm{av}}(t) = K\bar{G}^{\mathrm{av}}(t) + \epsilon K\phi^{\mathrm{av}}(\mathcal{D},t), \tag{17.66}$$

where the matrix $K := \mathrm{diag}\{k_1, k_2\}$ with entries $k_1 > 0$, $k_2 > 0$ and the auxiliary variable $\phi(\mathcal{D},t)$ is defined as

$$\phi(\mathcal{D},t) := -\begin{bmatrix} H_{12}^1\int_0^{D_2}(D_2-\tau)u_2(\tau,t)d\tau \\ H_{21}^2\int_0^{D_1}u_1(\tau,t)d\tau \end{bmatrix},$$

$$\phi(1,t) := -\begin{bmatrix} H_{12}^1\int_0^1 D_2^2(1-\xi)u_2(\xi,t)d\xi \\ H_{21}^2\int_0^1 D_1u_1(\xi,t)d\xi \end{bmatrix}, \tag{17.67}$$

since $\int_0^{D_j}(D_j-\tau)u_j(\tau,t)d\tau = \int_0^1 D_j^2(1-\xi)u_j(\xi,t)d\xi$ for $j \in \{1,2\}$. Then, it is possible to find a compact form for the overall average game from (17.63) and (17.66), such as

$$\dot{\bar{G}}^{\mathrm{av}}(t) = HK\bar{G}^{\mathrm{av}}(t) + \epsilon HK\phi^{\mathrm{av}}(1,t), \tag{17.68}$$

$$\partial_t u_1^{\mathrm{av}}(x,t) = D_1^{-1}\partial_x u_1^{\mathrm{av}}(x,t), \quad x \in (0,1), \tag{17.69}$$

$$\partial_t u_2^{\mathrm{av}}(x,t) = D_2^{-2}\partial_{xx} u_2^{\mathrm{av}}(x,t), \quad x \in (0,1), \tag{17.70}$$

$$\partial_x u^{\mathrm{av}}(0,t) = 0, \tag{17.71}$$

$$u^{\mathrm{av}}(1,t) = K\bar{G}^{\mathrm{av}}(t) + \epsilon K\phi^{\mathrm{av}}(1,t). \tag{17.72}$$

From (17.68), if HK is Hurwitz, it is clear that the dynamics of the ODE state variable $\bar{G}^{\mathrm{av}}(t)$ is exponentially Input-to-State Stable (ISS) [99] with respect to the PDE state $u^{\mathrm{av}}(x,t) = [u_1^{\mathrm{av}}(x,t), u_2^{\mathrm{av}}(x,t)]^T$ by means of the function $\phi^{\mathrm{av}}(1,t)$. Moreover, the PDE subsystem (17.69) and (17.70) is ISS [99] with respect to $\bar{G}^{\mathrm{av}}(t)$ in the boundary condition $u^{\mathrm{av}}(1,t)$.

17.3.3 ▪ Stability Analysis

In this section, we will show that this hyperbolic-parabolic PDE-ODE loop (17.68)–(17.72) contains a small-parameter ϵ which can lead to closed-loop stability if it is chosen sufficiently small. To this end, we assume the following condition for noncooperative games [70].

Assumption 17.1. *The Hessian matrix H given by* (17.17) *is strictly diagonal dominant, i.e.,*

$$\begin{cases} |\epsilon H_{12}^1| < |H_{11}^1|, \\ |\epsilon H_{21}^2| < |H_{22}^2|. \end{cases} \tag{17.73}$$

By Assumption 17.1, the Nash equilibrium Θ^* exists and is unique since strictly diagonally dominant matrices are nonsingular by the Levy–Desplanques theorem [90].

The next theorem provides the stability/convergence properties of the closed-loop error system of the proposed ES feedback for the 2-player noncooperative game with transport-heat PDEs.

Theorem 17.1. *Consider the closed-loop system* (17.40)–(17.47) *under transport-heat PDEs* (17.1)–(17.8) *of distinct transport-diffusion coefficients D_1 and D_2 for a heterogeneous duopoly quadratic game with payoff functions* (17.29) *given in* (17.12), (17.13) *satisfying Assumption* 17.1 *and control laws $U_i(t)$ defined in* (17.50) *or* (17.51), *with $\hat{G}_1(t)$, $\hat{G}_2(t)$ and $\hat{H}_1(t)$, $\hat{H}_2(t)$ given by* (17.34) *and* (17.37), *respectively. The ES perturbation signals are given in* (17.30) *and* (17.31). *Thus, there exist $c_1>0$, $c_2>0$, and $\omega>0$ sufficiently large as well as $\epsilon>0$ sufficiently small such that the closed-loop error system with state $\vartheta_i(t)$, $u_i(x,t)$, $\forall i\in\{1,2\}$, has a unique locally exponentially stable periodic solution in t of period $\Pi:=2\pi\times LCM^{11}\{1/\omega_i\}$, with ω_i in* (17.32) *of order $\mathcal{O}(\omega)$, denoted by $\vartheta_i^{\Pi}(t)$, $u_i^{\Pi}(x,t)$ and satisfying, $\forall t\geq 0$,*

$$\left(\sum_{i=1}^{2}\left[\vartheta_i^{\Pi}(t)\right]^2+\int_0^{D_i}\left[u_i^{\Pi}(x,t)\right]^2dx\right)^{1/2}\leq\mathcal{O}(1/\omega). \tag{17.74}$$

Furthermore,

$$\limsup_{t\to+\infty}|\Theta(t)-\Theta^*|=\mathcal{O}(|a|+1/\omega), \tag{17.75}$$

$$\limsup_{t\to+\infty}|\theta_1(t)-\theta_1^*|=\mathcal{O}\left(a_1+1/\omega\right), \tag{17.76}$$

$$\limsup_{t\to+\infty}|\theta_2(t)-\theta_2^*|=\mathcal{O}\left(a_2e^{D_2\sqrt{\omega/2}}+1/\omega\right), \tag{17.77}$$

where $a=[a_1\ a_2]^T$ and $\theta^*=\Theta^*$ is the unique (unknown) Nash equilibrium given by (17.18).

Proof. First, consider the equivalent hyperbolic/parabolic PDE-ODE representation (17.68)–(17.72) rewritten for each Player P_i, $i\in\{1,2\}$:

$$\dot{\bar{G}}_i^{\rm av}(t)=H_{ii}^ik_i\bar{G}_i^{\rm av}(t)+\epsilon H_{ii}^ik_i\phi_i^{\rm av}(1,t), \tag{17.78}$$

$$\partial_t u_1^{\rm av}(x,t)=D_1^{-1}\partial_x u_1^{\rm av}(x,t),\quad x\in(0,1), \tag{17.79}$$

$$\partial_t u_2^{\rm av}(x,t)=D_2^{-2}\partial_{xx}u_2^{\rm av}(x,t),\quad x\in(0,1), \tag{17.80}$$

$$\partial_x u_i^{\rm av}(0,t)=0, \tag{17.81}$$

$$u_i^{\rm av}(1,t)=k_i\bar{G}_i^{\rm av}(t)+\epsilon k_i\phi_i^{\rm av}(1,t). \tag{17.82}$$

For Player P_1, the average closed-loop system, given by (17.78)–(17.79) and (17.81)–(17.82), satisfies both assumptions **(H1)** and **(H2)** of the Small-Gain Theorem [99, Theorem 8.1, p. 198] (Appendix C) for the hyperbolic PDE-ODE loop with $n=1$, $x(t)=\bar{G}_1^{\rm av}(t)$, $F(\bar{G}_1^{\rm av}(t),u_1^{\rm av}(x,t),v(t))=H_{11}^1k_1\bar{G}_1^{\rm av}(t)+\epsilon v(t)$ with $v(t)=H_{11}^1k_1\phi_1^{\rm av}(1,t)$, $c=1/D_1$, $a(x)=0$, $g(x,\bar{G}_1^{\rm av}(t),u_1^{\rm av}(x,t))=0$, $f(x,t)=0$, $\varphi(d(t),u_1^{\rm av}(x,t),\bar{G}_1^{\rm av}(t))=k_1\bar{G}_1^{\rm av}(t)+\epsilon d(t)$, where $d(t)=k_1\phi_1^{\rm av}(1,t)$, $N=k_1$, $L=|H_{11}^1|k_1$, $B=0$, $\gamma_2=k_1$, $b_2=\epsilon$, and $A=\gamma_1=0$. Notice that assumption **(H1)** holds with $M=1$, $\gamma_3=0$, $\sigma=|H_{11}^1|k_1$, and $b_3>0$ being an appropriate constant of order $\mathcal{O}(\epsilon)$, as it can be readily verified by means of the variation-of-constants formula (for $i=1$)

$$\bar{G}_i^{\rm av}(t)=\exp(-|H_{ii}^i|k_it)\bar{G}_i^{\rm av}(0)+\int_0^1\exp(-|H_{ii}^i|k_i(t+s))\epsilon H_{ii}^ik_i\phi_i^{\rm av}(1,s)ds, \tag{17.83}$$

[11] The acronym LCM stands for the least common multiple.

and from the application of the Cauchy–Schwarz inequality to the term $\phi_1^{\rm av}(1,t)$ in (17.67):

$$\begin{aligned}\phi_1^{\rm av}(1,t) &\leq |H_{12}^1| D_2^2 \left(\int_0^1 (1-\tau)^2 d\tau\right)^{\frac{1}{2}} \times \left(\int_0^1 [u_2^{\rm av}(\xi,t)]^2 d\xi\right)^{\frac{1}{2}} \\ &\leq \frac{1}{\sqrt{3}} k_H D_2^2 \left(\int_0^1 [u_2^{\rm av}(\xi,t)]^2 d\xi\right)^{\frac{1}{2}}, \end{aligned} \tag{17.84}$$

since $|H_{ij}^i| < k_H < \frac{1}{\epsilon}|H_{ii}^i|$ according to Assumption 17.1, where k_H is a positive constant of order $\mathcal{O}(1)$. Therefore, if $0 < \epsilon < 1$ is sufficiently small, then [99, Theorem 8.1, p. 198] (Appendix C) allows us to conclude that there exist constants $\delta_1, \Delta_1 > 0$ such that for every $u_{1,0}^{\rm av} \in C^0([0,1])$, $\bar{G}_{1,0}^{\rm av} \in \mathbb{R}$, the unique generalized solution of this initial-boundary value problem, with $u_1^{\rm av}(x,0) = u_{1,0}^{\rm av}$ and $\bar{G}_1^{\rm av}(0) = \bar{G}_{1,0}^{\rm av}$, satisfies the following estimate:

$$\begin{aligned}|\bar{G}_1^{\rm av}(t)| + \|u_1^{\rm av}(t)\|_\infty &\leq \Delta_1(|\bar{G}_{1,0}^{\rm av}| + \|u_{1,0}^{\rm av}\|_\infty)\exp(-\delta_1 t) \\ &\quad + \bar{\gamma}_1 \epsilon \max_{0\leq s\leq t}(\|u_2^{\rm av}(s)\|_\infty) \quad \forall t \geq 0 \end{aligned} \tag{17.85}$$

for some adequate constant $\bar{\gamma}_1 > \max(|H_{11}^1| k_1, k_1)$.

For Player P_2, the average closed-loop system (17.78) and (17.80)–(17.82) satisfies all the assumptions **(A1)** to **(A7)** of the Small-Gain Theorem [99, Theorem 8.2, p. 205] for the parabolic PDE-ODE loop with $n=1$, $p(x)=1$, $r(x)=D_2^2$, $q(x)=0$, $F(\bar{G}_2^{\rm av}(t), u_2^{\rm av}(x,t), v(t)) = H_{22}^2 k_2 \bar{G}_2^{\rm av}(t) + \epsilon v(t)$, $v(t) = H_{22}^2 k_2 \phi_2^{\rm av}(1,t)$, $g(x, \bar{G}_2^{\rm av}(t), u_2^{\rm av}(x,t)) = 0$, $f(x,t) = 0$, $\varphi_0(d(t), u_2^{\rm av}(x,t), \bar{G}_2^{\rm av}(t)) = b_1 u_2^{\rm av}(0,t)$, $b_1 < 0$, $b_2 = 1$, $\varphi_1(d(t), u_2^{\rm av}(x,t), \bar{G}_2^{\rm av}(t)) = k_2 \bar{G}_2^{\rm av}(t) + \epsilon d(t)$, $d(t) = k_2 \phi_2^{\rm av}(1,t)$, $a_1 = 1$, $a_2 = 0$, $L = |H_{22}^2| k_2$, $K_0 = |b_1|$, $B_0 = C_0 = 0$, γ_0 is of order $\mathcal{O}(1)$, $K_1 = 0$, $B_1 = k_2$, $C_1 = \epsilon$, γ_1 is of order $\mathcal{O}(1)$, and $K_2 = B_2 = 0$. Assumption **(A6)** of [99, Theorem 8.2, p. 205] holds with $M=1$, $\gamma_3 = 0$, $\sigma = |H_{22}^2| k_2$, and $b_3 > 0$ being of order $\mathcal{O}(\epsilon)$, as it can be readily verified by means of the variation-of-constants formula (17.83), with $i=2$, and from the application of the Cauchy–Schwarz inequality to the term $\phi_2^{\rm av}(1,t)$ in (17.67):

$$\begin{aligned}\phi_2^{\rm av}(1,t) &\leq |H_{21}^2| \left(\int_0^1 D_1^2 d\tau\right)^{\frac{1}{2}} \times \left(\int_0^1 [u_1^{\rm av}(\xi,t)]^2 d\xi\right)^{\frac{1}{2}} \\ &\leq k_H D_1 \left(\int_0^1 [u_1^{\rm av}(\xi,t)]^2 d\xi\right)^{\frac{1}{2}}, \end{aligned} \tag{17.86}$$

with the same $k_H > 0$ defined just after (17.84). Hence, it follows that the small-gain condition in [99, inequality (8.3.24)] holds provided $0 < \epsilon < 1$ is sufficiently small. Thus, [99, Theorem 8.2, p. 205] allows us to conclude that there exist constants $\delta_2, \Delta_2 > 0$ such that

$$\begin{aligned}|\bar{G}_2^{\rm av}(t)| + \|u_2^{\rm av}(t)\|_\infty &\leq \Delta_2(|\bar{G}_{2,0}^{\rm av}| + \|u_{2,0}^{\rm av}\|_\infty)\exp(-\delta_2 t) \\ &\quad + \bar{\gamma}_2 \epsilon \max_{0\leq s\leq t}(\|u_1^{\rm av}(s)\|_\infty) \quad \forall t \geq 0 \end{aligned} \tag{17.87}$$

for some adequate constant $\bar{\gamma}_2 > \max(|H_{22}^2| k_2, k_2)$, $u_2^{\rm av}(x,0) = u_{2,0}^{\rm av}$, and $\bar{G}_2^{\rm av}(0) = \bar{G}_{2,0}^{\rm av}$. Since inequalities (17.85) and (17.87) are similar to those found in [99, Theorem 11.2, p. 269]—see inequalities (11.2.23) and (11.2.24)—we can finally invoke [99, Theorem 11.5, p. 277], under the condition of $0 < \epsilon < 1$ sufficiently small, to conclude that

$$\begin{aligned}&|\bar{G}_1^{\rm av}(t)| + |\bar{G}_2^{\rm av}(t)| + \|u_1^{\rm av}(t)\|_\infty + \|u_2^{\rm av}(t)\|_\infty \\ &\leq \Delta(|\bar{G}_{1,0}^{\rm av}| + |\bar{G}_{2,0}^{\rm av}| + \|u_{1,0}^{\rm av}\|_\infty + \|u_{2,0}^{\rm av}\|_\infty)\exp(-\delta t) \quad \forall t \geq 0 \end{aligned} \tag{17.88}$$

for some $\delta > 0$ and $\Delta > 0$. Therefore, we conclude that the origin of the average closed-loop

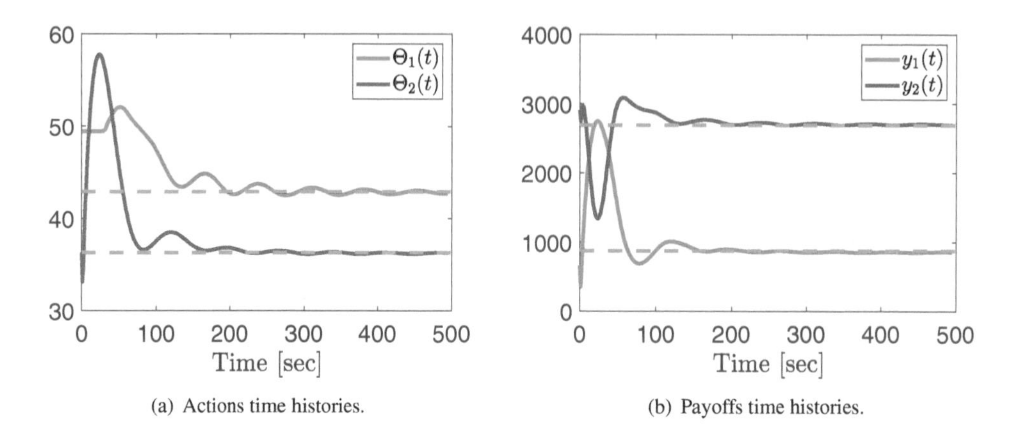

(a) Actions time histories.

(b) Payoffs time histories.

Figure 17.3. (a) *Actions time histories and* (b) *payoffs time histories for P_1 and P_2 for $\epsilon = 1$. The dashed lines denote the values at the Nash equilibrium, $\Theta_1^* = 43.33$ and $\Theta_2^* = 36.67$ (with $J_1(\Theta^*) = 889$ and $J_2(\Theta^*) = 2722$).*

system (17.68)–(17.72) is exponentially stable under the assumption of $0<\epsilon<1$ being sufficiently small. Then, from (17.48), (17.61), and (17.62), we conclude the same results in the norm

$$\left(\sum_{i=1}^{2}[\vartheta_i^{\mathrm{av}}(t)]^2 + \int_0^{D_i}[u_i^{\mathrm{av}}(x,t)]^2\,dx\right)^{1/2} \tag{17.89}$$

since H is nonsingular, i.e., $|\vartheta_i^{\mathrm{av}}(t)| \leq |H^{-1}||\hat{G}^{\mathrm{av}}(t)|$.

As developed in Theorem 10.1 (Chapter 10), the next steps of the proof would be the application of the local averaging theory for infinite dimensional systems in [83, Section 2] (Appendix B), showing that the periodic solutions satisfy (17.74) for ω sufficiently large, and then the conclusion of the attractiveness of the Nash equilibrium Θ^* according to (17.75). The final residual sets for the errors $\theta_i(t) - \theta_i^*$ in (17.76) and (17.77) depend on a_1 and $a_2 e^{D_2\sqrt{\frac{\omega}{2}}}$ due to the amplitude of the additive dithers $S_i(t)$ in (17.30) for $i \in \{1,2\}$. □

17.4 ▪ Simulations with a Heterogeneous Duopoly Game under Transport-Heat PDEs

For an example of a heterogeneous noncooperative game with 2 players that employ the proposed ES strategy for PDE compensation, we revisit the example in [185] and consider the following payoff functions (17.12) and (17.13) subject to transport-heat PDEs (17.1)–(17.4) and (17.5)–(17.8) with distinct transport-diffusion coefficients $D_1 = 30$ and $D_2 = 3$ in the players' decisions, $i \in \{1,2\}$:

$$J_1(\Theta(t)) = -5\,\Theta_1^2(t) + 5\,\epsilon\Theta_1(t)\Theta_2(t) + 250\,\Theta_1(t) - 150\,\Theta_2(t) - 3000, \tag{17.90}$$

$$J_2(\Theta(t)) = -5\,\Theta_2^2(t) + 5\,\epsilon\Theta_1(t)\Theta_2(t) - 150\,\Theta_1(t) + 150\,\Theta_2(t) + 2500, \tag{17.91}$$

which, according to (17.18), yield the unique Nash equilibrium

$$\Theta_1^* = \theta_1^* = \frac{100+30\epsilon}{4-\epsilon^2}, \quad \Theta_2^* = \theta_2^* = \frac{60+50\epsilon}{4-\epsilon^2}. \tag{17.92}$$

In order to attain (17.92), the players implement the non-model-based real-time optimization strategy of Section 17.3 acting through the transport-heat PDE dynamics (see Figure 17.2). For

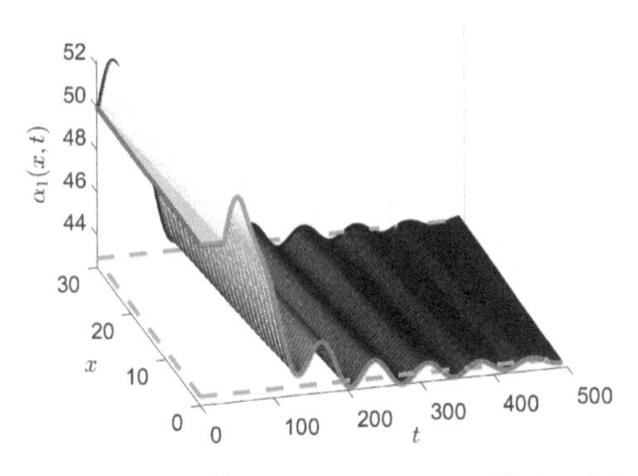

(a) Parameter $\Theta_1(t)$ (red) converges to a $\mathcal{O}(|a|+1/\omega)$-neighborhood of Θ_1^* (dashed-green).

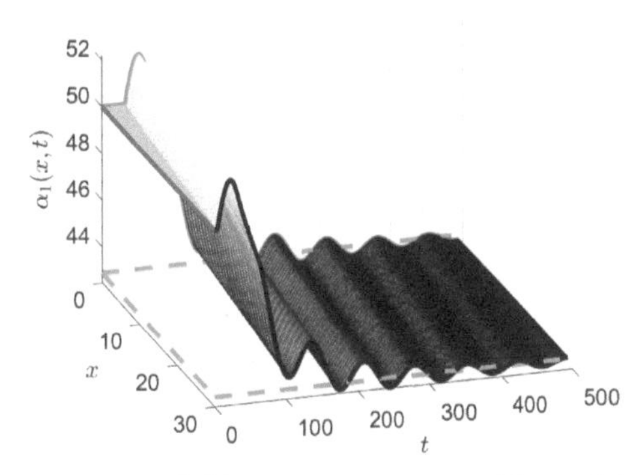

(b) Parameter $\theta_1(t)$ (black) converges to a $\mathcal{O}(a_1+1/\omega)$-neighborhood of θ_1^* (dashed-green).

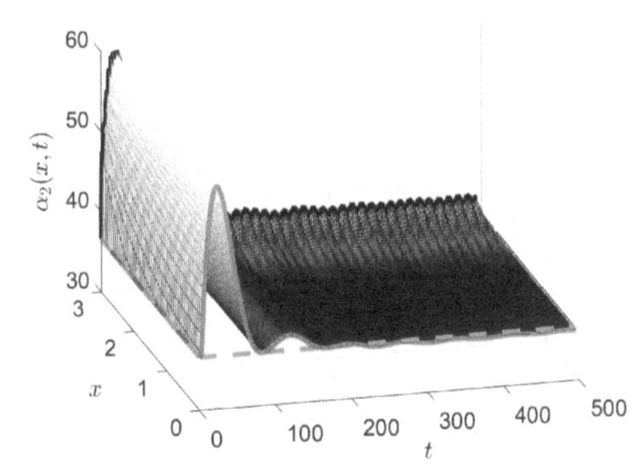

(c) Parameter $\Theta_2(t)$ (red) converges to a $\mathcal{O}(|a|+1/\omega)$-neighborhood of Θ_2^* (dashed-green).

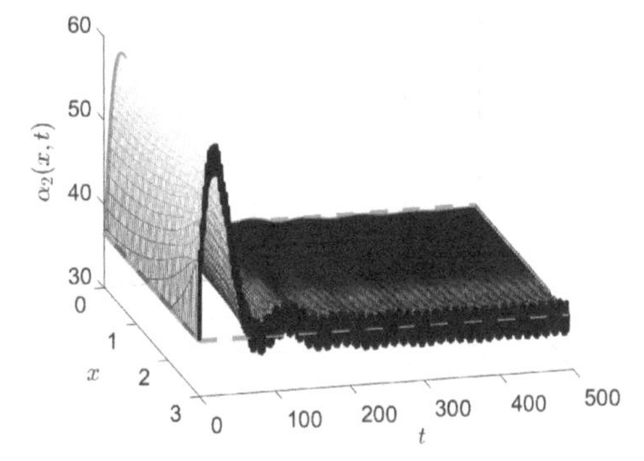

(d) Parameter $\theta_2(t)$ (black) converges to a $\mathcal{O}(a_2 e^{D_2\sqrt{\omega/2}} + 1/\omega)$-neighborhood of θ_2^* (dashed-green).

Figure 17.4. *Evolution of the infinite-dimensional states $\alpha_1(x,t)$ and $\alpha_2(x,t)$ of the transport-heat PDEs in a heterogeneous duopoly game with boundary Dirichlet actuation, according to* (17.75)–(17.77)*: from $\alpha_1(D_1,t)=\theta_1(t)$ to $\alpha_1(0,t)=\Theta_1(t)$, with $D_1=30$ for Player P_1, and from $\alpha_2(D_2,t)=\theta_2(t)$ to $\alpha_2(0,t)=\Theta_2(t)$, with $D_2=3$ for Player P_2.*

comparison purposes, except for the transport-heat PDEs in the players' input signals, the plant and the controller parameters were chosen similarly to [70] in the simulation tests: $\epsilon=1$, $a_1=0.075$, $a_2=0.05$, $k_1=2$, $k_2=5$, $\omega_1=26.75$ rad/s, $\omega_2=22$ rad/s and $\theta_1(0)=\hat{\theta}_1(0)=50$, $\theta_2(0)=\hat{\theta}_2(0)=\theta_2^*=110/3$. In addition, the time constants of the boundary control filters were set to $c_1=c_2=100$.

We can check that the ES approach proposed in [70] is effective when transport-heat PDEs are not present in the decision variables. However, in the presence of the transport-heat PDEs in the input signals θ_1 and θ_2, but without considering any kind of PDE compensation, the game collapses with the explosion of its variables (curves not shown). On the other hand, Figures 17.3(a) and 17.3(b) show that the proposed boundary control based scheme fixes this with a remarkable evolution in searching the Nash equilibrium and simultaneously compensates the effect of the transport-heat PDEs in our heterogeneous noncooperative game.

The evolution of the infinite-dimensional states $\alpha_1(x,t)$ and $\alpha_2(x,t)$ modeled by the transport-heat PDEs (17.1)–(17.4) and (17.5)–(17.8) is shown in Figures 17.4(a) to 17.4(d). The values

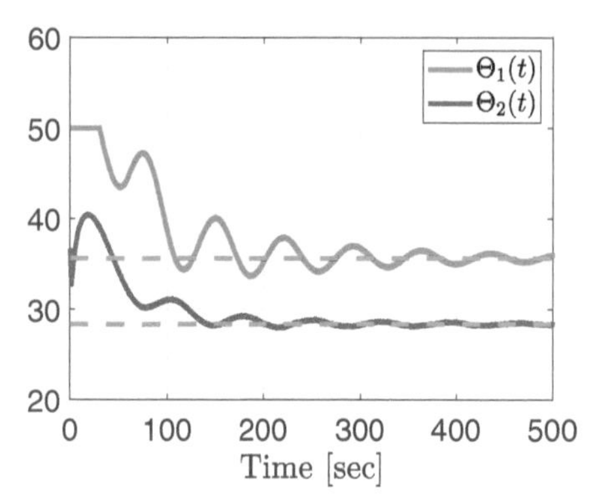

(a) Actions plots for $\epsilon = 0.75$. The dashed lines denote $\Theta_1^* = 35.64$ and $\Theta_2^* = 28.36$.

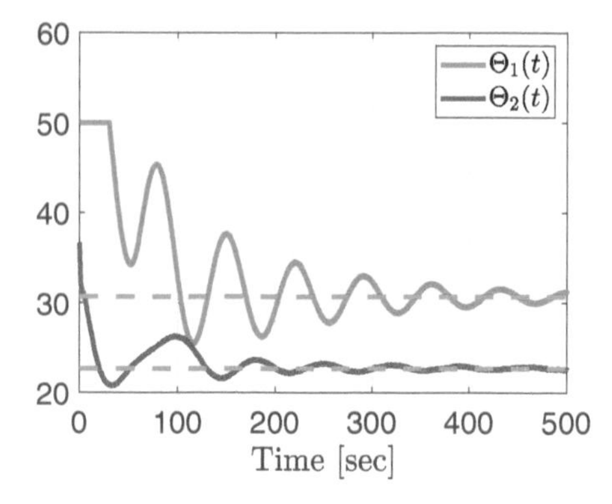

(b) Actions plots for $\epsilon = 0.5$. The dashed lines denote $\Theta_1^* = 30.67$ and $\Theta_2^* = 22.67$.

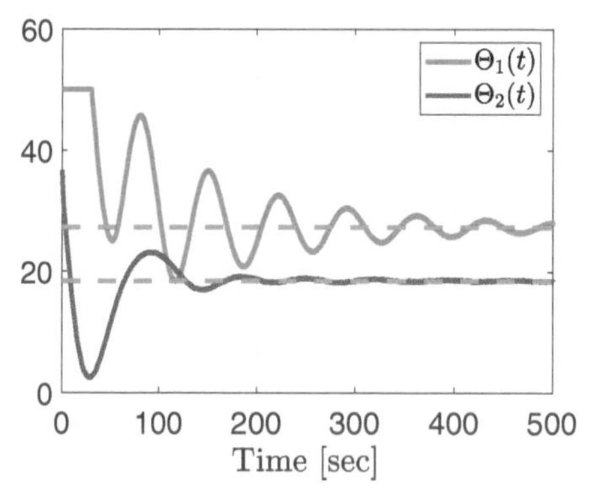

(c) Actions plots for $\epsilon = 0.25$. The dashed lines denote $\Theta_1^* = 27.30$ and $\Theta_2^* = 18.41$.

Figure 17.5. *Actions time histories for P_1 and P_2 for distinct values of the coupling coefficient ($\epsilon = 0.75$, $\epsilon = 0.5$, and $\epsilon = 0.25$).*

of the boundary inputs $\theta_1(t)$ and $\theta_2(t)$ as well as the boundary outputs $\Theta_1(t)$ and $\Theta_2(t)$ are highlighted with colors black and red, respectively. The initial condition is in blue color.

This first set of simulations indicates that even under an adversarial scenario of strong coupling between the players with $\epsilon = 1$, the proposed approach has behaved successfully. As in Chapters 15 and 16, this suggests that our stability analysis may be conservative and the theoretical assumption $0 < \epsilon < 1$ may be relaxed given the performance of the control system. In Figures 17.5(a) to 17.5(c), different values of $\epsilon = 0.75$, $\epsilon = 0.5$, and $\epsilon = 0.25$ are considered to evaluate the robustness of the proposed scheme under different levels of coupling between the two players and the corresponding impact on the transient responses.

17.5 ▪ Notes and References

This chapter generalized the ES results obtained in [184, 185] (Chapters 15 and 16) for a wider class of infinite-dimensional systems governed by heterogeneous PDEs [186], rather than being restricted to only pure delays or just one kind of PDE dynamics in the players' decision variables. This is particularly challenging since in the literature for multi-input PDE compensation [128], most authors consider a centralized approach for the PDE boundary control design (all vectors multiplying the control inputs need to be known) and homogeneous PDEs of one single type, while games are decentralized and heterogeneous—each player has his own particular dynamics, possibly of different nature.

In this scenario, we were able to develop a result for heterogeneous games with distinct transport and diffusion (heat) PDEs to be simultaneously compensated in the action paths of each player, where the players estimate only the diagonal entries of the Hessian matrix due to the players' payoffs in the corresponding real-time Nash seeking problem. We were able to dominate sufficiently small off-diagonal terms using a small-gain argument [99] for the average system.

Our analysis presented a properly designed sequence of steps of averaging in infinite dimensions [83] and Small-Gain Theorem for multiple Input-to-State Stable (ISS) cascades of hyperbolic/parabolic PDE-ODE systems [99] in the heterogeneous duopoly game in order to prove closed-loop stability with nonlinearities and periodic perturbations. Convergence to a small neighborhood of the Nash equilibrium was achieved, even in the presence of simultaneous but heterogeneous transport and heat PDEs.

APPENDICES

Appendix A

Averaging Theorem for Functional Differential Equations

Theorem A.1 (Averaging Theorem for FDEs [83]). *Consider the delay system*

$$\dot{x}(t) = f(t/\epsilon, x_t) \quad \forall t \geq 0, \tag{A.1}$$

where ϵ is a real parameter, $x_t(\Theta) = x(t+\Theta)$ for $-r \leq \Theta \leq 0$, and $f : \mathbb{R}_+ \times \Omega \to \mathbb{R}^n$ is a continuous functional from a neighborhood Ω of 0 of the supremum-normed Banach space $X = C([-r,0];\mathbb{R}^n)$ of continuous functions from $[-r,0]$ to $\mathbb{R}^n$. Assume that $f(t,\varphi)$ is periodic in t uniformly with respect to φ in compact subsets of Ω and that f has a continuous Fréchet derivative $\partial f(t,\varphi)/\partial\varphi$ in φ on $\mathbb{R}_+ \times \Omega$. If $y = y_0 \in \Omega$ is an exponentially stable equilibrium for the average system

$$\dot{y}(t) = f_0(y_t) \quad \forall t \geq 0, \tag{A.2}$$

where $f_0(\varphi) = \lim_{T\to\infty} \frac{1}{T}\int_0^T f(s,\varphi)ds$, then, for some $\epsilon_0 > 0$ and $0 < \epsilon \leq \epsilon_0$, there is a unique periodic solution $t \mapsto x^*(t,\epsilon)$ of (A.1) *with the properties of being continuous in t and ϵ, satisfying $|x^*(t,\epsilon) - y_0| \leq \mathcal{O}(\epsilon)$, for $t \in \mathbb{R}_+$, and such that there is $\rho > 0$ so that if $x(\cdot;\varphi)$ is a solution of* (A.1) *with $x(s) = \varphi$ and $|\varphi - y_0| < \rho$, then $|x(t) - x^*(t,\epsilon)| \leq Ce^{-\gamma(t-s)}$ for $C > 0$ and $\gamma > 0$.*

Appendix B

Averaging Theorem for General Infinite-Dimensional Systems

The key step of the proof of Theorem 10.1 in Section 10.4 is the Averaging Theorem for PDEs by [83], which we summarize here.

Theorem B.1. *Consider the infinite-dimensional system, defined in the Banach space* $\mathcal{X}$

$$\dot{z} = \mathcal{A}z + J(\omega t, z) \tag{B.1}$$

with $z(0) = z_0 \in \mathcal{X}$ *and the operator* $\mathcal{A} : \mathcal{D}(\mathcal{A}) \to \mathcal{X}$ *generating an analytic semigroup. Moreover, the nonlinearity* $J : \mathbb{R}_+ \times \mathcal{X} \to \mathcal{X}$ *with* $t \mapsto J(\omega t, z)$ *is Fréchet differentiable in* z*, strongly continuous and periodic in* t *uniformly with respect to* z *in a compact subset of* $\mathcal{X}$*. Along with* (B.1)*, the average system*

$$\dot{z}_{av} = \mathcal{A}z_{av} + J_0(z_{av}) \tag{B.2}$$

with $J_0(z_{av}) = \lim_{T\to\infty} \frac{1}{T}\int_0^T J(\tau, z_{av})d\tau$ *is considered. Suppose that* $z_{av} = 0 \in D \subset \mathcal{X}$ *is an exponentially stable equilibrium point of the average system* (B.2)*. Then for some* $\bar{\omega} > 0$ *and* $\omega > \bar{\omega}$*, the following hold:*

(a) *there exists a unique exponentially stable periodic solution* $t \mapsto \bar{z}(t, 1/\omega)$*, continuous in* t *and* $1/\omega$*, with* $\|\bar{z}(t, 1/\omega)\| \leq \mathcal{O}(1/\omega)$ *for* $t > 0$*;*

(b) *with* $\|z_0 - z_{av}(0)\| \leq \mathcal{O}(1/\omega))$*, the solution estimate of* (B.1) *is given by*

$$\|z(t) - z_{av}\| \leq \mathcal{O}(1/\omega), \quad t > 0; \tag{B.3}$$

(c) *for* $\|z_0\| \leq \mathcal{O}(1/\omega)$ *and the stable manifold theorem, it holds that*

$$\|z(t) - \bar{z}(t, 1/\omega)\| \leq Ce^{-\gamma t}, \quad t > 0, \tag{B.4}$$

for some C*,* $\gamma > 0$*.*

Appendix C
Small-Gain Theorems

Theorem C.1 (Small-Gain Theorem for ODE and Hyperbolic PDE Loops [99, Theorem 8.1, p. 198]). *Consider generalized solutions of the following initial-boundary value problem:*

$$\dot{x}(t) = F(x(t), u(z,t), v(t)) \quad \forall t \geq 0, \tag{C.1}$$

$$u_t(z,t) + cu_z(z,t) = a(z)u(z,t) + g(z, x(t), u(z,t)) + f(z,t) \quad \forall (z,t) \in [0,1] \times \mathbb{R}_+, \tag{C.2}$$

$$u(0,t) = \varphi(d(t), u(z,t), x(t)) \quad \forall t \geq 0, \quad u(z,0) = u_0, \quad x(0) = x_0. \tag{C.3}$$

The state of the system (C.1)–(C.3) *is* $(u(z,t), x(t)) \in C^0([0,1] \times \mathbb{R}_+) \times \mathbb{R}^n$, *while the other variables* $d \in C^0(\mathbb{R}_+; \mathbb{R}^q)$, $f \in C^0([0,1] \times \mathbb{R}_+)$ *and* $v \in C^0(\mathbb{R}_+; \mathbb{R}^m)$ *are external inputs. We assume that* $(0,0) \in C^0([0,1]) \times \mathbb{R}^n$ *is an equilibrium point for the input-free system, i.e.,* $F(0,0,0) = 0$, $g(z,0,0) = 0$, *and* $\varphi(0,0,0) = 0$. *Now, we assume that the ODE subsystem satisfies the ISS property.*
(H1) *There exist constants* $M, \sigma > 0$, $b_3, \gamma_3 \geq 0$, *such that for every* $x_0 \in \mathbb{R}^n$, $u \in C^0([0,1] \times \mathbb{R}_+)$, *and* $v \in C^0(\mathbb{R}_+; \mathbb{R}^m)$ *the unique solution* $x \in C^1(\mathbb{R}_+; \mathbb{R}^n)$ *of* (C.1) *with* $x(0) = x_0$ *satisfies the following estimate:*

$$|x(t)| \leq M|x_0| \exp(-\sigma t) + \max_{0 \leq s \leq t} (\gamma_3 \|u(s)\|_\infty + b_3 |v(s)|) \quad \forall t \geq 0. \tag{C.4}$$

We next need to estimate the static gain of the interconnections. To this purpose, we employ the following further assumption.
(H2) *There exist constants* $b_2, \gamma_1, \gamma_2, A, B \geq 0$ *such that the following growth conditions hold for every* $x \in C^1(\mathbb{R}_+; \mathbb{R}^n)$, $u \in C^0([0,1] \times \mathbb{R}_+)$, *and* $d \in C^0(\mathbb{R}_+; \mathbb{R}^q)$:

$$|g(z, x, u)| \leq A\|u\|_\infty + \gamma_1 |x| \quad \forall z \in [0,1], \tag{C.5}$$

$$|\varphi(d, u, x)| \leq B\|u\|_\infty + \gamma_2 |x| + b_2 |d|. \tag{C.6}$$

Let $c > 0$[12] *be a given constant and* $a \in C^0([0,1])$ *be a given function. Consider the mappings as* $F : \mathbb{R}^n \times C^0([0,1]) \times \mathbb{R}^m \to \mathbb{R}^n$, $g : [0,1] \times \mathbb{R}^n \times C^0([0,1]) \to \mathbb{R}$, $\varphi : \mathbb{R}^q \times C^0([0,1]) \times \mathbb{R}^n \to \mathbb{R}$ *being continuous mappings with* $F(0,0,0) = 0$ *for which there exist constants* $L > 0$, $\bar{N} \in [0,1[$ *such that the inequalities* $\max_{0 \leq z \leq 1}(|g(z,x,u) - g(z,y,w)|) + |F(x,u,v) - F(y,w,v)| \leq L|x-y| + L\|u-w\|_\infty$, $|\varphi(d,u,x) - \varphi(d,w,y)| \leq \bar{N}|x-y| + N\|u-w\|_\infty$ *hold for all* $u, w \in$

[12] If the scalar $c < 0$ is considered, the direction of convection must be reversed such that the boundary $u(0,t)$ is replaced by $u(1,t)$ and vice versa.

$C^0([0,1])$, $x,y \in \mathbb{R}^n$, $v \in \mathbb{R}^m$, $d \in \mathbb{R}^q$. *Suppose that assumptions* **(H1)** *and* **(H2)** *hold and that the following small-gain condition is satisfied:*

$$(\gamma_1\gamma_3+A)c^{-1}\max_{0\leq z\leq 1}\left(p(z)\int_0^z \frac{1}{p(l)}dl\right)+(\gamma_2\gamma_3+B)\max_{0\leq z\leq 1}(p(z))$$
$$+2\sqrt{(\gamma_1\gamma_3+A)c^{-1}(\gamma_2\gamma_3+B)\max_{0\leq z\leq 1}(p(z))\max_{0\leq z\leq 1}\left(p(z)\int_0^z \frac{1}{p(l)}dl\right)}<1 \quad \text{(C.7)}$$

with $p(z){:=}\exp\left(c^{-1}\int_0^z a(w)dw\right)$ *for* $z\in[0,1]$ *(recall* (8.2.11) *and* (8.2.14)*) in* [99, *Section* 8.2]. *Then, there exist constants* $\delta,\Theta,\gamma>0$ *such that for every* $u_0\in C^0([0,1])$, $x_0\in\mathbb{R}^n$, $d\in C^0(\mathbb{R}_+;\mathbb{R}^q)$ *with* $u_0(0)=\varphi(d(0),u_0,x_0)$, $f\in C^0([0,1]\times\mathbb{R}_+)$, *and* $v\in C^0(\mathbb{R}_+;\mathbb{R}^m)$ *the unique generalized solution of the initial-boundary value problem* (C.1), (C.2), (C.3) *satisfies the following estimate:*

$$|x(t)|+\|u(t)\|_\infty \leq \Theta(|x_0|+\|u_0\|_\infty)\exp(-\delta t)$$
$$+\gamma\left[\max_{0\leq s\leq t}(|v(s)|)+\max_{0\leq s\leq t}(\|f(s)\|_\infty)+\max_{0\leq s\leq t}(|d(s)|)\right] \quad \forall t\geq 0. \quad \text{(C.8)}$$

Analogous small-gain results for parabolic PDE-ODE loops and parabolic-hyperbolic PDE loops can be found in [99, Theorem 8.2, p. 205] and [99, Theorem 11.2(11.5), p. 269(277)], respectively. Such theorems are used in Chapters 16 and 17.

Appendix D

Auxiliary Proofs and Derivations

D.1 ▪ Computation of Constants α_1 and α_2 in Inequality (2.62)

For convenience, let us now repeat the backstepping transformation and its inverse in an re-arranged form as

$$w(x,t) = u_{\text{av}}(x,t) - kH\int_0^x u_{\text{av}}(\sigma,t)d\sigma - kH\tilde{\vartheta}_{\text{av}}(t), \tag{D.1}$$

$$u_{\text{av}}(x,t) = w(x,t) + kH\int_0^x e^{kH(x-\sigma)}w(\sigma,t)d\sigma + kHe^{kHx}\tilde{\vartheta}_{\text{av}}(t). \tag{D.2}$$

To derive a stability bound, we need to relate the norm

$$\left(|\tilde{\vartheta}_{\text{av}}(t)|^2 + \int_0^D u_{\text{av}}^2(x,t)dx\right)^{1/2} \text{ to the norm } \left(|\tilde{\vartheta}_{\text{av}}(t)|^2 + \int_0^D w^2(x,t)dx\right)^{1/2} \tag{D.3}$$

and then the norm

$$\left(|\tilde{\vartheta}_{\text{av}}(t)|^2 + \int_0^D w^2(x,t)dx\right)^{1/2} \text{ to } \sqrt{V} = \left(\frac{\tilde{\vartheta}_{\text{av}}^2(t)}{2} + \frac{a}{2}\int_0^D (1+x)w^2(x,t)dx\right)^{1/2}. \tag{D.4}$$

We start from the latter, as it is easier, and obtain

$$\Phi_1\left(|\tilde{\vartheta}_{\text{av}}(t)|^2 + \int_0^D w^2(x,t)dx\right) \leq V(t) \leq \Phi_2\left(|\tilde{\vartheta}_{\text{av}}(t)|^2 + \int_0^D w^2(x,t)dx\right), \tag{D.5}$$

where

$$\Phi_1 = \min\left\{\frac{1}{2}, \frac{a}{2}\right\}, \tag{D.6}$$

$$\Phi_2 = \max\left\{\frac{1}{2}, \frac{a(1+D)}{2}\right\}. \tag{D.7}$$

It is easy to show, using (D.1) and (D.2), that

$$\int_0^D w^2(x,t)dx \leq \alpha_1 \int_0^D u_{\mathrm{av}}^2(x,t)dx + \alpha_2|\tilde{\vartheta}_{\mathrm{av}}(t)|^2, \tag{D.8}$$

$$\int_0^D u_{\mathrm{av}}^2(x,t)dx \leq \beta_1 \int_0^D w^2(x,t)dx + \beta_2|\tilde{\vartheta}_{\mathrm{av}}(t)|^2, \tag{D.9}$$

where

$$\alpha_1 = 3(1+D|kH|^2), \tag{D.10}$$

$$\alpha_2 = 3|kH|^2, \tag{D.11}$$

$$\beta_1 = 3(1+D\|kHe^{kHx}\|^2), \tag{D.12}$$

$$\beta_2 = 3\|kHe^{kHx}\|^2, \tag{D.13}$$

and $\|\cdot\|$ denotes the $L_2[0,D]$ norm. Hence, we obtain

$$\phi_1\left(|\tilde{\vartheta}_{\mathrm{av}}(t)|^2 + \int_0^D u_{\mathrm{av}}^2(x,t)dx\right) \leq |\tilde{\vartheta}_{\mathrm{av}}(t)|^2 + \int_0^D w^2(x,t)dx, \tag{D.14}$$

$$|\tilde{\vartheta}_{\mathrm{av}}(t)|^2 + \int_0^D w^2(x,t)dx \leq \phi_2\left(|\tilde{\vartheta}_{\mathrm{av}}(t)|^2 + \int_0^D u_{\mathrm{av}}^2(x,t)dx\right), \tag{D.15}$$

where

$$\phi_1 = \frac{1}{\max\{\beta_1, \beta_2+1\}}, \tag{D.16}$$

$$\phi_2 = \max\{\alpha_1, \alpha_2+1\}. \tag{D.17}$$

Combining the above inequalities, we get

$$\alpha_1\Psi(t) \leq V(t) \leq \alpha_2\Psi(t), \tag{D.18}$$

with $\Psi(t) = |\tilde{\vartheta}_{\mathrm{av}}(t)|^2 + \int_0^D u_{\mathrm{av}}^2(x,t)dx + u_{\mathrm{av}}^2(D,t)$.

D.2 ▪ Alternative Stability Proof for State-Dependent Delays

Alternatively, the following PDE representation can be adopted for (8.109)–(8.110):

$$\partial_t u_{\mathrm{av}}(x,t) = -\partial_x u_{\mathrm{av}}(x,t), \quad x \in [0, D(G_{\mathrm{av}}(t))], \tag{D.19}$$

$$u_{\mathrm{av}}(0,t) = U_{\mathrm{av}}(t), \tag{D.20}$$

$$\dot{G}_{\mathrm{av}}(t) = Hu_{\mathrm{av}}\left(D(G_{\mathrm{av}}(t)),t\right). \tag{D.21}$$

Introducing the change of variables

$$z = \frac{D(G_{\mathrm{av}}(t)) - x}{D(G_{\mathrm{av}}(t))}, \tag{D.22}$$

the system (D.19)–(D.21) can be rewritten as

$$\partial_t v_{\mathrm{av}}(z,t) = \lambda_v(z,t)\partial_z v_{\mathrm{av}}(z,t), \quad z \in [0,1], \tag{D.23}$$

$$v_{\mathrm{av}}(1,t) = U_{\mathrm{av}}(t), \tag{D.24}$$

$$\dot{G}_{\mathrm{av}}(t) = Hv_{\mathrm{av}}\left(0,t\right), \tag{D.25}$$

where

$$\lambda_v(z,t)=\frac{1-z\dot{D}(G_{\text{av}}(t))}{D(G_{\text{av}}(t))}=\frac{1-zH\nabla D(G_{\text{av}}(t))v_{\text{av}}(0,t)}{D(G_{\text{av}}(t))}. \tag{D.26}$$

As expected, the transport speed is a function of the spatial variable z, the average state G_{av}, and the delayed input $u_{\text{av}}(0,t)$. Moreover, the system (D.23)–(D.25) is well-posed if the transport speed $\lambda_v(z,t)$ is strictly positive, that is,

$$\lambda_v(z,t)=1-zH\nabla D(G_{\text{av}}(t))v_{\text{av}}(0,t)>0 \tag{D.27}$$

$\forall t>0, z\in(0,1), D(\cdot)>0$. Noting that the function (D.27) is a linear function of z, its extrema are given by $\lambda_v(0,t)$ and $\lambda_v(1,t)$. Thus,

$$\lambda_v(0,t)=1, \tag{D.28}$$

$$\lambda_v(1,t)=1-H\nabla D(G_{\text{av}}(t))v_{\text{av}}(0,t), \tag{D.29}$$

leading to the following feasibility condition:

$$F_v(t)=H\nabla D(G_{\text{av}}(t))v_{\text{av}}(0,t), \tag{D.30}$$

$$F_v(t)<\bar{c}, \quad 0<\bar{c}<1. \tag{D.31}$$

Defining the predictor feedback control law of $p(x,t)$ as

$$U_{\text{av}}(t)=k\,p(D(G_{\text{av}}(t)),t), \tag{D.32}$$

$$p(x,t)=G_{\text{av}}(t)+H\int_0^x u_{\text{av}}(s,t)\,ds, \quad x\in[0,D(G_{\text{av}}(t))], \tag{D.33}$$

and applying again the change of variables (D.22), the following equivalent predictor feedback control law is obtained for the normalized system (D.23)–(D.25):

$$U_{\text{av}}(t)=k\,p_v(1,t), \tag{D.34}$$

$$p_v(z,t)=G_{\text{av}}(t)-D(G_{\text{av}}(t))H\int_0^z v_{\text{av}}(s,t)\,ds, \quad z\in[0,1]. \tag{D.35}$$

Here, $p(x,t)$ and $p_v(z,t)$ are the predictor states of the systems (D.19)–(D.21) and (D.23)–(D.25), respectively, and $k>0$ is the linear feedback gain.

For the stability analysis, let us consider the normalized PDE representation of (8.109)–(8.110) given by (D.23)–(D.25). We introduce the backstepping transformation [28]

$$w_{\text{av}}(z,t)=v_{\text{av}}(z,t)-kp_v(z,t), \tag{D.36}$$

where $p_v(z,t)$ is defined in (D.35) and its associated inverse transformation is [28]

$$v_{\text{av}}(z,t)=w_{\text{av}}(z,t)+k\pi_v(z,t), \tag{D.37}$$

where

$$\pi_v(z,t)=G_{\text{av}}(t)-D(G_{\text{av}}(t))H\int_0^z (w_{\text{av}}(s,t)+k\pi_v(s,t))\,ds, \quad z\in[0,1]. \tag{D.38}$$

Applying the backstepping transformation (D.36) to (D.23)–(D.25), the target system is given by

$$\partial_t w_{\rm av}(z,t) = \lambda_w(z,t)\partial_z w_{\rm av}(z,t), \quad z \in [0,1], \tag{D.39}$$

$$w_{\rm av}(1,t) = 0, \tag{D.40}$$

$$\dot{G}_{\rm av}(t) = kHG_{\rm av}(t) + Hw_{\rm av}(0,t), \tag{D.41}$$

where

$$\lambda_w(z,t) = \frac{1 - zH\nabla D(G_{\rm av}(t))\left(kG_{\rm av}(t) + w_{\rm av}(0,t)\right)}{D(G_{\rm av}(t))}. \tag{D.42}$$

Considering the Lyapunov functional

$$L(t) = \frac{1}{2}\int_0^1 e^{gz} w_{\rm av}(z,t)^2 dz, \tag{D.43}$$

we can calculate its time derivative as follows:

$$\begin{aligned}
\dot{L}(t) &= \int_0^1 e^{gz} w_{\rm av}(z,t)\partial_t w_{\rm av}(z,t)dz \\
&= \int_0^1 e^{gz} w_{\rm av}(z,t)\lambda_w(z,t)\partial_z w_{\rm av}(z,t)dz \\
&= \left[e^{gz}\lambda_w(z,t)w_{\rm av}(z,t)^2\right]_0^1 \\
&\quad - \int_0^1 \left(ge^{gz}\lambda_w(z,t) + e^{gz}\partial_z\lambda_w(z,t)\right) w_{\rm av}(z,t)^2 dz \\
&= e^{g}\lambda_w(1,t)w_{\rm av}(1,t)^2 - \lambda_w(0,t)w_{\rm av}(0,t)^2 \\
&\quad - \int_0^1 e^{gz}\left(g\lambda_w(z,t) + \partial_z\lambda_w(z,t)\right) w_{\rm av}(z,t)^2 dz.
\end{aligned} \tag{D.44}$$

By definition,

$$\partial_z\lambda_w(z,t) = \frac{-\dot{D}(G_{\rm av}(t))}{D(G_{\rm av}(t))} \tag{D.45}$$

and

$$\begin{aligned}
\Phi(z) &= g\lambda_w(z,t) + \partial_z\lambda_w(z,t) \\
&= \frac{g - (1+gz)\dot{D}(G_{\rm av}(t))}{D(G_{\rm av}(t))}.
\end{aligned} \tag{D.46}$$

Accounting for the derivative of the state-dependent delay, we can rewrite (D.46) into

$$\Phi(z) = \frac{g - (1+gz)F_w(t)}{D(G_{\rm av}(t))}, \tag{D.47}$$

where the feasibility condition is ensured if

$$F_w(t) = H\nabla D(G_{\rm av}(t))\left(kG_{\rm av}(t) + w_{\rm av}(0,t)\right), \tag{D.48}$$

$$F_w(t) < \bar{c}, \quad 0 < \bar{c} < 1. \tag{D.49}$$

Noting that the function (D.47) is a linear function of z, its minimum is given by $\Phi(0)$ or $\Phi(1)$. Thus,

$$\Phi(x) \geq \min\{\Phi(0), \Phi(1)\}, \tag{D.50}$$

$$\min\{\Phi(0), \Phi(1)\} = \frac{\min\{g+F_w(t), (1+g)F_w(t)\}}{D(G_{\text{av}}(t))}. \tag{D.51}$$

Recalling the prediction time $\sigma(t) = t + D(G_{\text{av}}(t))$ is always greater than the current time t, we can also show

$$\Phi(0) \geq \frac{g+\bar{c}}{D(G_{\text{av}}(t))}, \tag{D.52}$$

$$\Phi(1) \geq \frac{(1+g)\bar{c}}{D(G_{\text{av}}(t))} \quad \forall g > 0. \tag{D.53}$$

To preserve the causality of the system, the state-dependent delay function $D \in C^1(\mathbb{R}; \mathbb{R}_+)$ must be positive and therefore the function $\frac{1}{D(G_{\text{av}}(t))}$ for all $G_{\text{av}} \in \mathbb{R}$ admits a positive minimum g_0^* which implies that

$$g_0^* \Phi^* \leq \min\{\Phi(0), \Phi(1)\}, \tag{D.54}$$

$$\Phi^* = \min\{(g+\bar{c}), (1+g)\bar{c}\} = (1+g)\bar{c} \quad \forall \bar{c} < 1. \tag{D.55}$$

Choosing $g > 0$, from (D.40) and (D.44), the following inequality can be established:

$$\dot{L}(t) \leq -g_0^* w_{\text{av}}(0,t)^2 - \int_0^1 g_0^* \Phi^* e^{gx} w_{\text{av}}(x,t)^2 dx < 0. \tag{D.56}$$

Hence, the exponential stability of the complete average closed-loop system (D.23)–(D.25) and (D.34)–(D.35) can be guaranteed with the Lyapunov–Krasovskii functional

$$V(t) = \frac{G_{\text{av}}^2(t)}{2} + L(t) + \frac{1}{2} w_{\text{av}}^2(1,t), \tag{D.57}$$

reminding one that $G_{\text{av}}(t) = H\tilde{\theta}_{\text{av}}(t)$ from (8.107). It means that $G_{\text{av}}(t) \to 0$ implies $\tilde{\theta}_{\text{av}}(t) \to 0$, as desired.

D.3 ▪ Proof of (10.34)

First, by plugging (10.25) into (10.7), such that the output of the static map is given in terms of $\vartheta(t)$, we have

$$y(t) = y^* + \frac{H}{2}(\vartheta(t) + a\sin(\omega t))^2. \tag{D.58}$$

Then, by plugging (D.58) into (10.27), one has

$$\begin{aligned} G(t) &= \frac{2}{a}\sin(\omega t)\left(y^* + \frac{H}{2}(\vartheta(t) + a\sin(\omega t))^2\right) \\ &= \frac{2}{a} y^* \sin(\omega t) + \frac{H}{a}\left(\vartheta^2(t)\sin(\omega t)\right. \\ &\quad \left. + 2a\vartheta(t)\sin^2(\omega t) + a^2\sin^3(\omega t)\right). \end{aligned} \tag{D.59}$$

Furthermore, by plugging (D.58) into (10.26), one has

$$\begin{aligned}\hat{H}(t) =& -\frac{8}{a^2}\cos(2\omega t)\left(y^*\frac{H}{2}(\vartheta(t)+a\sin(\omega t))^2\right)\\ =& -\frac{8}{a^2}y^*\cos(2\omega t)-\frac{4H}{a^2}\,(\vartheta(t)\cos(2\omega t)\\ & +2a\vartheta(t)\sin(\omega t)\cos(2\omega t)+a^2\sin^2(\omega t)\cos(2\omega t)\big).\end{aligned} \tag{D.60}$$

Averaging (D.59) and (D.60) with ω large, hence $\vartheta(t)\approx$ const, results in

$$G_{\text{av}}(t)=H\vartheta_{\text{av}}(t),\quad \hat{H}_{\text{av}}(t)=H, \tag{D.61}$$

since

$$\begin{aligned}&\frac{\omega}{2\pi}\int_0^{\frac{2\pi}{\omega}}\sin(\omega\tau)d\tau=0,\quad \frac{\omega}{2\pi}\int_0^{\frac{2\pi}{\omega}}\sin^2(\omega\tau)d\tau=\frac{1}{2},\\ &\frac{\omega}{2\pi}\int_0^{\frac{2\pi}{\omega}}\sin^3(\omega\tau)d\tau=0,\quad \frac{\omega}{2\pi}\int_0^{\frac{2\pi}{\omega}}\sin^2(\omega\tau)d\tau=0,\\ &\frac{\omega}{2\pi}\int_0^{\frac{2\pi}{\omega}}\sin(\omega\tau)\cos(2\omega\tau)d\tau=0,\\ &\frac{\omega}{2\pi}\int_0^{\frac{2\pi}{\omega}}\sin^2(\omega\tau)\cos(2\omega t)d\tau=-\frac{1}{4}.\end{aligned} \tag{D.62}$$

D.4 ▪ Derivation of (10.69)

Starting with (10.68), we apply Young's inequality on $\vartheta_{\text{av}}(t)w(0,t)$ and Cauchy–Schwarz's and Young's inequalities on $-b\lambda^2 w_x(D,t)\left[\int_0^D\left(e^{-\lambda(D-r)}-1\right)w(y,t)dr\right]$ and

$$-d\lambda^2 w(D,t)\left[\int_0^D\left(e^{-\lambda(D-r)}-1\right)w(r,t)dr+e^{-\lambda D}\vartheta_{\text{av}}(t)\right],$$

where $\gamma_1,\gamma_2,\gamma_5>0$. Furthermore we choose $b=\frac{a}{c-\lambda}$, such that

$$\begin{aligned}\dot{\Upsilon}(t)\leq&\left(\frac{1}{2\gamma_1}-\lambda\right)\vartheta_{\text{av}}^2(t)+\frac{\gamma_1}{2}w^2(0,t)-a\|w_x(t)\|^2\\ &+\frac{1}{2\gamma_2}\|w(t)\|^2+\frac{\gamma_2}{2}b^2\lambda^4\left(\int_0^D(e^{-\lambda(D-r)}-1)^2dr\right)w_x^2(D,t)\\ &-b\lambda^2e^{-\lambda D}w_x(D,t)\vartheta_{\text{av}}(t)-b\|w_{xx}(t)\|^2\\ &+\frac{1}{2\gamma_5}\|w(t)\|^2+\frac{\gamma_5}{2}d^2\lambda^4\left(\int_0^D(e^{-\lambda(D-r)}-1)^2dr\right)w^2(D,t)\\ &-d\lambda^2e^{-\lambda D}w(D,t)\vartheta_{\text{av}}(t)-d(c-\lambda)w^2(D,t).\end{aligned} \tag{D.63}$$

By Agmon's and Poincaré's inequalities, we find the upper bound $w_x^2(D,t)\leq 4D\|w_{xx}^2(t)\|^2$ and again with Poincaré's inequality we can bound $\|w(t)\|^2\leq 2Dw^2(D,t)+4D^2\|w_x(t)\|^2$. Hence,

(D.63) results in

$$
\begin{aligned}
\dot{\Upsilon}(t) \leq & \left(\frac{1}{2\gamma_1} - \lambda\right)\vartheta_{\text{av}}^2(t) + \frac{\gamma_1}{2}w^2(0,t) \\
& + \left(\frac{D}{\gamma_2} + \frac{\gamma_5}{2}d^2\lambda^4\left(\int_0^D (e^{-\lambda(D-r)} - 1)^2 dr\right)\right. \\
& \left. + \frac{D}{\gamma_5} - d(c-\lambda)\right) w^2(D,t) \\
& + \left(\frac{2D^2}{\gamma_2} + \frac{2D^2}{\gamma_5} - a\right)\|w_x(t)\|^2 \\
& + \left(2D\gamma_2 b^2\lambda^4\left(\int_0^D (e^{-\lambda(D-r)} - 1)^2 dr\right) - b\right)\|w_{xx}(t)\|^2 \\
& - d\lambda^2 e^{-\lambda D} w(D,t)\vartheta_{\text{av}}(t) - b\lambda^2 e^{-\lambda D} w_x(D,t)\vartheta_{\text{av}}(t).
\end{aligned}
\tag{D.64}
$$

Finally, apply Young's inequality on $-\lambda^2 b w_x(D,t)e^{\lambda D}\vartheta_{\text{av}}(t)$, and apply Agmon's, Poincaré's, Young's inequalities to find an upper bound of $w^2(0,t)$, where $\gamma_3, \gamma_4 > 0$:

$$
\begin{aligned}
\dot{\Upsilon}(t) \leq & \left(\frac{1}{2\gamma_1} + \frac{\gamma_3}{2} + \frac{\gamma_6}{2} - \lambda\right)\vartheta_{\text{av}}^2(t) \\
& + \left(\frac{D}{\gamma_2} + \frac{\gamma_5}{2}d^2\lambda^4\left(\int_0^D (e^{-\lambda(D-r)} - 1)^2 dr\right)\right. \\
& \left. + \frac{D}{\gamma_5} - d(c-\lambda) + \frac{\gamma_1}{2}\left(1 + \frac{2D}{\gamma_4}\right) + \frac{d^2\lambda^4}{2\gamma_6}\right) w^2(D,t) \\
& + \left(\frac{\gamma_1}{2}\left(\frac{4D^2}{\gamma_4} + \gamma_4\right) + \frac{2D^2}{\gamma_2} + \frac{2D^2}{\gamma_5} - a\right)\|w_x(t)\|^2 \\
& + \left(2D\gamma_2 b^2\lambda^4\left(\int_0^D (e^{-\lambda(D-r)} - 1)^2 dr\right)\right. \\
& \left. + \frac{2Db^2\lambda^4}{\gamma_3} - b\right)\|w_{xx}(t)\|^2.
\end{aligned}
\tag{D.65}
$$

Choosing $\lambda_2 = \lambda_6 = \lambda/2$, $\gamma_1 = 2/\lambda$, $\gamma_4 = 1$, and $\gamma_2 = \gamma_5 = \lambda^{-1}\int_0^D (e^{-\lambda(D-r)} - 1)^{-2} dr$ leads to (10.69).

D.5 ▪ Constants of Step 4 in Proof of Theorem 10.1

It can easily be shown with (10.59) and (10.64) that

$$
\begin{aligned}
&\|w(t)\|^2 \leq \alpha_1\|u_{\text{av}}(t)\|^2 + \alpha_2\vartheta_{\text{av}}^2(t), \\
&\|\partial_x w(t)\|^2 \leq \beta_1\|u_{\text{av}}(t)\|^2 + \beta_2\|\partial_x u_{\text{av}}(t)\|^2, \\
&w^2(D,t) \leq \gamma_1 u_{\text{av}}^2(D,t) + \gamma_2\|u_{\text{av}}(t)\|^2 + \gamma_3\vartheta_{\text{av}}^2(t), \\
&\|u_{\text{av}}(t)\|^2 \leq \delta_1\|w(t)\|^2 + \delta_2\vartheta_{\text{av}}^2(t), \\
&\|\partial_x u_{\text{av}}(t)\|^2 \leq \epsilon_1\|w(t)\|^2 + \epsilon_2\|\partial_x w(t)\|^2 + \epsilon_3\vartheta_{\text{av}}^2(t), \\
&u_{\text{av}}^2(D,t) \leq \zeta_1 w^2(D,t) + \zeta_2\|w(t)\|^2 + \zeta_3\vartheta_{\text{av}}^2(t),
\end{aligned}
\tag{D.66}
$$

with

$$\alpha_1 = \left(3+\lambda^2 D^4\right),\ \alpha_2 = 3D\lambda^2,$$
$$\beta_1 = 2\lambda^2 D^2,\ \beta_2 = 2,$$

$$\gamma_1 = 3, \quad \gamma_2 = D^3 dx,\ \gamma_3 = 3\lambda^2,$$
$$\delta_1 = 3+3\lambda^2 D\|e^{-\lambda x}-1\|^2,\ \delta_2 = \frac{3\lambda}{2}(e^{-2\lambda D}-1),$$
$$\epsilon_1 = \frac{3\lambda^3 D}{2}(e^{-2\lambda D}-1),\ \epsilon_2 = 3,\ \epsilon_3 = \epsilon_2/D,$$
$$\zeta_1 = 3,\ \zeta_2 = 3\lambda^2\|e^{-\lambda x}-1\|^2,\ \zeta_3 = 3\lambda^2 e^{-2\lambda D},$$

such that

$$\Upsilon \le \bar{\tau}\bar{\sigma}\Psi \quad \text{and} \quad \underline{\tau}\underline{\sigma}\Psi \le \Upsilon$$

with

$$\underline{\tau} = 1/\max\{\delta_2+\epsilon_3+\zeta_3+1, \delta_1+\epsilon_1+\zeta_2, \epsilon_2, \zeta_1\},$$
$$\bar{\tau} = \max\{\alpha_2+\gamma_3+1, \alpha_1+\beta_1+\gamma_2, \beta_2, \gamma_1\}.$$

D.6 ▪ Deriving the Perturbation Signal (11.21) for the ES Control Loop with RAD PDEs

Considering the system (11.17)–(11.20) and transforming it, as in Section 11.4, with $\bar{\beta}(x,t) = \beta(x,t)e^{\frac{b}{2\epsilon}x}$ into an reaction-diffusion system, we have

$$S(t) = \bar{\beta}(1,t)e^{-\frac{b}{2\epsilon}}, \tag{D.67}$$
$$\bar{\beta}_t(x,t) = \epsilon\bar{\beta}_{xx}(x,t) - \xi\bar{\beta}(x,t), \quad x\in[0,1], \tag{D.68}$$
$$\bar{\beta}(0,t) = \frac{b}{2\epsilon}a\sin(\omega t), \tag{D.69}$$
$$\bar{\beta}(0,t) = a\sin(\omega t), \tag{D.70}$$

with

$$\xi = \frac{b^2}{4\epsilon} - \lambda. \tag{D.71}$$

As in Example 12.2 in [128], we postulate the full-state reference trajectory in the form

$$\bar{\beta}(x,t) = \sum_{k=0}^{\infty} a_k(t)\frac{x^k}{k!}. \tag{D.72}$$

From (D.69) and (D.70), we have

$$a_0(t) = a\sin(\omega t), \qquad a_1(t) = \frac{ab}{2\epsilon}\sin(\omega t), \tag{D.73}$$

and with the PDE (D.68), we get

$$a_{k+2}(t) = \frac{1}{\epsilon}\dot{a}_k(t) + \xi a_k(t). \tag{D.74}$$

Additionally, we observe the relation

$$a_{2k+1}(t) = \frac{b}{2\epsilon} a_{2k}. \tag{D.75}$$

Calculating a_{2k}, $k = 1, \ldots, 4$, only depending on $a_0^{(k)}$, $k = 1, \ldots, 4$, we have

$$\epsilon a_2 = \dot{a}_0 + \xi a_0, \tag{D.76}$$
$$\epsilon^2 a_4 = \ddot{a}_0 + 2\xi \dot{a}_0 + \xi^2 a_0, \tag{D.77}$$
$$\epsilon^3 a_6 = \dddot{a}_0 + 3\xi \ddot{a}_0 + 3\xi^2 \dot{a}_0 + \xi^3 a_0, \tag{D.78}$$
$$\epsilon^4 a_8 = \ddddot{a}_0 + 4\xi \dddot{a}_0 + 6\xi^2 \ddot{a}_0 + 4\xi^3 \dot{a}_0 + \xi^4 a_0. \tag{D.79}$$

By the derivative-law of the sine we can write

$$a_0^{(2k)} = (-1)^k \omega^{2k} a_0, \qquad a_0^{(2k+1)} = (-1)^k \omega^{2k} \dot{a}_0 \tag{D.80}$$

and (D.76)–(D.79) can be rewritten as

$$\frac{\epsilon}{a} a_2 = +\xi \sin(\omega t) + \omega \cos(\omega t), \tag{D.81}$$
$$\frac{\epsilon^2}{a} a_4 = (\xi^2 - \omega^2) \sin(\omega t) + 2\omega\xi \cos(\omega t), \tag{D.82}$$
$$\frac{\epsilon 3}{a} a_6 = (3\xi\omega^2 + \xi^3) \sin(\omega t) + (-\omega^3 + 3\xi^2 \omega) \cos(\omega t), \tag{D.83}$$
$$\frac{\epsilon^4}{a} a_8 = (\omega^4 + 6\xi^2\omega^2 + \xi^4) \sin(\omega t) + (4\xi\omega^3 + 4\xi^3). \tag{D.84}$$

Observation reveals that the coefficients of the sine and cosine terms of (D.81)–(D.84) are the same as of the modified Pascal triangle

$k=0$:					1				
$k=1$:				ω		ξ			
$k=2$:			ω^2		$2\xi\omega$		ξ^2		
$k=3$:		ω^3		$3\xi\omega^2$		$3\xi^2\omega$		ξ^3	
$k=4$:	ω^4		$4\xi\omega^3$		$6\xi^2\omega^2$		$4\xi^3\omega$		ξ^4,

where the bold (red) coefficients belongs to the sine and the other coefficients to the cosine terms. Since we have an iterative law to calculate a_{2k}, the expansion for $k > 4$ continues in the same way. Hence, with

$$P_m(k) = \omega^m \prod_{j=m}^{k} \binom{j}{m} \xi^j, \tag{D.85}$$

we can determine the value in the mth diagonal and kth row, where the diagonal $m = 0$ is $\{1, \xi, \xi^2, \xi^3, \ldots\}$. Combining this result and defining $\binom{y}{z} := 0$ for $y < z$ we arrive at

$$a_{2k} := \frac{a}{\epsilon^k} \sin(\omega t) \sum_{n=0}^{k} \binom{k}{2n} \xi^{k-2n} \omega^{2n} + \frac{a}{\epsilon^k} \cos(\omega t) \sum_{n=0}^{k} \binom{k}{2n+1} \xi^{k-2n-1} \omega^{2n+1} \tag{D.86}$$

$$\text{with} \quad \xi := \frac{b^2}{4\epsilon} - \lambda.$$

Finally with (D.72) and (D.67), we get

$$S(t) = e^{-\frac{b}{2\epsilon}} \sum_{k=0}^{\infty} \frac{a_{2k}(t)}{(2k)!} + \frac{b}{2\epsilon} \frac{a_{2k}(t)}{(2k+1)!}. \tag{D.87}$$

D.7 ▪ Inverse Backstepping Transformation (11.64) of the RAD-PDE Case

In this part of the appendix, we derive the inverse backstepping transformation of the backstepping transformation (11.29). Consider a candidate for the inverse backstepping transformation and its time and spatial derivatives of the backstepping transformation (11.29),

$$\bar{z}(x,t) = w(x,t) - \int_0^x p(x,y)w(y,t)dy - \eta(x)\vartheta(t), \tag{D.88}$$

$$\begin{aligned}\bar{z}_t(x,t) &= \epsilon w_{xx}(x,t) + w(x,t)\left[-\xi + \epsilon p_y(x,x)\right] - \epsilon p(x,x)w_x(x,t) \\ &\quad + w(0,t)\left[\frac{b}{2}p(x,0) - \epsilon p_y(x,0) - \eta(x)\right] - \bar{K}\eta(x)\vartheta(t) \\ &\quad - \int_0^x w(y,t)\left[\epsilon p_{yy}(x,t) - \xi p(x,y)\right]dy,\end{aligned} \tag{D.89}$$

$$\bar{z}_x(x,t) = w_x(x,t) - p(x,x)w(x,t) - \int_0^x p_x(x,y)w(y,t)dy - \eta'(x)\vartheta(t), \tag{D.90}$$

$$\begin{aligned}\bar{z}_{xx}(x,t) &= w_{xx}(x,t) - w(x,t)\frac{d}{dx}p(x,x) - p(x,x)w_x(x,t) - p_x(x,x)w(x,t) \\ &\quad - \int_0^x p_{xx}(x,y)w(y,t)dy - \eta''(x)\vartheta(t),\end{aligned} \tag{D.91}$$

where $p(x,y)$ is the gain kernel and $\eta(x)$ is a function in x. Note that the boundary value (11.27) and the integration by parts were used to derive (D.89). Inserting (D.88), (D.89), and (D.91) into the PDE (11.26)

$$\begin{aligned}\bar{z}_t(x,t) - \epsilon\bar{z}_{xx}(x,t) + \xi\bar{z}(x,t) &= 2\epsilon w(x,t)\frac{d}{dx}p(x,x) \\ &\quad + w(0,t)\left[\frac{b}{2}p(x,0) - \epsilon p_y(x,0) - \eta(x)\right] \\ &\quad + \vartheta(t)\left[-\eta(x)(\bar{K}+\xi) - \eta''(x)\right] \\ &\quad + \int_0^x \epsilon w(y,t)\left[p_{xx}(x,y) - p_{yy}(x,y)\right]dy\end{aligned} \tag{D.92}$$

and along with (D.88), (D.90), evaluated at $x=0$ plus the boundary value (11.27), leads to the following conditions for the gain kernel $p(x,y)$ and the function $\eta(x)$:

$$\eta(0) = -\bar{K}, \tag{D.93}$$

$$\eta'(0) = -\bar{K}\frac{b}{2\epsilon}, \tag{D.94}$$

$$\eta''(x) = \eta(x)(\bar{K}+\xi), \tag{D.95}$$

$$p(x,0) = \frac{2\epsilon}{b}p_y(x,0) + \frac{2}{b}\eta(x), \tag{D.96}$$

$$p(0,0) = 0, \tag{D.97}$$

$$p_{xx}(x,y) - p_{yy}(x,y) = 0. \tag{D.98}$$

The second-order ODE system (D.93)–(D.95) can be solved with the ansatz $\eta(x) = A\cosh(\mu x) + B\sinh(\mu x)$ to

$$\eta(x) = -\bar{K}\left(\cosh\left(\sqrt{\bar{K}+\xi}x\right) + \frac{b}{2\epsilon\sqrt{\bar{K}+\xi}}\sinh\left(\sqrt{\bar{K}+\xi}x\right)\right). \tag{D.99}$$

The hyperbolic PDE system of Goursat type (D.96)–(D.98) can be solved by the ansatz $p(x,y) = m(x-y)$, where $m(\cdot)$ is a scalar function. Additionally, by applying the variation-of-constants formula we get

$$p(x,y) = -\frac{\bar{K}}{\sqrt{\bar{K}+\xi}}\sinh\left(\sqrt{\bar{K}+\xi}(x-y)\right). \tag{D.100}$$

Hence, we have the backstepping and its inverse transformation in the variables (u,w),

$$w(x,t) = e^{\frac{b}{2\epsilon}x}u(x,t) - \int_0^x q(x,y)e^{\frac{b}{2\epsilon}y}u(y,t)dy - \gamma(x)\vartheta(t), \tag{D.101}$$

$$u(x,t) = e^{-\frac{b}{2\epsilon}x}w(x,t) - e^{-\frac{b}{2\epsilon}x}\int_0^x p(x,y)w(y,t)dy - e^{-\frac{b}{2\epsilon}x}\eta(x)\vartheta(t), \tag{D.102}$$

with $q(x,y)$ in (11.30) and $\gamma(x)$ in (11.31).

Appendix E
Important Inequalities

In this book, the inequalities of Young, Poincaré, Agmon, and Cauchy–Schwarz are used very frequently. Hence, we state these inequalities in the fashion we use them, namely, for a spatial domain $[0, D]$. In the literature [119], these inequalities are stated for the normed spatial domain $[0, 1]$.

Definition E.1. *Poincaré's inequality:*

$$\|w(t)\|^2 \leq 2Dw(D,t)^2 + 4D^2\|w_x(t)\|^2, \tag{E.1}$$

$$\|w(t)\|^2 \leq 2Dw(0,t)^2 + 4D^2\|w_x(t)\|^2. \tag{E.2}$$

Definition E.2. *Agmon's inequality (Case 1):*

$$\max_{x\in[0,D]} |w(x,t)|^2 \leq w(0,t)^2 + 2\|w(t)\|\,\|w_x(t)\|, \tag{E.3}$$

$$\max_{x\in[0,D]} |w(x,t)|^2 \leq w(D,t)^2 + 2\|w(t)\|\,\|w_x(t)\|. \tag{E.4}$$

Definition E.3. *Agmon's inequality (Case 2):*

$$w(0,t)^2 \leq \frac{D+1}{D}\|w(t)\|^2 + \|w_x(t)\|^2, \tag{E.5}$$

$$w(D,t)^2 \leq \frac{D+1}{D}\|w(t)\|^2 + \|w_x(t)\|^2. \tag{E.6}$$

Definition E.4. *Young's inequality:*

$$ab \leq \frac{\gamma}{2}a^2 + \frac{1}{2\gamma}b^2, \quad \gamma > 0. \tag{E.7}$$

Definition E.5. *Cauchy–Schwarz inequality:*

$$\int_0^D u(x,t)w(x,t)dx \leq \|u(t)\|\,\|w(t)\|. \tag{E.8}$$

Bibliography

[1] U. J. AARSNES, O. M. AAMO, AND M. KRSTIĆ, *Extremum seeking for real-time optimal drilling control*, in IEEE American Control Conference, Philadelphia, 2019, pp. 5222–5227.

[2] V. ADETOLA AND M. GUAY, *Parameter convergence in adaptive extremum-seeking control*, Automatica, 43 (2007), pp. 105–110.

[3] R. AIREAPETYAN, *Continuous Newton method and its modification*, Applicable Analysis Journal, 73 (1999), pp. 463–484.

[4] C. ALASSEUR, I. B. TAHER, AND A. MATOUSSI, *An extended mean field game for storage in smart grids*, Journal of Optimization Theory and Applications, 184 (2020), pp. 644–670.

[5] N. ALIBEJI, N. KIRSCH, S. FARROKHI, AND N. SHARMA, *Further results on predictor-based control of neuromuscular electrical stimulation*, IEEE Transactions on Neural Systems and Rehabilitation Engineering, 23 (2015), pp. 1095–1105.

[6] N. ALIBEJI AND N. SHARMA, *A PID-type robust input delay compensation method for uncertain Euler–Lagrange systems*, IEEE Transactions on Control Systems Technology, 25 (2017), pp. 2235–2242.

[7] T. ALPCAN AND T. BAŞAR, *Network Security: A Decision and Game Theoretic Approach*, Cambridge University Press, 2011.

[8] S. AMIN, G. A. SCHWARTZ, AND S. S. SASTRY, *Security of interdependent and identical networked control systems*, Automatica, 49 (2013), pp. 186–192.

[9] T. M. APOSTOL, *Mathematical Analysis*, Addison-Wesley, 1974.

[10] K. ARIYUR AND M. KRSTIĆ, *Real-Time Optimization by Extremum-Seeking Control*, John Wiley & Sons, 2003.

[11] ———, *Slope seeking: A generalization of extremum seeking*, International Journal of Adaptive Control and Signal Processing, 18 (2004), pp. 1–22.

[12] Z. ARTSTEIN, *Linear systems with delayed controls: A reduction*, IEEE Transactions on Automatic Control, 27 (1982), pp. 869–979.

[13] Z. ARTSTEIN AND M. SLEMROD, *On singularly perturbed retarded functional differential equations*, Journal of Differential Equations, 171 (2001), pp. 88–109.

[14] K. T. ATTA, A. JOHANSSON, AND T. GUSTAFSSON, *Extremum seeking control based on phasor estimation*, Systems & Control Letters, 85 (2015), pp. 37–45.

[15] ———, *On the stability analysis of phasor and classic extremum seeking control*, Systems & Control Letters, 91 (2016), pp. 55–62.

[16] J. AURIOL AND F. D. MEGLIO, *Robust output feedback stabilization for two heterodirectional linear coupled hyperbolic PDEs*, Automatica, 115 (2020), 108896.

[17] D. AUSSEL AND A. SVENSSON, *Towards tractable constraint qualifications for parametric optimisation problems and applications to generalised Nash games*, Journal of Optimization Theory and Applications, 182 (2019), pp. 404–416.

[18] F. E. AZAR AND M. PERRIER, *Slope seeking control using multi-units*, in IEEE Conference on Control Applications (CCA), 2014, pp. 1041–1045.

[19] A. BACCOLI, A. PISANO, AND Y. ORLOV, *Boundary control of coupled reaction–diffusion processes with constant parameters*, Automatica, 54 (2015), pp. 80–90.

[20] T. BAŞAR, *Relaxation techniques and the on-line asynchronous algorithms for computation of noncooperative equilibria*, Journal of Economic Dynamics and Control, 11 (1987), pp. 531–549.

[21] T. BAŞAR AND G. J. OLSDER, *Dynamic Noncooperative Game Theory*, SIAM, 1998.

[22] T. BAŞAR AND G. ZACCOUR (editors), *Handbook of Dynamic Game Theory*, Volume I (Theory of Dynamic Games), Springer, 2018.

[23] ———, *Handbook of Dynamic Game Theory*, Volume II (Applications of Dynamic Games), Springer, 2018.

[24] N. BEKIARIS-LIBERIS AND M. KRSTIĆ, *Lyapunov stability of linear predictor feedback for distributed input delays*, IEEE Transactions on Automatic Control, 56 (2011), pp. 655–660.

[25] ———, *Nonlinear Control under Nonconstant Delays*, SIAM, 2013.

[26] ———, *Compensation of wave actuator dynamics for nonlinear systems*, IEEE Transactions on Automatic Control, 59 (2014), pp. 1555–1570.

[27] ———, *Control of nonlinear systems with delays*, in Encyclopedia of Systems and Control, T. Samad and J. Baillieul (editors), Springer, 2014, pp. 1–6.

[28] ———, *Compensation of transport actuator dynamics via input-dependent moving uncontrolled boundary*, IEEE Transactions on Automatic Control, 63 (2018), pp. 3889–3896.

[29] M. BENOSMAN, *Learning-Based Adaptive Control: An Extremum Seeking Approach—Theory and Applications*, Butterworth-Heinemann, 2016.

[30] M. BENOSMAN, F. LEWIS, M. GUAY AND D. OWENS, *Editorial for the special issue on learning-based adaptive control: Theory and applications*, International Journal of Adaptive Control and Signal Processing, 33 (2019), pp. 225–228.

[31] D. BRESCH-PIETRI AND M. KRSTIĆ, *Adaptive trajectory tracking despite unknown input delay and plant parameters*, Automatica, 9 (2009), pp. 2075–2081.

[32] ———, *Delay-adaptive predictor feedback for systems with unknown long actuator delay*, IEEE Transactions on Automatic Control, 55 (2010), pp. 2106–2112.

[33] ———, *Compensation of state-dependent input delay for nonlinear systems*, IEEE Transactions on Automatic Control, 58 (2013), pp. 275–289.

[34] ———, *Output-feedback adaptive control of a wave PDE*, Automatica, 50 (2014), pp. 1407–1415.

[35] ———, *Delay-adaptive control for nonlinear systems*, IEEE Transactions on Automatic Control, 59 (2014), pp. 1203–1218.

[36] D. BRESCH-PIETRI AND N. PETIT, *Robust compensation of a chattering time-varying input delay*, in IEEE Conference on Decision and Control, 2014, pp. 457–462.

[37] R. A. BROWN, *Theory of transport processes in single crystal growth from the melt*, American Institute of Chemical Engineers Journals, 34 (1988), pp. 881–911.

[38] H. BUTLER, *Adaptive feedforward for a wafer stage in a lithographic tool*, IEEE Transactions on Control Systems Technology, 21 (2013), pp. 875–881.

[39] X. CAI, N. BEKIARIS-LIBERIS, AND M. KRSTIĆ, *Input-to-state stability and inverse optimality of linear time-varying-delay predictor feedbacks*, IEEE Transactions on Automatic Control, 63 (2018), pp. 233–240.

[40] R. C. CARLSON, I. PAPAMICHAIL, M. PAPAGEORGIOU, AND A. MESSMER, *Optimal motorway traffic flow control involving variable speed limits and ramp metering*, Transportation Science, 44 (2010), pp. 238–253.

[41] R. CARMONA, J-. P. FOUQUE, S. M. MOUSAVI, AND L.- H. SUN, *Systemic risk and stochastic games with delay*, Journal of Optimization Theory and Applications, 179 (2018), pp. 366–399.

[42] M. D. CILETTI, *Differential games with information time lag: Norm-invariant systems*, Journal of Optimization Theory and Applications, 9 (1972), pp. 293–301.

[43] J. COCHRAN, N. GHODS, A. SIRANOSIAN, AND M. KRSTIĆ, *3D source seeking for underactuated vehicles without position measurement*, IEEE Transactions on Robotics, 25 (2009), pp. 245–252.

[44] J. CORON, *Control and Nonlinearity*, American Mathematical Society, 2006.

[45] J. COTRINA AND J. ZUNIGA, *Time-dependent generalized Nash equilibrium problem*, Journal of Optimization Theory and Applications, 179 (2018), pp. 1054–1064.

[46] P. COUGNON, D. DOCHAIN, M. GUAY, AND M. PERRIER, *Real-time optimization of a tubular reactor with distributed feed*, American Institute of Chemical Engineers Journal, 52 (2006), pp. 2120–2128.

[47] R. DATKO, J. LAGNESE, AND M. P. POLIS, *An example on the effect of time delays in boundary feedback stabilization of wave equations*, SIAM Journal on Control and Optimization, 24 (1986), pp. 152–156.

[48] H. DENG AND M. KRSTIC, *Stochastic nonlinear stabilization—Part* II*: Inverse optimality*, Systems and Control Letters, 32 (1997), pp. 151–159.

[49] H. DENG, M. KRSTIC, AND R. WILLIAMS, *Stabilization of stochastic nonlinear systems driven by noise of unknown covariance*, IEEE Transactions on Automatic Control, 46 (2001), pp. 1237–1253.

[50] W. DENG, J. YAO, AND D. MA, *Time-varying input delay compensation for nonlinear systems with additive disturbance: An output feedback approach*, International Journal of Robust and Nonlinear Control, 28 (2018), pp. 31–52.

[51] G. DERVISOGLU, G. GOMES, J. KWON, R. HOROWITZ, AND P. VARAIYA, *Automatic calibration of the fundamental diagram and empirical observations on capacity*, in Transportation Research Board 88th Annual Meeting, 15 (2009), pp. 1–14.

[52] M. DIAGNE, N. BEKIARIS-LIBERIS, AND M. KRSTIĆ, *Time- and state-dependent input delay-compensated bang-bang control of a screw extruder for* 3D *printing*, International Journal of Robust and Nonlinear Control, 27 (2017), pp. 3727–3757.

[53] M. DIAGNE, N. BEKIARIS-LIBERIS, A. OTTO, AND M. KRSTIĆ, *Control of transport PDE/nonlinear ODE cascades with state-dependent propagation speed*, IEEE Transactions on Automatic Control, 62 (2017), pp. 6278–6293.

[54] D. Dochain, M. Perrier, and M. Guay, *Extremum seeking control and its application to process and reaction systems: A survey*, Mathematics and Computers in Simulation, 82 (2011), pp. 369–380.

[55] V. Dragan and A. Ionita, *Exponential stability for singularly perturbed systems with state delays*, Electronic Journal of Qualitative Theory of Differential Equations, 6 (2000), pp. 1–8.

[56] C. S. Draper and Y. T. Li, *Principles of Optimalizing Control Systems and an Application to the Internal Combustion Engine*, American Society of Mechanical Engineers, 1951.

[57] W. B. Dunbar, N. Petit, P. Rouchon, and P. Martin, *Motion planning for a nonlinear Stefan problem*, ESAIM: Control, Optimisation and Calculus of Variations, 9 (2003), pp. 275–296.

[58] H. B. Durr, M. Krstić, A. Scheinker, and C. Ebenbauer, *Extremum seeking for dynamic maps using Lie brackets and singular perturbations*, Automatica, 83 (2017), pp. 91–99.

[59] H. B. Durr, M. S. Stankovic, C. Ebenbauer, and K. H. Johansson, *Lie bracket approximation of extremum seeking systems*, Automatica, 49 (2013), pp. 1538–1552.

[60] L. Edelstein-Keshet, *Mathematical Models in Biology*, SIAM, 1988.

[61] H. Ehtamo and R. P. Hämäläinen, *Incentive strategies and equilibria for dynamic games with delayed information*, Journal of Optimization Theory and Applications, 63 (1989), pp. 355–369.

[62] K.-J. Engel and R. Nagel, *One-Parameter Semigroups for Linear Evolution Equations*, Springer Science & Business Media, 1999.

[63] S. Evesque, A. M. Annaswamy, S. Niculescu, and A. P. Dowling, *Adaptive control of a class of time-delay systems*, ASME Journal of Dynamic Systems, Measurement, and Control, 125 (2003), pp. 186–193.

[64] S. Fan and B. Seibold, *Data-fitted first-order traffic models and their second-order generalizations: Comparison by trajectory and sensor data*, Transportation Research Record, 2391 (2013), pp. 32–43.

[65] J. Feiling, S. Koga, M. Krstić, and T. R. Oliveira, *Gradient extremum seeking for static maps with actuation dynamics governed by diffusion PDEs*, Automatica, 95 (2018), pp. 197–206.

[66] Z. Feng and W. X. Zheng, *Improved stability condition for Takagi-Sugeno fuzzy systems with time-varying delay*, IEEE Transactions on Cybernetics, 47 (2017), pp. 661–670.

[67] E. Fridman, *Effects of small delays on stability of singularly perturbed systems*, Automatica, 38 (2002), pp. 897–902.

[68] ———, *Introduction to Time-Delay Systems—Analysis and Control*, Birkhäuser Basel, 2014.

[69] E. Fridman and J. Zhang, *Averaging of linear systems with almost periodic coefficients: A time-delay approach*, Automatica, 122 (2020), pp. 1–12 (109287).

[70] P. Frihauf, M. Krstic, and T. Başar, *Nash equilibrium seeking in noncooperative games*, IEEE Transactions on Automatic Control, 57 (2012), pp. 1192–1207.

[71] ———, *Finite-horizon LQ control for unknown discrete-time linear systems via extremum seeking*, European Journal of Control, 19 (2013), pp. 399–407.

[72] D. Fudenberg and J. Tirole, *Game Theory*, The MIT Press, 1991.

[73] A. Ghaffari, M. Krstić, and D. Nešić, *Multivariable Newton-based extremum seeking*, Automatica, 48 (2012), pp. 1759–1767.

[74] A. GHAFFARI, S. MOURA, AND M. KRSTIĆ, *PDE-based modeling, control, and stability analysis of heterogeneous thermostatically controlled load populations*, ASME Journal of Dynamic Systems, Measurement, and Control, 137 (2015), pp. 1–9.

[75] D. GILBARG AND N. S. TRUDINGER, *Elliptic Partial Differential Equations of Second Order*, Springer, 2015.

[76] V. Y. GLIZER AND J. SHINAR, *Optimal evasion from a pursuer with delayed information*, Journal of Optimization Theory and Applications, 111 (2001), pp. 7–38.

[77] V. GRUSHKOVSKAYA, A. ZUYEV, AND C. EBENBAUER, *On a class of generating vector fields for the extremum seeking problem: Lie bracket approximation and stability properties*, Automatica, 94 (2018), pp. 151–160.

[78] K. GU AND S. I. NICULESCU, *Survey on recent results in the stability and control of time-delay systems*, ASME Journal of Dynamic Systems, Measurement, and Control, 125 (2003), pp. 158–165.

[79] M. GUAY, *A perturbation-based proportional integral extremum-seeking control approach*, IEEE Transactions on Automatic Control, 61 (2016), pp. 3370–3381.

[80] M. GUAY AND D. DOCHAIN, *A time-varying extremum-seeking control approach*, Automatica, 51 (2015), pp. 356–363.

[81] M. GUAY, D. DOCHAIN, AND M. PERRIER, *Adaptive extremum seeking control of continuous stirred tank bioreactors with unknown growth kinetics*, Automatica, 40 (2004), pp. 881–888.

[82] M. GUAY AND T. ZHANG, *Adaptive extremum seeking control of nonlinear dynamic systems with parametric uncertainties*, Automatica, 39 (2003), pp. 1283–1293.

[83] J. K. HALE AND S. M. V. LUNEL, *Averaging in infinite dimensions*, Journal of Integral Equations and Applications, 2 (1990), pp. 463–494.

[84] ———, *Introduction to Functional Differential Equations*, Springer, 1993.

[85] Z. HAN, D. NIYATO, W. SAAD, AND T. BAŞAR, *Game Theory for Next Generation Wireless and Communication Networks: Modeling, Analysis, and Design*, Cambridge University Press, 2009.

[86] R. E. HAYES, *Dynamical analysis of distributed parameter tubular reactors*, Canadian Journal of Chemical Engineering, 80 (2002), pp. 334–335.

[87] A. HEGYI, B. DE SCHUTTER, AND H. HELLENDOORN, *Model predictive control for optimal co-ordination of ramp metering and variable speed limits*, Transportation Research Part C: Emerging Technologies, 13 (2005), pp. 185–209.

[88] L. HENNING, R. BECKER, G. FEUERBACH, R. MUMINOVIA, R. KING, A. BRUNN, AND W. NITSCHE, *Extensions of adaptive slope-seeking for active flow control*, Journal of Systems and Control Engineering, 222 (2008), pp. 309–322.

[89] C. D. HILL, *Parabolic equations in one space variable and the non-characteristic Cauchy problem*, Communications on Pure and Applied Mathematics, 20 (1967), pp. 619–633.

[90] R. A. HORN AND C. R. JOHNSON, *Matrix Analysis*, Cambridge University Press, 1985.

[91] N. HUDON, M. GUAY, M. PERRIER, AND D. DOCHAIN, *Adaptive extremum-seeking control of convection-reaction distributed reactor with limited actuation*, Computers and Chemical Engineering, 32 (2008), pp. 2994–3001.

[92] P. A. IOANNOU AND J. SUN, *Robust Adaptive Control*, PTR Prentice Hall, 1996.

[93] M. JANKOVIC, *Control Lyapunov-Razumikhin functions and robust stabilization of time delay systems*, IEEE Transactions on Automatic Control, 46 (2001), pp. 1048–1060.

[94] ———, *Forwarding, backstepping, and finite spectrum assignment for time delay systems*, Automatica, 45 (2009), pp. 2–9.

[95] ———, *Cross-term forwarding for systems with time delay*, IEEE Transactions on Automatic Control, 54 (2009), pp. 498–511.

[96] R. E. KALMAN, *When is a linear control system optimal?*, ASME Journal of Basic Engineering, 86 (1964), pp. 51–61.

[97] I. KARAFYLLIS AND M. KRSTIĆ, *Numerical schemes for nonlinear predictor feedback*, Mathematics of Control, Signals, and Systems, 26 (2014), pp. 519–546.

[98] ———, *Predictor Feedback for Delay Systems: Implementations and Approximations*, Birkhaüser, 2017.

[99] ———, *Input-to-State Stability for PDEs*, Springer, 2018.

[100] ———, *On the relation of delay equations to first-order hyperbolic partial differential equations*, ESAIM. Control, Optimization and Calculus Variations, 20 (2014), pp. 894–923.

[101] ———, *Stability of integral delay equations and stabilization of age-structured models*, ESAIM. Control, Optimization and Calculus Variations, 23 (2017), pp. 1667–1714.

[102] ———, *ISS in different norms for 1-D parabolic PDEs with boundary disturbances*, SIAM Journal on Control and Optimization, 55 (2017), pp. 1716–1751.

[103] ———, *Small-gain-based boundary feedback design for global exponential stabilization of 1-D semilinear parabolic PDEs*, SIAM Journal on Control and Optimization, 57 (2019), pp. 2016–2036.

[104] I. KARAFYLLIS AND M. PAPAGEORGIOU, *Feedback control of scalar conservation laws with application to density control in freeways by means of variable speed limits*, Automatica, 105 (2019), pp. 228–236.

[105] B. KASKOSZ AND T. TADUMADZE, *A differential game of evasion with delays*, Journal of Optimization Theory and Applications, 44 (1984), pp. 231–268.

[106] L. KATAFYGIOTIS AND Y. TSARKOV, *Averaging and stability of quasilinear functional differential equations with Markov parameters*, Journal of Applied Mathematics and Stochastic Analysis, 12 (1999), pp. 1–15.

[107] V. KAZAKEVICH, *Technique of automatic control of different processes to maximum or to minimum*, Avtorskoe svidetelstvo, USSR Patent, 1943.

[108] H. K. KHALIL, *Nonlinear Systems*, Prentice-Hall, 2002.

[109] S. Z. KHONG, D. NEŠIĆ, Y. TAN, AND C. MANZIE, *Unified frameworks for sampled-data extremum seeking control: Global optimisation and multi-unit systems*, Automatica, 49 (2013), pp. 2720–2733.

[110] S. KOGA, M. DIAGNE, AND M. KRSTIĆ, *Control and state estimation of the one-phase Stefan problem via backstepping design*, IEEE Transactions on Automatic Control, 64 (2019), pp. 510–525.

[111] S. KOGA, I. KARAFYLLIS, AND M. KRSTIĆ, *Input-to-state stability for the control of Stefan problem with respect to heat loss at the interface*, in Proceedings of the 2018 American Control Conference, Milwaukee, 2018, pp. 1740–1745.

[112] S. KOGA AND M. KRSTIĆ, *Materials Phase Change PDE Control & Estimation: From Additive Manufacturing to Polar Ice*, Birkhaüser, 2020.

[113] ——, *Control of the Stefan system and applications: A tutorial*, Annual Reviews in Control, Robotics, and Autonomous Systems, 5 (2022), pp. 547–577.

[114] M. KRSTIĆ, *Performance improvement and limitations in extremum seeking control*, Systems & Control Letters, 39 (2000), pp. 313–326.

[115] ——, *Lyapunov tools for predictor feedbacks for delay systems: Inverse optimality and robustness to delay mismatch*, Automatica, 44 (2008), pp. 2930–2935.

[116] ——, *Compensating actuator and sensor dynamics governed by diffusion PDEs*, Systems & Control Letters, 58 (2009), pp. 372–377.

[117] ——, *Compensating a string PDE in the actuation or sensing path of an unstable ODE*, IEEE Transactions on Automatic Control, 54 (2009), pp. 1362–1368.

[118] ——, *Control of an unstable reaction-diffusion PDE with long input delay*, Systems & Control Letters, 58 (2009), pp. 773–782.

[119] ——, *Delay Compensation for Nonlinear, Adaptive, and PDE Systems*, Birkhaüser, 2009.

[120] ——, *Input delay compensation for forward complete and strict-feedforward nonlinear systems*, IEEE Transactions on Automatic Control, 55 (2010), pp. 287–303.

[121] ——, *Lyapunov stability of linear predictor feedback for time-varying input delay*, IEEE Transactions on Automatic Control, 55 (2010), pp. 554–559.

[122] ——, *Dead-time compensation for wave/string PDEs*, Journal of Dynamic Systems, Measurement, and Control, 133 (2011), pp. 1–13.

[123] ——, *Extremum seeking control*, in Encyclopedia of Systems and Control, T. Samad and J. Baillieul (editors), Springer, 2014.

[124] M. KRSTIĆ AND H. DENG, *Stabilization of Nonlinear Uncertain Systems*, Springer, 1999.

[125] M. KRSTIĆ, I. KANELLAKOPOULOS, AND P. V. KOKOTOVIĆ, *Nonlinear and Adaptive Control Design*, Wiley, 1995.

[126] M. KRSTIC AND Z. H. LI, *Inverse optimal design of input-to-state stabilizing nonlinear controllers*, IEEE Transactions on Automatic Control, 43 (1998), pp. 336–351.

[127] M. KRSTIĆ AND A. SMYSHLYAEV, *Backstepping boundary control for first-order hyperbolic PDEs and application to systems with actuator and sensor delays*, Systems & Control Letters, 57 (2008), pp. 750–758.

[128] ——, *Boundary Control of PDEs: A Course on Backstepping Designs*, SIAM, 2008.

[129] M. KRSTIĆ AND H. H. WANG, *Stability of extremum seeking feedback for general nonlinear dynamic systems*, Automatica, 36 (2000), pp. 595–601.

[130] M. LEBLANC, *Sur l'electrification des chemins de fer au moyen de courants alternatifs de frequence elevee*, Revue Generale de l'Electricite, 1992.

[131] B. LEHMAN, *The influence of delays when averaging slow and fast oscillating systems: Overview*, IMA Journal of Mathematical Control and Information, 19 (2002), pp. 201–215.

[132] M. LEWICKA AND J. J. MANFREDI, *Game theoretical methods in PDEs*, Bollettino dell'Unione Matematica Italiana, 7 (2014), pp. 211–216.

[133] S. Li and T. Başar, *Distributed learning algorithms for the computation of noncooperative equilibria*, Automatica, 23 (1987), pp. 523–533.

[134] Z. -H. Li and M. Krstic, *Optimal design of adaptive tracking controllers for non-linear systems*, Automatica, 33 (1997), pp. 1459–1473.

[135] C. Li, Z. Qu, and M. A. Weitnauer, *Distributed extremum seeking and formation control for nonholonomic mobile network*, Systems & Control Letters, 75 (2015), pp. 27–34.

[136] Z. Li, K. You, S. Song, and S. Dong, *Distributed extremum seeking with stochastic perturbations*, in International Conference on Control and Automation (ICCA), 2018, pp. 367–372.

[137] C. K. Liao, C. Manzie, A. Chapman, and T. Alpcan, *Constrained extremum seeking of a MIMO dynamic system*, Automatica, 108 (2019), pp. 1–9.

[138] J. Lin, S.-M. Lee, H.-J. Lee, and Y.-M. Koo, *Modeling of typical microbial cell growth in batch culture: Modeling of typical microbial cell growth in batch culture*, Biotechnology and Bioprocess Engineering, 5 (2000), pp. 382–385.

[139] J. B. Lin and S. J. Song, K. Y. You, and M. Krstić, *Overshoot-free nonholonomic source seeking in* 3*D*, International Journal of Adaptive Control and Signal Processing, 31 (2017), pp. 1285–1295.

[140] S.-J. Liu and M. Krstić, *Stochastic averaging in continuous time and its applications to extremum seeking*, IEEE Transactions on Automatic Control, 55 (2010), pp. 2235–2250.

[141] ——, *Stochastic source seeking for nonholonomic unicycle*, Automatica, 46 (2010), pp. 1443–1453.

[142] ——, *Stochastic Averaging and Stochastic Extremum Seeking*, Springer, 2012.

[143] ——, *Newton-based stochastic extremum seeking*, Automatica, 50 (2014), pp. 952–961.

[144] X. Y. Lu, P. Varaiya, R. Horowitz, D. Su, and S. E. Shladover, *Novel freeway traffic control with variable speed limit and coordinated ramp metering*, Transportation Research Record, 2229 (2011), pp. 55–65.

[145] A. Lunardi, *Analytic Semigroups and Optimal Regularity in Parabolic Problems*, Birkhäuser, 1995.

[146] M. Malisoff and M. Krstić, *Delayed multivariable extremum seeking with sequential predictors*, in IEEE American Control Conference, Denver, 2020, pp. 2649–2653.

[147] A. Manitius and A. Olbrot, *Finite spectrum assignment problem for systems with delays*, IEEE Transactions on Automatic Control, 24 (1979), pp. 541–553.

[148] F. Mazenc, S.-I. Niculescu, and M. Krstić, *Lyapunov-Krasovskii functionals and application to input delay compensation for linear time-invariant systems*, Automatica, 48 (2012), pp. 1317–1323.

[149] M. Merad, R. J. Downey, S. Obuz, and W. E. Dixon, *Isometric torque control for neuromuscular electrical stimulation with time-varying input delay*, IEEE Transactions on Control Systems Technology, 24 (2016), pp. 971–978.

[150] T. Meurer, *Control of Higher-Dimensional PDEs: Flatness and Backstepping Designs*, Springer, 2013.

[151] G. Mills and M. Krstić, *Maximizing map sensitivity and higher derivatives via extremum seeking*, IEEE Transactions on Automatic Control, 63 (2018), pp. 3232–3247.

[152] L. MIRKIN, *On the approximation of distributed-delay control laws*, Systems & Control Letters, 51 (2004), pp. 331-342.

[153] A. MIRONCHENKO, *Local input-to-state stability: characterizations and counterexamples*, Systems & Control Letters, 87 (2016), pp. 23–28.

[154] A. MIRONCHENKO, I. KARAFYLLIS, AND M. KRSTIC, *Monotonicity methods for input-to-state stability of nonlinear parabolic PDEs with boundary disturbances*, SIAM Journal on Control and Optimization, 57 (2019), pp. 510–532.

[155] A. MOHAMMADI, C. MANZIE, AND D. NESIĆ, *Online optimization of spark advance in alternative fueled engines using extremum seeking control*, Control Engineering Practice, 29 (2014), pp. 201–211.

[156] S. MONDIÉ AND W. MICHIELS, *Finite spectrum assignment of unstable time-delay systems with a safe implementation*, IEEE Transactions on Automatic Control, 48 (2003), pp. 2207–2212.

[157] K. MORI AND E. SHIMEMURA, *Linear differential games with delayed and noisy information*, Journal of Optimization Theory and Applications, 13 (1974), pp. 275–289.

[158] S. J. MOURA AND Y. A. CHANG, *Lyapunov-based switched extremum seeking for photovoltaic power maximization*, Control Engineering Practice, 21 (2013), pp. 971–980.

[159] S. J. MOURA, N. A. CHATURVEDI, AND M. KRSTIĆ, *Adaptive partial differential equation observer for battery state-of-charge/state-of-health estimation via an electrochemical model*, ASME Journal of Dynamic Systems, Measurement, and Control, 136 (2014), pp. 1–11.

[160] P. MOYLAN AND B. ANDERSON, *Nonlinear regulator theory and an inverse optimal control problem*, IEEE Transactions on Automatic Control, 18 (1973), pp. 460-465.

[161] R. NAGEL, *Towards a "matrix theory" for unbounded operator matrices*, Mathematische Zeitschrift, 201 (1989), pp. 57–68.

[162] A. K. NAIMZADA AND L. SBRAGIA, *Oligopoly games with nonlinear demand and cost functions: Two boundedly rational adjustment processes*, Chaos, Solitons, Fractals, 29 (2006), pp. 707–722.

[163] J. F. NASH, *Noncooperative games*, Annals of Mathematics, 54 (1951), pp. 286–295.

[164] D. NEŠIĆ, *Extremum seeking control: Convergence analysis*, European Journal of Control, 15 (2009), pp. 331–347.

[165] D. NESIĆ, A. MOHAMMADI, AND C. MANZIE, *A framework for extremum seeking control of systems with parameter uncertainties*, IEEE Transactions on Automatic Control, 58 (2013), pp. 435–448.

[166] D. NEŠIĆ, Y. TAN, W. H. MOASE, AND C. MANZIE, *A unifying approach to extremum seeking: Adaptive schemes based on estimation of derivatives*, in IEEE Conference on Decision and Control, Atlanta, 2010, pp. 4625–4630.

[167] N. OLGAC AND R. SIPAHI, *Analysis of a system of linear delay differential equations*, ASME Journal of Dynamic Systems, Measurement, and Control, 125 (2003), pp. 215–223.

[168] ———, *The cluster treatment of characteristic roots and the neutral type time-delayed systems*, ASME Journal of Dynamic Systems, Measurement, and Control, 127 (2005), pp. 88–97.

[169] T. R. OLIVEIRA, J. FEILING, S. KOGA, AND M. KRSTIĆ, *Scalar Newton-based extremum seeking for a class of diffusion PDEs*, in IEEE Conference on Decision and Control, Miami Beach, 2018, pp. 2926–2931.

[170] ——, *Multivariable extremum seeking for PDE dynamic systems*, IEEE Transactions on Automatic Control, 65 (2020), pp. 4949–4956.

[171] ——, *Extremum seeking for unknown scalar maps in cascade with a class of parabolic partial differential equations*, International Journal of Adaptive Control and Signal Processing, 35 (2021), pp. 1162–1187.

[172] T. R. OLIVEIRA, J. FEILING, AND M. KRSTIĆ, *Extremum seeking for maximizing higher derivatives of unknown maps in cascade with reaction-advection-diffusion PDEs*, in IFAC Workshop on Adaptive and Learning Control Systems, 2019, pp. 210–215.

[173] T. R. OLIVEIRA, L. HSU, AND A. J. PEIXOTO, *Output-feedback global tracking for unknown control direction plants with application to extremum-seeking control*, Automatica, 47 (2011), pp. 2029–2038.

[174] T. R. OLIVEIRA AND M. KRSTIĆ, *Gradient extremum seeking with delays*, IFAC-PapersOnLine, 48 (2015), pp. 227–232.

[175] ——, *Newton-based extremum seeking under actuator and sensor delays*, IFAC-PapersOnLine, 48 (2015), pp. 304–309.

[176] ——, *Compensation of wave PDEs in actuator dynamics for extremum seeking feedback*, IFAC Workshop on Adaptive and Learning Control Systems, Winchester, 2019, pp. 134–139.

[177] T. R. OLIVEIRA, M. KRSTIĆ, AND D. TSUBAKINO, *Extremum seeking subject to multiple and distinct input delays*, IEEE Conference on Decision and Control, 2015, pp. 5635–5641.

[178] ——, *Multiparameter extremum seeking with output delays*, in IEEE American Control Conference, 2015, pp. 152–158.

[179] ——, *Extremum seeking for static maps with delays*, IEEE Transactions on Automatic Control, 62 (2017), pp. 1911–1926.

[180] T. R. OLIVEIRA AND M. KRSTIĆ, *Extremum seeking feedback with wave partial differential equation compensation*, ASME Journal of Dynamic Systems, Measurement, and Control, 143 (2021), 041002.

[181] T. R. OLIVEIRA, A. J. PEIXOTO, AND L. HSU, *Global real-time optimization by output-feedback extremum-seeking control with sliding modes*, Journal of the Franklin Institute, 349 (2012), pp. 1397–1415.

[182] T. R. OLIVEIRA, D. RUSITI, M. DIAGNE, AND M. KRSTIĆ, *Gradient extremum seeking with time-varying delays*, in IEEE American Control Conference (ACC), Milwaukee, 2018, pp. 3304–3309.

[183] T. R. OLIVEIRA, D. TSUBAKINO, AND M. KRSTIĆ, *A simplified multivariable gradient extremum seeking for distinct input delays with delay-independent convergence rates*, in IEEE American Control Conference, 2020, pp. 608–613.

[184] T. R. OLIVEIRA, V. H. P. RODRIGUES, M. KRSTIĆ, AND T. BAŞAR, *Nash equilibrium seeking with arbitrarily delayed player actions*, in IEEE Conference on Decision and Control, 2020, pp. 150–155.

[185] ——, *Nash equilibrium seeking with players acting through heat PDE dynamics*, IEEE American Control Conference, 2021, pp. 684–689.

[186] ——, *Nash equilibrium seeking in heterogeneous noncooperative games with players acting through heat PDE dynamics and delays*, in IEEE Conference on Decision and Control, Austin, 2021, pp. 1167–1173.

[187] O. M. PAMEN, *Optimal control for stochastic delay systems under model uncertainty: A stochastic differential game approach*, Journal of Optimization Theory and Applications, 167 (2015), pp. 998–1031.

[188] P. D. PARASKEVOPOULOS, *A singular perturbation analysis with applications to delay differential equations*, Journal of Dynamics and Differential Equations, 7 (1995), pp. 263–285.

[189] P. PAZ, T. R. OLIVEIRA, A. V. PINO, AND A. P. FONTANA, *Model-free neuromuscular electrical stimulation by stochastic extremum seeking*, IEEE Transactions on Control Systems Technology, 28 (2020), pp. 238–253.

[190] A. PAZY, *Semigroups of Linear Operators and Applications to Partial Differential Equations*, Springer Science & Business Media, 2012.

[191] B. PETROVIC AND Z. GAJIC, *Recursive solution of linear-quadratic Nash games for weakly interconnected systems*, Journal of Optimization Theory and Applications, 55 (1988), pp. 463–477.

[192] B. PETRUS, J. BENTSMAN, AND B. G. THOMAS, *Enthalpy-based feedback control algorithms for the Stefan problem*, in IEEE Conference on Decision and Control, Maui, 2012, pp. 7037–7042.

[193] M. S. RADENKOVIC AND J.-D. PARK, *Almost sure convergence of extremum seeking algorithm using stochastic perturbation*, Systems & Control Letters, 94 (2016), pp. 133-141.

[194] B. REN, P. FRIHAUF, R. J. RAFAC, AND M. KRSTIĆ, *Laser pulse shaping via extremum seeking*, Control Engineering Practice, 20 (2012), pp. 674–683.

[195] D. J. RIGGS AND O. HAUGAN, *Advanced Laser Wavelength Control*, United States Patent Application Publication, No. US 2011/0116522 A1, May 19, 2011.

[196] P. ROUCHON, *Motion planning, equivalence, infinite dimensional systems*, International Journal of Applied Mathematics and Computer Science, 11 (2001), pp. 165–188.

[197] E. O. ROXIN, *Differential games with partial differential equations*, in Differential Games and Applications, P. Hagedorn, H. W. Knobloch, and G. J. Olsder (editors), Springer, 1977, pp. 157–168.

[198] D. RUSITI, G. EVANGELISTI, T. R. OLIVEIRA, M. GERDTS AND M. KRSTIĆ, *Stochastic extremum seeking for dynamic maps with delays*, IEEE Control Systems Letters, 3 (2019), pp. 61–66.

[199] D. RUŠITI, T. R. OLIVEIRA, G. MILLS, AND M. KRSTIĆ, *Newton-based extremum seeking for higher derivatives of unknown maps with delays*, in IEEE Conference on Decision and Control, Las Vegas, 2016, pp. 1249–1254.

[200] ——, *Deterministic and stochastic Newton-based extremum seeking for higher derivatives of unknown maps with delays*, European Journal of Control, 41 (2018), pp. 72–83.

[201] D. RUŠITI, T. R. OLIVEIRA, M. KRSTIĆ, AND M. GERDTS, *Newton-based extremum seeking of higher-derivative maps with time-varying delays*, International Journal on Adaptive Control and Signal Processing, 35 (2021), pp. 1202–1216.

[202] G. C. SANTOS AND T. R. OLIVEIRA, *Gradient extremum seeking with nonconstant delays*, IEEE Access, 8 (2020), pp. 120429–120446.

[203] A. SCHEINKER AND M. KRSTIĆ, *Non-C^2 Lie bracket averaging for nonsmooth extremum seekers*, Journal of Dynamic Systems, Measurement, and Control, 136 (2014), pp. 011010.1–011010.10.

[204] ——, *Extremum seeking with bounded update rates*, Systems & Control Letters, 63 (2014), pp. 25–31.

[205] A. Sezgin and M. Krstić, *Boundary backstepping control of flow-induced vibrations of a membrane at high Mach numbers*, ASME Journal of Dynamic Systems, Measurement, and Control, 137 (2015), pp. 1–8.

[206] R. Sepulchre, M. Jankovic, and P. V. Kokotovic, *Constructive Nonlinear Control*, Springer, New York, 1997.

[207] N. Sharma, C. Gregory, and W. E. Dixon, *Predictor-based compensation for electromechanical delay during neuromuscular electrical stimulation*, IEEE Transactions on Neural Systems and Rehabilitation Engineering, 19 (2011), pp. 601–611.

[208] A. Siranosian, M. Krstić, A. Smyshlyaev, and M. Bement, *Motion planning and tracking for tip displacement and deflection angle for flexible beams*, ASME Journal of Dynamic Systems, Measurement, and Control, 131 (2009), pp. 1–10.

[209] A. Smyshlyaev and M. Krstić, *Closed form boundary state feedbacks for a class of* 1*-D partial integro-differential equations*, IEEE Transactions on Automatic Control, 49 (2004), pp. 2185–2202.

[210] ———, *Adaptive Control of Parabolic PDEs*, Princeton University Press, 2010.

[211] A. Smyshlyaev, Y. Orlov, and M. Krstić, *Adaptive identification of two unstable PDEs with boundary sensing and actuation*, International Journal of Adaptive Control and Signal Processing, 23 (2009), pp. 131–149.

[212] E. D. Sontag and Y. Wang, *On characterizations of the input-to-state stability property*, Systems & Control Letters, 24 (1995), pp. 351–359.

[213] R. Srikant and T. Başar, *Iterative computation of noncooperative equilibria in nonzero-sum differential games with weakly coupled players*, Journal of Optimization Theory and Applications, 71 (1991), pp. 137–168.

[214] A. W. Starr and Y. C. Ho, *Nonzero-sum differential games*, Journal of Optimization Theory and Applications, 3 (1969), pp. 184–206.

[215] G. A. Susto and M. Krstić, *Control of PDE–ODE cascades with Neumann interconnections*, Journal of the Franklin Institute, 347 (2010), pp. 284–314.

[216] I. A. M. Swinnen, K. Bernaerts, E. J. J. Dens, A. H. Geeraerd, and J. F. Van Impe, *Predictive modelling of the microbial lag phase: A review*, International Journal of Food Microbiology, 94 (2004), pp. 137–159.

[217] Y. Tan, D. Nesić, and I. Mareels, *On non-local stability properties of extremum seeking control*, Automatica, 42 (2006), pp. 889–903.

[218] ———, *On the choice of dither in extremum seeking systems: A case study*, Automatica, 44 (2008), pp. 1446–1450.

[219] Y. Tan, D. Nesić, I. Mareels, and A. Astolfi, *On global extremum seeking in the presence of local extrema*, Automatica, 45 (2009), pp. 245–251.

[220] Y. Tan, W. H. Moase, C. Manzie, D. Nešić, and I. M. Y. Mareels, *Extremum seeking from* 1922 *to* 2010, Proceedings of the 29th Chinese Control Conference, Beijing, 2010, pp. 14–26.

[221] O. Taussky, *A recurring theorem on determinants*, American Mathematical Monthly, 56 (1949), pp. 672–676.

[222] A. R. Teel, J. Peuteman, and D. Aeyels, *Semi-global practical asymptotic stability and averaging*, Systems & Control Letters, 37 (1999), pp. 329–334.

[223] A. R. TEEL AND D. POPOVIĆ, *Solving smooth and nonsmooth multivariable extremum seeking problems by the methods of nonlinear programming*, in IEEE American Control Conference, Arlington, 2001, pp. 2394–2399.

[224] M. TITICA, D. DOCHAIN, AND M. GUAY, *Adaptive extremum seeking control of fed-batch bioreactors*, European Journal of Control, 9 (2003), pp. 618–631.

[225] M. TREIBER AND A. KESTING, *Traffic Flow Dynamics*, Springer, 2013.

[226] D. TSUBAKINO, T. R. OLIVEIRA, AND M. KRSTIĆ, *Predictor-feedback for multi-input LTI systems with distinct delays*, in IEEE American Control Conference, Chicago, 2015, pp. 571–576.

[227] ———, *Extremum seeking under distributed input delay*, IFAC-PapersOnLine, 53 (2020), pp. 5423–5428.

[228] I. VANDERMEULEN, M. GUAY, AND P. J. MCLELLAN, *Distributed control of high-altitude balloon formation by extremum-seeking control*, IEEE Transactions on Control Systems Technology, 23 (2018), pp. 857–873.

[229] K. VINTHER, C. H. LYHNE, E. B. SORENSEN, AND H. RASMUSSEN, *Evaporator superheat control with one temperature sensor using qualitative system knowledge*, in IEEE American Control Conference (ACC), Montreal, 2012, pp. 374–379.

[230] K. VINTHER, H. RASMUSSEN, R. IZADI-ZAMANABADI, AND J. STOUSTRUP, *Single temperature sensor superheat control using a novel maximum slope-seeking method*, International Journal of Refrigeration, 36 (2013), pp. 1118–1129.

[231] H. YU, M. DIAGNE, L. ZHANG, AND M. KRSTIĆ, *Bilateral boundary control of moving shockwave in LWR model of congested traffic*, IEEE Transactions on Automatic Control, 66 (2020), pp. 1429–1436.

[232] H. YU, S. KOGA, T. R. OLIVEIRA, AND M. KRSTIĆ, *Extremum seeking for traffic congestion control with a downstream bottleneck*, ASME Journal of Dynamic Systems, Measurement, and Control, 143 (2021), pp. 1–10.

[233] H. YU AND M. KRSTIĆ, *Traffic congestion control for Aw-Rascle-Zhang model*, Automatica, 100 (2019), pp. 38–51.

[234] H. YU, L. ZHANG, M. DIAGNE, AND M. KRSTIC, *Bilateral boundary control of moving traffic shockwave*, IFAC-PapersOnLine, 52 (2019), pp. 48–53.

[235] L. WANG, S. CHEN, AND K. MA, *On stability and application of extremum seeking control without steady-state oscillation*, Automatica, 68 (2016), pp. 18–26.

[236] J. WANG, S. KOGA, Y. PI, AND M. KRSTIĆ, *Axial vibration suppression in a partial differential equation model of ascending mining cable elevator*, ASME Journal of Dynamic Systems, Measurement, and Control, 140 (2018), pp. 1–13.

[237] H.-H. WANG, M. KRSTIĆ, AND G. BASTIN, *Optimizing bioreactors by extremum seeking*, International Journal of Adaptive Control and Signal Processing, 13 (1999), pp. 651–669.

[238] W. WANG, H. SUN, R. VAN DEN BRINK, AND G. XU, *The family of ideal values for cooperative games*, Journal of Optimization Theory and Applications, 180 (2018), pp. 1065–1086.

[239] H.-H. WANG, S. YEUNG, AND M. KRSTIĆ, *Experimental application of extremum seeking on an axial-flow compressor*, IEEE Transactions on Control Systems Technology, 8 (1999), pp. 300–309.

[240] F. WANG, H. WANG, K. XU, J. WU, AND X. JIA, *Characterizing information diffusion in online social networks with linear diffusive model*, in International Conference on Distributed Computing Systems, 2013, pp. 307–316.

[241] J. S. WETTLAUFER, *Heat flux at the ice-ocean interface*, Journal of Geophysical Research: Oceans, 96 (1991), pp. 7215–7236.

[242] J. J. WINKIN, D. DOCHAIN, AND P. LIGARIUS, *Dynamical analysis of distributed parameter tubular reactors*, Automatica, 36 (2000), pp. 349 – 361.

[243] Z. ZAHEDI, M. M. AREFI, AND A. KHAYATIAN, *Fast convergence to Nash equilibria without steady-state oscillation*, Systems & Control Letters, 123 (2019), pp. 124–133.

[244] C. ZHANG, N. G. D. ARNOLD, A. SIRANOSIAN, AND M. KRSTIĆ, *Source seeking with nonholonomic unicycle without position measurement and with tuning of forward velocity*, Systems & Control Letters, 56 (2007), pp. 245–252.

[245] Y. ZHANG AND P. A. IOANNOU, *Coordinated variable speed limit, ramp metering and lane change control of highway traffic*, IFAC-PapersOnLine, 50 (2017), pp. 5307–5312.

[246] C. ZHANG, A. SIRANOSIAN, AND M. KRSTIĆ, *Extremum seeking for moderately unstable systems and for autonomous vehicle target tracking without position measurements*, Automatica, 43 (2007), pp. 1832–1839.

[247] Z. ZHANG AND E. ZUAZUA, *Polynomial decay and control of an* 1*-D hyperbolic-parabolic coupled system*, Comptes Rendus Mathematique, 336 (2003), pp. 745–750.

[248] Y. ZHU, E. FRIDMAN, AND T. R. OLIVEIRA, *Sampled-data extremum seeking with constant delay: A time-delay approach*, IEEE Transactions on Automatic Control, IEEE Xplore early access (2022), https://doi.org/10.1109/TAC.2022.3140259.

[249] Q. ZHU, H. TEMBINE, AND T. BAŞAR, *Hybrid learning in stochastic games and its applications in network security*, in Reinforcement Learning and Approximate Dynamic Programming for Feedback Control, Series on Computational Intelligence, F. L. Lewis and D. Liu (editors), IEEE Press/Wiley, 2013, pp. 305–329.

Index